A TEXTBOOK OF
Hydraulics, Fluid Mechanics and Hydraulic Machines

A TEXTBOOK OF
Hydraulics, Fluid Mechanics and Hydraulic Machines

[*For the Students of Degree and Diploma Courses*]

(SI UNITS)

R.S. KHURMI

S. CHAND
AN ISO 9001: 2000 COMPANY

S. CHAND & COMPANY LTD.
RAM NAGAR, NEW DELHI - 110 055

S. CHAND & COMPANY LTD.
(An ISO 9001 : 2000 Company)
Head Office: 7361, RAM NAGAR, NEW DELHI - 110 055
Phones : 23672080-81-82, 9899107446, 9911310888;
Fax : 91-11-23677446
Shop at: **schandgroup.com;** E-mail: **schand@vsnl.com**

Branches:

- 1st Floor, Heritage, Near Gujarat Vidhyapeeth, Ashram Road,
  Ahmedabad-380 014. Ph. 27541965, 27542369, ahmedabad@schandgroup.com
- No. 6, Ahuja Chambers, 1st Cross, Kumara Krupa Road,
  Bangalore-560 001. Ph: 22268048, 22354008, bangalore@schandgroup.com
- 238-A M.P. Nagar, Zone 1, **Bhopal** - 462 011. Ph : 4274723. bhopal@schandgroup.com
- 152, Anna Salai, **Chennai**-600 002. Ph: 28460026, chennai@schandgroup.com
- S.C.O. 6, 7 & 8, Sector 9D, **Chandigarh**-160017, Ph-2749376, 2749377,
  chandigarh@schandgroup.com
- 1st Floor, Bhartia Tower, Badambadi, **Cuttack**-753 009, Ph-2332580; 2332581,
  cuttack@schandgroup.com
- 1st Floor, 52-A, Rajpur Road, **Dehradun**-248 001. Ph : 2740889, 2740861,
  dehradun@schandgroup.com
- Pan Bazar, **Guwahati**-781 001. Ph: 2514155, guwahati@schandgroup.com
- Sultan Bazar, **Hyderabad**-500 195. Ph: 24651135, 24744815, hyderabad@schandgroup.com
- Mai Hiran Gate, **Jalandhar** - 144008 . Ph. 2401630, jalandhar@schandgroup.com
- A-14 Janta Store Shopping Complex, University Marg, Bapu Nagar, **Jaipur** - 302 015,
  Phone : 2719126, jaipur@schandgroup.com
- 613-7, M.G. Road, Ernakulam, **Kochi**-682 035. Ph: 2381740, cochin@schandgroup.com
- 285/J, Bipin Bihari Ganguli Street, **Kolkata**-700 012. Ph: 22367459, 22373914,
  kolkata@schandgroup.com
- Mahabeer Market, 25 Gwynne Road, Aminabad, **Lucknow**-226 018. Ph: 2626801, 2284815,
  lucknow@schandgroup.com
- Blackie House, 103/5, Walchand Hirachand Marg , Opp. G.P.O., **Mumbai**-400 001.
  Ph : 22690881, 22610885, mumbai@schandgroup.com
- Karnal Bag, Model Mill Chowk, Umrer Road, **Nagpur**-440 032 Ph : 2723901, 2777666
  nagpur@schandgroup.com
- 104, Citicentre Ashok, Govind Mitra Road, **Patna**-800 004. Ph : 2300489, 2302100,
  patna@schandgroup.com

© *1970, R.S. Khurmi*
All rights reserved. No part of this publication may be reproduced, stored in a retrieval system or transmitted, in any form or by any means, electronic, mechanical, photocopying, recording or otherwise, without the prior permission of the Publishers.

First Edition 1970
Subsequent Editions and Reprints 1971, 72, 73, 74, 75 (Twice), 76, 77 (Twice), 1978, 79, 80, 81, 82, 84, 85, 86, 87, 88, 89, 90, 91, 92, 93, 94, 95, 96, 97, 98, 2000, 2001, 2003, 2004, 2005, 2006
Reprint 2007

ISBN : 81-219-0162-6
Code : 10 026
PRINTED IN INDIA

By Rajendra Ravindra Printers (Pvt.) Ltd., 7361, Ram Nagar, New Delhi-110 055
and published by S. Chand & Company Ltd. 7361, Ram Nagar, New Delhi-110 055

Preface to the Nineteenth Edition

I feel very much elevated in presenting the new 'edition of this standard treatise'. The favourable and warm reception, which the previous editions and reprints of this popular book has enjoyed all over India and abroad has been a matter of great satisfaction for me.

The present edition of this treatise has been rewritten in S.I. units and brought up-to-date. Lot of useful changes have been incorporated in this volume to widen its scope and utility. About 100 examples, which have been carefully selected from the recent examination papers of various universities and examining bodies, have been solved and systematically graded in this volume.

I wish to express my sincere thanks to numerous professors and students for their valuable suggestions and recommending the book to their students and friends. I hope, that they will continue to patronise this standard treatise in the future also.

Any errors, omissions and suggestions for the improvement of this volume, brought to my notice, will be thankfully acknowledged and incorporated in the next edition.

R. S. Khurmi

Preface to the Eleventh Edition

I feel very much satisfied in presenting the eleventh edition of this standard treatise. The favourable and warm reception, which the previous editions and reprints of this popular book has enjoyed all over India and abroad has been a matter of great satisfaction for me. The revised edition of this treatise has been rewritten in S.I. units and M.K.S. units.

The mistakes, which had crept in, have been eliminated. It is earnestly hoped that the book will continue to cater the long-felt need of the teachers and students of all the Indian and Foreign Universities.

Any errors, omissions and suggestions for the improvement of this volume, brought to my notice, will be thankfully acknowledged and incorporated in the next edition.

R. S. Khurmi

Preface to the First Edition

I take an opportunity to present this standard treatise entitled as 'A Text Book of Hydraulics, Fluid Mechanics and Hydraulic Machines' to the students of Degree and Diploma classes in M.K.S. units. The object of this book is to present the subject matter in a most concise, compact, to-the-point and lucid manner.

While writing the book, I have always kept in mind the examination requirements of the students of Civil Engineering as well as Mechanical Engineering. I have also constantly kept in mind the requirements of other engineering students, who are always interested to have an up-to-date and latest information, by incorporating the research work conducted in various countries on allied topics; available from numerous specialised technical books, papers and periodicals. Thus the book will provide the students a necessary insight into the subject, and will prepare them adequately for their examinations.

To make the book more useful at all levels, it has been written in an easy style and such a simple manner, that even an average student can grasp the subject independently. Every care has been taken to make the book as self-explanatory as possible throughout. The rigorous mathematical steps have been replaced by simple treatments. The subject matter has been amply illustrated by incorporating a good number of solved and well graded examples of almost every variety. Most of these examples are taken from the recent examination papers of Indian as well as foreign universities and professional examining bodies, to make the students familiar with the types of questions, usually, set in their examinations. At the end of each topic, a few exercises have been added, for the students to solve them independently. Answer, along with hints wherever necessary, have been provided. But it is too much to hope that these are entirely free from errors.

Although every care has been taken to check mistakes and misprints, yet it is difficult to claim perfection. Any errors omissions and suggestions, for the improvement of this volume brought to my notice, will be thankfully acknowledged and incorporated in the next edition.

R. S. Khurmi

Contents

S.No.	Chapter Name	Page No.
1.	Introduction	1
2.	Fluid Pressure and its Measurement	13
3.	Hydrostatics	37
4.	Applications of Hydrostatics	64
5.	Equilibrium of Floating Bodies	87
6.	Hydrokinematics	107
7.	Bernoulli's Equation and its Applications	116
8.	Flow Through Orifices (Measurement of Discharge)	138
9.	Flow Through Orifices (Measurement of Time)	153
10.	Flow Through Mouthpieces	172
11.	Flow Over Notches	194
12.	Flow Over Weirs	210
13.	Flow Through Simple Pipes	231
14.	Flow Through Compound Pipes	253
15.	Flow Through Nozzles	270
16.	Uniform Flow Through Open Channels	284
17.	Non-Uniform Flow Through Open Channels	310
18.	Viscous Flow	333
19.	Viscous Resistance	350
20.	Fluid Masses Subjected to Acceleration	363
21.	Vortex Flow	373
22.	Mechanics of Compressible Flow	385
23.	Compressible Flow of Fluids	394
24.	Flow Around Immersed Bodies	408
25.	Dimensional Analysis	420
26.	Model Analysis (Undistorted Models)	440
27.	Model Analysis (Distorted Models)	454
28.	Non-Dimensional Constants	463
29.	Impact of Jets	471
30.	Jet Propulsion	489
31.	Water Wheels	499
32.	Impulse Turbines	508
33.	Reaction Turbines	528
34.	Performance of Turbines	561
35.	Centrifugal Pumps	582
36.	Reciprocating Pumps	602
37.	Performance of Pumps	632
38.	Pumping Devices	645
39.	Hydraulic Systems	651
	Index	661

Introduction

---

1. Introduction. 2. Beginning and Development of the Subject. 3. Fundamental Units. 4. Derived Units. 5. S.I. (International System of Units). 6. Metre. 7. Kilogram. 8. Second. 9. Presentation of Units and their Values. 10. Rules for S.I. Units. 11. Liquids and their Properties. 12. Density of Water. 13. Specific Weight of Water. 14. Specific Gravity of Water. 15. Compressibility of Water. 16. Surface Tension of Water. 17. Capillarity of Water. 18. Viscosity of Water. 19. Useful Data. 20. Force. 21. Resultant Force. 22. Composition of Forces. 23. Parallelogram Law of Forces. 24. Triangle Law of Forces. 25. Polygon Law of Forces. 26. Laws of Motion. 27. Newton's First Law of Motion. 28. Newton's Second Law of Motion. 29. Newton's Third Law of Motion. 30. Work. 31. Power. 32. Energy. 33. Law of Conservation of Energy. 34. Scalars and Vectors.

1·1 Introduction

The word 'Hydraulics' has been derived from a Greek word 'Hudour', which means water. Thus the subject of Hydraulics may be defined as that branch of Engineering-science, which deals with water (at rest or in motion). The subject of Fluid Mechanics may be defined as the mechanics of fluids (including water). The subject Hydraulic Machines may be defined as that branch of Engineering-science, which deals with the machines run by water under some head, or raising the water to higher levels.

In the text of this book some other liquids, in addition to water, will also be dealt with. But, in general, the given liquid will be assumed to be water, unless specified otherwise.

1·2 Beginning and Development of the Subject

It will be interesting to know as to how early the man had been curious to know about the earth and the various processes going on it. As a matter of fact, since the dim ages of pre-historic times, the man has been practising the art of engineering. And he started erecting very simple and ordinary types of structures (*e.g.*, digging wells and making water courses for water supply and irrigation purposes) for his livelihood. With the passage of time the man, entirely on the basis of his practical experience and common-sense, created a few thumb rules to serve as a guide for erecting very ordinary types of structures.

It is only since the human progress had reached its evolutionary height, we call the dawn of civilization, the man actually started to design and build the structures according to his requirements. Since then, the man had been very eager to know more and more about the various aspects of his structures, especially regarding their uses and smooth workability. Experience taught him, and he continued to make progress in the art of engineering bit by bit.

In the ancient Indian literature, based on the *Shilpasamhita*, a mention has been made about *Jala-shastra* (*i.e.*, science of water). In this book, so many properties of water at rest, and also in motion have been mentioned, some of which are given below :

1. The water, at rest, keeps its surface level.
2. The water exerts a uniform pressure in all directions.
3. The pressure of water is equal to its weight, and thus the pressure increases with the depth.
4. The velocity of water is proportional to the slope of the channel bed.
5. The velocity of water, in a channel, depends upon the roughness of the bed and sides. The smoother the bed and sides, the greater is velocity.
6. The velocity of water, in a channel, is more at the centre than at the sides.

Thus we find that the ancient Indians had a considerable knowledge of Hydraulics. The excavations of Mohenjodaro and Harappa, which indicate the Indus-valley civilization dating back to three thousand years before Christ, have brought to light some interesting hydraulic structures of the time. *e.g.*, house baths with ceramic pipes for water supply and brick conduits under the streets for drainage. These structures definitely indicate the qualitative knowledge of Hydraulics, without which these structures could not have been constructed. The vast study of ancient Indian civilization has revealed the construction of small earthen dams and use of water wheels. The study of ancient civilization of Babylonians, Egyptians, Greeks and Romans also reveal some knowledge of Hydraulics.

As a matter of fact, the scientific history of Hydraulics starts with Archimedes, who gave the law of buoyancy[*] and the connected theory. In the Roman empire, some aquaducts were constructed after finding out a set of rules for flowing water. After the decline of Roman empire (A.D. 476) no scientific progress was made for nearly 1000 years. This development of Hydraulics once again became active, when Leonardo da Vinci (1452—1519) gave a number of fresh ideas while working on the excavation of a canal in Northern Italy. He is universally believed to be the first person, who gave some knowledge of river Hydraulics[**]. In 1643, Torricelli gave the relation for the mean velocity of flow (*i.e.*, $v = \sqrt{2gh}$) through the orifice.

In fact, the progress in one branch of science, enriches the other branches of the bordering sciences. Similarly, with the passage of time, the concepts of Hydraulics aided by Mathematics and other physical sciences started contributing and the development of Hydraulics gained momentum in the 18th century. In the beginning of 18th century, Chezy gave a relation for the mean velocity in open channel (*i.e.*, $v = C\sqrt{mi}$). A succession of eminent scientists like Bernoulli, Euler, Laplace, Rankine and others gave a number of theories and built up the science of Hydraulics. In the later half of the 18th century, notable contributions were made by Venturi, Darcy, Hagen, Poiseuille and others. These contributions led to the creation of new subjects, now popularly known as Fluid Mechanics and Hydraulic Machines. In the 19th century all the three fields (*i.e.*, Hydraulics, Fluid Mechanics and Hydraulic Machines) started growing independently. With the introduction of Mathematics into Engineering-science, even the theoretical Hydraulics was separated from practical Hydraulics. Theoretical Hydraulics was developed by physicists and scientists, whereas practical Hydraulic was developed by practical engineers. Prandtl gave his boundary layer theory, which was based on his practical as well as theoretical experience. Reynold, in 1883, successfully conducted an experiment

---

[*]The principle of buoyancy was given by the sage Vasishtha long before Archimedes. But it was drowned down, like so many other theories, which were originally given by Indian thinkers. However, the principle of buoyancy given by Vasishtha did not have mathematical concept, as the mathematics was in its infancy at that time.

[**]The basic principles of river hydraulics were first given by Vasishtha, many centuries earlier than Leonardo da Vinci.

Introduction

to distinguish between streamline flow and turbulent flow. Fourneyron designed a water turbine in 1827. Pelton designed another water turbine in 1880.

Since the beginning of 20th century, the various fields of Hydraulic were enriched by a number of scientists, engineers and mathematicians like Rayleigh, Buckingham, Nikuradse and so many others. Now-a-days, before any big size hydraulic structure is constructed its model is prepared and tested at some hydraulic research centre. As a result of this, the model testing has become a common feature of the design of hydraulic structures and machines. Prof. Khurmi[*] gave a number of methods for the comparison of a model to the prototype.

Since the study of a subject, at university level, is done by engineers with only theoretical experience (having little or no practical experience, *i.e.*, field experience) therefore the present-day fashion is to prefer the study of Fluid Mechanics to that of Hydraulics. Moreover, the study of Fluid Mechanics has been very useful in space research and rocket launching programmes.

1·3 Fundamental Units

The measurement of physical quantities is one of the most important operations in engineering. Every quantity is measured in terms of some arbitrary, but internationally accepted units, called fundamental units.

All physical quantities, met with in this subject, are expressed in terms of the following three fundamental quantities :

1. length.   2. mass.   3. time.

1·4 Derived Units

Some units are expressed in terms of other units (which are derived from fundamental units) known as derived units. *e.g.*, the unit of area, velocity, acceleration, pressure etc.

1·5 S.I. Units (International System of Units)

The 11th[**] General Conference of Weights and Measures have recommended a unified and systematically constituted system of fundamental and derived units for international use. This system of units is now being used in about 100 countries. In India, the Standards of Weights and Measures Act 1956 (vide which we switched over to M.K.S. units) has been revised to recognise all the S.I. units in industry and commerce.

In this system of units, the fundamental units are metre, kilogram and second respectively. But there is a slight variation in their derived units. The derived units, which will be used in this system, are given below :

Density (mass density)	...	kg/m^3
Force	...	N (Newton)
Pressure	...	N/m^2 or N/mm^2
Dynamic viscosity	...	Ns/m^2
Kinematic viscosity	...	m^2/s
Work done (in Joules)	...	J = N-m
Power (in watts)	...	W = J/s

The international metre, kilogram and second is discussed here.

---

[*] These methods were given by the author of this book. A further research on these methods is being done in numerous universities and hydraulic research stations all over the world. As a matter of fact, the author of this book was awarded the degree of Ph.D. in this topic.

[**] It is known as General Conference of Weights and Measures (C.G.P.M.). It is an international organisation, of which most of the advanced and developing countries (including India) are members. The conference has been entrusted with the task of prescribing definitions for the various units of weights and measures, which are the very basis of science and technology today.

1·6 Metre

The international metre may (m) be defined as the shortest distance (at 0°C) between the two parallel lines, engraved upon the polished surface of a platinum-iridum bar, kept at the International Bureau of Weights and Measures at Sevres near Paris.

1·7 Kilogram

The international kilogram (kg) may be defined as the mass of the platinum-iridum cylinder, which is also kept at the International Bureau of Weights and Measures at Sevres near Paris.

1·8 Second

The fundamental unit of time, for all the systems, is second (s), which is $1/24 \times 60 \times 60$ = 1/86 400th of the mean solar day. A solar day may be defined as the interval of time, between the instants, at which the sun crosses a meridian on two consecutive days. This value varies slightly throughout the year. The average of all the solar days, during one year, is called the mean solar day.

1·9 Presentation of Units and their Values

The frequent changes in the present-day life are facilitated by an international body known as International Standard Organisation (ISO) which makes recommendations regarding international standard procedures. The implementation of ISO recommendations, in a country, is assisted by its organisation appointed for the purpose. In India, Bureau of Indian Standards formerly known as Indian Standards Institution (ISI) has been created for this purpose.

We have already discussed in the previous articles the units of length, mass and time. It is not always necessary to express all lengths in metres, all masses in kilograms and all times in seconds. According to convenience, we also use larger multiples or smaller fractions. As a typical example, although metre is the unit of length yet a smaller length of one-thousandth of a metre proves to be more convenient unit, especially in the dimensioning of drawings. Such convenient units are formed by using a prefix in front of the basic unit to indicate the multiplier. The full set of these prefixes is given in Table 1·1.

Table 1·1

Factor by which the unit is multiplied	Standard form	Prefix	Abbreviation
1 000 000 000 000	10^{12}	tera	T
1 000 000 000	10^{9}	giga	G
1 000 000	10^{6}	mega	M
1 000	10^{3}	kilo	k
100	10^{2}	hecto*	h
10	10^{1}	deca*	da
0·1	10^{-1}	deci*	d
0·01	10^{-2}	centi*	c
0·001	10^{-3}	milli	m
0·000 001	10^{-6}	micro	μ
0·000 000 001	10^{-9}	nano	n
0·000 000 000 001	10^{-12}	pico	p

*These prefixes are gradually becoming obsolete, probably, due to possible confusion. Moreover, it is becoming a conventional practice to use only those powers of ten, which confirm to 10^{3n} where n is a positive or negative whole number.

1·10 Rules for S.I. Units

The eleventh General Conference of Weights and Measures recommended only the fundamental and derived units for S.I. system. But it did not elaborate the rules for the usage of the units. Later on, many scientists and engineers held a number of meetings for the style and usage of S.I. units. Some of the decisions of these meetings are as follows :

1. For numbers having five or more digits, the digits should be placed in groups of three separated by spaces (instead of commas) counting both to the left and right of the decimal point.
2. In a four digit number, the space is not required unless the four digit number is used in a column of numbers with five or more digits.
3. A dash is to be used to separate units that are multiplied together. For example, newton × metre is written as N-m. It should not be confused as mN which stands for millinewton.
4. All symbols are written in small numbers, *e.g.*, m (for metre), s (for second), kg (for kilogram) except the symbols, derived from the proper names. For example, we write N for newton and W for watt.
5. The units with the names of scientists should not start with capital letter, when written in full. For example, we write 90 newton and not 90 Newton.

At the time of writing this book, the author sought the advice of various international authorities, regarding the use of units and their values. Keeping in view the global reputation of the author, as well as his books, it was decided to present* units and their values as per the recommendations of ISO and ISI. Some of these values are given below :

We shall use :

14 500	not	14500	or	14, 500
75 890 000	not	75890000	or	7, 58, 90, 000
0·012 55	not	0·01255	or	·01255
30×10^6	not	3, 00, 00, 000	or	3×10^7

The above-mentioned figures are meant for numerical values only. Now let us discuss about the units. We have already discussed that the fundamental units in S.I. system for length, mass and time are metre, kilogram, and second respectively. While expressing the quantities, we find it time-consuming to write the units such as metres, kilograms and seconds, in full, every time we use them. As a result of this, we find it quite convenient to use some standard abbreviations.

We shall use :

m	for	metre or metres
km	for	kilometre or kilometres
kg	for	kilogram or kilograms
t	for	tonne or tonnes
s	for	second or seconds
min	for	minute or minutes
N-m	for	newton × metres (*e.g.*, work done)
kN-m	for	kilonewton × metres
rev	for	revolution or revolutions
rad	for	radian or radians

---

* In some of the questions of universities and other examining bodies standard values are not used. The author has tried to avoid such questions in the text of the book. However, at certain places the questions with sub-standard values have to be included, keeping in view the merits of the question from the reader's angle.

1·11 Liquids and their Properties

Ordinarily, there is no difficulty in distinguishing a liquid from a solid or a gas. A sloid has a definite shape, which it retains, until some external force is applied to alter it. On the contrary, a liquid takes the shape of a vessel, into which it is poured. On the other hand, a gas completely fills up the vessel which contains it.

Among the liquids, water will be mostly dealt with in this book, which has the following properties :

1. Density. 2. Specific weight. 3. Specific gravity. 4. Compressibility. 5. Surface tension. 6. Capillarity. 7. Viscosity.

1·12 Density of Water

The density (or mass density) of a liquid may be defined as the mass per unit volume at a standard temperature and pressure and is usually denoted as (ρ). Mathematically density,

$$\rho = \frac{\text{Mass}}{\text{Volume}}$$

Note : The variation in the density of water, with the variation of pressure and temperature, is so small, that for all practical purposes it is generally neglected.

Example 1·1. *If 2·5 m^3 of a certain oil has a mass of 20 tonnes, find its mass density.*

Solution. Given : Volume = 2·5 m^2 and mass = 2·0 tonne.

We know that mass density of the oil,

$$\rho = \frac{\text{Mass}}{\text{Volume}} = \frac{2\cdot 0}{2\cdot 5} = 0\cdot 8 \text{ t/m}^3 = 800 \text{ kg/m}^3 \text{ Ans.}$$

1·13 Specific Weight of Water

The specific weight (or weight density) of a liquid may be defined as the weight per unit volume at the standard temperature and pressure and is usually denoted as (w). Mathematically specific weight,

$$w = \frac{\text{Weight}}{\text{Volume}}$$

Note : The variation in the specific weight of water, with the variation of pressure and temperature, is also so small, that for all practical purposes, it is generally neglected. It is also known as weight density.

For the purposes of all calculations, relating to Hydraulics, Fluid Mechanics and Hydraulic Machines, the specific weight of water is taken as 9·81 kN/m^3. However, the following table gives the values of specific weights of water at different temperatures as per manual 25 of American's Society of Civil Engineers.

Table 1·2

Temp. of water, in degree F	40	50	60	70	80	90	100	110	120	150
Specific weight (w) in kN/m^3	9·81	9·81	9·80	9·79	9·78	9·76	9·74	9·72	9·70	9·61

Example 1·2. *In an experiment, the weight of 2·5 m^3 of a certain liquid was found to be 18·75 kN. Find the specific weight of the liquid. Also find its density.*

Solution. Given : Volume = 2·5 m^3 and weight = 18·75 kN.

Specific weight of the liquid

We know that specific weight of the liquid

$$w = \frac{\text{Weight}}{\text{Volume}} = \frac{18\cdot 75}{2\cdot 5} = 7\cdot 5 \text{ kN/m}^3 \text{ Ans.}$$

Introduction

Density of the liquid

We also know that mass of the liquid

$$= \frac{\text{Weight}}{\text{Acceleration due to gravity}} = \frac{18 \cdot 75}{9 \cdot 81} = 1 \cdot 91 \text{ t/m}^3$$

and density of the liquid $= \dfrac{\text{Mass}}{\text{Volume}} = \dfrac{1 \cdot 91}{2 \cdot 5} = 0 \cdot 764 \text{ t/m}^3 = 764 \text{ kg/m}^3$ **Ans.**

1·14 Specific Gravity of Water

The specific gravity (briefly written as sp. gr.) of a liquid may be defined as the ratio of its specific weight to that of a standard substance at a standard temperature. For liquids, pure water is taken as a standard substance and at 4°C. Mathematically specific gravity

$$= \frac{\text{Specific weight of liquid}}{\text{Specific weight of pure water}} = \frac{w_{\text{liquid}}}{w_{\text{water}}}$$

Notes : 1. The specific gravity of water, in the calculation of Hydraulics, Fluid Mechanics and Hydraulic Machines is taken as unity.

2. If density of a liquid is in tonne per m³ (t/m³), it is numerically equal to specific gravity of the liquid.

Example 1·3. *Find the specific gravity of an oil whose specific weight is 7·85 kN/m³.*

Solution. Given : Specific weight of oil = 7·85 kN/m³.

We know that specific gravity of oil

$$= \frac{\text{Specific weight of liquid}}{\text{Specific weight of pure water}} = \frac{7 \cdot 85}{9 \cdot 81} = 0 \cdot 8 \text{ \textbf{Ans.}}$$

Example 1·4. *A vessel of 4 m³ volume contains an oil, which weighs 30·2 kN. Determine the specific gravity of the oil.*

Solution. Given : Volume = 4 m³ and weight 30·2 kN.

We know that specific weight of the oil

$$= \frac{\text{Weight}}{\text{Volume}} = \frac{30 \cdot 2}{4} = 7 \cdot 55 \text{ kN/m}^3$$

and specific gravity of oil $= \dfrac{\text{Specific weight of liquid}}{\text{Specific weight of pure water}} = \dfrac{7 \cdot 55}{9 \cdot 81} = 0 \cdot 77$ **Ans.**

1·15 Compressibility of Water

The compressibility of a liquid may be defined as the variation in its volume, with the variation of pressure. The variation in the volume of water, with the variation of pressure, is so small that for all practical purposes it is neglected. Thus, the water is considered to be an incompressible liquid.

1·16 Surface Tension of Water

The surface tension of a liquid is its property, which enables it to resist tensile stress. It is due to the cohesion between the molecules at the surface of a liquid.

The effect of surface tension may be easily seen in the case of tubes of smaller diameters, open to the atmosphere. For example, when a glass tube of small diameter is dipped in water, the water rises up in the tube with an upward *concave* surface as shown in Fig. 1·1 (*a*). But when the same tube is dipped in mercury, the mercury depresses down in the tube with an upward *convex* surface as shown in Fig. 1·1 (*b*).

For the purposes of all calculations, relating to Hydraulics, Fluid Mechanics and Hydraulic Machines, the effect of surface tension is generally neglected. But the surface tension has got its own importance in a variety of ways. As a result of surface tension, the liquid surface has a tendency to

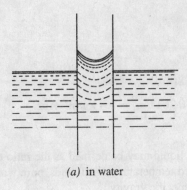
(a) in water

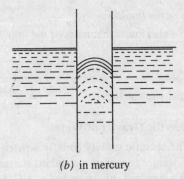
(b) in mercury

Fig. 1.1

reduce its surface as small as possible. That is why the falling drops of rain water become sphere. The property of surface tension is utilised in the manufacturing of lead shots. The molten lead is made to pass through a sieve from a high tower, and allowed to fall into water. The molten lead particles, while descending, assume a spherical shape and solidify in this form before falling into the water.

The following table gives the values of specific weight and surface tension of important liquids at 20°C (i.e., normal temperature).

Table 1.3

Liquid	Specific weight in kN/m^3	Surface tension in N/m
Water	9.80	0.0735
Glycerine	12.45	0.0490
Kerosene	7.85	0.0235
Castor oil	9.41	0.0392
Ethyl Alcohol	7.73	0.0216
Mercury	132.8	0.5100

1·17 Capillarity of Water

We have discussed in Art. 1·16 that when a tube of smaller diameter is dipped in water, the water rises up in the tube with an upward concave surface. This is due to the reason that the adhesion (between the tube and water molecules) is *more* than the cohesion between the water molecules. But when the same tube is dipped in mercury, the mercury depresses down in the tube with an upward convex surface. This is due to the reason that the adhesion (between the tube and mercury molecules) is *less* than the cohesion between the mercury molecules. Another example of surface tension is that the mercury does not wet the glass.

The phenomenon of rising water in the tube of smaller diameter is called the capillary rise as shown in Fig. 1·2.

Let h = Height of capillary rise,
 d = Diameter of the capillary tube,
 α = Angle of contact of the water surface, and

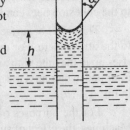
Fig. 1·2.
Effect of capillarity.

Introduction

σ = Force of surface tension per unit length of the periphery of the capillary tube in N/mm.

We know that weight of the water column in the tube above the water surface acting downwards

$$= wh \times \frac{\pi}{4}(d)^2 \qquad ...(i)$$

and vertical component of the force of surface tension

$$= \sigma \cdot \pi d \cos \alpha \qquad ...(ii)$$

Since the downward weight of the water column is balanced by the vertical component of the force of surface tension, therefore equating equations (i) and (ii),

$$wh \times \frac{\pi}{4}(d)^2 = \sigma \cdot \pi d \cos \alpha \quad \text{or} \quad h = \frac{4\sigma \cos \alpha}{wd}$$

Example 1·5. *Calculate the capillary effect in millimetres in a glass tube of 4 mm diameter, when immersed in (i) water and (ii) in mercury. The values of surface tension for water and mercury in contact with air are 0·0735 N/m and 0·5100 N/m respectively. The contact angle for water $\theta = 0°$ and for mercury $\theta = 130°$.*

Solution. Given : $d = 4$ mm $= 4 \times 10^{-3}$ m; Temperature $= 20°C$; $\sigma_w = 0·0735$ N/m; $\sigma_m = 0·5100$ N/m; $\alpha_w = 0°$ and $\alpha_m = 130°$.

Capillary effect in water

We know from Table 1·1, that the specific weight of water (w_w) at $20°C = 9·80$ kN/m^3 $= 9·80 \times 10^3$ N/m^3. Therefore capillary effect in water,

$$h_w = \frac{4\sigma_w \cos \alpha}{w_w \times d} = \frac{4 \times 0·0735 \times \cos 0°}{(9·80 \times 10^3) \times (4 \times 10^{-3})} \text{ m}$$

$$= 7·5 \times 10^{-3} \text{ m} = 7·5 \text{ mm} \quad \textbf{Ans.}$$

Capillary effect in mercury

We also know from Table 1·1, that the specific weight of mercury (w_m) at $20°C = 132·8$ kN/m^3. Therefore capillary effect in mercury,

$$h_m = \frac{4\sigma_m \cos \alpha}{w_m \times d} = \frac{4 \times 0·5100 \times \cos 130°}{(132·8 \times 10^3) \times (4 \times 10^{-3})} \text{ m}$$

$$= 3·84 \times 10^{-3} \, (-0·6428) \text{ mm}$$

$$= -2·47 \times 10^{-3} \text{ m} = 2·47 \text{ mm (depression)} \quad \textbf{Ans.}$$

1·18 Viscosity of Water

We see that the flow of thin liquids, such as alcohol or water, is much easier than thick liquids, such as syrup or heavy oil. It is thus obvious, that each liquid possesses some property, which controls its rate of flow. This property is termed as viscosity and is an essential property of the liquid. We shall study the viscosity in detail in a separate chapter.

1·19 Useful Data

The following data, which summarise the definition of previous memory and formulae, the revision of which is very essential at this stage.

1·20 Force

It is an important factor in the field of engineering-science, which may be defined as an agent, which produces or tends to produce, destroys or tends to destroy motion.

1·21 Resultant Force

If a number of forces P, Q, R ... etc. are acting simultaneously on a particle, then a single force, which will produce the same effect as that of all the given forces, then the single force is known as a resultant force. The forces P, Q, R ... etc. are called component forces. The resultant force of the component forces, or a point through which it acts, may be found out either mathematically or graphically.

1·22 Composition of Forces

It means the process of finding out the resultant force of the given component forces. A resultant force may be found out analytically, graphically or by the following laws :

1·23 Parallelogram Law of Forces

It states, *"If two forces acting simultaneously on a particle be represented in magnitude and direction by the two adjacent sides of a parallelogram, their resultant may be represented in magnitude and direction by the diagonal of parallelogram passing through the point."*

1·24 Triangle Law of Forces

It states, *"If two forces acting simultaneously on a particle be represented in magnitude and direction by the two sides of a triangle taken in order, their resultant may be represented in magnitude and direction by the third side of the triangle taken in opposite order."*

1·25 Polygon Law of Forces

It states, *"If a number of forces acting simultaneously on a particle be represented in magnitude and direction by the sides of a polygon taken in order, their resultant may be represented in magnitude and direction by the closing side of the polygon taken in opposite order."*

1·26 Laws of Motion

Newton has formulated three laws of motion, which are the basic postulates or assumptions on which the whole system of dynamics is based. Like other scientific laws, these are also justified as the results, so obtained, agree with the actual observations. These three laws of motion are popularly known as :

1. Newton's first law of motion,
2. Newton's second law of motion, and
3. Newton's third law of motion.

1·27 Newton's First Law of Motion

It states, *"Everybody continues in its state of rest or of uniform motion in a straight line, unless it is acted upon by some external force."*

1·28 Newton's Second Law of Motion

It states, *"The rate of change of momentum is directly proportional to the impressed force and takes place in the same direction in which the force acts."*

1·29 Newton's Third Law of Motion

It states, *"To every action, there is always an equal and opposite reaction."*

1·30 Work

Whenever a force acts on a body, and the body undergoes some displacement, work is said to be done.

Let P = Force acting on a body, and
S = Distance, through which the body has been displaced.

∴ Work done = Force × Distance = $P \times S$

In S.I. system of units, the work done is in N-m or Joule.

1·31 Power

The power may be defined as the rate of doing work. It is, thus, the measure of performance of engines and machines. *e.g.*, a machine doing a certain amount of work, in one second, will be twice as powerful as the machine doing the same amount of work in two seconds. In S.I. system of units, the unit of power is watt, which is equal to one Joule per second.

1·32 Energy

The energy may be defined as the capacity to do work. Since the energy of a body is measured by the work it can do, therefore the units of energy will also be the same as that of the work.

1·33 Law of Conservation of Energy

It states, *"The energy can neither be created nor destroyed, though it can be transformed from one form into another."*

1·34 Scalars and Vectors

1. Scalar quantities are those quantities, which have magnitude only. *i.e.*, mass, time, volume, density etc.

2. Vector quantities are those quantities, which have magnitude as well as direction. *i.e.*, velocity, acceleration, force etc.

3. As the vector quantities have both magnitude as well as direction, therefore while adding or subtracting vector quantities, their directions are also taken into account.

4. *Addition of vector quantities*. Consider two vector quantities *P* and *Q*, which are required to be added as shown in Fig. 1·3 (*a*).

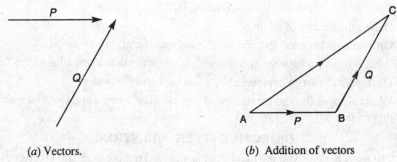

(*a*) Vectors. (*b*) Addition of vectors

Fig. 1·3

Take a point *A* and draw a line *AB* parallel to and equal in magnitude to the vector *P*. Through *B* draw *BC* parallel to and equal in magnitude to the vector *Q*. Join *AC*, which will give the required sum of the two vectors *P* and *Q* as shown in Fig. 1·3 (*b*).

5. *Subtraction of vector quantities*. Consider two vector quantities *P* and *Q*, whose difference is required to be fount out as shown in Fig. 1·4 (*a*).

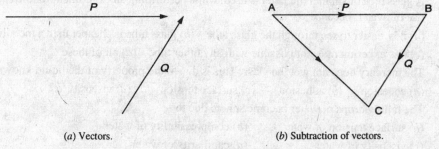

(*a*) Vectors. (*b*) Subtraction of vectors.

Fig. 1·4

Take a point *A* and draw a line *AB* parallel to and equal in magnitude to the vector *P*. Through *B*, draw *BC* parallel to and equal in magnitude to the vector *Q* but in opposite direction. Join *AC*, which will give the required difference of the two vectors *P* and *Q* as shown in Fig. 1·4 (*b*)

EXERCISE 1·1

1. Determine the mass density of an oil, if 3·0 tonnes of the oil occupies a volume of 4 m^3.

   [Ans. 750 kg/m^3]

2. A certain liquid, occupying a volume of 1·6 m^3, weighs 12·8 kN. What is the specific weight of the liquid?

   [Ans. 8 kN/m^3]

3. A container of volume 3·0 m^3 has 25·5 kN of an oil. Find the specific weight and mass density of the oil.   [Ans. 8·5 kN/m^2; 0·866 kg/m^3]

4. What is the specific gravity of a liquid, whose specific weight is 7·36 kN/m^3?   [Ans. 0·75]

5. A drum of 1 m^3 volume contains 8·5 kN an oil when full. Find its specific weight and specific gravity.

   [Ans. 8·5 kN/m^3; 0·866]

6. A 5 mm diameter glass tube is immersed vertically in water. If the contact angle is 5°, find the capillary rise. Take surface tension for the water as 0·074.

$$\text{Hint}: h = \frac{4\,\sigma \cos\alpha}{w \times d} = \frac{4 \times 0.074 \cos 5°}{(9.81 \times 10^3) \times (5 \times 10^{-3})} = \frac{0.296 \times 0.9962}{49.05} \text{ m}$$

$$= 6.0 \times 10^{-3} \text{ m} = 6 \text{ mm } \textbf{Ans.}$$

QUESTIONS

1. Define the density of a liquid.
2. Differentiate between specific weight and specific gravity of an oil.
3. What is the ratio between specific weight and mass density of a liquid?
4. Distinguish between compressibility and capillarity of water.
5. What role does the surface tension of a liquid play, when a glass tube of small diameter is dipped into it?

OBJECTIVE TYPE QUESTIONS

1. The mass per unit volume of a liquid at standard temperature and pressure is called
   (a) specific weight   (b) specific gravity
   (c) mass density   (d) none of the above

2. The ratio of specific weight of a liquid to the specific weight of pure water is known as
   (a) density of pure water   (b) density of liquid
   (c) specific gravity of water   (d) specific gravity of liquid

3. A glass tube of smaller diameter is used, while performing an experiment for the capillary rise of water, because
   (a) it is easier to see through the glass tube   (b) glass tube is cheaper than a metallic tube
   (c) this experiment is not possible with any other tube   (d) all of these

4. The mercury does not wet the glass. This is due to the property of the liquid known as
   (a) cohesion   (b) adhesion   (c) surface tension   (d) viscosity

5. The falling drops of water become sphere due to
   (a) surface tension of water   (b) compressibility of water
   (c) viscosity of water   (d) capillarity of water

ANSWERS

1. (c)   2. (d)   3. (a)   4. (c)   5. (a)

Fluid Pressure and its Measurement

1. Introduction. 2. Pressure Head. 3. Pascal's Law. 4. Atmospheric Pressure. 5. Gauge Pressure. 6. Absolute Pressure. 7. Measurement of Fluid Pressure. 8. Tube Gauges to Measure Fluid Pressure. 9. Piezometer Tube. 10. Manometer. 11. Simple Manometer. 12. Micromanometer. 13. Differential Manometer. 14. Inverted Differential Manometer. 15. Mechanical Gauges. 16. Bourdon's Tube Pressure Gauge. 17. Diaphragm Pressure Gauge. 18. Dead Weight Pressure Gauge.

2·1 Introduction

We see that whenever a liquid (such as water, oil etc.) is contained in a vessel, it exerts force at all points on the sides and bottom of the container. This force per unit area is called pressure. If P is the force acting on area a, then* intensity of pressure,

$$p = \frac{P}{a}$$

The direction of this pressure is always at right angles to the surface, with which the fluid at rest, comes in contact.

2·2 Pressure Head

Consider a vessel containing some liquid as shown in Fig. 2·1. We know that the liquid will exert pressure on all sides as well as bottom of the vessel. Now let a bottomless cylinder be made to stand in the liquid as shown in the figure.

Let
w = Specific weight of the liquid,
h = Height of liquid in the cylinder, and
A = Area of the cylinder base.

A little consideration will show, that there will be some pressure on the cylinder base due to weight of the liquid in it. Therefore pressure,

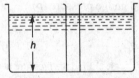

Fig. 2·1. Pressure head.

$$p = \frac{\text{Weight of liquid in the cylinder}}{\text{Area of the cylinder base}}$$

$$= \frac{whA}{A} = wh$$

This equation shows that the intensity of pressure at any point, in a liquid, is proportional to its depth, from the surface (as w is constant for the given liquid). It is thus obvious, that the pressure can be expressed in either of the following two ways :

1. As a force per unit area *i.e.*, N/m^2, kN/m^2 etc.
2. As a height of the equivalent liquid column.

*The intensity of pressure, in brief, is generally termed as *pressure*.

Note : The pressure is always expressed in pascal (briefly written as Pa) such that 1 Pa = 1 N/m^2, 1 kPa = 1 kN/m^2 and 1 MPa = 1 MN/m^2 = 1 N/mm^2.

Example 2·1. *Find the pressure at a point 4 m below the free surface of water.*

Solution. Given : $h = 4$ m.

We know that pressure at the point,

$$p = wh = 9·81 \times 4 = 39·24 \text{ kN/m}^2 = 39·24 \text{ kPa} \quad \text{Ans.}$$

Example 2·2. *A steel plate is immersed in an oil of specific weight 7·5 kN/m^3 upto a depth of 2·5 m. What is the intensity of pressure on the plate due to the oil ?*

Solution. Given : $w = 7·5$ kN/m^3 and $h = 2·5$ m.

We know that intensity of pressure on the plate,

$$p = wh = 7·5 \times 2·5 = 18·75 \text{ kN/m}^2 = 18·75 \text{ kPa} \quad \text{Ans.}$$

Example 2·3. *Calculate the height of a water column equivalent to a pressure of 0·15 MPa.*

Solution. Given : $p = 0·15$ MPa $= 0·15 \times 10^3$ kN/m^2

Let $h =$ Height of water column in metres.

We know that pressure of water column (p),

$$0·15 \times 10^3 = wh = 9·81 \times h$$

$$\therefore \quad h = (0·15 \times 10^3)/9·81 = 15·3 \text{ m} \quad \text{Ans.}$$

Example 2·4. *What is the height of an oil column of specific gravity 0·9 equivalent to a gauge pressure of 20·3 kPa?*

Solution. Given : Sp. gr. of oil = 0·9 and gauge pressure (p) = 20·3 kPa = 20·3 kN/m^2.

Let $h =$ Height of oil column in metres.

We know that specific weight of oil,

$$w = 0·9 \times 9·81 = 8·829 \text{ kN/m}^3$$

and gauge pressure (p) $\quad 20·3 = wh = 8·829 \times h$

$$\therefore \quad h = 20·3/8·829 = 2·3 \text{ m} \quad \text{Ans.}$$

EXERCISE 2·1

1. Find the pressure at a point 1·6 m below the free surface of water in a swiming pool. [**Ans.** 15·7 kPa]
2. A point is located at a depth of 1·6 m from the free surface of an oil of specific weight 8·0 kN/m^3. Calculate the intensity of pressure at the point. [**Ans.** 12·8 kPa]
3. Find the height of water column corresponding to a pressure of 5·6 kPa. [**Ans.** 0·57 m]
4. Determine the height of an oil column of specific gravity 0·8, which will cause a pressure of 25 kPa. [**Ans.** 3·19 m]
5. Calculate the height of mercury column equivalent to a gauge pressure of 150 kPa. [**Ans.** 1·12 m]

2·3 Pascal's Law

It states, "*The intensity of pressure at any point in a fluid at rest, is the same in all directions.*"

Proof. Consider a very small right-angled triangular element *ABC* of a liquid as shown in Fig. 2·2.

Let $\quad p_X =$ Intensity of horizontal pressure on the element of the liquid,

$p_Y =$ Intensity of vertical pressure on the element of the liquid,

p_z = Intensity of pressure on the diagonal of the tri-angular element of the liquid, and

θ = Angle of the triangular element of the liquid.

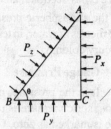

Fig. 2·2.
Element of liquid.

Now pressure on the vertical side AC of the liquid,
$$P_X = p_X \times AC \qquad ...(i)$$

Similarly, pressure on the horizontal side BC of the liquid,
$$P_Y = p_Y \times BC \qquad ...(ii)$$

and pressure on the diagonal AB of the liquid,
$$P_Z = p_Z \times AB \qquad ...(iii)$$

Since the element of the liquid is at rest, therefore sum of the horizontal and vertical components of the liquid pressures must be equal to zero.

Resolving the forces horizontally,
$$P_Z \sin\theta = P_X$$
or $\qquad p_Z . AB \sin\theta = p_X . AC \qquad\qquad ...(\because P_Z = p_Z . AB)$

From the geometry of the figure, we find that
$$AB \sin\theta = AC$$
$\therefore \qquad p_Z . AC = p_X . AC$

or $\qquad p_Z = p_X \qquad ...(iv)$

Now resolving the forces vertically,

i.e., $\qquad P_Z \cos\theta = P_Y - W \qquad$ ...(where W = Weight of the liquid element)

Since we are considering a very small triangular element of the liquid, therefore neglecting weight of the liquid (W), we find that
$$P_Z \cos\theta = P_Y$$
$\therefore \qquad p_Z AB . \cos\theta = p_Y . BC$

From the geometry of the figure, we find that
$$AB \cos\theta = BC$$
$\therefore \qquad p_Z . BC = p_Y . BC$

or $\qquad p_Z = p_Y \qquad ...(v)$

Now from equations (iv) and (v), we find that
$$p_X = p_Y = p_Z$$

Thus the intensity of pressure at any point in a fluid, at rest, is the same in all directions.

2·4 Atmospheric Pressure

It has been established, since long, that the air possesses some weight. Subsequently, it was also thought that the air, due to its weight, must exert some pressure on the surface of the earth. Since the air is compressible, therefore its density is different at different heights. The density of air has also been found to vary from time to time due to the changes in its temperature and humidity. It is thus obvious, that due to these difficulties, the atmospheric pressure (which is due to weight of the atmosphere or air above the surface of the earth) cannot be calculated, as is done in the case of liquids. However, it is measured by the height of the column of liquid that it can support.

It has been observed that at sea level, the pressure exerted by the column of air of 1 square metre cross-sectional area and of height equal to that of the atmosphere is 103 kN. Thus we may say that the atmospheric pressure at the sea level is 103 kN/m² (or 103 kPa). It can also be expressed as 10·3 metres of water, in terms of equivalent water column or 760 mm of mercury in terms of equivalent mercury column.

2·5 Gauge Pressure

It is the pressure, measured with the help of a pressure measuring instrument, in which the atmospheric pressure is taken as datum. Or in other words, the atmospheric pressure on the gauge scale is marked as zero. Generally, this pressure is above the atmospheric pressure.

2·6 Absolute Pressure

It is the pressure equal to the algebraic sum of atmospheric and gauge pressures. It may be noted that if the gauge pressure is minus (as in the case of vacuums or suctions), the absolute pressure will be atmospheric pressure *minus* gauge pressure. *e.g.*, if the absolute pressure at any point is 150 kN/m² and the atmospheric pressure is 103 kN/m², then the gauge pressure at that point will be 150 − 103 = 47 kN/m². A little consideration will show, that if the pressure intensity at a point is more than the local atmospheric pressure, the difference of these two pressures is called the positive gauge pressure. However, if the pressure intensity is less than the local atmospheric pressure, the difference of these two pressures is called the negative gauge pressure or vacuum pressure. Mathematically,

$$p_{absolute} = p_{atmospheric} + p_{gauge}$$

2·7 Measurement of Fluid Pressure

The principles, on which all the pressure measuring devices are based, are almost the same. However, for convenient sake, we may split up the same into the following two types :

1. By balancing the liquid column (whose pressure is to be found out by the same or another column. These are also called tube gauges to measure the pressure.
2. By balancing the liquid column (whose pressure is to be found out) by the spring or dead weight. These are also called mechanical gauges to measure the pressure.

2·8 Tube Gauges to Measure Fluid Pressure

The devices used for measuring the fluid pressure by tube gauges are :
1. Piezometer tube.   2. Manometer.

2·9 Piezometer Tube

A piezometer tube is the simplest form of instrument, used for measuring, moderate pressures. It consists of a tube, one end of which is connected to the pipeline in which the pressure is required to be found out. The other end is open to the atmosphere, in which the liquid can rise freely without overflow. The height, to which the liquid rises up in the tube, gives the pressure head directly.

If the pressure of a liquid flowing in a pipe is to be found out, the piezometer tube is connected to the pipe as shown in Fig. 2·3. While connecting the piezometer to a pipe, care should always be taken that the tube should not project inside the pipe beyond its surface. All burrs and roughness near the hole must be removed, and the edge of the hole should be rounded off.

It may be noted that piezometer tube is meant for measuring gauge pressure only as the surface of the liquid, in the tube, is exposed to the

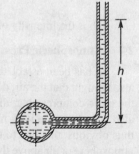

Fig. 2·3. Piezometer tube.

Fluid Pressure and its Measurement

atmosphere. A piezometer tube is also not suitable for measuring negative pressure; as in such a case the air will enter in the pipe through the tube.

2·10 Manometer

Strictly speaking, a manometer is an improved form of a piezometer tube. With the help of a manometer, we can measure comparatively high pressures and negative pressures also. Following are the few types of manometers :

1. Simple manometer,
2. Micromanometer,
3. Differential manometer, and
4. Inverted differential manometer.

2·11 Simple Manometer

A simple manometer is a slightly improved form of a piezometer tube for measuring high as well as negative pressures. A simple manometer, in its simplest form, consists of a tube bent in U-shape, one end of which is attached to the gauge point and the other is open to the atmosphere as shown in Fig. 2·4 (a) and (b).

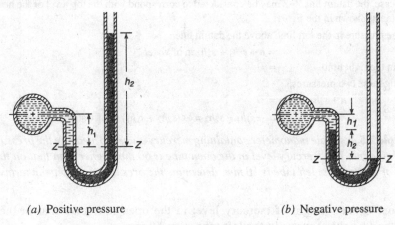

(a) Positive pressure　　　　　　　　　　　(b) Negative pressure

Fig. 2·4. Simple manometer.

The liquid used in the bent tube or simple manometer is, generally, mercury which is 13·6 times heavier than water. Hence it is suitable for measuring high pressures also.

Now consider a simple manometer connected to a pipe containing a light liquid under a high pressure. The high pressure in the pipe will force the heavy liquid, in the left limb of the U-tube, to move downward. This downward movement of the heavy liquid in the left limb will cause a corresponding rise of the heavy liquid in the right limb.

The horizontal surface, at which the heavy and light liquid meet in the left limb, is known as a common surface or datum line. Let Z–Z be the datum line as shown in Fig. 2·4 (a).

Let　　　　　　　　　h_1 = Height of the light liquid in the left limb above the commom surface in metres,

h_2 = Height of the heavy liquid in the right limb above the common surface in metres,

h = Pressure in the pipe, expressed in terms of head of water in metres,

s_1 = Specific gravity of the light liquid, and
s_2 = Specific gravity of the heavy liquid.

It will be interesting to know that the pressures in the left limb and right limb above the datum line are equal. We know that pressure in the left limb above the datum line Z–Z

$$= h + s_1 h_1 \text{ m of water} \qquad ...(i)$$

Similarly, pressure in the right limb above the datum line Z–Z

$$= s_2 h_2 \text{ m of water} \qquad ...(ii)$$

Since the pressure in both the limbs above the Z–Z datum is equal, therefore equating the pressures given by equations (*i*) and (*ii*),

$$h + s_1 h_1 = s_2 h_2$$

or $\qquad h = (s_2 h_2 - s_1 h_1) \text{ m of water}$

Note : If a negative pressure is to be measured by a simple manometer, the same can also be measured easily as discussed below :

In this case, negative pressure in the pipe will suck the light liquid which will pull up the heavy liquid in the left limb of the U-tube. This upward movement of the heavy liquid in the left limb will cause a corresponding fall of the liquid in the right limb as shown in Fig. 2·4 (*b*).

In this case, the datum line Z–Z may be considered to correspond with the top level of the heavy liquid in the right column as shown in the figure.

Now the pressure in the left limb above the datum line

$$= h + s_1 h_1 + s_2 h_2 \text{ m of water}$$

and pressure in the right limb $\qquad = 0$

Equating these two pressures,

$$h + s_1 h_1 + s_2 h_2 = 0$$

or $\qquad h = - s_1 h_1 - s_2 h_2 = - (s_2 h_2 + s_1 h_1) \text{ m of water}$

Example 2·5. *A simple manometer containing mercury is used to measure the pressure of water flowing in a pipeline. The mercury level in the open tube is 60 mm higher than that on the left tube. If the height of water in the left tube is 50 mm, determine the presure in the pipe in terms of head of water.*

Solution. Given : Height of mercury level in the open tube than that in the left tube h_2 = 60 mm and height of water in the left tube h_1 = 50 mm.

Let $\qquad h$ = Pressure in the pipe in terms of head of water

We know that pressure head in the left limb above Z–Z

$$= h + s_1 h_1 = h + (1 \times 50)$$
$$= h + 50 \text{ mm of water} \qquad ...(i)$$

and pressure head in the right limb above Z–Z

$$= s_2 h_2 = 13·6 \times 60$$
$$= 816 \text{ mm of water} \qquad ...(ii)$$

Since the pressure in both the limbs above the Z–Z datum equal, therefore equating (*i*) and (*ii*),

$$h + 50 = 816$$

or $\qquad h = 816 - 50$

$$= 766 \text{ mm of water} \qquad \textbf{Ans.}$$

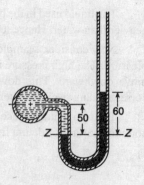

Fig. 2·5.

Fluid Pressure and its Measurement

Example 2·6. *A simple manometer is used to measure the pressure of oil (sp. gravity = 0·8) flowing in a pipeline. Its right limb is open to the atmosphere and the left limb is connected to the pipe. The centre of the pipe is 90 mm below the level of mercury (sp. gravity = 13·6) in the right limb. If difference of mercury levels in the two limbs is 150 mm, find the pressure of oil in the pipe.*

Solution. Given : $s_1 = 0·8$; Height of mercury above the centre of the pipeline = 90 mm; $s_2 = 13·6$ and $h_2 = 150$ mm.

From the geometry of the manometer, we find that height of oil column from the datum Z–Z,
$$h_1 = 150 - 90 = 60 \text{ mm}$$

Let $\quad h =$ Gauge pressure in the pipe line in terms of water head.

We know that pressure head in the left limb above Z–Z
$$= h + s_1 h_1 = h + (0·8 \times 60)$$
$$= h + 48 \text{ mm of water} \quad ...(i)$$

and pressure head in the right limb above Z–Z
$$= s_2 h_2 = 13·6 \times 150$$
$$= 2040 \text{ mm of water} \quad ...(ii)$$

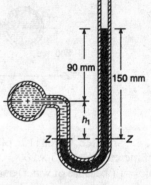

Since the pressure in both the limbs above the Z–Z datum is equal, therefore equating (*i*) and (*ii*),
$$h + 48 = 2040$$
$$\therefore \quad h = 2040 - 48 = 1992 \text{ mm}$$
$$= 1·992 \text{ m of water}$$

Fig. 2·6.

and gauge pressure (or pressure of oil in the pipe),
$$p = wh = 9·81 \times 1·992 = 19·54 \text{ kN/m}^2 = 19·54 \text{ kPa} \quad \textbf{Ans.}$$

Note : The value *h* may also be obtained directly by the relation :
$$h = s_2 h_2 - s_1 h_1 = (13·6 \times 150) - (0·8 \times 60)$$
$$= 1992 \text{ mm} = 1·992 \text{ mm of water}$$

Example 2·7. *A simple manometer containing mercury was used to find the negative pressure in the pipe containing water as shown in Fig. 2·7. The right limb of the manometer was open to atmosphere. Find the negative pressure, below the atmosphere in the pipe, if the manometer readings are given in the figure.*

Solution. Given : $h_2 = 50$ mm; $h_1 = 20$ mm; $s_1 = 1$ (because of water) and $s_2 = 13·6$ (because of mercury).

Let $\quad h =$ Gauge pressure in the pipe in terms of head of water.

We know that pressure head in the left limb above Z–Z
$$= h + s_1 h_1 + s_2 h_2$$
$$= h + (1 \times 20) + (13·6 \times 50)$$
$$= h + 700 \text{ mm} = h + 7 \text{ m of water} \quad ...(i)$$

and pressure head in the right limb above Z–Z
$$= 0$$

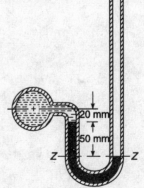

Since the pressure in both the limbs above the Z–Z datum is equal, therefore equating (*i*) and (*ii*),
$$h + 7 = 0 \quad \text{or} \quad h = -7 \text{ m of water}$$

Fig. 2·7.

and gauge pressure in the pipe,

$$p = wh = 9{\cdot}81 \times (-7) = -68{\cdot}67 \text{ kN/m}^2 = -68{\cdot}67 \text{ kPa}$$
$$= 68{\cdot}67 \text{ kPa (Vacuum)} \textbf{ Ans.}$$

Example 2·8. *The pressure of water flowing in a pipeline is measured by a manometer containing U-tubes as shown in Fig. 2·8.*

The measuring fluid is mercury in all the tubes and water is enclosed between the mercury columns. The last tube is open to the atmosphere. Find the pressure of oil in the pipeline.

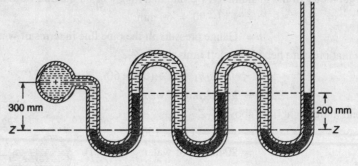

Fig. 2·8.

Solution. Given : No. of tubes = 3; Height of mercury in each column = 200 mm or total height of mercury (h_2) = 3 × 200 = 600 mm; Total height of water (h_1) = 300 + (2 × 200) = 700 mm; s_1 = 1 (because of water) and s_2 = 13·6 (because of mercury).

Let h = Gauge pressure in the pipe in terms of head of water.

We know that pressure head due to water above Z–Z

$$= h + s_1 h_1 = h + (1 \times 700) \text{ mm} = h + 0{\cdot}7 \text{ m of water} \quad ...(i)$$

and pressure head due to mercury above Z–Z

$$= s_2 h_2 = 13{\cdot}6 \times 600 = 8160 \text{ mm} = 8{\cdot}16 \text{ m of water} \quad ...(ii)$$

Since the pressure heads due to water and mercury are equal, therefore equating (*i*) and (*ii*),

$$h + 0{\cdot}7 = 8{\cdot}16 \quad \text{or} \quad h = 8{\cdot}16 - 0{\cdot}7 = 7{\cdot}46 \text{ m of water}$$

and gauge pressure (or pressure of oil in the pipeline),

$$p = wh = 9{\cdot}81 \times 7{\cdot}46 = 73{\cdot}2 \text{ kN/m}^2 = 73{\cdot}2 \text{ kPa} \textbf{ Ans.}$$

Example 2·9. *Fig. 2·9 shows a conical vessel having its outlet at A to which U tube manometer is connected.*

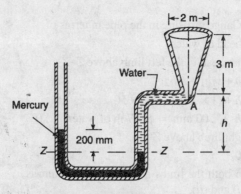

Fig. 2·9.

The reading of the manometer given in the figure shows when the vessel is empty. Find the reading of the manometer when the vessel is completely filled with water.

Solution. Given : Manometer reading when the vessel is empty $(h_2) = 200$ mm $= 0.2$ m; $s_1 = 1$ (because of water) and $s_2 = 13.6$ (because of mercury).

Let $\qquad h = $ Pressure head of oil in terms of head of water.

First of all, let us consider the vessel is to be empty and Z–Z be the datum line. We know that
pressure head in the right limb $= s_1 h_1 = 1 \times h = h$ m of water ...(i)
and pressure head in the left limb $= s_2 h_2 = 13.6 \times 0.2 = 2.72$ m of water ...(ii)

Since the pressure in both the limbs above the Z–Z datum is equal, therefore equating equations, (i) and (ii),

$$h = 2.72 \text{ m}$$

Now consider the vessel to be completely filled with water. As a result of this, let the mercury level go down by x metres in the right limb, and the mercury level go up by the same amount in the left limb. Therefore total height of water in the right limb

$$= x + h + 3 = x + 2.72 + 3 = x + 5.72 \text{ m}$$

and pressure head in the right limb $= 1 (x + 5.72) = x + 5.72$ m of water ...(iii)

We know that manometer reading in this case

$$= 0.2 + 2x \text{ m}$$

and pressure head in the left limb

$$= 13.6 \, (0.2 + 2x) = 2.72 + 27.2 \, x \qquad ...(iv)$$

Again equating the pressures of equations (iii) and (iv),

$$x + 5.72 = 2.72 + 27.2 \, x \quad \text{or} \quad 26.2 \, x = 3.0$$

$\therefore \qquad x = 3/26.2 = 0.115$ m

and manometer reading $\quad = 0.2 + (2 \times 0.115) = 0.43$ m $= 430$ mm **Ans.**

2·12 Micromanometer

It is a modified form of manometer, in which cross-sectional area of one of the limbs (say left limb) is made much larger (about 100 times) than that of the other limb as shown in Fig. 2·10. A micromanometer is used for measuring low pressures, where accuracy is of much importance. Though there are many types of micrometers, yet the following two types are important from the subject point of view :

1. Vertical tube micromanometer, and
2. Inclined tube micromanometer.

1. *Vertical tube micromanometer*

Now consider a vertical tube micromanometer connected to a pipe containing light liquid under a very high pressure. The pressure in the pipe will force the light liquid to push the heavy liquid in the basin downwards. Due to larger area of the basin, the fall of heavy liquid level will be very small. This downward movement of the heavy liquid, in the basin, will cause a considerable rise of the heavy liquid in the right limb. Let us consider our datum line Z–Z corresponding to heavy liquid level before the experiment.

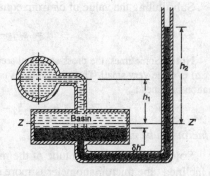

Fig. 2·10.
Vertical micromanometer.

Let
δh = Fall of heavy liquid level in the basin in m,
h_1 = Height of light liquid above the datum line in m
h_2 = Height of heavy liquid (after experiment) in the right limb above the datum line in m,
h = Pressure in the pipe, expressed in terms of head of water in m,
A = Cross-sectional area of the basin in m^2,
a = Cross-sectional area of the tube in m^2,
s_1 = Specific gravity of the light liquid, and
s_2 = Specific gravity of the heavy liquid.

We know that the fall of heavy liquid level, in the basin, will cause a corresponding rise of heavy liquid level.

$\therefore \qquad A.\delta h = ah_2 \qquad \text{or} \qquad \delta h = \dfrac{a}{A} \times h_2 \qquad \ldots(i)$

Now let us take the horizontal surface in the basin, at which the heavy and light liquid meet, as datum line. We also know that the pressures in the left limb and right limb, above the datum line are equal.

$\therefore$ Pressure in the left limb above the datum line Z–Z

$= h + s_1 h_1 + s_1 \delta h \text{ m of water} \qquad \ldots(ii)$

and pressure in the right limb above the datum line

$= s_2 h_2 + s_2 \delta h \text{ m of water} \qquad \ldots(iii)$

Since pressure in both the limbs above Z–Z datum is equal, therefore equating these two pressures,

$h + s_1 h_1 + s_1 \delta h = s_2 h_2 + s_2 \delta h$

or $\qquad h = s_2 h_2 + s_2 \delta h - s_1 h_1 - s_1 \delta h$

$\qquad\quad = s_2 h_2 - s_1 h_1 + \delta h (s_2 - s_1)$

Substituting the value of δh from equation (i),

$h = s_2 h_2 - s_1 h_1 + \dfrac{a}{A} \times h_2 (s_2 - s_1) \qquad \ldots(iv)$

Note : Sometimes, the cross-sectional area of the basin (i.e., A) is made very large and that of the tube (i.e., a) is made very small. Then the ratio a/A is extremely very small, and thus is neglected. Then the above equation becomes :

$h = s_2 h_2 - s_1 h_1 \qquad \ldots(i)$

2. Inclined tube micromanometer

Sometimes, the vertical tube of the micromanometer is made inclined as shown in Fig. 2·11. An inclined tube micromanometer is more sensitive than the vertical tube type. Due to inclination, the distance moved by the heavy liquid, in the narrow tube, will be comparatively more, and thus it gives a higher reading for the given pressure.

From the geometry of figure, we find

$\dfrac{h_2}{l} = \sin \alpha \qquad \text{or} \qquad h_2 = l \sin \alpha$

Fluid Pressure and its Measurement

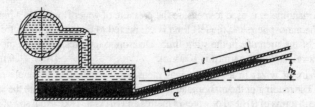

Fig. 2·11. Inclined tube micromanometer.

By substituting the value of h_2 in the micromanometer equation, we can find out the required pressure in the pipe.

Example 2·10. *In order to determine the pressure in a pipe, containing liquid of specific gravity 0·8, a micromanometer was used as shown in Fig. 2·12.*

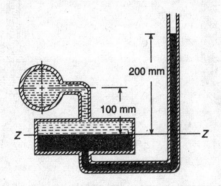

Fig. 2·12.

The ratio of area of the basin to that of the limb is 50. Find the intensity of pressure in the pipe for the manometer reading as shown in the figure.

Solution. Given: $s_1 = 0·8$; $A/a = 50$; $h_1 = 100$ mm; $h_2 = 200$ mm and $s_2 = 13·6$ (because of mercury).

We know that pressure head in the left limb above Z–Z

$$= h + s_1 h_1 + \frac{a}{A} \times s_1 h_2 = h + (0·8 \times 100) + \frac{1}{50} \times 0·8 \times 200$$

$$= h + 80 + 3·2 = h + 83·2 \text{ mm of water} \qquad \ldots(i)$$

and pressure head in the right limb above Z–Z

$$= s_2 h_2 + \frac{a}{A} \times s_2 h_2 = 13·6 \times 200 + \frac{1}{50} \times 13·6 \times 200$$

$$= 2720 + 54·4 = 2774·4 \text{ mm of water} \qquad \ldots(ii)$$

Since pressure in both the limbs above the Z–Z datum is equal, therefore equating (i) and (ii),

$$h + 83·2 = 2774·4$$

or
$$h = 2774·4 - 83·2 = 2691·2 \text{ mm} = 2·6912 \text{ m of water}$$

and intensity of pressure in the pipe

$$p = wh = 9·81 \times 2·6912 = 26·4 \text{ kN/m}^2 = 26·4 \text{ kPa} \quad \textbf{Ans.}$$

EXERCISE 2·2

1. A simple manometer is used to measure the pressure of water flowing in a pipeline. Its right limb is open to the atmosphere and the left limb is connected to the pipe. The centre of the pipe is in level with that of the mercury in the right limb. Determine the pressue in the pipe, if the difference of mercury levels in the two limbs is 100 mm. **[Ans.** 1·26 m of water; 12·36 kPa**]**

2. The pressure of water in a pipeline was measured by means of a simple manometer containing mercury. The reading of the manometer is shown in Fig. 2·13. Determine the static pressure of water in the pipe in terms of (i) head of water in metres and (ii) kPa. Take usual specific gravities of mercury and water. **[Ans.** 1·05 m; 10·3 kPa**]**

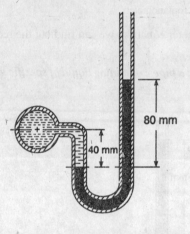

Fig. 2·13.

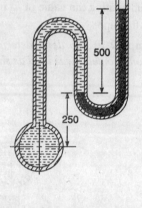

Fig. 2·14.

3. A U-tube containing mercury is used to measure the pressure of an oil (of specific gravity 0·8) as shown in Fig. 2·14. Calculate the pressure of the oil, if the difference of mercury level be 50 cm.

[Ans. 7·0 m of water**]**

[**Hint :** Pressure in the left limb = $h - (0·8 \times 250) = h - 200$ mm of water ...(i)
Pressure in the right limb = $s_2 h_2 = 13·6 \times 500 = 6800$ mm of water ...(ii)
Equating the equations (i) and (ii),
$$h = 200 = 6800 \quad \text{or} \quad h = 6800 + 200 = 7000 \text{ mm} = 7 \text{ m}$$

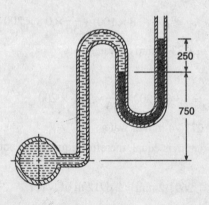

Fig. 2·15.

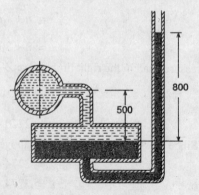

Fig. 2·16.

4. Fig. 2·15 shows a manometer connected to a pipeline, containing an oil of specific gravity 0·8. Find the pressure of oil in the pipe. [**Ans. 39·4 kPa**]

5. A micromanometer, having ratio of basin to limb areas as 40, was used to determine the pressure in a pipe containing water. Determine the pressure in the pipe for the manometer reading shown in Fig. 2·16. [**Ans. 104 kPa**]

2·13 Differential Manometer

It is a device used for measuring the difference of pressures, between two points in a pipe, or in two different pipes. A differential manometer, in its simplest form, consists of a U-tube containing a heavy liquid, whose two ends are connected to the points, whose difference of pressures is required to be found out.

Now consider a differential manometer whose two ends are connected with two different points A and B as shown in Fig. 2·17 (a) and (b). Let us assume that the pressure at point A is more than that at point B. A little consideration will show, that the greater pressure at A will force the heavy liquid in the U-tube to move downwards. This downward movement of the heavy liquid, in the left limb, will cause a corresponding rise of the heavy liquid in the right limb as shown in Fig. 2·17 (a).

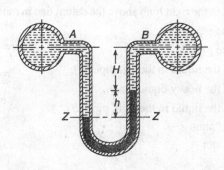

(a) A and B at the same level and containing same liquid.

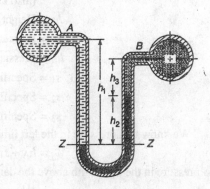

(b) A and B at different levels and containing different liquids.

Fig. 2·17.

Let us take the horizontal surface Z–Z, at which the heavy liquid and light liquid meet in the left limb, as the datum line.

Let
h = Difference of the levels of the heavy liquid in the right and left limb (also known as the reading of the differential manometer) in mm,

h_A = Pressure head in pipe A,

h_B = Pressure head in pipe B

s_1 = Specific gravity of the light liquid in the pipes, and

s_2 = Specific gravity of the heavy liquid.

We know that the pressure head in the left limb above Z–Z

$$= h_A + s_1(H+h) = h_A + s_1 H + s_1 h \text{ m of water} \qquad ...(i)$$

and pressure head in the right limb above Z–Z

$$= h_B + s_1 h_1 + s_2 h \text{ m of water} \qquad ...(ii)$$

Since the pressure heads in both the limbs above Z–Z datum are equal, therefore equating the equations (i) and (ii),

$$h_A + s_1 H + s_1 h = h_B + s_1 H + s_2 h$$
$$\therefore \qquad h_A - h_B = s_2 h - s_1 h = h(s_2 - s_1)$$

Sometimes, the two pipes or the two points, whose difference of pressures is required to be found out are not at the same level. And at the same time, the liquids flowing in the two pipes are different. In such a case, the same principle is applied to obtain the difference of pressure heads.

Now consider a differential manometer, whose two ends are connected to two different pipes A and B and containing different liquids at different levels as shown in Fig. 2·17 (b). Let us assume that the pressure at the point A is more than that at the point B. A little consideration will show that the greater pressure at A will force the heavy liquid to move downwards. This downward movement of the liquid, in the left limb, will cause a corresponding rise of the heavy liquid in the right limb as shown in Fig. 2·17 (b).

The horizontal surface at which the heavy and light liquids meet in the left limb is taken as a datum line. In this case, let Z–Z be the datum line, as shown in Fig. 2·17 (b).

Let
h_1 = Height of liquid in the left limb above the datum line in mm,

h_2 = Difference of levels of the heavy liquid in the right and left limb (also known as reading of the differential manometer) in mm,

h_3 = Height of the liquid in the right limb above the datum line in mm,

h_A = Pressure head in the pipe A,

h_B = Pressure head in the pipe B,

s_1 = Specific gravity of the liquid in the left pipe (A),

s_2 = Specific gravity of the heavy liquid, and

s_3 = Specific gravity of the liquid in the right pipe (B).

We know that pressure in the left limb above the datum Z–Z
$$= h_A + s_1 h_1 \quad \text{m of water} \qquad \qquad ...(i)$$
and pressure in the right limb above the datum Z–Z
$$= s_2 h_2 + s_3 h_3 + h_B \quad \text{m of water} \qquad \qquad ...(ii)$$

Since pressure in both the limbs above the Z–Z datum is equal, therefore equating these two pressures,
$$h_A + s_1 h_1 = s_2 h_2 + s_3 h_3 + h_B$$

From the above equation, the value of h_A, h_B or their difference may be found out.

Note : If a differential manometer has got readings other than those assumed in the above equation (*i.e.*, pressure in pipe A is more than that in B), then it is advisable to start the problem from the fundamentals.

Example 2·10. *A differential manometer connected at the two points A and B at the same level in a pipe containing an oil of specific gravity 0·8, shows a difference in mercury levels as 100 mm. Determine the difference in pressures at the two points.*

Solution. Given : $s_1 = 0·8$; $h = 100$ mm and $s_2 = 13·6$ (because of mercury)

We know that pressure head in the left limb above Z–Z
$$= h_A + s_1 (H + 100) \text{ mm of water}$$
$$= h_A + s_1 H + 100 s_1 \text{ mm of water} \quad ...(i)$$
and pressure in the right limb above Z–Z
$$= h_B + s_1 H + s_2 \times 100 \text{ mm of water}$$

Since pressure heads in both the limbs above the Z–Z datum are equal, therefore equating (i) and (ii),

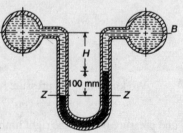

Fig. 2·18.

Fluid Pressure and its Measurement

$$h_A + s_1 H + 100 s_1 = h_B + s_1 H + 100 s_2$$

$$\therefore \quad h_A - h_B = 100 s_2 - 100 s_1 = 100(s_2 - s_1) = 100(13 \cdot 6 - 0 \cdot 8)$$

$$= 100 \times 12 \cdot 8 = 1280 \text{ mm} = 1 \cdot 28 \text{ m of water}$$

and difference of pressures,

$$p = w(h_A - h_B) = 9 \cdot 81 \times 1 \cdot 28 = 12 \cdot 56 \text{ kN/m}^3 = 12 \cdot 56 \text{ kPa} \quad \textbf{Ans.}$$

Example 2·11. *A manometer containing mercury is connected to two points 15 m apart, on a pipeline conveying water. The pipeline is straight and slopes at an angle of 15° with the horizontal. The manometer gives a reading of 150 mm. Determine the pressure difference between the two points of the pipeline. Take specific gravity of mercury as 13·6 and that of water as 1·0.*

Solution. Given : Pipe length $(l) = 15$ m; Inclination of pipe, $\alpha = 15°$; $h = 150$ mm $= 0 \cdot 15$ m; $s_2 = 13 \cdot 6$ and $s_1 = 1 \cdot 0$.

Let $\quad x = $ Height of water in the right limb.

We know that difference between the points A and B due to inclination,

$$AC = 15 \sin 15° = 15 \times 0 \cdot 2588 \text{ m}$$
$$= 3 \cdot 882 \text{ m}$$

We also know that pressure head in the left limb above Z–Z datum

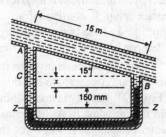

Fig. 2·19.

$$= h_A + 0 \cdot 15 \times s_1 + x \cdot s_1 + 3 \cdot 882 s_1$$
$$= h_A + (0 \cdot 15 \times 1) + (x \times 1) + 3 \cdot 882 \times 1$$
$$= h_A + x + 4 \cdot 032 \text{ m of water} \quad \ldots(i)$$

and pressure head in the right limb above Z–Z datum

$$= h_B + 0 \cdot 15 s_2 + x \cdot s_1$$
$$= h_B + (0 \cdot 15 \times 13 \cdot 6) + x_1 \times 1$$
$$= h_B + 2 \cdot 04 + x \text{ m of water} \quad \ldots(ii)$$

Since the pressure heads in both the limbs are equal, therefore equating (*i*) and (*ii*),

$$h_A + x + 4 \cdot 032 = h_B + 2 \cdot 04 + x$$

$$\therefore \quad h_B - h_A = 4 \cdot 032 - 2 \cdot 04 = 1 \cdot 992 \text{ m of water}$$

and difference of pressures,

$$p = w(h_B - h_A) = 9 \cdot 81 \times 1 \cdot 992 = 19 \cdot 54 \text{ kN/m}^2 = 19 \cdot 54 \text{ kPa} \quad \textbf{Ans.}$$

Example 2·12. *A U-tube differential manometer connects two pressure pipes A and B. The pipe A contains carbon tetrachloride having a specific gravity 1·6 under a pressure of 120 kPa. The pipe B contains oil of specific gravity 0·8 under a pressure of 200 kPa. The pipe A lies 2·5 m above pipe B. Find the difference of pressures measured by mercury as fluid filling U-tube.*

Solution. Given : $s_A = 1 \cdot 6$; $p_A = 120$ kPa; $s_B = 0 \cdot 8$; $p_B = 200$ kPa; $h_A = 2 \cdot 5$ m and $s = 13 \cdot 6$ (because of mercury).

Let $\quad h = $ Difference of pressure measured by mercury in terms of head of water.

We know that pressure head in pipe A,

$$\frac{p_A}{w} = \frac{120}{9 \cdot 81} = 12 \cdot 2 \text{ m of water}$$

and pressure head in pipe B, $\dfrac{p_B}{w} = \dfrac{200}{9 \cdot 81} = 20 \cdot 4$ m of water

We also know that pressure head in pipe A above Z–Z

$$= 12 \cdot 2 + (s_A \cdot h_A) + s \cdot h$$
$$= 12 \cdot 2 + (1 \cdot 6 \times 2 \cdot 5) + 13 \cdot 6 \times h$$
$$= 16 \cdot 2 + 13 \cdot 6\, h \qquad \ldots(i)$$

and pressure head in pipe B above Z–Z

$$= 20 \cdot 4 + s_B\, h = 20 \cdot 4 + (0 \cdot 8 \times h) \qquad \ldots(ii)$$

Since the pressure heads in both the pipes above the Z–Z datum are equal, therefore equating (i) and (ii)

$$16 \cdot 2 + 13 \cdot 6\, h = 20 \cdot 4 + 0 \cdot 8\, h \quad \text{or} \quad 12 \cdot 8\, h = 4 \cdot 2$$

$$\therefore \quad h = 4 \cdot 2/12 \cdot 8 = 0 \cdot 328 \text{ m} = 328 \text{ mm} \quad \textbf{Ans.}$$

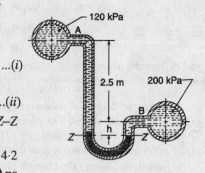

Fig. 2·20.

2·14 Inverted Differential Manometer

It is a particular type of differential manometer, in which an inverted U-tube is used. An inverted differential manometer is used for measuring difference of low pressures, where accuracy is the prime consideration. It consists of an inverted U-tube, containing a light liquid whose two ends are connected to the points whose difference of pressures is to be found out.

Now consider an inverted differential manometer, whose two ends are connected to two different points A and B as shown in Fig. 2·21. Let us assume that the pressure at point A is more than that at point B. A little consideration will show, that the greater pressure at A will force the light liquid in the inverted U-tube to move upwards. This upward movement of liquid in the left limb will cause a corresponding fall of the light liquid in the right limb as shown in Fig. 2·21. Let us take Z–Z as the datum line in this case.

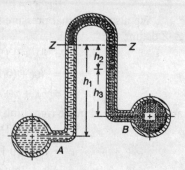

Fig. 2·21. Inverted differential manometer.

Let h_1 = Height of liquid in the left limb below the datum line in mm,

h_2 = Difference of levels of the light liquid in the right and left limbs (also known as manometer reading) in mm,

h_3 = Height of liquid in the right limb below the datum line in mm,

h_A = Pressure in the pipe A, expressed in terms of head of the liquid in mm,

h_B = Pressure in the pipe B, expressed in terms of head of the liquid in mm,

s_1 = Specific gravity of the liquid in the left limb,

s_2 = Specific gravity of the light liquid, and

s_3 = Specific gravity of the liquid in the right limb.

We know that pressure head in the left limb below the datum line

$$= h_A - s_1 h_1 \qquad \ldots(i)$$

and pressure head in the right limb below the datum line

$$= h_B - s_2 h_2 - s_3 h_3 \qquad \ldots(ii)$$

Since pressure heads in both the limbs below the Z–Z detum are equal, therefore equating (i) and (ii)

$$h_A - s_1 h_1 = h_B - s_2 h_2 - s_3 h_3$$

From the above equation, the value of h_A, h_B or their difference may be found out as usual.

Example 2·13. *An inverted differential manometer having an oil of specific gravity 0·75 was connected to two different pipes carrying water under pressure as shown in Fig. 2·22.*

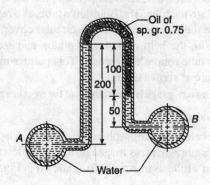

Fig. 2·22.

Determine the pressure in the pipe B in terms of kPa, if the manometer reads as shown in the figure. Take pressure in the pipe A as 1·5 metre of water.

Solution. Given : h_1 = 200 mm = 0·2 m; h_2 = 100 mm = 0·1 m; s_2 = 0·75, h_3 = 50 mm = 0·05 m ; h_A = 1·5 m and s_1 = s_3 (because of water).

Let h_B = Pressure in pipe B in terms of head of water.

We know that pressure head in the left limb below Z–Z

$$= h_A - s_1 h_1 = 1\cdot5 - (1 \times 0\cdot2) = 1\cdot3 \text{ m of water} \qquad ...(i)$$

and pressure head in the right limb below Z–Z

$$= h_B - s_2 h_2 - s_3 h_3 = h_B - (0\cdot75 \times 0\cdot1) - (1 \times 0\cdot05)$$

$$= h_B - 0\cdot125 \text{ m of water} \qquad ...(ii)$$

Since the pressures in both the limbs below Z–Z are equal, therefore equating (*i*) and (*ii*),

$$1\cdot3 = h_B - 0\cdot125$$

or $\qquad h_B = 1\cdot3 + 0\cdot125 = 1\cdot425 \text{ m of water}$

and pressure in the pipe B,

$$p_B = w.h_B = 9\cdot81 \times 1\cdot425 = 13\cdot98 \text{ kN/m}^2 = 13\cdot98 \text{ kPa} \quad \textbf{Ans.}$$

2·15 Mechanical Gauges

Whenever a very high fluid pressure is to be measured, a mechanical gauge is best suited for the purpose. A mechanical gauge is also used for the measurement of pressure in boilers or other pipes, where tube gauges cannot be conveniently used.

There are many types of gauges available in the market. But the principle, on which all these gauges work, is almost the same. Following three types of gauges are important from the subject point of view :

1. Bourdon's tube pressure gauge,
2. Diaphragm pressure gauge, and
3. Dead weight pressure gauge.

2·16 Bourdon's Tube Pressure Gauge

The pressure, above or below the atmospheric pressure, may be easily measured with the help of a Bourdon's tube pressure gauge. A Bourdon's tube pressure gauge, in its simplest form, consists of an elliptical tube *ABC* ; bent into an arc of a circle as shown in Fig. 2·23. This bent-up tube is called Bourdon's tube.

When the gauge tube is connected to the fluid (whose pressure is required to be found out) at *C*, the fluid under pressure flows into the tube. The Bourdon's tube, as a result of the increased pressure, tends to straighten itself. Since the tube is encased in a circular cover, therefore it tends to become circular instead of straight. With the help of a simple pinion and sector arrangement, the elastic deformation of the Bourdon's tube rotates the pointer. This pointer moves over a calibrated scale, which directly gives the pressure as shown in Fig. 2·23.

Note : A Bourdon's tube pressure gauge is generally used for measuring high pressures.

2·17 Diaphgram Pressure Gauge

The Pressure, above or below the atmospheric pressure, is also found out with the help of diaphragm pressure gauge. A diaphragm pressure gauge, in its simplest form, consists of a corrugated diaphragm (instead of Bourdon's tube as in Art. 2·16) as shown in Fig. 2·24.

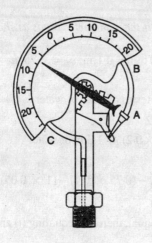

Fig. 2·23. Bourdon's tube pressure gauge.

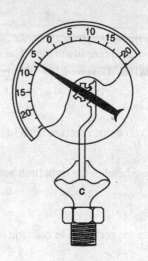

Fig. 2·24. Diaphragm pressure gauge.

When the gauge is connected to the fluid (whose pressure is required to be found out) at *C*, the fluid under pressure causes some deformation of the diaphragm. With the help of some pinion arrangement, the elastic deformation of the diaphragm rotates the pointer. This pointer moves over a calibrated scale, which directly gives the pressure as shown in Fig. 2·24.

A diaphragm pressure gauge is, generally, used to measure relatively low pressures.

2·18 Dead Weight Pressure Gauge

It is the most accurate pressure gauge, which is generally used for the calibration of the other pressure gauges in a laboratory. A dead weight pressure gauge, in its simplest form, consists of a piston and a cylinder of known area, which is connected to a fluid through a tube as shown in Fig. 2·25.

The pressure on the fluid, in the pipe, is calculated from the relation,

$$p = \frac{\text{Weight}}{\text{Area of the piston}}$$

A pressure gauge, to be calibrated, is fitted on the other end of the tube as shown in Fig. 2·25. By changing the weight, on the piston, the pressure on the fluid is calculated and marked on the gauge at the respective points, indicated by the pointer. A small error due to frictional resistance to the motion of the piston may come into play. But the same may be avoided by taking adequate precautions.

Fig. 2·25. Dead weight pressure gauge.

Example 2·14. *A closed tank fitted with a gauge and a manometer contains water as shown in Fig. 2·26.*

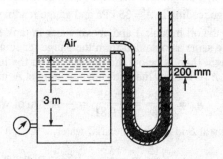

Fig. 2·26.

Find the gauge reading if the manometer, containing mercury, shows a reading of 200 mm.

Solution. Given : Manometer reading = 200 mm = 0·2 m.

Since the space above the water in the tank is full of air, therefore pressure of the air will be the same as that obtained from the manometer reading. Moreover, as the heavy liquid level in the right limb is below the liquid level in the left limb, therefore there is a negative pressure of the air.

We know that pressure of air in terms of head of water
$$= -0.2 \times 13.6 = -2.72 \text{ m}$$

Minus sign has been used for negative pressure in the tank. It has been done as the surface of heavy liquid in the right limb is lower than that in the left limb.

Now the gauge reading will be the pressure of air plus pressure due to water under a head of 3 m. Therefore gauge reading in terms of head of water,
$$h = -2.72 + 3.0 = 0.28 \text{ m}$$
and gauge reading,
$$= wh = 9.81 \times 0.28 = 2.75 \text{ kN/m}^2 = 2.75 \text{ kPa} \quad \textbf{Ans.}$$

Example 2·15. *A closed vessel is divided into two compartments. These compartments contain oil and water as shown in Fig. 2·27.*

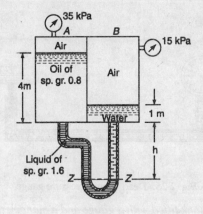

Fig. 2·27.

Determine the value of h, if the gauges show the readings as shown in the figure.

Solution. Given : Gauge reading in A = 35 kPa and gauge reading in B = 15 kPa.

Since the space above the oil in tank A and above water in tank B is full of air, therefore the pressure of the air will be the same as obtained from the gauge readings. Moreover, as the level of liquid of sp. gr. 1·6 in the right limb is below the liquid level in the left limb, therefore there is a negative pressure in the tank A. We know that the gauge reading at A in terms iof head of water,

$$h_A = -\frac{35}{w} = -\frac{35}{9\cdot 81} = -3\cdot 57 \text{ m of water}$$

Similarly, gauge reading at B in terms of head of water,

$$h_B = +\frac{15}{w} = \frac{15}{9\cdot 81} = 1\cdot 53 \text{ m of water}$$

From the geometry of the figure, we find that the pressure in the left limb above the datum line Z–Z

$$= -3\cdot 57 + (4 \times 0\cdot 8) + h \times 1\cdot 6 \text{ m of water}$$
$$= 1\cdot 6h - 0\cdot 37 \text{ m of water} \qquad \ldots(i)$$

and pressure in the right limb above the datum line Z–Z

$$= 1\cdot 53 + (1 \times 1) + h \times 1 \text{ m of water}$$
$$= h + 2\cdot 53 \text{ m of water} \qquad \ldots(ii)$$

Since pressures in both the limbs above the Z–Z datum are equal, therefore equating (*i*) and (*ii*),

$$1\cdot 6h - 0\cdot 37 = h + 2\cdot 53 \quad \text{or} \quad 0\cdot 6h = 2\cdot 53 + 0\cdot 37 = 2\cdot 9$$

$$\therefore \quad h = 2\cdot 9/0\cdot 6 = 4\cdot 83 \text{ m} \quad \textbf{Ans.}$$

Example 2·16. *A container with fluids, vacuum gauge, piezometers and manometer as shown in Fig. 2·8 is to be used in an experiment.*

Find the elevation of liquids in columns E and F and deflection of mercury in U-tube. Gauge reading in A = – 20·6 kPa.

Solution. Given : Gauge reading = – 20·6 kPa.

Elevation of liquid in column E

We know that pressure head due to gauge reading

$$= -\frac{20\cdot 6}{9\cdot 81} = -2\cdot 1 = 2\cdot 1 \text{ m (vacuum) of water}$$

Since the space above the elevation 20·7 m is full of air, therefore the pressure at elevation 20·7 will be the same as that of the gauge. The pressure in the column E will be the same as the bottom of the oil. *i.e.*, at elevation 17·4 m.

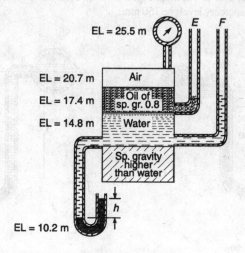

Fig. 2·28.

∴ Pressure at elevation 17·4 m

= Pressure at elevation 20·7 m
+ Weight of oil between 20·7 m and 17·4 m
= − 2·1 + (20·7 − 17·4) × 0·7 = 0·21 m of water
= 0·21 / 0·7 = 0·3 m of oil

and elevation of oil in column E = 17·4 + 0·3 = 17·7 m **Ans.**

Elevation of liquid in column F

From the geometry of the figure, we find that the pressure at elevation 14·8 m

= Weight of water between 17·4 m and 14·8 m
+ Pressure at elevation 17·4
= 0·21 + (17·4 − 14·8) = 2·81 m of water

∴ Elevation of water in column F

= 14·8 + 2·81 = 17·61 m **Ans.**

Deflection of mercury in U-tube

Let h = Deflection of mercury in U-tube.

Now let us consider the datum line at elevation 10·2 m. We know that the pressure head in the left limb above the datum line

= Pressure at elevation 14·8 m
+ Weight of water between 14·8 m and 10·2 m
= 2·81 + (14·8 − 10·2) = 7·41 m of water   ...(i)

and pressure head in the right limb above the common surface

= h m of mercury = $h \times 13·6$ m of water   ...(ii)

Since pressure heads in both the limbs above the elevation 10·2 m are equal, therefore equating (i) and (ii),

7·41 = 13·6h or h = 7·41/13·6 = 0·545 m = 545 mm **Ans.**

EXERCISE 2·3

1. A differential manometer was connected with two points at the same level in a pipe containing liquid of specific gravity 0·85 as shown in Fig. 2·29. Find the difference of pressures at the two points, if the difference of mercury levels be 150 mm. [**Ans.** 18·5 kPa]

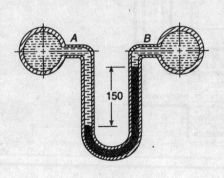

Fig. 2·29.   Fig. 2·30

2. A differential manometer containing mercury was used to measure the difference of pressures in two pipes containing water as shown in Fig. 2·30. Find the difference of pressures in the pipes, if the manometer reading is 0·8 m. [**Ans.** 9·88 m]

3. A differential manometer is connected to two pipes as shown in Fig. 2·31. The pipe A is containing water and the pipe B is containing an oil of specific gravity 0·8. Find the difference of mercury levels, the pressure difference in the two pipes be 80 kPa. [**Ans.** 426 mm]

[**Hint :** Pressure in the pipe B is more than that in pipe A.]

4. An inverted differential manometer containing an oil of sp. gr. 0·8 is connected to find the difference

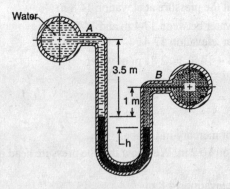 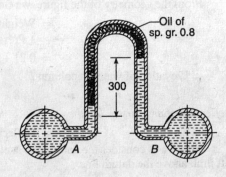

Fig. 2·31.   Fig. 2·32.

of pressures at two points of a pipe containing water as shown in Fig. 2·32. Find the difference of pressures, if the manometer reading be 300 mm. [**Ans.** 60 mm of water]

5. With the manometer reading as shown in Fig. 2·33, calculate the difference of pressures in the two tubes A and B containing water. [**Ans.** 9·07 kPa]

Fluid Pressure and its Measurement

6. An inverted differential manometer, when connected to two pipes A and B, gives the readings as shown in Fig. 2·34. Determine the pressure in the tube B, if the pressure in the pipe A be 50 kPa.

[**Ans.** 47·74 kPa]

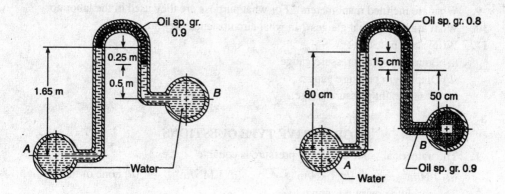

Fig. 2·33.   Fig. 2·34.

7. The compartments of the two tanks are closed and filled as shown in Fig. 2·35. Find the value of h, if the pressure in the left hand tank air is 0·2 m of mercury.   [**Ans.** 5·2 m]

[**Hint :** Pressure in the left hand tank above Z–Z datum

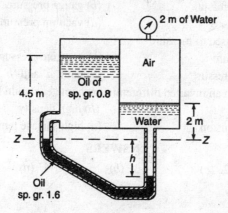

Fig. 2·35.

$$= (-0.2 \times 13.6) + (0.8 \times 4.5) + 1.6 \times h \text{ m of water} \quad \ldots(i)$$

and pressure in the right hand tank

$$= 2 + 2 + 1 \times h = h + 4 \text{ m of water} \quad \ldots(ii)$$

Equating equations (i) and (ii),

$$h = 5.2 \text{ m}$$

QUESTIONS

1. What do you understand by the term intensity of pressure ? State its units.
2. What is meant by pressure head ? Derive an expression for it.
3. State and prove Pascal's Law.
4. Distinguish between gauge pressure and absolute pressure.
5. State the different principles of measurement of pressure.

6. Distinguish between piezometer and pressure gauge. When and where are they used?
7. Describe the different types of manometers.
8. Distinguish between a simple manometer and a differential manometer.
9. What are inclined manometers? For what purpose are they used in the laboratory?
10. What are mechanical gauges? In what circumstance they are used?
11. Write short notes on :
    (a) Bourdon's tube pressure gauge.
    (b) Diaphragm pressure gauge.
    (c) Dead weight pressure gauge.

OBJECTIVE TYPE QUESTIONS

1. The numerical value of 1 Pa of pressure is equal to
   (a) 1 N/m^2 (b) 1 kN/m^2 (c) 1 MN/m^2 (d) none of these
2. The absolute pressure is equal to
   (a) Gauge pressure – Atmospheric pressure
   (b) Gauge pressure + Vacuum pressure
   (c) Atmospheric pressure + Gauge pressure
   (d) Atmospheric pressure – Gauge pressure
3. The pressure measured with the help of a piezometer tube is
   (a) atmospheric pressure   (b) gauge pressure
   (c) absolute pressure   (d) vacuum pressure
4. A manometer is used to measure
   (a) positive pressure   (b) negative pressure
   (c) atomspheric pressure   (d) both 'a' and 'b'
5. The liquid used in an inverted differential manometer should be of
   (a) low density   (b) high density
   (c) low surface tension   (d) high surface tension

ANSWERS

1. (a)   2. (c)   3. (b)   4. (d)   5. (a)

3

Hydrostatics

1. Introduction. 2. Total Pressure. 3. Total Pressure on an Immersed Surface. 4. Total Pressure on a Horizontally Immersed Surface. 5. Total Pressure on a Vertically Immersed Surface. 6. Total Pressure on an Inclined Immersed Surface. 7. Centre of Pressure. 8. Centre of Pressure of a Vertically Immersed Surface. 9. Centre of Pressure of an Inclined Immersed Surface. 10. Centre of Pressure of a Composite Section. 11. Pressure on a Curved Surface.

3·1 Introduction

We have discussed in the last chapter that a liquid, at rest, exerts some pressure on all sides of the container. We have been solving numerical problems to know the intensity of such pressures. In this chapter, we shall discuss the total pressure on a surface and its position (*i.e.*, the point, where the total pressure acts). The term hydrostatics means the study of pressure, exerted by a liquid at rest. It has been observed that the direction of such a pressure is always at right angles to the surface, on which it acts. In this chapter only the study of water will be dealt with, unless specified, otherwise.

3·2 Total Pressure

The total pressure, on an immersed surface, may be defined as the total pressure exerted by the liquid on it. Mathematically total pressure,

$$P = p_1 a_1 + p_2 a_2 + p_3 a_3 \ldots$$

where $p_1, p_2, p_3 \ldots$ = Intensities of pressure on different strips of the surface, and

$a_1, a_2, a_3, \ldots$ = Areas of the corresponding strips.

3·3 Total Pressure on an Immersed Surface

We see that whenever we dive in a swimming pool, we feel some uneasiness. As we dive deeper and deeper, we feel more and more uneasiness. This uneasiness is, in fact, due to the total weight (or in other words, total pressure) of water above us. Now we shall discuss the total pressure exerted by a liquid on an immersed surface. As a matter of fact, the position of an immersed surface may be :

1. horizontal, 2. vertical, and 3. inclined.

3·4 Total Pressure on a Horizontally Immersed Surface

Consider a plane horizontal surface immersed in a liquid as shown in Fig. 3·1.

Let w = Specific weight of the liquid,

A = Area of the immersed surface in m^2,

$\bar{x}$ = Depth of the horizontal surface from the liquid level in metres.

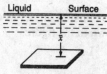

Fig. 3·1. Horizontally

We know that the total pressure on the surface,

$$P = \text{Weight of the liquid above the immersed surface}$$
$$= \text{Sp. wt. of liquid} \times \text{Volume of liquid}$$
$$= \text{Sp. wt. of liquid} \times \text{Area of Surface} \times \text{Depth of liquid}$$
$$= wA\bar{x} \text{ kN}$$

Where w is the specific weight of the liquid in kN/m^3.

Note : If the given liquid is water, then its specific weight taken as 9.81 kN/m^3.

Example 3·1. *A rectangular tank 4 metres long 2 metres wide contains water up to a depth of 2·5 metres. Calculate the pressure on the base of the tank.*

Solution. Given : $l = 4$ m ; $b = 2$ m and $\bar{x} = 2.5$ m

We know that base area of the tank,

$$A = l \times b = 4 \times 2 = 8 \text{ m}^2$$

and pressure on the base of the tank,

$$P = wA\bar{x} = 9.81 \times 8 \times 2.5 = 196.2 \text{ kN} \quad \textbf{Ans.}$$

Example 3·2. *A tank $3 m \times 4 m$ contains 1·2 m deep oil of specific gravity 0·8. Find (i) intensity of pressure at the base of the tank, and (ii) total pressure on the base of the tank.*

Solution. Given : Size of tank $(A) = 3$ m $\times 4$ m $= 12$ m^2; Depth of oil $(\bar{x}) = 1.2$ m and specific gravity of oil $= 0.8$ or specific weight of oil $(w) = 9.81 \times 0.8 = 7.85$ kN/m^3.

(*i*) *Intensity of pressure at the base of the tank*

We know that intensity of pressure at the base of the tank,

$$p = wh = 7.85 \times 1.2 = 9.42 \text{ kN/m}^2 = 9.42 \text{ kPa} \quad \textbf{Ans.}$$

(*ii*) *Total pressure on the base of the tank*

We also know that total pressure on the base of the tank,

$$p = wA\bar{x} = 7.85 \times 12 \times 1.2 = 113.4 \text{ kN} \quad \textbf{Ans.}$$

3·5 Total Pressure on a Vertically Immersed Surface

Consider a plane vertical surface immersed in a liquid as shown in Fig. 3·2.

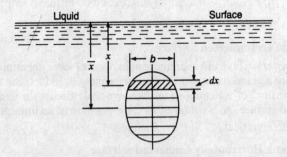

Fig. 3·2. Vertically immersed surface.

First of all, let us divide the whole immersed surface into a number of small parallel strips as shown in the figure.

Let
  w = Specific weight of the liquid,
  A = Total area of the immersed surface, and
  $\bar{x}$ = Depth of centre of gravity of the immersed surface from the liquid surface.

Hydrostatics

Let us consider a strip of thickness dx, width b and at a depth x from the free surface of the liquid as shown in Fig. 3·2.

We know that intensity of pressure on the strip $= wx$

and area of strip $= b.dx$

∴ Pressure on the strip,

$$p = \text{Intensity of pressure} \times \text{Area} = wx \cdot bdx$$

Now *total pressure on the surface,

$$P = \int wx \cdot bdx$$

$$= w \int x \cdot bdx$$

But $\int x \cdot bdx$ = Moment of the surface area about the liquid level.

$$= A\bar{x}$$

∴ $$P = wA\bar{x} \qquad \text{...(Same as in Art. 3·4)}$$

Example 3·3. *A circular door of 1 m diameter closes an opening in the vertical side of a bulkhead, which retains sea water. If the centre of the opening is at a depth of 2 m from the water level, determine the total pressure on the door. Take specific gravity of sea water as 1·03.*

Solution. Given : $d = 1$ m ; $\bar{x} = 2$ m and specific gravity of sea water = 1·03 or specific weight of sea water, $w = 9·81 \times 1·03 = 10·1$ kN/m³.

We know that surface area of the circular door,

$$A = \frac{\pi}{4}(d)^2 = \frac{\pi}{4} \times (1)^2 = 0·7854 \text{ m}^2$$

and total pressure on the door, $P = wA\bar{x} = 10·1 \times 0·7854 \times 2 = 15·9$ kN **Ans.**

---

*Total pressure may also be found out by dividing the whole surface into a number of small parallel strips.

Let $a_1, a_2, a_3...$ = Areas of the strips, and

$x_1, x_2, x_3....$ = Depths of the corresponding strips from the liquid surface.

We know that pressure on first strip

$$= wa_1x_1$$

Similarly, pressure on second strip

$$= wa_2x_2$$

and pressure on third strip $= wa_3x_3$

∴ Total pressure on the surface,

$$P = wa_1x_1 + wa_2x_2 + wa_3x_3 + ...$$

$$= w(a_1x_1 + a_2x_2 + a_3x_3 + ...)$$

$$= wA\bar{x}$$

Example 3.4. *A triangular lamina ABC is immersed in water with the side AB coinciding with the water surface as shown in Fig. 3·3.*

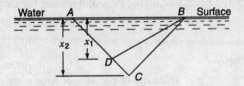

Fig. 3·3.

A point D is taken in AC, such that the water pressures on the two areas ABD and DBC are equal. Show that AD : AC = 1 : √2.

Solution. Given : Water pressure on area *ABD*,

$$P_{ABD} = P_{DBC} = \frac{P_{ABC}}{2}$$

Let w = Specific weight of the liquid,
a = Length of side *AB*,
x_1 = Depth of vertex *D* from the water surface, and
x_2 = Depth of vertex *C* from the water surface.

We know that the pressure on the area *ABD*,

$$P_{ABD} = wA\bar{x} = w \times \frac{a \times x_1}{2} \times \frac{x_1}{3} = \frac{wax_1^2}{6}$$

$$\therefore \quad x_1 = \sqrt{\frac{6 P_{ABD}}{wa}} \qquad \ldots(i)$$

and pressure on the area *ABC*,

$$P_{ABC} = wA\bar{x} = w \times \frac{a \times x_2}{2} \times \frac{x_2}{3} = \frac{wax_2^2}{6}$$

But

$$P_{ABD} = \frac{P_{ABC}}{2} = \frac{wax_2^2}{12}$$

$$\therefore \quad x_2 = \sqrt{\frac{12 P_{ABD}}{wa}} \qquad \ldots(ii)$$

Now from the geometry of the figure, we find that

$$AD : AC = x_1 : x_2 = \sqrt{\frac{6 P_{ABD}}{wa}} : \sqrt{\frac{12 P_{ABD}}{wa}} = 1 : \sqrt{2} \quad \textbf{Ans.}$$

3·6 Total Pressure on an Inclined Immersed Surface

Consider a plane inclined surface, immersed in a liquid as shown in Fig. 3·4.

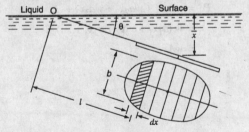

Fig. 3·4. Inclined Immersed surface.

Hydrostatics

First of all, let us divide the whole immersed surface into a number of small parallel strips as shown in the figure.

Let
- w = Specific weight of the liquid,
- A = Area of the surface,
- $\bar{x}$ = Depth of centre of gravity of the immersed surface from the liquid surface, and
- θ = Angle at which the immersed surface is inclined with the liquid surface.

Let us consider a strip of thickness, dx, width b and at a distance l from O (A point on the liquid surface where the immersed surface will meet, if produced).

We know that the intensity of pressure on the strip
$$= wl \sin \theta$$
and area of the strip $= b \cdot dx$

$\therefore$ Pressure on the strip, p = Intensity of pressure × Area
$$= wl \sin \theta \cdot b\,dx$$

Now *total pressure on the surface,
$$P = \int wl \sin \theta \cdot b\,dx$$
$$= w \sin \theta \int l \cdot b\,dx$$

But $\int l \cdot b\,dx$ = Moment of the surface area about O.
$$= \frac{A\bar{x}}{\sin \theta}$$

$\therefore \quad P = w \sin \theta \times \dfrac{A\bar{x}}{\sin \theta}$

$\qquad = wA\bar{x}$ ...(Same as in Arts. 3·4 and 3·5)

Example 3·5. *A rectangular plate 2 m × 3 m is immersed in water in such a way that its greatest and least depths are 6 m and 4 m respectively from the water surface. Calculate the total pressure on the plate.*

Solution. Given : Size of the plate = 2 m × 5 m and greatest and least depths of the plate = 6 m and 4 m.

We know that area of the plate,
$$A = 2 \times 3 = 6 \text{ m}^2$$

---

* The total pressure may also be found out by dividing the whole surface into a number of small parallel strips.

Let
- $a_1, a_2, a_3...$ = Areas of the strips, and
- $l_1, l_2, l_3...$ = Distances of corresponding strips from O.

We know that pressure on the first strip
$$= wa_1 l_1 \sin \theta$$
Similarly, pressure on the second strip
$$= wa_2 l_2 \sin \theta$$
and pressure on the third strip $= wa_3 l_3 \sin \theta$

$\therefore$ Total pressure on the surface $P = w_1 a_1 l_1 \sin \theta + wa_2 l_2 \sin \theta + wa_3 l_3 \sin \theta + ...$
$$= w \sin \theta \, (a_1 l_1 + a_2 l_2 + a_3 l_3 + ,...) = wAl \sin \theta$$
where l is the distance of centre of the surface from O.

$\therefore \qquad P = wA\bar{x}$ ... ($\because l \sin \theta = \bar{x}$)

and depth of centre of the plate,

$$\bar{x} = \frac{6+4}{2} = 5 \text{ m}$$

∴ Total pressure on the plate,

$$P = wA\bar{x} = 9.81 \times 6 \times 5 = 294.3 \text{ kN} \quad \textbf{Ans.}$$

Example 3·6. *A horizontal passage 1400 mm × 1400 mm has its outlet covered by a plane flap inclined at 60° with the horizontal and is hinged along the upper horizontal edge of the passage. If the depth of the flowing water is 500 mm in the passage, determine the thrust on the gate.*

Solution. Given : Width of passage = 1400 mm = 1·4 m; Depth of passage = 1400 mm = 1·4 m; Inclination of flap = 60° and depth of water = 500 mm = 0·5 m.

We know that area of the wetted flap,

$$A = 1.4 \times 0.5 \text{ cosec } 60° \text{ m}^2$$
$$= 0.7 \times 1.1547 \text{ m}^2$$
$$= 0.8083 \text{ m}^2$$

and depth of the centre of the watted flap

$$\bar{x} = 0.5/2 = 0.25 \text{ m}$$

∴ Thrust on the gate,

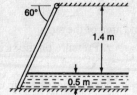

Fig. 3·5.

$$P = wA\bar{x} = 9.81 \times 0.8083 \times 0.25 = 1.98 \text{ kN} \quad \textbf{Ans.}$$

EXERCISE 3·1

1. A tank 10 m × 10 m contains water up to a height of 3 m. Determine the total pressure on the base of the tank. [**Ans.** 2943 kN]
2. A vessel 2 m × 1 m contains 1 m deep oil of specific gravity 0·84. What is the intensity of pressure and total pressure at the base of the tank ? [**Ans.** 8·24 kPa; 16·48 kN]
3. Find the total pressure on a rectangular plate 2 m wide and 4 m deep vertically immersed in water, in such a way that 2 m side is parallel to the water surface at 3 m below it. [**Ans.** 392·4 kN]
4. A circular plate 1·2 m diameter is immersed in oil of specific weight 8·5 kN/m³, such that its centre is 4 m below the oil surface. Find the intensity of pressure and total pressure on the plate. [**Ans.** 34 kPa; 38·42 kN]
5. Find the intensity of pressure on a rectangular plate 2 m wide and 2·5 m deep immersed vertically in water, such that its centre is 1·5 m deep from the water surface. Also find the total pressure on the plate. [**Ans.** 19·62 kPa; 98·1 kN]
6. A rectangular plate 1 m wide and 2 m deep is vertically immersed in water. Find the total pressure on the plate, when its upper edge is horizontal and (*i*) coincides with water and (*ii*) 1·5 m below the free surface of water. [**Ans.** 19·62 kN; 49·05 kN]
7. A circular plate of 1 m diameter is immersed in water in such a way that its plane makes an angle of 30° with the horizontal and its top edge is 1·25 m below the water surface. Determine the total pressure on the plate. [**Ans.** 11·56 kN]

3·7 Centre of Pressure

We have discussed in Art. 2·2 that the intensity of pressure on an immersed surface is not uniform, but increases with depth. As the pressure is greater over the lower portion of the figure, therefore the resultant pressure, on an immersed surface, will act at some point below the centre of gravity of the immersed surface and towards the lower edge of the figure. The point, through which this resultant pressure acts, is known as centre of pressure and is always expressed in terms of depth from the liquid surface.

Hydrostatics

In the following pages, we shall discuss the centres of pressure of vertically as well as inclined immersed surfaces.

3·8 Centre of Pressure of a Vertically Immersed Surface

Consider a plane surface immersed vertically in a liquid as shown in Fig. 3·6.

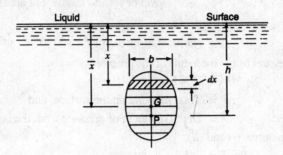

Fig. 3·6. Vertically immersed surface.

First of all, let us divide the whole immersed surface into a number of small parallel strips as shown in the figure.

Let
w = Specific weight of the liquid,
A = Area of the immersed surface, and
$\bar{x}$ = Depth of centre of gravity of the immersed surface from the liquid surface.

Let us consider a strip of thickness dx, width b and at a depth of x from the free surface of the liquid as shown in Fig. 3·6.

We know that intensity of pressure on the strip

$$= wx$$

and area of the strip $= b\,dx$

$\therefore$ Pressure on the strip, p = Intensity of pressure × Area

$$= wx \cdot b\,dx$$

Moment of this pressure about the liquid surface,

$$= (wx \cdot b\,dx)x = wx^2 \cdot b\,dx$$

Now the *sum of moments of all such pressures about the liquid surface,

---

*The sum of moments of water pressures about the liquid surface may also be found out by dividing the whole surface into a number of small parallel strips.

Let $a_1, a_2, a_3 \ldots$ = Areas of strips, and
$x_1, x_2, x_3 \ldots$ = Depths of the corresponding strips from the liquid surface.

We know that pressure on the first strip

$$= wa_1x_1$$

and moment of this pressure about the liquid surface

$$= wa_1x_1 \cdot x_1 = wa_1x_1^2$$

Similarly, moment of the pressure of second strip about liquid surface

$$= wa_2x_2^2$$

and moment of the pressure on the third strip about the liquid surface

$$= wa_3x_3^2$$

... (*Continued on next page*)

$$M = \int wx^2 \cdot b\,dx$$
$$= w \int x^2 \cdot b\,dx$$

But $\int x^2 \cdot b\,dx = I_0$ (*i.e.*, Moment of finertial of the surface about the liquid level or second moment of area)

∴ $M = w \cdot I_0$ ...(*i*)

where I_0 = Moment of inertia of the surface about the liquid level.

We know that the sum of the moments of the pressure
$$= P \times \bar{h}$$

where P = Total pressure on the surface, and ...(*ii*)
$\bar{h}$ = Depth of centre of pressure from the liquid surface.

Now equating equations (*i*) and (*ii*),
$$P \times \bar{h} = w \cdot I_0$$
$$wA\bar{x} \times \bar{h} = w \cdot I_0 \qquad \ldots(\because P = wA\bar{x})$$

∴ $$\bar{h} = \frac{I_0}{A\bar{x}} \qquad \ldots(iii)$$

We know from the *Theorem of Parallel Axis that
$$I_0 = I_G + Ah^2$$

where I_G = Moment of inertia of the figure, about horizontal axis through its centre of gravity, and
h = Distance between the liquid surface and the centre of gravity of the figure ($\bar{x}$ in this case).

Now rearranging the equation (*iii*),
$$\bar{h} = \frac{I_G + A\bar{x}^2}{A\bar{x}} = \frac{I_G}{A\bar{x}} + \bar{x}$$

Thus the centre of pressure is always below the centre of gravity of the area by a distance equal to $\dfrac{I_G}{A\bar{x}}$.

... (*Continued from last page*)

∴ Sum of moments of all such pressures about the liquid surface,
$$M = wa_1x_1^2 + wa_2x_2^2 + wa_3^2 + \ldots$$
$$= w(a_1x_1^2 + a_2x_2^2 + a_3x_3^2 +)\ldots = wI_0$$

where $I_0 = (a_1x_1^2 + a_2x_2^2 + a_3x_3^2 + \ldots)$ ...(*i*)
= Moment of inertia of the surface about the liquid surface (also known as second moment of area).

Further proceed from equation (*i*).

* For Theorem of Parallel Axis, please refer some standard book on Applied Mechanics or Author's book "A Textbook of Applied Mechanics" or "Strength of Materials."

Hydrostatics

Table 3·1

The centre of gravity (G) and moment of inertia (I) of some important geometrical figures is given below :

S.No.	Name of figure	C.G. from the base	I about an axis passing through C.G. and parallel to base	I about base
1.	(triangle, base b, height h)	$x = \dfrac{h}{3}$	$\dfrac{bh^3}{36}$	$\dfrac{bh^3}{12}$
2.	(inverted triangle, base b, height h)	$x = \dfrac{2h}{3}$	$\dfrac{bh^3}{36}$	$\dfrac{bh^3}{12}$
3.	(rectangle, base b, depth d)	$x = \dfrac{d}{2}$	$\dfrac{bd^3}{12}$	$\dfrac{bd^3}{3}$
4.	(circle, diameter d)	$x = \dfrac{d}{2}$	$\dfrac{\pi}{64} \times d^4$	—

Example 3·7. *A rectangular sluice gate is situated on the vertical wall of a lock. The vertical side of the sluice is (d) metres in length and depth of centriod of the area is (p) metres below the water surface. Prove that the depth of the centre of pressure is equal to*

$$\left(p + \dfrac{d^2}{12p}\right)$$

Solution. Given : Depth of the sluice gate = d and depth of the centroid of the area, $\bar{x} = p$.

Let b = Width of the sluice gate.

We know that area of the sluice gate,

$$A = b \times d$$

and moment of inertia of the rectangular section about its c.g. and parallel to water surface,

$$I_G = \dfrac{bd^3}{12}$$

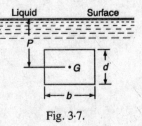

Fig. 3·7.

∴ Depth of the centre of pressure from the water surface,

$$\bar{h} = \frac{I_G}{A\bar{x}} + \bar{x} = \frac{\frac{bd^3}{12}}{bd \times p} + p = \frac{bd \times \frac{d^2}{12}}{bd \times p} + p = p + \frac{d^2}{12p} \quad \textbf{Ans.}$$

Example 3·8. *An isosceles triangular plate of base 3 metres and altitude 3 metres is immersed vertically in water as shown in Fig. 3·8.*

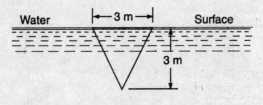

Fig. 3·8.

Determine the total pressure and centre of pressure of the plate.

Solution. Given : Base width $(b) = 3$ m and altitude $(h) = 3$ m.

Total pressure on the plate

We know that surface area of the triangular plate,

$$A = \frac{bh}{2} = \frac{3 \times 3}{2} = 4.5 \text{ m}^2$$

and depth of c.g. of the plate from the water surface,

$$\bar{x} = 3/3 = 1 \text{ m}$$

∴ Total pressure on the plate,

$$P = wA\bar{x} = 9.81 \times 4.5 \times 1 = 44.1 \text{ kN} \quad \textbf{Ans.}$$

Centre of pressure

We also know that moment of inertia of the triangular section about its c.g. and parallel to water surface,

$$I_G = \frac{bh^3}{36} = \frac{3 \times (3)^3}{36} = 2.25 \text{ m}^4$$

and depth of centre of pressure from the water surface,

$$\bar{h} = \frac{I_G}{A\bar{x}} + \bar{x} = \frac{2.25}{4.5 \times 1} + 1 = 1.5 \text{ m} \quad \textbf{Ans.}$$

Example 3·9. *A circular gate of 2 m diameter is immersed vertically in an oil of specific gravity 0·84 as shown in Fig. 3·9.*

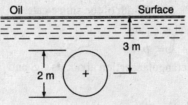

Fig. 3·9.

Find the oil pressure on the gate and position of the centre of pressure on the gate.

Hydrostatics

Solution. Given : $d = 2$ m; Specific gravity of oil = 0.84 or specific weight of oil $w = 9.81 \times 0.84 = 8.24$ kN/m^3 and $\bar{x} = 3$ m.

Pressure on the gate

We know that surface area of the circular gate,

$$A = \frac{\pi}{4} \times (d)^2 = \frac{\pi}{4} \times (2)^2 = 3.1416 \text{ m}^2$$

and pressure on the gate, $\quad P = wA\bar{x} = 8.24 \times 3.1416 \times 3 = 77.66$ kN **Ans.**

Position of the centre of pressure

We also know that moment of inertia of the circular section about its c.g. and parallel to the oil surface,

$$I_G = \frac{\pi}{64} \times (d)^4 = \frac{\pi}{64} \times (2)^4 = 0.7854 \text{ m}^4$$

and depth of centre of pressure from the surface,

$$\bar{h} = \frac{I_G}{A\bar{x}} + \bar{x} = \frac{0.7854}{3.1416 \times 3} + 3 = 3.08 \text{ m} \quad \textbf{Ans.}$$

Example 3·10. *A square plate ABCD 5 m × 5 m hangs in water from one of its corner as shown in Fig. 3·10.*

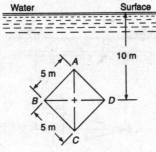

Fig. 3·10.

Determine the total pressure on the plate and the position of the centre of pressure.

Solution. Given : Side of the square plate = 5 m and $\bar{x}$ = 10 m.

Total pressure on the plate

We know that surface area of the square plate,

$$A = 5 \times 5 = 25 \text{ m}^2$$

and total pressure on the plate,

$$P = wA\bar{x} = 9.81 \times 25 \times 10 = 2452.5 \text{ kN} \quad \textbf{Ans.}$$

Centre of pressure

We know that length of the diagonal BD of the plate,

$$b = \sqrt{(5)^2 + (5)^2} = 5\sqrt{2} \text{ m}$$

First of all, let us find out the moment of inertia of square section $ABCD$ about its c.g. and parallel to water surface. Since the plate is lying with its diagonal BD parallel to the water surface, therefore the moment of inertia of the plate about BD will be obtained as discussed below :

1. Split up the plate into two triangles ABD and BCD.
2. Find out the moments of inertia of the two triangles ABD and BCD separately about BD.

3. Add the moments of inertia of the two triangles, which will give the required moment of inertia of the plate about the axis BD.

We know that moment of inertia of triangle ABD about BD

$$= \frac{bh^3}{12} = \frac{5\sqrt{2}\left(\frac{5\sqrt{2}}{2}\right)^3}{12} = \frac{625}{24} \text{ m}^4$$

Similarly, moment of inertia of the triangle BCD about BD,

$$= \frac{625}{24} \text{ m}^4$$

∴ Moment of inertia of the triangle BCD about BD

$$I_G = \frac{625}{24} + \frac{625}{24} = \frac{625}{12} = 52.1 \text{ m}^4$$

and depth of centre of pressure of the plate from the water level,

$$\bar{h} = \frac{I_G}{A\bar{x}} + \bar{x} = \frac{52.1}{25 \times 10} + 10 = 10.21 \text{ m} \quad \textbf{Ans.}$$

Example 3·11. *An isosceles triangle of base 3 metres, and altitude 6 metres, is immersed vertically in water, with its axis of symmetry horizontal as shown in Fig. 3·11.*

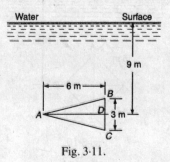

Fig. 3·11.

If head of water, on its axis, is 9 metres, calculate the total pressure on the plate. Also locate the centre of pressure both vertically and laterally.

Solution. Given : Base of the plate $(2h) = 3$ m or $h = 1.5$ m; Altitude of the plate $= 6$ m and $\bar{x} = 9$ m.

Total pressure on the plate

We know that surface area of the plate,

$$A = \frac{3 \times 6}{2} = 9 \text{ m}^2$$

and total pressure on the plate, $P = wA\bar{x} = 9.81 \times 9 \times 9 = 794.6 \text{ kN}$ **Ans.**

Vertical location of centre of pressure

First of all, let us find out the moment of inertia of the triangular section ABC about its c.g. and parallel to water surface. Since the plate is lying with its symmetrical axis horizontal, therefore the moment of inertia of the plate about AD may be obtained as discussed below :

1. Split up the plate into two triangles, ABD and ADC.
2. Find out the moments of inertia of the two triangles ABD and ADC separately about AD.
3. Add the moments of inertia of the two triangles, which will give the required moment of inertia of the triangle ABC about the axis AD.

Hydrostatics

We know that the moment of inertia of triangle ABD about AD

$$= \frac{bh^3}{12} = \frac{6 \times (1.5)^3}{12} = 1.6875 \text{ m}^4$$

Similarly, moment of inertia of triangle ADC about AD

$$= 1.6875 \text{ m}^4$$

∴ Moment of inertia of the triangle ABC about AD

$$I_G = 1.6875 + 1.6875 = 3.375 \text{ m}^4$$

and depth of centre of pressure of the plate from the water level,

$$\bar{h} = \frac{I_G}{A\bar{x}} + \bar{x} = \frac{3.375}{9 \times 9} + 9 = 9.04 \text{ m} \quad \textbf{Ans.}$$

Horizontal location of centre of pressure

The centre of pressure, in horizontal direction, will coincide with the centre of the triangle. Therefore centre of pressure will be at a distance of 6/3 = 2 m from BC. **Ans.**

3·9 Centre of Pressure of an Inclined Immersed Surface

Consider a plane inclined surface immersed in a liquid as shown in Fig. 3·12.

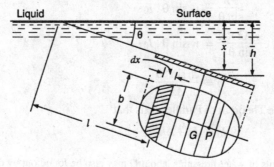

Fig. 3·12. Inclined immersed surface.

First of all, let us divide the whole immersed surface into a number of small parallel strips as shown in the figure.

Let w = Specific weight of the liquid,
  A = Area of the immersed face, and
  $\bar{x}$ = Depth of centre of gravity of the surface from the liquid surface.
  θ = Angle at which the immersed surface is inclined with the liquid surface.

Let us consider a strip of thickness of dx, width b and at a distance l from O (the point on the liquid surface where the immersed surface will meet, if extended).

We know that intensity of pressure on the strip

$$= wl \sin \theta$$

and area of strip $= b.dx$

∴   Pressure on the strip = Intensity of pressure × Area = $wl \sin \theta . b dx$

and moment of this pressure about O

$$= (wl \sin \theta \cdot dx) l = wl^2 \sin \theta \cdot bdx$$

Now sum of moments of all such pressures about O,

$$M = \int wl^2 \sin \theta \cdot bdx$$

$$= w \sin \theta \int l^2 \cdot bdx$$

But $\int l^2 \cdot bdx = I_0$ (i.e., moment of inertia of the surface about the point O or second moment of area)

$$\therefore \quad M = w \sin \theta \cdot I_0 \quad \quad ...(i)$$

We know that the sum of the *moments of all such pressures about O

$$= \frac{P\bar{h}}{\sin \theta} \quad \quad ...(ii)$$

where P = Total pressure on the surface, and
$\bar{h}$ = Depth of centre of pressure from the liquid surface.

Now equating equations (i) and (ii),

$$\frac{P\bar{h}}{\sin \theta} = w \sin \theta \cdot I_0$$

$$\frac{w A\bar{x} \times \bar{h}}{\sin \theta} = w \sin \theta \cdot I_0 \quad \quad ...(\because P = wA\bar{x})$$

$$\therefore \quad \bar{h} = \frac{I_0 \sin^2 \theta}{A\bar{x}} \quad \quad ...(iii)$$

We know from the Theorem of Parallel Axis that

$$I_0 = I_G + Ah^2$$

---

*The sum of moments of water pressures about O may also be found out by dividing the whole surface into a number of small parallel strips.

Let $a_1, a_2, a_3 ... $ = Areas of the strips, and
$l_1, l_2, l_3 ... $ = Distances of the corresponding strips from O.

Pressure on the first strip
$$= wa_1 l_1 \sin \theta$$
$\therefore$ Moment of this pressure about O
$$= wa_1 l_1 \sin \theta \times l_1 = wa_1 l_1^2 \sin \theta$$
Similarly, moment of the second strip about O
$$= wa_2 l_2^2 \sin \theta$$
and moment of the third strip about O
$$= wa_3 l_3^2 \sin \theta$$
The sum of moments of all such pressures about O,
$$M = wa_1 l_1^2 \sin \theta + wa_2 l_2^2 \sin \theta + wa_3 l_3^2 \sin \theta + ...$$
$$= w \sin \theta (a_1 l_1^2 + a_2 l_2^2 + a_3 l_3^2 + ...)$$
$$= w \sin \theta \cdot I_0$$
where $I_0 = (a_1 l_1^2 + a_2 l_2^2 + a_3 l_3^2 + ...)$
$$= \text{Moment of inertia of the surface about } O \text{ (also known as second moment of area).}$$
Further proceed from equation (i).

where
I_G = Moment of inertia of the figure about the horizontal axis through its centre of gravity, and
h = Distance between O and the centre of gravity of the figure (l in the case).

Now rearranging the equation (iii),

$$\bar{h} = \frac{\sin^2 \theta}{A\bar{x}} (I_G + Al^2) = \frac{\sin^2 \theta}{A\bar{x}} \left[I_G + A \left(\frac{\bar{x}}{\sin \theta} \right)^2 \right] \quad \left(\because l = \frac{\bar{x}}{\sin \theta} \right)$$

$$= \frac{I_G \sin^2 \theta}{A\bar{x}} + \bar{x}$$

Thus the centre of pressure is always below the centre of gravity of the area by a distance equal to $\dfrac{I_G \sin^2 \theta}{A\bar{x}}$.

Note : If the value of θ is substituted as 90°, the expression becomes the same as obtained in the last article.

Example 3·12. *A rectangular plate 2 m wide and 4 m deep is immersed in water in such a way that its plane makes an angle of 25° with the water surface as shown in Fig. 3·13.*

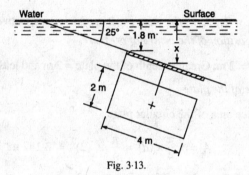

Fig. 3·13.

Determine the total pressure on one side of the plate and the position of the centre of pressure.

Solution. Given : $b = 2$ m; $d = 4$ m and $\theta = 25°$.

Total pressure on one side of the plate

We know that surface area of the rectangular plate,

$$A = b \times d = 2 \times 4 = 8 \text{ m}^2$$

and depth of c.g. of the plate from the water surface,

$$\bar{x} = 1·8 + \frac{4}{2} \sin 25° = 1·8 + (2 \times 0·4226) = 2·65 \text{ m}$$

∴ Total pressure on one side of the plate,

$$P = wA\bar{x} = 9·81 \times 8 \times 2·65 = 208·0 \text{ kN} \quad \textbf{Ans.}$$

Position of the centre of pressure

We also know that moment of inertia of the rectangular section about its c.g. and parallel to the water surface,

$$I_G = \frac{bd^3}{12} = \frac{2 \times (4)^3}{12} = 10·67 \text{ m}^4$$

and depth of centre of pressure from the water surface,

$$\bar{h} = \frac{I_G \sin^2 \theta}{A\bar{x}} + \bar{x} = \frac{10\cdot 67 \times \sin^2 25°}{8 \times 2\cdot 65} + 2\cdot 65 \text{ m}$$

$$= \frac{10\cdot 67 \times (0\cdot 4226)^2}{21\cdot 2} + 2\cdot 65 = 0\cdot 09 + 2\cdot 65 = 2\cdot 74 \text{ m Ans.}$$

Example 3·13. *A circular plate of 2 m diameter is submerged in water as shown in Fig. 3·14.*

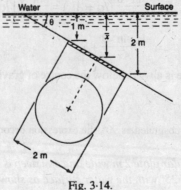

Fig. 3·14.

The greatest and least depths of the plate are 2 m and 1 m respectively. Find total pressure on the face of the plate and position of the centre of pressure.

Solution. Given : $d = 2$ m; Greatest depth of the plate = 2 m and least depth of the plate = 1 m.

Total pressure on the face of the plate

We know that surface area of the circular plate,

$$A = \frac{\pi}{4} \times (d)^2 = \frac{\pi}{4} \times (2)^2 = 3\cdot 142 \text{ m}^2$$

and depth of c.g. of the plate from the water surface,

$$\bar{x} = \frac{1+2}{2} = 1\cdot 5 \text{ m}$$

∴ Total pressure on the face of the plate,

$$P = wA\bar{x} = 9\cdot 81 \times 3\cdot 142 \times 1\cdot 5 = 46\cdot 2 \text{ kN Ans.}$$

Position of the centre of pressure

Let θ = Inclination of the plate.

We know that $\sin \theta = \dfrac{2-1}{2} = \dfrac{1}{2} = 0\cdot 5$

and moment of inertia of the circular plate about its c.g. and parallel to the water surface,

$$I_G = \frac{\pi}{64} \times (d)^4 = \frac{\pi}{64} \times (2)^4 = 0\cdot 7854 \text{ m}^4$$

∴ Depth of the centre of pressure from the water surface,

$$\bar{h} = \frac{I_G \sin^2 \theta}{A\bar{x}} + \bar{x} = \frac{0\cdot 7854 \times (0\cdot 5)^2}{3\cdot 142 \times 1\cdot 5} + 1\cdot 5 = 1\cdot 54 \text{ m Ans.}$$

Example 3·14. *A triangular plate of 1 metre base and 1·8 metre altitude is immersed in water. The plane of the plate is inclined at 30° with the free surface of water and the base is parallel to and at a depth of 2 metres from water surface as shown in Fig. 3·15.*

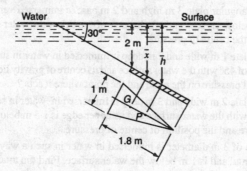

Fig. 3·15.

Find the total pressure on the plate and position of the centre of pressure.

Solution. Given : Base of the plate $(b) = 1$ m; Height $(h) = 1·8$ m and $\theta = 30°$.

Total pressure on the plate

We know that surface area of the triangular plate,

$$A = \frac{b \times h}{2} = \frac{1 \times 1·8}{2} = 0·9 \text{ m}^2$$

and depth of the c.g. of the plate from the water surface,

$$\bar{x} = 2 + \frac{1·8}{3} \sin 30° = 2 + (0·6 \times 0·5) = 2·3 \text{ m}$$

∴ Total pressure on the plate,

$$P = wA\bar{x} = 9·81 \times 0·9 \times 2·3 = 20·31 \text{ kN} \textbf{ Ans.}$$

Centre of pressure

We also know that moment of inertia of the triangular section about its c.g. and parallel to the water surface,

$$I_G = \frac{bh^3}{36} = \frac{1 \times (1·8)^3}{36} = 0·162 \text{ m}^4$$

and depth of centre of pressure from the water surface,

$$\bar{h} = \frac{I_G \sin^2 \theta}{A\bar{x}} + \bar{x} = \frac{0·162 \times \sin^2 30°}{0·9 \times 2·3} + 2·3 \text{ m}$$

$$= \frac{0·162 \times (0·5)^2}{2·07} + 2·3 = 0·02 + 2·3 = 2·32 \text{ m} \textbf{ Ans.}$$

EXERCISE 3·2

1. A circular plate of diameter d is submerged vertically in water in such a way that its centre is h below the water surface. Prove that the centre of the plate will be $\frac{d^2}{16h} + h$ below the water surface.

2. A rectangular plate 2 m side and 3 m deep is immersed vertically in water. Determine the total pressure and centre of pressure on the plate, when its upper edge is horizontal and 3·5 m below the free surface of water. [**Ans.** 294·3 kN; 5·15 m]

3. A square plate of 1 m side is immersed vertically with its centre is 4 m below the water surface. Find the total pressure and position of the centre of pressure. [**Ans.** 39·24 kN; 4·02 m]

4. A circular plate of 2·5 m diameter is vertically immersed in water, so that its centre is 4 m below the water surface. Find the total pressure on the plate and the point where it acts. [**Ans.** 192·7 kN; 4·1 m]

5. An isosceles triangular plate 3 m high and 2 m base is immersed vertically in water in such a way that its apex is below the base and at a distance of 4 m below the water surface. Find the total pressure and position of the centre of pressure. [**Ans.** 58·86 kN; 2·25 m]

6. A rectangular plate 1 m wide and 2 m deep is immersed in water in such a way that its plane surface makes an angle of 45° with the water surface and its centre of gravity lies 5 m below the water surface. What is the total pressure on the plate and the point where it acts? [**Ans.** 98·1 kN; 5·03 m]

7. A rectangular plate 2 m wide and 3 m deep is immersed in water in such a way that its plane makes an angle of 30° with the water surface and its upper edge is 1·5 m below the water surface. Calculate the total pressure and the position of centre of pressure. [**Ans.** 132·4 kN; 2·33 m]

8. A circular plate of 3 m diameter is immersed in water in such a way that it makes an angle of 30° with the horizontal and is 1 m below the water surface. Find the total pressure on the plate and the centre of pressure. [**Ans.** 121·4 kN; 1·8 m]

9. A gate 3 m wide and 2 m deep is fitted in a wall having a slope of 60° constructed across a channel. Determine the total pressure on the gate and its position, when the channel is full of water.
[**Ans.** 50·97 kN; 1·15 m]

3·10 Centre of Pressure of a Composite Section

The centre of pressure of a composite section (*i.e.*, a section with cut out hole or any other composite section) is obtained as discussed below :

1. Split up the composite section into convenient sections (*i.e.*, rectangles, triangles or circles).
2. Calculate the pressures. *i.e.*, P_1, P_2 ...... on all the sections.
3. Then calculate the total pressure P on the whole section by the algebraic sum of the pressures.
4. Now calculate the depths of centres of pressures *i.e.*, $\bar{h}_1$, $\bar{h}_2$........ for all the sections from the water surface.
5. Then equate $P\bar{h} = P_1\bar{h}_1 + P_2\bar{h}_2 + ...$

where $\bar{h}$ = Depth of centre of pressure of the section from the water level.

Example 3·15. *A hollow circular plate of 2 metres external and 1 metre internal diameter is immersed vertically in water, such that the centre of the plate is 4 metres deep from the water surface as shown in the figure given below.*

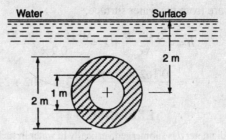

Fig. 3·16.

Calculate the total pressure and depth of the centre of pressure.

Solution. Given : d_1 = 2 m; d_2 = 1 m and $\bar{x}$ = 2 m.

Hydrostatics

Total pressure

We know that area of the hollow circular plate,

$$A = \frac{\pi}{4}\left[d_1^2 - d_2^2\right] = \frac{\pi}{4}\left[(2)^2 - (1)^2\right] = 2 \cdot 356 \text{ m}^2$$

and total pressure on the plate, $P = wA\bar{x} = 9 \cdot 81 \times 2 \cdot 356 \times 2 = 46 \cdot 2$ kN **Ans.**

Depth of centre of pressure

We also know that moment of inertia of the hollow circular section about its c.g. and parallel to water surface,

$$I_G = \frac{\pi}{64}\left[d_1^4 - d_2^4\right] = \frac{\pi}{64}\left[(2)^4 - (1)^4\right] = \frac{15\pi}{64} = 0 \cdot 736 \text{ m}^4$$

and depth of centre of pressure from the water surface,

$$\bar{h} = \frac{I_G}{A\bar{x}} + \bar{x} = \frac{0 \cdot 736}{2 \cdot 356 \times 2} + 2 = 2 \cdot 16 \text{ m} \quad \textbf{Ans.}$$

Note : Since the centres of the gravity of the main as well as that of the cut out circles coincide, therefore the depth of the centre of pressure, from the water, is obtained as usual.

Example 3·16. *A trapezoidal channel ABCD 2 m wide at the bottom, 5 m wide at the top and 1·5 m deep, has side slopes of 1 : 1 as shown in Fig. 3·17.*

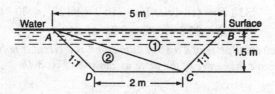

Fig. 3·17.

Determine (i) the total pressure, and (ii) the centre of pressure on the vertical gate closing the channel, when it is full of water.

Solution. Given : Bottom width $(b) = 2$ m; Top width $(a) = 5$ m; Water depth $(h) = 1 \cdot 5$ m and side slopes $= 1 : 1$.

(i) Total pressure on the plate

First of all, let us split up the trapezoidal section *ABCD* into two triangles 1 and 2 as shown in the figure. We know that area of triangle 1,

$$A_1 = \frac{5 \times 1 \cdot 5}{2} = 3 \cdot 75 \text{ m}^2$$

Similarly $\quad A_2 = \frac{2 \times 1 \cdot 5}{2} = 1 \cdot 5 \text{ m}^2$

From the geometry of the triangle 1, we know that depth of its c. g. from the water surface,

$$\bar{x}_1 = \frac{1 \cdot 5}{3} = 0 \cdot 5 \text{ m}$$

Similarly $\quad \bar{x}_2 = \frac{2 \times 1 \cdot 5}{3} = 1 \cdot 0 \text{ m}$

We know that pressure on triangular section 1,

$$P_1 = wA_1\bar{x}_1 = 9 \cdot 81 \times 3 \cdot 75 \times 0 \cdot 5 = 18 \cdot 4 \text{ kN} \qquad \ldots(i)$$

Similarly $\quad P_2 = wA_2\bar{x}_2 = 9 \cdot 81 \times 1 \cdot 5 \times 1 = 14 \cdot 7 \text{ kN} \qquad \ldots(ii)$

∴ Total pressure on the plate,
$$P = P_1 + P_2 = 18.4 + 14.7 = 33.1 \text{ kN} \quad \textbf{Ans.}$$

(ii) Centre of pressure of the gate

Let $\bar{h}$ = Depth of centre of pressure of the gate from the water surface.

We know that moment of inertia of triangular section 1 through its c.g. and parallel to the water surface,
$$I_{G1} = \frac{ah^3}{36} = \frac{5 \times (1.5)^3}{36} = 0.469 \text{ m}^4$$

Similarly, $$I_{G2} = \frac{bh^3}{36} = \frac{2 \times (1.5)^3}{36} = 0.188 \text{ m}^4$$

We also know that depth of centre of pressure of triangular section 1 from the water surface,
$$\bar{h}_1 = \frac{I_{G1}}{A_1 \bar{x}_1} + \bar{x}_1 = \frac{0.469}{3.75 \times 0.5} + 0.5 = 0.75 \text{ m}$$

Similarly, $$\bar{h}_2 = \frac{I_{G2}}{A_2 \bar{x}_2} + \bar{x}_2 = \frac{0.188}{1.5 \times 1} + 1 = 1.125 \text{ m}$$

Now taking moments of the pressures about the water surface and equating the same. *i.e.,*
$$P\bar{h} = P_1 \bar{h}_1 + P_2 \bar{h}_2$$
$$33.1 \bar{h} = (18.4 \times 0.75) + (14.7 \times 1.125) = 30.34$$
∴ $$\bar{h} = 30.34/33.1 = 0.92 \text{ m} \quad \textbf{Ans.}$$

Example 3·17. *A circular plate of 4 metres diameter has a circular hole of 2 metres diameter with its centre 1 metre above the centre of the plate as shown in Fig. 3·18.*

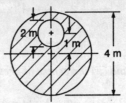

Fig. 3·18.

The plate is immersed in water at an angle of 30° to the horizontal and with its top edge 2 metres below the free surface. Find (i) the total pressure on the plate; and (ii) the depth of centre of pressure.

Solution. Given : $D = 4$ m; $d = 2$ m and $\theta = 30°$

(i) Total pressure on the plate

We know that area of the main plate,
$$A_1 = \frac{\pi}{4} \times (D)^2 = \frac{\pi}{4} \times (4)^2$$
$$= 12.57 \text{ m}^2$$

and area of the circular hole $A_2 = \frac{\pi}{4} \times (d)^2 = \frac{\pi}{4} \times (2)^2$
$$= 3.142 \text{ m}^2$$

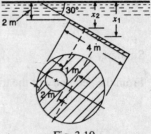

Fig. 3·19.

From the geometry of the main circular plate, we find that depth of its c.g. from the water surface,
$$\bar{x}_1 = 2 + 2 \sin 30° = 2 + (2 \times 0.5) = 3 \text{ m}$$

Similarly, $$\bar{x}_2 = 2 + 1 \sin 30° = 2 + (1 \times 0.5) = 2.5 \text{ m}$$

We know that pressure on the main circular plate,
$$P_1 = wA_1\bar{x}_1 = 9.81 \times 12.57 \times 3 = 369.9 \text{ kN} \qquad ...(i)$$
Similarly,
$$P_2 = wA_2\bar{x}_2 = 9.81 \times 3.142 \times 2.5 = 77.1 \text{ kN} \qquad ...(ii)$$
∴ Total pressure on the plate,
$$P = P_1 - P_2 = 369.9 - 77.1 = 292.8 \text{ kN} \quad \textbf{Ans.}$$

(ii) Depth of centre of pressure

Let $\bar{h}$ = Depth of the centre of pressure from the water surface.

We know that moment of inertia of the main circular plate through its c.g. and parallel to water surface,
$$I_{G1} = \frac{\pi}{64} \times (D)^4 = \frac{\pi}{64} \times (4)^4 = 12.57 \text{ m}^4$$
Similarly,
$$I_{G2} = \frac{\pi}{64} \times (d)^4 = \frac{\pi}{64} \times (2)^4 = 0.7854 \text{ m}^4$$

We also know that depth of centre of pressure of the main plate from the water surface,
$$\bar{h}_1 = \frac{I_{G1}}{A_1\bar{x}_1} + \bar{x}_1 = \frac{12.57}{12.57 \times 3} + 3 = 3.33 \text{ m}$$
Similarly,
$$\bar{h}_2 = \frac{I_{G2}}{A_2\bar{x}_2} + \bar{x}_2 = \frac{0.7854}{3.142 \times 2.5} + 2.5 = 2.6 \text{ m}$$

Now taking moments of the pressures about the water surface and equating the same. *i.e.*,
$$P\bar{h} = P_1\bar{h}_1 - P_2\bar{h}_2$$
$$292.8\,\bar{h} = (369.9 \times 3.33) - (77.1 \times 2.6) = 1031.3$$
∴
$$\bar{h} = 1031.3/292.8 = 3.52 \text{ m} \quad \textbf{Ans.}$$

EXERCISE 3·3

1. A hollow circular plate of 3 m external diameter and 1 m internal diameter is immersed in water such that its centre is at a depth of 2·5 m from the water surface. Find the total pressure and the point where it acts. [**Ans.** 154·1 kN; 2·75 m]

2. Find the total pressure and centre of pressure of a gate constructed across a channel 5 m wide at the top, 2 m at the bottom and 3 m deep when the channel is full of water. [**Ans.** 88·29 kN; 1·75 m]

3. A trapezoidal plate having its parallel sides $(2a)$ and (a) at a distance (h) apart is immersed vertically in water such that the $(2a)$ side is horizontal and at a depth of (h) below the water surface. Find the total thrust on the plate surface and position of the centre of pressure. [**Ans.** $2.167\ wah^2$; $1.5\ h$]

4. A composite section is made of a rectangle 4 m × 2 m and a triangle of base 2 m and height 3 m. The base of the triangle is connected to the 2 m side of the rectangle. The plate is immersed in water at an angle of 30° to the horizontal in such a way that the rectangular portion is above the triangular portion and its 2 m side is parallel to the water surface and 1 m below it. Find the total pressure on the plate and the position of the centre of pressure. [**Ans.** 260 kN; 2·71 m]

3·11 Pressure on a Curved Surface

The total pressure on a curved surface, when immersed in a liquid, cannot be found out*readily by the methods explained earlier. However, the same can be conveniently obtained by calculating the horizontal and vertical components of the resultant or total pressure, which is then combined together to give the total pressure on the curved surface.

---

*The reason for the same is that in case of a plane surface, the forces acting perpendicularly on all the strips have the same direction and form a system of parallel forces. But in case of a curved surface, all the strips do not lie in the same plane. Thus the forces (though perpendicular to their respective strips) do not form a system of parallel forces.

Consider a curved surface AB immersed in a liquid. Let BC be the vertical projection, and AC the horizontal projection of the curved surface as shown in Fig. 3·20.

The horizontal pressure, (P_H) will be the total horizontal pressure on the projection BC of the curved surface, and will act through the centre of pressure of the surface. The vertical pressure, (P_V), will be the total weight of the liquid in the portion ABC, and will act through the centre of gravity of the volume ABC.

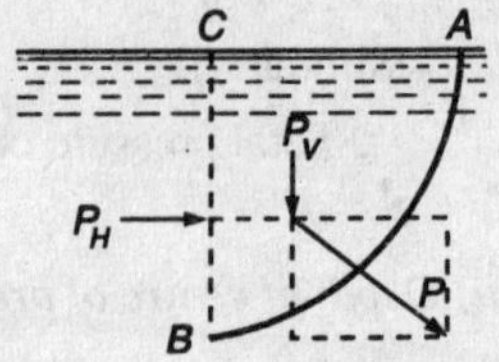

Now the total pressure or resultant pressure may be found out by the relation,

$$P = \sqrt{P_H^2 + P_V^2}$$

Fig. 3·20. Curved surface.

The inclination of the resultant pressure with the horizontal will by given by the equation,

$$\tan \alpha = \frac{P_V}{P_H}$$

where α is the angle, which the resultant pressure makes with the horizontal.

Note : Sometimes, the curved surface is subjected to a hydrostatic pressure on its lower side, while its upper side is not subjected to such pressure. In such cases, the vertical pressure P_V will be equal to the weight of the *imaginary* volume of liquid above this lower surface and up to the free surface of liquid. A little consideration will show that in such cases the direction of P_V will be *upwards*. The inclination of the resultant pressure may be found out as usual.

Example 3·18. *Determine the total pressure acting on the curved gate AB, per metre length, which is quadrant of a circular cylinder of radius 1 metre as shown in Fig. 3·21.*

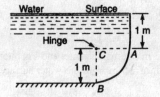

Fig. 3·21.

Also determine the angle at which the total pressure will act.

Solution. Given : Radius of the circular gate = 1 m

Total pressure acting on the curved gate

We know that vertical area of the curved gate (CB) per metre length,

$$A = 1 \times 1 = 1 \text{ m}^2$$

and depth of c.g. of the vertical area of the curved gate (CB),

$$\bar{x} = 1 + 0.5 = 1.5 \text{ m}$$

∴ Horizontal pressure due to water on the curved gate AB,

$$P_H = wA\bar{x} = 9.8 \times 1 \times 1.5 = 14.72 \text{ kN} \qquad ...(i)$$

and vertical pressure due to water on the curved gate AB per metre length,

$$P_V = \text{Weight of water in the curved portion } ACB$$
$$+ \text{Weight of water above } CA$$

$$P_V = 9.81\left[\frac{\pi}{4} \times (1)^2 \times 1 + (1 \times 1)\right] = 17.51 \text{ kN} \qquad ...(ii)$$

∴ Total pressure acting on the curved gate,

$$P = \sqrt{P_H^2 + P_V^2} = \sqrt{(14.72)^2 + (17.51)^2} = 22.87 \text{ kN} \quad \textbf{Ans.}$$

Angle at which the total pressure will act

Let $\quad\quad\quad\quad\quad\quad\quad \alpha =$ Angle which the total pressure makes with the horizontal.

We know that $\quad\quad \tan \alpha = \dfrac{P_V}{P_H} = \dfrac{17.51}{14.72} = 1.1895 \text{ or } \alpha = 49.95°$ **Ans.**

Example 3·19. *Find the resultant pressure due to water per metre length, acting on the circular gate of radius 3 metres as shown in Fig. 3·22.*

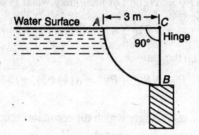

Fig. 3·22.

Also find the angle, at which the resultant pressure will act.

Solution. Given : Radius of the circular gate = 3 m

Resultant pressure due to water

We know that vertical area of the circular gate (AC) per metre length,

$$A = 3 \times 1 = 3 \text{ m}^2$$

and depth of c.g. of the vertical area of the circular gate (CB),

$$\bar{x} = 3/2 = 1.5 \text{ m}$$

∴ Horizontal pressure due to water on the circular gate AB,

$$P_H = wA\bar{x} = 9.81 \times 3 \times 1.5 = 44.15 \text{ kN} \qquad ...(i)$$

and vertical pressure due to weight of water on the circular gate AB,

$P_V =$ Weight of imaginary water lying in the circular surface AB

$$= 9.81\left[\frac{\pi}{4} \times (3)^2 \times 1\right] = 69.34 \text{ kN} \qquad ...(ii)$$

∴ Resultant pressure due to water,

$$P = \sqrt{P_H^2 + P_V^2} = \sqrt{(44.15)^2 + (69.34)^2} = 82.2 \text{ kN} \quad \textbf{Ans.}$$

Angle at which the resultant pressure will act

Let $\quad\quad\quad\quad\quad\quad\quad \alpha =$ Angle with the horizontal at which the resultant pressure will act.

We know that $\quad\quad \tan \alpha = \dfrac{P_V}{P_H} = \dfrac{69.34}{44.15} = 1.5706 \text{ or } \alpha = 57.52°$ **Ans.**

Example 3·20. *A roller gate of cylindrical form 3·0 metres in diameter has a span of 10 metres. Find the magnitude and direction of resultant force acting on the gate, when it is placed on the dam and the water level is such that it is going to spill.*

Solution. Given : Dia. of gate = 3 m; Span = 10 m and depth of water = 3 m.

Resultant force acting on the gate

We know that vertical area of the roller gate

$$A = 3 \times 10 = 30 \text{ m}^2$$

and depth of c.g. of the vertical area of the gate

$$\bar{x} = 3/2 = 1.5 \text{ m}$$

∴ Horizontal pressure on the roller gate

$$P_H = wA\bar{x} = 9.81 \times 30 \times 1.5 = 441.5 \text{ kN}$$

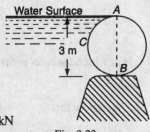

Fig. 3·23.

and vertical pressure acting on the gate

$$P_V = \text{Weight of imaginary water lying in the curved portion ACB}$$

$$= 9.81 \left[\frac{1}{2} \times \frac{\pi}{4} \times (3)^2 \times 10 \right] = 346.7 \text{ kN}$$

∴ Resultant force acting on the gate

$$P = \sqrt{P_H^2 + P_V^2} = \sqrt{(441.5)^2 + (346.7)^2} = 561.4 \text{ kN} \quad \textbf{Ans.}$$

Direction of the resultant force

Let α = Angle which the resultant force makes with the horizontal.

We know that $\tan \alpha = \dfrac{P_V}{P_H} = \dfrac{346.7}{441.5} = 0.7853$ or α = 38·1° **Ans.**

Example 3·21. *A tainter gate of 90° is subjected to water pressure as shown in Fig. 3·24.*

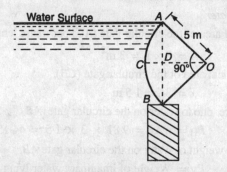

Fig. 3·24.

Determine horizontal and vertical pressures acting on the face of the gate.

Solution. Given : Angle of tainter gate sector = 90° and radius of the sector = 5 m.

Horizontal pressure acting on the face of the gate

We know that vertical area of the tainter gate (*AB*) per metre length,

$$A = (2 \times 5 \sin 45°) \times 1 = 2 \times 5 \times 0.707 = 7.07 \text{ m}^2$$

and depth of c.g. of the vertical area of the curved surface *AB*,

$$\bar{x} = 7.07/2 = 0.3535 \text{ m}$$

∴ Horizontal pressure acting on the face of the gate,

$$P_H = wA\bar{x} = 9.81 \times 7.07 \times 0.3535 = 245.2 \text{ kN} \quad \textbf{Ans.}$$

Vertical pressure acting on the face of the gate

We also know from the geometry of the gate that Area $ABCD$

$$= \text{Area of sector } ACBO - \text{Area of triangle } ABO$$

$$= \left[\frac{\pi}{4} \times (5)^2\right] - \left[\frac{5 \times 5}{2}\right] = 19 \cdot 635 - 12 \cdot 5 = 7 \cdot 135 \text{ m}^2$$

and vertical pressure acting on the face of the gate, per metre length,

P_V = Weight of imaginary water lying in the curved portion $ABCD$ of the tainter gate

$$= 9 \cdot 81 \times 7 \cdot 135 \times 1 = 70 \cdot 0 \text{ kN} \quad \textbf{Ans.}$$

Example 3·22. *The curved face of a dam, retaining water, is shaped according to the relationship $y = \dfrac{x^2}{4}$ as shown in Fig. 3·25.*

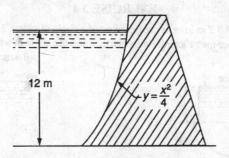

Fig. 3·25.

The height of water retained by the dam is 12 metres. Find the magnitude and direction of the resultant water pressure on the dam.

Solution. Given : Relation of water face curve, $y = \dfrac{x^2}{4}$ or $x^2 = 4y$ or $x = 2\sqrt{y}$ and depth of water = 12 m.

Magnitude of the resultant water pressure

We know that vertical area of the dam per metre length,

$$A = 12 \times 1 = 12 \text{ m}^2$$

and depth of c.g. of the vertical area of the dam,

$$\bar{x} = 12/2 = 6 \text{ m}$$

∴ Horizontal pressure due to water on the curved face of the dam,

$$P_H = wA\bar{x} = 9 \cdot 81 \times 12 \times 6 = 706 \cdot 3 \text{ kN} \qquad \ldots(i)$$

and vertical pressure due to water per metre length,

P_V = Weight of water in the curved portion

$$= \int_0^{12} 9 \cdot 81 \; x.dy$$

$$= 9 \cdot 81 \int_0^{12} 2y^{1/2} \; .dy \qquad \ldots (\because x = 2\sqrt{y})$$

$$= 9.81\left[2 \times \frac{2}{3} \times y^{3/2}\right]_0^{12} \text{ kN}$$

$$= 9.81\left[\frac{4}{3} \times 12\sqrt{12}\right] = 543.7 \text{ kN} \qquad ...(ii)$$

∴ Magnitude of the resultant water pressure,

$$P = \sqrt{P_H^2 + P_V^2} = \sqrt{(706.3)^2 + (543.7)^2} = 891.2 \text{ kN} \quad \textbf{Ans.}$$

Direction of the resultant water pressure

Let $\quad\quad\quad\quad\quad\quad\alpha$ = Angle, which the resultant pressure makes with the horizontal.

We know that $\quad\tan \alpha = \dfrac{P_V}{P_H} = \dfrac{543.7}{706.3} = 0.7698 \quad$ or $\quad \alpha = 37.6°$ **Ans.**

EXERCISE 3·4

1. A curved surface *AB* 2 m long is the quadrant of a circle of radius 2 m as shown in Fig 3·26. Find the total pressure per metre length on the surface and the angle at which it acts.
   [**Ans.** 183·0 kN; 50° with horizontal]

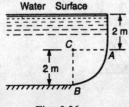

Fig. 3·26.

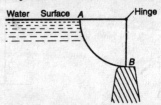

Fig. 3·27.

2. Find the resultant pressure due to weight per metre length of the gate as shown in Fig. 3·27. Also find the angle at which the total pressure will act. [**Ans.** 36·54 kN; 57·5° with horizontal]
3. A tainter gate 3 m long is mounted on a spillway as shown in Fig. 3·28. Find the total pressure on the gate and the angle at which it acts. [**Ans.** 184·8 kN; 39·3° with horizontal]
4. Fig. 3·29 shows the cross-section of a dam with a parabolic shape. Determine the force exerted by the water per metre length of the dam and its inclination with the vertical. [**Ans.** 8272·5 kN; 71·6°]

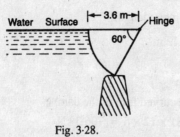

Fig. 3·28.

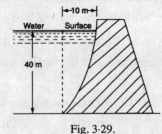

Fig. 3·29.

QUESTIONS

1. What do you understand by the term hydrostatic pressure ?
2. Derive an equation for the total pressure on a vertical immersed surface.
3. Define 'total pressure on a surface' and 'centre of pressure' of a surface.
4. Explain clearly with sketches the term centre of pressure.

Hydrostatics

5. From the first principles, derive a relation for the centre of pressure on a vertical immersed surface.
6. Prove that the centre of pressure of a fully submerged uniformly thick plane lamina is always below the centre of gravity of the lamina.
7. Derive an expression for the depth of centre of pressure of an inclined surface immersed in a liquid.
8. How would you find the total pressure and resultant pressure on a curved surface immersed in a liquid ?

OBJECTIVE TYPE QUESTIONS

1. The total pressure on a vertically immersed surface is
   (a) wA (b) $w\bar{x}$ (c) $wA\bar{x}$ (d) none of these

2. The total pressure on an immersed surface inclined at an angle θ with the liquid surface is
   (a) $wA\bar{x}$ (b) $\dfrac{wA\bar{x}}{\sin\theta}$ (c) $\dfrac{wa\bar{x}}{\cos\theta}$ (d) $\dfrac{wa\bar{x}}{\tan\theta}$

3. The centre of pressure acts .... the centre of gravity of the immersed surface
   (a) at   (b) above   (c) below   (d) can't say

4. The relationship of depth of centre of pressure for a vertically immersed surface is given by
   (a) $\dfrac{I_G}{A}+\bar{x}$ (b) $\dfrac{I_G}{\bar{x}}+\bar{x}$ (c) $\dfrac{I_G}{A\bar{x}}+\bar{x}$ (d) $\dfrac{I_G}{A\bar{x}}-\bar{x}$

   where I_G, A and $\bar{x}$ have usual meanings.

5. The distance between the centre of gravity and centre of pressure of a vertically immersed surface is equal to
   (a) $\dfrac{I_G}{\bar{x}}$ (b) $\dfrac{I_G}{A}$ (c) $\dfrac{I_G}{A\bar{x}}$ (d) $\dfrac{A\bar{x}}{I_G}$

   where I_G, A and $\bar{x}$ have usual meanings.

6. The depth of centre of pressure for an immersed surface inclined at an angle θ with the liquid surface lies at a distance .... below its centre of gravity.
   (a) $\dfrac{I_G \sin\theta}{A\bar{x}}$ (b) $\dfrac{I_G \sin^2\theta}{A\bar{x}}$ (c) $\dfrac{I_G \cos\theta}{A\bar{x}}$ (d) $\dfrac{I_G \cos^2\theta}{A\bar{x}}$

   where I_G, A and $\bar{x}$ have usual meanings.

ANSWERS

1. (c)   2. (a)   3. (b)   4. (c)   5. (c)   6. (b)

4

Applications of Hydrostatics

*1. Introduction.   2. Pressure Diagrams.   3. Pressure Due to One Kind of Liquid on One Side.
4. Pressure Due to One Kind of Liquid over Another on One Side.   5. Pressure Due to Liquids on Both Sides.
6. Practical Applications of Hydrostatics.   7. Water Pressure on Sluice Gates.   8. Water Pressure on Lock
Gates.   9. Water Pressure on Masonry Walls.   10. Water Pressure on Masonry Dams.   11. Water Pressure
on Rectangular Dams.   12. Water Pressure on Trapezoidal Dams.   13. Stability of a Dam.*

4·1 Introduction

In the previous chapter, we have been discussing the hydrostatic pressure on various types of immersed surfaces. We have also been discussing the centre of such pressure. In this chapter, we shall discuss the applications of hydrostatics *i.e.*, application of total pressure and centre of pressure on some engineering structures.

4·2 Pressure Diagrams

A pressure diagram may be defined as a graphical representation of the variation in the intensity of pressure over a surface. Such diagrams are very useful for finding out the total pressure and the centre of pressure of a liquid on a vertical surface (*i.e.*, wall or dam). A vertical surface may be subjected to the following types of pressures :

1. Pressure due to one kind of liquid on one side,
2. Pressure due to one kind of liquid, over another, on one side, and
3. Pressure due to liquids on both the sides.

Now we shall discuss the above three cases one-by-one.

4·3 Pressure Due to One Kind of Liquid on One Side

Consider a vertical wall, subjected to pressure due to one kind of liquid, on one of its sides as shown in Fig. 4·1.

Let H = Height of the liquid, and

w = Specific weight of the liquid.

We know that the pressure on the wall is zero at the liquid surface, and will increase by a straight line law to wH at the bottom. Therefore the pressure diagram will be a triangle ABC as shown in Fig. 4·1. The total pressure on the wall per unit length,

P = Area of triangle

$$= \frac{1}{2} ABC \times H \times wH = \frac{wH^2}{2}$$

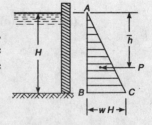

Fig. 4·1. Pressure due to one kind of liquid.

This pressure will act at the c.g. of the triangle. *i.e.*, at a depth of $2H/3$ from the liquid surface or in other words at a height of $H/3$ from the bottom of the liquid.

Applications of Hydrostatics

Example 4·1. *A water tank contains 1·3 m deep water. Find the pressure exerted by the water per metre length of the tank.*

Solution. Given : $H = 1·3$ m

We know that pressure exerted by the water per metre length of the tank,

$$P = \frac{wH^2}{2} = \frac{9·81 \times (1·3)^2}{2} = 8·29 \text{ kN} \quad \textbf{Ans.}$$

Example 4·2. *A swimming pool contains 3·6 m deep water in one of its sides. If the pool is 10 m wide on this side, calculate total pressure of the water on the vertical side of the pool. Also calculate the point where the total pressure acts on the wall.*

Solution. Given : $H = 3·6$ m and width of the swimming pool = 10 m

Total pressure on the wall

We know that water pressure on the vertical side of the swimming pool per metre length,

$$= \frac{wH^2}{2} = \frac{9·81 \times (3·6)^2}{2} = 63·57 \text{ kN}$$

and total pressure, $\quad P = 10 \times 63·57 = 635·7$ kN **Ans.**

Point where the total pressure acts on the wall

We also know that the total pressure will act at a point at a depth of $\dfrac{2H}{3} = \dfrac{2 \times 3·6}{3} = 2·4$ m from the water surface. **Ans.**

4·4 Pressure Due to One Kind of Liquid Over Another on One Side

Consider a vertical wall, subjected to pressure due to one kind of liquid, over another, on one side as shwon in Fig. 4·2. A little consideration will show, that this will happen, when one liquid is insoluble into the other.

Let $\quad H_1$ = Height of liquid 1,

w_1 = Specific weight of liquid 1,

H_2 = Height of liquid 2, and

w_2 = Specific weight of liquid 2.

We know that the pressure in such a case is zero at the liquid surface, and will increase by a straight line law to $w_1 H_1$ up to a depth of H_1. It will further increase, by a straight line law, to $w_1 H_1 + w_2 H_2$ as shown in Fig. 4·2.

The pressure P_1 on the surface AD, due to liquid 1, may be found out, as usual, from the area of triangle ADE $\left(i.e., P_1 = \dfrac{w_1 H_1^2}{2}\right)$.

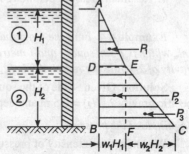

Fig. 4·2. Pressure due to one kind of liquid over another.

The pressure on the surface DB will consist of pressure P_2 due to superimposed liquid 1 as well as pressure P_3 due to liquid 2. This pressure will be given by the area of the trapezium $BCED$ *i.e.*, area of rectangle $DBFE$ due to liquid 1 ($P_2 = w_1 H_1 \times H_2$) and the area of triangle FCE due to liquid 2 $\left(i.e., P_3 = \dfrac{w_2 H_2^2}{2}\right)$.

The total pressure P will be sum of these three pressures (*i.e.*, $P = P_1 + P_2 + P_3$). The line of action of the total pressure may be found out by equating the moment of P, P_1, P_2 and P_3 about A.

Example 4·3. *A tank contains water for a height of 0·5 m and an immiscible liquid of specific gravity 0·8 above the water for a height of 1 m. Find the resultant pressure per metre length of the tank.*

Solution. Given : Height of water $(H_1) = 0·5$ m; Specific gravity of liquid = 0·8 and height of liquid $(H_2) = 1$ m.

We know that intensity of pressure at D (or B) due to oil of sp. gr. 0·8

$$= DE = BF = w_1 H_1$$

$$= (0·8 \times 9·81) \times 1·0 = 7·848 \text{ kN/m}^2$$

∴ Total Pressure at D due to oil of sp. gr. 0·8,

P_1 = Area of triangle ADE

$$= \frac{1}{2} \times 7·848 \times 1 = 3·924 \text{ kN} \quad ...(i)$$

and total pressure at B due to oil of sp. gr. 0·8,

P_2 = Area of rectangle $BDFE$

$$= 7·848 \times 0·5 = 3·924 \text{ kN} \quad ...(ii)$$

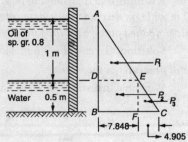

Fig. 4·3.

Similarly intensity of pressure at B due to water,

$$FC = W_2 H_2 = 9·81 \times 0·5 = 4·905 \text{ kN/m}^2$$

and total pressure at B due to water,

P_3 = Area of triangle EFC

$$= \frac{1}{2} \times 4·905 \times 0·5 = 1·226 \text{ kN/m}^2 \quad ...(iii)$$

∴ Resultant pressure per metre length of the tank

$$= P_1 + P_2 + P_3 = 3·924 + 3·924 + 1·226 = 9·074 \text{ kN} \quad \textbf{Ans.}$$

Example 4·4. *Find the magnitude and line of action of the pressure exerted on the side of a tank, which is 1·5 m square and 1 metre deep. The tank is filled half full with a liquid having specific gravity of 2, while the remainder is filled with a liquid having a specific gravity of 1.*

Solution. Given : Side of the square tank = 1·5 m; Depth of the tank = 1 m; Depth of liquid of specific gravity 2 $(H_2) = 0·5$ m and depth of liquid of specific gravity 1 $(H_1) = 0·5$ m.

Magnitude of the pressure

We know that intensity of pressure at D (or B) due to liquid of sp. gr. 1,

$$= DE = BF = w_1 H_1$$

$$= (1 \times 9·81) \times 0·5 = 4·905 \text{ kN/m}^2$$

∴ Total pressure at D due to liquid of sp. gr. 1,

P_1 = Area of triangle ADE
$\qquad \times$ Length of the tank wall

$$= \left(\frac{1}{2} \times 4·905 \times 0·5\right) \times 1·5 = 1·84 \text{ kN} \quad ...(i)$$

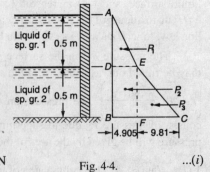

Fig. 4·4.

and total pressure at B due to liquid of sp. gr. 1,

P_2 = Area of rectangle $BDFE \times$ Length of the tank wall

$$= (4·905 \times 0·5) \times 1·5 = 3·68 \text{ kN} \quad ...(ii)$$

Similarly, intensity of pressure at B due to liquid of sp. gr. 2,

$$FC = w_2 H_2 = (2 \times 9\cdot 81) \times 0\cdot 5 = 9\cdot 81 \text{ kN/m}^2$$

total pressure from E to F or D to F (or B) due to liquid of sp. gr. 2,

P_3 = Area of triangle EFC × Length of the tank wall

$$= \left(\frac{1}{2} \times 9\cdot 81 \times 0\cdot 5\right) \times 1\cdot 5 = 3\cdot 68 \text{ kN} \qquad ...(iii)$$

∴ Magnitude of the pressure exerted on the side of the tank,

$$P = P_1 + P_2 + P_3 = 1\cdot 84 + 3\cdot 68 + 3\cdot 68 = 9\cdot 2 \text{ kN} \quad \textbf{Ans.}$$

Line of action of the resultant force (i.e., pressure)

Let $\bar{h}$ = Depth of the line of action of the resultant pressure from A.

Taking moments of all the pressures about A and equating the same,

$$P \times \bar{h} = \left[P_1 \times \frac{2 \times 0\cdot 5}{3}\right] + \left[P_2 \times \left(0\cdot 5 + \frac{0\cdot 5}{2}\right)\right] + \left[P_3 \times \left(0\cdot 5 + \frac{2 \times 0\cdot 5}{3}\right)\right]$$

$$9\cdot 2 \times \bar{h} = \left[1\cdot 84 \times \frac{1}{3}\right] + \left[3\cdot 68 \times \frac{3}{4}\right] + \left[3\cdot 68 \times \frac{5}{6}\right]$$

$$= 0\cdot 613 + 2\cdot 76 + 3\cdot 067 = 6\cdot 44$$

∴ $\bar{h} = 6\cdot 44/9\cdot 2 = 0\cdot 7$ m **Ans.**

Example 4·5. *A closed cylindrial tank of 1 m diameter and 2 m deep with vertical axis contains 1·5 m deep water. If air, at a pressure 5 kPa above atmosphere, is pumped into the cylinder, find the total pressure on the vertical wall of the tank and the distance of the total pressure from the bottom.*

Solution. Given : Dia. of tank = 1 m; Depth of tank = 2 m, Depth of water = 1·5 m and air pressure 5 kPa = 5 kN/m².

Total pressure on the vertical wall of the tank

We know that total pressure at A or B and to air (under pressure)

P_1 = Area of rectangle ADFB
 × Circumference of the tank
 = $(5 \times 2) \times (\pi \times 1)$
 = 31·42 kN

and pressure at B due to water

$$FC = wH = 9\cdot 81 \times 1\cdot 5$$
$$= 14\cdot 72 \text{ kN/m}^2$$

∴ Total pressure on the vertical wall of the tank,

P_2 = Area of triangle EFC
 × Circumference of tank

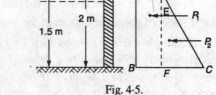

Fig. 4·5.

$$= \left(\frac{1}{2} \times 14\cdot 72 \times 1\cdot 5\right) \times (\pi \times 1) = 34\cdot 68 \text{ kN}$$

∴ Total pressure on the vertical wall of the tank,

$$P = P_1 + P_2 = 31\cdot 42 + 34\cdot 68 = 66\cdot 1 \text{ kN} \quad \textbf{Ans.}$$

Distance of the total pressure from the bottom

Let d = Distance of the total pressure from the bottom.

Taking moments of pressures about B and equating the same,

$$P \times d = \left[P_1 \times \frac{2}{2}\right] + \left[P_2 \times \frac{1.5}{3}\right]$$

$$66.1 \times d = \left[31.42 \times \frac{2}{2}\right] + \left[34.68 \times \frac{1.5}{3}\right] = 31.42 + 17.34 = 48.76$$

∴ $d = 48.76/66.1 = 0.74$ m **Ans.**

4.5 Pressure Due to Liquids on Both Sides

Consider a vertical wall subjected to pressures due to liquids on both sides as shown in Fig. 4.6.

Let H_1 = Height of liquid 1,
 w_1 = Specific weight of liquid 1,
 H_2 = Height of liquid 2,
 w_2 = Specific weight of liquid 2, and
 P = Resultant pressure on the wall per unit length.

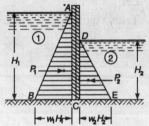

Fig. 4.6. Pressure due to liquids on both sides.

We know that the pressure of liquid 1 is zero at the liquid surface and will increase, by a straight line law, to $w_1 H_1$ at the bottom as shown in Fig. 4.6.

∴ Total pressure on the wall per unit length due to liquid 1,

$$P_1 = \frac{1}{2} \times H_1 \times w_1 H_1 = \frac{w_1 H_1^2}{2}$$

Similarly, total pressure on the wall per unit length due to liquid 2,

$$P_2 = \frac{1}{2} \times H_2 \times w_2 H_2 = \frac{w_2 H_2^2}{2}$$

A little consideration will show that as the two pressures are acting in the opposite directions, therefore the resultant pressure will be given by the difference of the two pressures (*i.e.*, $P = P_1 - P_2$). The line of action of the resultant pressure may be found out by equating the moments of P, P_1 and P_2 about the bottom of the wall.

Example 4.6. *A bulkhead 3 m long divides a storage tank. On one side, there is a petrol of specific gravity 0.78 stored to a depth of 1.8 m, while on the other side there is an oil of specific gravity 0.88 stored to a depth of 0.9 m. Determine the resultant pressure on the bulkhead and the position at which it acts.*

Solution. Given : Length of bulkhead = 3 m ; Specific gravity of petrol = 0.78; Depth of petrol (H_1) = 1.8 m; Sp. gr. of oil = 0.88 and depth of oil (H_2) = 0.9 m.

Resultant pressure on the bulkhead

We know that pressure due to petrol on the bulkhead per metre length

$$= \frac{w_1 H_1^2}{2} = \frac{(0.78 \times 9.81) \times (1.8)^2}{2} = 12.4 \text{ kN}$$

Applications of Hydrostatics

and total pressure due to petrol on the bulkhead,

$$P_1 = 3 \times 12.4 = 37.2 \text{ kN}$$

Similarly pressure due to oil on the bulkhead per metre length

$$= \frac{w_2 H_2^2}{2} = \frac{(0.88 \times 9.81) \times (0.9)^2}{2} = 3.5 \text{ kN}$$

and total pressure due to oil on the bulkhead

$$P_2 = 3 \times 3.5 = 10.5 \text{ kN}$$

∴ Resultant pressure on the bulkhead

$$P = P_1 - P_2 = 37.2 - 10.5 = 26.7 \text{ kN} \quad \textbf{Ans.}$$

Position of the resultant pressure

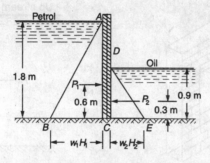

Fig. 4·7.

Let $\bar{h}$ = Depth of the point of the resultant pressure from C.

We know that the pressures P_1 and P_2 will act at height of $H/3$ from C, where H is the respective height of petrol and oil from C. Taking moment of all the pressure about C and equating the same,

$$P \times \bar{h} = P_1 \times 0.6 - P_2 \times 0.2$$
$$26.7 \bar{h} = (37.2 \times 0.6) - (10.5 \times 0.3) = 19.17$$
∴ $$\bar{h} = 19.17/26.7 = 0.72 \text{ m} \quad \textbf{Ans.}$$

EXERCISE 4·1

1. A water tank contains 0·8 m deep water. What is the total pressure exerted by the water per metre length of the tank ? [**Ans.** 3·14 kN]
2. A water tank of circular shape of diameter 1·2 m is fitted at the top of a building. Find the total pressure exerted by water, when it contains 1·6 m of water. [**Ans.** 47·35 kN]
3. A storage tank 30 m × 20 m contains 3 m deep water. Calculate the total pressure on the shorter side of the tank and point where it acts. [**Ans.** 882 kN; 2 m from water surface]
4. A tank contains 0·6 m deep water and an oil of sp. gr. 0·9 over the water for a depth of 0·9 m. Find the total pressure per metre length of the tank. [**Ans.** 5·35 kN]
5. A vertical wall contains water on its either sides up to 4 m and 2·5 m heights respectively. Find the resultant pressure on the wall per metre length. [**Ans.** 47·82 kN]
6. A vertical wall 5 m long contains water up to a height of 7 m on one side and 3 m on the other. Find the resultant pressure on the wall and position of its line of action.
[**Ans.** 981 kN; 2·63 m from the base]

4·6 Practical Applications of Hydrostatics

Strictly speaking, in the engineering-field, the hydrostatic pressure is either utilised in the working of a hydraulic structure or a structure is checked to withstand the hydrostatic pressure exerted on it. Thus the study of the subject hydrostatics is of much importance, while designing all sorts of hydraulic structures or hydraulic devices. In this chapter, we shall discuss the practical applications of the hydraulics on the following structures :

1. Sluice gates,
2. Lock gates,
3. Masonry walls, and
4. Dams.

4·7 Water Pressure on Sluice Gates

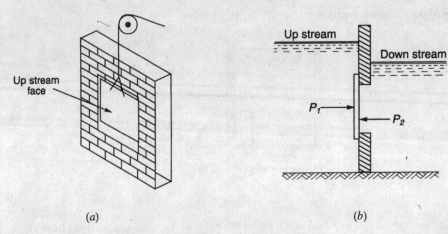

Fig. 4·8. Sluice gate.

A sluice gate is provided, in the path of a river or a stream, to regulate the flow of water. For doing so, the sluice gate is made to move up and down with the help of rollers fixed to the vertical plates (called skin plates) which travel on vertical rails called guides. These rails are fixed on piers or vertical walls as shown in Fig. 4·8 (a). In between these two skin plates, a number of I-beams are provided horizontally to withstand the water pressure. As the water pressure varies with the depth, the spacing between the I-beams is lesser at the bottom than that at the top of the sluice gate.

Water acts on both sides of a sluice gate as shown in Fig. 4·8 (b). The resultant pressure on the sluice gate is obtained as follows :

Let A_1 = Wetted area on the upstream of the gate.
 $\bar{x}_1$ = Depth of centre of gravity of the wetted area on the upstream side of the gate, and
 $A_2, \bar{x}_2$ = Corresponding values for the downstream side of the gate.

∴ Pressure on the upstream side of the gate,
$$P_1 = wA_1\bar{x}_1 \qquad \ldots(i)$$

Similarly, pressure on the downstream side of the gate,
$$P_2 = wA_2\bar{x}_2 \qquad \ldots(ii)$$

As the two pressures are acting in opposite directions, therefore the resultant pressure,
$$P = P_1 - P_2 \qquad \ldots(iii)$$

The point of application of the resultant pressure (P) may be found out by taking moments of the two pressures (P_1 and P_2) about the top or bottom of the gate.

Applications of Hydrostatics

Example 4·7. *A vertical sluice gate 3 m wide and 2·5 m deep contains water on both of its sides. On the upstream side, the water is 5 m deep and on the downstream side it is 2 m deep from the bottom of the sluice. What is the resultant pressure on the gate ?*

Solution. Given : Width of sluice gate $(b) = 3$ m; Depth of sluice gate $= 2·5$ m; Depth of water on the upstream side $= 5$ m and depth of water on the downstream side $= 2$ m.

We know that area of sluice gate submerged on the upstream side

$$A_1 = 2·5 \times 3 = 7·5 \text{ m}^2$$

and depth of c.g. of the sluice gate from the water level on the upstream side,

$$\bar{x}_1 = 2·5 + \frac{2·5}{2} = 3·75 \text{ m}$$

Similarly, area of sluice gate submerged on the downstream side,

$$A_2 = 2 \times 3 = 6 \text{ m}^2$$

and depth of c.g. of the sluice gate from the water level on the downstream side,

$$\bar{x}_2 = 2/2 = 1·0 \text{ m}$$

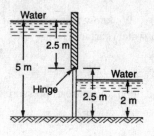

Fig. 4·9.

∴ Pressure due to water on the upstream side of the gate,

$$P_1 = wA_1\bar{x}_1 = 9·81 \times 7·5 \times 3·75 = 275·9 \text{ kN}$$

and pressure due to water on the downstream side of the gate,

$$P_2 = wA_2\bar{x}_2 = 9·81 \times 6 \times 1·0 = 58·9 \text{ kN}$$

∴ Resultant pressure on the gate,

$$P = P_1 - P_2 = 275·9 - 58·9 = 217 \text{ kN} \quad \textbf{Ans.}$$

Example 4·8. *A vertical sluice gate 4 metres wide and 2 metres deep is hinged at the top. A liquid of specific gravity 1·5 stands on the upstream side of the gate up to a height of 3·5 metres above the top edge of the gate and water on the downstream side up to the top edge of the gate. Find : (i) resultant pressure acting on the gate (ii) point at which the resultant pressure acts.*

Solution. Given : Width of sluice gate $(b) = 4$ m ; Depth of sluice gate $(d) = 2$ m; Specific gravity of liquid $= 1·5$; Height of liquid above the top edge of the gate on its upstream side $= 3·5$ m and height of water on the downstream side $= 2$ m.

(i) Resultant pressure acting on the gate

We know that area of the sluice gate,

$$A = 4 \times 2 = 8 \text{ m}^2$$

and depth of c.g. of the sluice gate from the liquid level on the upstream side,

$$\bar{x}_1 = 3·5 + \frac{2}{2} = 4·5 \text{ m}$$

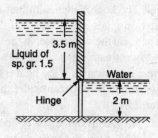

Fig. 4·10.

Similarly depth of c.g. of the sluice gate from the water level on the downstream side,

$$\bar{x}_2 = \frac{2}{2} = 1·0 \text{ m}$$

∴ Pressure due to liquid on upstream side of the gate,

$$P_1 = w_1 A \bar{x}_1 = (1·5 \times 9·81) \times 8 \times 4·5 \text{ kN} = 529·7 \text{ kN} \quad \text{...(i)}$$

and pressure due to water on downstream side of the gate,
$$P_2 = w_2 A \bar{x}_2 = 9.81 \times 8 \times 1.0 = 78.5 \text{ kN} \qquad ...(ii)$$
∴ Resultant pressure acting on the gate,
$$P = P_1 - P_2 = 529.7 - 78.5 = 451.2 \text{ kN} \quad \text{Ans.}$$

(ii) Point at which the resultant pressure acts

Let $\bar{h}$ = Depth of centre of the resultant pressure from the hinge (*i.e.*, top of the sluice gate).

We know that moment of inertia of the rectangular section about its c.g. and parallel to liquid (or water) surface,
$$I_G = \frac{bd^3}{12} = \frac{4 \times (2)^3}{12} = 2.67 \text{ m}^4$$
and depth of centre of pressure on the upstream side of the gate from the liquid surface,
$$\bar{h}_1 = \frac{I_G}{A\bar{x}_1} + \bar{x}_1 = \frac{2.67}{8 \times 4.5} + 4.5 = 4.574 \text{ m}$$

Similarly
$$\bar{h}_2 = \frac{I_G}{A\bar{x}_2} + \bar{x}_2 = \frac{2.67}{8 \times 1} + 1 = 1.334 \text{ m}$$

Now taking moments about the hinge and equating the same,
$$P \times \bar{h} = P_1 (4.574 - 3.5) - (P_2 \times 1.334)$$
$$451.2 \times \bar{h} = (529.7 \times 1.074) - (78.5 \times 1.334) = 464.2$$
∴ $\bar{h} = 464.2/451.2 = 1.03 \text{ m}$ **Ans.**

4·8 Water Pressure on Lock Gates

Whenever a dam or a weir is constructed across a river or a canal, the water levels on both the sides of the dam will be different. If it is desired to have navigation or boating in such a river or a canal, then a chamber, known as lock, is constructed between these two different water levels. Two sets of lock gates (one on the upstream side and the other on downstream side of the dam) are provided as shown in Fig. 4·11.

In order to transfer a boat from the upstream (*i.e.*, from a higher water level to the downstream (*i.e.*, to a lower water level) the upstream gates are opened (while the downstream gates are closed) and water level in the chamber rises up to the upstream water level. The boat is then admitted in the chamber. Then upstream gates are closed and downstream gates are opened and the water level in the chamber is lowered to the downstream water level. Now the boat can proceed further downwards. If the boat is to be transferred from downstream to upstream side, the above procedure is reversed.

Fig. 4·11. Lock gate.

Now consider a set of lock gates *AB* and *BC* hinged at the top and bottom at *A* and *C* respectively as shown in Fig. 4·12 (*a*). These gates will be held in contact at *B* by the water pressure, the water level being higher on the left hand side of the gates as shown in Fig. 4·12 (*b*).

Let P = Water pressure on the gate *AB* or *BC* acting at right angles on it,

Applications of Hydrostatics

F = Force exerted by the gate BC acting normally to the contact surface of the two gates AB and BC (also known as reaction between the two gates), and

R = Reaction at the upper and lower hinge.

Since the gate AB is in equilibrium, under the action of the above three forces, therefore they will meet at one point. Let P and F meet at O, then R must pass through this point.

Let α = Inclination of the lock gate with the normal to the walls of the lock.

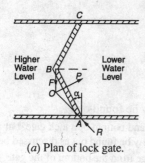 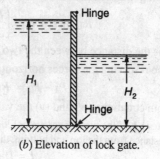

(a) Plan of lock gate.   (b) Elevation of lock gate.

Fig. 4·12.

From the geometry of the figure ABO, we find that it is an isosceles triangle having its angles OBA and OAB both equal to α.

Resolving the force, at O, parallel to AB,

$$R \cos \alpha = F \cos \alpha$$

$\therefore$ $R = F$ ...(i)

and now resolving the force at right angles to AB

$$P = R \sin \alpha + F \sin \alpha = 2R \sin \alpha \qquad ...(\because R = F)$$

$\therefore$ $R = \dfrac{P}{2 \sin \alpha}$

or $F = \dfrac{P}{2 \sin \alpha}$ ...(ii)

Now let us consider the water pressure on the top and bottom hinges of the gates.

Let H_1 = Height of water to the left side of the gate,

         A_1 = Wetted area (of one of the gates) on left side of the gate,

         P_1 = Total pressure of the water on the left side of the gate,

H_2, A_2, P_2 = Corresponding values for right side on the gate,

         R_T = Reaction of the top hinge, and

         R_B = Reaction of bottom hinge.

Since the total reaction (R) will be shared by the two hinges (R_T and R_B), therefore

$$R = R_T + R_B \qquad ...(iii)$$

and total pressure on the lock gate,

$$P = wA\bar{x} \text{ or } P_1 = wA_1 \times \frac{H_1}{2} = \frac{wA_1 H_1}{2} \qquad ...\left(\because \bar{x}_1 = \frac{H_1}{2}\right)$$

Similarly, $P_2 = \dfrac{wA_2 H_2}{2}$

Since the directions of P_1 and P_2 are in the opposite directions, therefore the resultant pressure,
$$P = P_1 - P_2$$
We know that the pressure P_1 will act through its centre of pressure, which is at a height of $H_1/3$ from the bottom of the gate. Similarly, the pressure P_2 will also act through its centre of pressure, which is also at a height of $H_2/3$ from the bottom of the gate.

A little consideration will show, that half of the resultant pressure (i.e., $P_1 - P_2$ or P) will be resisted by the hinges of one lock gate (as the other half will be resisted by the other lock gates).

Taking moments about the lower hinge,
$$R_T \sin \alpha \times h = \left(\frac{P_1}{2} \times \frac{H_1}{3}\right) - \left(\frac{P_2}{2} \times \frac{H_2}{3}\right) \qquad ...(iv)$$
where h is the distance between the two hinges.

Also resolving the forces horizontally,
$$P_1 - P_2 = R_B \sin \alpha + R_T \sin \alpha \qquad ...(v)$$
From equations (iv) and (v) the values of R_B and R_T may be found out.

Notes : 1. If the two hinges are not located at the extreme top and bottom of a lock gate, but are located at some distance from the top and bottom, then the hinge reactions may be found out as discussed below :

Let $\qquad d$ = Distance of the bottom hinge from the bottom of the gate.

Then by taking moments about the bottom hinge,
$$R_T \sin \alpha \times h = \left[\frac{P_1}{2} \times \left(\frac{H_1}{3} - d\right)\right] - \left[\frac{P_2}{2} \times \left(\frac{H_2}{3} - d\right)\right]$$

2. If there is water on one side of the lock gate only even then the above-mentioned relations are valid.

Example 4·9. *Two lock gates of 7·5 m height are provided in a canal of 16 metres width meeting at an angle of 120°. Calculate the force acting on each gate, when the depth of water on upstream side is 5 metres.*

Solution. Given : *Height of lock gates = 7·5 m; Width of lock gates = 16 m; Inclination of gates = 120° and H = 5 m.

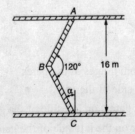 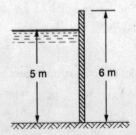

Fig. 4·13.

From the geometry of the lock gate, we find that inclination of the lock gates with the walls
$$\alpha = \frac{180° - 120°}{2} = 30°$$
and width of each gate $\qquad = \dfrac{16/2}{\cos \alpha} = \dfrac{8}{\cos 30°} = \dfrac{8}{0·866} = 9·24 \text{ m}$

∴ Wetted area of each gate $A = 5 \times 9·24 = 46·2 \text{ m}^2$

and force acting on each gate, $\qquad P = wA \times \dfrac{H}{2} = 9·81 \times 46·2 \times \dfrac{5}{2} = 1133 \text{ kN}$ **Ans.**

---

*Superfluous data.

Applications of Hydrostatics

Example 4·10. *The locks of a gate 10 m wide make an angle of 140° when closed. The water levels on the upstream and downstream sides are 6 m and 4 m respectively as shown in Fig. 4·14.*

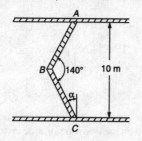
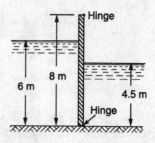

Fig. 4·14.

Each gate is carried by two hinges placed at the top and bottom of the gate. Find (i) resultant force on each gate, and (ii) force acting on each hinge.

Solution. Given : Width of the lock gate = 10 m; Angle which the gates make when closed $\theta = 140°$; $H_1 = 6$ m; $H_2 = 4·5$ m and distance between the hinges $(h) = 8$ m.

Resultant force on each gate

From the geometry of the lock gate, we find that inclination of the lock gates with the walls,

$$\alpha = \frac{180° - 140°}{2} = 20°$$

and width of each gate

$$= \frac{10/2}{\cos \alpha} = \frac{5}{\cos 20°} = \frac{5}{0·9397} = 5·32 \text{ m}$$

∴ Wetted area of each gate on the upstream side,

$$A_1 = 6 \times 5·32 =. 31·92 \text{ m}^2$$

and wetted area of each gate on the downstream side,

$$A_2 = 4·5 \times 5·32 = 23·94 \text{ m}^2$$

We know that force acting on each gate on the upstream side,

$$P_1 = wA_1 \times \frac{H_1}{2} = 9·81 \times 31·92 \times \frac{6}{2} = 939·4 \text{ kN} \qquad ...(i)$$

and force acting on each gate on the downstream side,

$$P_2 = wA_2 \times \frac{H_2}{2} = 9·81 \times 23·94 \times \frac{4·5}{2} = 528·4 \text{ kN} \qquad ...(ii)$$

∴ Resultant force on each gate

$$P = P_1 - P_2 = 939·4 - 528·4 = 411 \text{ kN } \textbf{Ans.}$$

(ii) Force acting on each hinge

Let R_T = Force acting on the top hinge, and
 R_B = Force acting on the bottom hinge.

We know that reaction between the gates,

$$F = \frac{P}{2 \sin \alpha} = \frac{411}{2 \times \sin 20°} = \frac{411}{2 \times 0·3420} = 600·9 \text{ kN}$$

We also know from the equation of forces acting on the hinges that

$$R_T \sin \alpha \times h = \left(\frac{P_1}{2} \times \frac{H_1}{3}\right) - \left(\frac{P_2}{2} \times \frac{H_2}{3}\right)$$

$$R_T \sin 20° \times 8 = \left(\frac{939 \cdot 4}{2} \times \frac{6}{3}\right) - \left(\frac{528 \cdot 4}{2} \times \frac{4 \cdot 5}{3}\right)$$

$$R_T \times 0 \cdot 3420 \times 8 = 939 \cdot 4 - 396 \cdot 3 = 543 \cdot 1$$

∴ $$R_T = \frac{543 \cdot 1}{0 \cdot 3420 \times 8} = \frac{543 \cdot 1}{2 \cdot 736} = 198 \cdot 5 \text{ kN} \quad \textbf{Ans.}$$

and $$R_B = F - R_T = 600 \cdot 9 - 198 \cdot 5 = 402 \cdot 4 \text{ kN} \quad \textbf{Ans.}$$

EXERCISE 4·2

1. A vertical sluice gate 2 m wide and 2 m deep contains water on its either side. The water on its upstream side is 1·5 m higher than the top of the gate and on the downstream side up to the top edge of the gate. Find the resultant froce on the gate. [**Ans.** 78·48 kN]

2. A vertical sluice gate 2 m wide and 1·2 m deep is containing a liquid of specific gravity 1·45 on its upstream side and up to height of 1·5 m above the top edge of the gate. There is water on the downstream side up to the top edge of the gate. Find the resultant force acting on the gate and its position. [**Ans.** 57·56 kN; 0·577 m from the bottom]

3. Two lock gates 4 m high meeting at an angle of 120° are provided in a canal of 6 m width. Each gate carries two hinges placed at the top and bottom of the gate. What is the reaction at the pivot, when the water on the upstream side is 3 m deep ? [**Ans.** 152·9 kN]

4. A pair of lock gates 5 m high are provided in a canal of 6·25 m width, which meet at an angle of 120°. These gates are provided with two hinges at the top and bottom. If the water levels on the upstream and downstream sides of the gate are 4 m and 2 m respectively, determine (*i*) resultant pressure on each gate, (*ii*) reaction between the gates, and (*iii*) reaction on both the hinges.
   [**Ans.** 212·5 kN; 212·5 kN; R_T = 66·1 kN; R_B = 146·4 kN]

4·9 Water Pressure on Masonry Walls

Consider a vertical masonry wall having water on one of its sides as shown in Fig. 4·15. Now consider a unit length of the wall. We know that the water pressure will act perpendicular to the wall. A little consideration will show, that the intensity of pressure, at the water level, will be zero, and will increase by a straight line law to wH at the bottom as shown in Fig. 4·15. Thus the pressure diagram will be a triangle.

The total pressure on the wall will be the area of the triangle. *i.e.*,

$$*P = \frac{wH}{2} \times H = \frac{wH^2}{2}$$

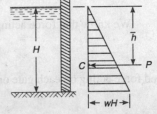

Fig. 4·15. Masonry wall.

This pressure will act through the centre of gravity of the pressure diagram.

---

*The value of total pressure (P) can also be found out by the relation :

$$P = wA\bar{x} = w\ (H \times 1) \times \frac{H}{2} = \frac{wH^2}{2}$$

and the value of $\bar{h}$ may also be found out by the relation,

$$= \frac{I_G}{\bar{x}} + \bar{x} = \frac{\frac{1 \times (H)^3}{12}}{(1 \times H) \times \frac{H}{2}} + \frac{H}{2} = \frac{2H}{3}$$

Applications of Hydrostatics

Let $\bar{h}$ = Depth of the centre of pressure from the water surface.

We know that the c.g. of triangle is at a height of $H/3$ from the base, where H is the height of the triangle. Therefore depth of centre of pressure from the water surface,

$$\bar{h} = H - \frac{H}{3} = \frac{2H}{3}$$

Thus the pressure of water on a vertical wall will act through a point at a distance $H/3$ from the bottom, where H is the depth of water.

Example 4·11. *One of the walls of a swimming pool contains 4 m deep water. Determine the total pressure on the wall, if it is 10 m wide.*

Solution. Given : Depth of water $H = 4$ m and width of wall 10 m.

We know that pressure on the wall per metre length

$$= \frac{wH^2}{2} = \frac{9 \cdot 81 \times (4)^2}{2} = 78 \cdot 48 \text{ kN}$$

and total pressure on the wall $P = 10 \times 78 \cdot 48 = 784 \cdot 8$ kN **Ans.**

Example 4·12. *A rectangular container 2·5 m wide has a vertical partition in the middle. It is filled with petrol of specific gravity 0·8 to a height of 1 m on one side and oil of specific gravity 0·9 to a height of 0·8 m on the other. Find the resultant thrust per metre length of the partition wall and its point of application.*

Solution, Given : Width of partition wall 2·5 m; Specific gravity of petrol = 0·7 ; Depth of petrol $(H_1) = 1$ m ; Specific gravity of oil = 0·9 and depth of oil $(H_2) = 0·8$ m.

Resultant thrust on the partition wall

We know that pressure of petrol per metre length of the partition wall,

$$P_1 = \frac{w_1 H_1^2}{2} = \frac{0 \cdot 7 \times 9 \cdot 81 \times (1)^2}{2} \text{ kN}$$

$$= 3 \cdot 434 \text{ kN}$$

and pressure of oil per metre length of the partition wall,

$$P_2 = \frac{w_2 H_2^2}{2} = \frac{0 \cdot 9 \times 9 \cdot 81 \times (0 \cdot 8)^2}{2} \text{ kN}$$

$$= 2 \cdot 825 \text{ kN}$$

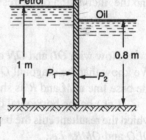

Fig. 4·16.

∴ Resultant thrust per metre length of the partition wall,

$$P = P_1 - P_2 = 3 \cdot 434 - 2 \cdot 825 = 0 \cdot 609 \text{ kN}$$

Point of application of the pressure

Let $\bar{h}$ = Height of the centre of the resultant pressure from the base.

We know that the pressures P_1 and P_2 will act at a height of $H/3$ from the base, where H is the respective height of petrol and oil from the base. Taking moments of the pressures about the base and equating the same,

$$P \times \bar{h} = P_1 \times \frac{1}{3} - P_2 \times \frac{0 \cdot 8}{3}$$

$$0 \cdot 609 \times \bar{h} = 3 \cdot 434 \times \frac{1}{3} - 2 \cdot 825 \times \frac{0 \cdot 8}{3} = 1 \cdot 145 - 0 \cdot 753 = 0 \cdot 392$$

∴ $\bar{h} = 0 \cdot 392 / 0 \cdot 609 = 0 \cdot 644$ m **Ans.**

4·10 Water Pressure on Masonry Dams

The dams are constructed in order to store large quantities of water, for the purpose of irrigation and power generation. A dam may be of any cross-section, but the following are important from the subject point of view :
1. Rectangular dams. 2. Trapezoidal dams.

4·11 Water Pressure on Rectangular Dams

Consider a rectangular dam retaining water on one of its sides as shown in Fig. 4·17. Now consider a unit length of the dam.

Let H = Height of water retained by the dam.

We know that total pressure on the dam due to water,

$$P = \frac{wH^2}{2}$$

and this pressure will act as height of $H/3$ above the base of the dam. Let W be the weight of the dam masonry per unit length of the dam. We know that W will act downwards through the centre of gravity of the dam section.

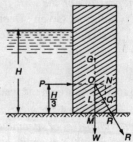

Fig. 4·17. Rectangular dam.

Now the resultant pressure of the force (P) and weight (W) will be given by the relation,

$$R = \sqrt{P^2 + W^2}$$

and the inclination of the resultant with the vertical (θ) will be given by the relation :

$$\tan \theta = \frac{P}{W}$$

Now with OL and ON or LQ (equal to W and P to some scale) complete the rectangle $OLQN$. We know that the diagonal OQ will give the resultant (R) to scale. Now extend OL and OQ to meet the base line at M and R as shown in Fig. 4·17.

Let x be the horizontal distance between the centre of gravity of the dam and the point through which the resultant cuts the base (i.e., MR). The distance x may be found out from the similar triangles OLQ and OMR. i.e.,

$$\frac{MR}{OM} = \frac{LQ}{OL} \quad \text{or} \quad \frac{x}{\frac{H}{3}} = \frac{P}{W} \quad \text{or} \quad x = \frac{P}{W} \times \frac{H}{3}$$

Example 4·13. *A retaining wall 6 m high and 2·5 m wide retains water up to its top. Find the total pressure per metre length of the wall and the point at which the resultant cuts the base. Also find the resultant thrust on the base of the wall per metre length. Assume weight of masonry as 23 kN/m³.*

Solution. Given : $H = 6$ m ; $b = 2.5$ m and weight of masonry = 23 kN/m³.

Total pressure per metre length of the wall

We know that total pressure per metre length of the wall,

$$P = \frac{wH^2}{2} = \frac{9 \cdot 81 \times (6)^2}{2} \text{ kN}$$

$$= 176 \cdot 6 \text{ kN} \quad \textbf{Ans.}$$

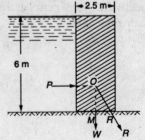

Fig. 4·18.

Applications of Hydrostatics

Point at which the resultant cuts the base

We also know that weight of masonry per metre length of the wall,

$$W = 23 \times 6 \times 2{\cdot}5 = 345 \text{ kN}$$

and distance between the mid-point (M) of the wall and the point where resultant cuts the base (R),

$$x = \frac{P}{W} \times \frac{H}{3} = \frac{176{\cdot}6}{345} \times \frac{6}{3} = 1{\cdot}02 \text{ m} \quad \text{Ans.}$$

Resultant thrust on the base of the wall per metre length

We know that resultant thrust on the base of the wall per metre length

$$R = \sqrt{P^2 + W^2} = \sqrt{(176{\cdot}6)^2 + (345)^2} = 387{\cdot}6 \text{ kN} \quad \text{Ans.}$$

Example 4·14. *A concrete dam of rectangular section, 15 metres high and 4 metres wide has water standing 3 metres below its top. Find :*

(a) water pressure on one metre length of dam,

(b) height of the centre of pressure above base, and

(c) the point at which the resultant cuts the base.

Assume the weight of concrete as 24 kN per cubic metre.

Solution. Given : Total height of the dam = 15 m; Width of dam (b) = 4 m ; Height of water (H) = 15 – 3 = 12 m and weight of concrete = 24 kN /m³.

Water pressure on one metre length of dam

We know that total water pressure on one metre length of the dam,

$$P = \frac{wH^2}{2} = \frac{9{\cdot}81 \times (12)^2}{2} \text{ kN}$$

$$= 706{\cdot}3 \text{ kN} \quad \text{Ans.}$$

Height of the centre of pressure above base

We also know that centre of pressure is at a height of $H/3$ above the base i.e., 12/3 = 4 m above the base. **Ans.**

The point at which the resultant cuts the base

We know that weight of concrete per metre length of the dam,

$$W = 24 \times 15 \times 4 = 1440 \text{ kN}$$

and distance between mid-point (M) of the dam and the point where resultant cuts the base (R).

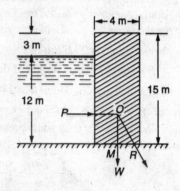

Fig. 4·19

$$x = \frac{P}{W} \times \frac{H}{3} = \frac{706{\cdot}3}{1440} \times \frac{12}{3} = 1{\cdot}96 \text{ m} \quad \text{Ans.}$$

4·12 Water Pressure on Trapezoidal Dams

A trapezoidal dam is more economical and also easier to construct than a rectangular dam. That is why, these days trapezoidal dams are preferred over the rectangular ones.

Consider a trapezoidal dam *ABCD* retaining water on one of its sides (say vertical side) as shown in Fig. 4·20. Now consider a unit length of the dam.

Let H = Height of water retained by the dam.

Like the rectangular dam, the total pressure on a trapezoidal dam per metre length will also be given by the relation,

$$P = \frac{wH^2}{2}$$

and the horizontal distance between the centre of gravity of the dam and the point, at which the resultant cuts the base will also be given by the relation :

$$x = \frac{P}{W} \times \frac{H}{3}$$

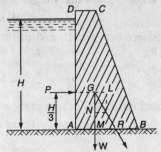

Fig. 4·20. Trapezoidal dam.

Notes : 1. The resultant thrust on the base of the dam per metre length may be found from the relation,

$$R = \sqrt{P^2 + W^2}$$

2. The point M at the base of the dam through which the centre of gravity of the dam passes may be found out either by taking moments about A or by the relation :

$$AM = \frac{a^2 + ab + b^2}{3\,(a+b)}$$

where a and b are the top and bottom widths of the dam, respectively.

3. If the water is retained on the inclined face, the water pressure will act normally to the face. In such a case, the horizontal and vertical components of the pressure are to be used for all calculations. This may be simplified by assuming the surface to be vertical. Now the weight of triangular wedge of water, over the inclined surface, is included in the weight of dam.

Example 4·15. *A concrete dam having water on vertical face is 16 metres high. The base of the dam is 8 metres wide and top 3 metres wide. Find the resultant thrust on the base per metre length of the dam and the point where it intersects the base, when it contains water 16 m deep. Take weight of the concrete as 23 kN/m³.*

Solution. Given : Height of dam = 20 m; Base width = 8 m; Top width = 3 m; Depth of water = 16 m and weight of concrete = 24 kN/m³.

Resultant thrust on the base per metre length of the dam

We know that total water pressure per metre length of the dam,

$$P = \frac{wH^2}{2} = \frac{9·81 \times (16)^2}{2} \text{ kN}$$

$$= 1255·7 \text{ kN}$$

and weight per metre length of the dam,

$$W = 24 \times \left[\frac{1}{2} \times (8+3) \times 20\right] \text{ kN}$$

$$= 2640 \text{ kN}$$

∴ Resultant thrust on the base per metre length of the dam,

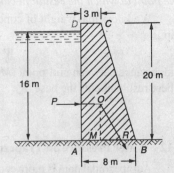

Fig. 4·21.

$$R = \sqrt{P^2 + W^2} = \sqrt{(1255·7)^2 + (2640)^2} = 2923·4 \text{ kN} \quad \textbf{Ans.}$$

The point where the resultant thrust intersects the base

First of all, let us find out the point (M) at the base of the dam, through which centre of gravity (G) passes. Taking moments of the area of trapezoidal section about A and equating the same,

Applications of Hydrostatics

$$*AM \times \left(20 \times \frac{3+8}{2}\right) = \left[20 \times 3 \times \frac{3}{2}\right] + \left[20 \times \frac{5}{2}\left(3 + \frac{5}{3}\right)\right]$$

or
$$110\,AM = 90 + 233 \cdot 3 = 323 \cdot 3$$

∴
$$AM = 323 \cdot 3/110 = 2 \cdot 94 \text{ m}$$

We know that distance between the mid-point (M) of the dam and point where the resultant cuts the base (R),

$$x = \frac{P}{W} \times \frac{H}{3} = \frac{1255 \cdot 7}{2640} \times \frac{16}{3} = 2 \cdot 54 \text{ m}$$

∴
$$AR = AM + x = 2 \cdot 94 + 2 \cdot 54 = 5 \cdot 48 \text{ m} \quad \textbf{Ans.}$$

Example 4·16. *A trapezoidal masonry dam having 2 m top width and 8 m bottom width is of 15 m height as shown in Fig. 4·22.*

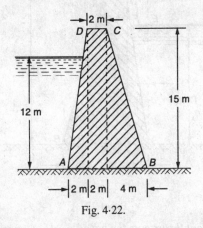

Fig. 4·22.

Find (i) total water pressure per metre length of the dam; (ii) resultant thrust on the base of the dam and (iii) the point where the resultant thrust cuts the base. Take weight of the masonry as 22·5 kN/m³.

Solution. Given : Top width = 2 m; Bottom width 8 m; Height of dam 15 m; Height of water $(H) = 12$ m and weight of the masonry $= 22 \cdot 5$ kN/m³.

(i) Total water pressure per metre length of the dam

We know that total water pressure per metre length of the dam,

$$P = \frac{wH^2}{2} = \frac{9 \cdot 81 \times (12)^2}{2} = 706 \cdot 3 \text{ kN} \quad \textbf{Ans.}$$

(ii) Resultant thrust on the base of the dam

We also know that weight per metre length of the dam,

$$W_1 = 22 \cdot 5 \times \frac{(2+8)}{2} \times 15 = 1687 \cdot 5 \text{ kN}$$

---

*The distance AM may also be found out from the relation :

$$AM = \frac{a^2 + ab + b^2}{3(a+b)} = \frac{(3)^2 + (3 \times 8) + (8)^2}{3(3+8)} = \frac{9 + 24 + 64}{33} = \frac{97}{33} = 2 \cdot 94 \text{ m.}$$

and weight of water wedge (*AEF*) per metre length of the dam,

$$W_2 = 9.81 \times \frac{1}{2} \times AE \times EF \text{*}$$

$$= 9.81 \times \left(\frac{1}{2} \times 12 \times 1.6\right) = 94.2 \text{ kN}$$

∴ Total weight $W = W_1 + W_2 = 1687.5 + 94.2 = 1781.7 \text{ kN}$

and resultant thrust on the base of the dam,

$$R = \sqrt{P^2 + W^2} = \sqrt{(706.3)^2 + (1781.7)^2} = 1916.6 \text{ kN} \quad \textbf{Ans.}$$

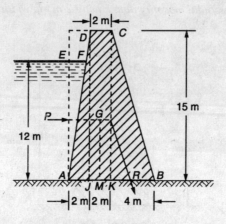

Fig. 4·23.

(iii) *The point where the resultant thrust cuts the base*

First of all, let us find out the point (*M*) at the base of the dam, through which centre of gravity (*G*) passes. Taking moments of the area of the water section of *AEF* and trapezoidal section about *A* and equating the same,

$$AM \times \text{Total weight} = \left[\text{Weight of water } AEF \times \frac{EF}{3}\right]$$
$$+ \left[\text{Weight of masonry } ADJ \times \frac{2AJ}{3}\right]$$
$$+ [\text{Weight of masonry } JKCD \times (2 + 1)]$$
$$+ \left[\text{Weight of masonry } KBC \times \left(4 + \frac{4}{3}\right)\right]$$

$$AM \times 1781.7 = \left[9.81 \times \left(\frac{1}{2} \times 12 \times 1.6\right) \times \frac{1.6}{3}\right]$$
$$+ \left[22.5 \times \left(\frac{1}{2} \times 15 \times 2\right) \times \frac{2 \times 2}{3}\right]$$
$$+ [22.5 \times (15 \times 2) \times 3]$$
$$+ \left[22.5 \times \left(\frac{1}{2} \times 15 \times 4\right) \times \frac{16}{3}\right]$$
$$= 50.2 + 450 + 2025 + 3600 = 6125.2$$

∴ $AM = 6125.2/1781.7 = 3.44 \text{ m}$

---

* Length *EF* may also be found out as :

$$\frac{12}{15} \times AJ = \frac{12}{15} \times 2 = 1.6 \text{ m}.$$

We know that distance between the mid-point (M) of the dam and point where the resultant cuts the base (R),

$$x = \frac{P}{W} \times \frac{H}{3} = \frac{706.3}{1781.7} \times \frac{12}{3} = 1.59 \text{ m}$$

∴ $AR = AM + x = 3.44 + 1.59 = 5.03$ m **Ans.**

4·13 Stability of a Dam

In Arts. 4·11 and 4·12, we have derived a relation which gives us the point, in the base of a dam, through which the resultant cuts the base. In these articles, we have not discussed the importance of this point. As a matter of fact, the point in the base, through which the resultant cuts gives us very important information and helps us in finding the stability of the dam. e.g.

1. To safeguard the dam against overturning, the resultant must pass within the base.
2. To avoid tension at the base, the resultant must pass through the middle third of the base.
3. The maximum stress developed at the bottom of the dam should be within the permissible stress of the site.
4. To prevent sliding, the maximum frictional force (i.e., $\mu \times W$, where μ is the coefficient of friction and W is the weight of the dam) should be *more than the horizontal force P.

Example 4·17. *A masonry wall 5 metres high and 1·8 metre wide is containing water up to a height of 4 metres. If the coefficient of friction between the wall and the soil is 0·6, check the stability of the wall. Take weight of the masonry as 22 kN/m³.*

Solution. Given : Height of wall = 5 m; Width of wall = 1·8 m; Height of water (H) = 4 m; Coefficient of friction (μ) = 0·6 and weight of masonry = 22 kN/m³.

First of all let us find out the point (R) where the resultant thrust of water pressure and weight of the wall cuts the base. We know that total water pressure per metre length of the wall,

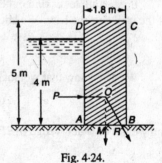

$$P = \frac{wH^2}{2} = \frac{9.81 \times (4)^2}{2} = 78.48 \text{ kN}$$

and weight of the wall,

$$W = 22 \times 5 \times 1.8 = 198 \text{ kN}$$

∴ Distance between the mid-point (M) of the wall and point where the resultant thrust cuts the base (R),

Fig. 4·24.

$$x = \frac{P}{W} \times \frac{H}{3} = \frac{78.48}{198} \times \frac{4}{3} = 0.53 \text{ m}$$

∴ $AR = AM + x = 0.9 + 0.53 = 1.43$ m

Now we shall check stability of the wall.

1. Check for tension at the base. Since the resultant thrust cuts the base beyond the middle third (because AR = 1·43 m and middle third of AB is from 0·6 m to 1·2 m), therefore the wall shall fail due to tension.

2. Check for overturning the wall. Since the resultant thrust is passing within the base, therefore the wall is safe against overturning. **Ans.**

3. Check for sliding the wall. We know that horizontal pressure due to water (P) = 78·48 kN. And the frictional force

$$= \mu W = 0.6 \times 198 = 118.8 \text{ kN}$$

---

*Certain authorities on the subject are of the opinion that magnitude of the weight should preferably be 1.5 times the horizontal pressure due to water.

Since the frictional force (118·8 kN) is more than the horizontal pressure (78·48 kN), therefore the wall is safe against sliding. **Ans.**

Example 4·18. *A masonry dam 12 metres high trapezoidal in section has top width 1 metre and bottom width 7·2 metres. The face exposed to water has a slope of 1 horizontal to 10 vertical. Check the stability of the dam, when the water level rises 10 m high. The coefficient of friction between the bottom of the dam and the soil as 0·6. Take the weight of the masonry as 22 kN/m³.*

Solution. Given : Height of dam = 12 m; Top width = 1 m; Bottom width = 7·2 m; Height of water (H) = 10 m; Coefficient of friction (μ) = 0·6 and weight of masonry = 22 kN/m³.

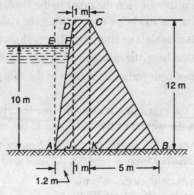

Fig. 4·25.

First of all, let us find out the point (R) where the resultant thrust of water pressure and weight of the wall cuts the base. We know that total water pressure per metre length of the dam,

$$P = \frac{wH^2}{2} = \frac{9 \cdot 81 \times (10)^2}{2} = 490 \cdot 5 \text{ kN}$$

We also know that weight per metre length of the concrete dam,

$$W_1 = 22 \times \frac{1 + 7 \cdot 2}{2} \times 12 = 1082 \cdot 4 \text{ kN}$$

and weight of water wedge (*AEF*) per metre length of the dam

$$W_2 = 9 \cdot 81 \times \left(\frac{1}{2} \times 10 \times 1\right) = 49 \cdot 05 \text{ kN}$$

∴ Total weight $\quad W = W_1 + W_2 = 1082 \cdot 4 + 49 \cdot 05 = 1131 \cdot 45$ kN

Now let us find out the point (M) at the base of the dam, through which centre of gravity (G) passes. Taking moments about *A* and equating the same,

$$AM \times \text{Total weight} = \left[\text{Weight of water } AEF \times \frac{EF}{3}\right]$$
$$+ \left[\text{Weight of concrete } ADJ \times \frac{2AJ}{3}\right]$$
$$+ \left[\text{Weight of concerete } JKCD \times \left(1 \cdot 2 + \frac{1}{2}\right)\right]$$
$$+ \left[\text{Weight of concrete } KBC \times \left(2 \cdot 2 + \frac{5}{3}\right)\right]$$

$$AM \times 1131\cdot 45 = \left[9\cdot 81 \times \left(\frac{1}{2} \times 10 \times 1\right) \times \frac{1}{3}\right]$$

$$+ \left[22 \times \left(\frac{1}{2} \times 12 \times 1\right) \times \frac{2 \times 1\cdot 2}{3}\right]$$

$$+ [22 \times (12 \times 1) \times 1\cdot 7]$$

$$+ \left[22 \times \left(\frac{1}{2} \times 12 \times 5\right) \times 3\cdot 87\right]$$

$$= 16\cdot 35 + 105\cdot 6 + 448\cdot 8 + 2554\cdot 2 = 3124\cdot 95$$

∴ $AM = 3124\cdot 95/1131\cdot 45 = 2\cdot 76$ m

and distance between mid-point (M) of the dam and point where the resultant cuts the base (R),

$$x = \frac{P}{W} \times \frac{H}{3} = \frac{490\cdot 5}{1131\cdot 45} \times \frac{10}{3} = 1\cdot 45 \text{ m}$$

∴ $AR = AM + x = 2\cdot 76 + 1\cdot 45 = 4\cdot 21$ m

Now we shall check stability of the dam.

1. Check for tension at the base. Since the resultant thrust cuts the base in the middle third (because $AR = 4\cdot 21$ m and middle third is from 2·4 m to 4·8 m), therefore the dam is safe against tension in the base. **Ans.**

2. Check for overturning the dam. Since the resultant thrust is passing within the base, therefore the dam is safe against overturning. **Ans.**

3. Check for sliding the dam. We know that horizontal pressure due to water (P) = 490·5 kN. And the frictional force

$$= \mu W = 0\cdot 6 \times 1131\cdot 45 = 678\cdot 9 \text{ kN}$$

Since the frictional force (678·9 kN) is more than the horizontal pressure (490·5 kN), therefore the dam is safe against sliding. **Ans.**

EXERCISE 4·3

1. A wall 5 m long contains water which is 3 m deep. What is the total pressure on the wall?
   [**Ans.** 220·7 kN]

2. A partition wall contains water for a depth of 2·4 m on one side and oil of specific gravity 0·8 for a depth of 1·5 m on the other side. Find (*i*) resultant thrust on the wall per metre length, and (*ii*) the point where it acts. [**Ans.** 19·42 kN; 0·94 m from the base]

3. A masonry retaining wall 5·5 m high and 2·4 m wide contains water up in its top. Find (*i*) total pressure per metre length of the wall, (*ii*) resultant thrust on the base of the wall per metre length. Take weight of the masonry as 23 kN/m³. [**Ans.** 148·4 kN; 362·9 kN]

4. A concrete dam 12 m high and 3·5 m wide contains 9 m deep water. Calculate the water pressure per metre length of the dam and the point where the resultant cuts the base. Assume weight of concrete as 24 kN/m³. [**Ans.** 379·3 kN; 1·13 m from the mid-point]

5. A masonry dam 1·5 m wide at the top and 4 m wide at the bottom is 6 m high. The face exposed to water is vertical and water stands up to the top level of the dam. Find (*i*) total pressure due to water per metre length of the dam; (*ii*) resultant thrust per metre length of the dam and (*iii*) the point where it acts. Take weight of the concrete as 24 kN/m³. [**Ans.** 176·6 kN; 216·5 kN; 2·37 m from A]

6. A masonry wall 8 m high and 3 m wide contains water for a height of 7 m. Check the stability of the wall, if the coefficient of friction between the wall and the soil is 0·55. Take weight of masonry as 22·2 kN/m^3.

[**Ans.** 1. The wall shall fail due to tension. 2. Safe for overturning. 3. Safe against sliding]

7. A concrete trapezoidal dam 2·5 m wide at the top and 10 m wide at the bottom is 25 m high. It contains water on its vertical side. Check the stability of the dam, when it contains water for a depth of 20 m. Take coefficients of friction between the wall and soil as 0·6 and weight of the concrete as 24 kN/m^3.

[**Ans.** 1. The dam shall fail due to tension. 2. Safe against overturning. 3. Safe against sliding]

QUESTIONS

1. What is a pressure diagram ? Discuss its utility in the Hydraulics ?
2. Explain the phenomenon of pressure due to one kind of liquid over another on one side of a wall.
3. Name any two applications of Hydrostatics.
4. Explain the difference between sluice gates and lock gates.
5. Derive an expression for the pressure on top and bottom hinges of a lock gate.
6. Obtain a relation for the point where the resultant pressure cuts the base of a dam retaining water on one of its sides.
7. What do you understand by the term 'stability of a dam'? Name the factors, which are important for checking the stability of a trapezoidal dam.

OBJECTIVE TYPE QUESTIONS

1. The water pressure per metre length on a vertical wall is

    (a) wH (b) $\dfrac{wH}{2}$ (c) $\dfrac{wH^2}{2}$ (d) $\dfrac{wH^2}{4}$

    where
    w = Specific weight of the liquid, and
    H = Height of the water

2. A vertical wall is subjected to a pressure due to one kind of liquid for a depth H on one of its sides. The total pressure acts at a distance of .... from the liquid surface.

    (a) $\dfrac{H}{3}$ (b) $\dfrac{H}{2}$ (c) $\dfrac{2H}{3}$ (d) $\dfrac{3H}{4}$

3. In a lock gate, the reaction between the two gates is

    (a) $\dfrac{P}{\sin \alpha}$ (b) $\dfrac{P}{\sin 2\alpha}$ (c) $\dfrac{2P}{\sin \alpha}$ (d) $\dfrac{2P}{\sin 2\alpha}$

    where
    P = Resultant pressure on the lock gate, and
    α = Inclination of the gate with the normal to the walls of the lock.

4. The stability of a dam is checked for
    (a) tension at the base   (b) overturning of the dam
    (c) sliding of the dam   (d) all of these

ANSWERS

1. (c)   2. (c)   3. (b)   4. (d)

5

Equilibrium of Floating Bodies

1. Introduction. 2. Archimedes' Principle. 3. Buoyancy. 4. Centre of Buoyancy. 5. Metacentre. 6. Metacentric Height. 7. Analytical Method for Metacentric Height. 8. Conditions of Equilibrium of a Floating Body. 9. Stable Equilibrium. 10. Unstable Equilibrium. 11. Neutral Equilibrium. 12. Maximum Length of Vertically Floating Body. 13. Floating Bodies Anchored at the Base. 14. Conical Buoys Floating in Liquid. 15. Experimental Method for Metacentric Height. 16. Time of Rolling (Oscillation) of a Floating Body.

5·1 Introduction

We see that whenever a body is placed over a liquid, either it sinks down or floats on the liquid. If we analyse the phenomenon of floatation, we find that the body, placed over a liquid, is subjected to the following two forces :

1. Gravitational force, and   2. Upthrust of the liquid.

Since the two forces act opposite to each other, therefore we have to study the comparative effect of these forces. A little consideration will show, that if the gravitational force is more than the upthrust of the liquid, the body will sink down. But if the gravitational force is less than the upthrust of the liquid, the body will float. This may be best understood by the Archimedes' principle as discussed below.

5·2 Archimedes' Principle

The Archimedes' principle states, *"Whenever a body is immersed wholly or partially in a fluid, it is buoyed up (i.e., lifted up) by a force equal to the weight of fluid displaced by the body."* Or in other words, whenever a body is immersed wholly or partially in a fluid, the resultant force acting on it, is equal to the difference between the upward pressure of the fluid on its bottom and the downward force due to gravity.

5·3 Buoyancy

The tendency of a fluid to uplift a submerged body, because of the upward thrust of the fluid, is known as the force of buoyancy or simply buoyancy. It is always equal to the weight of the fluid displaced by the body. It will be interesting to know that if the force of buoyancy is greater than the weight of the body, it will be pushed up till the weight of the fluid displaced is equal to the weight of the body, and the body will float. But if the force of buoyancy is less than the weight of the body, it will sink down.

5·4 Centre of Buoyancy

The centre of buoyancy is the point, through which the force of buoyancy is supposed to act. It is always the centre of gravity of the volume of the liquid displaced. In other words, the centre of buoyancy is the centre of area of the immersed section.

Example 5·1. *A uniform body 3 m long, 2 m wide and 1 m deep floats in water. If the depth of immersion is 0·6 m, what is the weight of the body ?*

[87]

Solution. Given : Length = 3 m; Width = 2 m; Depth = 1 m and depth of immersion = 0·6 m.

We know that volume of water displaced

$$= 3 \times 2 \times 0.6 = 3.6 \text{ m}^3$$

and weight of the body = Weight of water displaced

$$= 9.81 \times 3.6 = 35.3 \text{ kN} \quad \textbf{Ans.}$$

Example 5·2. *A floating buoy in harbour is to be assisted in floating upright by a submerged weight of concrete attached to the bottom of the buoy. How many cubic metres of concrete weighing 23 kN/m³ must be provided to get a net downward pull of 3·25 kN from the weight? Take specific weight of sea water as 10 kN/m³.*

Solution. Given : Specific weight of concrete = 23 kN/m³; Downward pull = 3·25 kN and specific weight of water, *w = 10 kN/m³.

We know that the weight of 1 m³ of concrete in sea water

$$= 23 - 10 = 13 \text{ kN/m}^3$$

and the weight of concrete required,

$$= \frac{\text{Downward pull}}{\text{Weight of concrete in water}} = \frac{3.25}{13} = 0.25 \text{ m}^3 \quad \textbf{Ans.}$$

Example 5·3. *A block of wood 4 m long, 2 m wide and 1 m deep is floating horizontally in water. If the density of the wood be 6·87 kN/m³, find the volume of the water displaced and position of the centre of buoyancy.*

Solution. Given : Size of wooden block = 4 m × 2 m × 1 m and density of wood = 6·87 kN/m³.

Volume of water displaced

We know that the volume of the wooden block

$$= 4 \times 2 \times 1 = 8 \text{ m}^3$$

and its weight = $8 \times 6.87 = 55$ kN

∴ Volume of water displaced

$$= \frac{\text{Weight of block}}{\text{Density of water}} = \frac{55}{9.81} = 5.6 \text{ m}^3 \quad \textbf{Ans.}$$

Position of centre of buoyancy

We also know that the depth of immersion

$$= \frac{\text{Volume}}{\text{Sectional area}} = \frac{5.6}{4 \times 2} = 0.7 \text{ m}$$

and centre of buoyancy = 0·7/2 = 0·35 m from the base   **Ans.**

Example 5·4. *A wooden block of 4 m × 1 m × 0 ·5 m in size and of specific gravity 0·75 is floating in water. Find the weight of concrete of specific weight 24 kN/m³ that may be placed on the block, which will immerse the wooden block completely.*

Solution. Given : Size of wooden block = 4 m × 1 m × 0·5 m; Specific gravity of block = 0·75 and specific weight of concrete = 24 kN/m³.

Let W = Weight of the concrete required to be placed on the wooden block.

---

*Superfluous data.

We know that the volume of the wooden block
$$= 4 \times 1 \times 0.5 = 2 \text{ m}^3$$
and its weight
$$= 9.81 \times (0.75 \times 2) = 14.72 \text{ kN}$$
∴ Total weight of the block and concrete
$$= 14.72 + W \text{ kN} \qquad ...(i)$$
We also know that when the block is completely immersed in water, volume of water displaced
$$= 2 \text{ m}^3$$
∴ Upward thrust when the block is completely immersed in water
$$= 9.81 \times 2 = 19.62 \text{ kN} \qquad ...(ii)$$
Now equating the total weight of block and concrete with the upward thrust,
$$14.72 + W = 19.62 \quad \text{or} \quad W = 19.62 - 14.72 = 4.9 \text{ kN} \quad \textbf{Ans.}$$

EXERCISE 5·1

1. A rectangular block of 1 m × 0·6 m × 0·4 m floats in water with 1/5th of its volume being out of water. Find the weight of the block. [Ans. 1·88 kN]
2. A wooden block of 3 cubic metre volume floats in the sea water. If the density of the wood is 7 kN/m^3, find the volume of water displaced. [Ans. 2·1 m^3]
3. A solid cube of 0·5 m side and specific gravity 0·6 floats in water with one of its faces parallel to the water surface. What load may be placed on the cube, so that it may completely immerse in water? [Ans. 490 kN]
4. A wooden block of 6 m × 2·5 m × 1·5 m is floating horizontally in the water. If the density of the wood is 6 kN/m^3, find the volume of water displaced and position of the centre of buoyancy.
[Ans. 13·8 m^3; 0·46 m from the bottom]

5·5 Metacentre

Whenever a body, floating in a liquid, is given a small angular displacement, it starts oscillating about some point. This point, about which the body starts oscillating, is called metacentre.

In other words, the metacentre may also be defined as the inter-section of the line passing through the original centre of buoyancy (B) and c.g., (G) of the body, and the vertical line through the new centre of buoyancy (B_1) as shown in Fig. 5·1.

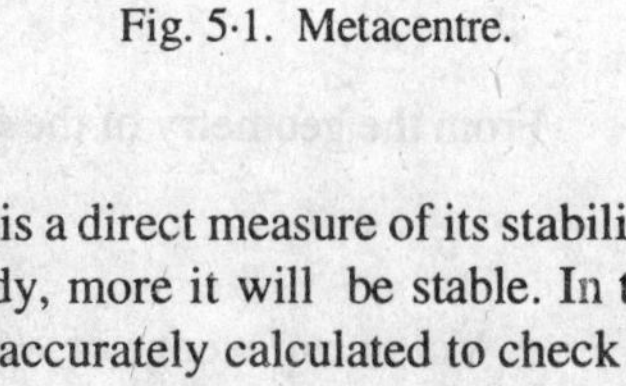

Fig. 5·1. Metacentre.

5·6 Metacentric Height

The distance between the centre of gravity of a floating body and the metacentre (i.e., distance GM as shown in Fig. 5·1) is called metacentric height.

As a matter of fact, the metacentric height of a floating body is a direct measure of its stability. Or in other words, more the metacentric height of a floating body, more it will be stable. In the modern design offices, the metacentric height of a boat or ship is accurately calculated to check its stability. Some values of metacentric height are given below :

        Merchant ships = up to 1·0 m
        Sailing ships   = up to 1·5 m
        Battle ships   = up to 2·0 m
        River craft   = up to 3·5 m

5·7. Analytical Method for Metacentric Height

Consider a ship floating freely in water. Let the ship be given a clockwise rotation through a small angle θ (in radians) as shown in Fig. 5·2. The immersed section has now changed from *acde* to acd_1e_1.

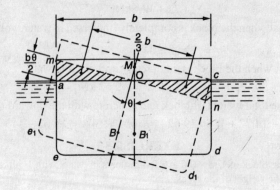

Fig. 5·2. Metacentric height.

The original centre of buoyancy B has now changed to a new position B_1. It may be noted that the triangular wedge *aom* has come out of water, whereas the triangular wedge *ocn* has gone under water. Since the volume of water displaced remains the same, therefore the two triangular wedges must have equal areas.

A little consideration will show that as the triangular wedge *aom* has come out of water, thus decreasing the force of buoyancy on the left, therefore it tends to rotate the vessel in an anti-clockwise direction about O. Similarly, as the triangular wedge *ocn* has gone under water, thus increasing the force of buoyancy on the right, therefore it again tends to rotate the vessel in an anti-clockwise direction. It is thus obvious, that these forces of buoyancy will form a couple, which will tend to rotate the vessel in anti-clockwise direction about O. If the angle (θ), through which the body is given rotation, is extremely small, then the ship may be assumed to rotate about M (*i.e.*, metacentre).

Let
l = Length of the ship,
b = Breadth of the ship,
θ = Very small angle (in radians) through which the ship is rotated, and
V = Volume of water displaced by the ship.

From the geometry of the figure, we find that

$$am = cn = \frac{b\theta}{2}$$

∴ Volume of wedge of water *aom*

$$= \frac{1}{2}\left(\frac{b}{2} \times am\right)l = \frac{1}{2}\left(\frac{b}{2} \times \frac{b\theta}{2}\right)l \qquad \ldots\left(\because am = \frac{b\theta}{2}\right)$$

$$= \frac{b^2 \theta l}{8}$$

∴ Weight of this wedge of water

$$= \frac{wb^2\theta l}{8} \qquad \ldots(w = \text{Sp. wt. of water})$$

and arm of the couple $= \dfrac{2}{3} b$

∴ Moment of the restoring couple

$$= \dfrac{wb^2 \theta l}{8} \times \dfrac{2}{3} b = \dfrac{wb^3 \theta l}{12} \qquad \ldots(i)$$

and moment of the disturbing force

$$= w \cdot V \times BB_1 \qquad \ldots(ii)$$

Equating these two moments,

$$\dfrac{wb^3 \theta l}{12} = w \times V \times BB_1$$

Substituting the values of $\dfrac{lb^3}{12} = I$ (*i.e.*, moment of inertia of the plan of the ship) and $BB_1 = BM \times \theta$ in the above equation,

$$w \cdot I \cdot \theta = w \times V (BM \times \theta)$$

∴ $BM = \dfrac{I}{V} = \dfrac{\text{Moment of inertia of the plan}}{\text{Volume of water displaced}}$

Now metacentric height *

$$GM = BM \pm BG$$

Note : +ve sign is to be used if G is lower than B, and –ve sign is to be used if G is higher than B.

Example 5·5. *A rectangular pontoon of 5 m long, 3 m wide and 1·2 m deep is immersed 0·8 m in sea water. If the density of sea water is 10 kN/m², find the metacentric height of the pontoon.*

Solution. Given : $l = 5$ m; $b = 3$ m; $d = 1·2$ m ; Depth of immersion = 0·8 m and **$w = 10$ kN/m³.

We know that distance of centre of buoyancy from the bottom of the block.

$$OB = 0·8/2 = 0·4 \text{ m}$$

and the distance of c.g. from the bottom of the block,

$$OG = 1·2/2 = 0·6 \text{ m}$$

∴ $BG = OG - OB = 0·6 - 0·4 = 0·2 \text{ m}$

We also know that moment of inertia of the rectangular plan about the central axis and parallel to the long side,

$$I = \dfrac{lb^3}{12} = \dfrac{5 \times (3)^3}{12} = 11·25 \text{ m}^4$$

and the volume of water displaced,

$$V = 5 \times 3 \times 0·8 = 12 \text{ m}^3$$

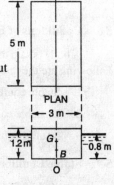

Fig. 5·3.

---

*If the moment of inertia of the section parallel to the short side is taken, then the metacentric height will be more than this. Since the metacentric height plays an important role in finding out the stability of a floating body (which will be discussed in succeeding pages), it is, therefore general practice to find out the smaller metacentric height of the two. For doing so, the moment of inertia of a rectangular section is always taken about the central axis and parallel to the long side. Such a moment of inertia is obtained by taking the cube of the breadth.

** Superfluous data.

∴ $BM = \dfrac{I}{V} = \dfrac{11 \cdot 25}{12} = 0 \cdot 94 \text{ m}$

and metacentric height, $GM = BM - BG = 0 \cdot 94 - 0 \cdot 2 = 0 \cdot 74 \text{ m}$ **Ans.**

Example 5·6. *A solid cylinder of 2 m diameter and 1 m height is made up of a material of specific gravity 0·7 and floats in water. Find its metacentric height.*

Solution. Given : Dia. of cylinder $(d) = 2$ m; Height of cylinder $(h) = 1$ m and sp. gr. of cylinder $= 0 \cdot 7$.

We know that the depth of immersion of the block
$$= 0 \cdot 7 \times 1 = 0 \cdot 7$$

∴ Distance of centre of buoyancy from the bottom of the block,
$$OB = 0 \cdot 7 / 2 = 0 \cdot 35 \text{ m}$$

and distance of c.g. from the bottom of the block,
$$OG = 1/2 = 0 \cdot 5 \text{ m}$$

∴ $BG = OG - OB = 0 \cdot 5 - 0 \cdot 35 = 0 \cdot 15 \text{ m}$

We also know that the moment of inertia of the circular plan about the central axis,

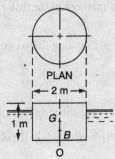

Fig. 5·4.

$$I = \dfrac{\pi}{64} \times (d)^4 = \dfrac{\pi}{64} \times (2)^4 = \dfrac{\pi}{4} = 0 \cdot 25 \, \pi$$

and volume of water displaced,
$$V = \dfrac{\pi}{4} \times (2)^2 \times 0 \cdot 7 = 0 \cdot 7 \, \pi$$

∴ $BM = \dfrac{I}{V} = \dfrac{0 \cdot 25 \, \pi}{0 \cdot 7 \, \pi} = 0 \cdot 36 \text{ m}$

and the metacentric height,
$$GM = BM - BG = 0 \cdot 36 - 0 \cdot 15 = 0 \cdot 21 \text{ m} \quad \textbf{Ans.}$$

5·8 Conditions of Equilibrium of a Floating Body

A body is said to be in equilibrium, when it remains in a steady state, while floating in a liquid. Following are the three conditions of equilibrium of a floating body :

1. Stable equilibrium,
2. Unstable equilibrium, and
3. Neutral equilibrium.

5·9 Stable Equilibrium

A body is said to be in a stable equilibrium, if it returns back to its original position, when given a small angular displacement. This happens when the metacentre (*M*) is higher than the centre of gravity (*G*) of the floating body.

5·10 Unstable Equilibrium

A body is said to be in an unstable equilibrium, if it does not return back to its original position and heels farther away, when given a small angular displacement. This happens when the metacentre (*M*) is lower than the centre of gravity (*G*) of the floating body.

5·11 Neutral Equilibrium

A body is said to be in a neutral equilibrium, if it occupies a new position and remains at rest in this new position, when given a small angular displacement. This happens when the metacentre (M) coincides with the centre of gravity (G) of the floating body.

Example 5·7. *A rectangular timber block 2 m long 1·8 m wide and 1·2 m deep is immersed in water. If the specific gravity of the timber is 0·65, prove that it is in a stable equilibrium.*

Solution. Given : $l = 2$ m; $b = 1.8$ m ; $d = 1.2$ m and sp. gr. of timber $= 0.65$.

We know that the depth of immersion of the block

$$= 0.65 \times 1.2 = 0.78 \text{ m}$$

∴ Distance of centre of buoyancy from the bottom of the block,

$$OB = 0.78/2 = 0.34 \text{ m}$$

and distance of c.g. from the bottom of the block :

$$OG = 1.2/2 = 0.6 \text{ m}$$

∴ $$BG = OG - OB = 0.6 - 0.34 = 0.26 \text{ m}$$

We know that the moment of inertia of the rectangular section about the central axis and parallel to the long side,

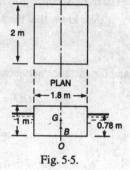

Fig. 5·5.

$$I = \frac{lb^3}{12} = \frac{2 \times (1.8)^3}{12} = 0.972 \text{ m}^4$$

and volume of water displaced, $V = 2 \times 1.8 \times 0.78 = 2.808 \text{ m}^3$

∴ $$BM = \frac{I}{V} = \frac{0.972}{2.808} = 0.35 \text{ m}$$

and the metacentric height, $GM = BM - BG = 0.35 - 0.26 = 0.09$

Since the metacentric height of the block (as obtained above) is positive, or in other words, metacentric (M) is above the centre of gravity (G), therefore, the block is in a stable equilibrium. **Ans.**

Example 5·8. *A cylindrical buoy of 3 m diameter and 4 m long is weighing 150 kN. Show that it cannot float vertically in water.*

Solution. Diameter of buoy (d) = 3 m ; Length of buoy (l) = 4 m and weight = 150 kN.

We know that the volume of water displaced by the buoy,

$$V = \frac{150}{9.81} = 15.3 \text{ m}^3$$

and depth of immersion $= \dfrac{150}{\dfrac{\pi}{4} \times (3)^2} = 2.16 \text{ m}$

∴ $$OB = 2.16/2 = 1.08 \text{ m}$$

and distance of c.g. from the bottom of the cylinder,

$$OG = 4/2 = 2 \text{ m}$$

∴ $$BG = OG - OB = 2 - 1.08 = 0.92 \text{ m}$$

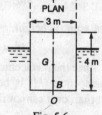

Fig. 5·6.

We also know that the moment of inertia of the circular plan about the central axis,

$$I = \frac{\pi}{64} \times (d)^4 = \frac{\pi}{64} \times (3)^4 = \frac{81 \pi}{64} = 3.98 \text{ m}^4$$

∴ $$BM = \frac{I}{V} = \frac{3.98}{15.3} = 0.26 \text{ m}$$

and metacentric height, $\quad GM = BM - BG = 0.26 - 0.92 = -0.66$ m

Since the metacentric height of the cylinder (as obtained above) is negative, or in other words, the metacentre (M) is below the centre of gravity (G), therefore it cannot float vertically in water. **Ans.**

Example 5·9. *A solid cylinder of 360 mm long and 80 mm diameter has its base 10 mm thick of specific gravity 7. The remaining part of the cylinder is of specific gravity 0·5. Determine, if the cylinder can float vertically in the water.*

Solution. Given : Total length of the cylinder = 360 mm; d = 80 mm; Thickness of base = 10 mm; Specific gravity of the base = 7 and specific gravity of the remaining part = 0·5.

We know that the area of the solid cylinder,

$$A = \frac{\pi}{4} \times (d)^2 = \frac{\pi}{4} \times (80)^2 = 1600\,\pi \text{ mm}^2$$

and the distance between combined centre of gravity (G) and bottom of the cylinder (O),

$$OG = \frac{[(0.5\,A \times 350) \times (10 + 175)] + [7A \times 10 \times 5]}{(0.5\,A \times 350) + (7A \times 10)}$$

$$= \frac{32\,725}{245} = 133.6 \text{ mm}$$

We also know that combined specific gravity of the cylinder

$$= \frac{(350 \times 0.5) + (10 \times 7)}{350 + 10} = 0.68$$

∴ Depth of immersion of the cylinder

$$= 360 \times 0.68 = 245 \text{ mm}$$

and distance of centre of buoyancy from the bottom of the cylinder,

$$OB = 245/2 = 122.5 \text{ mm}$$

∴ $\quad BG = OG - OB = 133.6 - 122.5 = 11.1$ mm

We know that the moment of inertia of the circular section,

$$I = \frac{\pi}{64} \times (d)^4 = \frac{\pi}{64} \times (80)^2 = 640\,000\,\pi \text{ mm}^4$$

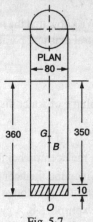

Fig. 5·7.

and volume of water displaced,

$$V = \frac{\pi}{4} \times d^2 \times \text{Depth} = \frac{\pi}{4} \times (80)^2 \times 245 = 392\,000\,\pi \text{ mm}^3$$

∴ $\quad BM = \dfrac{I}{V} = \dfrac{640\,000\,\pi}{392\,000\,\pi} = 1.6$ mm

and metacentric height,

$$GM = BM - BG = 1.6 - 11.1 = -9.5 \text{ mm}$$

Since the metacentric height of the cylinder (as obtained above) is negative or in other words, the metacentre (M) is below the centre of gravity (G), therefore the cylinder is in an unstable equilibrium or it cannot float vertically upon water. **Ans.**

EXERCISE 5·2

1. A wooden block of 1 m × 0·4 m × 0·3 m of specific gravity 0·8 floats in water with its 0·3 m side vertical. Determine its metacentric height, for tilt about its longitudinal axis. [**Ans.** 0·026 m]
2. A solid cube of 1200 mm side and of specific gravity 0·85 floats in water with one of its faces parallel to the water surface. Find the metacentric height of the block. [**Ans.** 27·6 mm]
3. A right circular solid cylinder of 1·5 m diameter and 0·5 m height having specific gravity 0·6 floats in water. What is its metacentric height ? [**Ans.** 0·37 m]

Equilibrium of Floating Bodies

4. A block of specific gravity 0·625 and size 1·28 m × 0·8 m × 0·8 m is floating in water with its length in vertical position. Prove that it is in a stable equilibrium.

5. A cylindrical buoy of 3 m diameter has a height of 3 m. It is made up of a material whose specific gravity is 0·8 and made to float on water with its axis vertical. State whether its equilibrium is stable or unstable. **[Ans. Unstable]**

6. A cylindrical buoy is of 2 m diameter and 3 m long. Determine the state of its equilibrium about its vertical axis, if it weighs 80 kN. **[Ans. Unstable]**

5·12 Maximum Length of a Vertically Floating Body

We see that a cube of wood (having specific gravity less than 1) can float in water in any position. If we maintain any two sides (say breadth and thickness) of the cube constant and go on increasing gradually the third side (say length) and try to float the block vertically in water, we see that the block can float vertically in water up to some length. If we increase the length of the block beyond this value, we find that it cannot float vertically in water, though it can float longitudinally.

This maximum permissible length of the block, floating vertically in water, may be found out by keeping the body in stable equilibrium. Or in other words, this can also be found out by avoiding the unstable equilibrium of the floating body. For doing so, the metacentre (M) should be above centre of gravity (G) of the floating body (a condition of stable equilibrium) or the metacentre (M) may coincide with the centre of gravity (G) of the floating body (a condition of neutral equilibrium. i.e., by avoiding the unstable equilibrium).

Example 5·10. *A block of wood of 120 mm × 120 mm in section is required to float in water. Find the maximum length of the block, so that it may float vertically in water. Take specific gravity of wood as 0·8.*

Solution. Given : Cross-section of the block = 120 mm × 120 mm and specific gravity of wood = 0·8.

Let l = Length of the block in mm.

We know that the depth of immersion of the block floating vertically in water
$$= 0.8 \times l = 0.8\, l$$

∴ Distance of centre of buoyancy from the bottom of the block,
$$OB = 0.8l/2 = 0.4\, l$$

and distance of c.g. from the bottom of the block,
$$OG = l/2 = 0.5\, l$$

∴ $$BG = OG - OB = 0.5\, l - 0.4\, l = 0.1\, l$$

We know that the moment of inertia of the square section,
$$I = \frac{120 \times (120)^3}{12} = 17.28 \times 10^6 \text{ mm}^4$$

and volume of water displaced,
$$V = 120 \times 120 \times 0.8\, l = 11.52 \times 10^3\, l \text{ mm}^3$$

∴ $$BM = \frac{I}{V} = \frac{17.28 \times 10^6}{11.52 \times 10^3\, l} = \frac{1.5 \times 10^3}{l} \text{ mm}$$

For stable equilibrium, the metacentre (M) should be above the centre of gravity (G) or may coincide with G. i.e.,

$$BG \leq BM \quad \text{or} \quad 0.1 l \leq \frac{1.5 \times 10^3}{l}$$

∴ $$l^2 \leq 1.5 \times 10^4 \quad \text{or} \quad l \leq 122.5 \text{ mm} \quad \textbf{Ans.}$$

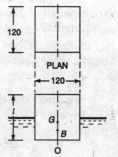

Fig. 5·8.

Example 5·11. *A uniform wooden circular cylinder of 400 mm diameter and of specific gravity 0·6 is required to float in an oil of specific gravity 0·8. Find the maximum length of the cylinder, in order that cylinder may float vertically in water.*

Solution. Given : $d = 40$ mm; Specific gravity of wood = 0·6 and specific gravity of oil 0·8.

Let $\qquad l$ = Length of the cylinder in mm.

We know that the depth of immersion of the cylinder floating vertically upon oil

$$= 0.6 \times \frac{1}{0.8} \times l = \frac{3l}{4} = 0.75\, l$$

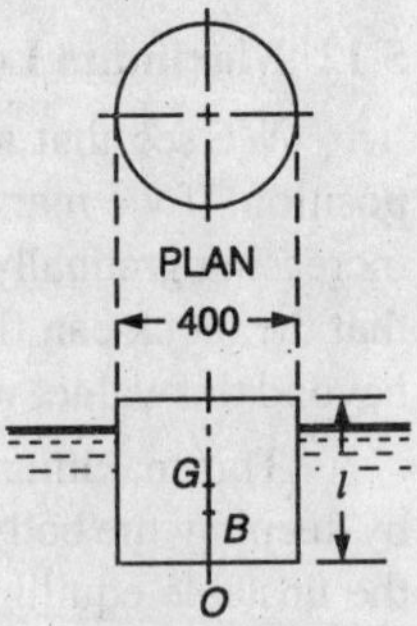

∴ Distance of centre of buoyancy from the bottom of the cylinder,

$$OB = 0.75\, l/2 = 0.375\, l$$

and distance of c.g. from the bottom face of the cylinder,

$$OG = l/2 = 0.5\, l$$

∴ $\qquad BG = OG - OB = 0.5\, l - 0.375\, l = 0.125\, l$

We know that moment of inertia of the circular section,

$$I = \frac{\pi}{64} \times (d)^4 = \frac{\pi}{64} \times (400)^4 = 400 \times 10^6\, \pi\, \text{mm}^4$$

and the volume of water displaced,

$$V = \frac{\pi}{4} \times (d)^2 \times 0.75\, l = \frac{\pi}{4} \times (400)^2 \times 0.75\, l = 30 \times 10^3\, \pi l\, \text{mm}^4$$

∴ $\qquad BM = \dfrac{I}{V} = \dfrac{400 \times 10^6\, \pi}{30 \times 10^3\, \pi l} = \dfrac{40 \times 10^3}{3l}\, \text{mm}$

Fig. 5·9.

For stable equilibrium, the metacentre (M) should be above centre of gravity (G) or may coincide with it. *i.e.*,

$$BG \leq BM \quad \text{or} \quad \frac{l}{8} \leq \frac{40 \times 10^3}{3l}$$

∴ $\qquad l^2 \leq \dfrac{320 \times 10^3}{3} = 10.67 \times 10^4 \quad \text{or} \quad l \leq 326.6\, \text{mm} \quad$ **Ans.**

Example 5·12. *A solid cylinder 1 m long, 200 mm diameter has its base 25 mm thick and of specific gravity 8. The remaining portion is of specific gravity 0·5. Find its maximum length for stable equilibrium.*

Solution. Given : Total length of the cylinder = 1 m; $d = 200$ mm; Base thickness = 25 mm; Specific gravity of base = 8 and specific gravity of remaining portion = 0·5.

Let $\qquad l$ = Length of cylinder *excluding* metal portion in mm.

We know that distance between the combined centre of gravity (G) and the bottom of the cylinder (O),

$$OG = \frac{\left[(0.5A \times l) \times \left(25 + \dfrac{l}{2}\right) + [(8A \times 25)\, 12.5]\right]}{(0.5\, A \times l) + (8A \times 25)}$$

$$= \frac{0.5l\,(25 + 0.5l) + 2500}{0.5l + 200} = \frac{12.5\, l + 0.25\, l^2 + 2500}{0.5\, l + 200}$$

$$= \frac{l^2 + 50l + 10\,000}{2l + 800}\, \text{mm}$$

Equilibrium of Floating Bodies

and the combined centre of gravity of the cylinder

$$= \frac{(0{\cdot}5 \times l) + (25 \times 8)}{l+25} = \frac{0{\cdot}5\,l + 200}{l+25}$$

∴ Depth of immersion of the cylinder,

$$= \text{Total length} \times \text{Combined sp. gr.}$$

$$= (l+25) \times \frac{0{\cdot}5\,l + 200}{l+25} = 0{\cdot}5\,l + 200 \text{ mm}$$

and the distance of centre of buoyancy from the bottom of the buoy,

$$OB = \frac{1}{2}(0{\cdot}5\,l + 200) = 0{\cdot}25\,l + 100 \text{ mm}$$

We know that the moment of inertia of the circular section,

$$I = \frac{\pi}{64} \times (d)^4 = \frac{\pi}{64} \times (200)^4 = 25 \times 10^6\,\pi \text{ mm}^4$$

Fig. 5·10.

and the volume of water displaced,

$$V = \frac{\pi}{4} \times (200)^2 \times (0{\cdot}5\,l + 200) = 10\,000\,\pi\,(0{\cdot}5\,l + 200)$$

∴ $$BM = \frac{I}{V} = \frac{25 \times 10^6\,\pi}{10\,000\,\pi\,(0{\cdot}5\,l + 200)} = \frac{2500}{0{\cdot}5\,l + 200} = \frac{5000}{l + 400}$$

and $$OM = OB + BM = (0{\cdot}25\,l + 100) + \frac{5000}{l + 400}$$

For stable equilibrium, the metacentric (M) should be above the centre of gravity (G) or may coincide with it. i.e.,

$$OM \leq OG$$

$$(0{\cdot}25\,l + 100) + \frac{5000}{l+400} \leq \frac{l^2 + 50l + 10\,000}{2l + 800}$$

$$0{\cdot}25\,(l + 400) + \frac{5000}{l+400} \leq \frac{l^2 + 50l + 10\,000}{2(l+400)}$$

$$\frac{0{\cdot}25\,(l+400)^2 + 5000}{l+400} \leq \frac{l^2 + 50\,l + 10\,000}{2\,(l+400)}$$

$$0{\cdot}5(l^2 + 160\,000 + 800\,l) + 10\,000 \leq l^2 + 50l + 10\,000$$

$$0{\cdot}5 l^2 + 80\,000 + 400\,l + 10\,000 \leq l^2 + 50\,l + 10\,000$$

$$0{\cdot}5\,l^2 - 350\,l - 80\,000 \geq 0$$

$$l^2 - 700\,l - 160\,000 \geq 0$$

This is a quadratic equation for l. Therefore

$$l \geq \frac{+700 + \sqrt{(700)^2 + 4 \times 160\,000}}{2} \geq \frac{700 + 1063}{2} \text{ mm}$$

$$\geq 881{\cdot}5 \text{ mm}$$

∴ Maximum length of the cylinder (including base)

$$= 881{\cdot}5 + 25 = 906{\cdot}5 \text{ mm} \quad \textbf{Ans.}$$

5·13 Floating Bodies Anchored at the Base

We have seen in Art. 5·12 that there is always a certain limit of length upto which a body can float vertically in a liquid. If the length of the body exceeds this limit, it cannot float vertically, though it can float horizontally. But, sometimes, due to certain reasons, the body is required to float vertically. For doing so, the body is anchored by means of a chain from the centre of its base. The tension, in the anchor chain, puts an additional downward force on the body, which will cause a larger volume of water to be displaced.

Let W = Weight of the body, and
T = Tension in the anchor chain.

Total downward force, which will displace the water
$$= W + T$$

The remaining procedure for solving the problems is the same.

Example 5·13. *A cylindrical buoy is 2 metres in diameter, 2·5 metres long weighs 20 kN. The density of sea water is 10 kN/m³. Show that the buoy cannot float with its axis vertical.*

What minimum pull should be applied to a chain attached to the centre of the base to keep it vertical ?

Solution. Given : Diameter of buoy $(d) = 2$ m; $l = 2·5$ m; Weight of buoy = 20 kN and density of sea water $(w) = 10$ kN/m³.

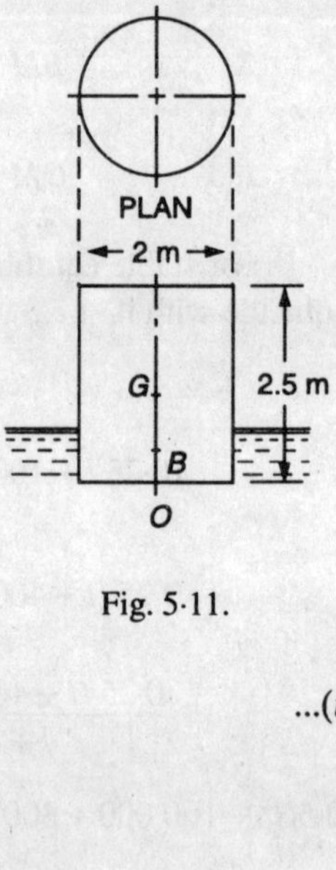

When the buoy is not anchored

We know that the volume of water displaced,

$$V_1 = \frac{\text{Weight of buoy}}{\text{Sp. wt. of water}} = \frac{20}{10} = 2·0 \text{ m}^3$$

and the depth of immersion of the cylinder

$$= \frac{\text{Volume}}{\text{Area}} = \frac{2·0}{\frac{\pi}{4} \times (2)^2} = 0·64 \text{ m}$$

∴ Distance of centre of buoyancy from the bottom of buoy,

$$OB_1 = 0·64/2 = 0·32 \text{ m}$$

and the distance of c.g. from the bottom of the buoy,

$$OG_1 = 2·5/2 = 1·25 \text{ m}$$

∴ $$B_1G_1 = OG_1 - OB_1 = 1·25 - 0·32 = 0·93 \text{ m} \quad \ldots(i)$$

We know that the moment of inertia of the circular section,

$$I = \frac{\pi}{64} \times (d)^4 = \frac{\pi}{64} \times (2)^4 = 0·7854 \text{ m}^4$$

∴ $$B_1M_1 = \frac{I}{V_1} = \frac{0·7854}{2·0} = 0·39 \text{ m}$$

and metracentric height $G_1M_1 = B_1M_1 - B_1G_1 = 0·39 - 0·93 = -0·54 \text{ m}$

Since the metracentric height of the cylinder (as obtained above) is negative, or in other words, the metacentre (M_1) is below the centre of gravity (G_1), therefore the buoy is an unstable equilibrium or it cannot float with its axis vertical. **Ans.**

When the buoy is anchored at the base

Let T = Tension in the base in kN.

We know that the total downward force
$$= \text{Weight of buoy} + \text{Tension} = 20 + T \text{ kN}$$

Fig. 5·11.

Equilibrium of Floating Bodies

∴ Volume of water displayed
$$V_2 = \frac{20 + T}{10}$$

and the depth of immersion of the cylinder

$$= \frac{\text{Volume}}{\text{Area}} = \frac{\frac{20 + T}{10}}{\frac{\pi}{4} \times (2)^2} = \frac{20 + 2T}{10\pi}$$

∴ Distance of centre of buoyancy from the bottom of the buoy,
$$OB_2 = \frac{1}{2} \times \frac{20 + T}{10\pi} = \frac{20 + T}{20\pi}$$

Now for finding out the combined centre of gravity, let us take moments of the two weights (W and $W + T$) and equate the same,

$$(20 + T) \times OG_2 = 20\, OG = 20 \times 1.25 = 25 \quad \ldots(\because OG = 1.25 \text{ m})$$

or
$$OG_2 = \frac{25}{20 + T}$$

∴
$$B_2G_2 = OG_2 - OB_2 = \frac{25}{20 + T} - \frac{20 + T}{20\pi}$$

and
$$B_2M_2 = \frac{I}{V_2} = \frac{0.7854}{\frac{20 + T}{10}} = \frac{7.854}{20 + T}$$

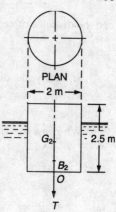

Fig. 5·12.

For stable equilibrium, the metacentre (M_2) should be above the centre of gravity (G_2) or may coincide with it. i.e.,

$$B_2G_2 \leq B_2M_2$$

$$\frac{25}{20 + T} - \frac{20 + T}{20\pi} \leq \frac{7.854}{20 + T}$$

$$\frac{20 + T}{20\pi} \geq \frac{25}{20 + T} - \frac{7.854}{20 + T} \geq \frac{17.15}{20 + T}$$

$$(20 + T)^2 \geq 17.15 \times 20\pi \geq 1078$$

Taking square root of both sides of the above equation,
$$20 + T \geq 32.8$$

∴
$$T \geq 32.8 - 20 = 12.8 \text{ kN} \quad \textbf{Ans.}$$

5·14 Conical Buoys Floating in Liquid

Consider a conical buoy with apex O floating in some liquid as shown in Fig. 5·13.

Let $\qquad D$ = Diameter of the cone,

2α = Apex angle of the cone,

L = Length of the cone, and

l = Length of the cone immersed in liquid.

From the geometry of the figure, we find that the diameter of the cone at liquid level,

$$d = 2l \tan \alpha$$

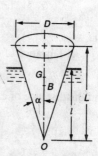

Fig. 5·13.

and the distance of centre of buoyancy from the apex O,

$$OB = \frac{3l}{4}$$

Similarly, the distance of centre of gravity from the apex O,

$$OG = \frac{3L}{4}$$

We know that the volume of liquid displaced,

$$V = \frac{1}{3} \times \frac{\pi}{4} \times (d)^2 \times l = \frac{1}{3} \times \frac{\pi}{4} \times (2l \tan \alpha)^2 \, l$$

$$= \frac{1}{3} \pi l^3 \tan^2 \alpha$$

and the moment of inertia of the circular section at the liquid level,

$$I = \frac{\pi}{64} \times (d)^4 = \frac{\pi}{64} (2l \tan \alpha)^4 = \frac{\pi}{4} l^4 \tan^4 \alpha$$

The metacentric height of the conical buoy may be found out as usual by the relation,

$$BM = \frac{I}{V} = \frac{\frac{\pi}{4} \times (l^4 \tan^4 \alpha)}{\frac{1}{3} \pi l^3 \tan^2 \alpha} = \frac{3}{4} \times l \tan^2 \alpha$$

Example 5·14. *A solid cone weighing 6·9 kN is floating in an oil of specific weight 9·3 kN/m³. Find the minimum apex angle, so that the cone may float with its apex downwards.*

Solution. Given : Sp. wt. of cone = 6·9 kN/m³ and sp. wt. of oil = 9·3 kN/m³.

Let $\quad L = $ Length of the cone,

$\quad l = $ Length of cone immersed in oil,

$\quad 2\alpha = $ Apex angle of the cone.

We know that the weight of the cone

$$= \frac{1}{3} \times \pi L^3 \tan^2 \alpha \times 6\cdot 9 \text{ kN}$$

and the weight of oil displaced $\quad = \frac{1}{3} \times \pi l^3 \tan^2 \alpha \times 9\cdot 3 \text{ kN}$

Since the cone is floating, therefore weight of the cone is equal to the weight of the liquid displaced. *i.e.*,

$$\frac{1}{3} \times \pi L^3 \tan^2 \alpha \times 6\cdot 9 = \frac{1}{3} \times \pi l^3 \tan^2 \alpha \times 9\cdot 3$$

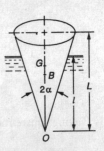

Fig. 5·14.

or
$$l = L \left(\frac{6\cdot 9}{9\cdot 3}\right)^{\frac{1}{3}} = L\,(0\cdot 742)^{1/3} = 0\cdot 905 \, L$$

∴ Distance of centre of buoyancy from the apex,

$$OB = 0\cdot 75 \, l = (0\cdot 75 \times 0\cdot 905 \, L) = 0\cdot 68 \, L$$

and the distance of c.g. from the apex,

$$OG = \frac{3}{4} L = 0\cdot 75 \, L$$

∴ $\quad BG = OG - OB = 0\cdot 75 \, L - 0\cdot 68 \, L = 0\cdot 07 \, L$

Equilibrium of Floating Bodies

For stable equilibrium, the metacentre (M) should be above the centre of gravity (G) or may coincide with it. i.e.,

$$BG \leq BM$$
$$0.07 L \leq 0.75\, l \tan^2 \alpha \leq 0.75 \times (0.905)\, L \tan^2 \alpha$$
$$\leq 0.68\, L \tan^2 \alpha$$

∴ $\tan^2 \alpha \geq 0.07 / 0.68 \geq 0.1029$

or $\tan \alpha \geq 0.3208$

$\alpha \geq 17.8°$

and least apex angle, $2\alpha \geq 35.6°$ **Ans.**

Example 5·15. *A conical buoy 1 metre long, and base diameter 1·2 metre, floats in water with its apex downwards. Determine the minimum weight of the buoy, for stable equilibrium.*

Solution. Given : $L = 1$ m and $D = 1.2$ m.

Let $l =$ Length of the cone immersed in water.

From the geometry of the buoy, we find that

$$\tan \alpha = \frac{0.6}{1} = 0.6$$

∴ Volume of water displaced,

$$V = \frac{1}{3} \pi\, l^3 \tan^2 \alpha = \frac{1}{3} \times \pi \times (l)^3 (0.6)^2$$
$$= 0.377\, l^3 \text{ m}^3$$

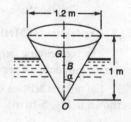

Fig. 5·15.

and moment of inertia of circular section at the liquid level,

$$I = \frac{\pi}{64} \times (D)^4 = \frac{\pi}{64} (1.2\, l)^4 = 0.1018\, l^4$$

∴ $$BM = \frac{I}{V} = \frac{0.1018\, l^4}{0.377\, l^3} = 0.27\, l$$

We know that distance of centre of buoyancy from the apex

$$OB = \frac{3}{4} \times l = 0.75\, l$$

and the distance of c.g. from the apex,

$$OG = \frac{3}{4} \times L = \frac{3}{4} \times 1 = 0.75 \text{ m}$$

∴ $BG = OG - OB = 0.75 - 0.75\, l$

For stable equilibrium the metacentre (M) should be above the centre of gravity (G) or may coincide with it. i.e.,

$$BG \geq BM$$
$$0.75 - 0.75\, l \geq 0.27\, l, \quad 1.02\, l \leq 0.75$$

or $l \leq 0.75/1.02 \geq 0.735$ m

∴ Volume of water displaced (substituting the value of l in equation for volume),

$$V = 0.377\, l^3 = 0.377 \times (0.735)^3 = 0.149 \text{ m}^3$$

This should be equal to the weight of the buoy. Therefore weight of the buoy

$$= 9.81 \times 0.149 = 1.46 \text{ kN} \textbf{ Ans.}$$

EXERCISE 5·3

1. A wooden block of 0·6 m × 0·6 m in section and specific gravity 0·7 floats vertically in water. Find its maximum length, so that it may float with stable equilibrium. [Ans. 0·54 m]
2. A wooden cylinder of circular section and of specific gravity 0·6 is required to float in an oil of specific gravity 0·9. If the cylinder has diameter (d), then show that the length of the cylinder cannot exceed 0·75 d for floating with its longitudinal axis vertical.
3. A cylindrical wooden block of specific gravity 0·6 has a diameter of 0·4 m. What is the maximum permissible length of the block, in order that it may float vertically ? [Ans. 0·29 m]
4. A cylindrical block of wood has 240 mm diameter and 0·8 specific gravity. Find the maximum permissible length of the block, so that it can float in water with its axis vertical. [Ans. 212 mm]
5. A cylindrical buoy of 1 m diameter and 2 m height is anchored at the centre of its base. Find the tension in the anchor chain, which will keep the buoy in upright position in sea water. Take the weight of the buoy as 15 kN. [Ans. 6 kN]
6. A wooden cone of specific gravity 0·8 is required to float vertically in water. Determine the least apex angle, which shall enable the cone to float in stable equilibrium. [Ans 31·2°]
7. A cone made of a material of specific gravity 0·7 is required to float in water with its apex vertical. Find the least apex angle, so that the cone may float in stable equilibrium. [Ans. 30·1°]

5·15 Experimental Method for Metacentric Height

The metacentric height of a floating body, like a ship, may also be found out experimentally provided the centre of gravity of the floating body is known.

Let all the articles on the ship be arranged in such a way that the ship is perfectly horizontal as shown in Fig. 5·16 (a).

Let
W = Weight of the ship, and
G = Centre of gravity of the ship.

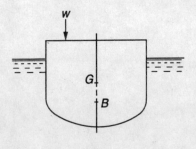
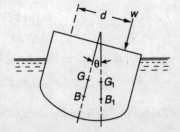

(a) Floating body.   (b) Tilted body.

Fig. 5·16.

Now let a movable weight w be moved right across the ship through a distance d as shown in Fig. 5·16 (b).

Due to this movement of load w, the boat will tilt. Let this angle of tilt be θ. Let the centre of gravity (G) move to a new position G_1 and the centre of buoyancy (B) to a new position B_1. Let us join B_1 and G_1 and extend it upwards to meet the line through B and G to meet at M, which is the metacentre and GM is the metacentric height.

The effect of moving the load w to the right, through a distance d, will cause a clockwise couple, whose moment

$$= w \cdot d \qquad \qquad ...(i)$$

The weight of the ship W and the force of buoyancy will form an anti-clockwise couple, whose moment

$$= W . GM . \tan \theta \qquad ...(ii)$$

Since these two moments are equal but opposite in directions, therefore equating (i) and (ii),

$$W. GM . \tan \theta = w . d$$

or

$$GM = \frac{wd}{W \tan \theta}$$

where GM is the metacentric height.

Example 5·16. *A weight of 20 kN moved through a distance of 9 metres across the deck of a pontoon of 1500 kN displacement, floating in water. This makes a pendulum 27 mm long, move through 1·3 mm horizontally. Calculate the metacentric height of the pontoon.*

Solution. Given : $w = 20$ kN; $d = 9$ m and $W = 1500$ kN

We know that the angle of heel,

$$\tan \theta = \frac{1 \cdot 3}{27} = 0 \cdot 048$$

and the metacentric height, $\quad GM = \dfrac{w.d}{W \tan \theta} = \dfrac{20 \times 9}{1500 \times 0 \cdot 048} = 2 \cdot 5$ m **Ans.**

Example 5·17. *A vessel has a length of 60 m, width 8·4 m and a displacement of 15 MN. A weight of 150 kN moved through a distance of 6 m across the deck causes the ship to heel through 3°. The moment of area at the water level is 72% of that of the circumscribing rectangle, and the position of the centre of buoyancy is 1·5 m below the water line.*

Determine the metacentric height and the position of centre of gravity of the ship. Take the density of sea water as 10 kN/m³.

Solution. Given : $l = 60$ m; $b = 8 \cdot 4$ m ; $W = 15$ MN $= 15\,000$ kN; $w = 150$ kN; $d = 6$ m; $\theta = 3°$; Second moment of area $= 72\%$ of circumscribing rectangle and position of B is 1·5 m below the water line.

Metacentric height

We know that the metacentric height of the vessel,

$$GM = \frac{w . d}{W \tan \theta} = \frac{150 \times 6}{15\,000 \times \tan 3°}$$

$$= \frac{900}{15\,000 \times 0 \cdot 0524} = 1 \cdot 15 \text{ m} \quad \textbf{Ans.}$$

Position of the centre of gravity of the ship

We also know that the moment of inertia of the vessel at the water level,

$$I = 0 \cdot 72 \times \frac{lb^3}{12} = 0 \cdot 72 \times \frac{60 \times (8 \cdot 4)^3}{12} \text{ m}^4$$

$$= 2134 \text{ m}^4$$

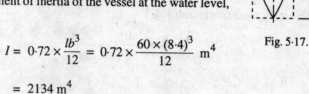

Fig. 5·17.

and the volume of water displaced,

$$V = \frac{\text{Weight of buoy}}{\text{Density of sea water}} = \frac{15\,000}{10} = 1500 \text{ m}^3$$

∴

$$BM = \frac{I}{V} = \frac{2134}{1500} = 1 \cdot 42 \text{ m}$$

Since the centre of buoyancy is 1·5 m below the water line, therefore, metacentre is at a distance of (1·5 − 1·42 = 0·08 m) below the water line. Or in other words, centre of gravity of the vessel is at a distance of 0·08 + 1·15 = 1·23 m below the water line.   **Ans.**

5·16 Time of Rolling (Oscillation) of a Floating Body

We have seen in Art. 5·5 that whenever a body, floating in a liquid, is given a small angular displacement, it starts oscillating about its metacentre in the same manner as a pendulum oscillates about its point of suspension. Now consider a body floating upon a liquid.

Let
- W = Weight of the floating body,
- θ = Angle (in radians) through which the body is depressed (*i.e.*, given a small angular displacement),
- GM = Metacentric height of the body,
- α = Angular acceleration of the ship in radians/s^2,
- T = Time of rolling (*i.e.*, one complete oscillation) in seconds,
- k = Radius of gyration about G, and
- I = Moment of inertia of the body about a line through the centre of gravity, whose value is equal to $\dfrac{W}{g} k^2$

A little consideration will show that the motion of the body will be simple harmonic. When the force, which has caused angular displacement, is removed the only force acting on the body, which will tend to bring the body in equilibrium

$$= W \cdot GM \cdot \tan\theta$$

For small angles, $\tan\theta$ can be taken to be equal to θ. Therefore the force acting on the body

$$= W \cdot GM \cdot \theta \qquad \ldots(i)$$

Angular acceleration of the ship,

$$\alpha = -\frac{d^2\theta}{dt^2}$$

*(Minus sign indicates that the force is acting in such a way that it tends to decrease the angle θ). We know that

$$\text{Inertia torque} = \text{Moment of inertia} \times \text{Angular acceleration}$$
$$= I \cdot \alpha$$
$$= \frac{W}{g} k^2 \times \left(-\frac{d^2\theta}{dt^2}\right) = -\frac{W}{g} k^2 \times \frac{d^2\theta}{dt^2} \qquad \ldots(ii)$$

Equating equations (*i*) and (*ii*),

$$W \cdot GM \cdot \theta = -\frac{W}{g} k^2 \times \frac{d^2\theta}{dt^2}$$

$$\frac{W}{g} k^2 \times \frac{d^2\theta}{dt^2} + W \cdot GM \cdot \theta = 0$$

$$\frac{k^2}{g} \times \frac{d^2\theta}{dt^2} + GM \cdot \theta = 0 \qquad \ldots\text{(Dividing both sides by } W\text{)}$$

---

*For details, please consult any standard book or Author's book, Text book of Engineering Mechanics.

Equilibrium of Floating Bodies

$$\frac{d^2\theta}{dt^2} + \frac{GM \cdot g\,\theta}{k^2} = 0 \qquad \ldots\left(\text{Dividing both sides by } \frac{k^2}{g}\right)$$

This is a differential equation of the second degree whose solution is :

$$\theta = A \sin\left(\sqrt{\frac{GM.g}{k^2}} \times t\right) + B \cos\left(\sqrt{\frac{GM.g}{k^2}} \times t\right)$$

where A and B are the constants of integration.

We know when $t = 0$, $\theta = 0$. Therefore $B = 0$.

$\therefore$
$$\theta = A \sin\left(\sqrt{\frac{GM.g}{k^2}} \times t\right)$$

We also know when $t = \dfrac{T}{2}$, then $\theta = 0$

$\therefore$
$$\theta = A \sin\left(\sqrt{\frac{GM.g}{k^2}} \times \frac{T}{2}\right)$$

Since A cannot be equal to zero, therefore

$$\sin\left(\sqrt{\frac{GM.g}{k^2}} \times \frac{T}{2}\right) = 0$$

or
$$\sqrt{\frac{GM.g}{k^2}} \times \frac{T}{2} = \pi \qquad \text{(When sin of some angle is zero, then that angle} = \pi\text{)}$$

$\therefore$
$$T = 2\pi\sqrt{\frac{k^2}{GM.g}}$$

Example 5·18. *Calculate the time of rolling of a ship whose metacentric height is 0·6 m and least radius of gyration is 4 m.*

Solution. Given : $GM = 0.6$ m and $k = 4$ m.

We know that the time of rolling,

$$T = 2\pi\sqrt{\frac{k^2}{GM.g}} = 2\pi\sqrt{\frac{(4)^2}{0.6 \times 9.81}} = 10.4 \text{ s} \quad \textbf{Ans.}$$

Example 5·19. *The metacentric height of a ship is 0·75 m and period time of rolling 22 seconds. Calculate the value of relevant radius of gyration.*

Solution. Given : $GM = 0.75$ m and $T = 22$ s.

Let $\qquad k =$ Relevant radius of gyration.

We know that time of rolling (T)

$$22 = 2\pi\sqrt{\frac{k^2}{GM.g}} = 2\pi\sqrt{\frac{k^2}{0.75 \times 9.81}}$$

Squaring both sides of the above equation,

$$484 = (2\pi)^2 \times \frac{k^2}{0.75 \times 9.81} = \frac{k^2}{5.366}$$

$\therefore \qquad k^2 = 484/5.366 = 90.2 \quad$ or $\quad k = 9.5$ m **Ans.**

EXERCISE 5.4

1. A ship weighing 20 MN heels over 2 degrees, when a load of 150 kN is moved across its deck through a distance of 5 m. Find the metacentric height of the ship. [**Ans.** 1·07 m]
2. A ship weighs 32 MN. If a load of 200 kN is moved through a distance of 6 m across the deck, it causes 3 m long pendulum to move 75 mm horizontally. Find the metacentric height of the ship. [**Ans.** 1·5 m]
3. A ship has 5·5 mm radius of gyration, while travelling in sea water. If its metacentric height is 0·8 m, find its rolling time. [**Ans.** 12·3 s]
4. A ship weighing 40 MN has centre of buoyancy 2 m below its centre of gravity. The moment of inertia of the ship area at the water level is 10 400 m^4. Find the period of rolling, when floating in sea water, if radius of gyration of the ship is 4 m. Take the density of sea water as 10 kN/m^3. [**Ans.** 10·4 s]

QUESTIONS

1. State the Archimedes' principle.
2. Explain the terms centre of buoyancy and metacentre.
3. Describe the meaning of metacentric height.
4. Obtain an expression for the metacentric height of a floating body.
5. What are the conditions of equilibrium of a floating body?
6. Explain the stability of a floating body with reference to its metacentric height.
7. Explain the necessity of anchoring a floating body.

OBJECTIVE TYPE QUESTIONS

1. The centre of gravity of the volume of liquid displaced by a floating body is called
   (*a*) centre of pressure   (*b*) centre of buoyancy
   (*c*) metacentre   (*d*) all of these
2. When a body floating in a liquid is given a small angular displacement, it starts oscillating about a point. This point is known as
   (*a*) centre of pressure   (*b*) centre of gravity
   (*c*) centre of buoyancy   (*d*) metacentre
3. The metacentric height of a floating body is the distance between
   (*a*) centre of gravity and centre of buoyancy   (*b*) centre of gravity and metacentre
   (*c*) centre of buoyancy and metacentre   (*d*) none of these
4. The metacentric height of two bodies *A* and *B* are 1 m and 1·5 m respectively. Select the correct statement
   (*a*) both *A* and *B* have equal stability   (*b*) both *A* and *B* are unstable
   (*c*) body *A* is more stable than *B*   (*d*) body *B* is more stable than *B*.
5. If a body floating in a liquid returns to its original position, when given a small angular displacement, the body is said to be in
   (*a*) neutral equilibrium   (*b*) stable equilibrium
   (*c*) unstable equilibrium   (*d*) either '*a*' or '*c*'

ANSWERS

1. (*b*)   2. (*d*)   3. (*b*)   4. (*d*)   5. (*b*)

6

Hydrokinematics

1. Introduction. 2. Rate of Discharge. 3. Equation of Continuity of a Liquid Flow. 4. Motion of Fluid Particles. 5. Types of Flow Lines. 6. Path Lines. 7. Stream Lines. 8. Streak Lines or Filament Lines. 9. Potential Lines or Equipotential Lines. 10. Flow Net. 11. Uses of Flow Nets. 12. Types of Flows in a Pipe. 13. Uniform Flow. 14. Non-uniform Flow. 15. Streamline Flow. 16. Turbulent Flow. 17. Steady Flow. 18. Unsteady Flow. 19. Compressible Flow. 20. Incompressible Flow. 21. Rotational Flow. 22. Irrotational Flow. 23. One-dimensional Flow. 24. Two-dimensional Flow. 25. Three-dimensional Flow. 26. Stream Function.

6·1 Introduction

In the previous chapters, we have been discussing the effect of force on the liquids at rest. But in this chapter, we shall study the motion of liquids without any reference to the force causing motion. The subject of hydrokinematics deals with the study of velocity and acceleration of the liquid particles without taking into consideration any force or energy.

6·2 Rate of Discharge

The quantity of a liquid, flowing per second through a section of a pipe or a channel, is known as the rate of discharge or simply discharge. It is generally denoted by Q. Now consider a liquid flowing through a pipe.

Let a = Cross-sectional area of the pipe, and
v = *Average velocity of the liquid,

∴ Discharge, Q = Area × Average velocity = $a.v$

Notes : 1. If the area is in m^2 and velocity in m/s, then the discharge,

$$Q = m^2 \times m/s = m^3/s = \text{cumecs}$$

2. Remember that 1 m^3 = 1000 litres.

6·3 Equation of Continuity of a Liquid Flow

If an incompressible liquid is continuously flowing through a pipe or a channel (whose cross-sectional area may or may not be constant) the quantity of liquid passing per second is the same at all sections. This is known as the equation of continuity of a liquid flow. It is the first and fundamental equation of flow.

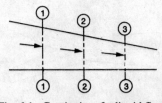

Fig. 6·1. Continuity of a liquid flow.

Consider a tapering pipe through which some liquid is flowing as shown in Fig. 6·1.

---

*In actual practice, the velocity of a liquid is maximum at the centre of a pipe and is minimum near the walls. For all calculations in Hydraulics, the average velocity of flow at a section is taken.

[107]

Let a_1 = Cross-sectional area of the pipe at section 1-1, and
v_1 = Velocity of the liquid at section 1-1,
Similarly a_2, v_2 = Corresponding values at section 2-2,
and a_3, v_3 = Corresponding values at section 3-3.

We know that the total quantity of liquid passing through section 1-1,
$$Q_1 = a_1 \cdot v_1 \qquad \ldots(i)$$
Similarly, total quantity of liquid passing through section 2-2,
$$Q_2 = a_2 \cdot v_2 \qquad \ldots(ii)$$
and total quantity of the liquid passing through section 3-3,
$$Q_3 = a_3 \cdot v_3 \qquad \ldots(iii)$$
From the law of conservation of matter, we know that the total quantity of liquid passing through the sections 1-1, 2-2 and 3-3 is the same. Therefore
$$Q_1 = Q_2 = Q_3 = \ldots\ldots \text{ or } a_1 \cdot v_1 = a_2 \cdot v_2 = a_3 \cdot v_3 \ldots\ldots \text{ and so on.}$$

Example 6·1. *Water is flowing through a pipe of 100 mm diameter with an average velocity of 10 m/s. Determine the rate of discharge of the water in litres/s. Also determine the velocity of water at the other end of the pipe, if the diameter of the pipe is gradually changed to 200 mm.*

Solution. Given : $d_1 = 100$ mm $= 0·1$ m; $v_1 = 10$ m/s and $d_2 = 200$ mm $= 0·2$ m.

Rate of discharge

We know that the cross-sectional area of the pipe at point 1,
$$a_1 = \frac{\pi}{4} \times (d_1)^2 = \frac{\pi}{4} \times (0·1)^2 = 7·854 \times 10^{-3} \text{ m}^2$$
and rate of discharge, $Q = a_1 \cdot v_1 = (7·854 \times 10^{-3}) \times 10 = 78·54 \times 10^{-3} \text{ m}^3/\text{s}$
$$= 78·54 \text{ litres/s } \textbf{Ans.}$$

Velocity of water at the other end of the pipe

We also know that cross-sectional area of the pipe at point 2,
$$a_2 = \frac{\pi}{4} \times (d_2)^2 = \frac{\pi}{4} \times (0·2)^2 = 31·42 \times 10^{-3} \text{ m}^2$$
and velocity of water at point 2, $v_2 = \dfrac{Q}{a_2} = \dfrac{78·54 \times 10^{-3}}{31·42 \times 10^{-3}} = 2·5$ m/s **Ans.**

Example 6·2. *A pipe AB branches into two pipes C and D as shown in Fig. 6·2.*

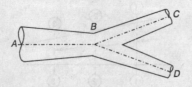

Fig. 6·2.

The pipe has diameter of 0·45 m at A, 0·3 m at B, 0·2 m at C and 0·15 m at D. Find the discharge at A, if the velocity of water at A is 2 m/s. Also find out the velocities at B and D, if velocity at C is 4 m/s.

Hydrokinematics

Solution. Given : $d_A = 0.45$ m; $d_B = 0.3$ m ; $d_C = 0.2$ m; $d_D = 0.15$ m; $v_A = 2$ m/s and $v_C = 4$ m/s.

Discharge at A
We know that cross-sectional area of the pipe at A,
$$a_A = \frac{\pi}{4} \times (d_A)^2 = \frac{\pi}{4} \times (0.45)^2 = 0.159 \text{ m}^2$$

and discharge at A, $Q_A = a_A \cdot v_A = 0.159 \times 2 = 0.318 \text{ m}^3/\text{s}$ **Ans.**

Velocity at B
We also know that cross-sectional area of the pipe at B,
$$a_B = \frac{\pi}{4} \times (d_B)^2 = \frac{\pi}{4} \times (0.3)^2 = 0.0707 \text{ m}^2$$

and velocity at B, $v_B = \dfrac{Q}{a_B} = \dfrac{0.318}{0.0707} = 4.5 \text{ m/s}$ **Ans.**

Velocity at D
We know that cross-sectional area of the pipe at C,
$$a_C = \frac{\pi}{4} \times (d_C)^2 = \frac{\pi}{4} \times (0.2)^2 = 0.0314 \text{ m}^2$$

and discharge at D, $Q_C = a_C \cdot v_C = 0.0314 \times 4 = 0.1256 \text{ m}^3/\text{s}$

$\therefore$ Discharge at D, $Q_D = Q_A - Q_C = 0.318 - 0.1256 = 0.1924 \text{ m}^3/\text{s}$

We also know that cross-sectional area of the pipe at D,
$$a_D = \frac{\pi}{4} \times (d_D)^2 = \frac{\pi}{4} \times (0.15)^2 = 0.0177 \text{ m}^2$$

$\therefore$ Velocity at D, $v_D = \dfrac{Q_D}{a_D} = \dfrac{0.1924}{0.0177} = 10.9 \text{ m/s}$ **Ans.**

6·4 Motion of Fluid Particles

A fluid consists of an innumerable number of particles, whose relative positions are never fixed. Whenever a fluid is in motion, these particles move along certain lines, depending upon the characteristic of the fluid and the shape of the passage through which the fluid particles move. For complete analysis of the fluid motion, it is necessary to observe the motion of the fluid particles at various points and times. For the mathematical analysis of the fluid motion, following two methods are generally used :

1. *Lagrangian method.* It deals with the study of flow pattern of the individual particles. In this method, the paths traced by a particle under consideration with the passage of time is studied in detail.
2. *Eulerian method.* It deals with the study of flow pattern of all the particles simultaneously at one section. In this method, the paths traced by all the particles at one section and one time are studied in detail.

Notes : 1. The general example, to explain both the methods, is the study of movement of a number of vehicles on a busy road. The Lagrangian method deals with the study of the movement of only one vehicle through a specified distance. The Eulerian method deals with the study of movement of all the vehicles on the road at one section and at one instant.

2. In the study of Hydraulics, Fluid Mechanics and Astrophysics, the Eulerian method is commonly used, because of its mathematical simplicity. Moreover, in Fluid Mechanics the movement of an individual fluid particle is not of much importance.

6·5 Types of Flow Lines

In the last article, we have discussed that whenever a fluid is in motion, its innumerable particles move along certain lines depending upon the conditions of flow. Though there are many types of flow lines, yet the following are important from the subject point of view :

6·6 Path Lines

The path followed by a fluid particle in motion is called a path line. Thus the path line shows the direction of a particle for a certain period of time or between two given sections.

6·7 Stream Lines

The imaginary line, drawn in the fluid in such a way that the tangent to any point gives the direction of motion at the point, is called the stream line. Thus the stream line shows the direction of motion of a number of particles at the same time.

An element of fluid, bonded by a number of streamlines, which confine the flow, is called stream tube. As there is no movement of fluid across a streamline, therefore no fluid can enter or leave the stream tube except at the ends. It is, thus, obvious that a stream tube behaves like a solid tube.

6·8 Steak Lines or Filament Lines

The instantaneous pictures of the position of all fluid particles, which have passed through a given point at some previous time, are called streak lines or filament lines. For example, the line formed by smoke particles ejected from a nozzle is a streak line.

6·9 Potential Lines or Equipotential Lines

We know that there is always a loss of head of the fluid particles, as we proceed along the flow lines. If we draw lines joining the points of equal potential on adjacent flow lines, we get potential lines or equipotential lines.

The lines AB, CD, EF, GH... etc., are streamlines and LM, NO, PO are potential lines as shown in Fig. 6·3.

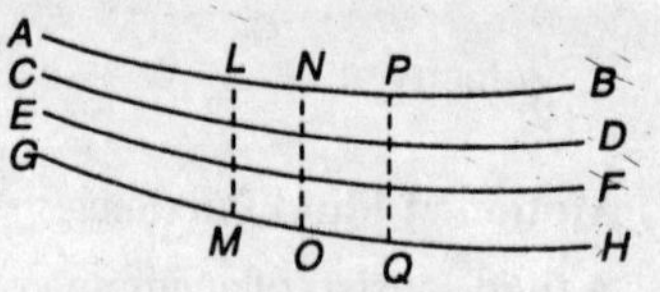

Fig. 6·3. Potential lines

6·10 Flow Net

We have already discussed in previous article the relation between streamlines and potential lines. If we draw both the lines for a flow, the pattern, obtained by the inter-section of the two sets of lines, is called flow net.

6·11 Uses of Flow Nets

The flow net helps us in depicting and analysing the behaviour of certain flow phenomenon, which cannot be easily analysed by mathematical means. Such a phenomenon is generally analysed and studied by drawing flow nets. A flow net may be constructed by drawing a system of stream lines between the boundaries by judgement and then a system of equipotential lines, so as to form a square mesh net. However, in a region where the boundaries converge, diverge or bend, the flow net does not contain squares.

Now consider a flow net when the stream lines are diverging as shown in Fig. 6·4. Now consider two sections of the flow nets 1 and 2.

Hydrokinematics

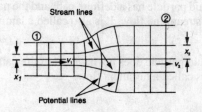

Fig. 6·4. Flow nets

Let
x_1 = Spacing between two streamlines at section 1,
v_1 = Velocity of liquid particles at section 1,
x_2, v_2 = Corresponding values at section 2.

Since there can be no flow across the streamlines, therefore discharge per unit width, between two consecutive streamlines, will be equal. Therefore discharge per unit width,

$$q = x_1 \cdot v_1 = x_2 \cdot v_2$$

It is, thus, obvious that the velocity of liquid particles varies inversely with the spacing between the streamlines. Or in other words, the velocity decreases with the increase in spacing and vice versa.

Example 6·3. *From a flow net diagram, it was found that the distance between two consecutive streamlines at two successive sections are 10 mm and 6 mm respectively. If the velocity at the first section is 1 m/s, find the velocity at the other section. Also find the discharge between these two streamlines.*

Solution. x_1 = 10 mm = 0·01 m; x_2 = 6 mm = 0·006 m and x_1 = 1 m/s.

Velocity at section 2

We know that the velocity at section 2,

$$v_2 = \frac{x_1 \cdot v_1}{x_2} = \frac{0 \cdot 01 \times 1}{0 \cdot 06} = 1 \cdot 67 \text{ m/s} \quad \textbf{Ans.}$$

Discharge between two streamlines

We also know that the discharge between two streamlines,

$$q = x_1 \cdot v_1 = 0 \cdot 01 \times 1 = 0 \cdot 01 \text{ m}^3/\text{s} = 10 \text{ litres/s} \quad \textbf{Ans.}$$

6·12 Types of Flows in a Pipe

When a fluid is flowing in a pipe, the innumerable small particles get together and form a flowing stream. These particles, while moving, group themselves in a variety of ways. *e.g.*, they may move in regular formation, just as disciplined soldiers do; or they may swirl or jostle, like the individuals, in a disorderly mob. The type of flow of a liquid depends upon the manner in which the particles unite and move. Though there are many types of flows, yet the following are important from the subject point of view :

6·13 Uniform Flow

A flow, in which the velocities of liquid particles at all sections of a pipe or channel are equal, is called a uniform flow. This term is generally applied to flow in channels.

6·14 Non-Uniform Flow

A flow, in which the velocities of liquid particles at all sections of a pipe or channel are not equal, is called a non-uniform flow.

6·15 Streamline Flow

A flow, in which each liquid particle has a definite path and the paths of individual particles do not cross each other, is called a streamline flow. It is also called a laminar flow.

6·16 Turbulent Flow

A flow, in which each liquid particle does not have a definite path, and the paths of individual particles also cross each other, is called a turbulent flow.

6·17 Steady Flow

A flow, in which the quantity of liquid flowing per second is constant, is called a steady flow. A steady flow may be uniform or non-uniform.

6·18 Unsteady Flow

A flow, in which the quantity of liquid flowing per second is not constant, is called unsteady flow.

6·19 Compressible Flow

A flow, in which the volume of a fluid and its density changes during the flow, is called a compressible flow. All the gases are, generally, considered to have compressible flows.

6·20 Incompressible Flow

A flow, in which the volume of the flowing fluid and its density does not change during the flow, is called an incompressible flow. All the liquids are, generally, considered to have incompressible flow.

6·21 Rotational Flow

A flow, in which the fluid particles also rotate (*i.e.*, have some angular velocity) about their own axes while flowing, is called a rotational flow. In a rotational flow, if a match stick is thrown on the surface of the moving fluid, it will rotate about its axis as shown in Fig. 6·5 (*a*).

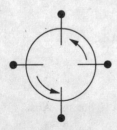
(*a*) Rotational flow

(*b*) Irrotational flow

Fig. 6·5

6·22 Irrotational Flow

A flow, in which the fluid particles do not rotate about their own axes and retain their original orientations, is called an irrotational flow. In an irrotational flow, if a match stick is thrown on the surface of the moving fluid, it does not rotate about its axis but retains its original orientation as shown in Fig. 6·5 (*b*).

6·23 One-dimensional Flow

A flow, in which the streamlines of its moving particles may be represented by straight line, is called one dimensional flow. It is because of the reason that a straight streamline, being a mathematical line, possesses one dimension only. *i.e.*, either x–x or y–y or z–z direction.

6·24 Two-dimensional Flow

A flow, whose streamlines may be represented by a curve, is called a two dimensional flow. It is because of the reason that a curved streamline will be along any two mutually perpendicular directions.

6·25 Three-dimensional Flow

A flow, whose streamlines may be represented in space *i.e.*, along three mutually perpendicular directions, is called three-dimensional flow.

6·26 Stream Function

It is a function, which describes the form of pattern of flow. Or in other words, it is the discharge per unit thickness. Mathematically stream function,

$$\psi = f(x, y)$$

where f = Coefficient of stream function, and

x, y = Co-ordinates of the point, where stream function is required to be found out.

Consider a point P along a streamline as shown in Fig. 6·6.

Let u = Velocity component in x–x direction at P,

v = Velocity component in y–y direction at P,

and ψ = Stream function at P.

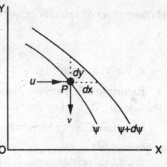

Fig. 6·6. Stream function.

Now let us consider another streamline, such that point P is displaced through a small distance dy in y–y direction and dx in x–x direction as shown in Fig. 6·6. Let the stream function of this new position be $\psi + d\psi$. The flow rate across dy will be given by

$$d\psi = u \cdot dy \qquad \text{or} \qquad u = \frac{d\psi}{dy} \qquad \qquad ...(i)$$

Similarly, the flow rate across dx will be given by

$$d\psi = -v \cdot dx \qquad \text{or} \qquad v = -\frac{d\psi}{dx} \qquad \qquad ...(i)$$

... (Minus sign indicates that the velocity (v) acts downwards)

Now with dx and dy as the components of differential distance, and u and v as the velocity components in x and y directions, we find that

$$\frac{v}{u} = \frac{dy}{dx}$$

or $v \cdot dx = u \cdot dy$

$u \cdot dy - v \cdot dx = 0$

Now substituting the values of u and v from equations (*i*) and (*ii*) in the above equation,

$$\frac{d\psi}{dy} + dy + \frac{d\psi}{dx} \times dx = 0$$

or $d\psi = 0 =$ Constant

The above expression shows that the discharge between the two streamlines is the difference in the two stream functions.

Note : The resultant velocity at any point P (with x and y coordinates) will be given by the relation :
$$V = \sqrt{u^2 + v^2}$$

Example 6·4. *If for two dimensional flow, the stream function is given by $\psi = 2xy$. Calculate the resultant velocity at the point (3, 6).*

Solution. Given : Stream function, $\psi = 2xy$ and co-ordinate of point P, $x = 3$ and $y = 6$.

We know that component of velocity in x–x direction,
$$u = \frac{d\psi}{dy} = \frac{d}{dy}(2xy) = 2x = 2 \times 3 = 6$$

and component of velocity in y–y direction,
$$v = -\frac{d\psi}{dx} = -\frac{d}{dx}(2xy) = -2y = -2 \times 6 = -12$$

∴ Resultant velocity, $V = \sqrt{u^2 + v^2} = \sqrt{(6)^2 + (-12)^2} = 13.4$ **Ans.**

Example 6·5. *A stream function is given by the expression : $\psi = 2x^2 - y^3$. Find components of the velocity, as well as the resultant velocity at a point $P(3, 1)$.*

Solution. Given : Stream function, $\psi = 2x^2 - y^3$ and co-ordinates of point P, $x = 3$ and $y = 1$.

Components of the velocity

We know that component of the velocity in x–x direction,
$$u = \frac{d\psi}{dy} = \frac{d}{dy}(2x^2 - y^3) = -3y^2 = -3 \times (1)^2 = -3 \text{ **Ans.**}$$

and component of the velocity in y–y direction,
$$v = -\frac{d\psi}{dx} = -\frac{d}{dx}(2x^2 - y^3) = -4x = -4 \times 3 = -12 \text{ **Ans.**}$$

Resultant velocity at the point

We also know that the resultant velocity at the point P,
$$V = \sqrt{u^2 + v^2} = \sqrt{(-3)^2 + (-12)^2} = 12.4 \text{ **Ans.**}$$

EXERCISE 6·1

1. The water is flowing through a pipe line of 100 mm diameter with a velocity of 1·5 m/s. Determine the discharge through the pipe in litres/s. [**Ans.** 12·8 litres/s]

2. Find the size of a pipe, which has to discharge an oil, at the rate of 2 m³/s and of specific gravity 0·8 with a velocity of 3 m/s. [**Ans.** 0·92 m]

3. A pipe of 100 mm diameter branches into two pipes of diameters 100 mm and 50 mm respectively. The flow in the larger branch pipe is 2/3 of the main pipe and the remaining discharge is through the smaller branch pipe. Determine the rate of flow in the main pipe, if average velocity of flow in any of the pipes is not to exceed 3 m/s. [**Ans.** 17·66 litres/s]

4. A stream function is given by the relation $\psi = x + y^2$. Find the velocity at a point $P(1, 3)$.
[**Ans.** 6·01 m]

[**Hint :** $u = \dfrac{d\psi}{dy} = \dfrac{d}{dy}(x + y^2) = 2y = 2 \times 3 = 6; \; v = -\dfrac{d\psi}{dx} = -\dfrac{d}{dy}(x + y^2) = -x = -1$

∴ $V = \sqrt{6^2 + (-1)^2} = 6\cdot01$] **Ans.**

Hydrokinematics

QUESTIONS

1. Give the equation for continuity of flow.
2. Distinguish clearly between streamlines and streak lines.
3. What is a flow net ? Describe its uses.
4. State the difference between
   (a) Uniform flow and non-uniform flow.
   (b) Steady and unsteady flow.
   (c) Rotational and irrotational flow.
   (d) One-dimensional, two-dimensional and three-dimensional flow.
5. What is a stream function ? How will you find out the resultant velocity at a point with the given co-ordinates,
6. Show that $\psi = x^2 - y^2$ represents a case of two dimensional fluid flow.

OBJECTIVE TYPE QUESTIONS

1. The principle of continuity is based on
   (a) law of conservation of energy
   (b) law of conservation of matter
   (c) law of conservation of momentum
   (d) all of these
2. A flow in which the velocities of liquid particles at all sections of the pipe are equal is called
   (a) uniform flow
   (b) streamline flow
   (c) steady flow
   (d) compressible flow
3. A flow in which each liquid particle has a definite path and the paths of individual particles do not cross each other is called
   (a) streamline flow
   (b) irrotational flow
   (c) laminar flow
   (d) both 'a' and 'c'
4. A flow whose streamline may be represented by a straightline flow is known as
   (a) stream function
   (b) one-dimensional flow
   (c) incompressible flow
   (d) unsteady flow

ANSWERS

1. (b)   2. (a)   3. (d)   4. (b)

7

Bernoulli's Equation and its Applications

1. Introduction. 2. Energy of a Liquid in Motion. 3. Potential Energy of a Liquid Particle in Motion. 4. Kinetic Energy of a Liquid Particle in Motion. 5. Pressure Energy of a Liquid Particle in Motion. 6. Total Energy of a Liquid Particle in Motion. 7. Total Head of a Liquid Particle in Motion. 8. Bernoulli's Equation. 9. Euler's Equation for Motion. 10. Limitations of Bernoulli's Equation. 11. Practical Applications of Bernoulli's Equation. 12. Venturimeter. 13. Discharge through a Venturimeter. 14. Inclined Venturimeter. 15. Orifice Meter. 16. Pitot Tube.

7·1 Introduction

In the previous chapter, we have discussed the motion of liquid particles without taking into consideration any force or energy causing the flow. But in this chapter, we shall discuss the motion of liquids and the forces causing the flow. This topic is also known as Hydrodynamics.

7·2 Energy of a Liquid in Motion

The energy, in general, may be defined as the capacity to do work. Though the energy exists in many forms, yet the following are important from the subject point of view :

1. Potential energy,
2. Kinetic energy, and
3. Pressure energy.

7·3 Potential Energy of a Liquid Particle in Motion

It is energy possessed by a liquid particle by virtue of its position. If a liquid particle is Z metres above the horizontal datum (arbitrarily chosen), the potential energy of the particle will be Z metre-kilogram (briefly written as mkg) per kg of the liquid. The potential head of the liquid, at that point, will be Z metres of the liquid.

7·4 Kinetic Energy of a Liquid Particle in Motion

It is the energy, possessed by a liquid particle, by virtue of its motion or velocity. If a liquid particle is flowing with a mean velocity of v metres per second, then the kinetic energy of the particle will be $\frac{v^2}{2g}$ mkg per kg of the liquid. Velocity head of the liquid, at that velocity, will be $\frac{v^2}{2g}$ metres of the liquid.

7·5 Pressure Energy of a Liquid Particle in Motion

It is the energy, possessed by a liquid particle, by virtue of its existing pressure. If a liquid particle is under a pressure of p kN/m² (*i.e.*, kPa), then the pressure energy of the particle will be $\frac{p}{w}$ mkg per kg of the liquid, where w is the specific weight of the liquid. Pressure head of the liquid under that pressure will be $\frac{p}{w}$ metres of the liquid.

7.6 Total Energy of a Liquid Particle in Motion

The total energy of a liquid, in motion, is the sum of its potential energy, kinetic energy and pressure energy. Mathematically total energy,

$$E = Z + \frac{v^2}{2g} + \frac{p}{w} \text{ m of liquid.}$$

Note : As a matter of fact, the units of energy are in N-m (or joule). But, according to the subject point of view, the units of energy are taken in terms of m of the liquid.

7.7 Total Head of a Liquid Particle in Motion

The total head of a liquid particle, in motion, is the sum of its potential head, kinetic head and pressure head. Mathematically, total head,

$$H = Z + \frac{v^2}{2g} + \frac{p}{w} \text{ m of liquid.}$$

Example 7·1. *Water is flowing through a tapered pipe having end diameters of 150 mm and 50 mm respectively. Find the discharge at the larger end and velocity head at the smaller end, if the velocity of water at the larger end is 2 m/s.*

Solution. Given : $d_1 = 150$ mm $= 0.15$ m ; $d_2 = 50$ mm $= 0.05$ m and $v_1 = 2.5$ m/s.

Discharge at the larger end

We know that the cross-sectional area of the pipe at the larger end,

$$a_1 = \frac{\pi}{4} \times (d_1)^2 = \frac{\pi}{4} \times (0.15)^2 = 17.67 \times 10^{-3} \text{ m}^2$$

and discharge at the larger end,

$$Q_1 = a_1 \cdot v_1 = (17.67 \times 10^{-3}) \times 2.5 = 44.2 \times 10^{-3} \text{ m}^3/\text{s}$$
$$= 44.2 \text{ litres/s Ans.}$$

Velocity head at the smaller end

We also know that the cross-sectional area of the pipe at the smaller end,

$$a_2 = \frac{\pi}{4} \times (d_2)^2 = \frac{\pi}{4} \times (0.05)^2 = 1.964 \times 10^{-3} \text{ m}^2$$

Since the discharge through the pipe is continuous, therefore

$$a_1 \cdot v_1 = a_2 \cdot v_2$$

or

$$v_2 = \frac{a_1 \cdot v_1}{a_2} = \frac{(17.67 \times 10^{-3}) \times 2.5}{1.964 \times 10^{-3}} = 22.5 \text{ m/s}$$

∴ Velocity head at the smaller end

$$= \frac{v_2^2}{2g} = \frac{(22.5)^2}{2 \times 9.81} = 25.8 \text{ m Ans.}$$

Example 7·2. *A circular pipe of 250 mm diameter carries an oil of specific gravity 0·8 at the rate of 120 litres/s and under a pressure of 2 kPa. Calculate the total energy in metres at a point which is 3 m above the datum line.*

Solution. Given : $d = 250$ mm $= 0.25$ m; Specific gravity oil $= 0.8$; $Q = 120$ litres/s $= 120 \times 10^{-3}$ m^3/s ; $p =$ kPa $= 20$ kN/m^2 and $Z = 3$ m.

We know that the area of the pipe,

$$a = \frac{\pi}{4} \times (d)^2 = \frac{\pi}{4} \times (0.25)^2 = 49.09 \times 10^{-3} \text{ m}^2$$

and velocity of oil,

$$v = \frac{Q}{a} = \frac{120 \times 10^{-3}}{49.09 \times 10^{-3}} = 2.44 \text{ m/s}$$

∴ Total energy
$$= Z + \frac{v^2}{2g} + \frac{p}{w} = 3 + \frac{(2 \cdot 44)^2}{2 \times 9 \cdot 81} + \frac{20}{0 \cdot 8 \times 9 \cdot 81} \text{ m}$$
$$= 3 + 0 \cdot 3 + 2 \cdot 5 = 5 \cdot 8 \text{ m} \quad \textbf{Ans.}$$

Example 7·3. *Water is flowing through a pipe of 70 mm diameter under a gauge pressure of 50 kPa, and with a mean velocity of 2·0 m/s. Neglecting friction, determine the total head, if the pipe is 7 metres above the datum line.*

Solution. Given : $d = 70$ mm $= 0 \cdot 07$ m; $p = 50$ kPa $= 50$ kN/m^2; $v = 2$ m/s and $Z = 7$ m.

We know that the total head of water,
$$H = Z + \frac{v^2}{2g} + \frac{p}{w} = 7 + \frac{(2)^2}{2 \times 9 \cdot 81} + \frac{50}{9 \cdot 81} \text{ m}$$
$$= 7 + 0 \cdot 2 + 5 \cdot 1 = 12 \cdot 3 \text{ m} \quad \textbf{Ans.}$$

7·8 Bernoulli's Equation

It states, "For a perfect incompressible liquid, flowing in a continuous stream, the total energy of a particle remains the same, while the particle moves from one point to another." This statement is based on the assumption that there are no **losses due to friction in the pipe. Mathematically,

$$Z + \frac{v^2}{2g} + \frac{p}{w} = \text{Constant}$$

where
Z = Potential energy,

$\frac{v^2}{2g}$ = Kinetic energy, and

$\frac{p}{w}$ = Pressure energy.

Proof

Consider a perfect incompressible liquid, flowing through a non-uniform pipe as shown in Fig. 7·1.

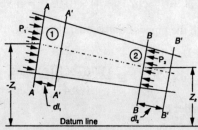

Fig. 7·1. Bernoulli's equation.

Let us consider two sections AA and BB of the pipe. Now let us assume that the pipe is running full and there is a continuity of flow between the two sections.

Let
Z_1 = Height of AA above the datum,

p_1 = Pressure at AA,

v_1 = Velocity of liquid at AA,

a_1 = Cross-sectional area of the pipe at AA, and

---

*Bernoulli Daniel was a Swiss engineer, who belonged to a renowned mathematical family and gave this equation in 1738.

**As a matter of fact, there is always some loss of head of the water while flowing through a pipe. For details, please refer to chapter 13.

Z_2, p_2, v_2, a_2 = Corresponding values at BB.

Let the liquid between the two sections AA and BB move to $A'A'$ and $B'B'$ through very small lengths dl_1 and dl_2 as shown in Fig. 7·1. This movement of the liquid between AA and BB is equivalent to the movement of the liquid between AA and $A'A'$ to BB and $B'B'$, the remaining liquid between $A'A'$ and BB being uneffected.

Let W be the weight of the liquid between AA and $A'A'$. Since the flow is continuous, therefore

$$W = wa_1 dl_1 = wa_2 dl_2$$

or $\qquad a_1 \cdot dl_1 = \dfrac{W}{w}$ ...(i)

Similarly $\qquad a_2 \cdot dl_2 = \dfrac{W}{w}$

$\therefore \qquad a_1 \cdot dl_1 = a_2 \cdot dl_2$ ...(ii)

We know that work done by pressure at AA, in moving the liquid to $A'A'$

$$= \text{Force} \times \text{Distance} = p_1 \cdot a_1 \cdot dl_1$$

Similarly, work done by pressure at BB, in moving the liquid to $B'B'$

$$= -p_2 a_2 \cdot dl_2$$

...(Minus sign is taken as the direction of p_2 is opposite to that of p_1)

$\therefore$ Total work done by the pressure

$$= p_1 a_1 \cdot dl_1 - p_2 a_2 \cdot dl_2$$
$$= p_1 \cdot a_1 dl_1 - p_2 \cdot a_1 dl_1 \qquad \ldots (\because a_1 \cdot dl_1 = a_2 \cdot dl_2)$$
$$= a_1 dl_1 (p_1 - p_2) = \dfrac{W}{w}(p_1 - p_2) \qquad \left(\because a_1 dl_1 = \dfrac{W}{w}\right)$$

Loss of potential energy $\qquad = W(Z_1 - Z_2)$

and again in kinetic energy $\qquad = W\left(\dfrac{v_2^2}{2g} - \dfrac{v_1^2}{2g}\right) = \dfrac{W}{2g}\left(v_2^2 - v_1^2\right)$

We know that loss of potential energy + Work done by pressure

$$= \text{Gain in kinetic energy}$$

$\therefore \qquad W(Z_1 - Z_2) + \dfrac{W}{w}(p_1 - p_2) = \dfrac{W}{2g}(v_2^2 - v_1^2)$

$\qquad (Z_1 - Z_2) + \dfrac{p_1}{2} - \dfrac{p_2}{w} = \dfrac{v_2^2}{2g} - \dfrac{v_1^2}{2g}$

or $\qquad Z_1 + \dfrac{v_1^2}{2g} + \dfrac{p_1}{w} = Z_2 + \dfrac{v_2^2}{2g} + \dfrac{p_2}{w}$

which proves the Bernoulli's equation.

7·9 Euler's Equation for Motion

The *Euler's equation for steady flow of an ideal fluid along a streamline is based on the Newton's Second Law of Motion. The integration of the equation gives Bernoulli's equation in the form of energy per unit weight of the flowing fluid. It is based on the following assumptions :

1. The fluid is non-viscous (*i.e.*, the frictional losses are zero).
2. The fluid is homogeneous and incompressible (*i.e.*, mass density of the fluid is constant).

---

*Euler Leonhard was a Swiss mathematician and a great learned man of his time.

3. The flow is continuous, steady and along the streamline.
4. The velocity of flow is uniform over the section.
5. No energy or force (except gravity and pressure forces) is involved in the flow.

Consider a steady flow of an ideal fluid along a streamline. Now consider a small element AB of the flowing fluid as shown in Fig. 7·2.

Let
dA = Cross-sectional area of the fluid element,
ds = Length of the fluid element,
dW = Weight of the fluid element,
p = Pressure on the element at A,
$p + dp$ = Pressure on the element at B, and
v = Velocity of the fluid element.

We know that the external forces tending to accelerate the fluid element in the direction of the streamline

$$= p \cdot dA - (p + dp) \, dA$$
$$= - dp \cdot dA \qquad \ldots(i)$$

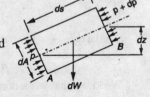

Fig. 7·2. Euler's equation.

We also know that the weight of the fluid element,
$$dW = \rho g \cdot dA \cdot ds$$

From the geometry of the figure, we find that the component of the weight of the fluid element in the direction of flow

$$= - \rho g \cdot dA \cdot ds \cos \theta$$
$$= - \rho g \cdot dA \cdot ds \left(\frac{dz}{ds} \right) \qquad \ldots \left(\because \cos \theta = \frac{dz}{ds} \right)$$
$$= - \rho g \cdot dA \cdot dz \qquad \ldots(ii)$$

∴ Mass of the fluid element $= \rho \cdot dA \cdot ds \qquad \ldots(iii)$

We see that the acceleration of the fluid element
$$= \frac{dv}{dt} = \frac{dv}{ds} \times \frac{ds}{dt} = v \cdot \frac{dv}{ds} \qquad \ldots(iv)$$

Now, as per Newton's Second Law of Motion, we know that
Force = Mass × Acceleration

$$(- dp \cdot dA) - (\rho g \cdot dA \cdot dz) = \rho \cdot dA \cdot ds \times v \cdot \frac{dv}{ds}$$

$$\frac{dp}{\rho} + g \cdot dz = v \cdot dv \qquad \ldots(\text{Dividing both sides by} - \rho dA)$$

or $\quad \dfrac{dp}{\rho} + g \cdot dz + v \cdot dv = 0 \qquad \ldots(v)$

This is the required Euler's equation for motion and is in the form of a differential equation. Integrating the above equation,

$$\frac{1}{\rho} \int dp + \int g \cdot dz + \int v \cdot dv = \text{Constant}$$

$$\frac{p}{\rho} + gZ + \frac{v^2}{2} = \text{Constant}$$

$$p + wZ + \frac{wv^2}{2g} = \text{Constant} \qquad \left(\text{Dividing by } \rho = \frac{w}{g} \right)$$

$$\frac{p}{w} + Z + \frac{v^2}{2g} = \text{Constant} \quad \text{(Dividing by } w\text{)}$$

or in other words, $\quad \dfrac{p_1}{w} + Z_1 + \dfrac{v_1^2}{2g} = \dfrac{p_2}{w} + Z_2 + \dfrac{v_2^2}{2g}$

which proves the Bernoulli's equation.

7·10 Limitations of Bernoulli's Equation

The Bernoulli's theorem or Bernoulli's equation has been derived on certain assumptions, which are rarely possible. Thus the Bernoulli's theorem has the following limitations :

1. The Bernoulli's equation has been derived under the assumption that the velocity of every liquid particle, across any cross-section of a pipe, is uniform. But, in actual practice, it is not so. The velocity of liquid particle in the centre of a pipe is maximum and gradually decreases towards the walls of the pipe due to the pipe friction. Thus, while using the Bernoulli's equation, only the mean velocity of the liquid should be taken into account.

2. The Bernoulli's equation has been derived under the assumption that no external force, except the gravity force, is acting on the liquid. But, in actual practice, it is not so. There are always some external forces (such as pipe friction etc.) acting on the liquid, which effect the flow of the liquid. Thus, while using the Bernoulli's equation, all such external forces should be neglected. But, if some energy is supplied to, or, extracted from the flow, the same should also be taken into account.

3. The Bernoulli's equation has been derived, under the assumption that there is no loss of energy of the liquid particle while flowing. But, in actual practice, it is rarely so. In a turbulent flow, some kinetic energy is converted into heat energy. And in a viscous flow, some energy is lost due to shear forces. Thus, while using Bernoulli's equation, all such losses should be neglected.

4. If the liquid is flowing in a curved path, the energy due to centrifugal force should also be taken into account.

Example 7·4. *The diameter of a pipe changes from 200 mm at a section 5 metres above datum to 50 mm at a section 3 metres above datum. The pressure of water at first section is 500 kPa. If the velocity of flow at the first section is 1 m/s, determine the intensity of pressure at the second section.*

Solution. Given : $d_1 = 200$ mm $= 0.2$ m; $Z_1 = 5$ m; $d_2 = 50$ mm $= 0.05$ m; $Z_2 = 3$ m; $p = 500$ kPa $= 500$ kN/m^2 and $v_1 = 1$ m/s.

Let $\quad v_2 = $ Velocity of flow at section 2, and

$\quad p_2 = $ Pressure at section 2.

We know that area of the pipe at section 1,

$$a_1 = \frac{\pi}{4} \times (d_1)^2 = \frac{\pi}{4} \times (0.2)^2 = 31.42 \times 10^{-3} \text{ m}^2$$

and area of pipe at section 2, $\quad a_2 = \dfrac{\pi}{4} \times (d_2)^2 = \dfrac{\pi}{4} \times (0.05)^2 = 1.964 \times 10^{-3}$ m^2

Since the discharge through the pipe is continuous, therefore

$$a_1 \cdot v_1 = a_2 \cdot v_2$$

or $\quad v_2 = \dfrac{a_1 \cdot v_1}{a_2} = \dfrac{(31.42 \times 10^{-3}) \times 1}{(1.964 \times 10^{-3})} = 16$ m/s

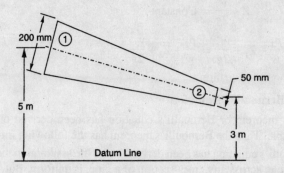

Fig. 7·3

Applying Bernoulli's equation for both the ends of the pipe,

$$Z_1 + \frac{v_1^2}{2g} + \frac{p_1}{w} = Z_2 + \frac{v_2^2}{2g} + \frac{p_2}{w}$$

$$5 + \frac{(1)^2}{2 \times 9\cdot81} + \frac{500}{9\cdot81} = 3 + \frac{(16)^2}{2 \times 9\cdot81} + \frac{p_2}{9\cdot81}$$

$$5 + 0\cdot05 + 51 = 3 + 13\cdot05 + \frac{p_2}{9\cdot81}$$

$$56\cdot05 = 16\cdot05 + \frac{p_2}{9\cdot81}$$

or $\qquad \frac{p_2}{9\cdot81} = 56\cdot05 - 16\cdot05 = 40$

∴ $\qquad p_2 = 40 \times 9\cdot81 = 392\cdot4 \text{ kN/m}^2 = 392\cdot4 \text{ kPa}$ **Ans.**

Example 7·5. *A pipe 300 metres long has a slope of 1 in 100 and tapers from 1 metre diameter at the higher end to 0·5 metre at the lower end. The quantity of water flowing is 900 litres/second. If the pressure at the higher end is 70 kPa, find the pressure at the lower end.*

Solution. Given : $l = 300$ m; Slope = 1 in 100; $d_2 = 1$ m; $d_1 = 0\cdot5$ m; $Q = 900$ litres/s = 0·9 m^3/s and $p_2 = 70$ kPa = 70 kN/m^2.

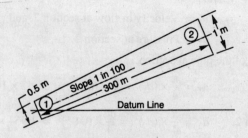

Fig. 7·4.

Let $\qquad p_1 = $ Pressure at section 1.

We know that area of the pipe at section 1,

$$a_1 = \frac{\pi}{4} \times (d_1)^2 = \frac{\pi}{4} \times (0\cdot5)^2 = 0\cdot1964 \text{ m}^2$$

and area of the pipe at section 2, $a_2 = \frac{\pi}{4} \times (d_2)^2 = \frac{\pi}{4} \times (1)^2 = 0.7854 \text{ m}^2$

First of all, let us assume the datum line to coincide with the centre of the pipe at section 1 as shown in the figure. Thus $Z_1 = 0$. Now from the geometry of the pipe, we find that the height of higher end from the datum line,

$$Z_2 = \frac{1}{100} \times 300 = 3 \text{ m}$$

We also know that velocity of water at section 1,

$$v_1 = \frac{Q}{a_1} = \frac{0.9}{0.1964} = 4.58 \text{ m/s}$$

and velocity of water at section 2,

$$v_2 = \frac{Q}{a_2} = \frac{0.9}{0.7854} = 1.15 \text{ m/s}$$

Now using Bernoulli's equation for both ends of the pipe,

$$Z_1 + \frac{v_1^2}{2g} + \frac{p_1}{w} = Z_2 + \frac{v_2^2}{2g} + \frac{p_2}{w}$$

$$0 + \frac{(4.58)^2}{2 \times 9.81} + \frac{p_1}{9.81} = 3 + \frac{(1.15)^2}{2 \times 9.81} + \frac{70}{9.81}$$

$$0 + 1.07 + \frac{p_1}{9.81} = 3 + 0.07 + 7.14 = 10.21$$

or

$$\frac{p_1}{9.81} = 10.21 - 1.07 = 9.14$$

∴ $p = 9.14 \times 9.81 = 89.7 \text{ kN/m}^2 = 89.7 \text{ kPa}$ **Ans.**

Example 7·6. *A pipe 5 metre long is inclined at an angle of 15° with the horizontal. The smaller section of the pipe, which is at a lower level, is of 80 mm diameter and the larger section of the pipe is of 240 mm diameter as shown in Fig. 7·5.*

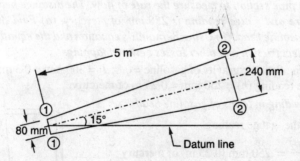

Fig. 7·5.

Determine the difference of pressures between the two sections, if the pipe is uniformly tapering and the velocity of water at the smaller section is 1 m/s.

Solution. Given : $l = 5$ m; $\alpha = 15°$; $d_1 = 80$ mm $= 0.08$ m; $d_2 = 240$ mm $= 0.24$ m and $v_1 = 1$ m/s.

Let
p_1 = Pressure at section 1, and
p_2 = Pressure at section 2.

We know that area of pipe at section 1,

$$a_1 = \frac{\pi}{4} \times (d_1)^2 = \frac{\pi}{4} \times (0.08)^2 = 5.027 \times 10^{-3} \text{ m}^2$$

and area of pipe at section 2, $a_2 = \frac{\pi}{4} \times (d_2)^2 = \frac{\pi}{4} \times (0.24)^2 = 45.24 \times 10^{-3} \text{ m}^2$

First of all, let us assume the datum line to coincide with the centre of the pipe at section 1 as shown in the figure. Thus $Z_1 = 0$. Now from the geometry of the pipe, we find that the height of the higher end from the datum line,

$$Z_2 = 0 + 5 \sin \alpha = 5 \sin 15° = 5 \times 0.2588 = 1.294 \text{ m}$$

Since the discharge through the pipe is continuous, therefore

$$a_1 \cdot v_1 = a_2 \cdot v_2$$

or $v_2 = \frac{a_1 \cdot v_1}{a_2} = \frac{(5.027 \times 10^{-3}) \times 1}{45.24 \times 10^{-3}} = 0.11 \text{ m/s}$

Applying Bernoulli's equation for both the sections of the pipe,

$$Z_1 + \frac{v_1^2}{2g} + \frac{p_1}{w} = Z_2 + \frac{v_2^2}{2g} + \frac{p_2}{w}$$

$$0 + \frac{(1)^2}{2 \times 9.81} + \frac{p_1}{9.81} = 1.294 + \frac{(0.11)^2}{2 \times 9.81} + \frac{p_2}{9.81}$$

$$0 + 0.05 + \frac{p_1}{9.81} = 1.294 + 0 + \frac{p_2}{9.81}$$

$$\frac{p_1}{9.81} - \frac{p_2}{9.81} = 1.294 + 0 - 0.05 = 1.244$$

∴ $(p_1 - p_2) = 1.244 \times 9.81 = 12.2 \text{ kN/m}^2 = 12.2 \text{ kPa}$ **Ans.**

Example 7·7. *An oil of sp. gr. 0·8 is flowing upwards through a vertical pipe line, which tapers from 300 mm to 150 mm diameter. A gasoline mercury differential monometer is connected between 300 mm and 150 mm pipe section to measure the rate of flow. The distance between the monometer the tapping is 1 metre and gauge reading is 250 mm of mercury. (a) Find the differential gauge reading in terms of gasoline head. (b) Using Bernoulli's equation and the equation of continuity, find the rate of flow. Neglect friction and other losses between tappings.*

Solution. Given : Specific gravity of gasoline = 0·8; d_1 = 300 mm = 0·3 m; d_2 = 150 mm = 0·15 m; l = 1 m and gauge reading (h) = 250 mm = 0·25 m of mercury.

Differential gauge reading in terms of gasoline head

We know that the gauge reading,

$$\frac{p_1}{W} - \frac{p_2}{w} = 250 \text{ mm } (0.25 \text{ m}) \text{ of mercury}$$

$$= 0.25 \times \left(\frac{13.6 - 0.8}{0.8}\right) \text{m of gasoline}$$

$$= 4 \text{ m of gasoline Ans.}$$

Rate of flow

Let v_1 = Velocity of oil at section 1, and

v_2 = Velocity of oil at section 2.

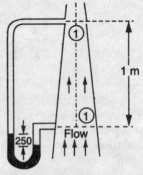

Fig. 7·6.

We know that the area of pipe at section 1,

$$a_1 = \frac{\pi}{4} \times (d_1)^2 = \frac{\pi}{4} \times (0.3)^2 = 70.69 \times 10^{-3} \text{ m}^2$$

and the area of pipe at section 2,

$$a_2 = \frac{\pi}{4} \times (d_2)^2 = \frac{\pi}{4} \times (0.15)^2 = 17.67 \times 10^{-3} \text{ m}^2$$

Since the discharge through the pipe is continuous, therefore

$$a_1 \cdot v_1 = a_2 \cdot v_2$$

or
$$v_2 = \frac{a_1 \cdot v_1}{a_2} = \frac{70.69 \times 10^{-3} \times v_1}{17.67 \times 10^{-3}} = 4v_1$$

Applying Bernoulli's equation for inlet or outlet of the pipe,

$$Z_1 + \frac{v_1^2}{2g} + \frac{p_1}{w} = Z_2 + \frac{v_2^2}{2g} + \frac{p_2}{w}$$

$$Z_1 + \left(\frac{p_1}{w} - \frac{p_2}{w}\right) = Z_2 + \frac{v_2^2}{2g} - \frac{v_1^2}{2g}$$

$$0 + 4 = 1 + \frac{(4v_1)^2}{2g} - \frac{v_1^2}{2g} = 1 + \frac{15v_1^2}{2g}$$

$$\frac{15v_1^2}{2g} = 4 - 1 = 3$$

or
$$v_1 = \sqrt{\frac{3 \times 2 \times 9.81}{15}} = 1.98 \text{ m/s}$$

∴ Rate of flow, $Q = a_1 \cdot v_1 = (70.69 \times 10^{-3}) \times 1.98 = 140 \times 10^{-3} \text{ m}^3/\text{s}$

= 140 litres/s **Ans.**

EXERCISE 7·1

1. A uniformly tapering pipe has a 120 mm and 80 mm diameters at its ends. If the velocity of water at the larger end is 2 m/s, find the discharge at the larger end and the velocity head at the smaller end. **[Ans. 22·62 litres/s; 1·03 m]**

2. Find the total head of water flowing with a velocity of 8 m/s under a pressure of 80 kPa. The centre line of the pipe is 5 m above the datum line. **[Ans. 16·41 m]**

3. A horizontal pipe 100 m long uniformly tapers from 300 mm diameter to 200 mm diameter. What is the pressure head at the smaller end, if the pressure at the larger end is 100 kPa and the pipe is discharging 50 litres of water per second? **[Ans. 99·1 kPa]**

4. Water is flowing through a pipe at the rate of 35 litres/s having diameters 200 mm and 100 mm at sections 1 and 2 respectively. The section 1 is 4 m above the datum and section 2 is 2 m above the datum. Find the pressure at section 2, if the pressure at section 1 is 40 kPa. **[Ans. 50·3 kPa]**

5. A 200 m long pipe slopes down at 1 in 100 and tapers from 0·25 m diameter to 0·15 m diameter at the lower end. If the pipe carries 100 litres of oil of specific gravity 0·85, find the pressure at the lower end. The upper end gauge reads 50 kPa. **[Ans. 54·9 kPa]**

6. A tapering pipe is used to carry water as shown in Fig. 7·7. The discharge through the pipe was observed to be 170 litres/s.

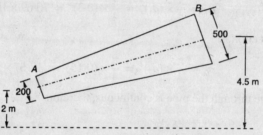

Fig. 7·7. Venturimeter.

If the pressures at A and B are 100 kPa and 75 kPa respectively, determine the direction in which the water will flow through the pipe. [**Ans.** from A to B]

7·11 Practical Applications of Bernoulli's Equation

The Bernoulli's theorem or Bernoulli's equation is the basic equation which has the widest applications in Hydraulics and Applied Hydraulics. Since this equation is applied for the derivation of many formulae, therefore its clear understanding is very essential. Though the Bernoulli's equation has a number of practical applications, yet in this chapter we shall discuss its applications on the following hydraulic devices :

1. Venturimeter. 2. Orificemeter. 3. Pitot tube.

7·12 Venturimeter

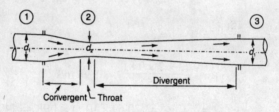

Fig. 7·8

A *venturimeter is an apparatus for finding out the discharge of a liquid flowing in a pipe. A venturimeter, in its simplest form, consists of the following three parts :

(*a*) Convergent cone. (*b*) Throat. (*c*) Divergent cone.

(*a*) *Convergent cone*

It is a short pipe which converges from a diameter d_1 (diameter of the pipe in which the venturimeter is fitted) to a smaller diameter d_2. The convergent cone is also known as inlet of the venturimeter. The slope of the converging sides is between 1 in 4 or 1 in 5 as shown in Fig. 7·7.

(*b*) *Throat*

It is a small portion of circular pipe in which the diameter d_2 is kept constant as shown in Fig. 7·7.

(*c*) *Divergent cone*

It is a pipe, which diverges from a diameter d_2 to a large diameter d_1. The divergent cone is also known as outlet of the venturimeter. The length of the divergent cone is about 3 to 4 times than that of the convergent cone as shown in Fig. 7·7.

---

*Venturi was an Italian engineer who discussed the phenomenon of pressure reduction at throats in pipes in 1791.

A little consideration will show that the liquid, while flowing through the venturimeter, is accelerated between the sections 1 and 2 (*i.e.*, while flowing through the convergent cone). As a result of the acceleration, the velocity of liquid at section 2 (*i.e.*, at the throat) becomes higher than that at section 1. This increase in velocity results in considerably decreasing the pressure at section 2. If the pressure head at the throat falls below the separation head (which is 2·5 metres of water), then there will be a tendency of separation of the liquid flow. In order to avoid the tendency of separation at throat, there is always a fixed ratio of the diameter of throat and the pipe (*i.e.*, d_2/d_1). This ratio varies from 1/4 to 3/4, but the most suitable value is 1/3 to 1/2.

The liquid, while flowing through the venturimeter, is decelerated (*i.e.*, retarded) between the sections 2 and 3 (*i.e.*, while flowing through the divergent cone). As a result of this retardation, the velocity of liquid decreases which, consequently, increases the pressure. If the pressure is rapidly recovered, then there is every possibility for the stream of liquid to break away from the walls of the metre due to boundary layer effects. In order to avoid the tendency of breaking away the stream of liquid, the divergent cone is made sufficiently longer. Another reason for making the divergent cone longer is to minimise the frictional losses. Due to these reasons, the divergent cone is 3 to 4 times longer than convergent cone as shown in Fig. 7·7.

7·13 Discharge through a Venturimeter

Consider a venturimeter through which some liquid is flowing as shown in Fig. 7·9.

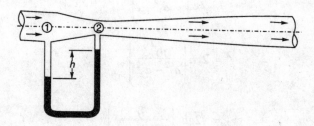

Fig. 7·9. Discharge through a venturimeter.

Let
p_1 = Pressure at section 1,
v_1 = Velocity of water at section 1,
Z_1 = Datum head at section 1,
a_1 = Area of the venturimeter at section 1, and
p_2, v_2, Z_2, a_2 = Corresponding values at section 2.

Applying Bernoulli's equation at sections 1 and 2. *i.e.*,

$$Z_1 + \frac{v_1^2}{2g} + \frac{p_1}{w} = Z_2 + \frac{v_2^2}{2g} + \frac{p_2}{w} \qquad \ldots(i)$$

Let us pass our datum line through the axis of the venturimeter as shown in Fig. 7·8.
Now $Z_1 = 0$ and $Z_2 = 0$.

∴
$$\frac{v_1^2}{2g} + \frac{p_1}{w} = \frac{v_2^2}{2g} + \frac{p_2}{w}$$

or
$$\frac{p_1}{w} - \frac{p_2}{w} = \frac{v_2^2}{2g} - \frac{v_1^2}{2g} \qquad \ldots(ii)$$

Since the discharge at sections 1 and 2 is continuous, therefore
$$v_1 = \frac{a_2 v_2}{a_1} \qquad (\because a_1 \cdot v_1 = a_2 \cdot v_2)$$

$$\therefore \quad v_1^2 = \frac{a_2^2 v_2^2}{a_1^2} \qquad \ldots(iii)$$

Substituting the above value of v_1^2 in equation (ii),

$$\frac{p_1}{w} - \frac{p_2}{w} = \frac{v_2^2}{2g} - \left(\frac{a_2^2}{a_1^2} \times \frac{v_2^2}{2g}\right)$$

$$= \frac{v_2^2}{2g}\left(1 - \frac{a_2^2}{a_1^2}\right) = \frac{v_2^2}{2g}\left(\frac{a_1^2 - a_2^2}{a_1^2}\right)$$

We know that $\frac{p_1}{w} - \frac{p_2}{w}$ is the difference between the pressure heads at sections 1 and 2. When the pipe is horizontal, this difference represents the venturi head and is denoted by h.

or
$$h = \frac{v_2^2}{2g}\left(\frac{a_1^2 - a_2^2}{a_1^2}\right)$$

or
$$v_2^2 = 2gh\left(\frac{a_1^2}{a_1^2 - a_2^2}\right)$$

$\therefore$
$$v_2 = \sqrt{2gh}\left(\frac{a_1}{\sqrt{a_1^2 - a_2^2}}\right)$$

We know that the discharge through a venturimeter,
$$Q = \text{Coefficient of venturimeter}^* \times q_2 \cdot v_2$$

$$= C \cdot a_2 \cdot v_2 = \frac{C a_1 a_2}{\sqrt{a_1^2 - a_2^2}} \sqrt{2gh}$$

Note : The venturi head (h), in the above equation, has been taken in terms of the liquid head. But, in actual practice, this head is given as the mercury head. In such a case, the mercury head should be converted into the liquid head. If the given liquid is water, the mercury head can be converted into water head by multiplying the mercury head by the specific gravity of mercury, minus the specific gravity of water (i.e., 13·6 – 1 = 12·6).

Sometimes, an oil is being discharged through the venturimeter. In such a case, the venturi head should be taken in terms of oil head. e.g.,

$$h = \frac{13 \cdot 6 - w}{w} \times \text{Head of mercury}$$

where
13·6 = Specific gravity of mercury, and
w = Specific weight of the oil.

Example 7·8. *A venturimeter with a 150 mm diameter at inlet and 100 mm at throat is laid with its axis horizontal and is used for measuring the flow of oil specific gravity 0·9. The oil-mercury differential monometer shows a gauge difference of 200 mm. Assume coefficient of the metre as 0·98. Calculate the discharge in litres per minute.*

Solution. Given : d_1 = 150 mm = 0·15 m; d_2 = 100 mm = 0·1 m; Specific gravity of oil = 0·9; h = 200 mm = 0·2 m of mercury and C = 0·98.

---

*Sometimes it is called as coefficient of discharge. For details, please refer to Art. 9·8.

We know that the area at inlet,

$$a_1 = \frac{\pi}{4} \times (d_1)^2 = \frac{\pi}{4} \times (0.15)^2 = 17.67 \times 10^{-3} \text{ m}^2$$

and the area at throat,

$$a_2 = \frac{\pi}{4} \times (d_2)^2 = \frac{\pi}{4} \times (0.1)^2 = 7.854 \times 10^{-3} \text{ m}^2$$

We also know that the difference of pressure head,

$$h = 0.2 \left(\frac{13.6 - 0.9}{0.9} \right) = 2.82 \text{ m of oil}$$

and the discharge through the venturimeter,

$$Q = \frac{C \cdot a_1 a_2}{\sqrt{a_1^2 - a_2^2}} \sqrt{2gh}$$

$$= \frac{0.98 \times (17.67 \times 10^{-3}) \times (7.854 \times 10^{-3})}{\sqrt{(17.67 \times 10^{-3})^2 - (7.854 \times 10^{-3})^2}} \times \sqrt{2 \times 9.81 \times 2.82} \text{ m}^3/\text{s}$$

$$= \frac{136 \times 10^{-6}}{15.83 \times 10^{-3}} \times 7.44 = 63.9 \times 10^{-3} \text{ m}^3/\text{s} = 63.9 \text{ litres/s}$$

$$= 63.9 \times 60 = 3834 \text{ litres/min} \quad \textbf{Ans.}$$

Example 7·9. *A venturimetre has an area ratio of 9 to 1, the larger diameter being 300 mm. During the flow, the recorded pressure head in the large section is 6·5 metres and that at the throat 4·25 metres. If the metre coefficient, (C) = 0·99, compute the discharge through the metre.*

Solution. Given : $a_1/a_2 = 9$; $d_1 = 300$ mm $= 0.3$ m; Pressure head at large section $(h_1) = 6.5$ m; Pressure head at smaller section $(h_2) = 4.25$ m and $C = 0.99$.

We know that the area of the large section (inlet),

$$a_1 = \frac{\pi}{4} \times (d_1)^2 = \frac{\pi}{4} \times (0.3)^2 = 70.69 \times 10^{-3} \text{ m}^2$$

and the area of the smaller section (throat),

$$a_2 = \frac{a_1}{9} = \frac{70.69 \times 10^{-3}}{9} = 7.854 \times 10^{-3} \text{ m}^2$$

We also know that the difference of pressure heads,

$$h = h_1 - h_2 = 6.5 - 4.25 = 2.25 \text{ m}$$

and the discharge through the metre,

$$Q = \frac{C \cdot a_1 a_2}{\sqrt{a_1^2 - a_2^2}} \sqrt{2gh}$$

$$= \frac{0.99 \times (70.69 \times 10^{-3}) \times (7.854 \times 10^{-3})}{\sqrt{(70.69 \times 10^{-3})^2 - (7.854 \times 10^{-3})^2}} \times \sqrt{2 \times 9.81 \times 2.25} \text{ m}^3/\text{s}$$

$$= \frac{549.6 \times 10^{-6}}{70.25 \times 10^{-3}} \times 6.644 = 52 \times 10^{-3} \text{ m}^3/\text{s} = 52 \text{ litres/s} \quad \textbf{Ans.}$$

Example 7·10. *A horizontal venturimeter 160 mm × 80 mm is used to measure the flow of an oil of specific gravity 0·8. Determine the deflection of the oil-mercury gauge, if the discharge of the oil is 50 litres/s. Take coefficient of venturimeter as 1.*

Solution. Given : $d_1 = 160$ mm $= 0.16$ m; $d_2 = 80$ mm $= 0.08$ m; Specific gravity of oil $= 0.8$; $Q = 50$ litres/s $= 50 \times 10^{-3}$ m^3/s and $C = 1$.

Let $\quad h =$ Deflection of oil-mercury gauge in m.

We know that the area at inlet,
$$a_1 = \frac{\pi}{4} \times (d_1)^2 = \frac{\pi}{4} \times (0.16)^2 = 20.11 \times 10^{-3} \text{ m}^2$$

and area at throat, $\quad a_2 = \frac{\pi}{4} \times (d_2)^2 = \frac{\pi}{4} \times (0.08)^2 = 5.027 \times 10^{-3} \text{ m}^2$

We also know that the deflection of oil mercury gauge,
$$h = \left(\frac{13.6 - 0.8}{0.8} \times h\right) = 16\, h \text{ m of oil}$$

and discharge of the oil through venturimetre, (Q),

$$50 \times 10^{-3} = \frac{C \cdot a_1 a_2}{\sqrt{a_1^2 - a_2^2}} \sqrt{2gh}$$

$$= \frac{1 \times (20.11 \times 10^{-3}) \times (5.027 \times 10^{-3})}{\sqrt{(20.11 \times 10^{-3})^2 - (5.027 \times 10^{-3})^2}} \times \sqrt{2 \times 9.81 \times 16h}$$

$$= \frac{101.1 \times 10^{-6}}{19.46 \times 10^{-3}} \times 17.7 \sqrt{h} = 91.96 \times 10^{-3} \sqrt{h}$$

$$\therefore \quad \sqrt{h} = \frac{50 \times 10^{-3}}{91.96 \times 10^{-3}} = 0.544$$

or $\quad h = (0.544)^2 = 0.296 \text{ m} = 296 \text{ mm}$ **Ans.**

Example 7·11. *A venturimeter is to be fitted to a 250 mm diameter pipe, in which the maximum flow is 7200 litres per minute and the pressure head is 6 metres of water. What is the minimum diameter of throat, so that there is no negative head in it ?*

Solution. Given : $d_1 = 250$ mm; $Q = 7200$ litres/min = 120 litres/s = 120×10^{-3} m³/s and *pressure head $(h) = 6$ m.

Let $\quad a_2 =$ Area of the throat.

We know that the area at inlet,
$$a_1 = \frac{\pi}{4} (d_1)^2 = \frac{\pi}{4} \times (0.25)^2 = 49.09 \times 10^{-3} \text{ m}^2$$

and flow through the venturimeter (Q),

$$120 \times 10^{-3} = \frac{C \cdot a_1 a_2}{\sqrt{a_1^2 - a_2^2}} \sqrt{2gh} = \frac{Ca_1}{\sqrt{\frac{a_1^2 - a_1^2}{a_2^2}}} \times \sqrt{2gh} = \frac{Ca_1}{\sqrt{\left(\frac{a_1}{a_2}\right)^2 - 1}} \times \sqrt{2gh}$$

$$= \frac{1 \times (49.09 \times 10^{-3})}{\sqrt{\left(\frac{a_1}{a_2}\right)^2 - 1}} \times \sqrt{2 \times 9.81 \times 6} = \frac{532.6 \times 10^{-3}}{\sqrt{\left(\frac{a_1}{a_2}\right)^2 - 1}}$$

---

*Pressure head, $\quad h = \dfrac{p_1}{w} - \dfrac{p_2}{w}$

Since the pressure head at the throat is not to be negative, or maximum it can be zero. Therefore venturihead
$$h = 6 - 0 = 6 \text{ m of water}. \quad \ldots\left(\because \frac{p_1}{w} = 6 \text{ m given}\right)$$

$$\sqrt{\left(\frac{a_1}{a_2}\right) - 1} = \frac{532 \cdot 6 \times 10^{-3}}{120 \times 10^{-3}} = 4 \cdot 44$$

Squaring both sides of the above equation,

$$\left(\frac{a_1}{a_2}\right)^2 - 1 = 19 \cdot 7 \quad \text{or} \quad \left(\frac{a_1}{a_2}\right)^2 = 19 \cdot 7 + 1 = 20 \cdot 7$$

Taking square root of both sides,

$$\frac{a_1}{a_2} = \sqrt{20 \cdot 7} = 4 \cdot 55$$

or
$$a_2 = \frac{a_1}{4 \cdot 55} = \frac{49 \cdot 09 \times 10^{-3}}{4 \cdot 55} = 10 \cdot 79 \times 10^{-3} \text{ m}^2$$

∴ Diameter of throat
$$d_2 = \sqrt{\frac{4 a_2}{\pi}} = \sqrt{\frac{4 \times 10 \cdot 79 \times 10^{-3}}{\pi}} \text{ m} \quad \ldots [\because \text{Area} = \frac{\pi}{4} \times (d_2)^2]$$

$$= 0 \cdot 117 \text{ m} = 117 \text{ mm} \quad \textbf{Ans.}$$

7·14 Inclined Venturimeter

Sometimes, a venturimeter is fitted to an inclined (or even a vertical) pipe as shown in Fig. 7·10 (*a*) and (*b*).

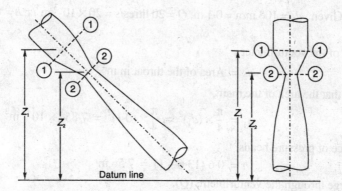

Fig. 7·10.

The same formula for discharge through the venturimeter, (as we derived in Art. 7·13) holds good. The discharge through an inclined venturimeter may also be found out, first by finding out the velocity at either section (by using Bernoulli's equation) and then by multiplying the velocity with the respective area of flow.

Example 7·12. *200 mm × 100 mm venturimeter is mounted in a vertical pipe carrying water, the flow being upwards. The throat section is 200 mm above the entrance section of the venturimeter. For a certain flow through the meter, the differential gauge between the throat and entrance indicates a gauge of deflection of 250 mm. Assuming the venturi coefficient as 0·98, find the discharge.*

Solution. Given : $d_1 = 200$ mm $= 0·2$ m; $d_2 = 100$ mm $= 0·1$ m; $h = 250$ mm $= 0·25$ m of mercury and $C = 0·98$.

We know that the area at inlet,

$$a_1 = \frac{\pi}{4} \times (d_1)^2 = \frac{\pi}{4} \times (0 \cdot 2)^2 = 31 \cdot 42 \times 10^{-3} \text{ m}^2$$

and the area at throat,
$$a_2 = \frac{\pi}{4} \times (d_2)^2 = \frac{\pi}{4} \times (0.1)^2 = 7.854 \times 10^{-3} \text{ m}^2$$

We also know that manometer reading,
$$h = 0.25(13.6 - 1) = 3.15 \text{ m of water}$$

and discharge through the venturimeter,
$$Q = \frac{C a_1 a_2}{\sqrt{(a_1^2 - a_2^2)}} \sqrt{2gh}$$

$$= \frac{0.98 \times (31.42 \times 10^{-3}) \times (7.854 \times 10^{-3})}{\sqrt{(31.42 \times 10^{-3})^2 - (7.854 \times 10^{-3})^2}} \times \sqrt{2 \times 9.81 \times 3.15} \text{ m}^3/\text{s}$$

$$= \frac{241.8 \times 10^{-6}}{30.42 \times 10^{-3}} \times 7.861 = 62.5 \times 10^{-3} \text{ m}^3/\text{s} = 62.5 \text{ litres/s} \text{ Ans.}$$

Example 7·13. *Find the throat diameter of a venturimeter, when fitted to a horizontal main 100 mm diameter having a discharge of 20 litres/s. The differential U-tube mercury manometer shows a deflection giving a reading of 0·6 m. Venture coefficient is 0·95.*

In case, this venturimeter is introduced in a vertical pipe with the water flowing upwards, find the difference in the readings of mercury gauge. The dimensions of pipe and venturimeter remain unaltered as well as the discharge through the pipe.

Solution. Given : $d_1 = 100 \text{ mm} = 0.1 \text{ m}$; $Q = 20 \text{ litres/s} = 20 \times 10^{-3} \text{ m}^3/\text{s}$; $h = 0.6 \text{ m of mercury}$ and $C = 0.95$.

Throat diameter of the venturimeter

Let a_2 = Area of the throat in m^2

We know that the area of the main,
$$a_1 = \frac{\pi}{4} \times (d_1)^2 = \frac{\pi}{4} \times (0.1)^2 = 7.854 \times 10^{-3} \text{ m}^2$$

and the difference of pressure heads,
$$h = 0.6(13.6 - 1) = 7.56 \text{ m}$$

∴ Discharge through the venturimetre (Q),

$$20 \times 10^{-3} = \frac{C a_1 a_2}{\sqrt{a_1^2 - a_2^2}} \sqrt{2gh} = \frac{C a_1}{\sqrt{\frac{a_1^2 - a_2^2}{a_2^2}}} \sqrt{2gh} = \frac{C a_1}{\sqrt{\left(\frac{a_1}{a_2}\right)^2 - 1}} \sqrt{2gh}$$

$$= \frac{0.95 \times (7.854 \times 10^{-3})}{\sqrt{\left(\frac{a_1}{a_2}\right)^2 - 1}} \times \sqrt{2 \times 9.81 \times 7.56} = \frac{90.87 \times 10^{-3}}{\sqrt{\left(\frac{a_1}{a_2}\right)^2 - 1}}$$

$$\sqrt{\left(\frac{a_1}{a_2}\right)^2 - 1} = \frac{90.87 \times 10^{-3}}{20 \times 10^{-3}} = 4.544$$

Squaring both sides of the above equation,
$$\left(\frac{a_1}{a_2}\right)^2 - 1 = 20.64 \quad \text{or} \quad \left(\frac{a_1}{a_2}\right)^2 = 20.64 + 1 = 21.64$$

Bernoulli's Equation and its Applications

and now taking square root of both sides,

$$\frac{a_1}{a_2} = \sqrt{21.64} = 4.652$$

$$a_2 = \frac{a_1}{4.652} = \frac{7.854 \times 10^{-3}}{4.652} = 1.688 \times 10^{-3} \text{ m}^2$$

or dia. of throat, $\quad d = \sqrt{\frac{4a_2}{\pi}} = \sqrt{\frac{4 \times (1.688 \times 10^{-3})}{\pi}}$ m ...[$\because$ Area $= \frac{\pi}{4} \times (d_2)^2$]

$$= 0.046 \text{ m} = 46 \text{ mm} \quad \textbf{Ans.}$$

Difference in the readings of mercury gauge, when the venturimeter is introduced in a vertical pipe

The difference in the readings of mercury gauge will be the same as that when the venturimeter is introduced in a horizontal pipe. *i.e.*, 600 mm of mercury. **Ans.**

Example 7·14. *A 300 mm × 150 mm venturimeter is provided in a vertical pipeline carrying oil of specific gravity 0·9, the flow being upwards. The difference in elevations of the throat section and entrance section of the venturimeter is 300 mm. The differential U-tube mercury manometer shows a gauge deflection of 250 mm. Calculate*

(i) discharge of the oil, and

(ii) pressure difference between the entrance and throat section.

Take the coefficient of meter as 0·98 and the specific gravity of the mercury as 13·6.

Solution. Given : $d_1 = 300$ mm $= 0.3$ m ; $d_2 = 150$ mm $= 0.15$ m; Specific gravity of oil $= 0.9$; Difference of elevations of throat section and entrance section $= 300$ mm $= 0.3$ m; $h = 250$ mm $= 0.25$ m of mercury and $C = 0.98$.

(i) Discharge of the oil

We know that the area of inlet,

$$a_1 = \frac{\pi}{4} \times (d_1)^2 = \frac{\pi}{4} \times (0.3)^2 \text{ m}^2$$

$$= 70.69 \times 10^{-3} \text{ m}^2$$

and the area at throat,

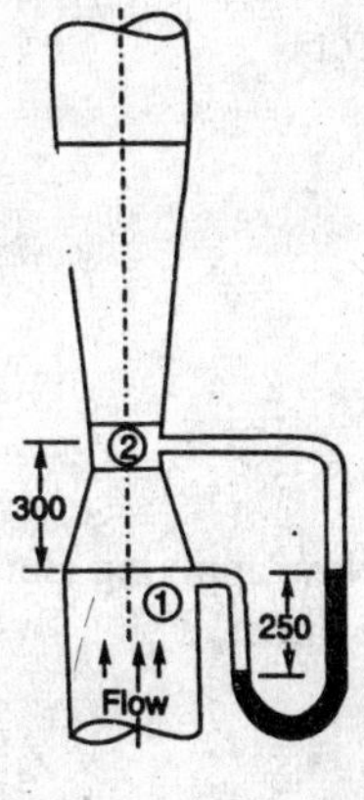

$$a_2 = \frac{\pi}{4} \times (d_2)^2 = \frac{\pi}{4} \times (0.15)^2 \text{ m}^2$$

$$= 17.67 \times 10^{-3} \text{ m}^2$$

We also know that the difference of pressure head,

$$h = 0.25 \left(\frac{13.6 - 0.9}{0.9} \right) = 3.53 \text{ m}$$

Fig. 7.11

and discharge of the oil, $Q = \dfrac{Ca_1 a_2}{\sqrt{a_1^2 - a_2^2}} \sqrt{2gh}$

$$= \frac{0.98 \times (70.69 \times 10^{-3}) \times (17.67 \times 10^{-3})}{\sqrt{(70.69 \times 10^{-3})^2 - (17.67 \times 10^{-3})^2}} \times \sqrt{2 \times 9.81 \times 3.53} \text{ m}^3/\text{s}$$

$$= \frac{1224 \times 10^{-6}}{68.45 \times 10^{-3}} \times 8.322 = 149 \times 10^{-3} \text{ m}^3/\text{s} = 149 \text{ litres/s} \quad \textbf{Ans.}$$

(ii) Pressure difference between the entrance and throat section

Let $\quad p_1 =$ Pressure at the entrance, and

$\quad\quad\quad p_2 =$ Pressure at the throat section.

We know that the velocity of oil at entrance,
$$v_1 = \frac{Q}{a_1} = \frac{149 \times 10^{-3}}{70.69 \times 10^{-3}} = 2.11 \text{ m/s}$$

and the velocity of oil at throat, $v_2 = \dfrac{Q}{a_2} = \dfrac{149 \times 10^{-3}}{17.67 \times 10^{-3}} = 8.43 \text{ m/s}$

Applying Bernoulli's equation for the entrance and throat section,

$$Z_1 + \frac{v_1^2}{2g} + \frac{p_1}{w} = Z_2 + \frac{v_2^2}{2g} + \frac{p_2}{w}$$

$$0 + \frac{(2.11)^2}{2 \times 9.81} + \frac{p_1}{w} = 0.3 + \frac{(8.43)^2}{2 \times 9.81} + \frac{p_2}{w}$$

$$0 + 0.227 + \frac{p_1}{w} = 0.3 + 3.622 + \frac{p_2}{w} = 3.922 + \frac{p_2}{w}$$

$$\therefore \quad \frac{p_1}{w} - \frac{p_2}{w} = 3.922 - 0.227 = 3.695 \text{ m of oil} \quad \textbf{Ans.}$$

7.15 Orifice Metre

An orifice metre is used to measure the discharge in a pipe. An orifice metre, in its simplest form, consists of a plate having a sharp edged circular hole known as an orifice. This plate is fixed inside a pipe as shown in Fig. 7.12.

A mercury manometer is inserted to know the difference of pressures between the pipe and the throat (*i.e.*, orifice).

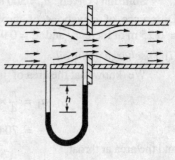

Let
h = Reading of the mercury manometer,
p_1 = Pressure at inlet,
v_1 = Velocity of liquid at inlet,
a_1 = Area of pipe at inlet, and
p_2, v_2, a_2 = Corresponding values at the throat.

Now applying Bernoulli's equation for inlet of the pipe and the throat,

Fig. 7.12. Orifice metre.

$$Z_1 + \frac{v_1^2}{2g} + \frac{p_1}{w} = Z_2 + \frac{v_2^2}{2g} + \frac{p_2}{w} \qquad \ldots(i)$$

$$\frac{p_1}{w} - \frac{p_2}{w} = \frac{v_2^2}{2g} - \frac{v_1^2}{2g} \qquad \ldots(\because Z_1 = Z_2)$$

or $\qquad h = \dfrac{v_2^2}{2g} - \dfrac{v_1^2}{2g} = \dfrac{1}{2g}(v_2^2 - v_1^2) \qquad \ldots(ii)$

Since the discharge is continuous, therefore $a_1 \cdot v_1 = a_2 \cdot v_2$

$$v_1 = \frac{a_2}{a_1} \times v_2 \quad \text{or} \quad v_1^2 = \frac{a_2^2}{a_1^2} \times v_2^2$$

Substituting the above value of v_1^2 in equation (ii),

$$h = \frac{1}{2g}\left(v_2^2 - \frac{a_2^2}{a_1^2} \times v_2^2\right) = \frac{v_2^2}{2g}\left(1 - \frac{a_2^2}{a_1^2}\right) = \frac{v_2^2}{2g}\left(\frac{a_1^2 - a_2^2}{a_1^2}\right)$$

∴ $\quad v_2^2 = 2gh\left(\dfrac{a_1^2}{a_1^2 - a_2^2}\right)\;$ or $\;v_2 = \sqrt{2gh}\left(\dfrac{a_1}{\sqrt{a_1^2 - a_2^2}}\right)$

We know that the discharge,

Q = Coefficient of orifice metre $\times a_2 \cdot v_2$

$$= \frac{C\, a_1 a_2}{\sqrt{a_1^2 - a_2^2}} \sqrt{2gh} \qquad \text{...(Same as in Art. 8·14)}$$

Note : The value of the coefficient of orifice metre will be such, which will also include the factor for coefficient of contraction for the orifice (for details of coefficient of contraction, please refer to Art. 8·6).

Example 7·15. *An orifice metre consisting of 100 mm diameter orifice in a 250 mm diameter pipe has coefficient equal to 0·65. The pipe delivers oil (sp. gr. 0·8). The pressure difference on the two sides of the orifice plate is measured by a mercury oil differential manometer. If the differential gauge reads 80 mm of mercury, calculate the rate of flow in litres/s.*

Solution. Given : $d_2 = 100$ mm $= 0\cdot 1$ m; $d_1 = 250$ mm $= 0\cdot 25$ m; $C = 0\cdot 65$; Specific gravity of oil $= 0\cdot 8$ and $h = 0\cdot 8$ m of mercury.

We know that the area of pipe,

$$a_1 = \frac{\pi}{4} \times (d_1)^2 = \frac{\pi}{4} \times (0\cdot 25)^2 = 49\cdot 09 \times 10^{-3}\ \text{m}^2$$

and area of throat, $\quad a_2 = \dfrac{\pi}{4} \times (d_2)^2 = \dfrac{\pi}{4} \times (0\cdot 1)^2 = 7\cdot 854 \times 10^{-3}\ \text{m}^2$

We also know that the pressure difference,

$$h = 0\cdot 8\left(\frac{13\cdot 6 - 0\cdot 8}{0\cdot 8}\right) = 12\cdot 8\ \text{m of oil}$$

and rate of flow, $\quad Q = \dfrac{C \cdot a_1 a_2}{\sqrt{a_1^2 - a_2^2}} \sqrt{2gh}$

$$= \frac{0\cdot 65 \times (49\cdot 09 \times 10^{-3}) \times (7\cdot 854 \times 10^{-3})}{\sqrt{(49\cdot 09 \times 10^{-3})^2 - (7\cdot 854 \times 10^{-3})^2}} \times \sqrt{2 \times 9\cdot 81 \times 12\cdot 8}\ \text{m}^3/\text{s}$$

$$= \frac{250\cdot 6 \times 10^{-6}}{48\cdot 45 \times 10^{-3}} \times 15\cdot 85 = 82 \times 10^{-3}\ \text{m}^3/\text{s} = 82\ \text{litres/s}\ \ \textbf{Ans.}$$

7·16 Pitot Tube

A Pitot[*] tube is an instrument to determine the velocity of flow at the required point in a pipe or a stream. In its simplest form, a pitot tube consists of a glass tube bent a through 90° as shown in Fig. 7·13.

The lower end of the tube faces the direction of the flow as shown in Fig. 7·13. The liquid rises up in the tube due to the pressure exerted by the flowing liquid. By measuring the rise of liquid in the tube, we can find out the velocity of the liquid flow.

Let $\quad h$ = Height of the liquid in the pitot tube above the surface,

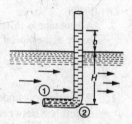

Fig. 7·13. Pitot tube.

---

[*] Pitot, Henri (1695—1771) was a French scientist and an engineer. He conducted a series of experiments on the flow of water in channels and pipes.

H = Depth of tube in the liquid, and
v = Velocity of the liquid.

Applying Bernoulli's equation for the sections 1 and 2,

$$H + \frac{v^2}{2g} = H + h \qquad \ldots(\because Z_1 = Z_2)$$

or
$$h = \frac{v^2}{2g}$$

$$\therefore \quad v = \sqrt{2gh}$$

Note : It has been experimentally found that if the pitot tube is placed, with its nose facing side way, in the flow, there will be no rise of the liquid in the tube. But if a pitot tube is placed, with its nose facing down stream, the liquid level in the tube will be depressed.

Example 7·16. *A pitot tube was inserted in a pipe to measure the velocity of water in it. If the water rises the tube is 200 mm, find the velocity of water.*

Solution. Given : $h = 200$ mm $= 0.2$ m.

We know that the velocity of water in the pipe,

$$v = \sqrt{2gh} = \sqrt{2 \times 9.81 \times 0.2} = 1.98 \text{ m/s} \quad \textbf{Ans.}$$

EXERCISE 7·2

1. A 200 mm × 120 mm venturimeter is installed in a pipe carrying water. If the mercury differential manometer shows a reading of 200 mm, find the discharge through the pipe. Take coefficient for the venturimeter as 0·98. **[Ans. 83·6 litres/s]**

2. A venturimeter has 400 mm diameter at the main and 150 mm at the throat. If the difference of pressures is 250 mm of mercury and the metre coefficient is 0·97, calculate the discharge of oil through the venturimeter. Take specific gravity of oil as 0·75. **[Ans. 158·6 litres/s]**

3. A venturimeter is inserted in a 150 mm diameter pipe carrying an oil of sp. gr. 0·85. If the differential head indicated by the mercury manometer is 0·25 m, and the rate of flow is 80 litres/s, find the diameter of the venturimeter at its throat. Take coefficient for the venturimeter as 0·97. **[Ans. 92 mm]**

4. In a laboratory, a 100 mm × 50 mm venturimeter was used, which recorded a discharge of 18 litres of water per second, when the mercury reading was 300 mm. What is the value of venturimeter coefficient ? **[Ans. 0·972]**

5. A 150 mm × 75 mm venturimeter is connected in a pipe discharging water, which is inclined at an angle of 45° with the horizontal. Find the discharge through the venturimeter, if the mercury gauge shows a deflection of 175 mm and coefficient for the venturimeter as 0·95. **[Ans. 28·5 litres/s]**

6. A 200 mm non-standard orifice is installed in a 250 mm pipe carrying water. When the flow is 165 litres/s, the mercury differential gauge reads 50 mm. Compute the value of coefficient for the orifice metre. **[Ans. 0·872]**

7. A pitot tube is installed in the centre of a pipe 80 mm diameter. Find the velocity of water in the centre of the pipe, if the water rises 300 mm in the tube. **[Ans. 2·43 m/s]**

QUESTIONS

1. What do you understand by the term total head of a moving fluid ? Explain clearly the difference between the total energy and total head of a moving fluid.
2. Derive Bernoulli's equation by any method.
3. State the limitations of the Bernoulli's theorem.
4. Name some practical applications of Bernoulli's theorem.

5. Derive an equation to measure the quantity of water flowing through a venturimeter.
6. Sketch a venturimeter and state why a certain angle of divergence is to be maintained.
7. What is an orifice metre ? Derive an expression for the discharge through an orifice metre.
8. Sketch a Pitot tube and explain how it is used to measure the velocity of a flowing liquid.

OBJECTIVE TYPE QUESTIONS

1. For a perfect incompressible liquid flowing in a continuous stream, the total energy of a particle remains the same, while the particle moves from one position to the other. This statement is called
   (a) continuity equation
   (b) Bernoulli's equation
   (c) Pascal's law
   (d) Archimedes' principle

2. According to Bernoulli's equation
   (a) $Z + p + v = C$
   (b) $Z + \dfrac{p}{w} + \dfrac{v}{g} = C$
   (c) $Z + \dfrac{p}{w} + \dfrac{v^2}{g} = C$
   (d) $Z + \dfrac{p}{w} + \dfrac{v^2}{2g} = C$

3. Bernoulli's equation is applied for
   (a) venturimeter
   (b) orifice metre
   (c) pitot tube
   (d) all of these

4. A venturimeter is used to measure
   (a) velocity of a flowing liquid
   (b) pressure of a flowing liquid
   (c) discharge of a flowing liquid
   (d) all of these

5. In order to avoid the tendency of separation at the throat in a venturimeter, the ratio of diameter at throat to that of the pipe should be
   (a) $\dfrac{1}{16}$ to $\dfrac{1}{8}$
   (b) $\dfrac{1}{8}$ to $\dfrac{1}{4}$
   (c) $\dfrac{1}{4}$ to $\dfrac{1}{3}$
   (d) $\dfrac{1}{3}$ to $\dfrac{1}{2}$

ANSWERS

1. (b) 2. (d) 3. (d) 4. (c) 5. (d)

8

Flow Through Orifices
(Measurement of Discharge)

---

1. Introduction 2. Types of Orifices 3. Jet of Water. 4. Vena Contracta. 5. Hydraulic Coefficients. 6. Coefficient of Contraction. 7. Coefficient of Velocity. 8. Coefficient of Discharge. 9. Coefficient of Resistance. 10. Experimental Method for Hydraulic Coefficients. 11. Discharge through a Rectangular Orifice. 12. Discharge through a small Rectangular Orifice. 13. Discharge through a Large Rectangular Orifice. 14. Discharge through a Submerged or Drowned Orifice. 15. Discharge through a Wholly Drowned Orifice. 16. Discharge through a Partially Drowned Orifice. 17. Discharge through a Drowned Orifice under Pressure.

8·1 Introduction

An opening, in a vessel, through which the liquid flows out is known as an orifice. This hole or opening is called an orifice, so long as the level of the liquid on the upstream side is above the top of the orifice. The usual purpose of an orifice is the measurement of discharge.

An orifice may be provided in the vertical side of a vessel or in the base. But the former is more common.

8·2 Types of Orifices

There are many types of orifices, depending upon their size, shape and nature of discharge. But the following are important from the subject point of view :

1. According to size :
   (a) Small orifice, and
   (b) Large orifice.
2. According to shape :
   (a) Circular orifice,
   (b) Rectangular orifice, and
   (c) Triangular orifice.
3. According to shape of the edge :
   (a) Sharp-edged orifice, and
   (b) Bell-mouthed orifice.
4. According to nature of discharge :
   (a) Fully submerged orifice, and
   (b) Partially submerged orifice.

All the above-mentioned orifices will be discussed in details at their appropriate places in the book. Before entering into details of the flow, through all types or orifices, following definitions should be clearly understood at this stage :

8·3 Jet of Water

The continuous stream of a liquid, that comes out or flows out of an orifice, is known as the jet of water.

8·4 Vena Contracta

Consider a tank, fitted with an orifice, as shown in Fig. 8·1. The liquid particles, in order to flow out through the orifice, move towards the orifice from all directions. A few of the particles first move downward, then take a turn to enter into the orifice and then finally flow through it. It may be noted, that the liquid particles lose some energy, while taking the turn to enter into the orifice. It has been, thus, observed that the jet, after leaving the orifice, gets contracted. The maximum contraction takes place at a section* slightly on the downstream side of the orifice, where the jet is more or less horizontal. Such a section is known as vena contracta as shown by the section C-C in Fig. 8·1.

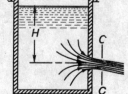

Fig. 8·1. Vena contracta.

8·5 Hydraulic Coefficients

The following four coefficients are known as hydraulic coefficients or orifice coefficients :

1. Coefficient of contraction.
2. Coefficient of velocity.
3. Coefficient of discharge, and
4. Coefficient of resistance.

8·6 Coefficient of Contraction

The ratio of area of the jet, at vena contracta, to the area of the orifice is known as coefficient of contraction. Mathematically coefficient of contraction,

$$C_c = \frac{\text{Area of jet at vena contracta}}{\text{Area of orifice}}$$

The value of coefficient of contraction varies slightly with the available head of the liquid, size and shape of the orifice. An average value of C_c is about 0·64.

8·7 Coefficient of Velocity

The ratio of actual velocity of the jet, at vena contracta, to the theoretical velocity is known as coefficient of velocity. Mathematically coefficient of velocity,

$$C_v = \frac{\text{Actual velocity of the jet at vena contracta}}{\text{Theoretical velocity of the jet}}$$

The difference between the velocities is due to friction of the orifice. The value of coefficient of velocity varies slightly with the different shapes of the edges of the orifice. *i.e.*, it is very small for sharp-edged orifices. For a sharp edged orifice, the value of C_v increases with the head of water. The following table gives the values of C_v for an orifice of 10 mm diameter with the corresponding head as given by **Weisbach.

H	20 mm	500 mm	3·5 m	20·0 m	100 m
C_v	0·959	0·967	0·975	0·991	0·994

Notes : 1. An average value of C_v is about 0·97.

2. The theoretical velocity of jet at vena contracta is given by the relation $v = \sqrt{2gh}$ where h is the head of water at vena contracta.

---

*As a matter of fact, the distance of this section depends upon the size of orifice and the head of water. Generally, this section is at a distance of about $d/2$ from the plane of orifice, where d is the diameter of the orifice.

** Julius Weisbach was a German Engineer, who made significant contribution in the field of Hydraulics and Geodesy.

8·8 Coefficient of Discharge

The ratio of a actual discharge through an orifice to the theoretical discharge is known as coefficient of discharge. Mathematically coefficient of discharge,

$$C_d = \frac{\text{Actual discharge}}{\text{Theoretical discharge}}$$

$$= \frac{\text{Actual velocity} \times \text{Actual area}}{\text{Theoretical velocity} \times \text{Theoretical area}}$$

$$= \frac{\text{Actual velocity}}{\text{Theoretical velocity}} \times \frac{\text{Actual area}}{\text{Theoretical area}}$$

$$= C_v \times C_c$$

The value of coefficient discharge varies with the values of C_c and C_v. An average of coefficient of discharge varies from 0·60 to 0·64.

8·9 Coefficient of Resistance

The ratio of loss of head in the orifice to the head of water available at the exit of the orifice is known as coefficient of resistance. Mathematically coefficient of resistance,

$$C_r = \frac{\text{Loss of head in the orifice}}{\text{Head of water}}$$

The loss of head in the orifice takes place, because the walls of the orifice offer some resistance to the liquid as it comes out. The coefficient of resistance is generally neglected, while solving numerical problems.

Example 8·1. *A jet of water issues from an orifice of diameter 20 mm under a head of 1 m. What is the coefficient of discharge for the orifice, if actual discharge is 0·85 litres/second.*

Solution. Given : $d = 20$ mm $= 0·02$ m; $H = 1$ m and $Q_{ac} = 0·85$ litres/s $= 0·85 \times 10^{-3}$ m³/s.

We know that cross-sectional area of the orifice,

$$a = \frac{\pi}{4} \times (d)^2 = \frac{\pi}{4} \times (0·02)^2 = 0·314 \times 10^{-3} \text{ m}^2$$

and theoretical discharge through the orifice,

$$Q_{th} = a\sqrt{2gh} = 0·314 \times 10^{-3} \times \sqrt{2 \times 9·81 \times 1} = 1·39 \times 10^{-3} \text{ m}^3/\text{s}$$

∴ Coefficient of discharge,

$$C_d = \frac{Q_{ac}}{Q_{th}} = \frac{0·85 \times 10^{-3}}{1·39 \times 10^{-3}} = 0·61 \text{ Ans.}$$

Example 8·2. *A 60 mm diameter orifice is discharging water under a head of 9 metres. Calculate the actual discharge through the orifice in litres per second and actual velocity of the jet in metres per second at vena contracta, if $C_d = 0·625$ and $C_v = 0·98$.*

Solution. Given : $d = 60$ mm $= 0·06$ m; $H = 9$ m; $C_d = 0·625$ and $C_v = 0·98$.

Actual discharge through the orifice

We know that cross-sectional area of the orifice,

$$a = \frac{\pi}{4} \times (d)^2 = \frac{\pi}{4} \times (0·06)^2 = 2·83 \times 10^{-3} \text{ m}^2$$

and theoretical discharge through the orifice,

$$Q_{th} = a\sqrt{2gh} = 2·83 \times 10^{-3} \times \sqrt{2 \times 9·81 \times 9} \text{ m}^3/\text{s}$$

$$= 37·6 \times 10^{-3} \text{ m}^3/\text{s}$$

∴ Actual discharge through the orifice,
$$Q_{ac} = C_d \cdot Q_{th} = 0.625 \times (37.6 \times 10^{-3}) \text{ m}^3/\text{s}$$
$$= 23.5 \times 10^{-3} \text{ m}^3/\text{s} = 23.5 \text{ litres/s} \quad \textbf{Ans.}$$

Actual velocity of the jet at vena contracta

We also know that theoretical velocity of the jet at vena contracta,
$$v_{th} = \sqrt{2gh} = \sqrt{2 \times 9.81 \times 9} = 13.3 \text{ m/s}$$
and actual velocity of the jet at vena contracta,
$$v_{ac} = C_v \cdot v_{th} = 0.98 \times 13.3 = 13.0 \text{ m/s} \quad \textbf{Ans.}$$

8·10 Experimental Method for Hydraulic Coefficients

Consider a tank containing water at a constant level, maintained by a constant supply, as shown in Fig. 8·2.

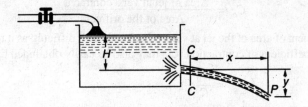

Fig. 8·2. Experiment for hydraulic coefficients.

Let the water flow out of the tank through an orifice, fitted in one side of the tank. Consider a particle of water in the jet at P. Let the section C-C represent the point of vena contracta.

Let
- H = Constant water head,
- x = Horizontal distance between C-C and P,
- y = Vertical distance between C-C and P,
- v = Velocity of the jet, and
- t = Time taken, in seconds, by the particle to reach from C-C to P

We know that
$$y = \frac{1}{2} gt^2 \qquad \ldots(i)$$

...(Where g is the gravitational acceleration)

and
$$x = v \times t \quad \text{or} \quad t = x/v \qquad \ldots(ii)$$

Substituting this value of t in equation (i),
$$y = \frac{1}{2} gt^2 = \frac{1}{2} g \left(\frac{x}{v}\right)^2 = \frac{gx^2}{2v^2}$$

or
$$v = \sqrt{\frac{gx^2}{2y}} \qquad \ldots(iii)$$

This is an equation of a parabola. It is, thus, obvious that the path of a jet is also a parabola.

We also know that the theoretical velocity of a particle,
$$v_{th} = \sqrt{2gh} \qquad \ldots(iv)$$

∴ Coefficient of velocity,

$$C_v = \frac{v_{ac}}{v_{th}} = \frac{\sqrt{\frac{gx^2}{2y}}}{\sqrt{2gh}} \sqrt{\frac{x^2}{4yh}} = \frac{x}{\sqrt{4yh}} \qquad ...(v)$$

The simplest method of determining the coefficient of discharge (C_d) is by measuring the actual quantity of discharge through the orifice in a given time t. This actual discharge may, then, be divided by the theoretical discharge, which will give the required value of the coefficient discharge. Mathematically coefficient of discharge,

$$C_d = \frac{Q_{ac}}{Q_{th}} = \frac{Q_{ac}}{\text{Area of orifice} \times \sqrt{2gh}}$$

Now the coefficient of contraction may be found out by measuring the actual area of the jet at vena contracta and then by dividing by the area of the jet. Mathematically coefficient of contraction,

$$C_c = \frac{\text{Area of jet at vena contracta}}{\text{Area of the orifice}}$$

The measurement of area of the jet at vena contracta is a bit difficult, as it requires a lot of time and patience. The coefficient of contraction (C_c) may also be easily obtained by the relation,

$$C_c = \frac{C_d}{C_v}$$

Example 8·3. *A jet of water issuing from a vertical orifice under a constant head of 500 mm crosses a point with horizontal and vertical co-ordinates as 1200 mm and 800 mm from the vena contracta. Find the value of coefficient of velocity for the orifice.*

Solution. Given : $h = 500$ mm $= 0.5$ m; $x = 1200$ mm $= 1.2$ m and $y = 800$ mm $= 0.8$ m.

We know that the value of coefficient of velocity for the orifice,

$$C_v = \frac{x}{\sqrt{4yh}} = \frac{1.2}{\sqrt{4 \times 0.8 \times 0.5}} = 0.95 \quad \textbf{Ans.}$$

Example 8·4. *Water discharges at the rate of 98·2 litres/second through a 120 mm diameter vertical sharp edged orifice under a constant head of 10 m. A point, on the jet, measured from the vena contracta has co-ordinates of 4·5 m horizontal and 0·54 m vertical. Find the values of C_v, C_c and C_d of the orifice.*

Solution. Given : $Q_{ac} = 98.2$ litres/s $= 0.0982$ m³/s; $d = 120$ mm $= 0.12$ m; $h = 10$ m; $x = 4.5$ m and $y = 0.54$ m.

We know that the value of coefficient of velocity,

$$C_v = \frac{x}{\sqrt{4yh}} = \frac{4.5}{\sqrt{4 \times 0.54 \times 10}} = 0.968 \quad \textbf{Ans.}$$

We also know that cross-sectional area of the orifice,

$$a = \frac{\pi}{4} \times (d)^2 = \frac{\pi}{4} \times (0.12)^2 = 0.0113 \text{ m}^2$$

and theoretical discharge through the orifice,

$$Q_{th} = a\sqrt{2gh} = 0.0113 \times \sqrt{2 \times 9.81 \times 10} = 0.158 \text{ m}^3/\text{s}$$

∴ Value of coefficient of discharge,

$$C_d = \frac{Q_{ac}}{Q_{th}} = \frac{0.0982}{0.158} = 0.62 \quad \textbf{Ans.}$$

We know that the value of coefficient of contraction,
$$C_c = \frac{C_d}{C_v} = \frac{0.62}{0.968} = 0.64 \text{ Ans.}$$

Example 8·5. *A tank has two identical orifices in one of its vertical sides. The upper orifice is 2 metres below the water surface and the lower one is 4 metres below the water surface as shown in Fig. 8·3.*

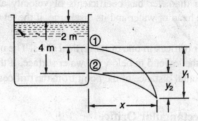

Fig. 8·3.

Find the point at which the two jets will intersect, if the coefficient of velocity is 0·98 for both the orifices.

Solution. Given : $h_1 = 2$ m; $h_2 = 4$ m and $C_v = 0.98$.

Let $\quad x = $ Horizontal distance between the orifices and the point where the two jets intersect.

We know that coefficient of velocity for orifice 1,
$$0.98 = \frac{x}{\sqrt{4y_1 h_1}} = \frac{x}{\sqrt{4y_1 \times 2}} = \frac{x}{\sqrt{8y_1}} \qquad \ldots(i)$$

and coefficient of velocity for orifice 2,
$$0.98 = \frac{x}{\sqrt{4y_2 h_2}} = \frac{x}{\sqrt{4y_2 \times 4}} = \frac{x}{\sqrt{16y_2}} \qquad \ldots(ii)$$

Since, both the orifices are identical and have the same values of coefficients of velocity, therefore equating equations (*i*) and (*ii*),
$$\frac{x}{\sqrt{8y_1}} = \frac{x}{\sqrt{16y_2}} \quad \text{or} \quad \sqrt{8y_1} = \sqrt{16y_2}$$
$$8y_1 = 16y_2 \quad \text{or} \quad y_1 = 2y_2$$

From the geometry of the figure, we find that
$$y_1 = y_2 + 2 \quad \text{or} \quad 2y_2 = y_2 + 2 \qquad \ldots(\because y_1 = 2y_2)$$
$$\therefore \quad y_1 = 4 \text{ m} \quad \text{and} \quad y_2 = 2 \text{ m}$$

Substituting this value of y_1 in equation (*i*),
$$0.98 = \frac{x}{\sqrt{8y_1}} = \frac{x}{\sqrt{8 \times 4}} = \frac{x}{\sqrt{32}} = \frac{x}{5.657}$$
$$\therefore \quad x = 0.98 \times 5.657 = 5.54 \text{ m Ans.}$$

EXERCISE 8·1

1. A jet of water issues from an orifice of diameter 16 mm under a constant head of 1·5 m. Find the coefficient of discharge for the orifice, when the actual discharge is 0·65 litres/s. **[Ans. 0·6]**

2. The head of water over an orifice of diameter 40 mm is 10 m. What is the actual discharge and actual velocity of the jet ? Take $C_d = 0.6$ and $C_v = 0.98$. [**Ans.** 10·6 litres/s ; 13·73 m/s]

3. In an experiment, water issues horizontally from an orifice under a head of 160 mm. Determine the coefficient of velocity of the jet, if the horizontal distance travelled by a point on the jet is 320 mm and vertical distance is 170 mm. [**Ans.** 0·97]

4. A jet of water issues from an orifice 1250 mm² in area under a constant head of 1·125 m. It falls vertically 1 m before striking the ground at a distance of 2 m measured horizontally from the vena contracta. Calculate the coefficients of discharge, velocity and contraction. [**Ans.** 0·622; 0·943; 0·66]

5. An orifice of 25 mm diameter has coefficients of velocity and contraction as 0·98 and 0·62 respectively. Find the head of water and its discharge, if the jet drops 1 m in a horizontal distance of 2·65 m. [**Ans.** 1·82 m; 1·78 litres/s]

6. A tank has two similar orifices in one of its vertical sides. The upper orifice is situated 3 m below the water surface and the lower 5 m below the water surface. If the value of C_v for both the orifices is 0·97, find the horizontal distance of the point from the orifices where the two jets intersect.
[**Ans.** 7·51 m]

8·11 Discharge through a Rectangular Orifice

Though there are many types of rectangular orifices, yet the following two types are important from the subject point of view :

1. Small rectangular orifices.   2. Large rectangular orifices

Now we shall discuss the discharge through both types of orifices one-by-one in the following pages.

8·12 Discharge through a Small Rectangular Orifice

An orifice is considered to be small, if the head of water is more than 5 times the height (or depth) of the orifice. It will be interesting to know that the width of the rectangular orifice has nothing to do with the type of orifice.

In a small rectangular orifice, the velocity of water in the entire cross-section of the jet is considered to be constant. The discharge through such an orifice is given by the relation,

$$Q = C_d \times a \times \sqrt{2gh} = C_d \times (b \times d) \times \sqrt{2gh}$$

where
C_d = Coefficient of discharge for the orifice,
a = Cross-sectional area of the orifice,
h = Height of the liquid above the centre of the orifice,
b = Width of the orifice, and
d = Depth of the orifice.

Example 8·6. *A small rectangular orifice 200 mm deep and 500 mm wide is discharging water under a constant head of 400 mm. What will be the discharge through the orifice in litres/s , if the coefficient of discharge for the orifice is 0·6.*

Solution. Given : $d = 200$ mm = 0·2 m; $b = 500$ mm = 0·5 m; $h = 400$ mm = 0·4 m and $C_d = 0.6$.

We know that area of the orifice,
$$a = b \times d = 0.5 \times 0.2 = 0.1 \text{ m}^2$$

and discharge through the orifice,
$$Q = C_d \times a \times \sqrt{2gh} = 0.6 \times 0.1 \times \sqrt{2 \times 9.81 \times 0.4} \text{ m}^3/\text{s}$$
$$= 0.168 \text{ m}^3/\text{s} = 168 \text{ litres/s} \quad \textbf{Ans.}$$

Example 8·7. *In a laboratory, 15 litres of water was collected through a small orifice of 100 mm diameter under a constant head of 500 mm. Determine the coefficient of discharge for the orifice.*

Flow Through Orifices (Measurement of Discharge)

Solution. Given : $Q = 15$ litre/s $= 15 \times 10^{-3}$ m^3/s ; $d = 100$ mm $= 0.1$ m; $h = 500$ mm $= 0.5$ m.

Let C_d = Coefficient of discharge for the orifice.

We know that the area of the orifice,

$$a = \frac{\pi}{4} \times (d)^2 = \frac{\pi}{4} \times (0.1)^2 = 7.854 \times 10^{-3} \text{ m}^2$$

and discharge through a small orifice (Q),

$$15 \times 10^{-3} = C_d \cdot a \cdot \sqrt{2gh} = C_d \times (7.854 \times 10^{-3}) \times \sqrt{2 \times 9.81 \times 0.5}$$

$$= 24.6 \times 10^{-3} C_d$$

∴ $C_d = 15/24.6 = 0.61$ **Ans.**

8·13 Discharge through a Large Rectangular Orifice

In the last article, we have discussed that in case of small rectangular orifices, the velocity of water in the entire cross-section of the jet is considered to be constant. But in case of large rectangular orifices, the velocity of various liquid particles will not be constant, because there is a considerable variation of head along the height of an orifice. And the velocity of liquid varies with the available head of the liquid.

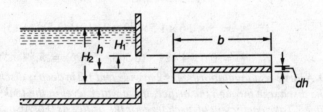

Fig. 8·4. Large rectangular orifice.

Now consider a large rectangular orifice, in one side of the tank, discharging water as shown in Fig. 9·4.

Let
H_1 = Height of liquid above the top of the orifice,
H_2 = Height of liquid above the bottom of the orifice,
b = Breadth of the orifice, and
C_d = Coefficient of discharge.

Consider a horizontal strip of thickness (dh) at a depth (h) from the water level as shown in Fig. 8·4. Therefore area of the strip

$$= b \cdot dh \qquad \ldots(i)$$

We know that the theoretical velocity of water through the strip

$$= \sqrt{2gh} \qquad \ldots(ii)$$

and discharge through the strip,

$$dq = C_d \times \text{Area} \times \text{Theoretical velocity}$$
$$= C_d \cdot b \cdot dh \sqrt{2gh} \qquad \ldots(iii)$$

Total discharge through the whole orifice may be found out by integrating the above equation between the limits H_1 and H_2. i.e.,

$$\therefore \quad Q = \int_{H_1}^{H_2} C_d \cdot b \cdot dh \sqrt{2gh} = C_d \cdot b \sqrt{2g} \int_{H_1}^{H_2} \sqrt{h} \cdot dh$$

$$= C_d \cdot b \sqrt{2g} \left[\frac{h^{3/2}}{\frac{3}{2}} \right]_{H_1}^{H_2} = \frac{2}{3} C_d \cdot b \sqrt{2g} \left[h^{3/2} \right]_{H_1}^{H_2}$$

$$= \frac{2}{3} C_d \cdot b \sqrt{2g} \left(H_2^{3/2} - H_1^{3/2} \right)$$

Example 8·8. *A rectangular orifice of 1·5 m wide and 0·5 m deep is discharging water from a tank. If the water level in the tank is 3 m above the top edge of the orifice, find the discharge through the orifice. Take coefficient of discharge for the orifice as 0·6.*

Solution. Given : $b = 1.5$ m; $d = 0.5$ m; $H_1 = 3$ m and $C_d = 0.6$.

We know that height of water above the bottom of the orifice,

$$H_2 = 3 + 0.5 = 3.5 \text{ m}$$

and discharge through the orifice,

$$Q = \frac{2}{3} C_d \cdot b \sqrt{2g} \left(H_2^{3/2} - H_1^{3/2} \right)$$

$$= \frac{2}{3} \times 0.6 \times 1.5 \times \sqrt{2 \times 9.81} \times \left[(3.5)^{3/2} - (3)^{3/2} \right] \text{m}^3/\text{s}$$

$$= 0.6 \times 4.429 \times (6.548 - 5.196) = 3.59 \text{ m}^3/\text{s} \quad \textbf{Ans.}$$

Example 8·9. *A large rectangular orifice of 2 m wide and 1·2 m deep is discharging water from a vessel. What is the discharge through the orifice, if the water level in the tank is 0·6 m above the top edge of the orifice ? Take coefficient of discharge for the orifice as 0·6. Also find the error in the discharge, if the orifice is treated as a small one.*

Solution. Given : $b = 2$ m; $d = 1.2$ m; $H_1 = 0.6$ m and $C_d = 0.6$.

Discharge through the orifice

We know that the head of water above the bottom of the orifice,

$$H_2 = 0.6 + 1.2 = 1.8 \text{ m}$$

and the discharge through the orifice (when treated as large),

$$Q_1 = \frac{2}{3} C_d \cdot b \cdot \sqrt{2g} \left(H_2^{3/2} - H_1^{3/2} \right)$$

$$= \frac{2}{3} \times 0.6 \times 2 \times \sqrt{2 \times 9.81} \left[(1.8)^{3/2} - (0.6)^{3/2} \right] \text{m}^3/\text{s}$$

$$= 0.8 \times 4.429 \times (2.415 - 0.465) = 6.91 \text{ m}^3/\text{s} \quad \textbf{Ans.}$$

Error in the discharge, if the orifice is treated as a small one

We also know that the head of water, when the orifice is treated as a small one,

$$H = H_1 + \frac{d}{2} = 0.6 + \frac{1.2}{2} = 1.2 \text{ m}$$

and the discharge through the orifice (when treated as small),

$$Q_2 = C_d \cdot (b \times d) \sqrt{2gH}$$

$$= 0.6 \times (2 \times 1.2) \times \sqrt{2 \times 9.81 \times 1.2} \text{ m}^3/\text{s}$$

$$= 1.44 \times 4.852 = 6.99 \text{ m}^3/\text{s}$$

∴ Error in the discharge $= \dfrac{Q_2 - Q_1}{Q_1} = \dfrac{6 \cdot 99 - 6 \cdot 91}{6 \cdot 91} = \dfrac{0 \cdot 08}{6 \cdot 91} = 0 \cdot 0116 = 1 \cdot 16\%$ **Ans.**

8·14 Discharge through a Submerged or Drowned Orifice

Sometimes, an orifice does not discharge the liquid freely into the atmosphere, but discharges into some other vessel containing the liquid. Such an orifice is known as submerged or drowned orifice. Following are the two types of submerged orifices :

1. Wholly drowned orifices. 2. Partially drowned orifices.

8·15 Discharge through a Wholly Drowned Orifice

Sometimes the entire outlet side of an orifice is under the liquid. It is known as a wholly submerged orifice as shown in Fig. 8·5. In such orifice, the coefficient of contraction is equal to one.

Consider a wholly drowned orifice discharging water as shown in Fig. 8·5.

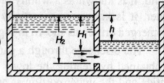

Let H_1 = Height of water (on the upstream side) above the top of the orifice,

H_2 = Height of water (on the upstream side) above the bottom of the orifice,

h = Difference between the two water levels on either side of the orifice,

Fig. 8·5. Wholly drowned orifice.

b = Width of the orifice, and

C_d = Coefficient of discharge.

∴ Area of orifice $= b (H_2 - H_1)$

We know that the theoretical velocity of water through the orifice

$$= \sqrt{2gh}$$

∴ Actual velocity of water $= C_v \sqrt{2gh}$

Since coefficient of contraction is 1 in this case therefore taking C_d equal to C_v, we find that the actual velocity of water

$$= C_d \sqrt{2gh}$$

Now the discharge through the orifice,

Q = Area of orifice × Actual velocity

$= b (H_2 - H_1) \times C_d \sqrt{2gh}$

$= C_d \cdot b (H_2 - H_1) \times \sqrt{2gh}$

Note : Sometimes, depth of the drowned orifice (d) is given instead of H_1 and H_2. In such cases, the discharge through the wholly drowned orifice is given by the relation :

$$Q = C_d \cdot b.d \sqrt{2gh}$$

Example 8·10. *A drowned orifice 1·5 metre wide and 0·5 metre deep is provided in one side of a tank. Find the discharge in litres/s through the orifice, if the difference of water levels on both the sides of the orifice be 4 metres. Take C_d = 0·64.*

Solution. Given : $b = 1 \cdot 5$ m; $d = 0 \cdot 5$ m; $h = 4$ m and $C_d = 0 \cdot 64$.

We know that discharge through the orifice,

$Q = C_d \cdot b \cdot d \sqrt{2gh} = 0 \cdot 64 \times 1 \cdot 5 \times 0 \cdot 5 \times \sqrt{2 \times 9 \cdot 81 \times 4}$ m³/s

$= 0 \cdot 48 \times 8 \cdot 859 = 4 \cdot 25$ m³/s $= 4250$ litres/s **Ans.**

Example 8·11. *A drowned orifice 1 metre wide has heights of water from the bottom and top of the orifice as 2·25 metres and 2·0 metres respectively. Find the discharge through the orifice, if the difference of water levels on both the sides of the orifice be 375 mm. Take coefficient of discharge as 0·62.*

Solution. Given : $b = 1$ m; $H_2 = 2·25$ m ; $H_1 = 2·0$ m; $h = 375$ mm $= 0·375$ m and $C_d = 0·62$.

We know that discharge through the orifice,

$$Q = C_d \cdot b (H_2 - H_1) \sqrt{2gh} \text{ m}^3/\text{s}$$
$$= 0·62 \times 1 (2·25 - 2·0) \times \sqrt{2 \times 9·81 \times 0·375} \text{ m}^3/\text{s}$$
$$= 0·155 \times 2·712 = 0·42 \text{ m}^3/\text{s} \quad \textbf{Ans.}$$

8·16 Discharge through a Partially Drowned Orifice

In the previous article, we have discussed that if the entire outlet side of an orifice is under the liquid, it is known as a wholly drowned orifice. But if the outlet side of an orifice is partly under water, it is known as a partially drowned or partially submerged orifice as shown in Fig. 8·6.

The discharge through a partially drowned orifice is obtained by treating the lower portion as a drowned orifice and the upper portion as an orifice running free, and then by adding the two discharges thus obtained.

We know that discharge through the free portion,

$$Q_1 = \frac{2}{3} C_d \cdot b \sqrt{2g} \left(H_2^{3/2} - H_1^{3/2} \right)$$

...(As in Art. 8·11)

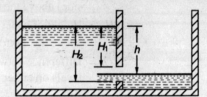

Fig. 8·6. Partially drowned orifice.

and discharge through the drowned orifice,

$$Q_2 = C_d \cdot b (H_2 - H) \sqrt{2gh} \qquad \text{...(As in Art. 8·13)}$$

Now total discharge,

$$Q = Q_1 + Q_2$$

Example 8·12. *An orifice in one side of a large tank is rectangular in shape, 2 metres broad and 1 metre deep. The water level on one side of the orifice is 4 metres above its top edge. The water level on the other side of the orifice is 0·5 metre below its top edge as shown in Fig. 8·7.*

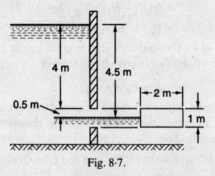

Fig. 8·7.

Calculate the discharge through the orifice per second, if $C_d = 0·63$.

Flow Through Orifices (Measurement of Discharge)

Solution. Given : $b = 2$ m; $d = 1$ m; $H_1 = 4$ m; $H_2 = 4 + 1 = 5$ m; $H = 4 + 0.5 = 4.5$ m and $C_d = 0.63$.

Since the orifice is partially drowned, therefore let us split up the orifice into two portions. The upper portion will be treated as a free orifice and the lower portion as a drowned orifice.

We know that the discharge through the free portion of the orifice,

$$Q_1 = \frac{2}{3} C_d \cdot b \sqrt{2g} \, (H_2^{3/2} - H_1^{3/2})$$

$$= \frac{2}{3} \times 0.63 \times \sqrt{2 \times 9.81} \, [(4.5)^{3/2} - (4)^{3/2}] \text{ m}^3/\text{s}$$

$$= 1.86 \times (9.546 - 8.0) = 2.88 \text{ m}^3/\text{s} \qquad ...(i)$$

and the discharge through the drowned portion of the orifice,

$$Q_2 = C_d \cdot b \, (H_2 - H) \times \sqrt{2gh}$$

$$= 0.63 \times 2(5 - 4.5) \sqrt{2 \times 9.81 \times 4.5} \text{ m}^3/\text{s}$$

$$= 0.63 \times 9.396 = 5.92 \text{ m}^3/\text{s} \qquad ...(ii)$$

∴ Total discharge, $\quad Q = Q_1 + Q_2 = 2.88 + 5.92 = 8.8 = \text{m}^3/\text{s}$ **Ans.**

8·17 Discharge through a Drowned Orifice under Pressure

Sometimes, flow from one vessel to the other, takes place under pressure through an orifice connecting the two vessels. In such a case, the head of liquid, causing flow, will be the algebraic sum of the actual difference of liquid levels and the pressure heads.

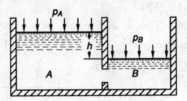

Fig. 8·8. Flow from one vessel to the other under pressure.

Consider two vessels A and B connected through an orifice. Let the liquid flow from vessel A to B as shown in Fig. 8·8.

Let
$\quad h = $ Difference of liquid levels,
$\quad p_A = $ Intensity of pressure over the liquid in vessel A,
$\quad p_B = $ Intensity of pressure over the liquid in vessel B,
$\quad w = $ Specific weight of the liquid, and
$\quad a = $ Area of the orifice.

We know that the pressure head over the vessel A,

$$H_A = \frac{p_A}{w}$$

Similarly, the pressure head over the vessel B,

$$H_B = \frac{p_B}{w}$$

∴ The net effective head of water, which causes flow,

$$H = h + H_A - H_B$$

and theoretical velocity of water,
$$v = \sqrt{2gH}$$

Now rate of discharge through the orifice,
$$Q = \text{Coefficient of discharge} \times \text{Area} \times \text{Theoretical velocity}$$
$$= C_d \cdot a \sqrt{2gH}$$

Notes. 1. The flow of liquid will take place from the vessel in which the head is more.
2. The orifice will behave like a drowned orifice.

Example 8·13. *Two tanks A and B of surface areas 800 m² and 500 m² are connected by an orifice of 0·15 m² in cross-section. If the pressures in the tank A and B are 40 kPa and 30 kPa respectively, find the rate of flow in litres/s at the instant when the water levels in the tank A is 1 m higher than that in the tank B. Take C_d for the orifice as 0·6.*

Solution. Given : $A_A = 800$ m²; $A_B = 500$ m²; $a = 0·15$ m²; $p_A = 40$ kPa $= 40$ kN/m²; $p_B = 30$ kPa $= 30$ kN/m²; $h = 1$ m and $C_d = 0·6$.

We know that the net effective head, which causes flow of water,
$$H = h + \frac{p_A}{w} - \frac{p_B}{w}$$
$$= 1 + \frac{40}{9·81} - \frac{30}{9·81} \text{ m}$$
$$= 1 + 4·08 - 3·06 = 2·02 \text{ m}$$

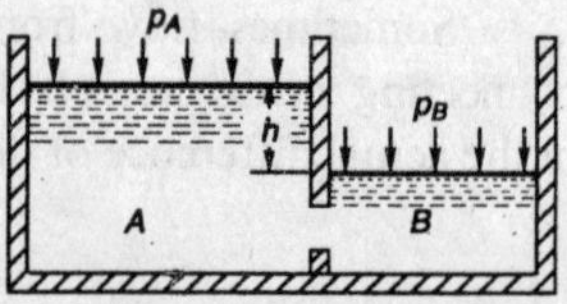

and the rate of flow through the orifice,
$$Q = C_d \cdot a \sqrt{2gH}$$
$$= 0·6 \times 0·15 \times \sqrt{2 \times 9·81 \times 2·02} \text{ m}^3/\text{s}$$
$$= 0·567 \text{ m}^3/\text{s} = 567 \text{ litres/s} \quad \textbf{Ans.}$$

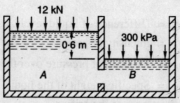

Fig. 8·9.

Example 8·14. *Two tanks A and B, circular in cross-section, are connected by a small pipe as shown in Fig. 8·10.*

Fig. 8·10.

The tank B is closed and contains entrapped air at a pressure of 300 kPa. Both the tanks contain oil of specific gravity 0·85, the level of A being 0·6 m higher than that in B. A load of 12 kN is placed on the piston A, whose area is 0·6 square metre. Find the discharge through the pipe at this instant, if diameter of the pipe is 75 mm. Take C_d for the pipe as 0·6.

Solution. Given : $p_B = 300$ kPa $= 300$ kN/m²; Specific gravity of oil $= 0·85$; $h = 0·6$ m; Load on piston of tank $A = 12$ kN; Area of piston $= 0·6$ m² : Diameter of pipe $(d) = 75$ mm $= 0·075$ m and $C_d = 0·6$.

We know that the specific weight of oil,
$$w = 0.85 \times 9.81 = 8.34 \text{ kN/m}^3$$
and the pressure in the tank A, $p_A = \dfrac{\text{Load}}{\text{Area}} = \dfrac{12}{0.6} = 20 \text{ kN/m}^2$

∴ Net effective head causing flow,
$$H = h + \frac{p_A}{w} - \frac{p_B}{w} = 0.6 + \frac{20}{8.34} - \frac{300}{8.34}$$
$$= 0.6 + 2.4 - 36.97 = -33.97 \text{ m}$$

...(Minus sign means that the flow of oil will be from tank B to A)

We also know that the area of the pipe connecting the tanks,
$$a = \frac{\pi}{4} \times (d)^2 = \frac{\pi}{4} \times (0.075)^2 = 4.418 \times 10^{-3} \text{ m}^2$$

and discharge through the orifice,
$$Q = C_d \cdot a \cdot \sqrt{2gH}$$
$$= 0.6 \times (4.418 \times 10^{-3}) \times \sqrt{2 \times 9.81 \times 33.97} \text{ m}^3/\text{s}$$
$$= 68.4 \times 10^{-3} \text{ m}^3/\text{s} = 68.4 \text{ litres/s} \quad \textbf{Ans.}$$

EXERCISE 8·2

1. Find the discharge in litres/s through a small orifice of 150 mm deep and 400 mm wide under a constant head of 250 mm. Take C_d as 0·625. [Ans. 83 litres/s]

2. In a laboratory, 53·5 litres of water per second is collected through a small orifice of 100 mm deep and 250 mm wide under a constant head of 600 mm. Find the value of coefficient of discharge. [Ans. 0·622]

3. A large rectangular orifice of 1·2 m wide and 0·6 m deep is discharging water from a tank, where the level is 0·6 m above the upper edge of the orifice. Find the discharge through the orifice, if coefficient of discharge for the orifice is 0·6. [Ans. 1·81 m³/s]

4. Find the discharge through a fully submerged orifice of 2 m wide and 1 m deep, if the difference of water levels on both sides of the orifice is 3 m. Take C_d = 0·6. [Ans. 9·21 m³/s]

5. Find the discharge through a drowned orifice of width 3 m, if the difference of water levels on both sides of the orifice is 0·5 m. The heights of water levels from the top and bottom of the orifice are 2·5 m and 2·75 m respectively. Take coefficient of discharge for the orifice as 0·6. [Ans. 1·41 m³/s]

6. A rectangular orifice of 2 m wide and 1·2 m deep is fitted in one side of a tank. The water level on one side of the orifice is 3 m above the top edge of the orifice. On the other side of the orifice, the water level is 0·5 m below its top edge. If C_d for the orifice is 0·64, find the discharge through the orifice. [Ans. 12·53 m³/s]

7. Two tanks P and Q are connected by an orifice of coefficient 0·6 at their bottom. At an instant, the pressures above the water surfaces of the tanks P and Q were recorded to be 25 kPa and 20 kPa respectively. If the water level in the tank P is 0·5 m higher than that of the tank Q, find the quantity of water flowing through the orifice. Take the area of the orifice as 0·6. [Ans 72 litres/s]

QUESTIONS

1. What is an orifice ? Discuss its classification.
2. Explain the term 'vena contracta' as applied to the flow of water through a sharp-edged orifice.

3. Define the terms coefficient of contraction, coefficient of velocity and coefficient of discharge. State the relation among them.
4. Derive an expression for the discharge through a rectangular orifice.
5. Define a submerged orifice. Give the names of various types of submerged orifices.
6. Obtain an expression for the discharge through a partially drowned orifice.

OBJECTIVE TYPE QUESTIONS

1. Theoretical velocity of the jet at vena contracta is
   (a) $\sqrt{2gh}$ (b) $2\sqrt{gh}$ (c) $2g\sqrt{h}$ (d) $2gh$
   where h = Head of water at vena contracta.

2. The relation among coefficients of discharge (C_d), coefficient of velocity (C_v) and coefficient of contraction (C_c) is
   (a) $C_d = C_v + C_c$ (b) $C_d = C_v - C_c$
   (c) $C_d = C_v / C_c$ (d) $C_d = C_v \times C_c$

3. The coefficient of velocity is determined experimentally by using the relation :
   (a) $C_v = \dfrac{y}{\sqrt{4 \times h}}$ (b) $C_v = \dfrac{y^2}{\sqrt{4 \times h}}$ (c) $C_v = \dfrac{x}{\sqrt{4yh}}$ (d) $C_v = \dfrac{x^2}{\sqrt{4yh}}$
   where x, y and h have usual meanings.

4. An orifice is said to be large, if
   (a) size of orifice is large
   (b) velocity of flow is large
   (c) available head of water is more than 5 times its height
   (d) available head of water is less than 5 times its height

5. The discharge through a wholly drowned orifice is given by the relation
   (a) $Q = C_d \cdot b \cdot H_1 \sqrt{2gh}$ (b) $Q = C_d \cdot b \cdot H_2 \sqrt{2gh}$
   (c) $Q = C_d \cdot H \sqrt{2gh}$ (d) $Q = C_d \cdot b (H_2 - H_1) \sqrt{2gh}$
   where H_1 = Height of water above the top of orifice on the upstream side
   H_2 = Height of water above the bottom of the orifice on the downstream side
   h = Difference between the water levels on both sides of the orifice.

ANSWERS

1. (a)   2. (d)   3. (c)   4. (d)   5. (d)

9

Flow Through Orifices
(Measurement of Time)

---

1. Introduction. 2. Time of Emptying a Square, Rectangular or Circular Tank through an Orifice at its Bottom. 3. Time of Emptying a Hemispherical Tank through an Orifice at its Bottom. 4. Time of Emptying a Circular Horizontal Tank through an Orifice at its Bottom. 5. Time of Emptying a Tank of Variable Cross-section through an Orifice. 6. Time of Emptying a Tank through Two Orifices. 7. Time of Flow of a Liquid from One Vessel into Another. 8. Flow of Liquid from One Vessel to Another under Pressure.

9·1 Introduction

In the last chapter, we have discussed the flow through orifices under various conditions. We have also discussed the discharge per second through these orifices. But in this chapter, we shall discuss the discharge through the orifices in a given time. Though this type of discharge has many applications, yet the time required for emptying a tank is important from the subject point of view. In this chapter, we shall discuss the time of emptying the following types of tanks :

1. Square or rectangular tank. 2. Hemispherical tank.
3. Circular horizontal tank. 4. Tank of variable cross-section.

9·2 Time of Emptying a Square, Rectangular or Circular Tank through an Orifice at its Bottom

Consider a square, rectangular or circular tank, of uniform cross-sectional area, containing some liquid and having an orifice at its bottom as shown in Fig. 9·1.

Let A = Surface area of the tank,
 H_1 = Initial height of the liquid,
 H_2 = Final height of the liquid, and
 a = Area of the orifice.

At some instant, let the height of the liquid be h above the orifice. We know that the theoretical velocity of the liquid at this instant,
$$v = \sqrt{2gh}$$

After a small interval of time (dt), let the liquid level fall down by an amount dh. Therefore volume of the liquid that has passed in time dt,

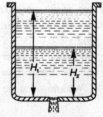

Fig. 9·1. Tank with an orifice at its bottom.

$$dq = A \times (-dh) = -A \cdot dh \qquad \ldots(i)$$

...(The value of dh is taken as negative, as its value will decrease with the increase in discharge)

We know that the volume of liquid that has passed through the orifice in time dt,

dq = Coefficient of discharge × Area × Theoretical velocity × Time
$$= C_d \cdot a \sqrt{2gh} \cdot dt \qquad \ldots(ii)$$

[153]

Equating equations (*i*) and (*ii*),
$$-A \cdot dh = C_d \cdot a \sqrt{2gh} \cdot dt$$
$$\therefore \quad dt = \frac{-A \cdot dh}{C_d \cdot a \sqrt{2gh}} = \frac{-A(h^{-1/2}) dh}{C_d \cdot a \sqrt{2g}}$$

Now the total time (T) required to bring the liquid level from H_1 to H_2 may be found out by integrating the above equation between the limits H_1 to H_2. i.e.,

$$T^* = \int_{H_1}^{H_2} \frac{-A(h^{-1/2}) dh}{C_d \cdot a \sqrt{2g}}$$

$$= \frac{-A}{C_d \cdot a \sqrt{2g}} \int_{H_1}^{H_2} h^{-1/2} dh$$

$$= \frac{-A}{C_d \, a \sqrt{2g}} \left[\frac{h^{1/2}}{1/2} \right]_{H_1}^{H_2}$$

$$= \frac{-2A}{C_d \cdot a \sqrt{2g}} \left[h^{1/2} \right]_{H_1}^{H_2}$$

$$= \frac{-2A}{C_d \cdot a \sqrt{2g}} [\sqrt{H_2} - \sqrt{H_1}]$$

Taking minus out from the bracket (as H_1 is greater than H_2),

$$T = \frac{2A(\sqrt{H_1} - \sqrt{H_2})}{C_d \cdot a \sqrt{2g}}$$

Note : If the tank is to be completely emptied, then putting $H_2 = 0$ in this equation, we get

$$T = \frac{2A \sqrt{H_1}}{C_d \cdot a \sqrt{2g}}$$

Example 9·1. *A water tank 10 metres long and 6 metres wide holds water to a depth of 1·25 metres. If the water is discharged through an opening at the bottom of the pool of an area of 0·23 square metre, find the time taken to empty it completely. Take coefficient of discharge for the opening as 0·62.*

---

*The time (T) may also be found out as follows :

We know that the volume of water that has passed through the orifice in T seconds
$$= A(H_1 - H_2)$$
and the initial theoretical velocity
$$= \sqrt{2gH_1}$$
$\therefore$ Initial rate of flow $= C_d \cdot a \sqrt{2gH_1}$
and the final rate of flow $= C_d \cdot a \sqrt{2gH_2}$
$\therefore$ Average rate of flow
$$= \frac{C_d \cdot a \sqrt{2gH_1} + C_d \cdot a \sqrt{2gH_2}}{2} = \frac{1}{2} C_d \cdot a \sqrt{2g} \, (\sqrt{H_1} + \sqrt{H_2})$$

Now the total time of flow,
$$T = \frac{\text{Volume of Water}}{\text{Average rate of flow}} = \frac{A(H_1 - H_2)}{\frac{1}{2} C_d \cdot \sqrt{2g} \, (\sqrt{H_1} + \sqrt{H_2})}$$

$$= \frac{2A(\sqrt{H_1} - \sqrt{H_2})}{C_d \cdot \sqrt{2g}}.$$

Solution. Given : Size of water tank $(A) = 10$ m $\times$ 6 m = 60 m^2; $H_1 = 1{\cdot}25$ m; $a = 0{\cdot}23$ m^2 and $C_d = 0{\cdot}62$.

We know that the time taken to empty the swimming pool completely,

$$T = \frac{2A\sqrt{H_1}}{C_d \cdot a\sqrt{2g}} = \frac{2 \times 60 \times \sqrt{1{\cdot}25}}{0{\cdot}62 \times 0{\cdot}23\,\sqrt{2 \times 9{\cdot}81}}\ \text{s}$$

$$= 212\ \text{s} = 3\ \text{min}\ 32\ \text{s} \quad \textbf{Ans.}$$

Example 9·2. *A circular water tank of 4 m diameter contains 5 m deep water. An orifice of 400 mm diameter is provided at its bottom. Find the time taken for water level to fall from 5 m to 2 m. Take $C_d = 0{\cdot}6$.*

Solution. Given : Diameter of circular tank $(D) = 4$ m; $H_1 = 5$ m; Diameter of orifice $(d) = 400$ mm $= 0{\cdot}4$ m; $H_2 = 2$ m and $C_d = 0{\cdot}6$.

We know that the surface area of the circular tank,

$$A = \frac{\pi}{4} \times (D)^2 = \frac{\pi}{4} \times (4)^2 = 12{\cdot}57\ \text{m}^2$$

and area of the orifice, $\quad a = \frac{\pi}{4} \times (d)^2 = \frac{\pi}{4} \times (0{\cdot}4)^2 = 0{\cdot}1257\ \text{m}^2$

∴ Time taken to fall the water level,

$$T = \frac{2A\,(\sqrt{H_1} - \sqrt{H_2})}{C_d \cdot a \cdot \sqrt{2g}} = \frac{2 \times 12{\cdot}57 \times (\sqrt{5} - \sqrt{2})}{0{\cdot}6 \times 0{\cdot}1257 \times \sqrt{2 \times 9{\cdot}81}} = 61{\cdot}9\text{s}\ \textbf{Ans.}$$

Example 9·3. *A cistern of cross-sectional area 1 square metre contains water 4 metres deep. An orifice of 60 mm diameter is provided at its bottom. Find the fall of water level after 2 minutes. Take coefficient of discharge as 0·6.*

Solution. Given : $A = 1$ m^2; $H_1 = 4$ m; Dia. of orifice $(d) = 60$ mm $= 0{\cdot}06$ m; $T = 2$ min $= 120$ s and $C_d = 0{\cdot}6$.

Let $\qquad H_2 = $ Height of water level in the cistern after 2 minutes (120 s).

We know that the area of the orifice,

$$a = \frac{\pi}{4} \times (d)^2 = \frac{\pi}{4} \times (0{\cdot}06)^2 = 0{\cdot}0028\ \text{m}^2$$

and time required to fall the water level (T),

$$120 = \frac{2A\,(\sqrt{H_1} - \sqrt{H_2})}{C_d \cdot a\,\sqrt{2g}} = \frac{2 \times 1(\sqrt{4} - \sqrt{H_2})}{0{\cdot}6 \times 0{\cdot}0028 \times \sqrt{2 \times 9{\cdot}81}} = \frac{4 - 2\sqrt{H_2}}{0{\cdot}0075}$$

$$120 \times 0{\cdot}0075 = 4 - 2\sqrt{H_2} \quad \text{or} \quad 0{\cdot}9 = 4 - 2\sqrt{H_2}$$

or $\qquad \sqrt{H_2} = \dfrac{4 - 0{\cdot}9}{2} = 1{\cdot}55$ or $H_2 = 2{\cdot}4$ m

∴ Fall of water level $= H_1 - H_2 = 4 - 2{\cdot}4 = 1{\cdot}6$ m **Ans.**

EXERCISE 9·1

1. A water tank of 8 m $\times$ 5 m contains 2 m deep water. If the tank has an orifice of area $0{\cdot}1$ m^2 in its bottom, find the time taken to empty the tank completely. Take C_d for the orifice as 0·6.
   [**Ans.** 7 min 6 s]

2. A rectangular tank of 5 m $\times$ 2 m is provided an orifice of 50 mm diameter at its bottom. Find the time taken to reduce the head of water from 1·8 m to 1 m. Take $C_d = 0{\cdot}625$. [**Ans.** 20 min 58 s]

3. A circular tank of 5 m diameter is provided with an orifice of 200 mm diameter at its bottom. Find the time taken for the water level to fall from 4 m to 2 m. Take coefficient of discharge for the orifice as 0·6. [Ans. 4 min 36 s]

4. A square tank of 2 m × 1·5 m, containing water up to a level of 2·25 m, contains an orifice of 60 mm diameter at its bottom. Find the quantity of water, which will flow in 2·5 minutes through the orifice, if C_d for the orifice is 0·6. [Ans. 1·59 m³]

9·3 Time of Emptying a Hemispherical Tank through an Orifice at its Bottom

Consider a hemispherical tank, containing some liquid, and having an orifice at its bottom as shown in Fig. 9·2.

Let
R = Radius of the tank,
H_1 = Initial height of the liquid,
H_2 = Final height of the liquid, and
a = Area of the orifice.

At some instant, let the height of the liquid be (h) above the orifice. We know that the theoretical velocity of the liquid at this instant,

$$v = \sqrt{2gh}$$

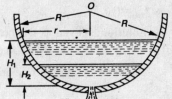

Fig. 9·2. Hemispherical tank with an orifice at its bottom.

At this instant, let (r) be the radius of the liquid surface. Then the surface area of the liquid,

$$A = \pi r^2$$

After a small interval of time dt, let the liquid level fall down by an amount dh. Therefore volume of the liquid that has passed in time dt,

$$dq = A \times (-dh) = -A \cdot dh = -\pi r^2 \cdot dh \quad \ldots(i)$$

...(The value of dh is taken as negative, as its value will decrease with the increase in discharge)

We know that the volume of liquid that has passed through the orifice in time dt,

$$dq = \text{Coefficient of discharge} \times \text{Area} \times \text{Theoretical velocity} \times \text{Time}$$
$$= C_d \cdot a \sqrt{2gh} \cdot dt \quad \ldots(ii)$$

Equating equations (i) and (ii),

$$-\pi r^2 \cdot dh = C_d \cdot a \sqrt{2gh} \cdot dt \quad \ldots(iii)$$

From the geometry of the tank, we find that

$$r^2 = R^2 - (R-h)^2 = R^2 - (R^2 + h^2 - 2Rh)$$
$$= R^2 - R^2 - h^2 + 2Rh = 2Rh - h^2$$

Substituting this value of r^2 in equation (iii),

$$-\pi (2Rh - h^2) dh = C_d \cdot a \sqrt{2gh} \cdot dt$$

$$\therefore dt = \frac{-\pi (2Rh - h^2) \times dh}{C_d \cdot a \sqrt{2gh}} = \frac{-\pi (2Rh - h^2) \times h^{-1/2} \, dh}{C_d \cdot a \sqrt{2g}}$$

Now the total time (T) required to bring the liquid level from H_1 to H_2 may be found out by integrating the above equation between the limits H_1 and H_2. i.e.,

Flow Through Orifices (Measurement of Time)

$$T = \int_{H_1}^{H_2} \frac{-\pi(2Rh - h^2) \times h^{-1/2} dh}{C_d \cdot a \sqrt{2g}}$$

$$= \frac{-\pi}{C_d \cdot a \sqrt{2g}} \int_{H_1}^{H_2} \left(2Rh^{1/2} - h^{3/2}\right) dh$$

$$= \frac{-\pi}{C_d \cdot a \sqrt{2g}} \left[\frac{2 \times 2}{3} Rh^{3/2} - \frac{2}{5} h^{5/2}\right]_{H_1}^{H_2}$$

$$= \frac{-\pi}{C_d \cdot a \sqrt{2g}} \left[\frac{4}{3} R\left(H_2^{3/2} - H_1^{3/2}\right) - \frac{2}{5}\left(H_2^{5/2} - H_1^{5/2}\right)\right]$$

Taking minus out from the bracket (as H_1 is greater than H_2),

$$T = \frac{\pi}{C_d \cdot a \sqrt{2g}} \left[\frac{4}{3} R\left(H_1^{3/2} - H_2^{3/2}\right) - \frac{2}{5}\left(H_1^{5/2} - H_2^{5/2}\right)\right]$$

$$= \frac{2\pi}{C_d \cdot a \sqrt{2g}} \left[\frac{2}{3} R\left(H_1^{3/2} - H_2^{3/2}\right) - \frac{1}{5}\left(H_1^{5/2} - H_2^{5/2}\right)\right]$$

Notes : 1. If the vessel is to be completely emptied, then putting $H_2 = 0$ in this equation,

$$T = \frac{2\pi}{C_d \cdot a \sqrt{2g}} \left[\frac{2}{3} RH_1^{3/2} - \frac{1}{5} H_1^{5/2}\right]$$

2. If the vessel was full at the time of the commencement and is to be completely emptied, then putting $H_1 = R$ in the above equation,

$$T = \frac{2\pi}{C_d a \sqrt{2g}} \left(\frac{2}{3} R \times R^{3/2} - \frac{1}{5} R^{5/2}\right) = \frac{2\pi}{C_d a \sqrt{2g}} \left(\frac{2}{3} R^{5/2} - \frac{1}{5} R^{5/2}\right) = \frac{14 \pi R^{5/2}}{15 C_d a \sqrt{2g}}$$

Example 9·4. *A hemispherical tank of 2 metres radius contains water up to a depth 1 metre. Find the time taken to discharge the tank completely through an orifice of 1000 sq mm provided at the bottom. Take coefficient of discharge as 0·62.*

Solution. Given : $R = 2$ m; $H_1 = 1$ m; $a = 1000$ mm$^2 = 0.001$ m^2 and $C_d = 0.62$.

We know that the time taken to discharge the tank completely,

$$T = \frac{2\pi}{C_d a \sqrt{2g}} \left(\frac{2}{3} RH_1^{3/2} - \frac{1}{5} H_1^{5/2}\right)$$

$$= \frac{2\pi}{0.62 \times 0.001 \times \sqrt{2 \times 9.81}} \left(\frac{2}{3} \times 2 \times (1)^{3/2} - \frac{1}{5} \times (1)^{5/2}\right) \text{s}$$

$$= 2593 \text{ s} = 43 \text{ min } 13 \text{ s} \quad \textbf{Ans.}$$

Example 9·5. *A hemispherical cistern of 6 metres radius is full of water. It is fitted with a 75 mm diameter sharp edged orifice at the bottom. Calculate the time required to lower the level in the cistern by 2 metres. Assume coefficient of discharge for the orifice as 0·6.*

Solution. Given : Dia. of cistern = 6 m or $R = 3$ m; Dia. of orifice (d) = 75 mm = 0·075 m; $H_1 = 6$ m; $H_2 = 6 - 2 = 4$ m and $C_d = 0.6$.

We know that the area of the orifice,

$$a = \frac{\pi}{4} \times (d)^2 = \frac{\pi}{4} \times (0.075)^2 = 0.0044 \text{ m}^2$$

and the time required to lower the level in the cistern,

$$T = \frac{2\pi}{C_d a \sqrt{2g}} \left[\frac{2}{3} R \left(H_1^{3/2} - H_2^{3/2} \right) - \frac{1}{5} \left(H_1^{5/2} - H_2^{5/2} \right) \right]$$

$$= \frac{2\pi}{0.6 \times 0.0044 \sqrt{2 \times 9.81}} \times \left[\frac{2}{3} \times 6 \left((6)^{3/2} - (4)^{3/2} \right) - \frac{1}{5} \left((6)^{5/2} - (4)^{5/2} \right) \right] \text{s}$$

$$= 8794 \text{ s} = 2 \text{ hrs } 26 \text{ min } 34 \text{ s} \quad \textbf{Ans.}$$

Example 9·6. *A tank has an upper cylindrical portion of 3 m diameter and 4 m high with a hemispherical base. The cylinder is full of water. Determine the time taken to empty it through an orifice of cross-sectional area of 8000 mm^2 at its bottom. Take $C_d = 0.62$.*

Solution. Given : Dia. of cylindrical portion $(d) = 3$ m or $R = 1.5$ m; Height of water above the spherical base = 4 m; $a = 8000$ mm$^2 = 0.008$ m^2 and $C_d = 0.62$.

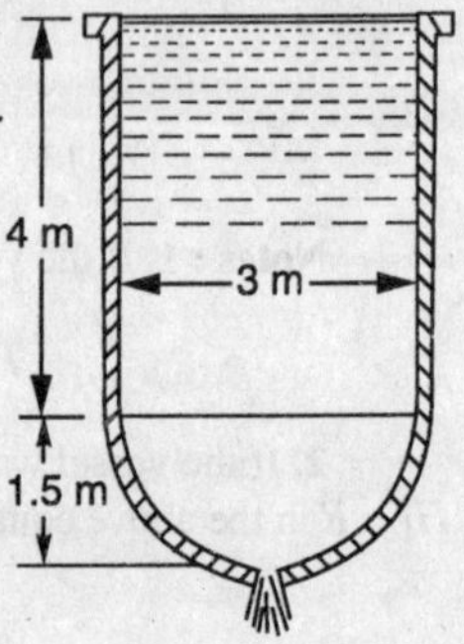

First of all, let us split up the tank into two portions. *i.e.*, cylindrical and hemispherical, and then find out the time required to empty both of them independently.

We know that the initial height of water above the orifice,

$$H_1 = 4 + 1.5 = 5.5 \text{ m}$$

and surface area of the tank,

$$A = \frac{\pi}{4} \times (D)^2 = \frac{\pi}{4} \times (3)^2 = 7.069 \text{ m}^2$$

We also know that the time required to empty the cylindrical portion (from $H_1 = 5.5$ m to $H_2 = 1.5$ m),

Fig. 9·3.

$$T_1 = \frac{2A (\sqrt{H_1} - \sqrt{H_2})}{C_d a \sqrt{2g}} = \frac{2 \times 7.069 (\sqrt{5.5} - \sqrt{1.5})}{0.62 \times 0.008 \sqrt{2 \times 9.81}} = 721 \text{ s} \qquad ...(i)$$

and the time required to empty the hemispherical portion when it is completely filled with water,

$$T_2 = \frac{14 \pi R^{5/2}}{15 C_d a \sqrt{2g}} = \frac{14 \times \pi \times (1.5)^{5/2}}{15 \times 0.62 \times 0.008 \sqrt{2 \times 9.81}} = 368 \text{ s} \qquad ...(ii)$$

∴ Total time, $T = T_1 + T_2 = 721 + 368 = 1089$ s = 18 min 9 s **Ans.**

9·4 Time of Emptying a Circular Horizontal tank through an Orifice at its Bottom

Consider a circular horizontal tank, containing liquid and having an orifice at its bottom as shown in Fig. 9·4 (*a*) and (*b*).

Let R = Radius of the tank,
H_1 = Initial height of the liquid,
H_2 = Final height of the liquid,
a = Area of the orifice, and
l = Length of the tank.

At some instant, let the height of the liquid be h above the orifice. We know that theoretical velocity of the liquid at this instant,

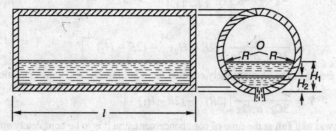

Fig. 9·4. Circular tank with an orifice at its bottom.

$$v = \sqrt{2gh}$$

At this instant, let b be the breadth of the liquid in the tank. We know that surface area of the liquid in the tank,

$$A = l \cdot b$$

After a small interval of time dt, let the liquid level fall down by an amount dh. Therefore volume of liquid that has passed in time dt,

$$dq = A \times (-dh) = -A \cdot dh = -l \cdot b \cdot dh \qquad \ldots(i)$$

... (The value of dh is taken as negative as it will decrease with the increase in discharge)

We know that the volume of liquid that has passed through the orifice in time dt,

$$dq = \text{Coefficient of discharge} \times \text{Area} \times \text{Theoretical Velocity} \times \text{Time}$$
$$= C_d \cdot \sqrt{2gh} \cdot dt \qquad \ldots(ii)$$

Equating equations (i) and (ii),

$$-l \cdot b \cdot dh = C_d \cdot a \sqrt{2gh} \cdot dt \qquad \ldots(iii)$$

From the geometry of the tank, we find that

$$\frac{b}{2} = \sqrt{R^2 - (R-h)^2} = \sqrt{R^2 - (R^2 + h^2 - 2Rh)}$$
$$= \sqrt{R^2 - R^2 - h^2 + 2Rh} = \sqrt{2Rh - h^2}$$
$$\therefore \quad b = 2\sqrt{2Rh - h^2}$$

Substituting the value of b in equation (iii),

$$-l \times 2\sqrt{2Rh - h^2} \cdot dh = C_d \cdot a \sqrt{2gh} \cdot dt$$

or

$$dt = \frac{-2l \sqrt{2Rh - h^2}\, dh}{C_d \cdot a \sqrt{2gh}} = \frac{-2l \sqrt{h(2R-h)}\, dh}{C_d \cdot a \sqrt{2gh}}$$

Dividing numerator and denominator by $\sqrt{h}$,

$$dt = \frac{-2l \sqrt{(2R-h)}\, dh}{C_d \cdot a \sqrt{2g}} = \frac{-2l (2R-h)^{\frac{1}{2}}\, dh}{C_d \cdot a \sqrt{2g}}$$

Now the total time (T) required to bring the liquid level from H_1 to H_2 may be found out by integrating the above equation between the limits H_1 to H_2. Therefore

$$T = \int_{H_1}^{H_2} \frac{-2l \times (2R-h)^{1/2}\, dh}{C_d \cdot a \sqrt{2g}} = \frac{-2l}{C_d \cdot a \sqrt{2g}} \int_{H_1}^{H_2} (2R-h)^{\frac{1}{2}}\, dh$$

$$= \frac{-2l}{C_d \cdot a \sqrt{2g}} \left[\frac{(2R-h)^{3/2}}{\frac{3}{2} \times (-1)} \right]_{H_1}^{H_2} = \frac{4l}{3C_d \cdot a \sqrt{2g}} \left[(2R-h)^{3/2} \right]_{H_1}^{H_2}$$

$$= \frac{4l}{3C_d \cdot a \sqrt{2g}} \left[(2R-H_2)^{3/2} - (2R-H_1)^{3/2} \right]$$

Notes : 1. If the tank is to be completely emptied, then putting $H_2 = 0$ in this equation,

$$T = \frac{4l}{3C_d \cdot a \sqrt{2g}} \left[(2R)^{3/2} - (2R-H_1)^{3/2} \right]$$

2. If the tank was half full at the time of commencement, and it is to be completely emptied, then putting $H_1 = R$ in the above equation,

$$T = \frac{4l}{3C_d \cdot a \sqrt{2g}} \left[(2R)^{3/2} - (2R-R)^{3/2} \right] = \frac{4l}{3C_d \cdot a \sqrt{2g}} \left[(2R)^{3/2} - (R)^{3/2} \right]$$

$$= \frac{4lR^{3/2}}{3 \, C_d \cdot a \sqrt{2g}} = (2\sqrt{2} - 1) = \frac{0.55 \times lR^{3/2}}{C_d \cdot a}$$

Note : The above simplified formula holds good, only if all the data is taken in metres.

Example 9·7. *An orifice is fitted at the bottom of a boiler drum for the purpose of emptying it. The drum, is horizontal and half full of water. It is 10 m long and 2 m in diameter. Find the time required to empty the boiler, if the diameter of the orifice is 150 mm. Assume coefficient of discharge as 0·6.*

Solution. Given : Length of boiler $(l) = 10$ m; Dia. of boiler $(D) = 2$ m or $r = 1$ m; Dia. of orifice $(d) = 150$ mm $= 0.15$ m and $C_d = 0.6$.

We know that the area of the orifice,

$$a = \frac{\pi}{4} \times (d)^2 = \frac{\pi}{4} \times (0.15)^2 = 0.0177 \text{ m}^2$$

and time required to empty the boiler,

$$T = \frac{0.55 \times l \cdot R^{3/2}}{C_d \cdot a} = \frac{0.55 \times 10 \times (1)^{3/2}}{0.6 \times 0.0177} = 518 \text{ s} = 8 \text{ min } 38 \text{ s} \quad \textbf{Ans.}$$

Example 9·8. *A horizontal boiler of 3 metres diameter and 15 metres long contains water to a height of 2·5 metres. Find the time taken for emptying the boiler through an orifice of 180 mm diameter at the bottom of the boiler. Take $C_d = 0.625$.*

Solution. Given : Dia. of boiler $(D) = 3$ m or radius $(r) = 1.5$ m; Length of boiler $(l) = 15$ m; $H_1 = 2.5$ m; Diameter of orifice $(d) = 180$ mm $= 0.18$ m and $C_d = 0.625$.

We know that the area of the orifice,

$$a = \frac{\pi}{4} \times (d)^2 = \frac{\pi}{4} \times (0.18)^2 = 0.0254 \text{ m}^2$$

and time taken for emptying the boiler,

$$T = \frac{4\,l}{3C_d \, a \sqrt{2g}} \left[(2R)^{3/2} - (2R-H_1)^{3/2} \right]$$

$$= \frac{4 \times 15}{3 \times 0.625 \times 0.0254 \times \sqrt{2 \times 9.81}} \left[(2 \times 1.5)^{3/2} - (3-2.5)^{3/2} \right] \text{ s}$$

$$= 472 \text{ s} = 7 \text{ min } 32 \text{ s} \quad \textbf{Ans.}$$

EXERCISE 9·2

1. A hemispherical tank of 4 m diameter is full of water. Find the time of emptying the tank through a 200 mm diameter orifice at the bottom of the tank. Take $C_d = 0·6$. **[Ans. 3 min 20 s]**

2. A tank with its upper cylindrical portion of 4 m diameter is of 6 m height with lower portion of hemispherical form. The tank is fitted with a 75 mm diameter orifice at its bottom. Find the time required to empty the tank completely, if it was initially full of water. Take coefficient of discharge for the orifice as 0·62. **[Ans. 12 min 40 s]**

3. A horizontal boiler of 2 m diameter and 3 m long is fitted with an orifice of 50 mm diameter at its bottom. Find the time required during which half full boiler will be emptied through the orifice. Take $C_d = 0·625$. **[Ans. 22 min 24 s]**

4. A horizontal boiler of 3 m diameter and 8 m long contains water up to a height of 2·4 m. Find the time taken to empty the boiler through an orifice of 150 mm diameter fitted at its bottom. Take coefficient of discharge for the orifice as 0·6. **[Ans. 17 min 53 s]**

9·5 Time of Emptying a Tank of Variable Cross-section through an Orifice

In the previous article, we have discussed the time of emptying of geometrical tanks (*i.e.*, rectangular, hemispherical and circular). But, sometimes, we come across tanks which have variable cross-section. In such cases, there are two variables instead of one, as in the case of tanks of uniform cross-section. Since a single relation cannot be derived for different cross-sections, it is therefore essential that such problems should be solved from the first principles. *i.e.*, from the equation :

$$dt = \frac{-A \cdot dh}{C_d \cdot a\sqrt{2gh}}$$

This can be best understood from the following examples.

Example 9·9. *A rectangular tank of 20 m × 12 m at the top and 10 m × 6 m at the bottom is 3 m deep as shown in Fig. 9·5.*

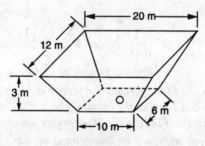

Fig. 9·5.

There is an orifice of 450 mm diameter at the bottom of the tank. Determine the time taken to empty the tank completely, if coefficient of discharge for the orifice is 0·64.

Solution. Given : Top length 20 m; Top width = 12 m; Bottom length = 10 m; Bottom width = 6 m; Depth of water = 3 m; Diameter of orifice (d) = 450 mm = 0·45 m and C_d = 0·64.

We know that the area of the orifice,

$$a = \frac{\pi}{4} \times (d)^2 = \frac{\pi}{4} \times (0·45)^2 = 0·159 \text{ m}^2$$

First of all, let us consider a small strip of water of thickness (dh) at a height (h) from the bottom of the tank. From the geometry of the figure, we find that the length of the strip of water,

$$l = 10 + \frac{10h}{3}$$

and the breadth $= 6 + \dfrac{6h}{3} = 6 + 2h$

∴ Area, $A = \left(10 + \dfrac{10h}{3}\right)(6 + 2h) = 60 + 20h + 20h + \dfrac{20h^2}{3}$

$= 6.67h^2 + 40h + 60$ ...(i)

Now let us use the general equation for the time at emptying a tank. i.e.,

$$dt = \dfrac{-A \cdot dh}{C_d \cdot a \sqrt{2gh}}$$

The total time required to empty the tank may be found by integrating the above equation between the limits 3 and 0 (because initial head of water is 3 m and final head of water is 0 m). i.e.,

$$T = \int_3^0 \dfrac{-A \cdot dh}{C_d \cdot a \sqrt{2gh}}$$

$$= \dfrac{-1}{C_d \cdot a \sqrt{2g}} \int_3^0 A \cdot h^{-\tfrac{1}{2}} \cdot dh$$

$$= \dfrac{1}{C_d \cdot a \sqrt{2g}} \int_0^3 (6.67h^2 + 40h + 60h)^{-\tfrac{1}{2}} dh$$

$$= \dfrac{1}{C_d \cdot a \sqrt{2g}} \int_0^3 (6.67h^{3/2} + 40h^{1/2} + 60)h^{-\tfrac{1}{2}} dh$$

$$= \dfrac{1}{0.64 \times 0.159 \sqrt{19.62}} \times \left[\dfrac{6.67h^{\tfrac{5}{2}}}{\tfrac{5}{2}} + \dfrac{40h^{\tfrac{3}{2}}}{\tfrac{3}{2}} + \dfrac{60h^{\tfrac{1}{2}}}{\tfrac{1}{2}}\right]_0^3 s$$

$$= \dfrac{1}{0.451}\left[2.668 \times (3)^{\tfrac{5}{2}} + 26.67 \times (3)^{\tfrac{3}{2}} + 120 \times (3)^{\tfrac{1}{2}}\right] s$$

$= 860 \text{ s} = 14 \text{ min } 20 \text{ s}$ **Ans.**

Example 9·10. *A tank in the form of frustum of a cone is 3 m high and filled with water. It has a diameter of 2·4 m at the top and 1·2 m at the bottom. There is an orifice in the bottom of the tank, whose coefficient of discharge is 0·6. Find the size of the orifice which can empty the tank in 6 minutes.*

Solution. Given : Depth of water = 3 m; Diameter at the top = 2·4 m; Diameter at the bottom = 1·2 m; C_d = 0·6 and time of emptying the tank = 6 min = 360 s.

Let d = Diameter of the orifice in metres.

First of all, consider a small strip of water of thickness (dh) at a height (h) from the bottom of the tank. From the geometry of the figure, we find that diameter of the strip

$$= 1.2 + \dfrac{1.2 h}{3} = (1.2 + 0.4 h) \text{ m}$$

∴ Area, $A = \dfrac{\pi}{4}(1.2 + 0.4h)^2 \text{ m}^2$

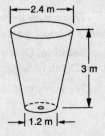

Fig. 9·6.

Flow Through Orifices (Measurement of Time)

Now let us use the general equation for the time of emptying a tank. *i.e.*,

$$dt = \frac{-A \cdot dh}{C_d \cdot a \sqrt{2gh}}$$

The total time required to empty the tank may be found out by integrating the above equation between the limits 3 and 0 (because initial head of water is 3 m and final head of water is 0 m). *i.e.*,

$$T = \int_3^0 \frac{-A \cdot dh}{C_d \cdot a \sqrt{2gh}}$$

$$360 = \frac{-1}{C_d \cdot a \sqrt{2g}} \int_3^0 \frac{\pi}{4}(1 \cdot 2 + 0 \cdot 4 h)^2 \times h^{-\frac{1}{2}} dh$$

$$= \frac{\pi}{4 C_d \cdot a \sqrt{2g}} \int_0^3 (1 \cdot 44 + 0 \cdot 16 h^2 + 0 \cdot 96 h) h^{-\frac{1}{2}} dh$$

$$= \frac{\pi}{4 \times 0 \cdot 6 \times a \sqrt{2 \times 9 \cdot 81}} \times \int_0^3 \left(1 \cdot 44 h^{-\frac{1}{2}} + 0 \cdot 16 \times h^{\frac{3}{2}} + 0 \cdot 96 h^{\frac{1}{2}}\right) dh$$

$$= \frac{0 \cdot 296}{a} \left[\frac{1 \cdot 44 h^{\frac{1}{2}}}{\frac{1}{2}} + \frac{0 \cdot 16 h^{\frac{5}{2}}}{\frac{5}{2}} + \frac{0 \cdot 96 h^{\frac{3}{2}}}{\frac{3}{2}} \right]_0^3$$

$$= \frac{0 \cdot 296}{a} \left[2 \cdot 88 (3)^{\frac{1}{2}} + 0 \cdot 064 (3)^{\frac{5}{2}} + 0 \cdot 64 (3)^{\frac{3}{2}} \right] = \frac{2 \cdot 756}{a}$$

or $\quad a = 2 \cdot 756 / 360 = 0 \cdot 0077 \text{ m}^2$

∴ Diameter of orifice, $\quad d = \sqrt{\dfrac{4a}{\pi}} = \sqrt{\dfrac{4 \times 0 \cdot 0077}{\pi}} \quad \left[\text{Area} = \dfrac{\pi}{4} \times (d)^2\right]$

$\qquad\qquad\qquad\qquad = 0 \cdot 099 \text{ m} = 99 \text{ mm}$ **Ans.**

9·6. Time of Emptying a Tank through Two Orifices

In the previous articles, we have discussed the cases of time of emptying the tank through one orifice at its bottom. But, sometimes, the tank has two (or even more) orifices. Here we shall discuss the following two cases :

1. When the orifices are at the same level, and
2. When the orifices are at different levels.

A little consideration will show that when both the orifices are at the same level, they will be discharging water under the same head. In such a case, both the orifices will behave as a single orifice of area equal to sum of the two orifices. But when the two orifices are at different levels, the water will flow through both the orifices (under their respective heads) so long as the water level reaches the upper orifice. After this, the water will flow through lower orifice only.

In such cases, it is convenient to divide the problem into two parts *i.e.*, first up to the upper orifice and then up to the bottom orifice.

Example 9·11. *A vertical circular tank of 600 mm diameter and 2·5 m height is full of water. It contains two orifices each of 1300 square mm area, one at the bottom of the tank and the other at a height of 1·25 m above the bottom as shown in Fig. 9·7.*

Determine the time required to empty the tank. Take coefficient of discharge for both the orifices as 0.62.

Solution. Given : Diameter of the tank = 600 mm = 0·6 m; $H = 2.5$ m ; Area of each orifice (a) = 1300 mm² = 0·0013 m² and $C_d = 0.62$.

For the sake of simplicity, let us divide the example into two parts. *i.e.*, first up to the centre of the top orifice, and then up to the bottom orifice. First of all, consider the first part of the example. In this case, the water is flowing through both the orifices.

Now consider an instant, when the height of water above the centre of the top orifice be (h) metres. At that instant, the height of water above the bottom orifice will be $(h + 1·25)$ metres.

We know that the surface area of the tank,

$$A = \frac{\pi}{4} \times (D)^2 = \frac{\pi}{4} \times (0.6)^2 = 0.2827 \text{ m}^2$$

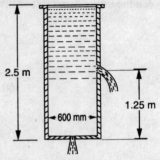

Fig. 9·7.

Now let us use the general equation for the time of emptying a tank. *i.e.*,

$$dt = \frac{-A \cdot dh}{C_d \cdot a \sqrt{2gh}}$$

$$= \frac{-A \cdot dh}{C_d \cdot a \sqrt{2g(h+1·25)} + C_d \cdot a \sqrt{2gh}}$$

$$= \frac{-A \cdot dh}{C_d \cdot a \sqrt{2g} \left[\sqrt{(h+1·25)} + \sqrt{h}\right]}$$

The total time (T_1) required to bring the water level up to the centre of the top orifice may now be found out by integrating the above equation between the limits 1·25 m and 0. Therefore

$$T_1 = \int_{1·25}^{0} \frac{-A \cdot dh}{C_d \cdot (2a) \sqrt{2g} \left[\sqrt{(h+1·25)} + \sqrt{h}\right]}$$

$$= \frac{A}{C_d (2a) \cdot \sqrt{2g}} \int_{0}^{1·25} \frac{dh}{\sqrt{(h+1·25)} + \sqrt{h}}$$

$$= \frac{0.2827}{0.62 \times (2 \times 0.0013) \times \sqrt{19.62}} \int_{0}^{1·25} \frac{\sqrt{(h+1·25)} - \sqrt{h}}{1·25} dh$$

...[Multiplying the numerator and denominator by $\left[\sqrt{(h+1·25)} - \sqrt{h}\right]$]

$$= \frac{79.2}{2.5} \times \left[\frac{(h+1·25)^{\frac{3}{2}}}{\frac{3}{2}} - \frac{h^{\frac{3}{2}}}{\frac{3}{2}}\right]_{0}^{1·25} \text{s}$$

$$= 31.7 \times 0.771 = 24.4 \text{ s} \qquad ...(i)$$

Now consider the flow of water below the centre of the top orifice. A little consideration will show that now the water will be flowing through the bottom orifice only. We know that time required to empty the tank,

$$T_2 = \frac{2A\sqrt{H_1}}{C_d \cdot a\sqrt{2g}} = \frac{2 \times 0.2827\sqrt{1.25}}{0.62 \times 0.0013 \times \sqrt{19.62}} = 177.1 \text{ s} \quad ...(ii)$$

∴ Total time, $T = T_1 + T_2 = 24.4 + 177.1 = 201.5$ s

$= 3$ min 21.5 s **Ans.**

Example 9·12. *A swimming pool 30 m long and 8 m wide has vertical sides, and the bottom slopes uniformly from a depth of 1 m of the shallow end to 2·5 m at and the deep end. There are two orifices each of 0·2 m² area, one at the deep end and the other of the shallow end as shown in the figure given below :*

Fig. 9·8.

Find the time required to empty the tank completely, if the coefficients of discharge for both the orifices is 0·64.

Solution. Given : Pool size (A) = 30 m × 8 m = 240 m²; Depth at one end = 1 m; Depth at the other end = 2·5 m; Area of each orifice = 0·25 m² and $C_d = 0.64$.

For the sake of simplicity, let us divide the problem into two parts. *i.e.*, first up to the centre of the top orifice and then up to the bottom orifice. First of all, consider the first part of the example. In this case, the water is flowing through both the orifices. Now consider an instant, when the height of water above the centre of the top orifice be (h) metre. At that instant, the height of water above the bottom orifice will be ($h + 1.5$) metres.

Now let us use the general equation for the time of emptying a tank. *i.e.*,

$$dt = \frac{-A \cdot dh}{C_d \cdot (2a)\sqrt{2gh}}$$

$$= \frac{-A \cdot dh}{C_d \cdot (2a)\sqrt{2g(h+1.5)} + C_d \cdot h\sqrt{2gh}}$$

$$= \frac{-A \cdot dh}{C_d \cdot (2a)\sqrt{2g}\,[\sqrt{(h+1.5)} + \sqrt{h}]}$$

The total time (T_1) may be found out by integrating the above equation between the limits 1 and 0. Therefore

$$T_1 = \int_1^0 \frac{-A \cdot dh}{C_d \cdot (2a)\sqrt{2g}\,\sqrt{(h+1.5)} + \sqrt{h}}$$

$$T_1 = \frac{A}{C_d \cdot (2a) \sqrt{2g}} \int_0^1 \frac{dh}{\sqrt{(h+1\cdot 5)} + \sqrt{h}}$$

$$= \frac{240}{0\cdot 64 \times 2 \times 0\cdot 25 \times \sqrt{19\cdot 62}} \int_0^1 \frac{\sqrt{(h+1\cdot 5)} - \sqrt{h}}{1\cdot 5} dh$$

... (Multiplying the numerator and denominator by $[\sqrt{(h+1\cdot 5)} - \sqrt{h}]$)

$$= \frac{169\cdot 3}{1\cdot 5} \times \left[\frac{(h+1\cdot 5)^{3/2}}{\frac{3}{2}} - \frac{h^{3/2}}{\frac{3}{2}} \right]_0^1 = 1129\cdot 7 \times 1\cdot 3 = 146\cdot 8 \text{ s} \quad ...(i)$$

Now consider the flow of water below the centre of the top orifice. Let us now consider an instant, when the height of water above the bottom orifice be (h) metre. At this instant, the surface area of water,

$$A = (30 \times 8) \times \frac{h}{1\cdot 5} = 160\, h \text{ m}^2$$

Again using the general equation for the time of emptying a tank,

$$dt = \frac{-A \cdot dh}{C_d \cdot a \sqrt{2g}} = \frac{-160\, h \cdot dh}{C_d \cdot a \sqrt{2gh}}$$

The total time (T_2) may be found out by integrating the above equation between the limits $1\cdot 5$ and 0. Therefore

$$T_2 = \int_{1\cdot 5}^0 \frac{-160\sqrt{h} \cdot dh}{C_d \cdot a \sqrt{2g}}$$

$$= \frac{-160}{C_d \cdot a \sqrt{2g}} \int_{1\cdot 5}^0 h^{1/2}\, dh$$

$$= \frac{160}{C_d \cdot a \cdot \sqrt{2g}} \int_0^{1\cdot 5} h^{1/2}\, dh$$

$$= \frac{160}{0\cdot 64 \times 0\cdot 25 \times \sqrt{19\cdot 62}} \left[\frac{h^{3/2}}{\frac{3}{2}} \right]_0^{1\cdot 5}$$

$$= 225\cdot 8 \times 1\cdot 225 = 276\cdot 5 \text{ s} \quad ...(ii)$$

∴ Total time, $T = T_1 + T_2 = 146\cdot 8 + 276\cdot 6 = 423\cdot 4$ s

$= 7$ min $53\cdot 4$ s **Ans.**

9·7 Time of Flow of Liquid from One Vessel into Another

When two vessels, containing liquid, are connected together by means of an orifice, the liquid will flow from the vessel with a higher level to the vessel with a lower level irrespective of their areas. In such a case, the liquid level will fall in one vessel with a corresponding rise in the other. The orifice, through which the flow takes place, is a drowned one and the liquid head causing flow will be the difference between the two liquid levels.

Consider two tanks connected at their bottom by a small orifice as shown in Fig. 9·9.
Let A_1 = Area of the larger vessel,

A_2 = Area of the smaller vessel,

a = Area of the orifice,

H_1 = Initial difference between the liquid levels of the two vessels,

H_2 = Final difference between the liquid levels of the two vessels,

and T = Time, in seconds, required to bring the difference of liquid levels from H_1 to H_2.

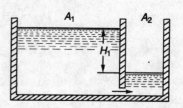

Fig. 9·9. Flow from one vessel into another.

At some instant, let the difference between the two liquid levels be h. We know that the theoretical velocity of the liquid at this instant,

$$v = \sqrt{2gh}$$

After a small interval of time dt, let the liquid level in the vessel A_1 fall down by an amount equal to x.

∴ The volume of liquid that has passed from the tank A_1

$$= A_1 \cdot x \qquad ...(i)$$

and the rise of liquid level in the other tank

$$= \frac{A_1}{A_2} x$$

If the change of liquid level in the tanks be dh in time dt, then

$$dh = -\left(x + \frac{A_1}{A_2} \times x\right)$$

...[Minus sign of dh is taken, as the value of h will decrease as the discharge will increase]

$$= -x\left(1 + \frac{A_1}{A_2}\right) = -x\left(\frac{A_1 + A_2}{A_2}\right)$$

or $\qquad x = \dfrac{-A_2 \cdot dh}{A_1 + A_2} \qquad ...(ii)$

In time dt, the volume of water that has passed from tank A_1

$$= \text{Coefficient of discharge} \times \text{Area} \times \text{Theoretical velocity} \times \text{Time}$$

$$= C_d \cdot a \sqrt{2gh} \times dt \qquad ...(iii)$$

Equating equations (i) and (iii),

$$A_1 \cdot x = C_d \cdot a \sqrt{2gh} \times dt$$

$$\frac{-A_1 A_2 \, dh}{A_1 + A_2} = C_d \cdot a \sqrt{2gh} \times dt$$

or $\qquad dt = \dfrac{-A_1 A_2 \, dh}{C_d \cdot a (A_1 + A_2) \sqrt{2gh}} = \dfrac{-A_1 A_2 \, h^{-1/2} \, dh}{C_d \cdot a (A_1 + A_2) \sqrt{2g}}$

Now the total time (T) required to bring the difference of liquid levels from H_1 to H_2 may be found out by integrating the above equation between the limits H_1 and H_2. i.e.,

$$^*T = \int_{H_2}^{H_1} \frac{-A_1 A_2 (h^{1/2}) \, dh}{C_d \cdot a \, (A_1 + A_2) \sqrt{2g}}$$

$$= \frac{-A_1 \times A_2}{C_d \cdot a \, (A_1 + A_2) \sqrt{2g}} \int_{H_1}^{H_2} h^{-1/2} dh$$

$$= \frac{-A_1 \, A_1}{C_d \cdot a \, (A_1 + A_2) \sqrt{2g}} \left[\frac{h^{1/2}}{\frac{1}{2}} \right]_{H_1}^{H_2}$$

$$= \frac{-2A_1 A_2 (\sqrt{H_2} - \sqrt{H_1})}{C_d \cdot a \, (A_1 + A_2) \sqrt{2g}}$$

Taking minus sign out of the bracket as H_1 is greater than H_2.

$$T = \frac{2A_1 A_2 (\sqrt{H_1} - \sqrt{H_2})}{C_d \cdot a (A_1 + A_2) \sqrt{2g}}$$

---

*The time (T) may also be found out as follows :
Let in T seconds the liquid level in the vessel A_1 fall down by an amount equal to l.
∴ Volume of liquid that has passed from the vessel A_1 in T seconds.

$$= A_1 \cdot l \qquad \qquad ...(i)$$

and volume of liquid that has passed into the vessel A_2 in T seconds

$$= A_2 (H_1 - H_2 - l) \qquad \qquad ...(ii)$$

Since both the volumes are equal, therefore

$$A_1 \cdot l = A_2 (H_1 - H_2 - l) = A_2 (H_1 - H_2) - A_2 l$$

$$l(A_1 + A_2) = A_2 (H_1 - H_2)$$

or
$$l = \frac{A_2 (H_1 - H_2)}{A_1 + A_2} \qquad \qquad ...(iii)$$

Substituting the value of l in equation (i),
Volume of liquid that has passed from the tank A_1 in T seconds

$$= \frac{A_1 A_2 (H_1 - H_2)}{A_1 + A_2} \qquad \qquad ...(iv)$$

Now initial theoretical velocity of liquid (when difference of liquid is H_1) = $\sqrt{2gH_1}$
∴ Initial rate of flow = $C_d \cdot a \sqrt{2gH_1}$
and final rate of flow = $C_d \cdot a \sqrt{2gH_2}$

∴ Average rate of flow $= \dfrac{C_d \cdot a \sqrt{2gH_1} + C_d \cdot a \sqrt{2gH_2}}{2} = \dfrac{1}{2} C_d \cdot a \sqrt{2g} (\sqrt{H_1} + \sqrt{H_2})$ ...(v)

and time of flow, $T = \dfrac{\text{Volume of water}}{\text{Average rate of flow}} = \dfrac{\dfrac{A_1 A_2 (H_1 - H_2)}{A_1 + A_2}}{\dfrac{1}{2} C_d \cdot a \sqrt{2g} (\sqrt{H_1} + \sqrt{H_2})}$

$$= \frac{2 A_1 A_2 (\sqrt{H_1} - \sqrt{H_2})}{C_d \cdot a (A_1 + A_2) \sqrt{2g}}$$

If the levels in the two tanks are to be brought down to the same level, then substituting $H_2 = 0$ in the above equation,

$$T = \frac{2 A_1 A_2 \sqrt{H_1}}{C_d \cdot a (A_1 + A_2) \sqrt{2g}}$$

Example 9·13. *A tank 6 metres long and 1·5 metre wide is divided into two parts, so that area of one part is 7·2 m^2 and that of the other is 1·8 m^2. The water level in the larger part is 3 metres higher than in the lower one. Find the time taken for the difference in water levels to reach 1 metre, if the water flows through a submerged orifice of 75 mm diameter, in the partition. Assume coefficient of discharge for orifice as 0·6.*

Solution. Given : Size of the tank = $6 \times 1\cdot 5 = 9$ m^2; $A_1 = 7\cdot 2$ m^2; $A_2 = 1\cdot 8$ m^2; $H_1 = 3$ m; $H_2 = 1$ m; $d = 75$ mm $= 0\cdot 075$ m and $C_d = 0\cdot 6$.

We know that the area of the orifice,

$$a = \frac{\pi}{4} \times (d)^2 = \frac{\pi}{4} \times (0\cdot 075)^2 = 4\cdot 418 \times 10^{-3} \text{ m}^2$$

and time taken for the difference of water levels,

$$T = \frac{2 A_1 A_2 (\sqrt{H_1} - \sqrt{H_2})}{C_d \cdot a (A_1 + A_2) \sqrt{2g}}$$

$$= \frac{2 \times 7\cdot 2 \times 1\cdot 8 (\sqrt{3} - \sqrt{1})}{0\cdot 6 \times (4\cdot 418 \times 10^{-3}) \times (7\cdot 2 + 1\cdot 8) \sqrt{2 \times 9\cdot 81}} \text{ s}$$

$$= 180 \text{ s} = 3 \text{ min} \quad \textbf{Ans.}$$

Example 9·14. *Two tanks of 9 m × 6 m and 6 m × 3 m are connected at their bottom by a circular orifice of 150 mm diameter. When the flow started, the water level in the larger tank was 4 metres higher than that in the smaller one. Find the time taken to flow 27 cubic metres of water from the larger tank into the smaller one, if coefficient of discharge for the orifice is 0·64.*

Solution. Given : $A_1 = 9 \times 6 = 54$ m^2 ; $A_2 = 6 \times 3 = 18$ m^2; $d = 150$ mm $= 0\cdot 15$ m; $H_1 = 4$ m; $Q = 27$ m^3 and $C_d = 0\cdot 64$.

We know that the fall in water level in the larger tank

$$= \frac{\text{Discharge}}{\text{Area}} = \frac{27}{54} = 0\cdot 5 \text{ m}$$

and the rise in water level in the smaller tank

$$= \frac{\text{Discharge}}{\text{Area}} = \frac{27}{18} = 1\cdot 5 \text{ m}$$

∴ The final difference in water levels,

$$H_2 = 4 - (0\cdot 5 + 1\cdot 5) = 2 \text{ m}$$

We also know that the area of the orifice,

$$a = \frac{\pi}{4} \times (d)^2 = \frac{\pi}{4} \times (0\cdot 15)^2 = 17\cdot 67 \times 10^{-3} \text{ m}^2$$

and time taken for the flow of water,

$$T = \frac{2 A_1 A_2 (\sqrt{H_1} - \sqrt{H_2})}{C_d \cdot a (A_1 + A_2) \sqrt{2g}}$$

$$T = \frac{2 \times 54 \times 18(\sqrt{4} - \sqrt{2})}{0 \cdot 64 \times (17 \cdot 67 \times 10^{-3})(54 + 18)\sqrt{2 \times 9 \cdot 81}} \text{ s}$$

$$= 316 \text{ s} = 5 \text{ min } 16 \text{ s} \quad \textbf{Ans.}$$

EXERCISE 9·3

1. A tank in the form of frustom of a cone of 2 m high has top and bottom diameters of 2 m and 0·8 m respectively. Find the time of emptying the tank through an orifice of 100 mm diameter provided at its bottom, when it is full of water. Take C_d for the orifice as 0·625. [**Ans.** 2 min 40 s]

2. A circular tank of 1 m² cross-section has two orifices each of 50 mm diameter in one of the vertical sides provided at 2·4 m and 6 m above the bottom of the tank. Find the time required to lower the depth of water from 9 m to 4·5 m. Take $C_d = 0·62$. [**Ans.** 5 min 35 s]

3. Two tanks having surface areas of 800 m² and 400 m² are connected by an orifice of 0·1 m² in area. If the initial difference of water levels in the tanks be 3 m, find the time taken to reduce it to 1 m. Take $C_d = 0·6$. [**Ans.** 24 min 29 s]

4. A tank of 3 m long and 1·5 m wide is divided into two parts by a partition, so that area of one part is double the area of the other. The partition contains a square orfice of 80 mm side. Find the time taken to bring the difference of the water levels from 3 m to 0·5 m. Take $C_d = 0·62$. [**Ans.** 1 min 57 s]

QUESTIONS

1. Derive an expression for the time taken to empty a tank through an orifice at its bottom.
2. What is a hemispherical tank ? Derive an equation for emptying a hemispherical tank through an orifice fitted at the bottom of the tank.
3. A circular boiler, kept in a horizontal position, is half full of water. Derive an equation for emptying it through an orifice fitted at its bottom.
4. Two tanks, containing water, are connected by an orifice at their bottom. Obtain a relation for the time required to reduce the difference of their water levels from H_1 to H_2.

OBJECTIVE TYPE QUESTIONS

1. A tank is discharging water through an orifice fitted at its bottom. The time taken for completely emptying the tank is
   (a) directly proportional to the surface area of the tank
   (b) inversely proportional to the area of the orifice
   (c) inversely proportional to the height of water in the tank
   (d) both 'a' and 'b'

2. A hemispherical tank of radius R is full of water. The time required to empty the tank depends upon
   (a) R (b) $R^{\frac{3}{2}}$ (c) $R^{\frac{5}{2}}$ (d) R^3

3. If a boiler, kept in horizontal position, is full of water, then the time taken to empty it completely depends upon
   (a) R (b) $R^{\frac{3}{2}}$ (c) R^2 (d) $R^{\frac{5}{2}}$

4. When two tanks having different water levels are connected by an orifice at their bottom, the time required to bring the water surfaces at the same levels is directly proportional to
   (a) difference of initial water levels
   (b) sum of the surface areas of both the tanks
   (c) coefficient of discharge for the orifice
   (d) all of these

ANSWERS

1. (d)   2. (c)   3. (b)   4. (a)

10

Flow Through Mouthpieces

1. Introduction. 2. Types of Mouthpieces. 3. Loss of a Head of a Liquid Flowing in a pipe. 4. Loss of Head due to Sudden Enlargement. 5. Loss of Head due to Sudden Contraction. 6. Loss of Head at the Entrance in a Pipe. 7. Loss of Head at the Exit of a Pipe. 8. Loss of Head due to an Obstruction in a Pipe. 9. Discharge through a Mouthpiece. 10. Discharge through an External Mouthpiece. 11. Discharge through an Internal Mouthpiece (Re-entrant or Borda's Mouthpiece). 12. Mouthpiece Running Free. 13. Mouthpiece Running Full. 14. Discharge through a Convergent Mouthpiece. 15. Discharge through a Convergent-divergent Mouthpiece (Bell-Mouthpiece). 16. Pressure in a Mouthpiece. 17. Pressure in an External Mouthpiece. 18. Pressure in an Internal Mouthpiece. 19. Pressure in a Convergent Mouthpiece. 20. Pressure in a Convergent-divergent Mouthpiece.

10·1 Introduction

We have discussed in Art. 9·11 that the discharge through an orifice depends upon its coefficient of discharge. It was felt by the engineers that the discharge through an orifice is too less (due to low value of the coefficient of discharge). Many scientists and engineers conducted experiments to improve the value of coefficient of discharge. For doing so, they introduced various types of orifices (as discussed in Art. 9·2).

It was, only after a series of experiments, the engineers found that if a short pipe be fitted to an orifice, it will increase the value of coefficient of discharge. We know that the increase in the value of coefficient of discharge will increase the discharge through the orifice. Such a pipe, whose length is generally more than 2 times the diameter of the orifice and is fitted (externally or internally) to the orifice is known as a mouthpiece.

10·2 Types of Mouthpieces

There are many types of mouthpieces, depending upon their size, shape or nature of discharge. But the following are important from the subject point of view :

1. *According to the position of the mouthpiece*
    (a) Internal mouthpiece, and
    (b) External mouthpiece.
2. *According to the shape of the mouthpiece*
    (a) Cylindrical mouthpiece,
    (b) Convergent mouthpiece, and
    (c) Convergent-divergent mouthpiece.
3. *According to the nature of discharge*
    (a) Mouthpiece running full, and
    (b) Mouthpiece running free.

Flow Through Mouthpieces

All the above mentioned mouthpieces will be discussed, in details, at the appropriate places in this book. Before entering into the details and various types of the mouthpieces, we shall study the loss of head of a flowing liquid, which takes place in the pipe.

10·3 Loss of Head of a Liquid Flowing in a Pipe

It has been experimentally found that when a liquid is flowing in a pipe, it loses its energy (or sometimes termed as head) due to friction* of the wall, change of cross-section or obstruction in the flow. All such losses are expressed in terms of velocity head. Though there are many types of losses of heads, yet the following losses, which occur in a flowing liquid, are important from the subject point of view :

1. Loss of head due to sudden enlargement,
2. Loss of head due to sudden contraction,
3. Loss of head at the entrance in a pipe,
4. Loss of head at the exit in a pipe, and
5. Loss of head due to an obstruction in a pipe.

All these losses of heads will be discussed in the following pages one-by-one.

10·4 Loss of Head due to Sudden Enlargement

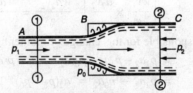

Fig. 10·1. Sudden enlargement.

Consider a liquid flowing in a pipe ABC having a sudden enlargement at B. As a result of this sudden enlargement, eddies will be formed in the corner of the pipe at B as shown in Fig. 10·1.

Let
a_1 = Area of pipe at section 1-1,
v_1 = Velocity of the liquid at section 1-1,
p_1 = Pressure of the liquid at section 1-1,
a_2, v_2, p_2 = Corresponding values at section 2-2,
p_0 = Pressure of the liquid eddies on area $(a_2 - a_1)$, and
h_e = Loss of head due to sudden enlargement.

It has been experimentally found that p_1 and p_0 are nearly equal to each other. And this is the assumption on which formula for loss of head due to sudden enlargement is derived. As a matter of fact, the loss of head, that takes place, is due to the eddies formed at the suddenly enlarged section as shown in Fig. 10·1.

Applying Bernoulli's equation to sections 1-1 and 2-2,

$$\frac{p_1}{w} + \frac{v_1^2}{2g} = \frac{p_2}{w} + \frac{v_2^2}{2g} + \text{Losses} \qquad \text{...(Taking } Z_1 = Z_2\text{)}$$

or losses,
$$h_e = \left(\frac{p_1}{w} - \frac{p_2}{w}\right) + \left(\frac{v_1^2}{2g} - \frac{v_2^2}{2g}\right) \qquad \text{...(i)}$$

---

* Loss of head due to friction will be discussed in Art. 13.3.

We know that the momentum of water flowing per second at section 1-1

$$= \text{Mass} \times \text{Velocity} = \frac{wa_1v_1}{g} \times v_1 = \frac{wa_1v_1^2}{g}$$

and the momentum of water flowing per second at section 2-2

$$= \frac{wa_2v_2^2}{g}$$

∴ Change of momentum per second

$$= \frac{wa_1v_1^2}{g} - \frac{wa_2v_2^2}{g} \qquad \ldots(ii)$$

Since the flow between the sections 1-1 and 2-2 is continuous, therefore

$$a_1v_1 = a_2v_2 \quad \text{or} \quad a_1 = \frac{a_2v_2}{v_1}$$

Substituting this value of a_1 in equation (ii),

Change of momentum per second

$$= \left(\frac{wv_1^2}{g} \times \frac{a_2v_2}{v_1}\right) - \frac{wa_2v_2^2}{g} = \frac{wa_2v_2 \times v_1}{g} - \frac{wa_2v_2^2}{g} \qquad \ldots(iii)$$

We know that the force responsible for this change of momentum

$$= p_2a_2 - p_1a_1 - p_0(a_2 - a_1) = p_2a_2 - p_1a_1 - p_0a_2 + p_0a_1$$
$$= p_2a_2 - p_0a_2 = a_2(p_2 - p_0) \qquad \ldots(\text{Taking } p_1 = p_0)$$
$$= a_2(p_2 - p_1) \qquad \ldots(\text{Taking } p_0 = p_1) \quad \ldots(iv)$$

Equating equations (iii) and (iv),

$$a_2(p_2 - p_1) = \frac{wa_2v_2 \times v_1}{g} - \frac{wa_2v_2^2}{g} = wa_2\left(\frac{v_2v_1}{g} - \frac{v_2^2}{g}\right)$$

$$\frac{p_2}{w} - \frac{p_1}{w} = \frac{v_2v_1}{g} - \frac{v_2^2}{g} \quad \text{or} \quad \frac{p_1}{w} - \frac{p_2}{w} = \frac{v_2^2}{g} - \frac{v_1v_2}{g}$$

Substituting this value of $\frac{p_1}{w} - \frac{p_2}{w}$ in equation (i) we find that the loss of head,

$$h_e = \left(\frac{v_2^2}{g} - \frac{v_1v_2}{g}\right) + \left(\frac{v_1^2}{2g} - \frac{v_2^2}{2g}\right) = \frac{1}{2g}(2v_2^2 - 2v_1v_2 + v_1^2 - v_2^2)$$

$$= \frac{1}{2g}(v_1^2 + v_2^2 - 2v_1v_2) = \frac{(v_1 - v_2)^2}{2g}$$

Example 10·1. *A pipe of 100 mm diameter is suddenly enlarged to 200 mm diameter. Find the loss of head, when the discharge is 60 litres/s.*

Solution. Given : $d_1 = 100$ mm $= 0·1$ m; $d_2 = 200$ mm $= 0·2$ m and $Q = 60$ litres/s $= 0·06$ m³/s.

We know that the area of pipe at section-1,

$$a_1 = \frac{\pi}{4} \times (d_1)^2 = \frac{\pi}{4} \times (0·1)^2 = 0·00785 \text{ m}^2$$

Similarly

$$a_2 = \frac{\pi}{4} \times (d_2)^2 = \frac{\pi}{4} \times (0·2)^2 = 0·03142 \text{ m}^2$$

∴ The velocity of water at section 1,
$$v_1 = \frac{Q}{a_1} = \frac{0.06}{0.007\ 85} = 7.64 \text{ m/s}$$

Similarly
$$v_2 = \frac{Q}{a_2} = \frac{0.06}{0.031\ 42} = 1.91 \text{ m/s}$$

We also know that the loss of head due to sudden enlargement,
$$h_e = \frac{(v_1 - v_2)^2}{2g} = \frac{(7.64 - 1.91)^2}{2 \times 9.81} = 1.67 \text{ m} \quad \textbf{Ans.}$$

Example 10·2. *A pipe of section 0.1 m^2 suddenly changes to 0.4 m^2 area. The quantity of water flowing in the pipe in 0.3 m^3/s and the pressure at the smaller part of pipe is 85 kPa. Find*
  (i) head loss due to sudden enlargement, and
  (ii) pressure at the larger part of the pipe.

Solution. Given : $a_1 = 0.1$ m²; $a_2 = 0.4$ m²; $Q = 0.3$ m³/s and $p_1 = 85$ kPa = 85 kN/m².

(i) Head loss due to sudden enlargement

We know that the velocity of water at section 1,
$$v_1 = \frac{Q}{a_1} = \frac{0.3}{0.1} = 3 \text{ m/s}$$

and the velocity of water at section 2,
$$v_2 = \frac{Q}{a_2} = \frac{0.3}{0.4} = 0.75 \text{ m/s}$$

∴ Head loss due to sudden enlargement,
$$h_e = \frac{(v_1 - v_2)^2}{2g} = \frac{(3 - 0.75)^2}{2 \times 9.81} = 0.258 \text{ m} \quad \textbf{Ans.}$$

(ii) Pressure at the larger part of the pipe

Let p_2 = Pressure at the larger part of the pipe (*i.e.*, at section 2) in kN/m².

Now using Bernoulli's equation for sections 1 and 2,
$$\frac{p_1}{w} + \frac{v_1^2}{2g} + Z_1 = \frac{p_2}{w} + \frac{v_2^2}{w} + Z_2 + h_e$$

$$\frac{85}{9.81} + \frac{(3)^2}{2 \times 9.81} = \frac{p_2}{9.81} + \frac{(0.75)^2}{2 \times 9.81} + 0.258 \quad \text{...(Taking } Z_1 = Z_2\text{)}$$

$$8.665 + 0.459 = \frac{p_2}{9.81} + 0.029 + 0.258$$

∴
$$\frac{p_2}{9.81} = 8.665 + 0.459 - 0.029 - 0.258 = 8.837$$

or
$$p_2 = 8.837 \times 9.81 = 86.7 \text{ kN/m}^2 = 86.7 \text{ kPa} \quad \textbf{Ans.}$$

10·5 Loss of Head due to Sudden Contraction

Consider a liquid flowing in a pipe *ABC*, having a sudden contraction at *B* as shown in Fig. 10·2. It will be interesting to know that the liquid, while flowing through a narrow pipe, will be further contracted at section 1-1 forming a vena contracta (in the same way as a jet issuing from an orifice). Strictly speaking, the loss of head due to sudden contraction is not due to the contraction itself, but it is due to the sudden enlargement, which takes place after the contraction.

Let
a_1 = Area of pipe at section 1-1,
v_1 = Velocity of liquid at section 1-1, and
a_2, v_2 = Corresponding values at section 2-2.

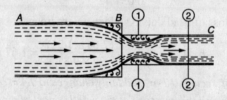

Fig. 10·2. Sudden contraction.

Since the discharge through the pipe *ABC* is continuous, therefore
$$a_1 v_1 = a_2 v_2 \qquad \ldots(i)$$
Assuming the value of C_c (*i.e.*, coefficient of contraction) to be 0·62, we find that
$$a_1 = C_c \times a_2 = 0.62\, a_2$$
Substituting this value of a_1 in equation (*ii*),
$$0.62\, a_2 v_1 = a_2 v_2 \quad \text{or} \quad v_1 = \frac{a_2}{0.62} \qquad \ldots(ii)$$
We know that the loss of head, due to sudden contraction, will actually be due to sudden enlargement from section 2-2. Therefore loss of head due to sudden enlargement after section 1-1,
$$= \frac{(v_1 - v_2)^2}{2g}$$
Substituting the value of v_1 from equation (*ii*),
$$h_e = \frac{\left(\dfrac{v_2}{0.62} - v_2\right)^2}{2g} = \frac{0.375\, v_2^2}{2g} = K \times \frac{v_2^2}{2g}$$

Notes : 1. The above relation has been derived after assuming the value of the coefficient of contraction as 0·62. Since the value of coefficient of contraction depends upon the type of orifice, therefore the exact relation will vary with the type of orifice.

2. It has been experimentally found that the actual value of loss of head also depends upon the ratio of the two diameters *i.e.*, d_1/d_2, where d_1 is diameter of pipe at section 1 and d_2 is the diameter of pipe at section 2. The following table gives the value of *K* for the corresponding value of d_1/d_2.

d_1/d_2	1·0	1·1	1·25	1·5	2·0	2·5	3·0	3·5	4·0
K	0	0·1	0·19	0·28	0·375	0·40	0·42	0·43	0·45

3. In general, if no data is given in the problem, the value of *K* is taken as 0·375. Though some authorities differ on this value, yet the internationally accepted figure is 0·375.

Example 10·3. *A horizontal pipe of 200 mm diameter suddenly enlarges to 300 mm diameter. After some length, it suddenly reduces to 150 mm diameter. If the water flowing in the pipe be 200 litres/s, find*
(a) loss of head due to sudden enlargement, and
(b) loss of head due to sudden contraction.

Solution. Given : $d_1 = 200$ mm $= 0.2$ m; $d_2 = 300$ mm $= 0.3$ m; $d_3 = 150$ mm $= 0.15$ m and $Q = 200$ litres/s $= 0.2$ m³/s.

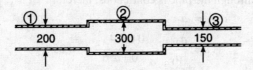

Fig. 10·3.

Loss of head due to sudden enlargement

We know that the area of pipe at section 1,

$$a_1 = \frac{\pi}{4} \times (d_1)^2 = \frac{\pi}{4} \times (0.2)^2 = 0.0314 \text{ m}^3$$

$$\therefore \quad v_1 = \frac{Q}{a_1} = \frac{0.2}{0.0314} = 6.37 \text{ m/s}$$

Similarly, the area of pipe at section 2,

$$a_2 = \frac{\pi}{4} \times (d_2)^2 = \frac{\pi}{4} \times (0.3)^2 = 0.0707 \text{ m}^3$$

$$\therefore \quad v_2 = \frac{Q}{a_2} = \frac{0.2}{0.0707} = 2.83 \text{ m/s}$$

and the area of pipe at section 3,

$$a_3 = \frac{\pi}{4} \times (d_3)^2 = \frac{\pi}{4} \times (0.15)^2 = 0.0177 \text{ m}^3$$

$$\therefore \quad v_3 = \frac{Q}{a_3} = \frac{0.2}{0.0177} = 11.3 \text{ m/s}$$

We know that the loss of head due to sudden enlargement,

$$h_e = \frac{(v_1 - v_2)^2}{2g} = \frac{(6.37 - 2.83)^2}{2 \times 9.81} = \frac{(3.54)^2}{19.62} = 0.639 \text{ m} \quad \textbf{Ans.}$$

Loss of head due to sudden contraction

We also know that the loss of head due to sudden contraction,

$$h_c = \frac{0.375 \, v_3^2}{2g} = \frac{0.375 \times (11.3)^2}{2 \times 9.81} = 2.44 \text{ m} \quad \textbf{Ans.}$$

Example 10·4. *When a sudden contraction is used in a horizantal pipe from 400 mm to 200 mm, the pressure changes from 100 kPa to 80 kPa. If the coefficient of contraction for the jet is 0·62, find the discharge through the pipe.*

Solution. Given $d_1 = 400$ mm $= 0.4$ m; $d_2 = 200$ mm $= 0.2$ m; $p_1 = 100$ kPa $= 100$ kN/m²; $p_2 = 80$ kPa $= 80$ kN/m² and $C_c = 0.62$.

Let $v_1 =$ Velocity of water in the pipe at section 1, and
$v_2 =$ Velocity of water in the pipe at section 2.

We know that area of the pipe at section 1,

$$a_1 = \frac{\pi}{4} \times (d_1)^2 = \frac{\pi}{4} \times (0.4)^2 = 0.1256 \text{ m}^2$$

and area of pipe at section 2, $a_2 = \dfrac{\pi}{4} \times (d_2)^2 = \dfrac{\pi}{4} \times (0\cdot 2)^2 = 0\cdot 0314 \text{ m}^2$

Since the discharge through the pipe is continuous, therefore

$$a_1 v_1 = a_2 v_2$$

or $\quad v_1 = \dfrac{a_2 v_2}{a_1} = \dfrac{0\cdot 0314 \, v_2}{0\cdot 1256} = 0\cdot 25 \, v_2$

We know that the loss of head due to sudden contraction,

$$h_e = \dfrac{0\cdot 375 \, v_2^2}{2g}$$

Now using Bernoulli's equation for sections 1 and 2,

$$\dfrac{p_1}{w} + \dfrac{v_1^2}{2g} + Z_1 = \dfrac{p_2}{w} + \dfrac{v_2^2}{2g} + Z_2 + h_e$$

$$\dfrac{100}{9\cdot 81} + \dfrac{(0\cdot 25 \, v_2)^2}{2g} = \dfrac{80}{9\cdot 81} + \dfrac{v_2^2}{2g} + \dfrac{0\cdot 375 \, v_2^2}{2g} \quad \ldots(\text{Taking } Z_1 = Z_2)$$

$$\dfrac{100}{9\cdot 81} - \dfrac{80}{9\cdot 81} = \dfrac{v_2^2}{2g} + \dfrac{0\cdot 375 \, v_2^2}{2g} - \dfrac{0\cdot 0625 \, v_2^2}{2g}$$

$$2\cdot 04 = \dfrac{1\cdot 3125 \, v_2^2}{2g} = \dfrac{1\cdot 3125 \, v_2^2}{2 \times 9\cdot 81} = 0\cdot 067 \, v_2^2$$

or $\quad v_2^2 = 2\cdot 04/0\cdot 067 = 30\cdot 45 \text{ or } v_2 = 5\cdot 52 \text{ m/s}$

$\therefore$ Discharge through the pipe,

$$Q = a_2 \cdot v_2 = 0\cdot 0314 \times 5\cdot 52 = 0\cdot 173 \text{ m}^3/\text{s} = 173 \text{ litres/s} \quad \textbf{Ans.}$$

10·6 Loss of Head at the Entrance in a Pipe

The loss of head, due to entrance in a pipe, is actually a loss due to sudden contraction and depends upon the form of entrance. Experimentally, the value of loss of head at entrance has been found to be equal to

$$= \dfrac{0\cdot 5 \, v^2}{2g}$$

where $\quad v =$ Velocity of the liquid in the pipe.

In case of long pipe, this loss of head is very small, as compared to the frictional loss. Thus it is sometimes neglected.

10·7 Loss of Head at the Exit of a Pipe

This loss of head, due to exit in a pipe, is actually a loss due to the energy of head, which the flowing liquid possesses, by virtue of its motion. Experimentally, the value of loss of head at exit has been found to be equal to

$$= \dfrac{v^2}{2g}$$

where $\quad v =$ Velocity of the liquid in the pipe.

In case of long pipes, this loss of head, like that at entrance, is also very small as compared to the frictional loss. Thus it also is sometimes neglected.

Example 10·5. *A pipe of 40 mm diameter is conveying water with a velocity of 2·5 m/s. Find the loss of head at the entrance and exit of the pipe.*

Solution. Given : $d = 40$ mm and $v = 2\cdot 5$ m/s.

Loss of head at the entrance of the pipe

We know that the loss of head at the entrance of the pipe

$$= \frac{0.5 v^2}{2g} = \frac{0.5 \times (2.5)^2}{2 \times 9.81} = 0.16 \text{ m} \quad \textbf{Ans.}$$

Loss of head at the exit of the pipe

We also know that the loss of head at the exit of the pipe,

$$= \frac{v^2}{2g} = \frac{(2.5)^2}{2 \times 9.81} = 0.32 \text{ m} \quad \textbf{Ans.}$$

10·8 Loss of Head due to an Obstruction in a Pipe

The liquid, while flowing around an obstruction, gets contracted (in the same way as a jet issuing from an orifice). Strictly speaking, this is a particular case of loss of head due to sudden contraction, which further depends upon the loss of head due to sudden enlargement.

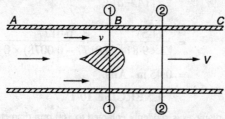

Fig. 10·4. Obstruction in a pipe.

Consider a liquid flowing in a pipe having an obstruction at B as shown in Fig. 10·4.

Let $\quad A$ = Area of the full pipe *i.e.*, at section 2-2,

a = Area of the obstruction,

∴ $(A - a)$ = Area of pipe at section 1-1 through which the liquid has to pass,

V = Velocity of liquid at section 2-2, and

v = Velocity of liquid at section 1-1.

We know from the equation of continuity,

$$A \times V = (A - a) \times v \times C_c \quad \text{...(where } C_c = \text{Coefficient of contraction)}$$

∴ $$v = \frac{A \times V}{(A - a) \times C_c} \quad \text{...(i)}$$

We have seen that the loss of head due to sudden enlargement,

$$h_e = \frac{(v_1 - v_2)^2}{2g} = \frac{(v - V)^2}{2g} \quad \text{...(ii)}$$

... (In this case, $v_1 = v$ and $v_2 = V$)

Substituting the value of v from equation (*i*),

$$h_0 = \frac{\left(\dfrac{A \times V}{A - a) \times C_c} - V\right)^2}{2g} \quad (\because h_0 = h_e)$$

$$= \frac{V^2}{2g} \left(\frac{A}{(A - a) \times C_c} - 1\right)^2$$

Example 10·6. *A horizontal pipe of 150 mm diameter is obstructed by a circular plate of 100 mm diameter. Find the loss of head due to the obstruction in the pipe, if the water is flowing with a velocity of 1·5 m/s in the pipe. Take C_c as 0·6.*

Solution. Given : D = 150 mm = 0·15 m; d_2 = 100 mm = 0·1 m and V = 1·5 m/s.

We know that area of the pipe,

$$A = \frac{\pi}{4} \times (D)^2 = \frac{\pi}{4} \times (0.15)^2 = 0.0177 \text{ m}^2$$

and area of the obstruction, $\quad a = \frac{\pi}{4} \times (d)^2 = \frac{\pi}{4} \times (0.1)^2 = 0.0078 \text{ m}^2$

∴ Loss of head due to obstruction,

$$h_0 = \frac{V^2}{2g}\left(\frac{A}{(A-a)C_c} - 1\right)^2$$

$$= \frac{(1.5)^2}{2 \times 9.81}\left(\frac{0.0177}{(0.0177 - 0.0078) \times 0.6} - 1\right)^2 \text{ m}$$

$$= 0.45 \text{ m} \quad \textbf{Ans.}$$

EXERCISE 10·1

1. A pipe of 80 mm in diameter is suddenly enlarged to 160 mm diameter. Find the loss of head due to sudden enlargement, if the velocity of water in 50 mm diameter section is 5 m/s. **[Ans.** 0·717 m**]**

2. A pipe of 150 mm diameter is suddenly enlarged to 300 mm diameter. Find the loss of head per kN of water, if the discharge through the pipe is 56 litres/s. **[Ans.** 0·287 m**]**

3. A horizontal pipe of 180 mm diameter suddenly reduces its diameter to 120 mm. If water flows in the pipe with a velocity of 2 m/s, find the loss of head due to sudden contraction and the discharge through the pipe. **[Ans.** 0·29 m; 50·8 litres/s**]**

4. A horizontal pipe of 100 mm diameter has its central portion enlarged to 200 mm. If discharge through the pipe is 1·2 m³/min, determine the loss of head at entrance, sudden enlargement and sudden contraction. **[Ans.** 0·166 m ; 0·186 m; 0·124 m**]**

5. A horizontal pipe of 100 mm diameter is obstructed by an obstruction of 5×10^{-3} m². Find the loss of head due to the obstruction, if the water is flowing with a velocity of 1 m/s through the pipe. Take $C_d = 0.6$. **[Ans.** 0·66 m**]**

10·9 Discharge through a Mouthpiece

We have discussed in Art. 10·1 that if a mouthpiece is fitted to an orifice, it will increase the value of coefficient of discharge for the orifice. The increase in the value of coefficient of discharge will increase the discharge through the orifice. In the succeeding pages, we shall find out the values of coefficient of discharge for the following types of mouthpieces :

1. Discharge through an external mouthpiece,
2. Discharge through an internal mouthpiece,
3. Discharge through a convergent mouthpiece,
4. Discharge through convergent-divergent mouthpiece.

10·10 Discharge through an External Mouthpiece

The discharge through an orifice may be increased by fitting a sufficient length of pipe to the outside of the orifice as shown in Fig. 10·5. Such a pipe, which is attached externally to an orifice, is known as an external mouthpiece.

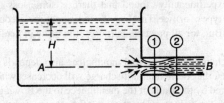

Fig. 10·5. External mouthpiece.

A little consideration will show that the jet, on entering the pipe will first contract, then expand and fill up the whole pipe as shown in Fig. 10·5.

Let H = Height of liquid above the mouthpiece,
a = Area of the orifice or mouthpiece,
a_c = Area of flow at vena contracta,
C_c = Coefficient of contraction,
v = Velocity of liquid at outlet, and
v_c = Velocity of liquid at vena contracta.

Assuming the coefficient of contraction at vena contracta to be 0·62, we have

$$a_c = C_c \times a = 0.62\, a$$

Since the liquid is flowing continuously, therefore,

$$v_c \cdot a_c = v \cdot a$$

or $\qquad v_c = \dfrac{v \cdot a}{a_c} = \dfrac{v}{0.62} \qquad \left(\because \dfrac{a}{a_c} = C_c = 0.62\right)$

We see that the jet after passing through the section 1-1, suddenly enlarges at section 2-2. Therefore due to the sudden enlargement the loss of head,

$$\therefore \quad h_c = \dfrac{(v_c - v)^2}{2g} = \dfrac{\left(\dfrac{v}{0.62} - v\right)^2}{2g} = \dfrac{0.375\, v^2}{2g} \qquad \ldots(i)$$

Applying Bernoulli's equation to points A and B,

$$H = \dfrac{v^2}{2g} + \text{Losses} = \dfrac{v^2}{2g} + \dfrac{0.375\, v^2}{2g} = \dfrac{1.375\, v^2}{2g}$$

$$\therefore \quad \dfrac{v^2}{2g} = \dfrac{H}{1.375} \qquad \text{or} \qquad v = \sqrt{\dfrac{2gH}{1.375}} = 0.855\, \sqrt{2gH} \qquad \ldots(ii)$$

We know that the theoretical velocity of liquid at the outlet

$$= \sqrt{2gH}$$

$\therefore$ Coefficient of velocity, $C_v = \dfrac{\text{Actual velocity}}{\text{Theoretical velocity}} = \dfrac{0.855\, \sqrt{2gH}}{\sqrt{2gH}} = 0.855 \qquad \ldots(iii)$

Thus for finding the discharge, through an external mouthpiece, we shall first find out the value of coefficient of discharge for the mouthpiece. We know that coefficient of discharge (assuming C_c equal to 1)

$$C_d = \text{Coefficient of velocity} \times \text{Coefficient of contraction}$$
$$= 0.855 \times 1 = 0.855$$

It shows that coefficient of discharge has considerably increased by fitting an external mouthpiece. The increase in coefficient of discharge means the increase in the discharge through the orifice.

∴ Discharge, $Q = C_d \cdot a \sqrt{2gH} = 0.855 \cdot a \sqrt{2gH}$ ...(∵ $C_d = 0.855$)

Notes : 1. It has been experimentally found that there is some loss of head at the entrance of the mouthpiece, depending upon the type of orifice. This loss of head, sometimes, reduces the coefficient of discharge by a small amount (up to 0.82). But, for all practical purposes, the value of coefficient of discharge is taken as 0.855.

2. The coefficient of discharge for an external mouthpiece also depends upon the length of the pipe. A little consideration will show, that the coefficient of discharge will decrease, with the increase in length, due to greater frictional resistance offered by the walls of the mouthpiece to the flowing water.

The following table gives the values of coefficient of discharge for the corresponding length of the mouthpiece :

Length of mouthpiece	3d	5d	10d	25d	50d
Coefficient of discharge	0.813	0.79	0.77	0.71	0.64

3. If length of the mouthpiece is not given, then the value of coefficient of discharge is taken as 0.855.

Example 10.7. *Find the discharge from a 100 mm diameter external mouthpiece, fitted to one side of a large vessel, if the head over the mouthpiece is 4 metres.*

Solution. Given : $d = 100$ mm $= 0.1$ m and, $H = 4$ m.

We know that the area of the mouthpiece,

$$a = \frac{\pi}{4} \times (d)^2 = \frac{\pi}{4} \times (0.1)^2 = 7.854 \times 10^{-3} \text{ m}^2$$

and discharge through an external mouthpiece,

$$Q = 0.855 \, a \sqrt{2gH} = 0.855 \times (7.584 \times 10^{-3}) \times \sqrt{2 \times 9.81 \times 4} \text{ m}^3/\text{s}$$
$$= 0.0595 \text{ m}^3/\text{s} = 59.5 \text{ litres/s} \textbf{ Ans.}$$

10.11 Discharge through an Internal Mouthpiece (Re-entrant or Borda's Mouthpiece)

An internal mouthpiece, extending into the fluid (*i.e.*, inside the vessel) is known as Re-entrant or *Borda's mouthpiece as shown in Figs. 10.6 and 10.7. Following are the two types of internal mouthpieces, depending upon their nature of discharge :

1. Mouthpiece running free, and
2. Mouthpiece running full.

If the jet, after contraction, does not touch the sides of the mouthpiece, it is said to be running free as shown in Fig. 10.6. But if the jet, after contraction, expands and fills up the whole mouthpiece, it is said to be running full as shown in Fig. 10.7.

It has been experimentally found that if the length of the mouthpiece extending into the fluid is less than 3 times the diameter of the orifice, it will run free. But, if the length of the mouthpiece is more than 3 times the diameter of the orifice, it will run full. The coefficient of discharge will be different in both the cases.

---

*Jean Charles Borda, a French Mathematician and Engineer, who conducted experiments on the flow of water through tubes.

10·12 Mouthpiece Running Free

We have discussed in Art. 10·11 that if the jet, after contraction does not touch the sides of the mouthpiece, it is said to be running free. Now consider a mouthpiece running free as shown in Fig. 10·6.

Let H = Height of the liquid above the mouthpiece,
a = Area of orifice or mouthpiece,
a_c = Area of the contracted jet,
v = Velocity of the liquid, and
and w = Specific weight of the liquid.

We know that pressure of the liquid on the mouthpiece,
$$p = wH$$
and force acting on the mouthpiece
$$= \text{Pressure} \times \text{Area} = wHa \qquad ...(i)$$

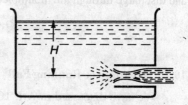

Fig. 10·6. Mouthpiece of running free.

Mass of the liquid flowing per second
$$= \frac{wa_c v}{g}$$

∴ Momentum of the flowing liquid
$$= \text{Mass} \times \text{Velocity} = \frac{wa_c v \times v}{g} = \frac{wa_c v^2}{g} \qquad ...(ii)$$

Since the water is initially at rest, therefore initial momentum
$$= 0$$
and change of momentum $= \dfrac{wa_c v^2}{g} \qquad ...(ii)$

According to Newton's *Second Law of Motion, the force is equal to the rate of change of momentum. Therefore equating equations (i) and (iii),
$$wHa = \frac{wa_c v^2}{g}$$
∴ $$Ha = \frac{a_c v^2}{g}$$
$$\frac{v^2}{2g} \times a = \frac{a_c v^2}{g} \qquad \left(\because H = \frac{v^2}{2g}\right)$$
or $$a = 2a_c \quad \text{or} \quad \frac{a_c}{a} = \frac{1}{2} = 0.5$$

∴ Coefficient of contraction, $C_c = \dfrac{a_c}{a} = 0.5$

We know that coefficient of discharge (assuming C_v equal to 1),
$$C_d = C_v \times C_c = 1 \times 0.5 = 0.5$$
∴ Discharge, $Q = C_d \cdot a \sqrt{2gH} = 0.5\, a \sqrt{2gH}$

Example 10·8. *A Borda's mouthpiece of 50 mm diameter is provided on one side of a tank containing water up to a height of 3 metres above the centre line or the orifice. Find the discharge through the mouthpiece, if the mouthpiece is running free.*

Solution. Given : $d = 50$ mm $= 0.05$ m and $H = 3$ m.

---

*Newton's Second Law of Motion states, "The rate of change of momentum is directly proportional to the impressed force and takes place in the same direction in which the force acts."

We know that the area of the mouthpiece,

$$a = \frac{\pi}{4} \times (d)^2 = \frac{\pi}{4} \times (0.05)^2 = 1.964 \times 10^{-3} \text{ m}^2$$

and discharge through the mouthpiece running free,

$$Q = 0.5 \, a \, \sqrt{2gH} = 0.5 \times (1.964 \times 10^{-3}) \times \sqrt{2 \times 9.81 \times 3} \text{ m}^3/\text{s}$$

$$= 7.53 \times 10^{-3} \text{ m}^3/\text{s } 7.53 \text{ litres/s. } \textbf{Ans.}$$

10·13 Mouthpiece Running Full

We have discussed in Art. 10·11 that if the jet, after contraction, expands and fills up the whole mouthpiece, it is said to be running full. Now consider a mouthpiece running full as shown in Fig. 10·7.

Let H = Height of the liquid above the mouthpiece,

a_c = Area of the flow at the vena contracta,

v = Velocity of the liquid at the outlet, and

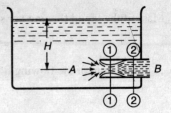

Fig. 10·7. Mouthpiece running full.

v_c = Velocity of the liquid at the vena contracta.

Since the liquid is flowing continuously, therefore

$$v_c \cdot a_c = v \cdot a \quad \text{or} \quad v_c = \frac{v \cdot a}{a_c}$$

We have seen in Art. 10·12 that the coefficient of contraction in an internal mouthpiece is 0·5. Therefore substituting the value of $\dfrac{a_c}{a} = 0.5$ or $\dfrac{a}{a_c} = 2$ in the above equation,

$$v_c = 2v \qquad \qquad \ldots(i)$$

We see that the jet after passing through section 1-1 suddenly enlarges at section 2-2. Therefore there will be a loss of head due to sudden enlargement. We know that the loss of head due to sudden enlargement,

$$h_c = \frac{(v_c - v)^2}{2g} = \frac{(2v - v)^2}{2g} = \frac{v^2}{2g} \qquad \ldots (\because v_c = 2v)$$

Applying Bernoulli's equation to points A and B,

$$H = \frac{v^2}{2g} + \text{Losses} = \frac{v^2}{2g} + \frac{v^2}{2g} = \frac{2v^2}{2g} = \frac{v^2}{g}$$

or

$$v = \sqrt{gH}$$

∴ Actual discharge $= a \times \sqrt{gH}$

We know that theoretical discharge,
$$= a\sqrt{2gH}$$

∴ Coefficient of discharge,
$$C_d = \frac{\text{Actual discharge}}{\text{Theoretical discharge}} = \frac{a\sqrt{gH}}{a\sqrt{2gH}} = \frac{1}{\sqrt{2}} = 0.707$$

and discharge, $Q = C_d a \sqrt{2gH} = 0.707 \, a \sqrt{2gH}$

Notes : 1. It has been experimentally found that in this case, the actual value of coefficient of discharge is slightly more than 0·707. But for all practical purposes, the value of coefficient of discharge is taken as 0·707.

2. We have seen in Arts. 10·12 and 10·13 that in the case of internal mouthpieces, the coefficients of discharge are less than that of the external mouthpiece. The reason, for the same, is that in the case of external mouthpieces the liquid particles have to deviate maximum through an angle of 90°. But in the case of internal mouthpieces the liquid particles have to deviate maximum through an angle of 180°. Due to more angle of deviation of the liquid particles, the contraction of the jet is more in the case of internal mouthpieces than that in the external mouthpieces.

Example 10·9. *An internal mouthpiece of 80 mm diameter is discharging water under a head of 4 metres. Find the discharge in litres/s through the mouthpiece, when (a) the mouthpiece is running free, and (b) the mouthpiece is running full.*

Solution. Given : $d = 80$ mm $= 0.8$ m and $H = 4$ m.

(a) Discharge through the mouthpiece when running free

We know that the area of the mouthpiece,
$$a = \frac{\pi}{4} \times (d)^2 = \frac{\pi}{4} \times (0.08)^2 = 5.027 \times 10^{-3} \text{ m}^2$$

and discharge through the internal mouthpiece when running free,
$$Q = 0.5 \, a \sqrt{2gH} = 0.5 \times (5.027 \times 10^{-3}) \times \sqrt{2 \times 9.81 \times 4} \text{ m}^3/\text{s}$$
$$= 0.0223 \text{ m}^3/\text{s} = 22.3 \text{ litres/s} \quad \textbf{Ans.}$$

(b) Discharge through the mouthpiece when running full

We also know that discharge through the internal mouthpiece when running full,
$$Q = 0.707 \, a \sqrt{2gH} = 0.707 \times (5.027 \times 10^{-3}) \times \sqrt{2 \times 9.81 \times 4} \text{ m}^3/\text{s}$$
$$= 0.0315 \text{ m}^3/\text{s} = 31.5 \text{ litres/s} \quad \textbf{Ans.}$$

10·14 Discharge through a Convergent Mouthpiece

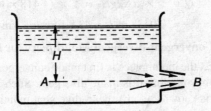

Fig. 10·8. Convergent mouthpiece.

We have seen in Arts. 10·10 and 10·13 that a jet of a liquid gets contracted at the entrance of a mouthpiece. There is always some loss of head due to this contraction of the jet. To counteract this loss of head, due to contraction of the jet up to vena contracta, the mouthpiece is given the same slope as that of the jet of liquid as shown in Fig. 10·8.

In such a mouthpiece, there will be no loss of head due to expansion. If such a mouthpiece terminates at the vena contracta of the jet, then the mouthpiece is called convergent mouthpiece.

Let
H = Height of the liquid above the mouthpiece
a = Area of the orifice at point B, and
v = Velocity of the jet.

Applying Bernoulli's equation to points A and B,

$$H = \frac{v^2}{2g}$$ ...($\because$ There is no loss of head)

or $v = \sqrt{2gH}$

$\therefore$ Actual discharge, Q = Area of orifice × Actual velocity = $a\sqrt{2gH}$ ...(i)

We know that the actual discharge
= Coefficient discharge × Area of orifice × Theoretical velocity
= $C_d \cdot a\sqrt{2gH}$ ...(ii)

Equating equations (i) and (ii),
$a\sqrt{2gH} = C_d \cdot a\sqrt{2gH}$

$\therefore$ Coefficient of discharge,

$$C_d = \frac{a\sqrt{2gH}}{a\sqrt{2gH}} = 1$$

and discharge $Q = a\sqrt{2gH}$

Note : It has been experimentally found that there is some loss of head at the entrance of the mouthpiece. This loss of head reduces the coefficient of discharge by a small amount (sometimes up to 0·975). But, for all practical purposes, the value of coefficient of discharge is taken as 1.

Example 10·10. *A convergent mouthpiece is discharging water under a constant head of 5 metres. Find the discharge, if diameter of the mouthpiece is 75 mm.*

Solution. Given : H = 5 m and d = 75 mm = 0·075 m.

We know that the area of the mouthpiece,

$$a = \frac{\pi}{4} \times (d)^2 = \frac{\pi}{4} \times (0.075)^2 = 4.418 \times 10^{-3} \text{ m}^2$$

and discharge through the convergent mouthpiece,

$Q = 1 \times a\sqrt{2gH} = 1 \times (4.418 \times 10^{-3}) \times \sqrt{2 \times 9.81 \times 5} \text{ m}^3/\text{s}$

= 0·0438 m³/s = 43·8 litres/s **Ans.**

10·15 Discharge through a Convergent-divergent Mouthpiece (or Bell-Mouthpiece)

In this type of mouthpiece, the mouthpiece is first made convergent up to the vena contracta of the jet (as in Art. 10·14) and beyond that it is made divergent. Such a mouthpiece, which is first convergent and then divergent is known as convergent–divergent mouthpiece as shown in Fig. 10·9.

In such a mouthpiece, there will be no loss of head due to sudden expansion. The *coefficient of discharge in the case of convergent-divergent mouthpiece is also 1. The diameter of the mouthpiece, for the purpose of calculating the discharge, is taken at the vena contracta *i.e.,* at B (or

---

*It may be derived in the same way as we did in Art. 10.14.

in other words where the convergent and divergent pieces meet). It is also known as throat diameter of the mouthpiece.

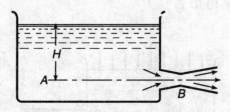

Fig. 10·9. Convergent-divergent mouthpiece.

Example 10·17. *A convergent-divergent mouthpiece having 80 mm throat diameter is discharging water under a constant head of 1·5 metre. Determine the discharge through the mouthpiece.*

Solution. Given : $d = 80$ mm $= 0·08$ m and $H = 4·5$ m.

We know that the area of the mouthpiece at its throat,

$$a = \frac{\pi}{4} \times (d)^2 = \frac{\pi}{4} \times (0·08)^2 = 5·027 \times 10^{-3} \text{ m}^2$$

and discharge through the convergent-divergent mouthpiece,

$$Q = 1 \times a \times \sqrt{2gH} = 1 \times (5·027 \times 10^{-3}) \times \sqrt{2 \times 9·81 \times 4·5} \text{ m}^3/\text{s}$$
$$= 47·2 \times 10^{-3} \text{ m}^3/\text{s} = 47·2 \text{ litres/s} \quad \textbf{Ans.}$$

EXERCISE 10·2

1. Find the discharge through an external mouthpiece of 80 mm diameter under a head of 2·5 m.
   [**Ans.** 30·1 litres/s]
2. An internal mouthpiece of 60 mm diameter is fitted to the vertical side of a tank containing water up to a height of 4 m. Find the discharge through the mouthpiece, when it is running free.
   [**Ans.** 12·5 litres/s]
3. A Borda's mouthpiece of 90 mm diameter is provided in one side of a tank and is running full. If water in the tank is 3 m above the centre line of the mouthpiece, find the discharge in litres/s.
   [**Ans.** 34·5 litres/s]
4. An internal mouthpiece of 100 mm diameter is discharging water under a head of 1·5 m. Determine the discharge through the mouthpiece, when it is running (*i*) free and (*ii*) full.
   [**Ans.** 30·1 litres/s ; 42·6 litres/s]
5. An external convergent-divergent mouthpiece is of 75 mm diameter. Find the discharge, if the head of water over the mouthpiece is 5 m. [**Ans.** 43·8 litres/s]

10·16 Pressure in a Mouthpiece

In the previous articles, we have discussed the discharge through the various types of mouthpieces. We have also seen that the discharge, through a mouthpiece, depends upon the coefficient of discharge for the particular mouthpiece; which is different for different types of mouthpieces. While deriving the value of coefficient of discharge, we have not taken into consideration the atmospheric pressure over the vessel to which the orifice is fitted. In the succeeding articles, we shall find out the pressure in the following mouthpieces, taking into consideration the atmospheric pressure also.

1. Pressure in an external mouthpiece,
2. Pressure in an internal mouthpiece,
3. Pressure in a convergent mouthpiece,
4. Pressure in a convergent-divergent mouthpiece.

10·17 Pressure in an External Mouthpiece

Consider a vessel, open to the atmosphere at its top, and having an orifice fitted with an external mouthpiece as shown in Fig. 10·10.

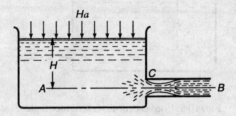

Fig. 10·10. Pressure in an external mouthpiece.

We know that the jet of the liquid, on entering the mouthpiece, will first contract up to the vena contracta C and then expand and fill up the whole orifice as shown in Fig. 10·10.

Let
H_a = Atmospheric pressure head,
H = Height of liquid above the mouthpiece,
H_c = Absolute pressure head at vena contracta,
v_c = Velocity of liquid at vena contracta, and
v = Velocity of liquid at outlet.

Applying Bernoulli's equation to points A and C,

$$H_a + H = H_c + \frac{v_c^2}{2g} \qquad \text{...(Taking } Z = Z_c\text{)}$$

or
$$H_c = H_a + H - \frac{v_c^2}{2g}$$

Assuming coefficient of contraction to be equal to 0·62,

$$H_c = H_a + H - \left(\frac{v}{0.62}\right)^2 \times \frac{1}{2g}$$

$$= H_a + H - \left(\frac{v^2}{2g} \times \frac{1}{(0.62)^2}\right) \qquad \text{...(}i\text{)}$$

We have seen in Art. 10·10 that

$$H = \frac{1.375 \, v^2}{2g} \qquad \text{or} \qquad \frac{v^2}{2g} = \frac{H}{1.375}$$

Substituting the value of $\frac{v^2}{2g}$ in equation (i),

$$H_c = H_a + H - \left(\frac{H}{1.375} \times \frac{1}{(0.62)^2}\right) = H_a + H - 1.89 \, H = H_a - 0.89 \, H$$

This shows that the absolute pressure head at vena contracta is less than the atmospheric pressure by an amount equal to 0·89 times the height of the liquid above the vena contracta. But, in actual practice, due to some loss of head at the entrance of the mouthpiece, the absolute pressure head at vena contracta reduces it by an amount equal to 0·74 times the height of the liquid at the vena contracta. But, for all practical purposes, the absolute pressure at vena contracta is taken as :

$$H_c = H_a - 0.89 \, H$$

A little consideration will show that absolute pressure head, at vena contracta, will be zero, when
$$0.89 H = H_a = 10.3 \text{ metres of water}$$
$$\therefore \quad H = \frac{10.3}{0.89} = 11.5 \text{ metres of water.}$$

It means that if this condition is reached (*i.e.*, height of water above the mouthpiece reaches 11.5 m), the flow of the liquid would be no longer steady. This means that the mouthpiece will not run full at its outlet, and the jet will flow straight without touching the walls of the mouthpiece. But, in actual practice, this condition reaches at 7.5 metres only.

Example 10.12. *Water under a constant head of 5 metres is discharging through an external mouthpiece of 100 mm diameter. Determine the absolute pressure head of water at the vena contracta. Take atmospheric pressure as 10.3 m of water.*

Solution. Given : $H = 5$ m ; *$d = 100$ mm $= 0.1$ m and $H_a = 10.3$ m.

We know that absolute pressure head of water at vena contracta of an external mouthpiece,
$$H_c = H_a - 0.89 H = 10.3 - (0.89 \times 5) = 5.88 \text{ m} \quad \textbf{Ans.}$$

10.18 Pressure in an Internal Mouthpiece

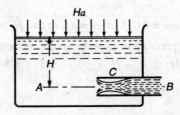

Fig. 10.11. Pressure in an internal mouthpiece.

Consider a vessel, open to atmosphere at its top, having an orifice with an internal mouthpiece, as shown in Fig. 10.11. We know that the jet of liquid, on entering the mouthpiece, will first contract up to vena contracta C and then expand and fill up the whole orifice as shown in Fig. 10.11.

Let
H_a = Atmospheric pressure head,
H = Height of liquid above the mouthpiece,
H_c = Absolute pressure head at vena contracta,
v_c = Velocity of liquid at vena contracta, and
v = Velocity of liquid at outlet.

Applying Bernoulli's equation to points A and C,
$$H_a + H = H_c + \frac{v_c^2}{2g} \qquad \qquad ...(\text{Taking } Z_a = Z_c)$$

or
$$H_c = H_a + H - \frac{v_c^2}{2g} = H_a + H - \left(\frac{v}{0.5}\right)^2 \times \frac{1}{2g} \qquad ...\left(\because v_c = \frac{v}{0.5}\right)$$
$$= H_a + H - \frac{4v^2}{2g} = H_a + H - \frac{2v^2}{g} \qquad \qquad ...(i)$$

---

*Superfluous data.

We have seen in Art. 10·13 that

$$\frac{v^2}{g} = H$$

Substituting the value of $\frac{v^2}{g}$ equation (i),

$$H_c = H_a + H - 2H = H_a - H$$

This shows that the absolute pressure head at vena contracta is less than the atmopheric pressure by an amount equal to height of the liquid above the vena contracta. A little consideration will show that the absolute pressure head at vena contracta will be zero, when

$$H = H_a = 10\cdot3 \text{ metres of water.}$$

It means that if this condition is reached (*i.e.*, height of water above the mouthpiece reaches 10·3 m), the flow of the liquid would be no longer steady. This means that the mouthpiece will not run full at its outlet and the jet will flow straight without touching the walls of the mouthpiece. But, in actual practice, this condition also reaches at 7·5 metres.

Example 10·13. *A re-entrant mouthpiece of 75 mm diameter is discharging water under a constant head of 3·5 metres. Determine the absolute pressure head at vena contracta, if the atmosphere pressure head be 10·3 metres of water.*

Solution. Given : *d = 75 mm = 0·075 m; H = 3·5 m and H_a = 10·3 m.

We know that the absolute pressure head at the vena contracta of an internal mouthpiece,

$$H_c = H_a - H = 10\cdot3 - 3\cdot5 = 6\cdot8 \text{ m} \quad \textbf{Ans.}$$

10·19 Pressure in a Convergent Mouthpiece

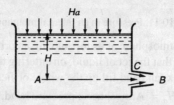

Fig. 10·12. Pressure in convergent mouthpiece.

Consider a vessel, open to atmosphere at its top, having an orifice fitted with a convergent mouthpiece as shown in Fig. 10·12. We know that the slope of the mouthpiece is the same as that of the jet up to vena contracta.

Let H_a = Atmospheric pressure head,

H = Height of liquid above the mouthpiece,

H_c = Absolute pressure head at vena contracta, and

v = Velocity of the jet.

Applying Bernoulli's equation to points *A* and *C*,

$$H_a + H = H_c + \frac{v^2}{2g} \qquad \ldots$$

...[Taking $Z_a = Z_c$]

---

*Superfluous data.

or $$H_c = H_a + H - \frac{v^2}{2g} \qquad ...(i)$$

We have seen in Art. 10·14 that
$$H = \frac{v^2}{2g}$$

Now substituting this value of H in equation (i),
$$H_c = H_a + H - H = H_a$$

This shows that the absolute pressure head at the vena contracta is the same as that of the atmosphere. But, in actual practice, the flow is no longer a steady one.

10·20 Pressure in a Convergent-divergent Mouthpiece

Consider a vessel open to atmosphere at its top, having an orifice fitted with a convergent-divergent mouthpiece as shown in Fig. 10·13. We know that the slope of the mouthpiece is the same as that of the jet up to vena contracta, and beyond that is it made divergent. The theoretical absolute pressure head at vena contracta is the same as that of atmospheric pressure head (as per Art. 10·19).

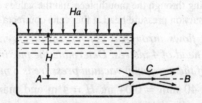

Fig. 10·13. Pressure in convergent-divergent mouthpiece.

The pressure at the outlet of the mouthpiece is atmospheric. We know that the jet will expand from vena contracta (*i.e.*, C) to outlet of the tube (*i.e.*, B). For a steady flow, through the outlet of the mouthpiece, the shape of the divergent portion is made according to the profile of the expanding jet. However, if the divergence is made too large, the jet will not touch the walls of the mouthpiece.

Let H_a = Atmospheric pressure head,

H = Height of liquid above the mouthpiece,

H_c = Absolute pressure head at vena contracta,

v_c = Velocity of liquid at vena contracta,

v = Velocity of liquid at outlet,

a_c = Area of mouthpiece at vena contracta, and

a = Area of mouthpiece at outlet.

We have seen in Art. 10·14 that
$$H = \frac{v^2}{2g} \qquad \text{or} \qquad v = \sqrt{2gH}$$

Now applying Bernoulli's equation to points C and B,
$$H_c + \frac{v_c^2}{2g} = H_a + \frac{v^2}{2g} \qquad ...(\text{Taking } Z_C = Z_B)...(i)$$

Substituting the value of $\frac{v^2}{2g}$ in equation (i),
$$H_c + \frac{v_c^2}{2g} = H_a + H$$

$$\therefore \quad \frac{v_c^2}{2g} = H_a + H - H_c$$

or $\quad v_c = \sqrt{2g(H_a + H - H_c)}$

Since the flow of the liquid is continuous, therefore

$$a_c \cdot v_c = a \cdot v$$

or $\quad \dfrac{a}{a_c} = \dfrac{v_c}{v} = \dfrac{\sqrt{2g(H_a + H - H_c)}}{\sqrt{2gH}} = \sqrt{1 + \left(\dfrac{H_a - H_c}{H}\right)}$

The above expression gives the ratio of areas of divergence to the convergence of the mouthpieces. If d and d_c be the diameters of the mouthpiece at outlet and vena contracta (i.e., convergence), then this expression can also be expressed as :

$$\frac{\frac{\pi}{4} \times d^2}{\frac{\pi}{4} \times d_c^2} = \sqrt{1 + \left(\frac{H_a - H_c}{H}\right)} \quad \text{or} \quad \frac{d^2}{d_c^2} = \sqrt{1 + \left(\frac{H_a - H_c}{H}\right)}$$

Note : If the water is flowing through the mouthpiece and the values of H_a or H_c are not given, then the value of $(H_a - H_c)$ i.e., value of suction pressure head at the vena contracta is taken as 7·8 m.

Example 10·14. *Water flows through a convergent-divergent mouthpiece of diameter at convergence 40 mm, under a head of 4 metres. Determine the maximum diameter of divergence to avoid separation of the flow, if the maximum vacuum pressure is 8 metres of water.*

Solution. Given : $d_c = 40$ mm $= 0.04$ m; $H = 4$ m and maximum vacuum pressure head $(H_a - H_c) = 8$ m.

Let $\quad d = $ Maximum diameter of the divergence in metres.

We know that

$$\frac{d^2}{d_c^2} = \sqrt{1 + \left(\frac{H_a - H_c}{H}\right)} \quad \text{or} \quad \frac{d^2}{(0.04)^2} = \sqrt{1 + \frac{8}{4}} = \sqrt{3} = 1.732$$

$\therefore \quad d^2 = 1.732 \times (0.04)^2 = 2.77 \times 10^{-3} = 27.7 \times 10^{-4}$

or $\quad d = \sqrt{27.7 \times 10^{-4}} = 5.26 \times 10^{-2}$ m $= 52.6$ mm **Ans.**

Example 11·15. *A convergent-divergent mouthpiece is fitted to the side of a tank. If the diameters at the outlet and vena contracta be 30 mm and 20 mm respectively, find the maximum head of water for steady flow. Take atmospheric pressure as 10 metres and separation pressure as 2·5 metres of water.*

Solution. Given : $d = 30$ mm $= 0.03$ m; $d_c = 20$ mm $= 0.02$ m; $H_a = 10$ m and $H_c = 2.5$ m.

Let $\quad H = $ Maximum head of water in metres.

We know that

$$\frac{d^2}{d_c^2} = \sqrt{1 + \frac{H_a - H_c}{H}} \quad \text{or} \quad \frac{(0.03)^2}{(0.02)^2} = \sqrt{1 + \frac{10 - 2.5}{H}}$$

$\therefore \quad 2.25 = \sqrt{1 + \dfrac{7.5}{H}}$

Squaring both sides of the above equation,

$$5.0625 = 1 + \frac{7.5}{H} \quad \text{or} \quad \frac{7.5}{H} = 5.0625 - 1 = 4.0625$$

$$H = 7.5 / 4.0625 = 1.85 \text{ m} \text{ **Ans.**}$$

EXERCISE 10.3

1. An external mouthpiece is discharging water under a constant head of 10 m. Determine the absolute pressure head at vena contracta, if the atmospheric pressure head is 10.3 m of water. [**Ans.** 1.4 m]

2. A re-entrant mouthpiece of 50 mm diameter is discharging water under a head of 7.5 m. Find the absolute pressure head at vena contracta, if the barometer reads 10.2 metres. [**Ans.** 2.7 m]
   [**Hint.** Barometer reading means atmospheric pressure head.]

3. A convergent-divergent mouthpiece having 25 mm diameter at the convergence is discharging water under a head of 4 m. Calculate the maximum diameter of divergence to avoid separation, which takes place at 2.1 m of water. Assume the atmospheric head as 10 m of water. [**Ans.** 32.8 mm]

4. A convergent-divergent mouthpiece discharges water under a constant head of 5 m. Find the throat and exit diametres, if the discharge through the mouthpiece is 7 litres/s and the maximum vacuum pressure is 7.5 m of water. [**Ans.** 30 mm; 42.4 mm]

   [**Hint**: $7 \times 10^{-3} = 1 \times a \times \sqrt{2g \times 5} = a \times \sqrt{98.1} = 9.9a$

   $a = \dfrac{7 \times 10^{-3}}{9.9} = 0.707 \times 10^{-3}$ or $d_c = 0.03$ m = 30 mm

∴ $\dfrac{d^2}{d_c^2} = \sqrt{1 + \dfrac{H_a - H_c}{H}} = \sqrt{1 + \dfrac{7.5}{2.5}} = \sqrt{4} = 2$

or $d^2 = (0.03)^2 \times 2 = 0.18 \times 10^{-4}$ or $d = 42.4$ mm]

QUESTIONS

1. Obtain an expression for the loss of head due to sudden enlargement of area.
2. Derive a relation for the head lost due to sudden contraction in the diameter of a pipe.
3. Sketch the different types of mouthpieces.
4. Obtain an expression for the coefficient of discharge for an external mouthpiece.
5. Explain Borda's mouthpiece with the help of a neat sketch.
6. What is a convergent-divergent mouthpiece? Where is it used?

OBJECTIVE TYPE QUESTIONS

1. The loss of head due to sudden enlargement in a pipe is equal to
   (a) $\dfrac{v_1 - v_2}{2g}$ (b) $\dfrac{(v_1 - v_2)^2}{2g}$ (c) $\dfrac{v_1^2 - v_2^2}{2g}$ (d) $\dfrac{v_1^2 + v_2^2}{2g}$

2. The ratio of loss of head at entrance to that at the exit of pipe is
   (a) 0.375 (b) 0.4 (c) 0.5 (d) 0.855

3. The value of coefficient of discharge for an external mouthpiece is
   (a) 0.375 (b) 0.5 (c) 0.855 (d) 1.0

4. Two internal mouthpieces of the same diameter are discharging water under the same head. One of them is running free and the other full. How much extra water will be discharged by the mouthpiece running full to that running free?
   (a) 20% (b) 30% (c) 40% (d) 50%

5. If a convergent mouthpiece is replaced by a convergent-divergent mouthpiece, the discharge through the mouthpiece will
   (a) decrease
   (b) remain the same
   (c) increase
   (d) depend upon the head of water

ANSWERS

1. (b)   2. (c)   3. (c)   4. (c)   5. (b)

11

Flow Over Notches

1. Introduction. 2. Types of Notches. 3. Discharge over a Rectangular Notch. 4. Discharge over a Triangular Notch. 5. Advantages of a Trinagular Notch over a Rectangular Notch. 6. Discharge over a Trapezoidal Notch. 7. Discharge over a Stepped Notch. 8. Time of Emptying a Tank over a Rectangular Notch. 9. Time of Emptying a Tank over a Triangular Notch. 10. Effect on the Discharge over a Notch, due to an Error in the Measurement of Head. 11. Effect on the Discharge over a Rectangular Notch, due to an Error in the Measurement of Head. 12. Effect on the Discharge over a Triangular Notch due to an Error in the Measurement of Head.

11·1 Introduction

A notch may be defined as an opening in one side of a tank or a reservoir, like a large orifice, with the upstream liquid level below the top edge of the opening as shown in Fig. 11·1.

Since the top edge of the notch above the liquid level serves no purpose, therefore a notch may have only the bottom edge and sides. The bottom edge, over which the liquid flows, is known as sill or crest of the notch and the sheet of liquid flowing over a notch (or a weir) is known as nappe or vein. A notch is, usually, made of a metallic plate and is used to measure the discharge of liquids.

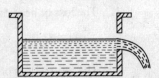

Fig. 11·1. A Notch.

11·2 Types of Notches

There are many types of notches, depending upon their shapes. But the following are important from the subject point of view :

1. Rectangular notch,
2. Triangular notch,
3. Trapezoidal notch, and
4. Stepped notch.

In this chapter, we shall discuss the discharge or the phenomenon associated with discharge over all these notches one-by-one.

11·3 Discharge over a Rectangular Notch

Consider a rectangular notch in one side of a tank over which water is flowing as shown in Fig. 11·2.

Let H = Height of water above sill of the notch,
b = Width or length of the notch, and
C_d = Coefficient of discharge.

[194]

Flow Over Notches

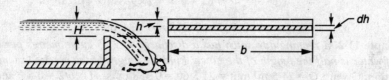

Fig. 11·2. Rectangular notch.

Let us consider a horizontal strip of water of thickness dh at a depth of h from the water level as shown in Fig. 11·2.

∴ Area of the strip $= b.dh$ ... (i)

We know that the theoretical velocity of water through the strip

$$= \sqrt{2gh} \quad \ldots (ii)$$

and discharge through the strip,

$$dq = C_d \times \text{Area of strip} \times \text{Theoretical velocity}$$
$$= C_d . b.dh \sqrt{2gh} \quad \ldots (iii)$$

The total discharge, over the whole notch, may be found out by integrating the above equation within the limits 0 and H.

∴
$$Q = \int_0^H C_d . b.dh \sqrt{2gh}$$

$$= C_d . b.dh \sqrt{2g} \int_0^H h^{1/2} \, dh$$

$$= C_d . b \sqrt{2g} \left[\frac{h^{3/2}}{\frac{3}{2}} \right]_0^H$$

$$= \frac{2}{3} C_d . b \sqrt{2g} \left[h^{3/2} \right]_0^H$$

$$= \frac{2}{3} C_d . b \sqrt{2g} \, (H)^{3/2}$$

Note : Sometimes, the limits of integration, in the above equation, are from H_1 to H_2 (i.e., the liquid level is at a height of H_1 above the top of the notch and H_2 above the bottom of the notch, instead of 0 to H. Then the discharge over such a notch will be given by the equation :

$$Q = \frac{2}{3} Cd.b \sqrt{2g} \left(H_2^{3/2} - H_1^{3/2} \right)$$

Example 11·1. *A rectangular notch 0·5 metres wide has a constant head of 400 mm. Find the discharge over the notch in litres per second, if the coefficient of discharge for the notch is 0·62.*

Solution. Given : $b = 0.5$ m; $H = 400$ mm $= 0.4$ m and $C_d = 0.62$.

We know that the discharge over the rectangular notch,

$$Q = \frac{2}{3} Cd.b \sqrt{2g} \, (H)^{3/2} \text{ m}^3/\text{s}$$

$$Q = \frac{2}{3} \times 0.62 \times 0.5 \sqrt{2 \times 9.81} (0.4)^{3/2} \text{ m}^3/\text{s}$$

$$= 0.915 \times 0.253 = 0.231 \text{ m}^3/\text{s} = 231 \text{ litres/s} \quad \textbf{Ans.}$$

Example 11·2. *A rectangular notch has a discharge of 21·5 cubic metres per minute, when the head of water is half the length of the notch. Find the length of the notch. Assume $C_d = 0.6$.*

Solution. Given : $Q = 21.5$ m³/min $= 21.5/60 = 0.358$ m³/s; $H = b/2 = 0.5\,b$ and $C_d = 0.6$.

We know that the discharge over the rectangular notch (Q),

$$0.358 = \frac{2}{3} C_d \cdot b \sqrt{2g}\, (H)^{3/2}$$

$$= \frac{2}{3} \times 0.6 \times b \sqrt{2 \times 9.81} \left(\frac{b}{2}\right)^{3/2} = 0.626\, b^{5/2}$$

∴ $\quad b^{5/2} = 0.358/0.626 = 0.572 \quad$ or $\quad b = 0.8$ m $\quad$ **Ans.**

11·4 Discharge over a Triangular Notch

A triangular notch is also called a *V*-notch. Consider a triangular notch, in one side of the tank, over which water is flowing as shown in Fig. 11·3.

Let $\qquad H = $ Height of the liquid above the apex of the notch,

$\theta = $ Angle of the notch, and

$C_d = $ Coefficient of discharge.

From the geometry of the figure, we find that the width of the notch at the water surface

$$= 2H \tan \frac{\theta}{2}$$

∴ $\quad$ Area of the strip $= 2\,(H-h) \tan \dfrac{\theta}{2} \cdot dh$

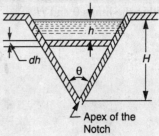

Fig. 11·3. Triangular notch.

We know that the theoretical velocity of water through the strip

$$= \sqrt{2gh}$$

and discharge over the notch,

$$dq = C_d \times \text{Area of strip} \times \text{Theoretical velocity}$$

$$= C_d \times 2\,(H-h) \tan \frac{\theta}{2}\, dh\, \sqrt{2gh}$$

The total discharge, over the whole notch, may be found out by integrating the above equation within the limits 0 and H.

∴ $$Q = \int_0^H C_d \cdot 2(H-h) \tan \frac{\theta}{2}\, dh \cdot \sqrt{2gh}$$

$$= 2 C_d \sqrt{2g} \tan \frac{\theta}{2} \int_0^H (H-h) \sqrt{h}\; dh$$

$$= 2 C_d \sqrt{2g} \tan \frac{\theta}{2} \int_0^H \left[(Hh^{1/2} - h)^{3/2}\right] dh$$

Flow Over Notches

$$Q = 2C_d \sqrt{2g} \tan\frac{\theta}{2} \left[\frac{H.h^{3/2}}{\frac{3}{2}} - \frac{h^{5/2}}{\frac{5}{2}}\right]_0^H$$

$$= \frac{8}{15} C_d \sqrt{2g} \tan\frac{\theta}{2} \times H^{5/2}$$

Cor. If $\theta = 90°$, $C_d = 0.6$ and $g = 9.81$ m/s², then $Q = 1.417\, H^{5/2}$

11·5 Advantages of a Triangular Notch over a Rectangular Notch

Following are the advantages of a triangular notch over a rectangular notch :
1. Only one reading i.e., head (H) is required to be taken for the measurement of discharge in a given triangular notch.
2. If, in a triangular notch the angle of the notch i.e., $\theta = 90°$, the formula becomes very simple (i.e., $Q = 1.417\, H^{5/2}$) to remember.
3. A traingular notch gives more accurate results for low discharges than a rectangular notch.
4. The same triangular notch can measure a wide range of flows accurately.

Example 11·3. *A right-angled V-notch was used to measure the discharge of a centrifugal pump. If the depth of water at V-notch is 200 mm, calculate the discharge over the notch in litres per minute. Assume coefficient of discharge as 0·62.*

Solution. Given $\theta = 90°$; $H = 200$ mm $= 0·2$ m and $C_d = 0·62$.
We know that the discharge over the triangular notch,

$$Q = \frac{8}{15} C_d \sqrt{2g} \tan\frac{\theta}{2} \times H^{5/2}$$

$$= \frac{8}{15} \times 0·62 \times \sqrt{2 \times 9·81}\ \tan 45° \times (0·2)^{5/2}\ \text{m}^3/\text{s}$$

$$= 1·465 \times 0·018 = 0·026\ \text{m}^3/\text{s}$$

$$= 26\ \text{litres/s} = 1560\ \text{litres/min}\quad \textbf{Ans.}$$

Example 11·4. *During an experiment in a laboratory, 280 litres of water flowing over a right-angled notch was collected in one minute. If the head of water over the sill is 100 mm, calculate the coefficient of discharge of the notch.*

Solution. Given : $Q = 280$ litres/min $= 0·28$ m³/min $= 0·0047$ m³/s and $H = 100$ mm $= 0·1$ m.
We know that the discharge over the triangular notch (Q),

$$0·0047 = \frac{8}{15} C_d \sqrt{2g} \tan\frac{\theta}{2} \times H^{5/2}$$

$$= \frac{8}{15} C_d \sqrt{2 \times 981}\ \tan 45° \times (0·1)^{5/2} = 0·0075\, C_d$$

∴ $C_d = 0·0047 / 0·0075 = 0·627$ **Ans.**

Example 11·5. *A rectangular channel of 1·5 metre width is used to carry 0·2 cubic metre of water. The rate of flow is measured by placing a 90° V-notch weir. If the maximum depth of water is not to exceed 1·2 metres, find the position of the apex of the notch from the bed of the channel. Assume $C_d = 0·6$.*

Solution. Given : *Width of rectangular notch (b) = 1·5 m; $Q = 0·2$ m³/s; $\theta = 90°$ and $C_d = 0·6$.
Let H = Height of water above the apex of notch.

---

* Superfluous data.

We know that discharge over the triangular notch (Q),

$$0 \cdot 2 = \frac{8}{15} C_d \sqrt{2g} \tan \frac{\theta}{2} \times H^{5/2}$$

$$= \frac{8}{15} \times 0 \cdot 6 \sqrt{2 \times 9 \cdot 81} \tan 45° \times H^{5/2} = 1 \cdot 417 \, H^{5/2}$$

$\therefore \quad H^{5/2} = 0 \cdot 2/1 \cdot 417 = 0141 \quad$ or $\quad H = 0 \cdot 46 \text{ m}$

Since maximum depth of water in the channel is 1·2 m, therefore apex of the triangular notch is to be kept at a height of 1·2 − 0·46 = 0·74 m = 740 mm from the bed of the channel. **Ans.**

Example 11·6. *Water flows over a rectangular notch of 1 metre length over a depth of 150 mm. Then the same quantity of water passes through a triangular right-angled notch. Find the depth of water through the notch.*

Take the coefficeints of discharges for the rectangular and triangular notch as 0·62 and 0·59 respectively.

Solution. Given : For rectangular notch : b = 1 m; H_1 = 150 mm = 0·15 m and C_d = 0·62. For triangular notch : θ = 90° and C_d = 0·59.

First of all, consider the flow of water over the rectangular notch. We know that the discharge over the rectangular notch,

$$Q = \frac{2}{3} C_d \cdot b \sqrt{2g} \, (H)^{5/2}$$

$$= \frac{2}{3} \times 0 \cdot 62 \times 1 \times \sqrt{2 \times 9 \cdot 81} \times (0 \cdot 15)^{5/2} \text{ m}^3/\text{s}$$

$$= 1 \cdot 831 \times 0 \cdot 058 = 0 \cdot 106 \text{ m}^3/\text{s}$$

Now consider the flow of water over the triangular notch. We know that discharge over the triangular notch, (Q),

$$0.106 = \frac{8}{15} C_d \sqrt{2g} \tan \frac{\theta}{2} \times H_2^{5/2}$$

$$= \frac{8}{15} \times 0 \cdot 59 \sqrt{2 \times 9 \cdot 81} \tan 45° \times H_2^{5/2} = 1 \cdot 394 \, H_2^{5/2}$$

$\therefore \quad H_2^{5/2} = 0 \cdot 106/1 \cdot 394 = 0 \cdot 076 \quad$ or $\quad H_2 = 0 \cdot 357 \text{ m} \quad$ **Ans.**

11·6 Discharge over a Trapezoidal Notch

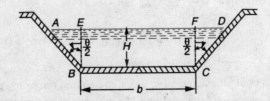

Fig. 11·4. Trapezoidal notch.

A trapezoidal notch is a combination of a rectangular notch and two triangular notches as shown in Fig. 11·4. It is, thus, obvious that the discharge over such a notch will be the sum of the discharges over the rectangular and triangular notches.

Flow Over Notches

Consider a trapezoidal notch *ABCD* as shown in Fig. 11·4. For analysis purpose, split up the notch into a rectangular notch *BCFE* and two triangular notches *ABE* and *DCF*. The discharge over these two triangular notches is equivalent to the discharge over a single triangular notch of angle θ.

Let
H = Height of the liquid above the sill of the notch,
C_{d1} = Coefficient of discharge for the rectangular portion,
C_{d2} = Coefficient of discharge for the triangular portions,
b = Breadth of the rectangular portion of the notch, and
$\dfrac{\theta}{2}$ = Angle, which the sides make with the vertical.

∴ Discharge over the trapezoidal notch,

Q = Discharge over the rectangular notch
+ Discharge over the triangular notch.

$$= \frac{2}{3} C_{d1} \cdot b \sqrt{2g} \, H^{3/2} + \frac{8}{15} C_{d2} \sqrt{2g} \tan\frac{\theta}{2} \times H^{5/2}$$

Example 11·7. *A trapezoidal notch of 1·2 wide at the top and 450 mm at the bottom is 300 mm high. Find the discharge through the notch, if the head of water is 225 mm. Take coefficient of discharge as 0·6.*

Solution. Given : Width of the notch = 1·2 m; b = 450 mm = 0·45; Height of the notch = 300 mm = 0·3 m; H = 225 mm = 0·225 m and C_d = 0·6.

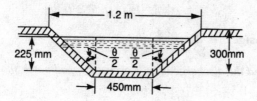

Fig. 11·5.

From the geometry of the notch, we find that

$$\tan\frac{\theta}{2} = \frac{1200 - 450}{2} \times \frac{1}{300} = \frac{750}{600} = 1·25$$

and the discharge over the trapezoidal notch,

$$Q = \frac{2}{3} C_d \cdot b \sqrt{2g} \times H^{3/2} + \frac{8}{15} C_d \sqrt{2g} \tan\frac{\theta}{2} \times H^{5/2}$$

$$= \frac{2}{3} \times 0·6 \times 0·45 \sqrt{2 \times 9·81} \times (0·225)^{3/2}$$

$$+ \frac{8}{15} \times 0·6 \sqrt{2 \times 9·81} \times 1·25 \times (0·225)^{5/2} \text{ m}^3/\text{s}$$

$$= 0·085 + 0·043 = 0·128 \text{ m}^3/\text{s} = 128 \text{ litres/s} \quad \textbf{Ans.}$$

11·7 Discharge over a Stepped Notch

A stepped notch is a combination of rectangular notches as shown in Fig. 11·6. It is, thus, obvious that the discharge over such a notch will be the sum of the discharges over the different rectangular notches.

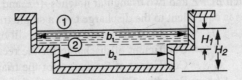

Fig. 11·6. Steeped notch.

Consider a stepped notch as shown in Fig. 11·6. For analysis purpose, let us split up the notch into two rectangular notches 1 and 2. The total discharge over the notch will be the sum of the discharges over the two rectangular notches.

Let H_1 = Height of the liquid above sill of notch 1,
b_1 = Breadth of notch 1,
H_2, b_2 = Corresponding values for notch 2, and
C_d = Coefficient of discharge for both the notches.

From the geometry of the notch, we find that the discharge over the notch 1,

$$Q_1 = \frac{2}{3} C_d \cdot b_1 \sqrt{2g} \times H_1^{3/2}$$

and discharge over the notch 2,

$$Q_2 = \frac{2}{3} C_d \cdot b_2 \sqrt{2g} \left[H_2^{3/2} - H_1^{3/2} \right]$$

Now the total discharge over the notch,
$$Q = Q_1 + Q_2 + \ldots$$

Example 11·8. *Find the discharge in m^3/s over a stepped notch shown in Fig. 11·7.*

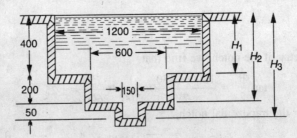

Fig. 11·7.

The level of water coincides with the top of the notch. Take C_d for all sections as 0·6. All dimensions are in mm.

Solution. Given : $C_d = 0.6$.
We know that the total discharge over the stepped notch,

$$Q = \frac{2}{3} C_d b_1 \sqrt{2g} \times H_1^{3/2} + \frac{2}{3} C_d b_2 \sqrt{2g} \left[H_2^{3/2} - H_1^{3/2} \right]$$

$$+ \frac{2}{3} C_d \cdot b_3 \sqrt{2g} \left[H_3^{3/2} - H_2^{3/2} \right]$$

$$Q = \frac{2}{3} \times 0.6 \times 1.2 \sqrt{2 \times 9.81} \times (0.4)^{3/2}$$

$$+ \frac{2}{3} \times 0.6 \times 0.6 \sqrt{2 \times 9.81} \left[(0.6)^{3/2} - (0.4)^{3/2} \right]$$

$$+ \frac{2}{3} \times 0.6 \times 0.15 \sqrt{2 \times 9.81} \left[(0.65)^{3/2} - (0.6)^{3/2} \right]$$

$$= 0.538 + 0.225 + 0.016 = 0.779 \text{ m}^3/\text{s} \quad \textbf{Ans.}$$

EXERCISES 11·1

1. A rectangular notch of 1 m wide is discharging water under a constant head of 0·6 m. What is the discharge over the notch, if its coefficient of discharge is 0·6? [**Ans.** 0·823 m³/s]
2. In a laboratory experiment, water is passing through a rectangular notch 500 mm wide under a constant head of 100 mm. Find the coefficient of discharge, if water is being collected in the tank at the rate of 29 litres/s. [**Ans.** 0·62].
3. A rectangular notch 0·8 m wide is discharging water at the rate of 200 litres/s. Calculate the head of water over the notch, if the coefficient of discharge for the notch is 0·62. [**Ans.** 0·226 m]
4. A right angled triangular notch is discharging water under a constant head of 300 mm. What will be the discharge, if C_d for the notch is 0·61 ? [**Ans.** 71 litres/s]
5. In a laboratory experiment, 14·8 litres of water was found to be discharged over a right-angled V-notch under a head of 160 mm. Determine the coefficient of discharge for the notch.[**Ans.** 0·61]
6. A trapezoidal notch having 0·8 m base has sides inclined at 30° to the horizontal. Find the discharge over the notch under a head of 0·4 m. [**Ans.** 0·627 m³/s]
7. A trapezoidal notch with sides slopes 1 : 1 is discharging 300 litres of water per second under a head of 500 mm as shown in Fig. 11·8. Find the width of the notch at its bottom. [**Ans.** 0·8 m]

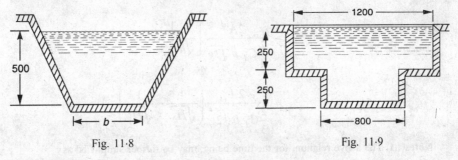

Fig. 11·8   Fig. 11·9

8. A stepped rectangular notch is shown in Fig. 11·9. Find the discharge through the notch, if the coefficient of discharge for both the portions is 0·62. [**Ans.** 0·61 m³/s]

11·8. Time of Emptying a Tank over a Rectangular Notch

Consider a tank of uniform cross-sectional area having a rectangular notch in one of its vertical sides discharging liquid contained in the tank.

Let
A = Surface area of the tank,
H_1 = Initial height of the liquid above the sill of the notch, and
H_2 = Final height of the liquid above the sill of the notch.

At some instant, let the height of the liquid be h above the sill of the rectangular notch. After a small interval of time (dt), let the liquid level fall by an amount equal to dh.

Therefore volume of liquid that has passed in this time,
$$dq = A \times (-dh) = -A \cdot dh \qquad \ldots(i)$$
... (The value of dh is taken as negative, as its value will decrease with the increase in discharge)

We know that the volume of liquid that has passed over the rectangular notch in a small interval of time dh,

$$dq = \text{Rate of discharge} \times \text{Time}$$
$$= \frac{2}{3} C_d \, b \, \sqrt{2g} \, h^{3/2} \times dt \qquad \ldots(ii)$$

Equating equations (i) and (ii),

$$-A \cdot dh = \frac{2}{3} C_d \cdot b \sqrt{2g} \times h^{2/3} \times dt$$

or
$$dt = \frac{-A \cdot dh}{\frac{2}{3} C_d \cdot b \sqrt{2g} \times h^{3/2}} = \frac{-A(h^{-3/2}) dh}{\frac{2}{3} C_d \cdot b \sqrt{2g}}$$

The total time taken (in seconds) to lower the height of liquid from H_1 to H_2 may be found out by integrating the above equation between the limits H_1 to H_2. i.e.,

$$T = \int_{H_1}^{H_2} \frac{-A(h^{-3/2}) \cdot dh}{\frac{2}{3} C_d \cdot b \sqrt{2g}} = \frac{-A}{\frac{2}{3} C_d \cdot b \sqrt{2g}} \int_{H_1}^{H_2} h^{-3/2} \cdot dh$$

$$= \frac{-A}{\frac{2}{3} C_d \cdot b \sqrt{2g}} \left[\frac{h^{-1/2}}{-\frac{1}{2}} \right]_{H_1}^{H_2}$$

$$= \frac{2A}{\frac{2}{3} C_d \cdot b \sqrt{2g}} \left[\frac{1}{\sqrt{h}} \right]_{H_1}^{H_2}$$

$$= \frac{2A}{\frac{2}{3} C_d \cdot b \sqrt{2g}} \left(\frac{1}{\sqrt{H_2}} - \frac{1}{\sqrt{H_1}} \right)$$

Notes : 1. The above relation, for the time being, may be further simplified as :

$$T = \frac{3A}{C_d \cdot b \sqrt{2g}} \left(\frac{1}{\sqrt{H_2}} - \frac{1}{\sqrt{H_1}} \right)$$

But, it has not been done intentionally, because it is not easier to remember the formula in this fashion.

2. If the tank is to be completely emptied, then we have to substitute the value of H_2 which is taken as zero in the relation. But by doing so, we find that the relation becomes infinity. It means that the relation does not hold good for extremely small values of H_2, as a thin layer of water will adhere to the bottom of the tank due to surface tension. But, for all practical purposes, relation for completely emptying the tank is used as :

$$T = \frac{2A}{\frac{2}{3} C_d \cdot b \sqrt{2gH_1}}$$

Flow Over Notches

Example 11·9. *A water tank of surface area 3000 m² contains 1·5 m wide notch in one of its sides. Determine the time required to empty the tank, when the height of water above the sill is 1 m. Take $C_d = 0.61$.*

Solution. : Given : $A = 3000$ m²; $b = 1.5$ m; $H_1 = 1$ m and $C_d = 0.61$.

We know that the time required to empty the tank,

$$T = \frac{2A}{\frac{2}{3} C_d . b . \sqrt{2gH_1}} = \frac{2 \times 3000}{\frac{2}{3} \times 0.61 \times 1.5 \times \sqrt{2 \times 9.81 \times 1}} \text{ s}$$

$= 2222$ s $= 37$ min 2 s **Ans.**

Example 11·10. *A reservoir 100 metres long and 100 metres wide is provided with a rectangular notch 2 metres long. Find the time required to lower the water level in the reservoir from 2 metres to 1 metre. Take $C_d = 0.6$.*

Solution. Given : Length of reservoir $= 100$ m ; Width of reservoir $= 100$ m; $b = 2$ m; $H_1 = 2$ m; $H_2 = 1$ m and $C_d = 0.6$.

We know that the surface area of the reservoir,

$$A = 100 \times 100 = 10\,000 \text{ m}^2$$

and the time required to lower the water level,

$$T = \frac{2A}{\frac{2}{3} C_d . b \sqrt{2g}} \left(\frac{1}{\sqrt{H_2}} - \frac{1}{\sqrt{H_1}} \right)$$

$$= \frac{2 \times 10\,000}{\frac{2}{3} \times 0.6 \times 2 \sqrt{2 \times 9.81}} \left(\frac{1}{\sqrt{1}} - \frac{1}{\sqrt{2}} \right) \text{ s}$$

$= 5645 \times 0.293 = 1654$ s $= 27$ min 34 s **Ans.**

11·9 Time of Emptying a Tank over a Triangular Notch

Consider a tank of uniform cross-sectional area having a triangular notch in one of its sides, discharging liquid contained in the tank.

Let A = Surface area of the tank,
 θ = Angle of the notch,
 H_1 = Initial height of the liquid, above the apex of the notch, and
 H_2 = Final height of the liquid above the apex of the notch.

At some instant, let the height of the liquid be h above the sill of triangular notch. After a small interval of time (dt), let the liquid level fall by an amount dh. Therefore volume of liquid that has passed in this time,

$$dq = A \times (-dh) = -A . dh \qquad \ldots (i)$$

... (The value of dh is taken as negative, as its value will decrease with the increase in discharge)

We know that the volume of liquid that has passed over the triangular notch in a small interval of time dt.

$$dq = \text{Rate of Discharge} \times \text{Time}$$

$$= \frac{8}{15} C_d \sqrt{2g} \tan \frac{\theta}{2} \times h^{5/2} \times dt$$

Equating equations (*i*) and (*ii*),

$$-A . dh = \frac{8}{15} C_d \sqrt{2g} \tan \frac{\theta}{2} \times h^{5/2} \times dt$$

or
$$dt = \frac{-A \cdot dh}{\frac{8}{15} C_d \sqrt{2g} \tan\frac{\theta}{2} \times h^{5/2}} = \frac{-A(dh^{-5/2})dh}{\frac{8}{15} C_d \sqrt{2g} \tan\frac{\theta}{2}}$$

The total time taken (in seconds) to lower the height of liquid from H_1 to H_2 may be found out by integrating the above equation between the limits H_1 to H_2. i.e.,

$$T = \int_{H_1}^{H_2} \frac{-A(h^{-5/2})dh}{\frac{8}{15} C_d \sqrt{2g} \tan\frac{\theta}{2}}$$

$$= \frac{-A}{\frac{8}{15} C_d \sqrt{2g} \tan\frac{\theta}{2}} \int_{H_1}^{H_2} h^{-5/2} dh$$

$$= \frac{-A}{\frac{8}{15} C_d \sqrt{2g} \tan\frac{\theta}{2}} \times \left[\frac{h^{-3/2}}{-\frac{3}{2}}\right]_{H_1}^{H_2}$$

$$= \frac{\frac{2}{3} A}{\frac{8}{15} C_d \sqrt{2g} \tan\frac{\theta}{2}} \times \left[\frac{1}{h^{3/2}}\right]_{H_1}^{H_2}$$

$$= \frac{\frac{2}{3} A}{\frac{8}{15} C_d \sqrt{2g} \tan\frac{\theta}{2}} \times \left[\frac{1}{H_2^{3/2}} - \frac{1}{H_1^{3/2}}\right]$$

Notes : 1. If the tank is to be completely emptied, then the value of H_2 is taken as zero in the above relation. And the relation used is :

$$T = \frac{\frac{2}{3} A}{\frac{8}{15} C_d \sqrt{2g} \tan\frac{\theta}{2} \times H_1^{3/2}}$$

2. The notes given at the end of the last article are also applicable to this article.

Example 11·11. *A water tank of 300 m² surface area is fitted with a notch of 120°. Calculate the time required to empty the tank, if it contains 1 m deep water above the apex of the notch. Take* $C_d = 0.625$.

Solution. Given : $A = 300$ m²; $\theta = 120°$; and $C_d = 0.625$.

We know that the time required to empty the tank,

$$T = \frac{\frac{2}{3} A}{\frac{8}{15} C_d \sqrt{2g} \tan\frac{\theta}{2} \times H_1^{3/2}}$$

$$= \frac{\frac{2}{3} \times 300}{\frac{8}{15} \times 0.625 \times \sqrt{2 \times 9.81} \tan 60° \times (1)^{3/2}} = 78.2 \text{ s} \quad \textbf{Ans.}$$

Example 11·12. *A tank of 25 metres long and 15 metres wide is provided with a right angled V-notch. Find the time required to lower the water level of the tank from 1·5 metres to 0·5 metre. Take coefficient of discharge as 0·62.*

Solution. Given : Length of tank 25 m; Width of tank = 15 m; $\theta = 90°$; $H_1 = 1\cdot 5$ m; $H_2 = 0\cdot 5$ m and $C_d = 0\cdot 62$.

We know that the surface area of the tank,
$$A = 25 \times 15 = 375 \text{ m}^2$$
and the time required to lower the water level of the tank,

$$T = \frac{\frac{2}{3}A}{\frac{8}{15} C_d \sqrt{2g} \tan \frac{\theta}{2}} \times \left[\frac{1}{H_2^{3/2}} - \frac{1}{H_1^{3/2}}\right]$$

$$= \frac{\frac{2}{3} \times 375}{\frac{8}{15} \times 0\cdot 62 \sqrt{2 \times 9\cdot 81} \tan 45°} \left[\frac{1}{(0\cdot 5)^{3/2}} - \frac{1}{(1\cdot 5)^{3/2}}\right] \text{s}$$

$$= 170\cdot 7 \times 2\cdot 284 = 390 \text{ s } 6 \min 30 \text{ s} \quad \textbf{Ans.}$$

EXERCISE 11·2

1. A water tank of 23 m × 15 m contains 1·2 m wide rectangular notch in one of its sides. Find the time taken to empty the tank, if the height of water above the sill of the notch is 600 mm. Take C_d for the notch as 0·6. **[Ans. 7 min 35 s]**
2. A reservoir of 50 m long and 40 m wide has been provided with a rectangular notch of 1 m wide in one of its sides at its base. Find the time required to lower the water level from 1·5 m to 1·0 m. Take $C_d = 0\cdot 63$. **[Ans. 6 min 36 s]**
3. A water tank of 30 m × 20 m is provided with a right-angled V-notch in one of its vertical sides. If the depth of water above the apex of the notch is 500 mm, find the time required to empty the tank. Take coefficient of discharge for the notch as 0·62. **[Ans. 12 min 53 s]**
4. A small tank 20 m × 15 m is provided a right-angled triangular notch in one of its sides. Determine the time required to reduce the water level, in the tank from 1 m to 0·5 m. Assume C_d for the notch as 0·62. **[Ans. 4 min 10 s]**
5. Two similar reservoirs of 100 mm × 60 mm are provided with V-notches of 90° and 120° respectively. Find the ratio of the times required to empty the reservoirs, if the heads of water and coefficients of discharge are same in both the cases. **[Ans. 1·732 : 1]**

11.10 Effect on the Discharge over a Notch due to an Error in the Measurement of Head

We have seen in Arts. 11·3 and 11·4 that the discharge over a rectangular notch is directly proportional to $H^{3/2}$. And the discharge over a triangular notch is directly proportional to $H^{5/2}$ (where H is the height of the liquid level above the sill of the notch). It is, thus, obvious that the accurate measurement of the height of the liquid, is very essential to know the accurate discharge over the notch. Sometimes, some error is induced, while measuring the height of the liquid above the sill of the notch. This error in the measurement of height (H) effects in the calculation of discharge over the notch and is generally expressed as the percentage of the discharge. Following two cases of error in the measurement of head are important from the subject point of view :

1. Over a rectangular notch. 2. Over a triangular notch.

11·11 Effect on the Discharge over a Rectangular Notch due to an Error in the Measurement of Head

Consider a rectangular notch, in one side of a tank, over which the liquid is flowing.
Let H = Height of liquid above the crest of the notch,

b = Width or length of the notch, and
C_d = Coefficient of discharge for the notch.

We have studied in Art. 11·3 that the discharge over a rectangular notch,

$$Q = \frac{2}{3} C_d \cdot b \sqrt{2g} \times H^{3/2}$$

$$= K \cdot H^{3/2} \qquad \ldots \text{(where } K = \frac{2}{3} C_d \cdot b \sqrt{2g}\text{)}$$

Differentiating the above equation with respect to H,

$$\frac{dQ}{dH} = K \times \frac{3}{2} \times H^{1/2}$$

or

$$dQ = \frac{3}{2} K \cdot H^{1/2} \cdot dH$$

$$\therefore \quad \frac{dQ}{Q} = \frac{\frac{3}{2} K \cdot H^{1/2} \cdot dH}{K \cdot H^{3/2}} = \frac{3}{2} \times \frac{dH}{H}$$

$$\frac{dQ}{Q} \times 100 = \frac{3}{2} \times \frac{dH}{H} \times 100 \qquad \ldots \text{(When expressed as per cent)}$$

where

$\dfrac{dQ}{Q} \times 100$ = Percentage error in discharge over the notch, and

$\dfrac{dH}{H} \times 100$ = Percentage error in the measurement of head over the crest of the notch.

Example 11.13. *In a laboratory, water is flowing over a rectangular notch under a head of 200 mm. If the possible error in the reading is 0·6 mm, find the error in discharge.*

Solution. Given : $H = 200$ mm and $dH = 0·6$ mm.

We know that error in discharge,

$$\frac{dQ}{Q} = \frac{3}{2} \times \frac{dH}{H} = \frac{3}{2} \times \frac{0·6}{200} = 0·0045 = 0·45\% \quad \textbf{Ans.}$$

Example 11·14. *A discharge of 85 litres per second was measured over a rectangular notch 300 mm wide. While measuring the head over the notch, an error of 1·25 mm was made. Calculate the percentage of error in the discharge, if the coefficient of discharge for the notch is 0·62.*

Solution. Given : $Q = 85$ litres/s $= 0·085$ m³/s; $b = 300$ mm $= 0·3$ m; $dH = 1·25$ mm and $C_d = 0·62$.

Let H = Height of water above the sill of the rectangular notch.

We know that the discharge over the rectangular notch (Q),

$$0·085 = \frac{2}{3} C_d \cdot b \sqrt{2g} \times H^{3/2}$$

$$= \frac{2}{3} \times 0·62 \times 0·3 \sqrt{2 \times 9·81} \times (H)^{3/2} = 0·55 \, H^{3/2}$$

$$\therefore \quad H^{3/2} = 0·085/0·55 = 0·1545 \qquad \text{or} \qquad H = 0·288 \text{ m} = 288 \text{ mm}$$

and the error in discharge, $\quad \dfrac{dQ}{Q} = \dfrac{3}{2} \times \dfrac{dH}{H} = \dfrac{3}{2} \times \dfrac{1·25}{288} = 0·0065 = 0·65\%$ **Ans.**

Flow Over Notches

11·12 Effect on the Discharge, over a Triangular Notch, due to an Error in the Measurement of Head

Consider a triangular notch, in one side of a tank, over which the liquid is flowing.

Let
  H = Height of the liquid over the apex of the notch
  θ = Angle of the notch, and
  C_d = Coefficient of discharge, for the notch.

We have seen in Art. 11·5 that the discharge over a triangular notch,

$$Q = \frac{8}{15} C_d \sqrt{2g} \tan\frac{\theta}{2} H^{5/2}$$

$$= KH^{5/2} \qquad \ldots \text{(where } K = \frac{8}{15} C_d \sqrt{2g} \tan\frac{\theta}{2}\text{)}$$

Differentiating the above equation with respect to H,

$$\frac{dQ}{dH} = K \times \frac{5}{2} \times H^{3/2} = \frac{5}{2} KH^{3/2}$$

or
$$dQ = \frac{5}{2} KH^{3/2} \, dH$$

$\therefore$
$$\frac{dQ}{Q} = \frac{\frac{5}{2} KH^{3/2} \, dH}{KH^{3/2}} = \frac{5}{2} \times \frac{dH}{H}$$

$$\frac{dQ}{Q} \times 100 = \frac{5}{2} \times \frac{dH}{H} \times 100 \qquad \ldots \text{(When expressed as per cent)}$$

where
$\frac{dQ}{Q} \times 100$ = Percentage error, in the discharge over the notch, and

$\frac{dH}{H} \times 100$ = Percentage error, in the measurement of head over the crest of the notch.

Example 11·15. *Water is flowing over a 90° V-notch under a head of 360 mm. If a probable error of 2 mm is made in reading the value of H, find the percentage accuracy of the value of discharge. Take C_d as 0·62.*

Solution. Given $\theta = 90°$; $H = 360$ mm; $dH = 2$ mm and $C_d = 0·62$.
We know that the error in discharge,

$$\frac{dQ}{Q} = \frac{5}{2} \times \frac{dH}{H} = \frac{5}{2} \times \frac{2}{360} = 0·014 = 1·4\%$$

$\therefore$ Accuracy = $100 - 1·4 = 98·6\%$ **Ans.**

Example 11·16. *A discharge of 60 litres per second was measured over a right-angled triangular notch. While measuring the head over the notch, an error of 1 mm was made. Determine the percentage of error in the discharge, if coefficient of discharge for the notch is 0·6.*

Solution. Given : $Q = 60$ litres/s = $0·06$ m³/s; $\theta = 90°$; $dH = 1$ mm and $C_d = 0·6$.
Let H = Height of water above the apex of the triangular notch.
We know that the discharge over the triangular notch (Q),

$$0·06 = \frac{8}{15} C_d \sqrt{2g} \tan\frac{\theta}{2} \times H^{5/2}$$

$$= \frac{8}{15} \times 0·6 \sqrt{2 \times 9·81} \times \tan 45° \times H^{5/2} = 1·417 \, H^{5/2}$$

$$\therefore \qquad H^{5/2} = 0.06/1.472 = 0.0423 \qquad \text{or} \qquad H = 0.282 \text{ m} = 282 \text{ mm}$$

and the error in discharge,

$$\frac{dQ}{Q} = \frac{5}{2} \times \frac{dH}{H} = \frac{5}{2} \times \frac{1}{282} = 0.0089 = 0.89\% \qquad \textbf{Ans.}$$

Example 11·17. *A right-angled triangular notch and a rectangular notch of 0·5 m width are used alternatively for measuring the discharge of about 20 litres/s.*

Find in each case, the percentage of error in finding the discharge, if an error of 2 mm is introduced in measuring the head of water in both the cases. Take C_d for the triangular and rectangular notches as 0·62 and 0·63 respectively.

Solution. Given : $Q = 20$ litres/s $= 0.02$ m³/s and $dH = 2$ mm.

First of all, consider the flow over the triangular notch. In this case, $\theta = 90°$ and $C_d = 0.62$.

Let $\qquad H_1 = $ Height of water above the apex of the notch.

We know that the discharge over the triangular notch (Q),

$$0.02 = \frac{8}{15} C_d \sqrt{2g} \tan\frac{\theta}{2} \times H_1^{5/2}$$

$$= \frac{8}{15} \times 0.62 \times \sqrt{2 \times 9.81} \times \tan 45° \times H_1^{5/2} = 1.465 \, H_1^{3/2}$$

$$\therefore \qquad H_1^{5/2} = 0.02/1.465 = 0.0137 \qquad \text{or} \qquad H_1 = 0.18 \text{ m} = 180 \text{ mm}$$

and error in the discharge, $\qquad \dfrac{dQ}{Q} = \dfrac{5}{2} \times \dfrac{dH}{H_1} = \dfrac{5}{2} \times \dfrac{2}{180} = 0.0278 = 2.78\% \qquad \textbf{Ans.}$

Now consider the flow over the rectangular notch. In this case, $b = 0.5$ m and $C_d = 0.63$.

Let $\qquad H_2 = $ Height of water above the sill of the rectangular notch.

We know that the discharge over the rectangular notch (Q),

$$0.02 = \frac{2}{3} C_d \cdot b \sqrt{2g} \times H^{3/2}$$

$$= \frac{2}{3} \times 0.63 \times 0.5 \sqrt{2 \times 9.81} \times H_2^{3/2} = 0.9301 \, H_2^{3/2}$$

$$H_2^{3/2} = 0.02/0.9301 = 0.0215 \qquad \text{or} \qquad H_2 = 0.077 \text{ m} = 77 \text{ mm}$$

and error in the discharge, $\qquad \dfrac{dQ}{Q} = \dfrac{3}{2} \times \dfrac{dH}{H} = \dfrac{3}{2} \times \dfrac{2}{77} = 0.039 = 3.9\% \qquad \textbf{Ans.}$

EXERCISE 11·3

1. In a research station, water was made to flow over a rectangular notch under a head of 400 mm. Find the percentage of error in measuring the discharge, if the possible error in observing the head of water is 1·5 mm. **[Ans. 0·56]**
2. A discharge of 150 litres of water per second was found over a rectangular notch of 500 mm width. What is the percentage of error in measuring the discharge, if an error of 2·5 mm was made while measuring the head of water ? Take C_d for the notch as 0·61. **[Ans. 1·2]**
3. A triangular notch is discharging water under a head of 500 mm. A possible error of 3·6 mm was made in reading the head of water. Find the percentage of error in measuring the discharge. **[Ans. 1·8%]**

Flow Over Notches

4. A discharge of 150 litres/s was measured over a V-notch under a head of 400 mm. Determine the percentage error in the measurement of discharge, if an error of 4 mm had taken place, while measuring the head of water. **[Ans. 2·5%]**

5. In an experiment, water was made to flow over two notches, one triangular and the other rectangular. If the heads of water and possible errors in reading the heads is same in both the cases, determine the ratio of errors in measuring the discharge in both the cases. **[Ans. 1 : 0·6]**

QUESTIONS

1. What is a notch? Drive an equation for the discharge over a rectangular notch.
2. Derive an expression for the discharge through a V-notch.
3. Show that the flow over a triangular notch is given by :

$$Q = \frac{8}{15} C_d \sqrt{2g} \tan \frac{\theta}{2} \times H^{5/2}$$

4. Explain, why the triangular notch is preferred to measure small quantities of flow of water in laboratory.
5. What are the advantages of V-notch over a rectangular notch?
6. Derive, from fundamentals, an equation for the discharge over a trapezoidal notch.
7. Explain clearly, how the error, involved in the measurement of head, effects the discharge over a rectangular notch and a triangular notch.

OBJECTIVE TYPE QUESTIONS

1. The discharge over a rectangular notch is
    (a) $\frac{2}{3} C_d \cdot b \sqrt{2gH}$
    (b) $\frac{2}{3} C_d \cdot b \sqrt{2g} \times H$
    (c) $\frac{2}{3} C_d \cdot b \sqrt{2g} (H)^{3/2}$
    (d) $\frac{2}{3} C_d \cdot b \sqrt{2g} (H)^{5/2}$

2. The discharge over a rectangular triangular notch is
    (a) directly proportional to $H^{3/2}$
    (b) inversely proportional to $H^{3/2}$
    (c) directly proportional to $H^{5/2}$
    (c) inversely proportional to $H^{5/2}$

3. The time of emptying of a tank over a rectangular notch depends upon
    (a) surface area of the tank
    (b) length of the notch
    (c) head of water
    (d) all of these

4. The time of emptying a tank over a triangular notch is
    (a) directly proportional to $H^{3/2}$
    (b) inversely proportional to $H^{3/2}$
    (c) directly proportional to $H^{5/2}$
    (c) inversely proportional to $H^{5/2}$

ANSWERS

1. (c)   2. (a)   3. (d)   4. (b)

12

Flow Over Weirs

1. Introduction. 2. Types of Weirs. 3. Discharge over a Rectangular Weir. 4. Francis' Formula for Discharge over a Rectangular Weir (Effect of End Contractions). 5. Bazin's Formula for Discharge over a Rectangular Weir. 6. Discharge over a Cippoletti Weir. 7. Velocity of Approach. 8. Determination of Velocity of Approach. 9. Types of Weirs. 10. Discharge over a Narrow-crested Weir. 11. Discharge over a Broad-crested Weir. 12. Discharge over a Sharp-crested Weir. 13. Discharge over an Ogee Weir. 14. Discharge over a Submerged or Drowned Weir. 15. Ventilation of Rectangular Weirs. 16. Free Nappe. 17. Depressed Nappe. 18. Clinging Nappe. 19. Practical Applications of Weirs. 20. Anicut (i.e., Raised Weir) and Barrages. 21. Bridge Openings. 22. Syphon Spillway. 23. Advantages of Syphon Spillway.

12·1 Introduction

A structure, used to dam up a stream or river, over which the water flows, is called a weir. The conditions of flow, in the case of a weir, are practically the same as those of a rectangular notch. That is why, a notch is, sometimes, called as a weir and *vice versa*.

The only difference between a notch and a weir is that the notch is of a small size and the weir is of a bigger one. Moreover, a notch is usually made in a plate, whereas a weir is usually made of masonry or concrete.

12·2 Types of Weirs

There are many types of weirs depending upon their shape, nature of discharge, width of crest and nature of crest. But the following are important from the subject point of view :

1. *According to the shape :*
    (a) Rectangular weir, and
    (b) Cippoletti weir.
2. *According to the nature of discharge :*
    (a) Ordinary weir, and
    (b) Submerged or drowned weir.
3. *According to the width of crest :*
    (a) Narrow-crested weir, and
    (b) Broad-crested weir.
4. *According to the nature of crest :*
    (a) Sharp-crested weir, and
    (b) Ogee weir.

All the above mentioned weirs will be discussed, in details, at the appropriate places in the book.

12·3 Discharge over a Rectangular Weir

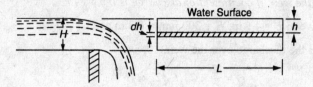

Fig. 12·1. Rectangular weir.

Consider a rectangular weir over which the water is flowing as shown in Fig. 12·1.

Let H = Height of the water above the crest of the weir,

L = Length of the weir, and

C_d = Coefficient of discharge.

Let us consider a horizontal strip of water of thickness dh at a depth h from the water surface as shown in Fig. 12·1.

∴ Area of strip = $L.dh$ ...(i)

We know that the theoretical velocity of water through the strip

$$= \sqrt{2gh} \qquad \ldots(ii)$$

∴ Discharge through the strip,

$dq = C_d \times$ Area of strip × Theoretical velocity

$= C_d . L. dh \sqrt{2gh}$ ...(iii)

The total discharge, over the weir, may be found out by integrating the above equation within the limits 0 and H.

∴
$$Q = \int_0^H C_d L.dh \sqrt{2gh}$$

$$= C_d . L \sqrt{2g} \int_0^H h^{1/2} \, dh$$

$$= C_d . L \sqrt{2g} \left[\frac{h^{3/2}}{\frac{3}{2}} \right]_0^H$$

$$= \frac{2}{3} \times C_d . L \sqrt{2g} \left[h^{3/2} \right]_0^H$$

$$= \frac{2}{3} \times C_d . L \sqrt{2g} \times H^{3/2}$$

Note : Sometimes, the limits of integration, in the above equation, are from H_1 to H_2 (*i.e.*, the liquid level is at a height of H_1 above the top of the weir and H_2 above the bottom of the weir) instead of 0 and H. Then the discharge over such a weir will be given by the equation :

$$Q = \frac{2}{3} \times C_d . L \sqrt{2g} \left(H_2^{3/2} - H_1^{3/2} \right).$$

Example 12·1. *A rectangular weir of 4·5 metres long has a 300 mm head of water. Determine the discharge over the weir, if coefficient of discharge is 0·6.*

Solution. Given : $L = 4.5$ m; $H = 300$ mm $= 0.3$ m and $C_d = 0.6$.

We know that the discharge over the weir,

$$Q = \frac{2}{3} \times C_d . L \sqrt{2g} \times H^{3/2}$$

$$= \frac{2}{3} \times 0.6 \times 4.5 \sqrt{2 \times 9.81} \times (0.3)^{3/2} \text{ m}^3/\text{s}$$

$$= 7.972 \times 0.164 = 1.31 \text{ m}^3/\text{s} = 1310 \text{ litres/s} \quad \textbf{Ans.}$$

Example 12·2. *A weir of 8 m long is to be built across a rectangular channel to discharge a flow of 9 m³/s. If the maximum depth of water on the upstream side of the weir is to be 2 m, what should be the height of the weir ? Adopt $C_d = 0.62$.*

Solution. Given : $L = 8$ m; $Q = 9$ m³/s; Depth of water $= 2$ m and $C_d = 0.62$.
Let $\quad\quad\quad\quad\quad\quad H = $ Height of water above the sill of the weir.

We know that the discharge over the weir (Q),

$$9 = \frac{2}{3} \times C_d . L \sqrt{2g} \; H^{3/2}$$

$$= \frac{2}{3} \times 0.62 \times 8 \sqrt{2 \times 9.81} \times H^{5/2} = 14.65 \, H^{5/2}$$

$$H^{3/2} = 9/14.65 = 0.614 \quad \text{or} \quad H = 0.72 \text{ m}$$

Therefore height of weir should be $2.0 - 0.72 = 1.28$ m **Ans.**

Example 12·2. *The daily record of rainfall over a catchment area is 0·2 million cubic metres. It has been found that 80% of the rain water reaches the storage reservoir and then passes over a rectangular weir. What should be the length of the weir, if the water is not to rise more than 400 mm above the crest ?*

Assume the value of coefficient of discharge for the weir as 0·61.

Solution. Given : Rainfall $= 0.2 \times 10^6$ m³/day; Discharge into the reservoir $= 80\%$ of rain water $H = 400$ mm $= 0.4$ m and $C_d = 0.61$.
Let $\quad\quad\quad\quad\quad\quad L = $ Length of the water in metres.

We know that the volume of water which reaches the reservoir from the catchment area,

$$Q = 80\% \text{ of rain water} = 0.8 \times (0.2 \times 10^6) \text{ m}^3/\text{day}$$

$$= 0.16 \times 10^6 \text{ m}^3/\text{day} = \frac{0.16 \times 10^6}{24 \times 60 \times 60} = 1.85 \text{ m}^3/\text{s}$$

We also know that the discharge into the reservoir over the rectangular weir (Q),

$$1.85 = \frac{2}{3} \times C_d . L \sqrt{2g} \times H^{3/2}$$

$$= \frac{2}{3} 0.61 \times L \sqrt{2g \times 9.81} \times (0.4)^{3/2} = 0.456 L$$

∴ $\quad\quad\quad\quad\quad L = 1.85/0.456 = 4.06$ m **Ans.**

12·4 Francis's Formula for Discharge over a Rectangular Weir (Effect of End Contractions)

Francis, after carrying out a series of experiments, proposed an empirical formula for the discharge over a rectangular weir. He found that the length of the stream of liquid, while flowing over a weir, gets contracted at the ends of the sill as shown in Fig. 12·2.

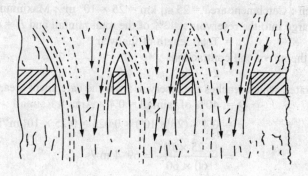

Fig. 12.2. Effect of end contractions.

This end contraction of the stream of liquid is known as lateral contraction or side contraction. Francis also found that the amount of the end contractions depend upon the conditions of sides of the channel and top of the sill as well as velocity of liquid. But an approximate value of end contraction at each end is 1/10 of the height of the liquid above the sill of the weir. Thus, if there are two end contractions only (as in the case of simple rectangular weir) the effective length of the weir is $(L - 0.2 H)$. Substituting this value of length in equation for discharge,

$$Q = \frac{2}{3} \times C_d (L - 0.2H) \sqrt{2g} \times H^{3/2}$$

Sometimes, the total length of a weir is divided into a number of bays or spans by vertical posts as shown in Fig. 12·2. In such a case, the number of end contractions will be twice the number of bays or spans, into which the weir is divided. Thus, in general, we may write the empirical formula proposed by Francis as :

$$Q = \frac{2}{3} \times C_d (L - 0.1 nH) \sqrt{2g} \times H^{3/2}$$

where n = No. of end contractions.

Now substituting $C_d = 0.623$ and $g = 9.81$ m/s² in the general equation for discharge,

$$Q = 1.84 (L - 0.1 nH)^{3/2}$$

Note : When the end contractions are suppressed, the value of n in the above equation is taken as zero.

Example 12·4. *A 30 metres long weir is divided into 10 equal bays by vertical posts each of 0.6 metre width. Using Francis' formula, calculate the discharge over the weir under an effective head of 1 metre.*

Solution. Given : Total length of weir = 30 m; No. of bays = 10; Width of each post = 0·6 m and $H = 1$ m.

We know that the no. of end contractions,

$$n = 10 \times 2 = 20 \quad \text{...(Each bay has two end contractions)}$$

and the effective length of the weir,

$$L = 30 - (9 \times 0.6) = 24.6 \text{ m}$$

∴ Discharge over the weir (by Francis' formula),

$$Q = 1.84 [L - 0.1 nH] H^{3/2}$$
$$= 1.84 [24.6 - (0.1 \times 20 \times 1)] \times (1)^{3/2} = 41.6 \text{ m}^3/\text{s} \quad \textbf{Ans.}$$

Example 12·5. *A reservoir has a catchment area of 25 square kilometres. The maximum rainfall over the area is 25 mm per hour, 40% of which flows to the reservoir over a weir. Using Francis' formula, find the length of the weir. The head of water over the weir should not exceed 0·8 m.*

Solution. Given : Catchment area = 25 sq km = 25×10^6 m² ; Maximum rainfall = 25 mm = 0·025 m/hr; Discharge into the reservoir = 40% of the total rainfall and $H = 0·8$ m.

Let L = Total length of the weir.

We know that the no. of end contractions :
$$n = 2 \qquad \ldots(\because \text{ of simple weir})$$

and the volume of water, which reaches the reservoir from the catchment area,
$$Q = 40\% \text{ of rainfall} = 0·4 \times \text{Area} \times \text{Rainfall}$$
$$= 0·4 \times (25 \times 10^6) \times 0·025 = 0·25 \times 10^6 \text{ m}^3/\text{hr}$$
$$= \frac{0·25 \times 10^6}{60 \times 60} = 69·4 \text{ m}^3/\text{s}$$

We also know that the discharge into the reservoir over the weir (by Francis' formula),
$$69·4 = 1·84 \, [L - 0·1 \, nH] \, H^{3/2} = 1·84 \, [L - (0·1 \times 2 \times 0·8)] \times (0·8)^{3/2}$$
$$= 1·32 \, (L - 0·16)$$
$$\therefore \qquad (L - 0·16) = 69·4/1·32$$
or $\qquad L = 52·58 + 0·16 = 52·74$ m **Ans.**

12·5 Bazin's Formula for Discharge over a Rectangular Weir

Bazin, after carrying out a series of experiments, proposed an empirical formula for the discharge over a rectangular weir. He found that the value of coefficient of discharge varies with the height of water over the sill of a weir. Thus, the proposed an amendment for the formula for discharge over a rectangular notch as given below :

We have seen in Art. 12·3 that the discharge over a rectangular weir,
$$Q = \frac{2}{3} \times C_d \, L \, \sqrt{2g} \times H^{3/2}$$

Bazin proposed that the discharge over a weir,
$$Q = mL\sqrt{2g} \times H^{3/2} \qquad \text{where } m = \frac{2}{3} C_d$$

He found that the value of m varies with the head of water, whose value may be obtained from the relation :
$$m = 0·405 + \frac{0·003}{H}$$

where $\qquad H$ = Height of water in metres.

He found the above relation by experiments in which he avoided the effect of end contractions.

Example 12·6. *Find the discharge over a rectangular weir 4·5 metre long under a head of 600 mm by using Bazin's formula.*

Solution. Given : $L = 4·5$ m and $H = 600$ mm = 0·6 m

We know from Bazin's relation that
$$m = 0·405 + \frac{0·003}{H} = 0·405 + \frac{0·003}{0·6} = 0·41$$

and discharge over the weir (by Bazin's formula),

$$Q = mL\sqrt{2g} \times H^{3/2} = 0\cdot 41 \times 4\cdot 5 \sqrt{2 \times 9\cdot 81} \times (0\cdot 6)^{3/2} \text{ m}^3/\text{s}$$

$$= 8\cdot 172 \times 0\cdot 465 = 3\cdot 8 \text{ m}^3/\text{s} \quad \textbf{Ans.}$$

Example 12·7. *A rectangular weir 6 m long is discharging water under a head of 300 mm. Calculate the discharge over the weir by using (i) Francis' formula and (ii) Bazin's formula.*

Solution. Given : $L = 6$ m and $H = 300$ mm $= 0\cdot 3$ m.

(i) *Discharge over the weir by using Francis' formula*

We know that the no. of end contractions,

$$n = 2 \qquad \qquad \text{...($\because$ of simple weir)}$$

and discharge ove the weir, $\quad Q = 1\cdot 84\,[L - 0\cdot 1\,nH]\,H^{3/2}$

$$= 1\cdot 84[6 - (0\cdot 1 \times 2 \times 0\cdot 3)] \times (0\cdot 3)^{3/2} = 1\cdot 796 \text{ m}^3/\text{s} \quad \textbf{Ans.}$$

(ii) *Discharge over the weir by using Bazin's formula*

We know from Bazin's relation that

$$m = 0\cdot 405 + \frac{0\cdot 003}{H} = 0\cdot 405 + \frac{0\cdot 003}{0\cdot 3} = 0\cdot 415$$

and discharge over the weir $\quad Q = mL\sqrt{2g} \times H^{3/2} = 0\cdot 415 \times 6 \times \sqrt{2 \times 9\cdot 81} \times (0\cdot 3)^{3/2} \text{ m}^3/\text{s}$

$$= 11\cdot 03 \times 0\cdot 164 = 1\cdot 81 \text{ m}^3/\text{s} \quad \textbf{Ans.}$$

12·6 Discharge over a Cippoletti weir

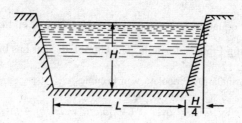

Fig. 12·3. Cippoletti weir.

The Cippoletti* weir is a trapezoidal weir, having 1 horizontal to 4 vertical side slopes, as shown in Fig. 12·3.

The purpose of slope, on the sides, is to obtain an increased discharge through the triangular portions of the weir, which, otherwise, would have been decreased due to end contractions in the case of rectangular weirs. Thus the advantage of a Cippoletti weir is that the factor of end contractions is not required, while using the Francis' formula. Strictly speaking, a Cippoletti weir is a theoretical weir, whose side slope (*i.e.,* 1 horizontal to 4 vertical) has been obtained mathemetically as follows :

Let us split up the trapezoidal weir into a rectangular weir and a triangular notch. We have seen in Art. 12·4 that the discharge over a rectangular weir, considering end contraction,

$$Q_1 = \frac{2}{3} \times C_d \sqrt{2g}\,(L - 0\cdot 2H)\,H^{3/2} \qquad \qquad \text{...(i)}$$

and the discharge over the triangular notch (as per Art. 10·4) :

$$Q_2 = \frac{8}{15} \times C_d \sqrt{2g}\,\tan\frac{\theta}{2} \times (H)^{3/2} \qquad \qquad \text{...(ii)}$$

---

*Cippoletti was an Italian engineer who did a lot of work on the discharge over weirs.

∴ Total discharge, $Q = Q_1 + Q_2$

$$= \frac{2}{3} \times C_d \sqrt{2g} \, (L - 0.2H)^{3/2} + \frac{8}{15} \times C_d \sqrt{2g} \, \tan\frac{\theta}{2} \times H^{5/2} \quad ...(iii)$$

Since the main idea of Cippoletti was to avoid the factor of end contractions, and as such he gave the formula for the discharge,

$$Q = \frac{2}{3} \times C_d \, L\sqrt{2g} \times H^{3/2} \qquad \qquad ...(iv)$$

Equating the equations (*iii*) and (*iv*),

$$\frac{2}{3} \times C_d L\sqrt{2g} \times H^{3/2} = \frac{2}{3} \times C_d\sqrt{2g} \, (L-0.2H) + H^{3/2} + \frac{8}{15} \times C_d \sqrt{2g} \, \tan\frac{\theta}{2} \times H^{5/2}$$

Dividing both sides by $\left[\frac{2}{3} \times C_d \sqrt{2g} \times H^{3/2}\right]$

$$L = L - 0.2H + \frac{4}{5} \tan\frac{\theta}{2} \times H$$

or $\qquad \frac{4}{5} \tan\frac{\theta}{2} \times H = 0.2H$

∴ $\qquad \tan\frac{\theta}{2} = 0.2 \times \frac{5}{4} = \frac{1}{5} \times \frac{5}{4} = \frac{1}{4}$

It is, thus, obvious that if a trapezoidal weir is having 1 horizontal to 4 vertical side slopes (as is clear from the above equation *i.e.*, $\tan\frac{\theta}{2} = \frac{1}{4}$), the factor of end contraction is not required for the discharge, while using Francis' formula. The Cippoletti weir has been designed on the above derivation as shown in Fig. 12·3.

Note : The Francis' formula for Cippoletti weir becomes

$$Q = 1.84 \, L \times H^{3/2}$$

Example 12·8. *Water is flowing over a Cippoletti weir of 4 metres long under a head of 1 metre. Compute the discharge, if the coefficient of discharge for the weir is 0·62.*

Solution. Given : $L = 4$ m ; $H = 1$ m and $C_d = 0.62$.

We know that the discharge over the Cippoletti weir,

$$Q = \frac{2}{3} \times C_d . L \sqrt{2g} \times H^{3/2}$$

$$= \frac{2}{3} \times 0.62 \times 4 \times \sqrt{2 \times 9.81} \times (1)^{3/2} \text{ m}^3/\text{s}$$

$$= 7.32 \times 1 = 7.32 \text{ m}^3/\text{s} \qquad \textbf{Ans.}$$

Example 12·9. *Find the length of a Cippoletti weir required for a flow of 425 litres per second, if the head of water is not to exceed 1/10th of its length. Use Francis's formula for the weir.*

Solution. Given : $Q = 425$ litres/s $= 0.425$ m³/s and head of water $(H) = 0.1 L$ (where L is the length of Cippoletti weir).

Let $\qquad L = $ Length of the cippoletti weir.

We know that the discharge over the Cippoletti weir (as per Francis' formula),

$$0.425 = 1.84 L \times H^{3/2} = 1.84 \times 10H \times H^{3/2} \times 18.4 H^{5/2}$$

∴ $\qquad H^{5/2} = 0.425/18.4 = 0.023 \qquad \text{or} \qquad H = 0.221 \text{ m}$

and $\qquad L = 10 \times 0.221 = 2.21 \text{ m} \qquad \textbf{Ans.}$

EXERCISE 12·1

1. A weir 12 m long has 0·5 m head of water. Find the discharge over the weir, if the coefficient of discharge is 0·61. [**Ans.** 7·64 m³/s]
2. A weir of 10 m long has a constant head of water as 300 mm. Taking coefficient of discharge as 0·62, determine the discharge over the weir in litres/s. [**Ans.** 3000 litres/s]
3. In a laboratory, a weir of 1·25 m length was discharging water at the rate of 200 litres/s under a head of 250 mm. What is the value of coefficient of discharge for the weir? [**Ans.** 0·61]
4. The daily rainfall over a catchment area was found to be 250 million litres. It was observed that 30% of the rain water is lost and the remaining reaches the reservoir, which passes over a weir. Find the length of the weir, if water over the weir shall never rise more than 350 mm. Take $C_d = 0·6$. [**Ans.** 5·53 m]
5. A 15 m long weir has 6 equal bays with 500 mm wide vertical posts. With the help of Francis' formula, find the discharge over the weir under a head of 0·8 m. [**Ans.** 10·1 m³/s]
6. A rectangular weir 8·5 m long is discharging water under a head of 500 mm. Find the discharge over the weir in m³/s with the help of Bazin's formula. [**Ans.** 1·235 m³/s]
7. Water is flowing over a rectangular weir 10 m long under a constant head of 200 mm. Find the discharge over the weir by using Bazin's formula and Francis' formula.
[**Ans.** 1·664 m³/s ; 1·639 m³/s]
8. Find the quantity of water flowing over a Cippoletti weir of 8·5 m length under a constant head of 250 mm. Take coefficient of discharge for the weir as 0·6. [**Ans.** 1·88 m³/s]

12·7 Velocity of Approach

Sometimes, a weir is provided in a stream or a river to measure the flow of water. In such a case, the water, approaching the weir, has got some velocity, known as velocity of approach. It is assumed to be uniform over the whole weir.

In the previous articles, we have altogether neglected the velocity of approach, and all the formulae have been derived on the assumption that the water on the upstream side of the weir is not in motion. A little consideration will show that this velocity of approach of the water is sure to increase the discharge over the weir.

Let A = Cross-sectional area of the channel on the upstream side of the weir, and
Q = Discharge over the weir.

∴ Velocity of approach,

$$v = \frac{Q}{A} \qquad \ldots(i)$$

Since the formulae for discharge, over the weir, involve the height of water above the crest of weir, therefore this velocity of approach should also be converted into an additional head of water acting over the whole weir.

∴ Additional height of water due to velocity of approach,

$$H_a = \frac{v^2}{2g}$$

Thus, if the velocity of approach is considered for the discharge over the weir, then the additional height of water should also be taken into account.

Let H = Height of water, over the crest of the weir, and
H_a = Height of water, due to velocity of approach.

∴ Total height of water above the weir,

$$H_1 = H + H_a$$

It is, thus, obvious that the limits of integration for the discharge over a rectangular weir in Art. 12·3 will be H_a and H_1 instead of 0 and H.

∴ Discharge over the weir with the velocity of approach,

$$Q = \frac{2}{3} \times C_d \, L \, \sqrt{2g} \left[H_1^{3/2} - H_a^{3/2} \right]$$

Notes : 1. Francis' formula for discharge with the velocity of approach,

$$Q = 1 \cdot 84 \, (L - 0 \cdot 1 \, nH_1) \left[H_1^{3/2} - H_a^{3/2} \right]$$

2. Similarly, Bazin's formula for discharge with the velocity of approach,

$$Q = mL\sqrt{2g} \times H_1^{3/2}$$

Example 12·10. *Find the discharge in m^3/min over a rectangular weir of 30 metres in length with a head of 1 metre. Take the velocity of approach as 1 metre/s and $C_d = 0 \cdot 58$.*

Solution. Given : $L = 30$ m; $H = 1$ m; $v = 1$ m/s and $C_d = 0 \cdot 58$.
We know that the head due to velocity of approach,

$$H_a = \frac{v^2}{2g} = \frac{(1)^2}{2 \times 9 \cdot 81} = 0 \cdot 05 \text{ m}$$

and total head, $\qquad H_1 = H + H_a = 1 + 0 \cdot 05 = 1 \cdot 05$ m

∴ Discharge over the rectangular weir,

$$Q = \frac{2}{3} \times C_d \, L \, \sqrt{2g} \left[H_1^{3/2} - H_a^{3/2} \right]$$

$$= \frac{2}{3} \times 0 \cdot 58 \times 30 \, \sqrt{2 \times 9 \cdot 81} \left[(1 \cdot 05)^{3/2} - (0 \cdot 05)^{3/2} \right] \text{m}^3/\text{s}$$

$$= 51 \cdot 38 \times 1 \cdot 065 = 54 \cdot 72 \text{ m}^3/\text{s} = 3283 \text{ m}^3/\text{min} \quad \textbf{Ans.}$$

Example 12·11. *A weir, 36 m long, is divided into 12 equal bays by vertical posts each of 600 mm width. Determine the discharge over the weir, if the head over the crest is $1 \cdot 2$ m and the velocity approach is 2 m/s.*

Solution. Total length of the weir = 36 m; No. of bays = 12; Width of each post = 600 mm = 0·6 m; $H = 1 \cdot 2$ m and $v = 2$ m/s.

We know that the no. of end contractions,

$$n = 12 \times 2 = 24 \qquad \ldots (\because \text{Each bay has two end contractions})$$

and the effective length of the weir,

$$L = 36 - (11 \times 0 \cdot 6) = 29 \cdot 4 \text{ m}$$

We also know that the head due to velocity of approach,

$$H_a = \frac{v^2}{2g} = \frac{(2)^2}{2 \times 9 \cdot 81} = 0 \cdot 2 \text{ m}$$

and the total head, $\qquad H_1 = H + H_a = 1 \cdot 2 + 0 \cdot 2 = 1 \cdot 4$ m

∴ Discharge over the weir,

$$Q = 1 \cdot 84 \, (L - 0 \cdot 1 \, nH_1) \times \left(H_1^{3/2} - H_a^{3/2} \right)$$

$$= 1 \cdot 84 \left[29 \cdot 4 - (0 \cdot 1 \times 24 \times 1 \cdot 4) \right] \times \left[(1 \cdot 4)^{3/2} - (0 \cdot 2)^{3/2} \right] \text{m}^3/\text{s}$$

$$= 47 \cdot 91 \times 1 \cdot 568 = 75 \cdot 1 \text{ m}^3/\text{s} \quad \textbf{Ans.}$$

12·8 Determination of Velocity of Approach

Sometimes, in a problem, the value of velocity of approach is not given. In such a case, the discharge over the weir is found out first by ignoring the velocity of approach. Then the velocity of approach is obtained by dividing the discharge by the cross-sectional area of the channel on the upstream side of the weir. This velocity of approach, so obtained, is then used for finding the discharge over the weir considering the velocity of approach.

Flow Over Weirs

If more accurate discharge is required, the above process can be repeated. But it has been, generally, seen that the discharge, obtained by substituting the first obtained value of the velocity of approach, does not differ appreciably from the further values. Thus, in actual practice, the repetition is not necessary.

Example 12·12. *In a laboratory experiment, a Cippoletti weir having a crest length of 400 mm is used to measure the flow of water in a rectangular channel 600 mm wide and 75 mm deep. If the water level in the channel is 50 mm above the weir crest, estimate the discharge in the channel in litres/minute by considering velocity of approach. Take coefficient of discharge as 0·63.*

Solution. Given : Length of Cappoletti weir (L) = 400 mm = 0·4 m; Width of channel = 600 mm = 0·6 m; Depth of channel = 75 mm = 0·075 m; H = 50 mm = 0·05 m and C_d = 0·63.

First of all, let us find out the discharge in the rectangular channel ignoring the velocity of approach. We know that the discharge in the channel,

$$Q = \frac{2}{3} \times C_d \cdot L \sqrt{2g} \times H^{3/2}$$

$$= \frac{2}{3} \times 0·63 \times 0·4 \times \sqrt{2 \times 9·81} \times (0·05)^{3/2} \; m^3/s$$

$$= 0·744 \times 0·011 = 0·0082 \; m^3/s$$

and the cross-sectional area of the water flowing in the channel,

$$A = 0·6 \times 0·05 = 0·03 \; m^2$$

∴ Velocity of approach, $\quad v = \dfrac{Q}{A} = \dfrac{0·0082}{0·03} = 0·28 \; m/s$

and head due to velocity of approach

$$H_a = \frac{v^2}{2g} = \frac{(0·28)^2}{2 \times 9·81} = 0·004 \; m$$

∴ Total head, $\quad H_1 = H + H_a = 0·05 + 0·004 = 0·054 \; m$

We also know that the discharge in the channel by considering the velocity of approach,

$$Q = \frac{2}{3} \times C_d \; L \times \sqrt{2g} \left(H_1^{3/2} - H_a^{3/2} \right)$$

$$= \frac{2}{3} \times 0·63 \times 0·4 \times \sqrt{2 \times 9·81} \left[(0·054)^{3/2} - (0·004)^{3/2} \right] \; m^3/s$$

$$= 0·744 \times 0·012 = 0·009 \; m^3/s$$

$$= 0.54 \; m^3/min = 540 \; litres/min \quad \textbf{Ans.}$$

Example 12·13. *A weir 2·0 metres long has 0·6 m head of water over its crest. Using Francis' formula, find the discharge over the weir, if the channel approaching the weir is 6 metres wide and 1·2 metres deep.*

Also determine the new discharge, considering the velocity of approach.

Solution. Given : L = 2·0 m; H = 0·6 m; Width of the channel = 6 m and depth of the channel = 1·2 m.

Discharge over the weir without considering the velocity of approach

We know that the discharge over the weir (by Francis' formula),

$$Q = 1·84 \, (L - 0·1 \, nH) \, H^{3/2}$$

$$= 1·84 \, [2·4 - (0·1 \times 2 \times 0·6)] \times (0·6)^{3/2} \; m^3/s$$

$$= 4·195 \times 0·465 = 1·95 \; m^3/s \quad \textbf{Ans.}$$

Discharge over the weir considering the velocity of approach

We know that the cross-sectional area of the water flowing in the channel,
$$A = 6 \times 0.6 = 3.6 \text{ m}^3$$

$\therefore$ Velocity of approach, $v = \dfrac{Q}{A} = \dfrac{1.95}{3.6} = 0.54$ m/s

and head due to velocity of approach,
$$H_a = \dfrac{v^2}{2g} = \dfrac{(0.54)^2}{2 \times 9.81} = 0.015 \text{ m}$$

$\therefore$ Total head $\quad H_1 = H + H_a = 0.6 + 0.015 = 0.615$ m

We also know that the discharge over the weir (by Francis' formula),
$$Q = 1.84\,(L - 0.1\,nH_1) \times (H_1^{3/2} - H_a^{3/2})$$
$$= 1.84\,[2.4 - (0.1 \times 2 \times 0.615)] \times [(0.615)^{3/2} - (0.015)^{3/2}]$$
$$= 4.19 \times 0.48 = 2.01 \text{ m}^3/\text{s} \quad \textbf{Ans.}$$

EXERCISE 12·2

1. Water flowing in a channel with a velocity of 0·6 m/s approaches a weir of 6 m long. If the head of water over the weir is 300 mm, find the discharge over the weir. Take coefficient of discharge as 0·6. **[Ans. 1·89 m³/s]**
2. A weir 2·3 m long is discharging water under a head of 270 mm. Using Bazin's formula, find the discharge over the weir, when the water approaches the weir with a velocity of 0·75 m/s. **[Ans. 0·67 m³/s]**
3. In a laboratory, a Cippoletti weir with a crest width of 400 mm is used to find the discharge under a head of 250 mm. If the channel is 600 mm wide and 450 mm deep, determine the discharge in litres/s over the weir (*i*) without considering the velocity of approach and (*ii*) with considering velocity of approach. Take $C_d = 0.62$. **[Ans. 91·5 litres/s; 94·5 litres/s]**
4. A rectangular channel 3·2 m wide has at the end a weir 2·5 m long with a sill 240 mm from the bottom. Find the discharge over the weir considering the velocity of approach with the help of Francis' formula. **[Ans. 0·61 m³/s]**

12·9 Types of Weirs

Though there are numerous types of weirs, yet the following are important from the subject point of view :
1. Narrow-crested weirs,
2. Broad-crested weirs,
3. Sharp-crested weirs,
4. Ogee weirs, and
5. Submerged or drowned weirs.

In the following pages, we shall discuss the discharge over all these types of weirs one-by-one.

12·10 Discharge over a Narrow-crested Weir

The weirs are generally classified according to the width of their crests into two types. *i.e.*, narrow-crested weirs and broad crested weirs.

Let $\quad b$ = Width of the crest of the weir, and
$\quad H$ = Height of water above the weir crest.

If, $2b$ is less than H, the weir is called a narrow-crested weir. But if $2b$ is more than H, it is called a broad-crested weir. Unless mentioned otherwise, the given weir is taken to be a narrow-crested weir.

A narrow-crested weir is hydraulically similar to an ordinary weir or to a rectangular weir. Thus, the same formula for discharge over a narrow-crested weir holds good, which we derived from an ordinary weir (as in Art. 12·3). *i.e.*

$$Q = \frac{2}{3} \times C_d . L \sqrt{2g} \times H^{3/2}$$

where
Q = Discharge over the weir,
C_d = Coefficient of discharge,
L = Length of the weir, and
H = Height of water level above the crest of the weir.

Example 12·14. *A narrow-crested weir of 10 metres long is discharging water under a constant head of 400 mm. Find discharge over the weir in litres/s. Assume coefficient of discharge as 0·623.*

Solution. Given : $L = 10$ m; $H = 400$ mm $= 0\cdot4$ m and $C_d = 0\cdot623$.
We know that the discharge over the weir,

$$Q = \frac{2}{3} \times C_d . L \sqrt{2g} \times H^{3/2}$$

$$= \frac{2}{3} \times 0\cdot623 \times 10 \sqrt{2 \times 9\cdot81} \times (0\cdot4)^{3/2} \text{ m}^3/\text{s}$$

$$= 18\cdot4 \times 2\cdot53 = 46\cdot55 \text{ m}^2/\text{s} = 4655 \text{ litres/s} \quad \textbf{Ans.}$$

12·11 Discharge over a Broad-crested Weir

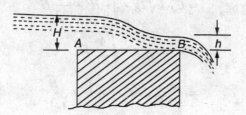

Fig. 12·4. Broad-crested weir.

Consider a broad-crested weir as shown in Fig. 12·4. Let A and B be the upstream and downstream ends of the weir.

Let
H = Head of water on the upstream side of the weir (*i.e.*, at A),
h = Head of water on the downstream side of the weir (*i.e.*, at B),
v = Velocity of the water on the downstream side of the weir (*i.e.*, at B),
C_d = Coefficient of discharge, and
L = Length of the weir.

Applying Bernoulli's equation at A and B,

$$0 + 0 + H = 0 + h + \frac{v^2}{2g}$$

∴
$$\frac{v^2}{2g} = H - h$$

or
$$v = \sqrt{2g(H-h)} \qquad \ldots(i)$$

∴ Discharge over the weir,

$$Q = C_d \times \text{Area of flow} \times \text{Velocity}$$
$$= C_d \cdot L \cdot h \cdot v \qquad \ldots(ii)$$
$$= C_d \cdot L \cdot h \cdot \sqrt{2g(H-h)}$$
$$= C_d \cdot L\sqrt{2g} \times \sqrt{Hh^2 - h^3} \qquad \ldots(iii)$$

From the above equation, we see that the discharge will be maximum, when $(Hh^2 - h^3)$ is maximum. Therefore differentiating the equation $(Hh^2 - h^3)$ and equating the same to zero,

$$\frac{dQ}{dh}(Hh^2 - h^3) = 0$$

$$2Hh - 3h^2 = 0$$
or $\qquad 2H - 3h = 0$

or $\qquad h = \dfrac{2}{3}H$

Substituting this value of h in equation (iii),

$$Q_{max} = C_d \cdot L\sqrt{2g} \times \sqrt{H\left(\frac{2}{3}H\right)^2 - \left(\frac{2}{3}H\right)^3}$$

$$= C_d \cdot L\sqrt{2g} \times \sqrt{\frac{4}{9}H^3 - \frac{8}{27}H^3}$$

$$= C_d \cdot L\sqrt{2g} \times \sqrt{\frac{4}{27}H^3}$$

$$= C_d \cdot L\sqrt{2g}\ \frac{2}{3} \times H \times \sqrt{\frac{H}{3}}$$

$$= \frac{2}{3\sqrt{3}} C_d \cdot L \sqrt{2g} \times H^{3/2} \qquad \ldots(iv)$$

$$= 0.384\ C_d \cdot L \sqrt{2 \times 9.81} \times H^{3/2} \qquad (v)$$

$$= 1.71\ C_d \cdot L \times H^{3/2} \qquad \ldots(vi)$$

Example 12·15. *A broad-crested weir 20 m long is discharging water from a reservoir into a channel. What will be the discharge over the weir, if the head of water on the upstream and downstream sides is 1 m and 0·5 m respectively? Take coefficient of discharge for the flow as 0·6.*

Solution. Given : $L = 20$ m; $H = 1$ m; $h = 0.5$ m and $C_d = 0.6$.

We know that the discharge over the weir,

$$Q = C_d \cdot L \cdot h \sqrt{2g(H-h)}$$
$$= 0.6 \times 20 \times 0.5 \times \sqrt{2 \times 9.81(1 - 0.5)}\ \text{m}^3/\text{s}$$
$$= 6 \times 3.13 = 18.8\ \text{m}^3/\text{s} \quad \textbf{Ans.}$$

Example 12·16. *A broad-crested weir 10 metres long has a maximum discharge of 10 000 litres of water per second. Determine the head of water on the upstream side of the weir for this discharge, if coefficient of discharge is 0·62.*

Solution. Given : $L = 10$ m; $Q_{max} = 10\,000$ litres/s $= 10$ m³/s and $C_d = 0.62$.

Let H = Head of water on the upstream side of the weir for maximum discharge.

We know that the maximum discharge over the weir (Q_{max})

$$10 = 1.71\, C_d \cdot L \times H^{3/2} = 1.71 \times 0.62 \times 10 \times H^{3/2} = 10.6\, H^{3/2}$$

$\therefore \quad H^{3/2} = 10/10.6 = 0.943$ or $H = 0.92$ m $= 920$ mm **Ans.**

Example 12·17. *Determine the maximum discharge over a broad-crested weir 60 metres long having 0·6 m height of water above its crest. Take coefficient of discharge as 0·595. Also determine the new discharge over the weir, considering the velocity of approach. The channel at the upstream side of the weir has a cross-sectional area of 45 sq metres.*

Solution. Given : $L = 60$ m; $H = 0.6$ m; $C_d = 0.595$ and $A = 45$ m².

Maximum discharge over the weir without considering the velocity of approach

We know that the maximum discharge over the weir,

$$Q_{max} = 1.71\, C_d \cdot L \times H^{3/2} = 1.71 \times 0.595 \times 60 \times (0.6)^{3/2} \text{ m}^3/\text{s}$$
$$= 28.4 \text{ m}^3/\text{s}$$

Maximum discharge over the weir considering the velocity of approach

We know that velocity of approach,

$$v = \frac{Q}{A} = \frac{28.4}{45} = 0.63 \text{ m/s}$$

and the head due to velocity of approach,

$$H_a = \frac{v^2}{2g} = \frac{(0.63)^2}{2 \times 9.81} = 0.02 \text{ m}$$

$\therefore$ Total head, $H_1 = H + H_a = 0.6 + 0.02 = 0.62$ m

We also know that the maximum discharge over the weir,

$$Q = 1.71\, C_d \cdot L\, (H_1^{3/2} - H_a^{3/2})_{3/2}$$
$$= 1.71 \times 0.595 \times 60\, (0.62) - (0.02)^{2/3} = 29.6 \text{ m}^3/\text{s} \textbf{ Ans.}$$

12·12 Discharge over a Sharp-crested Weir

It is a special type of weir, having a sharp-crest as shown in Fig. 12·5. The water flowing over the crest comes in contact with the crest line and then springs up from the crest and falls as a trajectory as shown in Fig. 12·5.

In a sharp-crested weir, the thickness of the weir is kept less than half of the height of water on the weir. *i.e.*,

$$b < \frac{H}{2}$$

where b = Thickness of the weir,

and H = Height of water, above the crest of the weir.

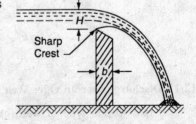

Fig. 12·5. Sharp-crested weir.

The discharge equation, for a sharp crested weir, remains the same as that of a rectangular weir. *i.e.*,

$$Q = \frac{2}{3}\, C_d \cdot L \sqrt{2g} \times H^{3/2}$$

where C_d = Coefficient of discharge, and

L = Length of sharp-crested weir.

Example 12·18. *In a laboratory experiment, water flows over a sharp-crested weir 200 mm long under a constant head of 75 mm. Find the discharge over the weir in litres/s, if $C_d = 0.6$.*

Solution. Given : $L = 200$ mm $= 0.2$ m; $H = 75$ mm $= 0.075$ m and $C_d = 0.6$.

We know that the discharge over the weir,

$$Q = \frac{2}{3} \times C_d . L \sqrt{2g} \times H^{3/2}$$

$$= \frac{2}{3} \times 0.6 \times 0.2 \times \sqrt{2 \times 9.81} \times (0.075)^{3/2} \text{ m}^3/\text{s}$$

$$= 0.354 \times 0.0205 = 0.0073 \text{ m}^3/\text{s} = 7.3 \text{ litres /s} \quad \textbf{Ans.}$$

Example 12·19. *A rectangular sharp-crested weir is to be constructed in a testing station with small stream in which the discharge varies from 50 litres/s and 1250 litres/s. Find the suitable length of the weir, if the minimum head to be measured is 50 mm and the maximum head on it does not exceed one-third of its length.*

Solution. $Q_{min} = 50$ litres/s $= 0.05$ m³/s; $Q_{max} = 1250$ litres/s $= 1.25$ m³/s and $H_{min} = 50$ mm $= 0.05$ m.

Let $\quad H = $ Length of weir in metres.

∴ Maximum head of water,

$$H_{max} = L/3$$

We know that the minimum discharge over the weir (Q_{min}),

$$0.05 = \frac{2}{3} C_d . L \sqrt{2g} \times H^{3/2}$$

$$0.05 = \frac{2}{3} C_d . L \sqrt{2g} \times (0.05)^{3/2} \quad \dots(i)$$

and maximum discharge over the weir (Q_{max})

$$1.25 = \frac{2}{3} \times C_d . L \sqrt{2g} \times \left(\frac{L}{3}\right)^{3/2} \quad \dots(ii)$$

Dividing equation (*ii*) by (*i*),

$$\frac{1.25}{0.05} = \frac{\frac{2}{3} \times C_d L \sqrt{2g} \times \left(\frac{L}{3}\right)^{3/2}}{\frac{2}{3} \times C_d L \sqrt{2g} (0.05)^{3/2}} = \frac{L^{3/2}}{(3 \times 0.05)^{3/2}}$$

$$25 = \frac{L^{3/2}}{0.058} \quad \text{or} \quad L^{3/2} = 25 \times 0.058 = 1.45$$

∴ $\quad L = 1.28$ m **Ans.**

12·13 Discharge over an Ogee Weir

It is a special type of weir, generally, used as a spillway of a dam as shown in Fig. 12·6.

The crest of an ogee weir slightly rises up from the point *A* (*i.e.*, crest of the sharp-crested weir) and after reaching the maximum rise of $0.115 H$ (where *H* is the height of a water above the point *A*) falls in a parabolic form as shown in Fig. 12·6.

The discharge equation for an ogee weir remains the same as that of a rectangular weir. *i.e.*,

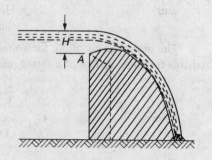

Fig. 12·6. Ogee weir.

$$Q = \frac{2}{3} \times C_d \cdot L \sqrt{2g} \times H^{3/2}$$

where
C_d = Coefficient of discharge, and
L = Length of an ogee weir.

Example 12·20. *An ogee weir 4 metres long has 500 mm head of water. Find the discharge over the weir, if $C_d = 0.62$.*

Solution. Given : $L = 4$ m; $H = 500$ mm $= 0.5$ m and $C_d = 0.62$.

We know that the discharge over the weir,

$$Q = \frac{2}{3} \times C_d \cdot L \sqrt{2g} \times H^{3/2}$$

$$= \frac{2}{3} \times 0.62 \times 4 \sqrt{2 \times 9.81} \times (0.5)^{3/2} \text{ m}^3/\text{s}$$

$$= 7.323 \times 0.354 = 2.59 \text{ m}^3/\text{s} = 2590 \text{ litres/s} \quad \textbf{Ans.}$$

Example 12·21. *An ogee weir 8 metres long, with suppressed end contractions, is discharging water under a head of 0·25 m. Using Francis' and Bazin's formula, determine the discharge over the weir in litres/s.*

Solution. Given : $L = 8$ m and $H = 0.25$ m.

Discharge over the weir by Francis' formula

We know that the discharge over the weir (by Francis' formula),

$$Q = 1.84 (L - 0.1nH) H^{3/2}$$

$$= 1.84 [8 - (0.1 \times 2 \times 0.25)] \times (0.25)^{3/2} \text{ m}^3/\text{s}$$

$$= 14.63 \times 0.125 = 1.83 \text{ m}^3/\text{s}$$

Discharge over the weir by Bazin's formula

We know from Bazin's relation that

$$m = 0.405 + \frac{0.003}{H} = 0.405 + \frac{0.003}{0.25} = 0.417$$

and discharge over the weir (by Bazin's formula),

$$Q = mL \sqrt{2g} \times H^{3/2}$$

$$= 0.417 \times 8 \times \sqrt{2 \times 9.81} \times (0.25)^{3/2} \text{ m}^3/\text{s}$$

$$= 14.78 \times 0.125 = 1.85 \text{ m}^3/\text{s} \quad \textbf{Ans.}$$

12·14 Discharge over a Submerged or Drowned Weir

When the water level on the downstream side of a weir is above the top surface of weir, it is known a submerged or drowned weir as shown in Fig. 12·7.

The total discharge, over such a weir, is found out by splitting up the height of water, above the sill of the weir, into two portions as discussed below :

Let
H_1 = Height of water on the upstream side of the weir, and
H_2 = Height of water on the downstream side of the weir.

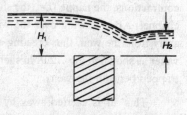

Fig. 12·7. Submerged weir.

The discharge over the upper portion may be considered as a free discharge under a head of water equal to $(H_1 - H_2)$. And the discharge over the lower portion may be considered as a submerged discharge under a head of H_2.

Thus discharge over the free portion (*i.e.,* upper portion),

$$Q_1 = \frac{2}{3} \times C_d . L \sqrt{2g} (H_1 - H_2)^{3/2}$$

and the discharge over the submerged (*i.e.,* lower portion),

$$Q_2 = C_d . L . H_2 \sqrt{2g (H_1 - H_2)}$$

∴ Total discharge, $Q = Q_1 + Q_2$

Example 12·22. *A submerged sharp crested weir 0·8 metre high stands clear across a channel having vertical sides and a width of 3 metres. The depth of water in the channel of approach is 1·25 metre. And 10 metres downstream from the weir, the depth of water is 1 metre. Determine the discharge over the weir in litres per second. Take C_d as 0·6.*

Solution. Given : $L = 3$ m and $C_d = 0.6$.

From the geometry of the weir, we find that the depth of water on the upstream side,

$$H_1 = 1.25 - 0.8 = 0.45 \text{ m}$$

and depth of water on the downstream side,

$$H_2 = 1 - 0.8 = 0.2 \text{ m}$$

We know that the discharge over the free portion of the weir,

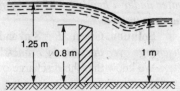

Fig. 12·8

$$Q_1 = \frac{2}{3} \times C_d L \sqrt{2g} (H_1 - H_2)^{3/2}$$

$$= \frac{2}{3} \times 0.6 \times 3 \sqrt{2 \times 9.81} (0.45 - 0.20)^{3/2}$$

$$= 5.315 \times 0.125 = 0.664 \text{ m}^3/\text{s} = 664 \text{ litres/s} \qquad ...(i)$$

and discharge over the submerged portion of the weir,

$$Q_2 = C_d . L . H_2 \sqrt{2g (H_1 - H_2)}$$

$$= 0.6 \times 3 \times 0.2 \sqrt{2 \times 9.81 (0.45 - 0.2)} \text{ m}^3/\text{s}$$

$$= 0.36 \times 2.215 = 0.797 \text{ m}^3/\text{s} = 797 \text{ litres/s} \qquad ...(ii)$$

∴ Total discharge : $Q = Q_1 + Q_2 = 664 + 797 = 1461$ litres/s **Ans.**

12·15 Ventilation of Rectangular Weirs

It has been observed that whenever water is flowing over a rectangular weir, having no end contractions, the nappe (*i.e.,* the sheet of water flowing over the weir) touches the side walls of the channel. After flowing over the weir, the nappe falls away from the weir, thus creating a space beneath the water as shown in Fig. 12·9. In such a case, some air is trapped beneath the weir.

This air is carried away by the flowing water, which results in creating a negative pressure beneath the nappe. This negative pressure drags the lower side of the nappe towards the surface of the weir wall. This

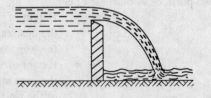

Fig. 12·9. Ventilation of weirs.

Flow Over Weirs

results in more discharge than the normal discharge* obtained in Art. 12·3. In order to keep the atmospheric pressure in the space below the nappe, holes are made through the channel walls which are connected through the pipes to the atmosphere as shown in Fig. 12·10. Such holes are called ventilation of a weir. Though there are many types of the nappes, yet the following are important from the subject point of view :
1. Free nappe,
2. Depressed nappe, and
3. Clinging nappe.

12·16 Free Nappe

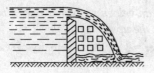
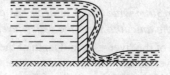
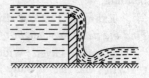

Fig. 12.10. Free nappe.   Fig. 12.11. Depressed nappe.   Fig. 12.12. Clinging nappe.

If the atmospheric pressure exists beneath the nappe, it is known as a free nappe as shown in Fig. 12·10. The discharge of a free nappe is the same as given in Art. 12·3. A free nappe is obtained by ventilating a weir (*i.e.*, by providing holes through the channel wall and connecting the same through pipes with the atmosphere) as shown in Fig. 12·10.

12·17 Depressed Nappe

Sometimes, a weir is not fully ventilated, but is partially ventilated as shown in Fig. 12·11. If the pressure below the nappe is negative, it is called a depressed nappe.

The discharge of the nappe, in this case, depends upon the amount of ventilation and the negative pressure. Generally, the discharge of a depressed nappe is 6% to 7% more than that of a free nappe.

12·18 Clinging Nappe

Sometimes, no air is left below the water, and the nappe adheres or clings to the downstream side of the weir as shown in Fig. 12·12. Such a nappe is called clinging nappe or an adhering nappe.

The discharge of a clinging nappe is 25% to 30% more than that of a free nappe.

12·19 Practical Applications of Weirs

Though there are numerous practical applications of weirs in different forms, yet the following are important from the subject point of view :
1. Anicut, raised weir or barrage,
2. Bridge openings, and
3. Siphon spillway.

12·20 Anicut (*i.e.*, Raised Weir) and Barrage

An anicut is a small masonry dam type structure, built across a river for raising the water level on the upstream side. It is, generally, constructed to regulate the supply of water to the canals.

---

*$Q = \dfrac{2}{3} \times C_d \, L \, \sqrt{2g} \times H^{3/2}$

As a matter of fact, anicut is nothing but a weir over which the water flows. During floods, an anicut behaves like a drowned weir (as the downstream water level is higher than the crest level).

A barrage is an improved form of weir in which gates are provided on the top of the weir crest. In a barrage, the weir crest is kept at a low level and ponding up of the river water is accomplished primarily by means of gates. These gates can be raised or lowered to clear off the high flood level. A barrage provides a perfect control of the river water on the upstream.

The difference between a barrage and a weir is only qualitative. In the former, the gates or shutters provide larger part of the ponding, while in the latter, the solid crest carries out most of the ponding.

Example 12·23. *A river 30 m wide and 3 m deep has a mean velocity of water as 1·2 m/s. Find the necessary height of the anicut to raise the water level on the upstream side by 1 m.*

Solution. Given : Width of the river = 30 m; Depth of the river = 3 m; Mean velocity of water = 1·2 m/s and rise of water level on the upstream side = 1 m.

Let H = Height of water above the anicut.

We know that the discharge in the river,

Q = Width of river × Depth of river × Mean velocity of water
$= 30 \times 3 \times 1 \cdot 2 = 108 \text{ m}^3/\text{s}$

and the depth of water after the construction of anicut
$= 3 + 1 = 4 \text{ m}$

∴ Area of flow, $A = 30 \times 4 = 120 \text{ m}^2$

Now velocity of approach after the construction of anicut,

$$v = \frac{Q}{A} = \frac{108}{120} = 0 \cdot 9 \text{ m/s}$$

and head due to velocity of approach,

$$H_a = \frac{v^2}{2g} = \frac{(0 \cdot 9)^2}{2 \times 9 \cdot 81} = 0 \cdot 041 \text{ m}$$

∴ Total head, $H_1 = H + H_a = H + 0 \cdot 041 \text{ m}$

We also know that the discharge in the river (Q),

$$108 = \frac{2}{3} \times C_d . L \sqrt{2g} \left[H_1^{3/2} - H_a^{3/2} \right]$$

$$= \frac{2}{3} \times 0 \cdot 6 \times 30 \times \sqrt{2 \times 9 \cdot 81} \times \left[(H + 0 \cdot 041)^{3/2} - (0 \cdot 041)^{3/2} \right]$$

$$= 53 \cdot 1 \left[(H + 0 \cdot 041)^{3/2} - 0 \cdot 008 \right]$$

$$(H + 0 \cdot 041)^{3/2} - 0 \cdot 008 = \frac{108}{53 \cdot 1} = 2 \cdot 03$$

By trial and error, $H = 1 \cdot 565 \text{ m}$

∴ Height of anicut $= 4 - 1 \cdot 565 = 2 \cdot 435 \text{ m}$ **Ans.**

12·21 Bridge Openings

A bridge consists of piers and abutments, which offer obstruction to the flow of river water. At a bridge site, the river water flows through the piers. As a result of this, the depth of water on the upstream side rises. Such a rise of water is known as afflux. This phenomenon will be discussed in detail in chapter non-uniform flow through open channels.

12·22 Syphon Spillway

A syphon spillway is a part of a dam through which the water flows. It consists of one or more conduits, usually, of rectangular section. These conduits communicate at the upper end with the upstream water level of the dam and at the lower end with the downstream as shown in Fig. 12·13.

As soon as the upstream water level rises above the inner crest of the spillway, the water begins to flow through the siphon. As the water flows down, it sucks the air from the crown of the siphon, which causes partial vacuum and increases the rate of flow. The increased rate of flow, over the crest, sweeps the air with a greater speed. As a result of this, the siphon runs full, while discharging water. The head of water, under which the flow takes place will be increased from h to H. As the head of water H is more than h, therefore the discharge can be considerably increased with the help of a syphon spillway.

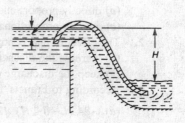

Fig. 12·13. Syphon spillway.

12·23 Advantages of Syphon Spillway

A syphon spillway has the following two advantages, which are important from the subject point of view :
1. The head of water, under which the flow takes place, can be increased as a result of which the discharge through the siphon can be increased.
2. The crest of a siphon spillway can be raised as the head of water above the crest (h), required for the commencement of flow, is only a few cm. As a result of raising the crest of siphon, a greater amount of water can be stored in the reservoir.

EXERCISE 12·3

1. A narrow crested weir of 6·5 m long is discharging water under a head of 300 mm. If the coefficient of discharge for the weir is 0·6, find the discharge in litres/s. [**Ans.** 1892 litres/s]
2. A broad crested weir 15 m long is discharging water with 500 mm and 300 mm heads of water on the upstream and downstream sides respectively. Determine the discharge over the weir, if the coefficient of discharge is 0·6. What will be the maximum discharge over the weir?
   [**Ans.** 5·35 m³/s; 5·44 m³/s]
3. Find the discharge over a sharp-crested weir of 5 m long under a head of 150 mm. Take C_d for the weir as 0·625. [**Ans.** 0·536 m³/s]
4. An ogee weir of 5·5 m long is discharging water under a constant head of 350 mm. If the coefficient of discharge is 0·6, calculate the discharge over the weir. [**Ans.** 2·02 m³/s]
5. A submerged weir 3·5 m long has upstream and downstream water levels of 800 mm and 500 mm respectively above its crest. Find the discharge over the weir, if the coefficients of discharge for the free portion is 0·61 and that for the drowned portion as 0·6. [**Ans.** 4·44 m³/s]

QUESTIONS

1. What is a weir ? Explain the difference between a weir and a notch.
2. Obtain an expression for the discharge over a rectangular weir.
3. What is Francis' formula for discharge over rectangular weirs?
4. What is a Cippoletti weir ? How does it differ from a rectangular weir ?
5. Define velocity of approach. How does it affect the discharge over a weir ?
6. Derive an expression for the theoretical discharge of a rectangular weir with velocity of approach.

7. What is nappe of a weir? Describe the free, depressed and clinging nappes with the help of sketches. State how do they effect the discharge measurement in case of weir.
8. Explain clearly the difference between a narrow and a broad-crested weir.
9. What is a drowned weir? Derive an equation for discharge over such a weir.

OBJECTIVE TYPE QUESTIONS

1. The discharge over a rectangular weir is
   (a) directly proportional to H
   (b) indirectly proportional to H
   (c) directly proportional to L
   (d) both 'a' and 'c'
2. The effect of end contractions over a rectangular weir is to
   (a) increase the discharge
   (b) decrease the discharge
   (c) keep the discharge constant
   (d) keep the head constant
3. According to Francis' formula, the discharge over a rectangular weir is
   (a) $1 \cdot 84 \, (L - 0 \cdot 1 H) \sqrt{2g} \times H^{3/2}$
   (b) $1 \cdot 84 \, (L - 0 \cdot 1 \, H)^{3/2}$
   (c) $1 \cdot 84 \, (L - 0 \cdot 2 \, H) \sqrt{2g} \, H^{3/2}$
   (d) $1 \cdot 84 \, (L - 0 \cdot 2 \, H)^{3/2}$
   where H = Height of water above the crest of the weir.
4. A weir is said to be narrow-crested, if its crest width is less than
   (a) height of water
   (b) half the height of water
   (c) length of the weir
   (d) half the length of weir
5. The maximum discharge over a broad crested weir is
   (a) $1 \cdot 71 \, C_d \, . \, L \, . \, \sqrt{2gH}$
   (b) $1 \cdot 71 \, C_d \, . \, L \, . \, \sqrt{2g} \times H$
   (c) $1 \cdot 71 \, C_d \, . \, L \, . \, \sqrt{2g} \times H^{3/2}$
   (d) $1 \cdot 71 \, C_d \, . \, L \, . \, H^{3/2}$

ANSWERS

1. (d)   2. (b)   3. (d)   4. (b)   5. (d)

13

Flow Through Simple Pipes

1. Introduction. 2. Loss of Head in Pipes. 3. Darcy's Formula for Loss of Head in Pipes. 4. Chezy's Formula for Loss of Head in Pipes. 5. Graphical Representation of Pressure Head and Velocity Head. 6. Hydraulic Gradient Line. 7. Total Energy Line. 8. Transmission of Power through Pipes. 9. Time of Emptying a Tank through a Long Pipe. 10. Time of Flow from One Tank into Another through a Long Pipe.

13·1 Introduction

A pipe is a closed conduit, generally of circular cross-section, used to carry water or any other fluid. When the pipe is running full, the flow is under pressure. But, if the pipe is not running full (as in the case of sewer pipes, culverts etc.), the flow is not under pressure. In such a case the atmospheric pressure exists inside the pipe. In this chapter, we shall discuss the flow in the pipes under pressure only.

13·2 Loss of Head in Pipes

When the water is flowing in a pipe, it experiences some resistance to its motion, whose effect is to reduce the velocity and ultimately the head of water available. Though there are many types of losses, yet the major loss is due to frictional resistance of the pipe only. The frictional resistance of a pipe depends upon the roughness of the inside surface of the pipe. It has been experimentally found that more the roughness of the inside surface of the pipe, greater will be the resistance. This friction is known as fluid friction and the resistance is known as frictional resistance.

The earlier experiments on the fluid friction were conducted by *Froude who concluded that :

1. The frictional resistance varies approximately with the square of the velocity of the liquid.
2. The frictional resistance varies with the nature of the surface.

Later on, some empirical formulae were derived for the loss of head due to friction, out of which the following two are important from the subject point of view :

1. Darcy's formula for loss of head in pipes, and
2. Chezy's formula for loss of head in pipes.

---

*Froude William was an English scientist and engineer who first performed his experiments on ship models in tanks. He gave the basic theory for the liquid flow through the pipes.

13·3 Darcy's Formula for Loss of Head in Pipes

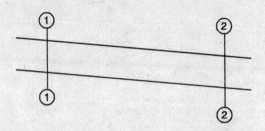

Fig. 13·1. Uniform long pipe.

Consider a uniform long* pipe through which water is flowing at a uniform rate as shown in Fig. 13·1.

Let
  l = Length of pipe,
  d = Diameter of the pipe,
  v = Velocity of water in the pipe,
  f' = Frictional resistance per unit area (of wetted surface) per unit velocity, and
  h_f = Loss of head due to friction.

Let us consider sections (1-1) and (2-2) of the pipe.

Now let
  p_1 = Intensity of pressure at section 1-1, and
  p_2 = Intensity of pressure at section 2-2.

A little consideration will show, that p_1 and p_2 would have been equal, if there would have been no frictional resistance. Now considering horizontal forces on water between sections (1-1) and (2-2) and equating the same,

$$p_1 A = p_2 A + \text{Frictional resistance}$$

or frictional resistance $= p_1 A - p_2 A$

$\therefore \quad \dfrac{\text{Frictional resistance}}{w} = \dfrac{p_1 A - p_2 A}{w}$ ...(Dividing both sides by w)

$\dfrac{\text{Frictional resistance}}{Aw} = \dfrac{p_1}{w} - \dfrac{p_2}{w}$

But $\dfrac{p_1}{w} - \dfrac{p_2}{w} = h_f$ = Loss of pressure head due to friction

$\therefore \quad h_f = \dfrac{\text{Frictional resistance}}{A w}$

$= \dfrac{\text{Frictional resistance}}{\dfrac{\pi}{4} \times d^2 \times w}$ ...$\left(\text{Area, } A = \dfrac{\pi}{4} \times d^2\right)$

---

*Strictly speaking, there is no hard and fast rule to define a long pipe. But a pipe is, generally, termed as a long pipe when its length is more than 1000 times its diameter. And it is termed as a short pipe when its length is less than 1000 times its diameter.

In a long pipe, the major loss of head is due to friction in the pipe only. The minor losses are so small, as compared to the friction loss, that they may be neglected altogether. But in case of a short pipe, the minor losses, as compared to the friction loss, are of appreciable amount and, thus, cannot be neglected.

Flow Through Simple Pipes

We know that as per Froude's* experiment, frictional resistance

h_f = Frictional resistance per unit area at unit velocity × Wetted area × (Velocity)²

$$f \times \pi dl \times v^2$$

Substituting the value of frictional resistance in the above equation,

$$h_f = \frac{f \, \pi dl \times v^2}{\frac{\pi}{4} \times d^2 w} = \frac{4f' lv^2}{wd}$$

Let us introduce another coefficient** (f') such that

$$f' = \frac{fw}{2g}$$

$$\therefore \quad h_f = \frac{4}{wd} \times \frac{fw}{2g} \times lv^2 = \frac{4flv^2}{2gd} \qquad \ldots(i)$$

We know that the discharge,

$$Q = \frac{\pi}{4} \times d^2 \times v \qquad \text{or} \qquad v = \frac{4Q}{\pi d^2}$$

$$\therefore \quad v^2 = \frac{16Q^2}{\pi^2 d^4}$$

Substituting the value of v^2 in equation (i),

$$h_f = \frac{4fl}{2gd} \times \frac{16Q^2}{\pi^2 d^4} = \frac{flQ^2}{3d^5} \qquad \ldots(ii)$$

Notes: 1. The equations (i) and (ii) give us the value of loss of head in pipes due to friction. The following points should be taken into consideration at the time of using the above equation :
(a) The equation (i) should be used when the velocity of water in the pipe is known.
(b) The equation (ii) should be used when the discharge in the pipe is known.
2. In addition to the major loss of head due to friction the following minor losses of head also take place +

(a) Loss of head at entrance $= \dfrac{0 \cdot 5 v^2}{g}$

(b) Loss of head due to velocity of water at outlet

$$= \frac{v^2}{2g}$$

---

*Prof. Froude conducted a series of experiments to find the resistance of vertical plane boards, when propelled in water. He conducted the experiments in a tank about 85 m long, 11 m wide and 3 m deep.

A truck was propelled by an endless wire rope, on a light railway track, with different velocities between 30 m/min to 300 m/min. He immersed boards of wood in water whose surfaces were covered with varnish, oil and sand. These boards were towed in water, so that its upper edge was 50 cm below the water surface, and the force required was measured.

**This coefficient is called Darcy's coefficient or friction coefficient. Darcy recommended the following values of f:

(i) $f = 0 \cdot 005 \left(1 + \dfrac{1}{12 \, d}\right)$ ...(For new and smooth pipes)

(ii) $f = 0 \cdot 01 \left(1 + \dfrac{1}{12 \, d}\right)$ ...(For old and rough pipes)

3. Thus, if all losses are considered, then

$$h_f = \frac{0.5v^2}{2g} + \frac{4flv^2}{2gd} + \frac{v^2}{2g}$$

But, if the minor losses (being very small as compared to the frictional loss) are neglected, then

$$h_f = \frac{4flv^2}{2gd}$$

4. In actual practice, the minor losses are neglected, until and unless mentioned in the example. One of the reasons, for this, is to make the calculation simple.

Example 13·1. *Find the loss of head, due to friction, in a pipe of 500 mm diameter and 1·5 kilometres long. The velocity of water in the pipe is 1 metre per second. Take coefficient of friction as 0·0 05.*

Solution. Given : $d = 500$ mm $= 0.5$ m; $l = 1.5$ km $= 1500$ m; $v = 1$ m/s and $f = 0.005$.

Since length of the pipe (1500 m) is more than 1000 d (1000 × 0·5 = 500 m), therefore it is taken as a long pipe. Now let us neglect all the minor losses except friction. We know that the loss of head due to friction,

$$h_f = \frac{4flv^2}{2gd} = \frac{4 \times 0.005 \times 1500 \times (1)^2}{2 \times 9.81 \times 0.5} = 3.01 \text{ m} \quad \textbf{Ans.}$$

Example 13·2. *Water is flowing through a pipe 1500 m long with a velocity of 0·8 m/s. What should be the diameter of the pipe, if the loss of head due to friction is 8·7 m. Take f for the pipe as 0·01.*

Solution. Given : $l = 1500$ m; $v = 0.8$ m/s; $h_f = 8.7$ m and $f = 0.01$.
Let $\quad\quad\quad d = $ Diameter of the pipe
We know that the loss of head due to friction (h_f),

$$8.7 = \frac{4flv^2}{2gd} = \frac{4 \times 0.01 \times 1500 \times (0.8)^2}{2 \times 9.8 \times d} = \frac{1.957}{d}$$

∴ $\quad\quad\quad d = 1.957/8.7 = 0.225$ m $= 225$ mm **Ans.**

Example 13·3. *It was observed that the difference of heads between the two ends of a pipe 250 metres long and 300 mm diameter is 1·5 metre. Taking Darcy's coefficient as 0·01 and neglecting minor losses, calculate the discharge flowing through the pipe.*

Solution. Given : $l = 250$ m; $d = 300$ mm $= 0.3$ m; $h_f = 1.5$ m and $f = 0.01$.
Let $\quad\quad\quad Q = $ Discharge through the pipe.
As a matter fact, the pipe should be taken as a short pipe as the length of the pipe (250 m) is less than 1000 d (1000 × 0·3 = 300 m). But, in the example, it has been mentioned that we have to calculate the discharge by neglecting, minor losses. So this point is not to be considered.

We know that the loss of head due to friction (h_f),

$$1.5 = \frac{flQ^2}{3d^5} = \frac{0.01 \times 250 \times Q^2}{3 \times (0.3)^5} = 342.9 Q^2$$

∴ $$Q = \sqrt{\frac{1.5}{342.9}} = 0.0661 \text{ m}^3/\text{s} = 66.1 \text{ litres/s} \quad \textbf{Ans.}$$

Example 13·4. *A pipe of 60 metres long and 150 mm in diameter is connected to a water tank at one end and flows freely into the atmosphere at the other end. The height of water level in the*

Flow Through Simple Pipes

tank is 2·6 metres above the centre of the pipe. The pipe is horizontal and f = 0·01. Determine the discharge through the pipe in litres/s, if all the minor losses are to be considered.

Solution. Given : $l = 60$ m; $d = 150$ mm $= 0·15$ m; $h_f = 2·6$ m and $f = 0·01$.

Let $\qquad v =$ Velocity of water through the pipe.

In the example, it has been mentioned that we have to determine the discharge by considering all the minor losses. We know that loss of head due to friction, considering minor losses (h_f),

$$2·6 = \frac{0·5v^2}{2g} + \frac{4flv^2}{2gd} + \frac{v^2}{2g}$$

$$= \frac{v^2}{2g}\left(0·5 + \frac{4 \times 0·01 \times 60}{0·15} + 1\right) = \frac{17·5\,v^2}{2g}$$

$$\therefore \qquad v = \sqrt{\frac{2·6 \times 2 \times 9·81}{17·5}} = 1·71 \text{ m/s}$$

and discharge through the pipe,

$$Q = a \times v = \frac{\pi}{4} \times (0·15)^2 \times 1·71 \text{ m}^3/\text{s}$$

$$= 0·0302 \text{ m}^3/\text{s} = 30·2 \text{ litres/s} \quad \textbf{Ans.}$$

Example 13·5. *A reservoir has been built 4 kilometres away from a college campus having 5000 inhabitants. Water is to be supplied from the reservoir to the campus. It is estimated that each inhabitant will consume 200 litres of water per day, and that half of the daily supply is pumped within 10 hours. Calculate the size of the supply main, if the loss of head due to friction in pipeline is 20 m. Assume coefficient of friction for the pipe line as 0·008.*

Solution. Given : $l = 4$ km $= 4000$ m; No. of habitants (n) $= 5000$; Consumption of water by each inhabitant per day $= 200$ litres $= 0·2$ m³; $h_f = 20$ m and $f = 0·008$.

Let $\qquad d =$ Diameter of the pipe.

We know that the total supply

$$= \text{No. of inhabitants} \times \text{Consumption by each inhabitant}$$
$$= 5000 \times 0·2 = 1000 \text{ m}^3/\text{day}$$

Since half of this supply is to be pumped in 10 hours, therefore maximum flow for which the pipe is to be designed,

$$Q = \frac{1000}{2 \times 10 \times (60 \times 60)} = 0·014 \text{ m}^3/\text{s}$$

We also know that the loss of head due to friction (h_f),

$$20 = \frac{flQ^2}{3d^5} = \frac{0·008 \times 4000 \times (0·014)^2}{3d^5} = \frac{0·0021}{d^5}$$

$$d^5 = 0·0021/20 = 0·0001$$

or $\qquad d = 0·158$ m $= 158$ mm say 160 mm $\quad$ **Ans.**

13·4 Chezy's Formula for Loss of Head in Pipes

Consider a uniform long pipe through which water is flowing at a uniform rate as shown in Fig. 13·2.

Let $\qquad l =$ Length of the pipe, and
$\qquad d =$ Diameter of the pipe,

∴ Area of pipe, $\qquad A = \dfrac{\pi}{2} \times d^2$

and perimeter of pipe, $\qquad P = \pi d$

Let $\qquad v =$ Velocity of water in pipe,

$\qquad f' =$ Frictional resistance, per unit area (of wetted surface) per unit velocity, and

$\qquad h_f =$ Loss of head due to friction.

Now let us consider sections (1-1) and (2-2) of the pipe.

Let $\qquad p_1 =$ Intensity of pressure at section 1-1, and

$\qquad p_2 =$ Intensity of pressure at section 2-2.

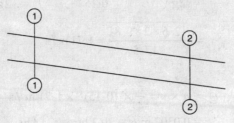

Fig. 13·2. Uniform long pipe.

A little consideration will show that p_1 and p_2 would have been equal, if there would have been no frictional resistance. Now considering horizontal forces on water between sections (1-1) and (2-2) and equating the same,

$$p_1 A = p_2 A + \text{Frictional resistance}$$

or $\qquad$ frictional resistance $= p_1 A - p_2 A$

$$\dfrac{\text{Frictional resistance}}{w} = \dfrac{p_1 A - p_2 A}{w} \qquad \text{...(Dividing both sides by } w\text{)}$$

or $\qquad \dfrac{\text{Frictional resistance}}{Aw} = \dfrac{p_1}{w} - \dfrac{p_2}{w}$

But $\qquad \dfrac{p_1}{w} - \dfrac{p_2}{w} = h_f =$ Loss of pressure head due to friction

∴ $\qquad h_f = \dfrac{\text{Frictional resistance}}{Aw}$

We know that as per Froude's experiment, frictional resistance

$\qquad$ = Frictional resistance per unit area at unit velocity

$\qquad \qquad \times$ Wetted area $\times$ (Velocity)^2

$\qquad = f' \times \pi d l \times v^2$

Substituting the value of frictional resistance in the above equation,

$$h_f = \dfrac{f' \times \pi d l \times v^2}{Aw} = \dfrac{f' \times P \times l v^2}{Aw} \qquad \text{...(} \because \pi d = P \text{ i.e., perimeter)}$$

$$= \dfrac{f' l v^2}{w} \times \dfrac{P}{A}$$

Flow Through Simple Pipes

Now substituting another term of hydraulic mean depth (also known as hydraulic radius) in the above equation, such that hydraulic mean depth,

$$m = \frac{\text{Area of flow}}{\text{Wetted perimeter}} = \frac{A}{P}$$

∴
$$h_f = \frac{f'lv^2}{w} \times \frac{1}{m}$$

or
$$v^2 = \frac{h_f \cdot w \cdot m}{f' \cdot l} = \frac{w}{f'} \times m \times \frac{h_f}{l}$$

∴
$$v = \sqrt{\frac{w}{f'} \times m \times \frac{h_f}{l}} \qquad ...(i)$$

Now substituting two more terms in the above equation, such that

∴
$$v = \sqrt{\frac{w}{f'}} = C \qquad ...(A \text{ constant known as Chezy's constant})$$

and
$$\frac{h_f}{l} = i \qquad ...(i.e., \text{ loss of head per unit length})$$

Now substituting the above two values in equation (i),

$$v = C\sqrt{mi}$$

Notes: 1. Darcy's formula for loss of head is generally used for the flow through pipes, until and unless mentioned in the example.

2. Chezy's formula for loss of head is, generally, used for the flow through open channels. It will be discussed in the next chapters.

3. The value of hydraulic mean depth for a circular pipe,

$$m = \frac{\text{Area}}{\text{Perimeter}} = \frac{\frac{\pi}{4} \times d^2}{4\pi} = \frac{d}{4}$$

Example 13·6. *Water flows through a pipe of 200 mm in diameter 60 m long with a velocity of 2·5 m/s. Find the head lost due to friction.*

(a) by using Darcy's formula,

$$h_f = \frac{4flv^2}{2gd}; \text{ assuming } f = 0.005$$

(b) by using Chezy's formula,

$$v = C\sqrt{mi} \text{ ; assuming } C = 55$$

Solution. Given: $d = 200$ mm $= 0·2$ m; $l = 60$ m; $v = 2·5$ m/s; $f = 0·005$ and $C = 55$.

Head lost due to friction by using the Darcy's formula.

We know that the head lost due to friction,

$$h_f = \frac{4flv^2}{2gd} = \frac{4 \times 0·005 \times 60 \times (2·5)^2}{2 \times 9·81 \times 0·2} = 1·91 \text{ m} \qquad \textbf{Ans.}$$

Head lost due to friction by using Chezy's formula

We know that the hydraulic mean depth of a circular pipe,

$$m = \frac{d}{4} = \frac{0·2}{4} = 0·05 \text{ m}$$

and the loss of head per unit length,

$$i = \frac{h_f}{l} = \frac{h_f}{60}$$

We also know that the velocity of water (v),

$$2 \cdot 5 = C\sqrt{mi} = 55\sqrt{0 \cdot 55 \times \frac{h_f}{60}}$$

Squaring both sides, $\quad 6 \cdot 25 = 3025 \times 0 \cdot 05 \times \dfrac{h_f}{60} = 2 \cdot 52\, h_f$

$\therefore \quad h_f = 6 \cdot 25/2 \cdot 52 = 2 \cdot 48$ m   **Ans.**

Example 13·7. *A town having a population of 100 000 is to be supplied with water from a reservoir at 5 kilometres distance. It is stipulated that one-half of the daily supply of 150 litres per head should be delivered within 8 hours. What must be the size of the pipe to furnish the supply, if the head available is 12 metres? Take C = 45 in Chezy's formula.*

Solution. Given : Population = 100 000; l = 5 km = 5000 m; Daily supply = 150 litres/head = 0·15 m³/head; h_f = 12 m and C = 45

Let $\qquad\qquad d$ = Diameter of the pipe.

We know that the total supply

$$= \text{Population} \times \text{Daily consumption}$$
$$= 100\,000 \times 0 \cdot 15 = 15\,000 \text{ m}^3/\text{day}$$

Since half of this quantity is to be delivered in 8 hours, therefore maximum flow,

$$Q = \frac{15\,000}{2 \times 8 \times 3\,600} = 0 \cdot 26 \text{ m}^3/\text{sec}$$

We know that the hydraulic mean depth for a circular pipe,

$$m = \frac{d}{4}$$

and loss of head per unit length,

$$i = \frac{h_f}{l} = \frac{12}{5000} = 0 \cdot 0024$$

$\therefore$ Velocity of water, $\quad v = C\sqrt{mi} = 45 \times \sqrt{\dfrac{d}{4} \times 0 \cdot 0024} = 45\sqrt{\dfrac{d}{0 \cdot 0096}}$

We also know that the discharge through the pipe (Q),

$$0 \cdot 26 = A \times v = \frac{\pi}{4} \times (d)^2 \times 45\sqrt{\frac{d}{0 \cdot 0096}}$$

Squaring both sides of the above equation,

$$0 \cdot 0676 = \frac{\pi^2}{16} \times d^4 \times 2025 \times \frac{d}{0 \cdot 0096} = 130\,100\, d^5$$

$\therefore \qquad\qquad d^5 = 0 \cdot 0676/130\,100 = 0 \cdot 0052 \times 10^{-5}$

or $\qquad\qquad d = 0 \cdot 35 \times 10^{-1} = 0 \cdot 035$ m = 35 mm   **Ans.**

Flow Through Simple Pipes

13·5 Graphical Representation of Pressure Head and Velocity Head

In design offices, sometimes, we are interested to know the pattern in which the pressure head or the velocity head changes at the pipe junctions. It becomes more interesting at points where the cross-section of the beam suddenly changes (*i.e.*, at sudden enlargement or sudden contraction). It is generally done by drawing the following lines :

1. Hydraulic gradient lines. 2. Total energy line.

13·6 Hydraulic Gradient Line

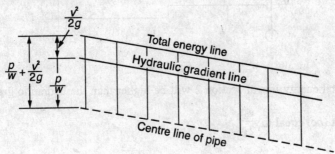

Fig. 13·3. Hydraulic gradient and total energy lines

If pressure heads (*i.e.*, p/w) of a liquid flowing in a pipe be plotted as vertical ordinates on the centre line of the pipe, then the line joining the tops of such ordinates is known as hydraulic gradient line (briefly written as H.G.L.) as shown in Fig. 13·3.

13·7 Total Energy Line

If the sum of pressure heads and velocity heads $\left(\dfrac{p}{w} + \dfrac{v^2}{2g}\right)$ of a liquid flowing in a pipe be plotted as vertical ordinates on the centre line of the pipe, then the line joining the tops of such ordinates is known as total energy line (briefly written as T.E.L.) as shown in Fig. 13·3.

Or, in other words, the total energy line lies over the hydraulic gradient by an amount equal to the velocity heads as shown in Fig. 13·3.

Note : The above mentioned two lines give very useful information regarding the flow of liquid in the pipeline. Some of them will be discussed in the following pages.

Example 13·8. *At a sudden enlargement of a water line from 240 mm to 480 mm diameter pipe, the hydraulic gradient rises by 10 mm. Estimate the rate of flow.*

Solution. Given : $d_1 = 240$ mm $= 0.24$ m; $d_2 = 480$ mm $= 0.48$ m and rise in hydraulic gradient line $= 10$ mm $= 0.01$ m.

Let $v_1 =$ Velocity of water at section 1, and

$v_2 =$ Velocity of water at section 2.

First of all, let us draw the hydraulic gradient line and the total energy line above the centre line of the pipe as discussed below and shown in Fig. 13·4 (*a*), (*b*) and (*c*).

1. The total energy line at section 1 will be higher than the hydraulic gradient line by an amount (*ad*) equal to $\dfrac{v_1^2}{2g}$.

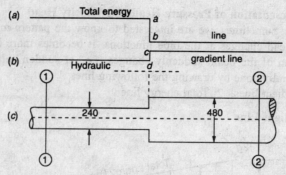

Fig. 13·4.

2. The total energy line at section 2 will be higher than the hydraulic gradient line by an amount (bc) equal to $\dfrac{v_2^2}{2g}$.

3. The total energy line, at sudden enlargement, will come down by an amount (ab) equal to $\dfrac{(v_1 - v_2)^2}{2g}$.

4. The hydraulic gradient line, at sudden enlargement, will rise by an amount (dc) equal to 10 mm = 0·01 m (as given).

From the equation of continuity, we know that

$$\frac{\pi}{4} \times (0 \cdot 24)^2 \times v_1 = \frac{\pi}{4} \times (0 \cdot 48)^2 \times v_2 \qquad \ldots (\because a_1 v_1 = a_2 v_2)$$

$\therefore \qquad v_1 = 4v_2$

Now from the geometry of the figure, we find that

$$ad = ab + bc + cd$$

$\therefore \qquad \dfrac{v_1^2}{2g} = \dfrac{(v_1 - v_2)^2}{2g} + \dfrac{v_2^2}{2g} + 0 \cdot 01$

$$\frac{16v_2^2}{2g} = \frac{9v_2^2}{2g} + \frac{v_2^2}{2g} + 0 \cdot 01 \qquad \ldots \text{(Substituting } v_1 = 4v_2\text{)}$$

$\therefore \qquad \dfrac{6v_2^2}{2g} = 0 \cdot 01$

or $\qquad v_2 = \sqrt{\dfrac{2 \times 9 \cdot 81 \times 0 \cdot 01}{6}} = \sqrt{0 \cdot 0327} = 0 \cdot 181 \text{ m/s}$

$\therefore$ Rate of flow, $Q = a_2 v_2 = \dfrac{\pi}{4} \times (0 \cdot 48)^2 \times 0 \cdot 181 \text{ m}^3/\text{s}$

$\qquad = 0 \cdot 0328 \text{ m}^3/\text{s} = 32 \cdot 8 \text{ litres/s}$ **Ans.**

EXERCISE 13·1

1. A pipe of 200 mm diameter and 300 m long is discharging water with a velocity of 1·2 m/s. Find the loss of head due to friction. Take coefficient of friction for the pipe as 0·01. [Ans. 4·4 m]

Flow Through Simple Pipes

2. A pipe 500 m long is conveying water with a velocity of 1 m/s. Find the suitable diameter of the pipe, if the loss of head due to friction is 3·4 m. Take $f = 0.01$. **[Ans. 300 mm]**

3. Water is flowing in a pipeline of 400 m long and 150 mm diameter at the rate of 20 litres/s. Determine the loss of head due to friction. Assume f for the pipe as 0·008. **[Ans. 5·62 m]**

4. A pipe of 600 m long and 200 mm diameter connects two reservoirs, whose surface levels differ by 10 m. What will be the discharge through the pipe ? Take $f = 0.008$. **[Ans. 45 litres/s]**
   [**Hint** : The loss of head due to friction is to be taken as 10 m]

5. A town with 4 lakh inhabitants is to be supplied with water from a reservoir 6·4 km away from a town with 25 m available head. Calculate the size of the pipeline, if half of the daily supply of 180 litres per day is to be pumped within 8 hours. Take coefficient of friction for the pipeline as 0·0075. **[Ans. 1 m]**

   [**Hint** : Total supply = $\dfrac{400\,000 \times 180}{1000}$ = 72 000 m³/day

   ∴ Discharge, $Q = \dfrac{72\,000}{2 \times 8 \times (60 \times 60)}$ = 1·25 m³/s]

6. Find the loss of head due to friction in a pipe of 100 m length and 180 mm diameter, when the velocity of water in the pipe is 1·4 m/s. Also find the discharge through the pipe. Take Chezy's constant as 55. **[Ans. 3·18 m; 35·6 litres/s]**

7. Water is flowing through a pipe 1·5 km long and 400 mm diameter with a velocity of 1 m/s. Find the loss of head by using
   (a) Darcy's equation with $f = 0.006$.
   (b) Chezy's equation with $C = 60$. **[Ans. 4·59 m; 4·17 m]**

13·8 Transmission of Power through Pipes

Whenever water is allowed to fall from a higher level to a lower level, we can always generate some power. As a matter of fact, whenever we come across a waterfall, we do not allow the water simply to fall. But it is made to flow through a pipe, so that the direction of the water may be set in some convenient way from which we may produce some power.

A little consideration will show that some head of water will be lost due to friction in the pipe through which the water is flowing.

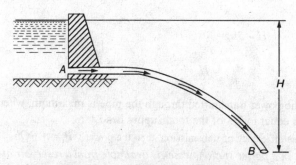

Fig. 13·5. Transmission of power.

Consider a high level storage tank. Let a pipe AB lead water from this tank from A to a power house at B as shown in Fig. 13·5.

Let
H = Head of water at the power house AB in metres,
l = Length of the pipe AB in metres,
v = Velocity of water in the pipe in m/s,
h_1 = Loss of head in the pipe AB due to friction in metres,
f = Coefficient of friction, and
d = Diameter of the pipe AB in metres,

∴ Cross-sectional area of the pipe,
$$a = \frac{\pi}{4} \times (d)^2 \text{ m}^2$$

We know that the weight of water flowing per second
$$= wQ = wav \text{ kN} \qquad \ldots(i)$$
and net head of water available at B (neglecting minor losses),
$$h = H - h_f$$

∴ Efficiency of transmission,
$$\eta = \frac{h}{H} = \frac{H - h_f}{H}$$

We also know that the power available,
$$P = \text{Weight of water flowing per second} \times \text{Head of water}$$
$$= wQ \cdot h = wav(H - h_f) = wav\left(H - \frac{4flv^2}{2gd}\right)$$
$$= wa\left(Hv - \frac{4flv^3}{2gd}\right)$$

A little consideration will show that, in the above equation, the power transmitted depends upon the velocity of water (v), as the other things are constant. Therefore the power transmitted will be maximum when
$$\frac{dP}{dv} = 0$$
or when the differential coefficient of the amount inside the bracket of equation (*iii*) is zero, *i.e.*,

$$\frac{d\left(Hv - \frac{4flv^3}{2gd}\right)}{dv} = 0 \quad \text{or} \quad H - 3\left(\frac{4flv^2}{2gd}\right) = 0$$

$$\therefore \quad H - 3h_f = 0 \qquad \ldots\left(\because \frac{4flv^2}{2gd} = h_f\right)$$

or
$$h_f = \frac{H}{3}$$

It means that the power transmitted through the pipe is maximum, when the head lost due to friction in the pipe is equal to 1/3 of the total supply head.

Note : If η is the efficiency of transmission, then the power : $P = \eta \cdot wQh$.

Example 13·9. *In a power station, water is available from a reservoir at a head of 75 m. If the efficiency of transmission is 60%, find the power available when* $1\cdot25$ m^3 *of water flows to the station in one second.*

Solution. Given : $H = 75$ m; $\eta = 60\% = 0\cdot6$ and $Q = 1\cdot25$ m³/s
Let $h_f = $ Loss of head due to friction in the pipeline.
We know that the efficiency of transmission (η),
$$0\cdot6 = \frac{H - h_f}{H} = \frac{H - h_f}{75} \quad \text{or} \quad H - h_f = 0\cdot6 \times 75 = 40 \text{ m}$$
and power available at the station,
$$P = wQ(H - h_f) = 9\cdot81 \times 1\cdot25 \times 40 = 490\cdot5 \text{ kW} \quad \textbf{Ans.}$$

Flow Through Simple Pipes

Example 13·10. *Find the maximum power that can be transmitted by a power station through a hydraulic pipe of 3 kilometres long and 200 mm diameter. The pressure of water at the power station is 1500 kPa. Take f = 0·01.*

Solution. Given : $l = 3$ km $= 3000$ m; $d = 200$ mm $= 0·2$ m, $p = 1500$ kPa and $f = 0·01$.
Let $\qquad Q$ = Discharge through the pipe.
We know that the pressure head at the power station,

$$H = \frac{p}{w} = \frac{1500}{9·81} = 153 \text{ m}$$

and for maximum transmission of power, the loss of head due to friction,

$$h_f = \frac{H}{3} = \frac{153}{3} = 51 \text{ m}$$

We also know that the loss of head due to friction (h_f),

$$51 = \frac{f l Q^2}{3 d^5} = \frac{0·01 \times 3000 \, Q^2}{3 \times (0·2)^5} = 31\,250 \, Q^2$$

$\therefore \qquad Q^2 = 51/31\,250 = 0·0016 \quad \text{or} \quad Q = 0·04 \text{ m}^3/\text{s}$

and maximum power that can be transmitted
$$= wQ \, (H - h_f) = 9·8 \times 0·04 \times (153 - 51) = 40 \text{ kW} \qquad \textbf{Ans.}$$

Example 13·11. *The pressure at the inlet of a pipeline is 1000 kPa and the pressure drop is 200 kPa. The pipeline is 1·5 kilometre long. If 100 kW is to be transmitted over this pipeline, find the diameter of the pipe and efficiency of transmission. Take f = 0·006.*

Solution. Given : Pressure head at inlet = 1000 kPa; Pressure drop = 200 kPa; $l = 1·5$ km $= 1500$ m; $P = 100$ kW and $f = 0·006$.

Diameter of the pipe
Let $\qquad d$ = Diameter of the pipe, and
$\qquad Q$ = Discharge through the pipe.
We know that the pressure head at inlet,

$$H = \frac{p_1}{w} = \frac{400}{9·81} = 40·8 \text{ m}$$

and pressure drop in terms of head (due to friction),

$$h_f = \frac{p_2}{w} = \frac{200}{9·81} = 20·4 \text{ m}$$

$\therefore$ Power to be transmitted (P)
$$100 = wQ \, (H - h_f) = 9·81 \times Q \, (40·8 - 20·4) = 200 \, Q$$

or $\qquad Q = 100/200 = 0·5 \text{ m}^3/\text{s}$

We also know that the loss of head due to friction (h_f),

$$20·4 = \frac{f l Q^2}{3(d)^5} = \frac{0·06 \times 1500 \times (0·5)^2}{3(d)^5} = \frac{7·5}{d^5}$$

$\therefore \qquad d^5 = 7·5/20·4 = 0·3676 \quad \text{or} \quad d = 0·82 \text{ m} \quad \textbf{Ans.}$

Efficiency of transmission
We also know that the efficiency of transmission,

$$\eta = \frac{H - h_f}{H} = \frac{40·8 - 20·4}{40·8} = 0·5 = 50\% \qquad \textbf{Ans.}$$

EXERCISE 13·2

1. Water flows from a reservoir to a hydraulic power station through a pipeline of 800 m length and 250 mm diameter with a velocity of 2 m/s. The power station is 45 m below the water level in the reservoir. Find the efficiency of transmission. Take $f = 0·01$. [Ans. 42%]

2. Find the maximum power available at the end of a pipeline of 3·5 km long and 200 mm diameter, when the difference of levels between the two ends is 120 m. Assume $f = 0·007$. [Ans. 175·8 kW]

3. In a hydraulic power station, 400 litres of water is supplied per second through a pipe 3 km long and 150 mm diameter. Find the maximum power supplied to the machine, if the pressure at the outlet of the pipe is recorded to be 883 kPa. [Ans. 235·4 kW]
   [Hint : Length of the pipe (3 km) and it diameter (150 mm) is superfluons data]

4. In a hydraulic power scheme, the water is available under a head of 120 m and is to be carried through a pipe of length 800 m. Determine the minimum diameter of the pipe, that will convey the water for an output of 1000 kW at 80% efficiency. Take $f = 0·01$. [Ans. 0·67 m]
   [Hint : For minimum diameter of the pipe, there should be maximum transmission of power]

13·9 Time of Emptying a Tank through a Long pipe

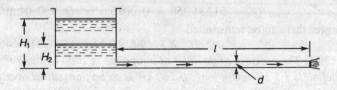

Fig. 13·6. Time of emptying a tank.

Consider a tank, which is to be emptied through a long pipe as shown in Fig. 13·6.
Let
H_1 = Initial head of water, in the tank, before opening the pipe,
H_2 = Final head of water, in the tank, after opening the pipe in T seconds,

∴ $(H_1 - H_2)$ = Fall of water level in the tank,
A = Surface area of the tank,

∴ $A(H_1 - H_2)$ = Volume of water discharged through the pipe,
l = Length of the pipe,
d = Diameter of the pipe, and
T = Time taken, in seconds, to fall the water level in the tank from H_1 to H_2.

Consider an instant, when the head of water in the tank is h. After a small interval of time dt, let the water level in the tank fall down by an amount equal to dh.

∴ Volume of water that has passed through the pipe in the time dt
$$= -A.dh \qquad \ldots(i)$$

(Minus value of dh is taken, as the value of h decreases as the discharge increases).

Since the water is being discharged in the atmosphere, therefore (ignoring all other losses, except friction) there will be some loss of head of water at outlet also.

∴
$$h = \frac{4flv^2}{2gd} + \frac{v^2}{2g} = \frac{v^2}{2g}\left(1 + \frac{4fl}{d}\right)$$

or
$$v = \sqrt{\frac{2gh}{\left(1 + \frac{4fl}{d}\right)}}$$

Flow Through Simple Pipes

In time dt, the volume of water, that has passed through the pipe
= Area of pipe × Velocity of water × Time

$$= \frac{\pi}{4} \times d^2 \times \sqrt{\frac{2gh}{\left(1+\frac{4fl}{d}\right)}} \times dt \qquad \ldots(ii)$$

Equating equations (*i*) and (*ii*),

$$-A.dh = \frac{\pi}{4} \times d^2 \times \sqrt{\frac{2gh}{\left(1+\frac{4fl}{d}\right)}} \times dt$$

or

$$dt = \frac{-4A\sqrt{1+\frac{4fl}{d}} \cdot dh}{\pi d^2 \sqrt{2gh}} = \frac{-4A\sqrt{1+\frac{4fl}{d}}\left(h^{-1/2}\right)dh}{\pi d^2 \sqrt{2g}}$$

Now the total time T, required to bring the water level from H_1 to H_2, may be found by integrating the above equation between the limits H_1 and H_2. i.e.,

$$T = \int_{H1}^{H_2} \frac{-4A\sqrt{1+\frac{4fl}{d}}\left(h^{-1/2}\right)dh}{\pi d^2 \sqrt{2g}}$$

---

*The time T may be found out as follows :
We have seen that the velocity of water,

$$v = \sqrt{\frac{2gh}{\left(1+\frac{4fl}{d}\right)}}$$

∴ Velocity of water when the head of water is H_1.

$$v_1 = \sqrt{\frac{2gH_1}{\left(1+\frac{4fl}{d}\right)}}$$

and the velocity of water when the head of water is H_2,

$$v_2 = \sqrt{\frac{2gH_2}{\left(1+\frac{4fl}{d}\right)}}$$

∴ Average velocity

$$= \frac{v_1 + v_2}{2} = \frac{\sqrt{\frac{2gH_1}{\left(1+\frac{4fl}{d}\right)}} + \sqrt{\frac{2gH_2}{\left(1+\frac{4fl}{d}\right)}}}{2}$$

$$= \sqrt{\frac{2g}{\left(1+\frac{4fl}{d}\right)}} \times \left(\sqrt{H_1}+\sqrt{H_2}\right)$$

...(Contd. on next page).

$$T = \frac{-4A\sqrt{1+\frac{4fl}{d}}}{\pi d^2 \sqrt{2g}} \int_{H_1}^{H2} h^{1/2}\, dh$$

$$= \frac{-4A\sqrt{1+\frac{4fl}{d}}}{\pi d^2 \sqrt{2g}} \left[\frac{h^{1/2}}{\frac{1}{2}}\right]_{H_1}^{H_2}$$

$$= \frac{-8A\sqrt{1+\frac{4fl}{d}}}{\pi d^2 \sqrt{2g}} \left(\sqrt{H_2} - \sqrt{H_1}\right)$$

Taking minus out of the bracket,

$$T = \frac{8A\sqrt{1+\frac{4fl}{d}}}{\pi d^2 \sqrt{2g}} \times \left(\sqrt{H_1} - \sqrt{H_2}\right)$$

If the tank is to be completely emptied, then substituting $H_2 = 0$, in the above equation,

$$T = \frac{8A\sqrt{1+\frac{4fl}{d}}}{\pi d^2 \sqrt{2g}} \times \sqrt{H_1}$$

Example 13·12. *A tank of 100 square metres in area contains water of 4 metres deep. Find the time taken to lower the water level to 2 metres through a pipe of 300 metres long and 150 mm diameter connected to the bottom of the tank. Take f = 0·01.*

Solution. Given : $A = 100$ m²; $H_1 = 4$ m; $H_2 = 2$ m; $l = 300$ m ; $d = 150$ mm $= 0·15$ m and $f = 0·01$.

...(Continued from last page)

∴ Average discharge = Area × Average velocity

$$= \frac{\pi}{4} \times d^2 \times \frac{1}{2} \sqrt{\frac{2g}{\left(1+\frac{4fl}{d}\right)}} \times \left(\sqrt{H_1} + \sqrt{H_2}\right)$$

Now the time taken to lower the water level from H_1 to H_2 in the tank,

$$T = \frac{\text{Total volume of water discharged}}{\text{Average rate of discharge}}$$

$$= \frac{A(H_1 - H_2)}{\frac{\pi}{4} \times d^2 \times \frac{1}{2} \sqrt{\frac{2g}{\left(1+\frac{4fl}{d}\right)}} \left(\sqrt{H_1} + \sqrt{H_2}\right)}$$

$$= \frac{8A\sqrt{1+\frac{4fl}{d}}}{\pi d^2 \sqrt{2g}} \left(\sqrt{H_1} - \sqrt{H_2}\right)$$

We know that time taken to lower the water level,

$$T = \frac{8A\sqrt{1+\frac{4fl}{d}}}{\pi d^2 \sqrt{2g}} \times \left(\sqrt{H_1} - \sqrt{H_2}\right)$$

$$= \frac{8 \times 100 \sqrt{1+\frac{4 \times 0.01 \times 300}{0.15}}}{\pi \times (0.15)^2 \sqrt{2 \times 9.81}} \times \left(\sqrt{4} - \sqrt{2}\right) \text{ s}$$

$= 23\,000 \times 0.5858 = 13\,470 = 3$ hrs 44 min 30 s **Ans.**

13·10 Time of from Flow One Tank into Another through a Long Pipe

Whenever two tanks, containing water, are connected together by a long pipe, the water will flow from the tank with a higher level to the tank with a lower level, irrespective of their areas. In such a case, the water level will fall in one tank, with a corresponding rise in the other. The head of water, causing flow in the pipe, will be the difference between the two water levels.

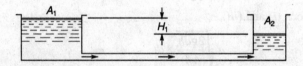

Fig. 13·7. Flow from one tank to another.

Consider two tanks connected at their bottom by a long pipe as shown in Fig. 13·7.
Let
A_1 = Area of the larger tank,
A_2 = Area of the smaller tank,
l = Length of the pipe,
d = Diameter of the pipe,
a = Area of the pipe,
f = Coefficient of friction for the pipe,
H_1 = Initial difference of water levels in the two tanks,
H_2 = Final difference of water levels in the two tanks,
T = Time, in seconds, required to bring the difference of water levels from H_1 to H_2.

At some instant, let the difference of water levels between the two tanks be h. After a small interval of time dt, let the water level in the tank A_1 fall down by an amount equal to x.

∴ The volume of water that has passed from the tank A_1

$$= A_1 . x \qquad \ldots(i)$$

and the rise of water level in the tank A_2

$$= \frac{A_1}{A_2} x$$

If the change of water levels in the two tanks be dh in time dt, then

$$dh = x + \frac{A_1}{A_2} x = x\left(1 + \frac{A_1}{A_2}\right)$$

$$dh = x\left(\frac{A_1 + A_2}{A_2}\right)$$

or $$x = \frac{-A_2\, dh}{A_1 + A_2} \qquad \ldots(ii)$$

(Minus sign of dh is taken, as the value of h decreases as the discharge increases)
At the instant when the difference of water levels in the two tanks is h, then

$$h = \frac{4flv^2}{2gd}$$

or $$v = \sqrt{\frac{2gdh}{4fl}}$$

The volume of water that has passed from tank A_1
= Area of pipe × Velocity × Time

$$= a\sqrt{\frac{2gdh}{4fl}} \times dt \qquad \ldots(iii)$$

Equating equations (*i*) and (*iii*),

$$A_1 \cdot x = a\sqrt{\frac{2gdh}{4fl}} \times dt$$

$$\frac{-A_1 A_2\, dh}{A_1 + A_2} = a\sqrt{\frac{2gdh}{4fl}} \times dt \qquad \ldots\left(\because x = \frac{-A_2 dh}{A_1 + A_2}\right)$$

or $$dt = \frac{-A_1 A_2\, dh}{a(A_1 + A_2)\sqrt{\frac{2gdh}{4fl}}}$$

$$= \frac{-A_1 A_2 \sqrt{\frac{4fl}{d}} \times h^{-\frac{1}{2}}\, dh}{a(A_1 + A_2)\sqrt{2g}}$$

Now the *total time (T), required to bring down the difference of water levels from H_1 to H_2 may be found out by integrating the above equation between the limits H_1 and H_2. i.e.,

$$T = \int_{H_2}^{H_1} \frac{-A_1 A_2 \sqrt{\frac{4fl}{d}} \times h^{-\frac{1}{2}}\, dh}{a(A_1 + A_2)\sqrt{2g}}$$

---

*The time T may also be found out as discussed below :
In T seconds, let the water level in tank A_1 fall down by an amount equal to h.
∴ Volume of water that has passed from the tank A_1 in T seconds
$$= A_1 h \qquad \ldots(i)$$
and volume of water that has passed into tank A_2 in T seconds.
$$= A_2(H_1 - H_2 - H) \qquad \ldots(ii)$$

...(Contd. on next page)

$$= \frac{-A_1 A_2 \sqrt{\frac{4fl}{d}}}{a(A_1+A_2)\sqrt{2g}} \int_{H_1}^{H_1} \times h^{-\frac{1}{2}} dh$$

$$T = \frac{-A_1 A_2 \sqrt{\frac{4fl}{d}}}{a(A_1+A_2)\sqrt{2g}} \times \left[\frac{h^{\frac{1}{2}}}{\frac{1}{2}}\right]_{H_1}^{H_2}$$

$$= \frac{-2A_2 A_2 \sqrt{\frac{4fl}{d}}}{a(A_1+A_2)\sqrt{2g}} \left(\sqrt{H_2} - \sqrt{H_1}\right)$$

...*(Contd. from previous page)*

Since both the volumes are equal, therefore equating equations (*i*) and (*ii*)
$$A_1 h = A_2(H_1 - H_2 - h) = A_2(H_1 - H_2) - A_2 h$$
$\therefore \quad h(A_1 + A_2) = A_2(H_1 - H_2)$

or
$$h = \frac{A_2(H_1 - H_2)}{A_1 + A_2} \qquad ...(iii)$$

Now the volume of water that has passed from the tank A_1 in T seconds may be found out by substituting the value of h in equation (*i*),

$$= \frac{A_1 A_2 (H_1 - H_2)}{A_1 + A_2} \qquad ...(iv)$$

Initially, when the difference of water level is H_1, then loss of head,

$$H_1 = \frac{4fl v_1^2}{2gd}$$

or
$$v_1 = \sqrt{\frac{2gdH_1}{4fl}}$$

Similarly,
$$v_2 = \sqrt{\frac{2gdH_2}{4fl}}$$

$\therefore$ Average velocity
$$= \frac{\sqrt{\frac{2gdH_1}{4fl}} + \sqrt{\frac{2gdH_2}{4fl}}}{2} = \frac{1}{2}\sqrt{\frac{2gd}{4fl}}\left(\sqrt{H_1} + \sqrt{H_2}\right) \qquad ...(v)$$

We also know that the time of flow,

$$T = \frac{\text{Volume of water}}{\text{Rate of flow}} = \frac{\text{Volume of water}}{\text{Area of pipe} \times \text{Average velocity}}$$

$$= \frac{\frac{A_1 A_2 (H_1 - H_2)}{A_1 + A_2}}{a \times \frac{1}{2}\sqrt{\frac{2gd}{4fl}}\left(\sqrt{H_1} + \sqrt{H_2}\right)} = \frac{2A_1 A_2 \sqrt{\frac{4fl}{d}}}{a(A_1+A_2)\sqrt{2g}}\left(\sqrt{H_1} - \sqrt{H_2}\right)$$

Taking minus out of the bracket,

$$T = \frac{2A_1 A_2 \sqrt{\frac{4fl}{d}}}{a(A_1 - A_2)} \sqrt{2g} \left(\sqrt{H_1} - \sqrt{H_2}\right)$$

Example 13·13. *Two tanks A and B, having surface dimensions as 4 m × 2 m and 4 m × 1 m respectively, are connected by a 150 m long pipe and of 50 mm diameter. Determine the time taken to reduce the difference of water levels from 2 metres to 0·5 metre. Take f = 0·0075.*

Solution. Given = $A_1 = 4 \times 2 = 8$ m²; $A_2 = 4 \times 1 = 4$ m²; $l = 150$ m; $d = 50$ mm = 0·05 m; $H_1 = 2$m; $H_2 = 0·5$ m and $f = 0·0075$.

We know that the area of the pipe,

$$a = \frac{\pi}{4} \times (d)^2 = \frac{\pi}{4} \times (0·05)^2 = 1·96 \times 10^{-3} \text{ m}^2$$

and time taken to reduce the difference of water levels,

$$T = \frac{2A_1 A_2 \sqrt{\frac{4fl}{d}}}{a(A_1 + A_2)\sqrt{2g}} \left(\sqrt{H_1} - \sqrt{H_2}\right)$$

$$= \frac{2 \times 8 \times 4 \sqrt{\frac{4 \times 0·0075 \times 150}{0·05}}}{1·96 \times 10^{-3}(8+4)\sqrt{2 \times 9·81}} \left(\sqrt{2} - \sqrt{0·5}\right) \text{ s}$$

$$= 5829 \times 0·707 = 4121 \text{ s} = 1 \text{ hr } 8 \text{ min } 41 \text{ s} \qquad \textbf{Ans.}$$

Exampl 13·14. *Two tanks of respectively 8 metres and 3 metres diameter are connected by a 300 metres long and 250 mm diameter pipe. The level of water in the bigger tank is 10 metres higher than that in the smaller tank. Determine the time taken to flow 1800 litres of water. Take f = 0·0075.*

Solution. Given : $D_1 = 8$ m; $D_2 = 3$ m; $l = 300$ m; $d = 250$ mm = 0·25 m; $H_1 = 10$; $Q = 18000$ litres = 18 m³ and $f = 0·0075$.

We know that the area of first tank,

$$A_1 = \frac{\pi}{4} \times (D_1)^2 = \frac{\pi}{4} \times (8)^2 = 16\pi \text{ m}^2$$

Similarly area of second tank,

$$A_2 = \frac{\pi}{4} \times (D_2)^2 = \frac{\pi}{4} \times (3)^2 = 2·25\, \pi \text{ m}^2$$

and area of pipe,

$$a = \frac{\pi}{4} \times (d)^2 = \frac{\pi}{4} \times (0·25)^2 = \frac{\pi}{64} \text{ m}^2$$

We know that the fall of water level in the bigger tank

$$= \frac{\text{Discharge}}{\text{Area}} = \frac{18}{16\pi} = 0.358 \text{ m}$$

and the rise of water level in the smaller tank

$$= \frac{\text{Discharge}}{\text{Area}} = \frac{18}{2.25\pi} = 2.546 \text{ m}$$

∴ Final difference of water levels,

$$H_2 = 10 - 0.358 - 2.546 = 7.096 \text{ m}$$

and the time taken to flow the water,

$$T = \frac{2A_1 A_2 \sqrt{\frac{4fl}{d}}}{a(A_1 + A_2)\sqrt{2g}} \left(\sqrt{H_1} - \sqrt{H_2}\right)$$

$$= \frac{2 \times 16\pi \times 2.25\pi \sqrt{\frac{4 \times 0.0075 \times 300}{0.25}}}{\frac{\pi}{64}(16\pi + 2.25\pi)\sqrt{2 \times 9.81}} \left(\sqrt{10} - \sqrt{7.096}\right) \text{ s}$$

$$= 342 \times 0.498 = 170 \text{ s} = 2 \text{ min } 50 \text{ s} \quad \textbf{Ans.}$$

EXERCISE 13·3

1. A tank 18 m long and 15 m wide contains 4 m deep water. Find the time required to empty, the tank through a pipe 25 m long and 120 mm diameter. Assume $f = 0.006$.
   [**Ans.** 14 hrs 40 min 10 s]

2. A tank 20 m × 20 m contains water 4 m deep. The tank is provided with a pipe 400 m long and 200 mm diameter at its bottom. If the value of f is 0·01, find the time taken
   (a) to fall the water level to 2 metres.   (b) in emptying the tank.
   [**Ans.** 8 hrs 25 min 10 s; 28 hrs 44 min 50 s]

3. Two tanks of surface areas of 20 m² and 15 m² respectively are connected by a 100 m long pipe and of 100 mm diameter. Find the time taken to reduce the difference of water levels from 4 m to 2·56 m. Take $f = 0.01$.
   [**Ans.** 20 min 46 s]

QUESTIONS

1. Derive Darcy's equation for the determination of loss of head due to friction in pipeline.
2. Derive from first principles, Chezy's formula for loss of head due to friction in pipe.
3. Explain with the help of a neat sketches
   (a) Hydraulic gradient, and
   (b) Total energy line.
4. Obtain the condition for transmission of maximum power, through a pipe, when the loss of head is due to friction only.

5. Obtain an expression for the time of emptying a tank through a long pipe.
6. Derive an expression for the time of flow from one tank to another through a long pipe.

OBJECTIVE TYPE QUESTIONS

1. The loss of head due to friction according to Darcy's formula is

   (a) $\dfrac{4flv^2}{gd}$ (b) $\dfrac{4flv^2}{2gd}$ (c) $\dfrac{4flv}{gd}$ (d) $\dfrac{4flv}{2gd}$

2. The hydraulic mean depth for a circular pipe of diameter (d) is

   (a) $\dfrac{d}{2}$ (b) $\dfrac{d}{3}$ (c) $\dfrac{d}{4}$ (d) $\dfrac{d}{6}$

3. The total energy line lies over the hydraulic gradient line by an amount equal to

   (a) $\dfrac{v^2}{2g}$ (b) $\dfrac{v^2}{g}$ (c) $\dfrac{v}{2g}$ (d) $\dfrac{v}{g}$

4. The power transmitted through a pipe is maximum, when the head lost due to friction is equal to

   (a) one-half of the supply head   (b) one-third of the supply head
   (c) one-fourth of the supply head   (d) two-third of the supply head

5. The maximum efficiency of transmission through a pipe is
   (a) 50%   (b) 60%   (c) 62.5%   (d) 67.7%

ANSWERS
1. (b)   2. (c)   3. (a)   4. (b)   5. (d)

14

Flow Through Compound Pipes

1. Introduction. 2. Discharge from One Reservoir to Another Through a Pipeline. 3. Discharge Through Compound Pipe (i.e., Pipes in Series). 4. Equivalent Size of a Pipe. 5. Discharge Through Pipes in Parallel. 6. Discharge through Branched Pipes from One Reservoir to Another. 7. Discharge through Branched Pipes from One Reservoir to Two or More Reservoirs. 8. Flow through Syphon Pipes.

14·1 Introduction

In the previous chapter, we have studied the flow of water through simple pipes or, in other words, discharge through a pipe in the atmosphere or through a pipe of the same diameter throughout its length and arranged in simple manner. But, in this chapter, we shall discuss the flow through compound pipes *i.e.,* pipes of different lengths and diameters arranged in different manners.

14·2 Discharge from One Reservoir to Another through a Pipe line

Sometimes, discharge from one reservoir to another takes place through a pipe line. In such a case, it is assumed that the difference between the water levels of the two reservoirs (or, in other words, head of water causing flow) is lost due to friction in the pipe line. In the following pages, we shall discuss the discharge from one reservoir to another through,

1. Compound pipes,
2. Pipes in parallel,
3. Branched pipes, and
4. Siphon pipes.

Note : The discharge through a branched pipe may be either from one reservoir to other, or from one reservoir to two or more reservoirs.

14·3 Discharge through a Compound Pipe (i.e., Pipes in Series)

Sometimes, while laying a pipeline, we have to connect pipes of different lengths and diameters with one another to form a pipeline. Such a pipe line is called a compound pipe or pipes in series. A little consideration will show that as the pipes are in series, therefore the discharge will be continuous.

Now consider a compound pipe discharging water from one tank with a higher water level to another with a lower water level as shown in Fig. 14·1.

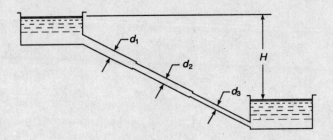

Fig. 14·1. Pipes in series.

Let
Q = Discharge through the pipeline,
H = Total loss of head,
d_1 = Diameter of pipe 1,
l_1 = Length of pipe 1,
f_1 = Coefficient of friction for pipe 1,
d_2, l_2, f_2 = Corresponding values of pipe 2,
d_3, l_3, f_3 = Corresponding values of pipe 3 and so on.

Neglecting minor losses except friction, we know that the total loss of head,
H = Loss of head in pipe 1, + Loss of head in pipe 2,
+ Loss of head in pipe 3, and so on

$$= \frac{4f_1 l_1 v_1^2}{2gd_1} + \frac{4f_2 l_2 v_2^2}{2gd_2} + \frac{4f_3 l_3 v_3^2}{2gd_3} + \ldots$$

$$= \frac{4}{2g}\left(\frac{f_1 l_1 v_1^2}{d_1} + \frac{f_2 l_2 v_2^2}{d_2} + \frac{f_3 l_3 v_3^2}{d_3} + \ldots\right)$$

Notes : 1. If the coefficient of friction is the same for all the pipes, then

$$H = \frac{4f}{2g}\left(\frac{l_1 v_1^2}{d_1} + \frac{l_2 v_2^2}{d_2} + \frac{l_3 v_3^2}{d_3} + \ldots\right)$$

2. If the discharge through the pipeline is given then the total loss of head,

$$H = \frac{f_1 l_1 Q^2}{3 d_1^5} + \frac{f_2 l_2 Q^2}{3 d_2^5} + \frac{f_3 l_3 Q^2}{3 d_2^5} + \ldots$$

$$= \frac{Q^2}{3}\left(\frac{f_1 l_1}{d_1^5} + \frac{f_2 l_2}{d_2^5} + \frac{f_3 l_3}{d_2^5} + \ldots\right)$$

3. If the coefficient of friction is the same for all the pipes, then

$$H = \frac{fQ^2}{3}\left(\frac{l_1}{d_1^5} + \frac{l_2}{d_2^5} + \frac{l_3}{d_3^5} + \ldots\right)$$

Example 14·1. *Water is discharged from a tank to another with 30 metres difference of water levels through a pipe 1200 metres long. The diameter for the first 600 metres length of the pipe is 400 mm and 250 mm for the remaining 600 metres long. Find the discharge in litres per second through the pipe, taking into consideration the frictional losses only. Assume the coefficient of friction as 0·009 for both the pipes.*

Flow Through Compound Pipes

Solution. Given : $H = 30$ m; $l = 1200$ m; $l_1 = 600$ m; $d_1 = 400$ mm $= 0.4$ m; $l_2 = 600$ m; $d_2 = 250$ mm $= 0.25$ m and $f = 0.009$.

Let $\qquad Q =$ Discharge through the pipe.

We know that the difference of water levels taking into account frictional losses only (H),

$$30 = \frac{fQ^2}{3}\left(\frac{l_1}{d_1^5} + \frac{l_2}{d_2^5}\right) = \frac{0.009\, Q^2}{3}\left(\frac{600}{(0.4)^5} + \frac{600}{(0.25)^5}\right)$$

$$= 0.003 \times 672\,700 = 2018\, Q^2$$

$\therefore \qquad Q^2 = 30/2018 = 0.015$ or $Q = 0.122$ m³/s $= 122$ litres/s **Ans.**

14·4 Equivalent Size of a Pipe

Sometimes, a compound pipe is required to be replaced by a pipe of a uniform diameter and of the same length as that of the compound pipe, so that the loss of head as well as the discharge is the same in both the cases. The new pipe of uniform diameter is called equivalent pipe and its diameter is called equivalent size of the pipe.

Now consider a compound pipe to be replaced by an equivalent pipe.

Let $\qquad H =$ Total loss of head,
$\qquad Q =$ Discharge through the pipes,
$\qquad l_1 =$ Length of pipe 1,
$\qquad d_1 =$ Diameter of pipe 1,
$\qquad l_2, d_2 =$ Corresponding values of pipe 2,
$\qquad l_3, d_3 =$ Corresponding values of pipe 3, and
$\qquad d =$ Diameter of the equivalent pipe.

$\therefore$ Length of the equivalent pipe :

$$l = l_1 + l_2 + l_3.$$

Neglecting minor losses, except friction and assuming the coefficient of friction to be the same for both the cases, we know that the total loss of head in the first case,

$$H = h_{f1} + h_{f2} + h_{f3} + ...$$

$$= \frac{fl_1 Q^2}{3d_1^5} + \frac{fl_2 Q^2}{3d_2^5} + \frac{fl_3 Q^2}{3d_3^5} + ...$$

$$= \frac{fQ^2}{3}\left(\frac{l_1}{d_1^5} + \frac{l_2}{d_2^5} + \frac{l_3}{d_3^5} + ...\right) \qquad ...(i)$$

Similarly, the total loss of head in the equivalent pipe,

$$H = \frac{flQ^2}{3d^5} \qquad ...(ii)$$

Since the loss of head is same in both the cases, therefore equating (i) and (ii),

$$\frac{flQ^2}{3d^5} = \frac{fQ^2}{3}\left(\frac{l_1}{d_1^5} + \frac{l_2}{d_2^5} + \frac{l_3}{d_3^5} + ...\right)$$

or $\qquad \dfrac{l}{d^5} = \dfrac{l_1}{d_1^5} + \dfrac{l_2}{d_2^5} + \dfrac{l_3}{d_3^5} + ...$

Example 14·2. *A compound pipeline 1650 metres long is made up of pipes of 450 mm diameter for 900 metres, 375 mm for 450 metres and 300 mm for 300 metres. If the compound pipe is to be replaced by a pipe of uniform diameter, find the diameter of the new pipe, assuming the length to remain the same.*

Solution. Given : $l = 1650$ m; $d_1 = 450$ mm $= 0.45$ m; $l_1 = 900$ m; $d_2 = 375$ mm $= 0.375$ m; $l_2 = 450$ m; $d_3 = 300$ mm $= 0.3$ m and $l_3 = 300$ m.

Let $\qquad d =$ Diameter of equivalent pipe.

We know that for equivalent diameter of a compound pipe,

$$\frac{l}{d^5} = \frac{l_1}{d_1^5} + \frac{l_2}{d_2^5} + \frac{l_3}{d_3^5}$$

$$\frac{1650}{d^5} = \frac{900}{(0.45)^5} + \frac{450}{(0.375)^5} + \frac{300}{(0.3)^5} = 232\ 900$$

∴ $\qquad d^5 = 1650/232\ 900 = 0.0071 \qquad$ or $\quad d = 0.372$ m **Ans.**

14·5 Discharge through Pipes in Parallel

Sometimes, in order to increase the discharge from one tank into another, a new pipe has to be laid alongwith the existing one. Such an arrangement is known as pipes in parallel as shown in Fig. 14·2.

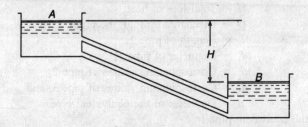

Fig. 14·2. Pipes in parallel.

A little consideration will show that as the pipes are parallel, therefore the loss of head for all the pipes will be the same and all the pipes will discharge water independently. The total discharge through all the pipes will be the sum of the discharges in the various pipes.

Example 14·3. *Two pipes are connected parallel to each other between two reservoirs with length $l_1 = 2400$ metres, $d_1 = 1·2$ m diameter, Darcy's coefficient $f_1 = 0.006$; $l_2 = 2400$ metres, $d_2 = 1·0$ m and $f_2 = 0·005$. Find the total flow, if the difference in elevation is 20 m metres.*

Solution. Given : $l_1 = 2400$ m; $d_1 = 1·2$ m; $f_1 = 0·006$; $l_2 = 2400$ m; $d_2 = 1·0$ m; $f_2 = 0·005$ and $H = 20$ m.

Let $\qquad Q_1 =$ Discharge through the pipe 1, and
$\qquad\qquad Q_2 =$ Discharge through the pipe 2.

Since the pipes are connected in parallel, therefore loss of head (or difference in water levels) will be the same in both the pipes.

We know that the difference of water levels for pipe 1 (H),

$$20 = \frac{f_1 l_1 Q_1^2}{3(d_1)^5} = \frac{0.006 \times 2400 \times Q_1^2}{3 \times (1·2)^5} = 1·93\, Q_1^2$$

or $\qquad Q_1^2 = 20/1·93 = 10·4 \quad$ or $\quad Q_1 = 3·22$ m³/s $\qquad\qquad$ ...(*i*)

Similarly, the difference of water levels for pipe 2 (H),

$$20 = \frac{f_2 l_2 Q_2^2}{3(d_2)^5} = \frac{0·005 \times 2400 \times Q_2^2}{3 \times (1·0)^5} = 4 Q^2$$

Flow Through Compound Pipes

or $\qquad Q_2^2 = 20/4 = 5 \quad$ or $\quad Q_2 = 2\cdot 24$ m³/s ...(ii)

∴ Total flow, $\qquad Q = Q_1 + Q_2 = 3\cdot 22 + 2\cdot 24 = 5\cdot 46$ m³/s **Ans.**

Example 14·4. *Two reservoirs are connected by three pipes, laid in parallel. Their diameters are d, 2d and 3d respectively, and they are of the same length l. Assuming f to be the same for all pipes, what will be the discharge through each of the larger pipes, if the smallest pipe is discharging 1 m³/s?*

Solution. Given : Dia. of pipe 1 = d; Dia. of pipe 2 = $2d$; Dia. of pipe 3 = $3d$ and $Q_1 = 1$ m³/s.

Since all the three pipes are arranged in parallel, therefore loss of head will be the same for all the pipes.

or
$$h_f = \frac{f_1 l_1 Q_1^2}{3d^5} = \frac{f_2 l_2 Q_2^2}{3(2d)^5} = \frac{f_3 l_3 Q_3^2}{3(3d)^5}$$

Moreover, as the lengths and coefficients of friction for all the pipes are the same, therefore $l = l_1 = l_2 = l_3$ and $f = f_1 = f_2 = f_3$. Thus

$$\frac{fl Q_1^2}{3d^5} = \frac{fl Q_2^2}{3(2d)^5} = \frac{fl Q_3^2}{3(3d)^5}$$

or $\qquad 1 = \dfrac{Q_2^2}{32} = \dfrac{Q_3^2}{243} \qquad$ ...(∵ $Q_1 = 1$ m³/s)

∴ $\qquad Q_2 = \sqrt{32} = 5\cdot 667$ m³/s **Ans.**

and $\qquad Q_3 = \sqrt{243} = 15\cdot 59$ m³/s **Ans.**

Example 14·5. *An old water supply distribution pipe of 250 mm diameter is to be replaced by two parallel pipes of equal diameter having equal lengths and identical values of coefficient friction. Find the diameter of the new pipes.*

Solution. Given : $d = 250$ mm $= 0\cdot 25$ m.

Let $\qquad d_1 = $ Diameter of the new pipes.

Since the new pipes are of the same length diameter and have identical value of coefficient of friction, thus they have equal discharges. Or in other words, each pipe will discharge half of water of the old pipe. Mathematically, the discharge through the new pipe,

$$Q_1 = \frac{Q}{2} = 0\cdot 5\, Q$$

We know that the loss of head in the old water supply pipe,

$$h_f = \frac{fl Q^2}{3d^5} = \frac{fl Q^2}{3 \times (0\cdot 25)^5} = 341\cdot 3\, fl Q^2 \qquad ...(i)$$

and loss of head in the new pipes,

$$hf_1 = \frac{fl Q_1^2}{3d_1^5} = \frac{fl(0\cdot 5Q)^2}{3d_1^5} = \frac{0\cdot 083\, fl Q^2}{d_1^5} \qquad ...(ii)$$

Since the loss of head is same in both the cases, therefore equating equations (*i*) and (*ii*),

$$341\cdot 3\, fl Q^2 = \frac{0\cdot 083\, fl Q^2}{d_1^5}$$

∴ $\qquad d_1^5 = 0\cdot 083 / 341\cdot 3 = 0\cdot 00024$

or $\qquad d = 0\cdot 19$ m $= 190$ mm **Ans.**

EXERCISE 14·1

1. Two tanks having difference of water levels as 50 m are connected by a pipeline. The first 1 km long pipe is of 600 mm diameter and the last 500 m long pipe is of 500 mm diameter. Find the discharge through the pipeline, if the coefficient of friction for both the pipes is 0·009.
   [Ans. 0·76 m³/s]

2. Two reservoirs whose surface levels are 30 m different in elevation are connected by a compound pipe consisting 0f 30 m of 75 mm diameter, 15 m of 150 mm diameter and 30 m of 100 mm diameter. Find the rate of discharge through the pipe in litres/s, if the coefficient of friction for all the pipes is 0·01.
   [Ans. 24 litres/s]

3. A compound pipe is made up of two pipes 450 m long and of diameters 300 mm and 400 mm respectively. What should be the diameter of an equivalent pipe, if the discharge, length and coefficient of friction is to be the same for both the systems?
   [Ans. 330 mm]

4. Three pipes of lengths of 800 m, 500 m and 400 m and of diameters 500 mm, 400 mm and 300 mm respectively are connected in series. If this compound pipe is to be replaced by a single pipe of length of 1700 m, find the diameter of this single pipe for the same discharge.
   [Ans. 372 mm]

5. Two pipes of length of 1 km each and of diameter 300 mm and 400 mm respectively are connected in parallel between two reservoirs. Find the ratio of their discharges, if the value of coefficients of friction is same for both the pipes.
   [Ans. 1 : 2·05]

14·6 Discharge through Branched Pipes from One Reservoir to Another

Sometimes, a pipe is laid partly alongwith an existing pipe to increase the discharge in the lower tank as shown in Fig. 14·3 (a) and (b). In case (a), the discharge through the pipe 1 (i.e., Q_1) is equal to the sum of discharges through the pipes 2 and 3 (i.e., $Q_2 + Q_3$). But in case (b) the sum of discharges through the pipes 1 and 2 (i.e., $Q_1 + Q_2$) is equal to the discharge through the pipe, 3 (i.e., Q_3).

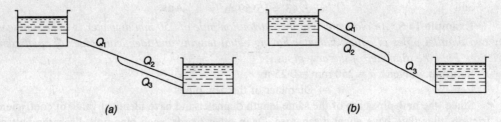

Fig. 14·3. Branched pipes.

The examples in such cases are solved by making discharge equations through pipes 1 and 2 and 3 and then solved as usual.

Example 14·5. *Two reservoirs having difference of water levels as 12 m are connected by a pipe system as shown in Fig. 14·4.*

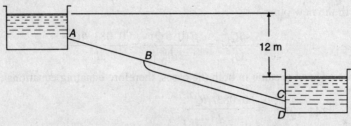

Fig. 14·4.

The lengths of the pipes AB, BC and BD are 1·5 km each of 300 mm diameter. Find the discharge from one tank to another, if the value of f is 0·01 for all the pipes.

Flow Through Compound Pipes

Solution. Given : $H = 12$ m; $l_1 = l_2 = l_3 = 1.5$ km $= 1500$ m; $d_1 = d_2 = d_3 = 300$ mm $= 0.3$ m and $f_1 = f_2 = f_3 = 0.01$.

Let
Q_1 = Discharge through the pipe AB,
Q_2 = Discharge through the pipe BC, and
Q_3 = Discharge through the pipe BD.

Since the lengths, diameters and values of f for the pipes BC and BD are equal, therefore Q_2 will be equal to Q_3. Or in other words,

$$Q_1 = Q_2 + Q_3 = 2Q_2 \qquad \ldots(\because Q_2 = Q_3)$$

Consider the flow of water through ABC. We know that the difference of water levels in the two reservoirs (H)

$$12 = \frac{fl Q_1^2}{3d_1^5} + \frac{fl_2 Q_2^2}{3d_2^5} = \frac{fl Q_1^2}{3d_1^5}\left[1 + \left(\frac{1}{2}\right)^2\right] \qquad \ldots(l_1 = l_2 \text{ and } d_1 = d_2)$$

$$= \frac{0.1 \times 1500\, Q_1^2}{3 \times (0.3)^5} \times \frac{5}{4} = 2572\, Q_1^2$$

∴ $Q_1^2 = 12/2572 = 0.0047$ or $Q = 0.068$ m³/s **Ans.**

Example 14·6. *Two reservoirs are connected by a pipe of 200 mm diameter and 3000 metres long, the difference in surface being 45 metres. Calculate the discharge through the pipe in litres per minute.*

If a loop line of 300 mm diameter and 1200 metres long is connected to the last 1200 metres of the pipe line, calculate the increase in discharge in litres per minute due to addition of loop line. Neglect all losses other than those due to friction. Assume $f = 0.008$.

Solution. Given : $d = 200$ mm $= 0.2$ m; $l = 3000$ m; $H = 45$ m; $d_3 = 300$ mm $= 0.3$ m; $l_3 = 1200$ m and $f = 0.008$.

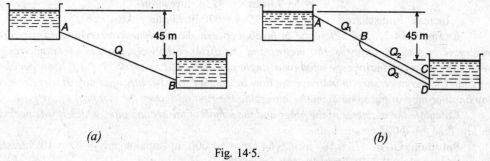

Fig. 14·5.

Ist Case

In this case, the reservoirs are connected by a single pipe only as shown in Fig. 14·5 (a).

Let Q = Discharge through the one pipe AB.

We know that the difference in the water surfaces (H),

$$45 = \frac{flQ^2}{3(d)^5} = \frac{0.008 \times 3000\, Q^2}{3 \times (0.2)^5} = 25\,000\, Q^2$$

∴ $Q^2 = 45/25\,000 = 0.0018$ or $Q = 0.0424$ m³/s
$= 42.4$ litres/s $= 2544$ litres/min **Ans.**

2nd Case

In this case, the reservoirs are connected by a branched pipe also as shown in Fig. 14·5 (b).

Let
Q_1 = Discharge through the pipe AB,
Q_2 = Discharge through the pipe BC, and
Q_3 = Discharge through the pipe BD.

First of all, consider the flow of water through ABC. We know that the difference in the water surface (H),

$$45 = \frac{f l_1 Q_1^2}{3 d_1^5} + \frac{f l_2 Q_2^2}{3 d_2^5} = \frac{0.008 \times 1800 \, Q_1^2}{3 \times (0.2)^5} + \frac{0.008 \times 1200 \, Q_2^2}{3 \times (0.2)^5}$$

$$= 15000 \, Q_1^2 + 10000 \, Q_2^2 \qquad \qquad \ldots(i)$$

Now consider the flow of water through ABD. We know that the difference in the water surfaces (H),

$$45 = \frac{f l_1 Q_1^2}{3 d_1^5} + \frac{f l_3 Q_3^2}{3 d_3^5} = \frac{0.008 \times 1800 \, Q_1^2}{3 \times (0.2)^5} + \frac{0.008 \times 1200 \, Q_3^2}{3 \times (0.3)^5}$$

$$= 15\,000 \, Q_1^2 + 1317 \, Q_3^2 \qquad \qquad \ldots(ii)$$

Subtracting equation (ii) from equation (i),

$0 = 10\,000 \, Q_2^2 - 1317 \, Q_3^2$ or $1317 Q_3^2 = 10\,000 \, Q_2^2$

∴ $Q_3^2 = 10\,000 \, Q_2^2 / 1317 = 7.59 \, Q_2$ or $1317 \, Q_3 = 2.75 \, Q_2$ ...(iii)

From the geometry of the pipes, we find that
$Q_1 = Q_2 + Q_3 = Q_2 + 2.75 \, Q_2 = 3.75 \, Q_2$
or $Q_2 = Q_1 / 3.75$

Substituting this value of Q_2 in equation (i),

$$45 = 15\,000 \, Q_1^2 + 10\,000 \times \left(\frac{Q_1}{3.75}\right)^2 = 15\,710 \, Q_1^2$$

$Q_1^2 = 45/15\,170 = 0.0029$ or $Q_1 = 0.0539 \text{ m}^3/\text{s}$
$= 53.9$ litres/s $= 3234$ litres/min

∴ Increase in discharge $= 3234 - 2544 = 690$ litres/min **Ans.**

Example 14·7. *The difference in levels between the catchment reservoir and the service reservoir of a town supply is 180 metres and the distance between them is 64 kilometres. The reservoirs were originally connected by a single pipe designed to carry 27×10^6 litres per day. It was later found necessary to increase the flow by another 9×10^6 litres per day. It was decided to lay another pipe of the same diameter alongside the first and over the last part of the pipe.*

Calculate the diameter of the pipes and the length of the second pipe, which it was necessary to lay. Take $f = 0·008$ for each pipe.

Solution. Given : $H = 180$ m; $l = 64$ km $= 64\,000$ m; Initial supply $= 27 \times 10^6$ litres/day; Increased supply $= 9 \times 10^6$ litres/day and $f = 0·008$.

Diameter of the pipe

Let d = Diameter of the pipe.

We know that initial discharge through the pipe,

$$Q = \frac{27 \times 10^6}{1000 \times 24 \times 60 \times 60} = 0.312 \text{ m}^3/\text{s}$$

and difference between the catchment reservoir level and service reservoir level (H),

$$180 = \frac{f l Q^2}{3 d^5} = \frac{0.008 \times 64\,000 \times (0.312)^2}{3 (d)^5} = \frac{16.6}{d^5}$$

∴ $d^5 = 16.6/180 = 0.0922$ or $d = 0.62$ m **Ans.**

Flow Through Compound Pipes

Length of the second pipe

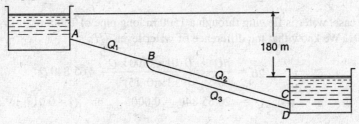

Fig. 14·6.

We know that increase in discharge,

$$q = \frac{9 \times 10^6}{1000 \times 24 \times 60 \times 60} = 0\cdot 104 \text{ m}^3/\text{s}$$

and total discharge, $Q_1 = Q + q = 0\cdot 312 + 0\cdot 104 = 0\cdot 416$ m³/s

Let $l_1 = $ Length of the pipe AB,

$Q_2, l_2 = $ Corresponding values for pipe BC, and

$Q_3, l_3 = $ Corresponding values for pipe BD.

From the geometry of the pipe system, we find that

$$l_2 = l_3 = 64\ 000 - l_1$$

Since the lengths, diameters and coefficients of friction for the pipes BC and BD are equal, therefore Q_2 will be equal to Q_3. Or in other words,

$$Q_2 = Q_3 = \frac{Q_1}{2} = \frac{0\cdot 416}{2} = 0\cdot 208 \text{ m}^3/\text{s}$$

Now consider flow through ABC. We know that the difference between catchment reservoir level and surface reservoir level (H).

$$180 = \frac{fl_1 Q_1^2}{3d_1^5} + \frac{fl_2 Q_2^2}{3d_2^2}$$

$$= \frac{0\cdot 008\, l_1 \times (0\cdot 416)^2}{3 \times (0\cdot 62)^5} + \frac{0\cdot 008\, (64\ 000 - l_1) \times (0\cdot 208)^2}{3 (0\cdot 62)^2}$$

$$= \frac{0\cdot 008 \times (0\cdot 208)^2}{3 (0\cdot 62)^2} (4l_1 + 64\ 000 - l_1)$$

$$= 0\cdot 0013\, (3l_1 + 64\ 000) = 0\cdot 0039\, l_1 + 83\cdot 2$$

or $l_1 = (180 - 83\cdot 2)/0\cdot 0039 = 24\ 820$ m $= 24\cdot 82$ km

∴ Length of second pipe, $l_2 = 64 - 24\cdot 82 = 39\cdot 18$ km **Ans.**

Example 14·8. *In order to improve water supply of a reservoir from another reservoir whose water level is 20 m higher and 1·5 km away, the following two different schemes have been proposed:*

The existing 150 mm pipe may be replaced by a new 200 mm pipe over the whole distance. Or alternatively, a new 200 mm pipe may be installed over the 750 m and the old pipes relaid in parallel over the remaining length. Find the increase in discharge in both the schemes.

Also determine which schemes will give more water supply. Take the values of f_1 for old pipes as 0·01 and that for new pipes as 0·008.

Solution. Given : $H = 20$ m and $l = 1.5$ km = 1500 m.

1st Case

In this case, water is flowing through a 1500 m long pipe of 150 mm (0·15 m) diameter with f equal to 0·01. We know that the difference of water levels (H),

$$20 = \frac{flQ^2}{3d^5} = \frac{0.01 \times 1500 \times Q^2}{3 \times (0.15)^5} = 65\,840\, Q^2$$

or $\qquad Q^2 = 20/65\,840 = 0.0003 \qquad$ or $\qquad Q = 0.017$ m³/s

2nd Case

In this case, water is flowing through a 1500 m long pipe of 200 mm (0·2 m) diameter with f equal to 0·008. We know that the difference of water levels (H),

$$20 = \frac{flQ^2}{3d^5} = \frac{0.008 \times 1500 \times Q^2}{3 \times (0.2)^5} = 12\,500\, Q^2$$

or $\qquad Q^2 = 20/12\,500 = 0.0016 \qquad$ or $\qquad Q = 0.04$ m³/s

∴ Increase in discharge $= 0.04 - 0.017 = 0.023$ m³/s **Ans.**

3rd Case

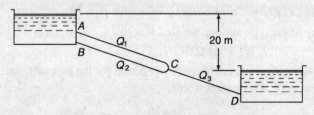

Fig. 14.7.

In this case, the old pipe of 1500 m length and 150 mm (0·15 m) diameter with f equal to 0·01 is relaid as AC and BC. In addition to this, a new pipe of 750 m length and 200 mm (0·2 m) diameter with f equal to 0·008 is installed as CD as shown in Fig. 14·7. Thus, in this case, $l_1 = l_2 = 750$ m; $d_1 = d_2 = 0.15$ m; $f_1 = f_2 = 0.01$; $l_3 = 750$ m; $d_3 = 0.2$ m and $f_3 = 0.008$.

Since the lengths, diameters and coefficients of friction for the pipes AC and BC are equal, therefore Q_1 will be equal to Q_2. Or, in other words,

$$Q_3 = Q_1 + Q_2 = 2Q_1 \qquad \qquad ...(Q_1 = Q_2)$$

Now consider the flow through ACD. We know that the difference of water levels (H),

$$20 = \frac{f_1 l_1 Q_1^2}{3d_1^5} + \frac{f_2 l_2 Q_3^2}{3d_3^5} = \frac{0.01 \times 750\, Q_1^2}{3 \times (0.15)^5} + \frac{0.008 \times 750\, (2Q_1)^2}{3 \times (0.2)^5}$$

$$= 32\,920\, Q_1^2 + 25\,000\, Q_1^2 = 57\,920\, Q_1^2$$

or $\qquad Q_1^2 = 20/57\,920 = 0.0003 \qquad$ or $\qquad Q = 0.017$ m³/s

and $\qquad Q_3 = 2Q_1 = 2 \times 0.017 = 0.034$ m³/s

∴ Increase in discharge $= 0.034 - 0.017 = 0.017$ m³/s **Ans.**

From the above two answers, we find that the discharge in the second case (0·034 m³/s) is more than that in the third case (0·034 m³/s). Therefore it gives more water supply. **Ans.**

Flow Through Compound Pipes

Example 14·9. *A pipeline of 6 km long and of 700 mm diameter connects two reservoirs A and B as shown in Fig. 14·8.*

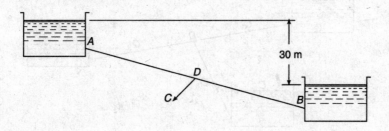

Fig. 14·8.

In the middle of the pipeline, there is a tap through which water can be supplied to a third reservoir C. Determine the rate of flow to the reservoir B, when (i) no water is discharged to the reservoir C, and (ii) the quantity of water discharged to the reservoir C is 150 litres/s. Neglect all losses except friction and take the value of f as 0·006.

Solution. Given : $l = 6$ km $= 6000$ m; $d = 700$ mm $= 0.7$ m; $H = 30$ m and $f = 0.006$.

Rate of flow to the reservoir B, when no water is discharged to the reservoir C

Let $\qquad Q = $ Rate of flow to the reservoir B.

We know that the difference of water levels in the reservoirs (H),

$$30 = \frac{flQ^2}{3(d)^5} = \frac{0.006 \times 6000\, Q^2}{3 \times (0.7)^2} = 71.4\, Q^2$$

$\therefore \qquad Q^2 = 30/71.4 = 0.42 \quad $ or $\quad Q = 0.648$ m³/s **Ans.**

Rate of flow to the reservoir B, when water is discharged to the reservoir C

We know that in this case $l_1 = 3000$ m; $l_2 = 3000$ m and discharge to the reservoir $= 150$ litres /s $= 0.15$ m³/s.

Let $\qquad Q_1 = $ Discharge in the pipe AD.

$\therefore$ Discharge in the pipe DB,

$$Q_2 = (Q_1 - 0.15) \text{ m}^3/\text{s}$$

Consider the flow through ADB. We know that difference of water levels in the reservoirs (H),

$$30 = \frac{fl_1Q_1^2}{3d^5} + \frac{fl_2Q_2^2}{3d^5} = \frac{fl}{3d^5}\left[Q_1^2 + (Q_1 - 0.15)^2\right]$$

$$= \frac{0.006 \times 3000}{3 \times (0.7)^5}\left[Q_1^2 + Q_1^2 + 0.0225 - 0.3\, Q_1\right]$$

$$= 71.4 Q_1^2 + 0.8 - 10.7 Q_1$$

or $\qquad 71.4 Q_1^2 - 10.7 Q_1 - 29.2 = 0$

This is a quadratic equation for Q_1. Therefore

$$Q_1 = \frac{+10.7 \pm \sqrt{(10.7)^2 + 4 \times 71.4 \times 29.2}}{2 \times 71.4} = 0.72 \text{ m}^3/\text{s}$$

$\therefore$ Flow to the reservoir $B = 0.72 - 0.15 = 0.57$ m³/s   **Ans.**

14·7. Discharge through Branched Pipes from One Reservoir to Two or More Reservoirs

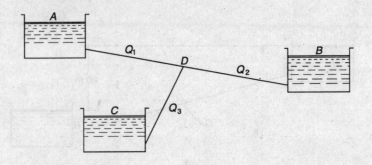

Fig. 14·9. Branched Pipes.

Sometimes, three (or more) reservoirs are connected through branched pipes as shown in Fig. 14·9. In such a case, the discharge through pipe 1 (*i.e.*, Q_1) is also equal to the sum of discharge through the pipes 2 and 3 (*i.e.*, $Q_2 + Q_3$).

The examples, in such cases, are solved either by making discharge equations through pipes 1 and 2, as well as 1 and 3 or by the application of Bernoulli's equation.

Example 14·10. *Water is being discharged, from a reservoir, through a pipe of 4 kilometres long and of 500 mm diameter to another reservoir having water level of 12·5 metres below the first reservoir.*

It is required to feed a third reservoir, whose water level is 15 metres below the first reservoir, through a pipe line of 1·5 kilometre long to be connected to the pipe at a distance of 1 kilometre from its entrance as shown in Fig. 14·10.

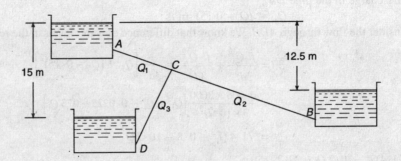

Fig. 14·10.

Find the diameter of this new pipe, so that the flow into both the reservoirs may be the same. Assume (f) for all the pipes as 0·008.

Solution. Given : $l_{AB} = 4$ km $= 4000$ m; $d_{AB} = 500$ mm $= 0.5$ m; $H_{AB} = 12.5$ m; $H_{AD} = 15$ m; $l_{CD} = 1.5$ km $= 1500$ m; $l_{AC} = 1$ km $= 1000$ m or $l_{CB} = 4 - 1 = 3$ km $= 3000$ m; $Q_{CB} = Q_{CD}$ and $f = 0.008$.

Flow Through Compound Pipes

We know that $Q_2 = Q_3$ and $Q_1 = Q_2 + Q_3$

$\therefore \quad Q_2 = Q_3 = \dfrac{Q_1}{2}$

Let d_{cd} = Diameter of the new pipe (*i.e.*, CD).

First of all, let us consider the flow of water through ACB. We know that the difference of water levels between the tanks A and B (H_{AB}),

$$12 \cdot 5 = \dfrac{fl_1 Q_1^2}{3d_1^5} + \dfrac{fl_2 Q_2^2}{3d_2^5} = \dfrac{fQ_1^2}{3d^5} = \left[l_1 + l_2 \times \left(\dfrac{1}{2}\right)^2\right]$$

$$= \dfrac{0 \cdot 008\, Q_1^2}{3 \times (0 \cdot 5)^5} \left[1000 + 3000 \times \dfrac{1}{4}\right] = 149 \cdot 3\, Q_1^2$$

or $\quad Q_1^2 = 12 \cdot 5/149 \cdot 3 = 0 \cdot 0837 \quad$ or $\quad Q_1 = 0 \cdot 29$ m³/s

$\therefore \quad Q_2 = Q_3 = 0 \cdot 5 Q_1 = 0 \cdot 29/2 = 0 \cdot 145$ m³/s

Now consider the flow of water through ACD. We know that the difference of water levels between the tanks A and D (H_{AD}),

$$15 = \dfrac{fl_1 Q_1^2}{3d_1^5} + \dfrac{fl_3 Q_3^2}{3d_2^5}$$

$$= \dfrac{0 \cdot 008 \times 1000 \times (0 \cdot 29)^2}{3 \times (0 \cdot 5)^2} + \dfrac{0 \cdot 008 \times 1500 \times (0 \cdot 145)^2}{3d_2^5}$$

$$= 7 \cdot 18 + \dfrac{0 \cdot 084}{d_3^5}$$

$\therefore \quad d_3^5 = \dfrac{0 \cdot 084}{15 - 7 \cdot 18} = 0 \cdot 0107 \quad$ or $\quad d_3 = 0 \cdot 4$ m   **Ans.**

EXERCISE 14·2

1. A pipe of 2 km length and 400 mm diameter is connected to a reservoir and a junction box. From the junction box, two parallel pipes of 1 km length and 300 mm diameter are connected to another reservoir whose water level is 10 m below the upper reservoir. What is the discharge from the upper reservoir in litres/s, if $f = 0 \cdot 015$? **[Ans. 82·2 litres/s]**

2. Two reservoirs with a difference of water levels as 12 m are connected by a pipe of 3·2 km long and 300 mm diameter. A second pipe of 300 mm diameter is laid alongside the first for the last 800 m, which connects first pipe and the reservoir. Find the increase in discharge. Take f as 0·01. **[Ans. 5·7 litres/s]**

3. A pipeline of 1·5 km long and 600 mm diameter is discharging water from a reservoir to another reservoir whose water is 30 m below the upper reservoir. To augment the discharge, another line of the same diameter is introduced in parallel to the first in the second half of the length. Neglecting minor losses, find the percentage increase in discharge, if $f = 0 \cdot 01$. **[Ans. 26·6%]**

4. Two reservoirs with 20 m difference in water levels are connected by a pipeline of 4 km long and 600 mm diameter. At a distance of 1 km from the upper reservoir, a small pipe is connected to the pipeline. Find the discharge to the lower reservoir, if (*i*) no water is taken from the small pipe, and (*ii*) 100 litres/s of water is taken from the small pipe. **[Ans. 485 litres/s; 458 litres/s]**

14·8 Flow through Syphon Pipes

Sometimes, while laying a pipeline between two reservoirs, an obstacle in the form of a high ridge or high ground comes up. In such a case, it is not advisable to cut the ridge and lay the pipe as the cost of laying the pipe will be more. Moreover, the repairs after laying the pipe will also be difficult. Thus the pipe is, usually, laid at the surface of the ridge. Sometimes, it so happens that the summit of the pipe is above the water level in the supply reservoir. Such a pipe is known as a syphon pipe as shown in Fig. 14·11.

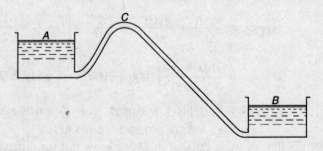

Fig. 14·11. Syphon pipe.

Water will flow through the pipeline, if the absolute pressure at the summit C is not below zero. If it falls below zero, the syphon will not run full as some air will be liberated from the water. For this reason, there is a limit for maximum height of summit C over the inlet A. Assuming the pipeline to be of uniform diameter, the height of summit above the inlet A

$$= \frac{p}{w} = 10.3 \text{ m} \qquad \text{...(Atmospheric pressure head)}$$

In actual practice water stops flowing, if the summit C is higher than 7·5 m from the inlet A (or in other words 10·3 − 7·5 = 2·8 m negative pressure head).

Note : The syphon can be made to work by exhausting air, thus creating vacuum in it, or by filling it with water.

Example 14·12. *Two reservoirs having difference of 30 metres in their levels are connected by a syphon pipe of 1200 metres long and 300 mm diameter. Determine the discharge in litres/s through the pipe.*

If the vertex of the pipe is 5 metres above water surface of the upper reservoir, find the maximum length of the inlet leg. Assume that the separation takes place at 3 metres of water. Take $f = 0·01$ and atmospheric pressure = 10·3 m of water.

Solution. Given : $H = 30$ m; $l = 1200$ m; $d = 300$ mm $= 0·3$ m; $Z_c = 5$ m; Separation head $= \left(\frac{p_c}{w}\right) 3$ m of water; $f = 0·01$ and atmospheric pressure head $\left(\frac{p_a}{w}\right) = 10·3$ m of water.

Discharge through the pipe

Let $\qquad Q =$ Discharge through the pipe

We know that the difference of water levels in the reservoirs (H)

$$30 = \frac{f l Q^2}{3(d)^5} = \frac{0·01 \times 1200 \times Q^2}{3 \times (0·3)^5} = 1646 \, Q^2$$

$\therefore \qquad Q^2 = 30/1646 = 0·0182 \qquad$ or $\qquad Q = 0·135$ m³/s   **Ans.**

Flow Through Compound Pipes

Length of the inlet leg

Let l = Length of the inlet leg.

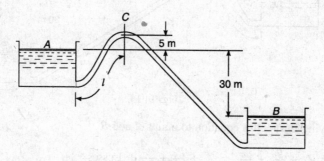

Fig. 14·12.

We know that the cross-sectional area of the pipe,

$$a = \frac{\pi}{4} \times (d)^2 = \frac{\pi}{4} \times (0.3)^2 = 0.0707 \text{ m}^2$$

and velocity of water in the pipe,

$$v = \frac{Q}{a} = \frac{0.135}{0.0707} = 1.91 \text{ m/s}$$

Applying Bernoulli's equation to points A and C,

$$Z_a + \frac{p_a}{w} + \frac{v_a^2}{2g} = Z_c + \frac{p_c}{w} + \frac{v_c^2}{2g} + \text{Losses}$$

$$0 + 10.3 + 0 = 5 + 3.0 \frac{(1.91)^2}{2 \times 9.81} + \left(\frac{l_1}{1200} \times 30\right)$$

$$10.3 = 8.19 + 0.025 \, l_1$$

$$\therefore \quad l_1 = \frac{10.3 - 8.19}{0.025} = 84.4 \text{ m} \quad \textbf{Ans.}$$

Example 14·13. *A pipe 1800 metres long having a diameter of 200 mm connects two reservoirs, the surface of water in one being 30 metres below the other. The pipe crosses a ridge, whose summit is 7·5 metres above the upper reservoir. Find the minimum depth of the pipe below the summit of the ridge, in order that the pressure at the apex does not fall 7·5 metres below the atmosphere. $f = 0.008$. The length of pipe from the upper reservoir to the summit ridge is 300 metres. Find also the discharge to the lower reservoir in litres/s.*

Solution. Given $l = 1800$ m; $d = 200$ mm $= 0.2$ m; $H = 30$ m; Pressure at the apex (p_c/w) $= 10.3 - 7.5 = 2.8$ m; $f = 0.008$ and length of pipe from the upper reservoir to the summit of the ridge $(l_1) = 300$ m.

Minimum depth of pipe below the summit of the ridge

Let Z_c = Height of apex above water level of the upper reservoir.

We know that the difference of water levels (H)

$$30 = \frac{4flv^2}{2gd} = \frac{4 \times 0.008 \times 1800 \times v^2}{2 \times 9.81 \times 0.2} = 14.68 \, v^2$$

$$\therefore \quad v^2 = 30/14.68 = 2.044 \quad \text{or} \quad v = 1.43 \text{ m/s}$$

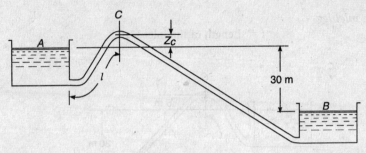

Fig. 14·13.

Now applying Bernoulli's equation to points A and B,

$$Z_a + \frac{p_a}{w} + \frac{v_a^2}{2g} = Z_c + \frac{p_c}{w} + \frac{v_c^2}{2g} + \text{Losses}$$

$$0 + 10\cdot3 + 0 = Z_c + 2\cdot8 + \frac{(1\cdot43)^2}{2\times9\cdot81} + \left(\frac{300}{1800}\times30\right)$$

$$10\cdot3 = Z_c + 7\cdot9$$

$$\therefore \quad Z_c = 10\cdot3 - 7\cdot9 = 2\cdot4 \text{ m}$$

and minimum depth of pipe below the summit of the ridge

$$= 7\cdot5 - 2\cdot4 = 5\cdot1 \text{ m} \quad \textbf{Ans.}$$

Discharge to the lower reservoir

We also know that the cross-sectional area of the pipe,

$$a = \frac{\pi}{4}\times(d)^2 = \frac{\pi}{4}\times(0\cdot2)^2 = 0\cdot0314 \text{ m}^2$$

and discharge to the lower reservoir,

$$Q = a.v = 0\cdot0314 \times 1\cdot43 = 0\cdot045 \text{ m}^3/\text{s} = 45 \text{ litres/s} \quad \textbf{Ans.}$$

EXERCISE 14·3

1. The difference of water levels in two reservoirs is connected by a syphon is 10 m. The pipe is 1200 m long and of 300 mm diameter. If the pipe is raised to a height of 3 m above the upper reservoir level at a point 400 m from the entry, calculate the absolute pressure head in the pipe at this point. Take atmospheric pressure head as 10·3 m of water and $f = 0\cdot0075$. [**Ans.** 3·9 m]

2. Two reservoirs, whose surface levels differ by 30 m are connected by a pipe of 600 mm diameter and 3 km long. The pipeline crosses a ridge, whose summit is 9 m above the higher reservoir. Find the minimum depth below the ridge at which the pipe must be laid, if the absolute pressure in the pipe is not to fall below 8 m of water. Also calculate the discharge through the pipe. Take $f = 0\cdot0075$. [**Ans.** 4·9 m; 560 litres/s]

3. Two reservoirs, whose surface levels differ by 100 m are connected by a pipe of 2 m diameter and 10 km long. The pipeline crosses a ridge, whose summit is 30 m above the level of and 500 m distant from the higher reservoir.
Find the minimum depth, below the ridge, at which the pipe should be laid if the absolute pressure in the pipe is not to fall below 3 m of water. Also find out the discharge in litres/s through the pipeline considering all the losses. [**Ans.** 28·3 m; 11 300 litres/s]

QUESTIONS

1. State the assumption for flow of water from one tank to another through a pipe.
2. What is a compound pipe? Under what circumstances is it used?

Flow Through Compound Pipes

3. Explain the reason for connecting two tanks with the pipes in parallel.
4. What do you understand by the term 'branched pipes'?
5. What is a syphon pipe? Explain the phenomenon of flow through a syphon pipe.

OBJECTIVE TYPE QUESTIONS

1. A compound pipe is required to be replaced by a new pipe. Both the pipes are said to be equivalent, if both of them have same
   (a) length and diameter
   (b) loss of head
   (c) discharge
   (d) both 'b' and 'c'

2. A compound pipe consisting of three pipes of lengths l_1, l_2 and l_3 having diameters d_1, d_2 and d_3 is to be replaced by an equivalent pipe of the same length (l) and diameter (d). The size of the equivalent pipe is given by the relation :

   (a) $\dfrac{l}{d^2} = \dfrac{l_1}{d_1^2} + \dfrac{l_2}{d_2^2} + \dfrac{l_3}{d_3^2}$

   (b) $\dfrac{l^2}{d^2} = \dfrac{l_1^2}{d_1^2} + \dfrac{l_2^2}{d_2^2} + \dfrac{l_3^2}{d_3^2}$

   (c) $\dfrac{l}{d^5} = \dfrac{l_1}{d_1^5} + \dfrac{l_2}{d_2^5} + \dfrac{l_3}{d_3^5}$

   (d) $\dfrac{l^2}{d^5} = \dfrac{l_1^2}{d_1^5} + \dfrac{l_2^2}{d_2^5} + \dfrac{l_3^2}{d_3^5}$

3. In case of flow through parallel pipes
   (a) loss of head for all the pipes is same
   (b) total discharge is equal to the sum of discharge in all the pipes
   (c) total loss of head is equal to the sum of loss of heads in all the pipes
   (d) both 'a' and 'b'

4. A syphon will work satisfactorily, if the pressure at its summit is .... than the atmospheric pressure
   (a) less than   (b) equal to   (c) more than   (d) either 'b' or 'c'

ANSWERS

1. (a)   2. (c)   3. (d)   4. (a)

15

Flow Through Nozzles

1. *Introduction.* 2. *Velocity of Water through a Nozzle.* 3. *Transmission of Power through a Nozzle.* 4. *Efficiency of Power Transmission through a Nozzle.* 5. *Diametre of Nozzle for Maximum Transmission of Power.* 6. *Uses of Nozzles.* 7. *Water Hammer.* 8. *Water Hammer when the Valve is Gradually Closed.* 9. *Water Hammer when the Valve is Suddenly Closed.* 10. *Effect of Pipe Elasticity on Hammer Blow.* 11. *Surge Tanks.*

15·1 Introduction

A nozzle is a tapering mouthpiece, which is fitted to the outlet end of a pipe. A nozzle is, generally, used to have a high velocity of water, as it converts pressure head into kinetic head at its outlet. A high velocity of water is required in fire fighting, mining power developments etc.

15·2 Velocity of Water through a Nozzle

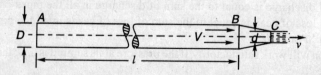

Fig. 15·1. A Nozzle.

Consider a nozzle *BC* fitted at the end of a pipeline *AB* through which water is flowing as shown in Fig. 15·1.

Let
- l = Length of the pipe *AB*,
- D = Diameter of the pipe *AB*,
- V = Velocity of water in the pipe *AB*,
- f = Darcy's coefficient of friction for the pipe *AB*,
- d = Diameter of the nozzle,
- v = Velocity of water through the nozzle, and
- H = Head of water under which the flow takes place.

From the geometry of the pipeline, we find that the area of pipe *AB*,

$$A = \frac{\pi}{4} \times D^2$$

and the area of nozzle at *C*,

$$a = \frac{\pi}{4} \times d^2$$

Since water is continuously flowing through the pipe and nozzle, therefore

$$AV = av \quad \text{or} \quad V = \frac{av}{A}$$

We know that the loss of head due to friction in the pipe *AB* ...(*i*)

$$= \frac{4flV^2}{2gD}$$

[270]

Flow Through Nozzles

and the loss of head due to velocity at outlet

$$= \frac{v^2}{2g} \qquad \ldots(ii)$$

Assuming that the total available head of water is lost, while flowing through the pipe and nozzle, therefore loss of head,

$$H = \text{Loss of head due to friction}$$
$$+ \text{Loss of head due to velocity at outlet}$$

$$= \frac{4flV^2}{2gD} + \frac{v^2}{2g} \qquad \ldots\text{(Neglecting minor losses)}$$

$$= \frac{4fl}{2gD}\left(\frac{a^2 v^2}{A^2}\right) + \frac{v^2}{2g} = \frac{v^2}{2g}\left(\frac{4fl}{D} \times \frac{a^2}{A^2} + 1\right)$$

$$\therefore \quad v = \sqrt{\frac{2gH}{1 + \frac{4fl}{D} \times \frac{a^2}{A^2}}}$$

Example 15·1. *A pipe 3·2 kilometres long and of 0·9 m diameter is fitted with a nozzle of 200 mm diameter at its discharge end. Find the velocity of water through the nozzle, if the head of water is 50 m. Take f = 0·006 for the pipe.*

Solution. Given : $l = 3·2$ km $= 3200$ m; $D = 0·9$ m; $d = 200$ mm $= 0·2$ m; $H = 50$ m and $f = 0·006$.

We know that the area of the pipe,

$$A = \frac{\pi}{4} \times (D)^2 = \frac{\pi}{4} \times (0·9)^2 = 0·636 \text{ m}^2$$

and area of nozzle,

$$a = \frac{\pi}{4} \times (d)^2 = \frac{\pi}{4} \times (0·2)^2 = 31·4 \times 10^{-3} \text{ m}^2$$

∴ Velocity of water through the nozzle,

$$v = \sqrt{\frac{2gH}{1 + \frac{4fl}{D} \times \frac{a^2}{A^2}}}$$

$$= \sqrt{\frac{2 \times 9·81 \times 50}{1 + \frac{4 \times 0·006 \times 3200}{0·9} \times \frac{(31·4 \times 10^{-3})}{(0·636)^2}}} \text{ m/s}$$

$$= \sqrt{\frac{981}{1·208}} = 28·5 \text{ m/s } \textbf{Ans.}$$

Example 15·2. *A pipe, 100 mm in diameter, has a nozzle attached to it at the discharge end. The diameter of the nozzle is 50 mm. The rate of discharge of water through the nozzle of 15 litres/s and pressure of the base of the nozzle is 60 kPa. Calculate the coefficient of discharge. Assume that the base of the nozzle and the outlet of the nozzle are of the same elevation.*

Solution. Given : $D = 100$ mm $= 0·1$ m; $d = 50$ mm $= 0·05$ m; $Q = 15$ litres/s $= 0·015$ m³/s and $p_1 = 60$ kPa $= 60$ kN/m².

Let $V =$ Theoretical velocity of water in the pipe, and
$v =$ Theoretical velocity of jet at the nozzle.

We know that area of the pipe,

$$A = \frac{\pi}{4} \times (D)^2 = \frac{\pi}{4} \times (0 \cdot 1)^2 = 7 \cdot 854 \times 10^{-3} \text{ m}^2$$

and the area of nozzle,

$$a = \frac{\pi}{4} \times (d)^2 = \frac{\pi}{4} \times (0 \cdot 05)^2 = 1 \cdot 964 \times 10^{-3} \text{ m}^2$$

Since the discharge is continuous, therefore

$$A \cdot V = a \cdot v \quad \text{or} \quad 7 \cdot 854 \times 10^{-3} \cdot V = 1 \cdot 964 \times 10^{-3} \cdot v$$

$$\therefore \quad V = \frac{1 \cdot 964 \times 10^{-3} \cdot v}{7 \cdot 854 \times 10^{-3}} = 0 \cdot 25 \, v$$

Using Bernoulli's equation at the base of the nozzle and outlet of the nozzle (Taking base of the nozzle as section 1, and outlet of the nozzle as section 2),

$$Z_1 + \frac{p_1}{w} + \frac{v_1^2}{2g} = Z_2 + \frac{p_2}{w} + \frac{v_2^2}{2g} \qquad \text{...(Neglecting minor losses)}$$

$$\frac{p_1}{w} + \frac{V^2}{2g} = \frac{v^2}{2g} \qquad \text{...}(\because Z_1 = Z_2, \, v_1 = V, \, v_2 = v \text{ and } p_2 = 0)$$

$$\frac{60}{9 \cdot 81} = \frac{v^2}{2g} - \frac{V^2}{2g} = \frac{v^2 - (0 \cdot 25 \, v)^2}{2g} = \frac{15 \, v^2}{16 \times 2g}$$

$$6 \cdot 12 = \frac{15 \, v^2}{32 \, g} = \frac{15 \, v^2}{32 \times 9 \cdot 81} = \frac{v^2}{20 \cdot 93}$$

$$\therefore \quad v^2 = 6 \cdot 12 \times 20 \cdot 93 = 128 \cdot 1 \quad \text{or} \quad v = 11 \cdot 3 \text{ m/s}$$

We know that the theoretical discharge

$$= a \cdot v = (1 \cdot 964 \times 10^{-3}) \times 11 \cdot 3 = 0 \cdot 022 \text{ m}^3/\text{s}$$

and coefficient of discharge, $\quad C_d = \dfrac{\text{Actual discharge}}{\text{Theoretical discharge}} = \dfrac{0 \cdot 015}{0 \cdot 022} = 0 \cdot 68$ **Ans.**

15·3 Transmission of Power through a Nozzle

In the previous article, we have discussed the velocity of water flowing through a nozzle. This flowing water, under the available head, may be used for the transmission of power. Now consider a nozzle fitted at one end of the pipeline through which water is flowing.

Let
- l = Length of the pipe in metres
- H = Head of water under which the flow takes place in metres
- f = Darcy's coefficient of friction for the pipe,
- D = Diameter of the pipe in metres
- V = Velocity of water in the pipe in m/s
- A = Area of the pipe in metres and
- d, v, a = Corresponding values for the nozzle at the outlet.

Assuming that the total available head of water is lost, while flowing through the pipe, therefore

$$H = \frac{4 f l V^2}{2gD} + \frac{v^2}{2g} \qquad \text{...(Neglecting minor losses)}$$

$$= \frac{4fl}{2gD}\left(\frac{a^2 \cdot v^2}{A^2}\right) + \frac{v^2}{2g} \qquad \text{...}\left(\because V = \frac{av}{A}\right)$$

Flow Through Nozzles

$$\therefore \quad \frac{v^2}{2g} = H - \frac{4fl}{2gD}\left(\frac{a^2 \cdot v^2}{A^2}\right)$$

and power available at the outlet of the jet,

$$P = \frac{wQv^2}{2g} = wQ \times \frac{v^2}{2g} = wav \times \frac{v^2}{2g} = \frac{v^2}{2g} \text{ kW}$$

Substituting the value of $\frac{v^2}{2g}$ from the last article,

$$P = wQ\left[H - \frac{4fl}{2gD}\left(\frac{a^2 \cdot v^2}{A^2}\right)\right] \text{ kW}$$

Example 15·3. *A hydro-electric plant is supplied water at the rate of 500 litres/s under a head of 250 metres through a pipeline 3·2 km long and 500 mm diameter. The pipeline terminates in a nozzle, which has a diameter of 200 mm. Find the power that can be transmitted, if the Darcy's coefficient for the pipe is 0·01.*

Solution. Given : $Q = 500$ litres/s $= 0.5$ m³/s; $H = 250$ m; $l = 3.2$ km $= 3200$ m; $D = 500$ mm $= 0.5$ m; $d = 200$ mm $= 0.2$ m and $f = 0.01$.

We know that the area of pipeline,

$$A = \frac{\pi}{4} \times (D)^2 = \frac{\pi}{4} \times (0.5)^2 = 0.1964 \text{ m}^2$$

and the area of nozzle

$$a = \frac{\pi}{4} \times (d)^2 = \frac{\pi}{4}(0.2)^2 = 0.0314 \text{ m}^2$$

∴ Velocity of water through the nozzle,

$$v = \frac{Q}{a} = \frac{0.5}{0.0314} = 15.9 \text{ m/s}$$

and power that can be transmitted,

$$P = wQ\left[H - \frac{4fl}{2gD}\left(\frac{a^2 \cdot v^2}{A^2}\right)\right]$$

$$= 9.81 \times 0.5 \times \left[250 - \frac{4 \times 0.01 \times 3200}{2 \times 9.81 \times 0.5} \times \left(\frac{(0.0314)^2 \times (15.9)^2}{(0.1964)^2}\right)\right] \text{ kW}$$

$$= 4.905 \times (250 - 84.3) = 812.8 \text{ kW} \quad \textbf{Ans.}$$

15·4 Efficiency of Power Transmission through a Nozzle

In the last article, we have discussed the amount of power transmitted through a nozzle. But in this article, we shall discuss the efficiency of power transmission. Now consider a nozzle fitted at one end of the pipeline through which water is flowing.

Let l = Length of the pipe,

H = Head of water under which the flow takes place,

f = Darcy's coefficient of friction for the pipe,

D = Diameter of the pipe,

V = Velocity of water in the pipe,

A = Area of the pipe,

d, v, a = Corresponding values for the nozzle at outlet.

We have seen in the previous article that the power available at the outlet of the nozzle,

$$P = \frac{wQv^2}{2g} = wav\left[H - \frac{4fl}{2gD}\left(\frac{a^2 \cdot v^2}{A^2}\right)\right] \qquad \ldots(i)$$

We know that the power available at the inlet of the nozzle

$$= wQH \qquad \ldots(ii)$$

∴ Efficiency of transmission,

$$\eta = \frac{\text{Power available at the outlet of the nozzle}}{\text{Power available at the inlet of the nozzle}}$$

$$= \frac{\frac{wQv^2}{2g}}{wQH} = \frac{v^2}{2gH}$$

Now the power transmitted will be maximum, when $\left(\frac{d(P)}{dv}\right) = 0$ or when the differential coefficient of equation (i) is zero. i.e.,

$$\frac{d\left[wav\left(H - \frac{4fl}{2gD} \times \frac{a^2 v^2}{A^2}\right)\right]}{dv} = 0$$

$$\frac{d\left[wa\left(Hv - \frac{4fl}{2gD} \times \frac{a^2 v^3}{A^2}\right)\right]}{dv} = 0 \quad \text{or} \quad H - 3\left(\frac{4fl}{2gD} \times \frac{a^2 v^2}{A^2}\right) = 0$$

$$H - 3\left(\frac{4flV^2}{2gD}\right) = 0 \qquad \ldots\left(\because V = \frac{av}{A}\right)$$

$$H - 3h_f = 0 \qquad \ldots\left(\because h_f = \frac{4flV^2}{2gD}\right)$$

∴ $$h_f = \frac{H}{3}$$

It means that the power transmitted through the nozzle is maximum, when the head lost due to friction in the pipe is equal to 1/3 of the total supply head.

Note : The general formula for efficiency $\left(i.e., \frac{H - h_f}{H}\right)$ remains applicable.

Example 15·4. *Find the transmission efficiency of a nozzle fitted to a pipe, if water through it is flowing with a velocity of 13 m/s under 15 m head of water.*

Solution. Given : $v = 13$ m/s and $H = 15$ m.

We know that the transmission efficiency of nozzle,

$$\eta = \frac{v^2}{2gH} = \frac{(13)^2}{2 \times 9\cdot81 \times 15} = 0\cdot574 = 57\cdot4\% \quad \textbf{Ans.}$$

Example 15·5. *A pipe of 75 mm diameter and 250 metres long has a nozzle of 25 mm fitted at the discharge end. If the total head of water is 48 metres, find the maximum power transmitted. Take f as 0.01 for the pipe.*

Solution. Given: $D = 75$ mm $= 75 \times 10^{-3}$ m; $l = 250$ m; *$d = 25$ mm $= 25 \times 10^{-3}$ m; $H = 48$ m and $f = 0.01$.

We know that for maximum transmission of power, the head lost due to friction is 1/3 of the total head. Therefore loss of head,

$$\therefore \qquad h_f = \frac{H}{3} = \frac{48}{3} = 16 \text{ m}$$

We also know that loss of head due to friction (h_f),

$$16 = \frac{flQ^2}{3D^5} = \frac{0.01 \times 250 \ Q^2}{3 \times (75 \times 10^{-3})^5} = 351\ 200\ Q^2$$

or $\qquad Q^2 = 16/351\ 200 = 0.4556 \times 10^{-4}$ or $Q = 0.0067$ m³/s

$\therefore$ Maximum power transmitted,

$$P_{max} = wQ\,(H - h_f) = 9.81 \times 0.0067\,(48 - 16) = 2.1 \text{ kW} \textbf{ Ans.}$$

15·5 Diameter of Nozzle for Maximum Transmission of Power

In the previous article, we have discussed the amount of power transmitted by a nozzle. But in this article, we shall discuss the condition for the diameter of the nozzle for the maximum transmission of power. Now consider a nozzle fitted at the end of a pipeline through which the water is flowing.

Let
l = Length of the pipe,
H = Head of water under which the flow takes place,
f = Darcy's coefficient of friction for the pipe,
D = Diameter of the pipe,
V = Velocity of water in the pipe,
A = Area of the pipe, and
d, v, a = Corresponding values for the nozzle at the outlet.

We have discussed in Art. 15·3 that the power transmitted through the nozzle,

$$P = wav \times \frac{v^2}{2g} = w \times \left(\frac{\pi}{4} \times d^2\right) \times \frac{1}{2g} \times v^3 \qquad \ldots(i)$$

We have also seen in the same article that the velocity of water through the nozzle,

$$v = \sqrt{\frac{2gH}{1 + \frac{4fl}{D^2} \times \frac{a^2}{A^2}}} \qquad \ldots(ii)$$

Now substituting the value of v in equation (i),

$$P = w \times \frac{\pi}{4} \times d^2 \times \frac{1}{2g} \left[\frac{2gH}{1 + \frac{4fl}{D} \times \frac{a^2}{A^2}}\right]^{3/2}$$

$$= w \times \frac{\pi}{4} \times d^2 \times \frac{1}{2g} \left\{\frac{2gH}{1 + \frac{4fl}{D} \frac{\left(\frac{\pi}{4} \times d^2\right)^2}{\left(\frac{\pi}{4} \times D^2\right)^2}}\right\}^{3/2}$$

---

*Superfluous data.

$$P = w \times \frac{\pi}{4} \times d^2 \times \frac{1}{2g}\left[\frac{2gH}{1 + \frac{4fld^4}{D^5}}\right]^{3/2}$$

$$= w \times \frac{\pi}{4} \times d^2 \times \frac{1}{2g}\left[\frac{2gHD^5}{D^5 + 4fld^4}\right]^{3/2}$$

$$= w \times \frac{\pi}{4 \times 2g} (2gHD^5)^{3/2} \times \frac{d^2}{(D^5 + 4fld^4)^{3/2}}$$

$$= K \times \frac{d^2}{(D^5 + 4fld^4)^{3/2}} \qquad \ldots(iii)$$

where K is constant, whose value $= w \times \frac{\pi}{4 \times 2g} (2gHD^5)^{3/2}$

Now power transmitted will be maximum, when $\frac{d(P)}{d(d)} = 0$ or when the differential coefficient of equation (iii) is zero. i.e.,

$$\frac{d\left[K \times x \frac{d^2}{(D^5 + 4fld^2)^{3/2}}\right]}{d(d)} = 0$$

or $\quad \dfrac{[(D^5 + 4fld^4)^{3/2} \times 2d] - [d^2 \times \frac{3}{2}(D^5 + 4fld^4)^{1/2} \times 4 \times 4fld^3]}{(D^5 + 4fld^4)^{3/2}} = 0$

$[(D^5 + 4fld^4)^{3/2} \times 2d] - [d^2 \times \frac{3}{2}(D^5 + 4fld^4)^{1/2} \times 4 \times 4fld^3] = 0$

$\therefore \quad [(D^5 + 4fld^4)^{3/2} \times 2d] - [(D^5 + 4fld^4)^{1/2} \times 24fld^5] = 0$

or $\quad (D^5 + 4fld^4)^{3/2} \times 2d = (D^5 + 4fld^4)^{1/2} \times 24fld^5$

$(D^5 + 4fld^4) \times 2d = 24fld^5$

$D^5 + 4fld^4 = 12fld^4$

$8fld^4 = D^5 \quad$ or $\quad d^4 = \dfrac{D^5}{8fl} \qquad \ldots(iv)$

$\therefore \quad d = \left(\dfrac{D^5}{8fl}\right)^{1/4} \qquad \ldots(v)$

Cor. Equation (iv) may be rewritten as:

$$\frac{D^5}{d^4} = 8fl \quad \text{or} \quad \frac{D^4}{d^4} = \frac{8fl}{D} \qquad \ldots(vi)$$

or $\quad \dfrac{D^2}{d^2} = \sqrt{\dfrac{8fl}{D}} \qquad \ldots$(Taking square root)

$$\frac{\frac{\pi}{4} \times D^2}{\frac{\pi}{4} d^2} = \sqrt{\frac{8fl}{D}} \qquad \ldots\left(\text{Multiplying and dividing by } \frac{\pi}{4}\right)$$

$\therefore \quad \dfrac{A}{a} = \sqrt{\dfrac{8fl}{D}}$

Flow Through Nozzles

Example 15·6. *A pipe 250 metres long and of 75 mm diameter has a nozzle fitted at the discharge end. Find the diameter of the nozzle, so that maximum power may be transmitted. Take f = 0·01.*

Solution. Given : $l = 250$ m; $D = 75$ mm $= 0·075$ m and $f = 0·01$.

We know that diameter of the nozzle, so that maximum power may be transmitted,

$$d = \left(\frac{D^5}{8fl}\right)^{1/4} = \left(\frac{0·075}{8 \times 0·01 \times 250}\right)^{1/4} = 0·0185 \text{ m} = 18·5 \text{ mm. Ans.}$$

Example 15·7. *Find the diameter of the nozzle and the maximum power transmitted by a jet of water discharging freely out of a nozzle, fitted to a pipe 300 m long and 100 mm diameter with coefficient of friction as 0·01. The available head at nozzle is 90 m.*

Solution. Given : $l = 300$ m; $D = 100$ mm $= 0·1$ m; $f = 0·01$ and total head $= 90$ m.

Diameter of the nozzle for maximum power

We know that diameter of the nozzle for maximum transmission of power,

$$d = \left(\frac{D^5}{8fl}\right)^{1/4} = \left(\frac{(0·1)^5}{8 \times 0·01 \times 300}\right)^{1/4} = 0·0254 \text{ m} = 25·4 \text{ mm Ans.}$$

Maximum power transmitted by the jet

We also know that area of the nozzle,

$$a = \frac{\pi}{4} \times (d)^2 = \frac{\pi}{4} \times (0·0254)^2 = 0·507 \times 10^{-3} \text{ m}^2$$

and for maximum transmission of power, the head lost due to friction is $\frac{1}{3}$ of the total head.

or $$h_f = \frac{\text{Total head}}{3} = \frac{90}{3} = 30 \text{ m}$$

∴ Available head, $H = 90 - 30 = 60$ m

and velocity of water through the nozzle,

$$v = \sqrt{2gH} = \sqrt{2 \times 9·81 \times 60} = 34·3 \text{ m/s}$$

∴ Maximum power transmitted by the jet,

$$P_{max} = \frac{wav^3}{2g} = \frac{9·81 \times (0·507 \times 10^{-3}) \times (34·3)^3}{2 \times 9·81} \text{ kW}$$

$$= 10·2 \text{ kW Ans.}$$

EXERCISE 15·1

1. A pipeline 1·5 km long and 500 mm diameter is fitted with a nozzle of 150 mm diameter at its discharge end. If the flow takes place under a head of 100 m, determine the velocity of water through the nozzle. Take f as 0·01 for the pipeline. **(Ans. 31·5 m/s)**

2. A cast iron pipe of 500 mm diameter and 2 km long is fitted with a nozzle of 100 mm diameter. The pipe is conveying water from a reservoir, which is 60 m above the nozzle level. Find the velocity and discharge of the water through the nozzle, if the value of the coefficient of friction for the pipe is 0·01. **(Ans. 30·6 m/s ; 0·24 m³/s)**

3. A pipeline 600 m long and 400 mm diameter is supplying water at the rate of 800 litres/s under a head of 180 m at a power plant. If the pipeline is fitted with a nozzle of 150 mm diameter, find the power which can be transmitted. Take f as 0·01. **(Ans. 174·2 kW)**

4. Water is flowing through a pipe fitted with a nozzle with a velocity of 10·5 m/s under a head of 12·8 m. What is the efficiency of transmission ? (Ans. 43·9%)
5. Find the diameter of a nozzle to be fitted at the end of a pipeline of 500 m long and of diameter 200 mm for the maximum transmission of power. Take $f = 0.008$. (Ans. 56·2 mm)

15·6 Uses of Nozzles

We have already discussed that a nozzle is used at the end of a pipeline in order to obtain the jet of water with a high velocity. Though the nozzles are widely used for numerous purposes, yet the water hammer is one of the problems in the use of nozzles,

15·7 Water Hammer

Water, while flowing in a pipe, possesses some momentum on account of its motion. It has been experienced that if the flowing water is suddenly brought to rest by closing the valve, its momentum is destroyed, which causes a very high pressure on the valve. This high pressure is followed by a series of pressure vibrations. These pressure vibrations set up noises in the pipe, known as knocking. Such a knocking is often heard in water pipes, if the tap is turned off quickly. The sudden rise of pressure has the effect of hammering action on the walls of the pipe and, thus, is known as hammer blow or water hammer. Sometimes, the hammer blow is so high, that it may even burst the pipe. It is, thus, obvious that the valves of the pipelines or penstocks should always be closed gradually.

Now we shall discuss the following two cases of water hammer, depending upon the time taken in closing the valve :

1. When the valve is gradually closed.
2. When the valve is suddenly* closed.

15·8 Water Hammer when the Valve is Gradually Closed

Consider a uniform pipeline, through which some liquid (say water) is flowing with a uniform velocity, whose valve is gradually closed.

Let
l = Length of the pipeline
a = Cross-sectional area of the pipeline,
v = Velocity of water in the pipeline, and
t = Time, in seconds, in which the water is brought to rest by the closure of the valve.

∴ Mass of water in the pipe $= \dfrac{wal}{g}$ ...(i)

and rate of retardation of the water $= \dfrac{\text{Velocity}}{\text{Time}} = \dfrac{v}{t}$ ...(ii)

We know that force = Mass × Acceleration

$$= \dfrac{wal}{g} \times \dfrac{v}{t} = \dfrac{walv}{gt} \qquad ...(iii)$$

∴ Intensity of pressure rise in the valve

$$p = \dfrac{\text{Force}}{\text{Area}} = \dfrac{\dfrac{walv}{gt}}{a} = \dfrac{wlv}{gt}$$

---

*Sometimes, valve closure is said to be gradual when time taken to close the valve, $t > \dfrac{2l}{v}$ and sudden when $t < \dfrac{2l}{v}$, where l is the length of the pipe and v is the velocity of water in the pipeline.

Flow Through Nozzles

and pressure head due to increase in pressure,

$$H = \frac{p}{w} = \frac{\frac{wlv}{gt}}{w} = \frac{lv}{gt} \qquad \ldots(iv)$$

Example 15·8. *Water is supplied to a turbine through a pipe of 3 kilometres long. Water flows in the pipe with a velocity of 2 m/s. A valve near the turbine end is closed in 30 seconds. Find the rise in pressure behind the valve.*

Solution. Given : $l = 3$ km $= 3000$ m; $v = 2$ m/s and $t = 30$ s.

We know that the rise in pressure,

$$p = \frac{lv}{gt} = \frac{3000 \times 2}{9\cdot 81 \times 30} = 20\cdot 4 \text{ kN/m}^2 = 20\cdot 4 \text{ kPa} \quad \textbf{Ans.}$$

15·9 Water Hammer when the Valve is Suddenly Closed

We have discussed, in the last article that the increase in pressure when the valve is gradually closed,

$$p = \frac{lv}{gt}$$

A little consideration will show that if the valve fitted to a pipeline is suddenly closed or instantaneously closed (*i.e.*, $t = 0$), the increase in pressure will be infinite. However, in actual practice, it is not possible to close the valve instantaneously, as it always takes some time. Thus, the infinite pressure is hypothetical. Moreover, the above equation has been derived on the assumption that the liquid flowing through the pipe is incompressible. As a matter of fact, this assumption is not always correct because at very high pressure, the liquids get compressed to some extent and, thus, behave like compressible fluids.

Now consider a uniform pipeline through which some liquid (say water) is flowing with a uniform velocity, whose valve is suddenly closed.

Let l = Length of the pipeline,

a = Cross-sectional area of the pipeline,

v = Velocity of water in the pipeline

K = Bulk modulus of water, and

p = Increase in pressure.

We know that when the flowing water is suddenly brought to rest (by suddenly closing the valve), the kinetic energy of the water is converted into strain energy, which is stored in water (assuming the pipe to be perfectly rigid and non-elastic, so that there is no radial expansion of its walls).

$\therefore$ Mass of water in the pipe $= \dfrac{wal}{g}$

and kinetic energy of water $= \dfrac{\text{Mass} \times (\text{Velocity})^2}{2} = \dfrac{\frac{wal}{g} \times v^2}{2} = \dfrac{walv^2}{2g} \qquad \ldots(i)$

We also know that the strain energy stored in water

$$= \frac{1}{2}\left(\frac{p^2}{K}\right) \times al \qquad \ldots(ii)$$

Equating equations (i) and (ii),

$$\frac{walv^2}{2g} = \frac{1}{2}\left(\frac{p^2}{K}\right)al \quad \text{or} \quad p^2 = \frac{wv^2K}{g}$$

$$\therefore \quad p = v\sqrt{\frac{Kw}{g}} = v\sqrt{K\rho} \quad \quad \ldots\left(\because \rho = \frac{w}{g}\right)$$

Example 15·9. *A valve at the outlet of a pipe is suddenly closed to bring the water to rest, which was flowing at 3 m/s. Find the pressure increase due to sudden closure of the valve. Take K as 2 GPa.*

Solution. Given : $v = 3$ m/s and $K = 2$ GPa $= 2000$ kPa $= 2000$ kN/m^2

We know that the pressure increase due to sudden closure of valve,

$$p = v\sqrt{\frac{Kw}{g}} = 3 \times \sqrt{\frac{2000 \times 9 \cdot 81}{9 \cdot 81}} \text{ kN/m}^2$$

$$= 134 \cdot 2 \text{ kN/m}^2 = 134 \cdot 2 \text{ kPa} \quad \textbf{Ans.}$$

15·10 Effect of Pipe Elasticity on Hammer Blow

In the previous article, we have obtained a relation for the increase in the pressure, when the valve is suddenly closed. In this article, we assumed the pipeline to be perfectly rigid and non-elastic. But, in actual practice, every pipe possesses some elasticity. As a result of this, the radial pressure of water on the pipe causes circumferential* and longitudinal * stresses in the pipe walls. It is, thus, obvious that the pipe also absorbs some strain energy alongwith the water.

Now consider a uniform pipeline, through which some liquid (say water) is flowing with a uniform velocity, whose valve is suddenly closed.

Let
  l = Length of the pipeline,
  a = Cross-sectional area of the pipeline,
  R = Radius of the pipeline,
  t = Thickness of the pipeline,
  v = Velocity of water in the pipeline,
  K = Bulk modulus of water,
  E = Modulus of elasticity for pipe material,
  m = Poisson's ratio for pipe material, and
  p = Pressure of water due to hammer blow.

We know that the circumferential stress in the pipe,

$$\sigma_c = \frac{pr}{t}$$

and the longitudinal stress in the pipe,

$$\sigma_l = \frac{pr}{2t}$$

We also know that the volume of pipe wall
$$= 2\pi r \cdot t \cdot l$$

and the strain energy stored in the pipe walls

$$= \left(\frac{1}{2}\frac{\sigma_c^2}{E} \times 2\pi rtl\right) + \left(\frac{1}{2} \times \frac{\sigma_l^2}{E} \times 2\pi rtl\right) - \left(\frac{\sigma_c \cdot \sigma_l}{mE} \times 2\pi rtl\right)$$

$$= \frac{1}{2E} \times 2\pi rtl \left(\sigma_c^2 + \sigma_l^2 - \frac{2\sigma_c \cdot \sigma_l}{m}\right)$$

---
*For details, please refer to some standard book or author's book *Strength of Materials*.

Flow Through Nozzles

$$= \frac{\pi r t l}{E}\left(\frac{p^2 r^2}{t^2} + \frac{p^2 r^2}{4t^2} - \frac{p^2 r^2}{4t^2}\right) \qquad \ldots \text{(Assuming } m = 4\text{)}$$

$$= \frac{p^2 \pi r^3 l}{Et} = \frac{p^2 (al) r}{Et} \qquad \ldots (\because \pi r^2 = a) \ldots (i)$$

We also know that the strain energy stored in water,

$$= \frac{1}{2}\left(\frac{p^2}{K}\right) \times al \qquad \ldots (ii)$$

and kinetic energy of water $= \dfrac{walv^2}{2g}$ \qquad ...(iii)

Now loss of kinetic energy of water

$$= \text{Strain energy stored in water}$$
$$+ \text{Strain energy stored in pipe walls}$$

$$\frac{walv^2}{2g} = \frac{1}{2}\left(\frac{p^2}{K}\right) \times al + \frac{p^2(al)r}{Et}$$

$$\frac{wv^2}{2g} = \frac{p^2}{2K} + \frac{p^2 r}{Et} = p^2\left(\frac{1}{2K} + \frac{r}{Et}\right)$$

$$\therefore \quad p^2 = \frac{wv^2}{2g} \times \frac{1}{\left(\dfrac{1}{2K} + \dfrac{r}{Et}\right)}$$

or $\quad p = v \times \sqrt{\dfrac{w}{g\left(\dfrac{1}{K} + \dfrac{2r}{Et}\right)}} = v \times \sqrt{\dfrac{w}{g\left(\dfrac{1}{K} + \dfrac{d}{Et}\right)}} \quad \ldots (2r = d = \text{Dia.})$

Notes : 1. After obtaining the value of p i.e., increased pressure, the values of circumferential stress (i.e., σ_c) or longitudinal stress (i.e., σ_l) may be found out as usual.

2. If the pipeline is considered to be rigid (i.e., E is infinity), the above equation is reduced to :

$$p = v\sqrt{\frac{Kw}{g}}$$

Example 15.10. *Water flows through a pipeline of 1·5 km long 150 mm diameter and 5 mm thick with a velocity of 3 m/s. Determine the increase in pressure when the valve at the outlet of the pipeline is suddenly closed. Take the values of E and m for pipe as 210 GPa and 4 respectively. Also take K for water as 2·1 GPa.*

Solution. Given : $l = 1\cdot 5$ km $= 1500$ m; $d = 150$ mm $= 0\cdot 15$ m; $t = 5$ mm $= 0\cdot 005$ m; $v = 3$ m/s; $E = 210$ GPa $= 210 \times 10^3$ kPa and $K = 2\cdot 1$ GPa $= 2\cdot 1 \times 10^3$ kPa.

We know that the increase in water pressure,

$$p = v\sqrt{\frac{w}{8\left(\dfrac{1}{K} + \dfrac{d}{Et}\right)}}$$

$$= 0\cdot 8 \times \sqrt{\frac{9\cdot 81}{9\cdot 81\left(\dfrac{1}{2\cdot 1 \times 10^3} + \dfrac{0\cdot 15}{(210 \times 10^3) \times 0\cdot 005}\right)}}$$

$$= 3 \times 40\cdot 2 = 120\cdot 6 \text{ kN/m}^2 = 120\cdot 6 \text{ kPa} \quad \textbf{Ans.}$$

15·11 Surge Tanks

We have already discussed in the previous articles that whenever a valve, fitted at the end of a pipe, is suddenly closed it causes hammer blow in the pipeline. Moreover, in hydro-electric power plants*, since the requirement of water goes on changing, it is, therefore, essential to increase or decrease the discharge flowing through the pipeline.

A little consideration will show that whenever the requirement of water is suddenly decreased, the valve is suddenly closed, as a result of which the entire pipe length between the reservoir and the turbine will experience an increased pressure.

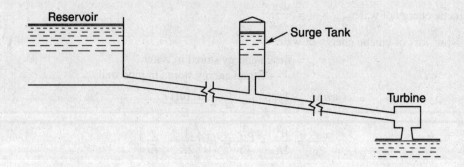

Fig. 15·2. Surge tank.

In order to overcome the above mentioned problems, a storage reservoir is fitted at some opening made on the pipeline (called pen-stock) in order to store water when the valve is suddenly closed, or to discharge water when increased discharge is required as shown in Fig. 15·2. Such a storage reservoir is called surge tank, which has the following two functions :

1. To control the pressure variations, due to rapid changes in the pipeline flow, thus eliminating water hammer possibilities.
2. To regulate the flow of water to the turbines by providing necessary retarding head of water.

The surge tanks are placed as close to the turbine as possible. The height of surge tank is generally kept above the maximum water level in the supply level reservoir. Though there are many types of surge tanks, yet the following are important from the subject point of view.
1. Simple surge tank,
2. Restricted orifice type surge tank, and
3. Differential surge tank.

EXERCISE 15·2

1. A pipeline of 1·2 km long and 200 mm diameter is conveying water with a velocity of 2·5 m/s. If the valve near the nozzle is gradually closed in 20 seconds, find the increase of pressure in the pipeline. **(Ans. 150 kPa)**
2. Water is flowing through a conduit pipe of 2 km long with a velocity of 1·8 m/s. Find the time in which the nozzle should be closed, if the pressure in the pipe should not exceed 225 kPa. **(Ans. 16 s)**
3. In a power plant, water is flowing through a pipeline with a velocity of 2·5 m/s. Water is required to be suddenly closed for repairs. Find the increase in pressure, if the value of bulk modulus for water is 210 GPa. **(Ans. 114·5 kPa)**

---

*These are the turbines, which convert the potential energy or kinetic energy of water into mechanical energy. These turbines will be discussed in the later part of this book.

4. A pipe 1 km long 400 mm diameter is 8 mm thick. If water through the pipe flows with a velocity of 1 m/s, find the increase in pressure when the valve at the outlet is suddenly closed. Take E for the pipe as 200 GPa and K for water as 2 GPa. **(Ans. 29·8 kPa)**

QUESTIONS

1. What is a nozzle ? State its uses.
2. Obtain relation for the transmission of power through a nozzle.
3. Write a short note on water hammer in pipes.
4. Establish an equation for the rise in pressure in a pipeline when the valve is closed gradually.
5. What is the difference in gradual and sudden closure of valve fitted at the discharge end of a pipeline running full of water ? Give the equation of pressure rise in the pipe due to sudden closure.
6. Explain the effect of hammer blow on pipes.

OBJECTIVE TYPE QUESTIONS

1. The shape of a nozzle is
   (a) cylindrical   (b) convergent   (c) divergent   (d) first 'b' then 'c'
2. A nozzle placed at the end of a water pipeline discharges water at a
   (a) low pressure   (b) high pressure   (c) low velocity   (d) high velocity
3. The power transmitted through the nozzle is maximum, when the head lost due to friction in the pipe is
   (a) one-fourth of the supply head   (b) one-third of the supply head
   (c) one-half of the supply head   (d) two-third of the supply head
4. The diameter of the nozzle (d) for maximum transmission of power is equal
   (a) $\left(\dfrac{D^5}{8fl}\right)^{1/4}$ (b) $\left(\dfrac{D^5}{8fl}\right)^{1/5}$ (c) $\left(\dfrac{D^4}{8fl}\right)^{1/4}$ (d) $\left(\dfrac{D^4}{8fl}\right)^{1/5}$

ANSWERS

1. (b)   2. (d)   3. (b)   4. (a)

16
Uniform Flow Through Open Channels

1. Introduction. 2. Chezy's Formula for Discharge through an Open Channel. 3. Values of Chezy's Constant in the Formula for Discharge through an Open Channel. 4. Bazin's Formula for Discharge. 5. Kutter's Formula for Discharge. 6. Manning's Formula for Discharge. 7. Discharge through a Circular Channel. 8. Channels of Most Economical Cross-sections. 9. Condition for Maximum Discharge through a Channel of Rectangular Section. 10. Condition for Maximum Discharge through Channel of Trapezoidal Section. 11. Condition for Maximum Velocity through a Channel of Circular Section. 12. Condition for Maximum Discharge through a Channel of Circular Section. 13. Measurement of River Discharge. 14. Area of Flow. 15. Simple Segments Method. 16. Simpson's Rule. 17. Average Velocity of Flow. 18. Floats. 19. Double Floats. 20. Rod Floats. 21. Pitot Tube. 22. Chemical Method for the Discharge of a River. 23. Current Metre. 24. Rating of Current Metres. 25. Precautions for the Rating Current Metres.

16·1 Introduction

An open channel is a passage through which the water flows under the force of gravity and atmospheric pressure. Or in other words, when the free surface of the flowing water is in contact with the atmosphere as in the case of a canal, a sewer or an aquaduct, the flow is said to be through an open channel. A channel may be covered or open at the top. As a matter of fact, the flow of water, in an open channel, is not due to any pressure as in the case of pipe flow. But it is due to the slope of the bed of the channel. Thus during the construction of a channel, a uniform slope in its bed is provided to maintain the flow of water.

It has been experimentally found that the velocity of flow is different at different points in the cross-section of a channel. But all calculations are based on the mean velocity of flow. In this chapter, while discussing the problems, we shall assume the flow to be steady and uniform, if specified, otherwise. It is, thus, obvious that we shall always refer to the condition in which the rate of discharge, depth of flow, velocity, slope of bed and cross-sectional area remain constant over the given length of the channel.

16·2 Chezy's Formula for Discharge through an Open Channel

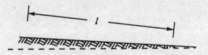

Fig. 16·1. Sloping bed of a channel.

Consider an open channel of uniform cross-section and bed slope as shown in Fig. 16·1.

Let
    l = Length of the channel,
    A = Area of flow,
    v = Velocity of water,
    P = Wetted perimeter of the cross-section,
    f = Frictional resistance per unit area at unit velocity, and
    i = Uniform slope in the bed.

Uniform Flow Through Open Channels

It has been experimentally found that the total frictional resistance in the length l of the channel follows a law,

Frictional resistance $= f \times$ Contact area $\times$ (Velocity)^n
$$= f \times Pl \times v^n \qquad ...(i)$$

The value of n has been experimentally found to be nearly equal to 2. But, for all practical purposes, its value is taken to be 2. Therefore frictional resistance
$$= f \times Pl \times v^2 \qquad ...(ii)$$

Since the water moves through a distance v in one second, therefore work done in overcoming the friction
$$= \text{Frictional resistance} \times \text{Distance}$$
$$= f \times Pl \times v^2 \times v = f \times Pl \times v^3 \qquad ...(iii)$$

We know that weight of water in the channel in a length of l metres
$$= wAl$$
where w = Specific weight of water.

This water will fall vertically down by a distance equal to $(v \cdot i)$ in one second. Therefore loss of potential energy
$$= \text{Weight of water} \times \text{Height}$$
$$= wAl.v.i \qquad ...(iv)$$

We also know that the work done in overcoming friction
$$= \text{Loss of potential energy}$$

$\therefore \qquad f \times Pl \times v^3 = wAl \cdot v \cdot i$

or $\qquad v^2 = \dfrac{w \cdot Ai}{fP}$

$\therefore \qquad v = \sqrt{\dfrac{w}{f}} \times \sqrt{\dfrac{A}{P} \times i} = C\sqrt{mi}$

where $\qquad C = \dfrac{w}{f} \qquad$ ...(Known as Chezy's constant)

and $\qquad m = \dfrac{A}{P} \qquad$ ...(Known as hydraulic mean depth or hydraulic radius)

$\therefore$ Discharge, $Q = A \times v = AC\sqrt{mi}$.

Example 16·1. *A rectangular channel is 1·5 metres deep and 6 metres wide. Find the discharge through channel, when it runs full. Take slope of the bed as 1 in 900 and Chezy's constant as 50.*

Solution. Given : $d = 1·5$ m; $b = 6$ m; $i = 1/900$ and $C = 50$.

We know that the area of the channel,
$$A = b \cdot d = 6 \times 1·5 = 9 \text{ m}^2$$
and wetted perimeter, $\qquad D = b + 2d = 6 + (2 \times 1·5) = 9 \text{ m}$

$\therefore$ Hydraulic mean depth, $m = \dfrac{A}{P} = \dfrac{9}{9} = 1$ m

and the discharge through the channel,
$$Q = A \cdot C\sqrt{mi} = 9 \times 50\sqrt{1 \times \dfrac{1}{900}} = 450 \times \dfrac{1}{30} = 15 \text{ m}^3/\text{s } \textbf{Ans.}$$

Example 16·2. *A rectangular channel having hydraulic mean depth of 0·5 metres discharges water with a velocity of 1 m/s. Find the value of Chezy's constant, if the bed slope of the channel is 1 in 2000.*

Solution. Given : $m = 0.5$ m; $v = 1$ m/s and $i = 1/2000 = 0.0005$

Let $\quad\quad\quad\quad C = $ Value of Chezy's constant.

We know that the velocity of water (v),

$$1 = C\sqrt{mi} = C\sqrt{0.5 \times 0.0005} = 0.0158\ C$$

∴ $\quad\quad\quad\quad C = 1/0.0158 = 63.2$ **Ans.**

Example 16·3. *Water is flowing at the rate of 16·5 cubic metres per second in an earthen trapezoidal channel with bed width 9 metres, water depth 1·2 metre and side slope 1 : 2. Calculate the bed slope, if the value of C in the Chezy's formula be 49·5.*

Solution. Given : $Q = 16.5$ m³/s; $b = 9$ m ; $d = 1.2$ m; Side slope = 1 : 2 (*i.e.*, one horizontal to two vertical) and $C = 49.5$

Let $\quad\quad\quad\quad i =$ Bed slope of the trapezoidal channel.

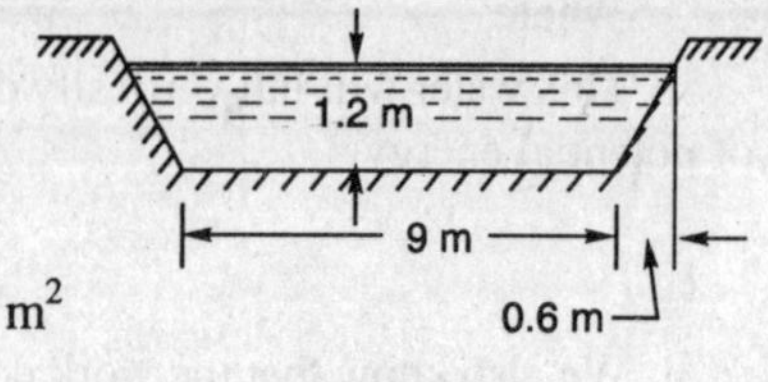

We know that area of flow,

$$A = \frac{1}{2} \times (9 + 10.2) \times 1.2\ \text{m}^2$$

$$= 11.52\ \text{m}^2$$

Fig. 16·2

and wetted perimeter, $\quad P = 9 + 2\sqrt{(1.2)^2 + (0.6)^2} = 11.68$ m

∴ Hydraulic mean depth, $m = \dfrac{A}{P} = \dfrac{11.52}{11.68} = 0.986$ m

We also know that the discharge through the pipe (Q),

$$16.5 = A \cdot C\sqrt{mi} = 11.52 \times 49.5 \sqrt{0.986\ i} = 566.2 \sqrt{i}$$

∴ $\quad \sqrt{i} = \dfrac{16.5}{566.2} = 0.0291 \quad$ or $\quad i = 8.47 \times 10^{-4} = \dfrac{1}{1181}$ **Ans.**

Example 16·4. *A channel has two sides vertical and semi-circular bottom of 2 metres diameter. Calculate the discharge of water through the channel, when the depth of flow is 2 metres. Take C = 70 and slope of bed as 1 in 1000.*

Solution. Given : Bottom dia. = 2 m; Depth of water = 2 m; $C = 70$ and $i = 1/1000$

We know that area of flow,

$$A = (2 \times 1) + \frac{\pi}{2} \times (1)^2$$

$$= 3.57\ \text{m}^2$$

and wetted perimeter, $\quad P = 1 + (\pi \times 1) + 1$

$$= 5.142\ \text{m}$$

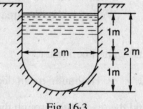

Fig. 16·3

∴ Hydraulic mean depth, $m = \dfrac{A}{P} = \dfrac{3.57}{5.14} = 0.695$ m

Uniform Flow Through Open Channels

and discharge of water through the channel,

$$Q = A \cdot C \sqrt{m \cdot i} = 3 \cdot 57 \times 70 \sqrt{0 \cdot 695 \times \frac{1}{1000}} \; m^3/s$$

$$= 249 \cdot 9 \times 0 \cdot 0264 = 6 \cdot 597 \; m^3/s \quad \textbf{Ans.}$$

16·3 Values of Chezy's Constant in the Formula for Discharge in Open Channel

The earliest formula for flow in open channels was given by Chezy in 1775 which is :

$$v = C \sqrt{mi}$$

where C is known as Chezy's constant, and its value depends upon the roughness of the inside surface of the channel. If the surface is smooth, there will be less frictional resistance to the motion of water, and as such the value of C will be more. It will result in more velocity or discharge and *vice versa*.

Later on, other scientists and engineers conducted a series of experiments and proposed many relations for the values of C in the Chezy's formula, out of which the following relations are important from the subject point of view :

1. Bazin's formula,
2. Kutter's formula, and
3. Manning's formula.

16·4 Bazin's Formula for Discharge

Bazin, after carrying out a series of experiments, deduced the following relation for the value of C in the Chezy's formula for discharge,

$$C = \frac{157 \cdot 6}{1 \cdot 81 + \dfrac{K}{\sqrt{m}}}$$

where K is a constant known as Bazin constant, whose value depends upon the roughness of the channel surface as given in Table 16·1 and m is the hydraulic mean depth.

Table 16·1. Values of K in the Bazin's formula

S. No.	Type of inside surface of channel	Value of K
1.	Smooth cement plaster or planed wood	0·11
2.	Brickwork, stone or unplaned wood	0·21
3.	Poor brickwork or rubble stone	0·83
4.	Earth of very good surface	1·54
5.	Earth of ordinary surface	2·35
6.	Earth of rough surface	3·17

Example 16·5. *A rectangular channel 1·2 metre wide and 1 metre deep has longitudinal slope of 1 in 3 000. Using Banzin's formula, find the discharge through the channel. Take K = 1·54.*

Solution. Given : $b = 1·2$ m; $d = 1$ m; $i = 1/3000$ and $K = 1·54$.

We know that the area of flow,

$$A = b \cdot d = 1·2 \times 1 = 1·2 \; m^2$$

and wetted perimeter, $P = 1 + 1{\cdot}2 + 1 = 3{\cdot}2$ m

$\therefore$ Hydraulic mean depth, $m = \dfrac{A}{P} = \dfrac{1{\cdot}2}{3{\cdot}2} = 0{\cdot}375$ m

We also know that Chezy's constant with Bazin's formula,

$$C = \dfrac{157{\cdot}6}{1{\cdot}81 + \dfrac{K}{\sqrt{m}}} = \dfrac{157{\cdot}6}{1{\cdot}81 + \dfrac{1{\cdot}54}{\sqrt{0{\cdot}375}}} = \dfrac{157{\cdot}6}{4{\cdot}325} = 36{\cdot}4$$

and discharge through the channel,

$$Q = A \cdot C\sqrt{mi} = 1{\cdot}2 \times 36{\cdot}4 \sqrt{0{\cdot}375 \times \dfrac{1}{3000}} \text{ m}^3/\text{s}$$

$$= 43{\cdot}68 \times 0{\cdot}0112 = 0{\cdot}489 \text{ m}^3/\text{s} \quad \textbf{Ans.}$$

16·5 Kutter's Formula for Discharge

Swiss engineers, Kutter and Ganguillet, after carrying out a series of experiments on canals and streams, gave the following relation for the value of C in the Chezy's formula for discharge. This formula gives more correct value, but is a bit elaborate.

$$C = \dfrac{23 + \dfrac{0{\cdot}001\,55}{i} + \dfrac{1}{N}}{1 + \left(23 + \dfrac{0{\cdot}001\,55}{i}\right) \times \dfrac{N}{\sqrt{m}}}$$

where N is a constant, known as Kutter's constant, whose value depends upon roughness of the channel surface as given in Table 16·2 and i is the bed slope and m the hydraulic mean depth.

Table 16·2. Values of N in the Kutter's formula

S. No.	Type of inside surface of channel	Value of N
1.	Smooth cement plaster or planed wood	0·010
2.	Concrete or unplaned wood	0·012
3.	Poor brick or rubble stone	0·017
4.	Earth of very good surface	0·020
5.	Earth of ordinary surface	0·025
6.	Earth of rough surface	0·030

Example 16·6. *An open canal of rectangular cross-section with horizontal base and vertical sides is built through rock, which is not lined. The canal is 6 metres deep and 9 metres wide, while the slope of the bed is 1 in 1 000. What will be the discharging capacity of the canal, if $N = 0{\cdot}0293$ in the Kutter's formula :*

$$C = \dfrac{23 + \dfrac{0{\cdot}001\,55}{i} + \dfrac{1}{N}}{1 + \left(23 + \dfrac{0{\cdot}001\,55}{i}\right) \times \dfrac{N}{\sqrt{m}}}$$

where i = slope and m = hydraulic mean depth.

Solution. Given : $d = 6$ m; $b = 9$ m; $i = 1/1000$ and $N = 0.0293$.

We know that the area of flow,
$$A = b \cdot d = 9 \times 6 = 54 \text{ m}^2$$
and wetted perimeter,
$$P = 6 + 9 + 6 = 21 \text{ m}$$
∴ Hydraulic mean depth, $m = \dfrac{A}{P} = \dfrac{54}{21} = 2.57$ m

We also know that Chezy's constant with Kutter's formula,

$$C = \dfrac{23 + \dfrac{0.001\,55}{i} + \dfrac{1}{N}}{1 + \left(23 + \dfrac{0.001\,55}{i}\right) \times \dfrac{N}{\sqrt{m}}} = \dfrac{23 + \dfrac{0.001\,55}{0.001} + \dfrac{1}{0.029\,3}}{1 + \left(23 + \dfrac{0.001\,55}{0.001}\right) \times \dfrac{0.029\,3}{\sqrt{2.57}}}$$

$$= \dfrac{58.68}{1.449} = 40.5$$

and discharge through the canal,

$$Q = A \cdot C \sqrt{mi} = 54 \times 40.5 \sqrt{2.57 \times \dfrac{1}{1000}} \text{ m}^3/\text{s}$$

$$= 2187 \times 0.0507 = 110.9 \text{ m}^3/\text{s} \quad \textbf{Ans.}$$

16·6 Manning Formula for Discharge

Manning, after carrying out a series of experiments, deduced the following relation for the value of C in Chezy's formula for discharge :

$$C = \dfrac{1}{N} \times m^{1/6}$$

where N is the Kutter's constant and has the same values as per Table 16·2.

Now we see that the velocity,

$$v = C\sqrt{mi} = \dfrac{1}{N} \times m^{1/6} \sqrt{mi} = \dfrac{1}{N} \times m^{1/6} \times m^{1/2} \times i^{1/2}$$

$$= \dfrac{1}{N} \times m^{2/3} \times i^{1/2} = M \times m^{2/3} \times i^{1/2}$$

where $M = \dfrac{1}{N}$ and is known as Manning's constant.

Now the discharge, $Q = \text{Area} \times \text{Velocity} = A \times \dfrac{1}{N} \times m^{2/3} \times i^{1/2}$

$$= A \times M \times m^{2/3} \times i^{1/2}$$

Example 16·7. *An earthen channel with a 3 m wide base and side slopes 1 : 1 carries water with a depth of 1 m. The bed slope is 1 in 1600. Estimate the discharge. Take value of N in Manning's formula as 0·04.*

Solution. Given : $b = 3$ m; Side slopes = 1 : 1; $d = 1$ m; $i = 1/1600$ and $N = 0.04$.

We know that the area of flow,
$$A = \dfrac{1}{2} \times (3 + 5) \times 1 = 4 \text{ m}^2$$
and wetted perimeter, $P = 3 + 2\sqrt{(1)^2 + (1)^2}$ m
$$= 5.83 \text{ m}$$

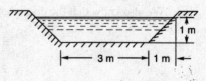

Fig. 16·4.

∴ Hydraulic mean depth $m = \dfrac{A}{P} = \dfrac{4}{5.83} = 0.686$ m

We know that the discharge through the channel,

$$Q = A \times \dfrac{1}{N} \times m^{2/3} \times i^{1/2}$$

$$= 4 \times \dfrac{1}{0.04} \times (0.686)^{2/3} \times \left(\dfrac{1}{1600}\right)^{1/2} \text{ m}^3/\text{s}$$

$$= 100 \times 0.778 \times 0.025 = 1.945 \text{ m}^3/\text{s} \quad \textbf{Ans.}$$

Example 16·8. *Water at the rate of 0·4 cumec flows through a 1 m diameter vitrified sewer, when the sewer pipe is half full. Find the slope of the water, if Manning's N is 0·013.*

Solution. Given : $Q = 0.4$ m³/s; Dia. $(d) = 1$ m or radius $(r) = 0.5$ m and $N = 0.013$.

Let $\qquad i =$ Slope of the water surface

We know that the area of flow,

$$A = \dfrac{\pi}{2} \times (r)^2 = \dfrac{\pi}{2} \times (0.5)^2 = 0.393 \text{ m}^2$$

and the wetted perimeter, $\qquad P = \dfrac{\pi \times d}{2} = \dfrac{\pi \times 1}{2} = 1.571$

∴ Hydraulic mean depth, $\quad m = \dfrac{A}{P} = \dfrac{0.393}{1.571} = 0.25$ m

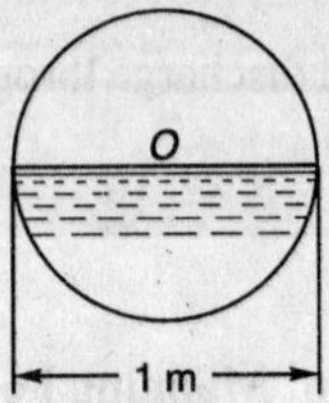

We also know that the discharge through the sewer (Q),

Fig. 16·5.

$$0.4 = A \times \dfrac{1}{N} \times m^{2/3} \times i^{1/2}$$

$$= 0.393 \times \dfrac{1}{0.013} \times (0.25)^{2/3} \times i^{1/2} = 12 \, i^{1/2}$$

∴ $\quad i^{1/2} = \dfrac{0.4}{12} = \dfrac{1}{30} \quad$ or $\quad i = \left(\dfrac{1}{30}\right)^2 = \dfrac{1}{900} \quad \textbf{Ans.}$

Example 16·9. *A cement-lined rectangular channel 6 metres wide carries water at the rate of 30 m³/s. Find the value of Manning's constant, if the slope required to maintain a depth of 1·5 m is 1/625.*

Solution. Given : $b = 6$ m; $Q = 30$ m³/s; $d = 1.5$ m and $i = 1/625$.

Let $\qquad N =$ Value of Manning's constant.

We know that the area of flow,

$$A = 6 \times 1.5 = 9 \text{ m}^2$$

and wetted perimeter, $\qquad P = 1.5 + 6 + 1.5 = 9$ m

∴ Hydraulic mean depth, $\quad m = \dfrac{A}{P} = \dfrac{9}{9} = 1$ m

We also know that the discharge through the channel (Q),

$$30 = A \times \dfrac{1}{N} \times m^{2/3} \times i^{1/2} = 9 \times \dfrac{1}{N} \times (1)^{2/3} \times \left(\dfrac{1}{625}\right)^{1/2} = \dfrac{0.36}{N}$$

∴ $\qquad N = 0.36/30 = 0.012 \quad \textbf{Ans.}$

16·7. Discharge through a Circular Channel

We have discussed in Art. 16·1 that if a circular channel (*i.e.*, sewer etc.) does not run full, it is also termed as an open channel. The discharge through such channels takes place due to slope in the bed of the channel.

Consider a circular channel, through which the water is flowing, as shown in Fig. 16·6.

Let
- r = Radius of the circular channel,
- d = Depth of water in the channel,
- 2θ = Total angle (in radians) subtended by the water surface AB at the centre,
- P = Wetted perimeter ACB of the channel, and
- A = Area of the section through which water is flowing.

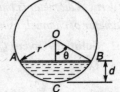

Fig. 16·6. Circular channel.

From the geometry of the figure, we find that the wetted perimeter,

$$P = 2r\theta$$

and area,

$$A = r^2\theta - \frac{r^2 \sin 2\theta}{2} = r^2\left(\theta - \frac{\sin 2\theta}{2}\right)$$

The remaining formulae for m (*i.e.*, hydraulic mean depth) and the discharge remain the same.

Example 16·10. *A circular pipe of 1 metre radius is laid at an inclination of 5° with the horizontal. Calculate the discharge through the pipe, if the depth of water in the pipe is 0·75 metre. Take C = 65.*

Solution. Given : $r = 1$ m; $i = \tan 5° = 0·0875$; $d = 0·75$ m and $C = 65$.

Let 2θ = Total angle (in radians) subtended at the centre by the water surface AB.

From the geometry of the figure, we find that

$$\cos \theta = \frac{OD}{OB} = \frac{(1 - 0·75)}{1} = 0·25$$

$$\therefore \theta = 75·5° = 75·5 \times \frac{\pi}{180} = 1·318 \text{ rad}$$

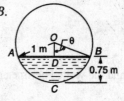

Fig. 16·7

We know that the area of flow,

$$A = r^2\left(\theta - \frac{\sin 2\theta}{2}\right) = (1)^2 \times \left[1·318 - \frac{\sin (2 \times 75·5°)}{2}\right] \text{ m}^2$$

$$= (1)^2 \times \left[1·318 - \frac{\sin 151°}{2}\right] = 1 \times \left[1·318 - \frac{\sin 29°}{2}\right] \text{ m}^2$$

...[$\because \sin (180 - \theta) = \sin \theta$]

$$= 1 \times \left(1·318 - \frac{0·4848}{2}\right) = 1·076 \text{ m}^2$$

and wetted perimeter, $P = 2r\theta = 2 \times 1 \times 1·318 = 2·636$ m

$\therefore$ Hydraulic mean depth, $m = \dfrac{A}{P} = \dfrac{1·076}{2·636} = 0·408$ m and discharge through the pipe, $Q = A \cdot C \sqrt{mi} = 1.076 \times 65 \sqrt{0.408 \times 0.0875}$ m³/s

$= 69.94 \times 0.1889 = 13.21$ m³/s   **Ans.**

Example 16·11. *A circular channel conveys 3·25 cubic metres of water per second, when 3/4 of vertical diameter is immersed. The slope of the channel is 0·8 metre per kilometre. Determine the diameter of channel, using Manning's formula $v = M \cdot m^{2/3} \; i^{1/2}$ Take M = 87·5.*

Solution. Given : $Q = 3.25$ m³/s; Depth of water $= 0.75\, d$; $i = 0.8/1000 = 0.0008$ and $M = 87.5$.

Let
- r = Radius of the channel, and
- 2θ = Total angle (in radians) subtended at the centre by the water surface AB.

From the geometry of the figure we find that

$$\cos\alpha = \frac{OD}{BO} = \frac{r - \frac{1}{2}r}{r} = \frac{1}{2}$$

or $\alpha = 60°$

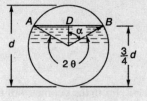

Fig. 16·8.

∴ $\theta = 180° - 60° = 120 = 120° \times \dfrac{\pi}{180} = 2.094$ rad

We know that the area of flow,

$$A = r^2 \left(\theta - \frac{\sin 2\theta}{2}\right) = r^2 \left(2.094 - \frac{\sin(2 \times 120°)}{2}\right)$$

$$= r^2 \left(2.094 - \frac{\sin 240°}{2}\right) = r^2 \left(2.094 + \frac{0.866}{2}\right)$$

...(sin 240° = − sin 60°)

$$= 2.527\, r^2$$

and the wetted perimeter,

$$P = 2r\theta = 2 \times r \times 2.094 = 4.188\, r$$

∴ Hydraulic mean depth, $m = \dfrac{A}{P} = \dfrac{2.527\, r^2}{4.188\, r} = 0.063\, r$

We know that the discharge through the channel (Q),

$3.25 = A \times M \cdot m^{3/2} \cdot i^{1/2}$

$= 2.527\, r^2 \times 87.5\, (0.603\, r)^{3/2} \times (0.0008)^{1/2}$

$= 221.1\, r^2 \times 0.7136\, r^{2/3} \times 0.0283 = 4.465\, r^{8/3}$

∴ $r^{8/3} = 3.25 / 4.465 = 0.728$ or $r = 0.89$ m

or diameter $d = 2r = 2 \times 0.89 = 1.78$ m   **Ans.**

EXERCISE 16·1

1. A rectangular channel 3 m wide and 1 m deep has a longitudinal slope of 1 in 1000. Determine the discharge through the channel, if Chezy's constant is 55.   (**Ans.** 4·04 m³/s)

2. In a laboratory, water is flowing through a rectangular channel of width of 600 mm under a constant head of 100 mm. If the discharge through the channel of bed slope of 1 in 750 is recorded as 31 litres/s, find the value of Chezy's constant. **(Ans. 51.7)**

3. A channel of rectangular cross-section is 4.5 m wide and 1.5 m deep. The discharge of water through the channel is 8.5 m^3/s. Find slope of the channel bed, if Chezy's constant is 52. **(Ans. 1/1538)**

4. A trapezoidal channel of 2.5 m wide at its bottom has side slopes of 1 : 1. Find the velocity of water through the channel flowing 1 m deep. Take $C = 52.5$ and bed slope 1 in 2000. **(Ans. 0.95 m/s)**

5. Find the discharge in a trapezoidal channel having base width of 3 m and side slopes of 1 : 1 and depth of flow as 400 mm. Take $C = 55$ and bed slope 1 in 1000. **(Ans. 1.354 m^3/s)**

6. A rectangular channel 4 m wide and 2 m deep has bed slope of 1 in 3600. Using Bazin's formula, find the discharge through the channel. Take $K = 0.83$. **(Ans. 7.96 m^3/s)**

7. A trapezoidal channel 3.5 m wide at the bottom has side and bed slopes of 1 : 1 and 1 in 1000 respectively. Using Manning's formula, find the discharge through the channel, if depth of water is 0.5 m. Take $N = 0.03$. **(Ans. 0.925 m^3/s)**

8. A circular sewer of 0.75 m radius is laid with a slope of 1 in 400. Find the discharge through the sewer, when it runs half full. Take Manning's constant as 65. **(Ans. 1.495 m^3/s)**

9. Water is flowing through a circular channel of 1.5 m radius having bed slope of 1 in 700. Find the velocity of water through the channel, if the depth of water in the channel is 1 m. Take Chezy's constant as 60. **(Ans. 1.7 m/s)**

16.8 Channels of Most Economical Cross-sections

A channel, which gives maximum discharge for a given cross-sectional area and bed slope is called a channel of most economical cross-section. Or in other words, it is a channel which involves least excavation for a designed amount of discharge. A channel of most economical cross-section is, sometimes, also defined as a channel which has a minimum wetted perimeter; so that there is a minimum resistance to flow and thus resulting in a maximum discharge. From the above definitions, it is obvious that while deriving the condition for a channel of most economical cross-section, the cross-sectional area is assumed to be constant. The relation between depth and breadth of the section is found out to give the maximum discharge.

The conditions for maximum discharge for the following sections will be dealt with in the succeeding pages :
1. Rectangular section,
2. Trapezoidal section, and
3. Circular section.

16.9 Condition for Maximum Discharge through a Channel of Rectangular Section

A rectangular section is, usually, not provided in channels except in rocky soils where the faces of rocks can stand vertically. Though a rectangular section is not of much practical importance, yet we shall discuss it for its theoretical importance only.

Consider a channel of rectangular section as shown in Fig. 16.9.

Let b = Breadth of the channel, and
d = Depth of the channel.

∴ Area of flow, $A = b \times d$

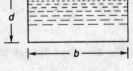

Fig. 16.9. A rectangular channel.

and discharge, $Q = A \times v = AC \sqrt{mi}$ ...($\because v = C\sqrt{mi}$)

$= AC \sqrt{\dfrac{A}{P} \times i}$ ...$\left(\because m = \dfrac{A}{P}\right)$

Keeping A, C and i constant in the above equation, the discharge will be maximum when A/P is maximum or the perimeter P is minimum. Or in other words,

$$\frac{dP}{dd} = 0$$

We know that perimeter of a rectangular section,

$$P = b + 2d = \frac{A}{d} + 2d \qquad \ldots(\because A = b \times d)$$

Differentiating the above equation with respect to d and equating the same to zero. *i.e.*,

$$\frac{dP}{dd} = -\frac{A}{d^2} + 2 \quad \text{or} \quad -\frac{A}{d^2} + 2 = 0$$

$$\therefore \quad A = 2d^2 \quad \text{or} \quad bd = 2d^2 \qquad \ldots(\because A = b \times d)$$

or $\quad b = 2d$

i.e., breadth is equal to double the depth. In this case, the hydraulic mean depth,

$$m = \frac{A}{P} = \frac{bd}{b + 2d} = \frac{2d \times d}{2d + 2d} \qquad \ldots(\because b = 2d)$$

$$= \frac{2d^2}{4d} = \frac{d}{2}$$

Hence, for maximum discharge or maximum velocity, these two conditions (*i.e.*, $b = 2d$ and $m = d/2$) should be used for solving the problems of channels of rectangular cross-sections.

Example 16·12. *A rectangular channel has a cross-section of 8 square metres. Find its size and discharge through the most economical section, if bed slope is 1 in 1000. Take C = 55.*

Solution. Given : $A = 8 \text{ m}^2$; $i = 1/1000 = 0\cdot001$ and $C = 55$.

Size of the channel

Let $\qquad b = $ Breadth of the channel, and
$\qquad d = $ Depth of the channel.

We know that for the most economical rectangular section,

$$b = 2d$$

$\therefore$ Area $(A) \qquad 8 = b \times d = 2d \times d = 2d^2$

or $\qquad d^2 = 8/2 = 4$ or $d = 2$ m **Ans.**

and $\qquad b = 2d = 2 \times 2 = 4$ m **Ans.**

Discharge through the channel

We also know that for the most economical rectangular section, hydraulic mean depth,

$$m = \frac{d}{2} = \frac{2}{2} = 1 \text{ m}$$

and the discharge through the channel,

$$Q = AC \sqrt{mi} = 8 \times 55 \sqrt{1 \times 0\cdot001} \text{ m}^3/\text{s}$$

$$= 440 \times 0\cdot0316 = 13\cdot9 \text{ m}^3/\text{s} \quad \textbf{Ans.}$$

Example 16·13. *Find the most economical cross-section of a rectangular channel to carry 0·3 m^3/s of water, when bed slope is 1 in 1000. Assume Chezy's C = 60.*

Solution. Given : $Q = 0.3$ m³/s; $i = 1/1000 = 0.001$ and $C = 60$.

Let b = Breadth of the channel, and
d = Depth of the channel.

We know that for the most economical rectangular section,
$$b = 2d$$

∴ Area $A = b \times d = 2d \times d = 2d^2$

and hydraulic mean depth, $m = \dfrac{d}{2} = 0.5\, d$

We also know that the discharge through the channel (Q),
$$0.3 = A \cdot C \sqrt{mi} = 2d^2 \times 60 \sqrt{0.5d \times 0.001}$$

Squaring both sides of the above equation,
$$0.09 = 4d^4 \times 3600 \times 0.0005\, d = 7.2\, d^5$$

∴ $d^5 = 0.09/7.2 = 0.0125$ or $d = 0.42$ m **Ans.**

and breadth, $b = 2d = 2 \times 0.42 = 0.84$ m **Ans.**

16·10 Condition for Maximum Discharge through a Channel of Trapezoidal Section

A trapezoidal section is always provided in the earthen channels. The side slopes, in a channel of trapezoidal cross-section are provided, so that the soil can stand safely. Generally, the side slope in a particular soil is decided after conducting experiments on that soil. In a soft soil, flatter side slopes should be provided whereas in a harder one, steeper side slopes may be provided.

Consider a channel of trapezoidal cross-section $ABCD$ as shown in Fig. 16·10.

Let b = Breadth of the channel at the bottom,
d = Depth of the channel, and
$\dfrac{1}{n}$ = Side slope (*i.e.*, 1 vertical to n horizontal)

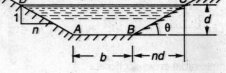

Fig. 16·10. A trapezoidal section.

∴ Area of flow, $A = d(b + nd)$

or $\dfrac{A}{d} = b + nd$ or $b = \dfrac{A}{d} - nd$ ...(*i*)

and discharge, $Q = A \times v = AC \sqrt{mi} = AC \sqrt{\dfrac{A}{P} \cdot i}$

Keeping A, C and i constant, in the above equation, the discharge will be maximum, when A/P is maximum or the perimeter P is minimum. Or in other words,
$$\dfrac{dP}{dd} = 0$$

We know that parameter of a trapezoidal section,
$$P = b + 2\sqrt{n^2 d^2 + d^2} = b + 2d\sqrt{n^2 + 1}$$

Substituting the value of b from equation (i),

$$P = \frac{A}{d} - nd + 2d\sqrt{n^2 + 1}$$

Differentiating the above equation with respect to d and equating the same to zero,

$$\frac{dP}{dd} = -\frac{A}{d^2} - n + 2\sqrt{n^2 + 1}$$

or $\qquad -\dfrac{A}{d^2} - n + 2\sqrt{n^2 + 1} = 0$

$$\frac{A}{d^2} + n = 2\sqrt{n^2 + 1}$$

$$\frac{d(b + nd)}{d^2} + n = 2\sqrt{n^2 + 1} \qquad\qquad ...[\because A = d(b + nd)]$$

$$\frac{b + 2nd}{d} = 2\sqrt{n^2 + 1}$$

or $\qquad \dfrac{b + 2nd}{2} = d\sqrt{n^2 + 1} \qquad\qquad\qquad\qquad ...(ii)$

i.e., Sloping side is equal to half of the top width. In this case, the hydraulic mean depth,

$$m = \frac{A}{P} = \frac{d(b + nd)}{b + 2d\sqrt{n^2 + 1}} = \frac{d(b + nd)}{b + (b + 2nd)}$$

$$= \frac{d(b + nd)}{2(b + nd)} = \frac{d}{2} \qquad\qquad\qquad ...(iii)$$

Hence, for maximum discharge or maximum velocity these two conditions $\left(i.e., \dfrac{b + 2nd}{2} = d\sqrt{n^2 + 1} \text{ and } m = \dfrac{d}{2}\right)$ should be used for solving problems on channels of trapezoidal cross-sections.

Example 16·14. *A most economical trapezoidal channel has an area of flow 3·5 m². Find the discharge in the channel, when running 1 metre deep. Take C = 60 and bed slope 1 in 800.*

Solution. Given : $A = 3·5$ m² ; $d = 1$ m; $C = 60$ and $i = 1/800$.
We know that for the most economical trapezoidal channel the hydraulic mean depth,

$$m = \frac{d}{2} = \frac{1}{2} = 0·5 \text{ m}$$

and discharge in the channel, $\quad Q = A \cdot C \sqrt{mi} = 3·5 \times 60 \sqrt{0·5 \times \dfrac{1}{800}} \text{ m}^3/\text{s}$

$$= 210 \times 0·025 = 5·25 \text{ m}^3/\text{s} \quad \textbf{Ans.}$$

Example 16·15. *A trapezoidal channel having side slopes of 1 : 1 and bed slope of 1 in 1200 is required to carry a discharge of 180 m³/min. Find the dimensions of the channel for minimum cross-section. Take Chezy's constant as 50.*

Solution. Given : Side slopes $(n) = 1$ (*i.e.*, 1 vertical to n horizontal); $i = 1/1200$; $Q = 180$ m³/min $= 3$ m³/s and $C = 50$.

Let $\qquad b =$ Breadth of the channel of its bottom, and

$\qquad d =$ Depth of the water flow.

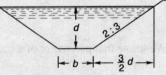

Fig. 16·11.

We know that for minimum cross-section, the channel should be most economical. And for a most economical trapezoidal section half of the top width is equal to the sloping side. *i.e.*,

$$\frac{b + 2nd}{2} = d\sqrt{n^2 + 1}$$

$$\frac{b + (2 \times 1d)}{2} = d\sqrt{(1)^2 + 1} = 1.414\,d$$

$$b + 2d = 2 \times 1.414\,d = 2.828\,d$$

$$b = 0.828\,d$$

∴ Area, $A = d(b + nd) = d(0.828\,d + 1 \times d) = 1.828\,d^2$

We know that in the case of a most economical trapezoidal section, the hydraulic mean depth,

$$m = \frac{d}{2}$$

and discharge through the channel (Q),

$$3 = A \cdot C\sqrt{mi} = 1.828\,d^2 \times 50\sqrt{\frac{d}{2} \times \frac{1}{1200}}$$

$$= 91.4\,d^2 \times \sqrt{\frac{d}{2400}} = 1.866\,d^{5/2}$$

∴ $d^{5/2} = 3/1.866 = 1.608$ or $d = 1.21$ m **Ans.**

and $\quad b = 0.828\,d = 0.828 \times 1.21 = 1$ m **Ans.**

Example 16·16. *A trapezoidal channel has side slope 2 vertical to 3 horizontal. It is discharging water at the rate of 20 cumecs with a bed slope 1 in 2 000. Design the channel for its best form. Use Manning's formula, taking N = 0·01.*

Solution. Given : Side slope (n) = 3/2 = 1·5 (*i.e.*, 1 vertical to n horizontal); $Q = 20$ m³/s ; $i = 1/1200 = 0·0005$ and $N = 0·01$.

Let $\quad b$ = Breadth of the channel at its bottom, and

d = Depth of the water flow.

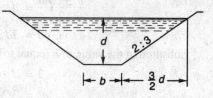

Fig. 16·12.

We know that a channel in its best form is the most economical one. And for a most economical trapezoidal section, half of the top width is equal to the sloping side. *i.e.*,

$$\frac{b + 2nd}{2} = d\sqrt{n^2 + 1} \quad \text{or} \quad \frac{b + (2 \times 1.5\,d)}{2} = d\sqrt{(1.5)^2 + 1} = 1.8\,d$$

or $\quad b + 3d = 2 \times 1.8\,d = 3.6\,d \quad$ or $\quad b = 0.6\,d$

∴ Area, $\quad A = d(b + nd) = d(0.6\,d + 1.5\,d) = 2.1\,d^2$

We know that in the case of a most economical trapezoidal section, the hydraulic mean depth,

$$m = \frac{d}{2}$$

and discharge through the channel (Q),

$$20 = A \times \frac{1}{N}\,m^{2/3} \cdot i^{1/2} = 2.1\,d^2 \times \frac{1}{0.01} \times \left(\frac{d}{2}\right)^{2/3} \times (0.0005)^{1/2}$$

$$= 210\,d^2 \times 0.63\,d^{2/3} \times 0.0224 = 2.96\,d^{8/3}$$

$$\therefore \quad d^{8/3} = 20/2.96 = 6.757 \quad \text{or} \quad d = 2.05 \text{ m} \quad \textbf{Ans.}$$
and
$$b = 0.6 \, d = 0.6 \times 2.05 = 1.23 \text{ m} \quad \textbf{Ans.}$$

Example 16·17. *Design a most economical earthen trapezoidal channel with velocity of flow as 1 m/s, and to discharge 3 m^3/s having side slope 1 in 2. Take C = 55.*

Solution. Given: $v = 1$ m/s; $Q = 3$ m³/s; Side slope $(n) = 2$ (1 in 2 means 1 vertical to 2 horizontal) and $C = 55$.

As a matter of fact, designing of a trapezoidal channel means finding out its bottom width and bed slope as well as depth of water.

Depth of water

Let $\quad d =$ Depth of water,

and $\quad b =$ Bottom width of the channel.

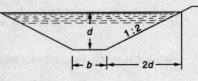

Fig. 16·13.

We know that the area of flow from the given data,

$$A = \frac{Q}{v} = \frac{3}{1} = 3 \text{ m}^2 \qquad \ldots(i)$$

and the area of flow from the assumptions,

$$A = d \, (b + nd)$$
$$= d \, (b + 2d) \qquad \ldots(ii)$$

Equating areas of flow from equations (*i*) and (*ii*)

$$3 = d \, (b + 2d) \qquad \ldots(iii)$$

We also know that when we have to design a channel, it should be the most economical one. And for a most economical trapezoidal section, half of the top width is equal to the sloping side. i.e.,

$$\frac{b + 2nd}{2} = d \sqrt{n^2 + 1} \quad \text{or} \quad \frac{b + (2 \times 2d)}{2} = d \sqrt{(2)^2 + 1} = d\sqrt{5}$$

$$\therefore \quad b + 4d = 2d \sqrt{5} = 4.47 \, d \quad \text{or} \quad b = 4.47d - 4 \, d = 0.47d$$

Substituting the value of b (equal to 0·47d) in equation (*iii*),

$$3 = d \, (b + 2d) = d \, (0.47d + 2d) = 2.47d^2$$

$$\therefore \quad d^2 = 3/2.47 = 1.21 \text{ m} \quad \text{or} \quad d = 1.1 \text{ m} \quad \textbf{Ans.}$$

Bottom width of the channel

Now susbtituting the value of d (equal to 1·1 m) in equation for bottom width,

$$b = 0.47 \, d = 0.47 \times 1.1 = 0.52 \text{ m} \quad \textbf{Ans.}$$

Slope of the channel bed

Let $\quad i =$ Slope of the channel bed.

We know that in the case of most economical trapezoidal section, hydraulic mean depth,

$$m = \frac{d}{2} = \frac{1.1}{2} = 0.55 \text{ m}$$

and velocity of flow (v),

$$1 = C \sqrt{mi} = 55 \sqrt{0.55 \, i} = 55 \times 0.742 \sqrt{i} = 40.8 \sqrt{i}$$

$$\sqrt{i} = \frac{1}{40.8} \quad \text{or} \quad i = \frac{1}{1665} \quad \textbf{Ans.}$$

Example 16·18. *A brick lined trapezoidal canal has side slopes of 1·5 horizontal to 1 vertical. It is required to carry 15 cubic metres of water per second. If the average velocity of flow is not to exceed 1 m/s, find*

(a) *the wetted perimeter for minimum amount of lining, and*

(b) *bed slope, assuming Manning's N = 0·015.*

Solution. Given : $n = 1·5$; $Q = 15$ m³/s; $v = 1$ m/s and $N = 0·015$.

Wetted perimeter for minimum amount of lining

Let b = Breadth at the bottom, and
d = Depth of water flow.

We know that the area of flow from the given data,

$$A = \frac{Q}{v} = \frac{15}{1} = 15 \text{ m}^2 \quad ...(i)$$

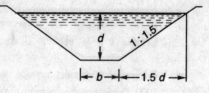

Fig. 16·14.

and the area of flow from the assumptions,

$$A = d(b + nd) = d(b + 1·5 d) \quad ...(ii)$$

Equating areas of flow from equations (i) and (ii),

$$15 = d(b + 1·5 d) \quad ...(iii)$$

We also know that a channel for minimum amount of living in the most economical one. And for a most economical trapezoidal section, half of the top width is equal to the sloping side. *i.e.*,

$$\frac{b + 2nd}{2} = d\sqrt{n^2 + 1}$$

or $\qquad \dfrac{b + (2 \times 1·5 d)}{2} = d \sqrt{(1·5)^2 + 1} = 1·8 d$

∴ $\qquad b + 3d = 1·8 d \times 2 = 3·6 d \qquad$ or $\qquad b = 0·6 d$

Substituting this value of b in equation (iii),

$$15 = d(0·6 d + 1·5 d) = 2·1 d^2 \text{ or } d^2 = 15/2·1 = 7·14$$

or $\qquad d = 2·67$ m $\qquad$ and $\qquad b = 0·6 d = 0·6 \times 2·67 = 1·6$ m

∴ Wetted perimeter,

$$P = b + 2d\sqrt{n^2 + 1} = 1·6 + [2 \times 2·67 \sqrt{(1·5)^2 + 1}] \text{ m}$$
$$= 11·2 \text{ m} \textbf{ Ans.}$$

Bed slope of the canal

Let i = Bed slope of the canal.

We know that in the case of the most economical trapezoidal section, hydraulic mean depth,

$$m = \frac{d}{2} = \frac{2·67}{2} = 1·335 \text{ m}$$

and average velocity of flow (v), $1 = \dfrac{1}{N} m^{2/3} \cdot i^{1/2} = \dfrac{1}{0·03} \times (1·335)^{2/3} \times i^{1/2} = 40·5 \ i^{1/2}$

∴ $\qquad i^{1/2} = \dfrac{1}{40·5} \qquad$ or $\qquad i = \dfrac{1}{1640}$ **Ans.**

EXERCISE 16·2

1. A rectangular channel of cross-sectional area of 18 square metres is to be laid with a bed slope of 1 in 1600. Design the channel for the most economical section. Also find the discharge through the channel, if Chezy's constant for the channel is 50. **(Ans.** $b = 6$ m; $d = 3$ m; $Q = 32·5$ m³/s)

2. Design the most economical cross-section of a rectangular channel to carry water at the rate of 2·75 m³/s. Take Chezy's constant and bed slope as 55 and 1 in 800 respectively.
(**Ans.** $b = 2$ m; $d = 1$ m)

3. A most economical rectangular channel is discharging water at the rate of 10 m³/s with a velocity of 1·25 m/s. Design the channel, if Chezy's constant is 50. (**Ans.** $b = 4$ m; $d = 2$ m; $i = 1/1600$)

4. A trapezoidal channel has side slopes of 2 horizontal to 1 vertical and bed slope of 1 in 800. Find the dimensions of the most economical section for a discharge of 50 m³/s. Take C as 60.
(**Ans.** $b = 1·34$ m; $d = 2·83$ m)

5. A trapezoidal channel of most economical section has side slopes of 1 : 1. It is required to discharge 10 cubic metres of water per second with a slope of 1 in 1500. Design the section using Manning's formula: $v = M \cdot m^{2/3} \cdot i^{1/2}$. Take $M = 50$. (**Ans.** $d = 2·04$ m; $b = 1·69$ m)

6. A trapezoidal channel has side slopes of 3 horizontal to 4 vertical and the slope of its bed is 1 in 2000. Determine the optimum dimensions of the channel, if it is to carry 0·5 m³ of water per second. Take $C = 80$. (**Ans.** $b = d = 0·552$ m)

16·11. Condition for Maximum Velocity through a Channel of Circular Section

Consider a channel of circular section, discharging water under the atmospheric pressure, as shown in Fig. 16·15.

Let r = Radius of the channel,
  h = Depth of water in the channel, and
  2θ = Total angle (in radians) subtended at the centre by the water surface AB.

From the geometry of the figure, we find that the wetted perimeter of the channels,
$$P = 2r\theta \qquad \ldots(i)$$

and area of the section, through which the water is flowing,
$$A = r^2\theta - \frac{r^2 \sin 2\theta}{2}$$
$$= r^2\left(\theta - \frac{\sin 2\theta}{2}\right) \qquad \ldots(ii)$$

We know that the velocity of flow in an open channel,
$$v = C\sqrt{mi} = C\sqrt{\frac{A}{P} \cdot i}$$

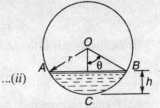

Fig. 16·15. A circular channel.

Keeping C and i constant, in the above equation, the velocity will be maximum when A/P is maximum. Or, in other words,
$$\frac{d\left(\frac{A}{P}\right)}{d\theta} = 0$$

Differentiating the above equation, with respect to θ, and equating the same to zero. *i.e.*,
$$\frac{P\frac{dA}{d\theta} - A\frac{dP}{d\theta}}{P^2} = 0 \quad \text{or} \quad P\frac{dA}{d\theta} - A\frac{dP}{d\theta} = 0.$$

Substituting the values of P and A from equations (i) and (ii); $\frac{dA}{d\theta}$ and $\frac{dP}{d\theta}$ by differentiating the equations (i) and (ii) in the above equation,

Uniform Flow Through Open Channels

$$2r\theta \left[r^2 (1 - \cos 2\theta) \right] - \left[r^2 \left(\theta - \frac{\sin 2\theta}{2} \right) \times 2r \right] = 0$$

$$2r^3 \theta (1 - \cos 2\theta) - 2r^3 \left(\theta - \frac{\sin 2\theta}{2} \right) = 0$$

$$\theta (1 - \cos 2\theta) - \left(\theta - \frac{\sin 2\theta}{2} \right) = 0 \qquad \text{...(Dividing by } 2r^3 \text{)}$$

$$\theta - \theta \cos 2\theta - \theta + \frac{\sin 2\theta}{2} = 0$$

$$\therefore \qquad \theta \cos 2\theta = \frac{\sin 2\theta}{2} \quad \text{or} \quad \tan 2\theta = 2\theta$$

Solving this equation by trial and error, we get

$$2\theta = 257.5° \quad \text{or} \quad \theta = 128.75° \qquad \text{...(iii)}$$

We know that the depth of water,

$$d = r - r \cos \theta = r (1 - \cos \theta) = r (1 - \cos 128.75°)$$
$$= r (1 + \cos 51.25°) \qquad \text{...}[\because \cos (180 - \theta) = - \cos \theta]$$
$$= r (1 + 0.62) = 1.62 \text{ Radius} = 0.81 \text{ Diameter}$$

It means that the maximum velocity will take place, when the depth of water is 0.81 times the diameter of the circular channel.

Example 16·19. *A circular pipe of 1·2 m diameter is conveying water from one point to another. Find the depth of water in the pipe, so that it flows with a maximum velocity. Also find the angle subtended by the water surface at the centre of the pipe when the water flows with a maximum velocity.*

Solution. Given : $d = 1.2$ m.

Depth of water in the pipe

We know that the depth of water in the pipe so that it flows with maximum velocity,

$$d = 0.81 \times d = 0.81 \times 1.2 = 0.972 \text{ m} \quad \textbf{Ans.}$$

Angle subtended by the water surface

We also know that the angle subtended by the water surface at the centre of the pipe when the water flows with a maximum velocity

$$= 2\theta = 2 \times 128.75° = 257.5° \quad \textbf{Ans.}$$

Example 16·20. *Find the maximum velocity of water in a circular channel of 500 mm radius, if the bed slope is 1 in 400. Take Manning's constant as 50.*

Solution. Given : $d = 500$ mm $= 0.5$ m or $r = 0.5/2 = 0.25$ m; $i = 1/400$ and $M = 50$.

Let $2\theta =$ Total angle (in radians) subtended by the water surface at the centre of the channel.

We know that for maximum velocity, the angle subtended by the water surface at the centre of the channel,

$$2\theta = 257° 30' \quad \text{or} \quad \theta = 128.75° = 128.75 \times \frac{\pi}{180} = 2.247 \text{ rad}$$

$\therefore$ Area of flow,

$$A = r^2 \left(\theta - \frac{\sin 2\theta}{2} \right) = (0.25)^2 \times \left(2.247 - \frac{\sin 257.5°}{2} \right) \text{ m}^2$$

$$= 0.0625 \left(2.247 - \frac{-\sin 102.5°}{2} \right) \qquad \text{...}[\because \sin (360° - \theta) = - \sin \theta]$$

$$= 0.0625 \left(2.247 - \frac{-0.9763}{2} \right) = 0.171 \text{ m}^2$$

and perimeter, $P = 2r\theta = 2 \times 0.25 \times 2.247 = 1.124$ m

∴ Hydraulic mean depth, $m = \dfrac{A}{P} = \dfrac{0.171}{1.124} = 0.152$ m

and velocity of water, $v = M \times m^{2/3} \times i^{1/2}$ m/s $= 50 \times (0.152)^{2/3} \times \left(\dfrac{1}{400}\right)^{1/2}$ m/s

$= 0.71$ m/s **Ans.**

16·12 Condition for Maximum Discharge through a Channel of Circular Section

Consider a channel of circular section, discharging water under the atmospheric pressure, as shown in Fig. 16·16.

Let
- $r =$ Radius of the channel,
- $h =$ Depth of water in the channel, and
- $2\theta =$ Total angle (in radians) subtended at the centre by the water surface AB.

From the geometry of the figure, we find that the wetted perimeter of the channel,

$$P = 2r\theta$$

and the area of section, through which the water is flowing,

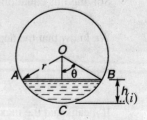

Fig. 16·16. A circular channel

$$A = r^2\theta - \dfrac{r^2 \sin 2\theta}{2} = r^2\left(\theta - \dfrac{\sin 2\theta}{2}\right) \qquad \ldots(ii)$$

We know that the discharge through an open channel,

$$Q = AC\sqrt{mi} = AC\sqrt{\dfrac{A}{P} \cdot i} = C\sqrt{\dfrac{A^3}{P} \cdot i}$$

Keeping C and i constant in the above equation, the discharge will be maximum when A^3/P is maximum. Or in other words,

$$\dfrac{d\left(\dfrac{A^3}{P}\right)}{d\theta} = 0$$

Differentiating the above equation with respect to θ and equating the same to zero.

$$\dfrac{P \times 3A^2 \dfrac{dA}{d\theta} - A^3 \dfrac{dP}{d\theta}}{P^2} = 0$$

Multiplying both sides of the above equation by P^2 and dividing by A^2

$$3P\dfrac{dA}{d\theta} - A\dfrac{dP}{d\theta} = 0 \qquad \text{or} \qquad 3P\dfrac{dA}{d\theta} = A\dfrac{dP}{d\theta}$$

Substituting the values of P and A from equations (i) and (ii); $\dfrac{dA}{d\theta}$ and $\dfrac{dP}{d\theta}$ by differentiating the equations (i) and (ii) in above equation,

$$3 \times 2r\theta \times r^2(1 - \cos 2\theta) = r^2\left(\theta - \dfrac{\sin 2\theta}{2}\right) \times 2r$$

$$3\theta \times 2r^3(1 - \cos 2\theta) = 2r^3\left(\theta - \dfrac{\sin 2\theta}{2}\right)$$

> # Uniform Flow Through Open Channels

$$3\theta(1 - \cos 2\theta) = \theta - \frac{\sin 2\theta}{2} \qquad \text{...(Dividing by } 2r^3)$$

$$3\theta - 3\theta \cos 2\theta = \theta - \frac{\sin 2\theta}{2}$$

or $\quad 2\theta - 3\theta \cos 2\theta + \dfrac{\sin 2\theta}{2} = 0$

Solving this equation by trial and error, we get

$$\theta = 154° \qquad \text{...(iii)}$$

We know that the depth of water,

$$d = r - r \cos \theta = r(1 - \cos \theta)$$
$$= r(1 - \cos 154°) = r(1 + \cos 26°)$$

...[$\because \cos(180 - \theta) = -\cos \theta$]

$= 1·898$ Radius $= 0·949$ Diameter say $0·95$ Diameter

It means that the maximum discharge will take place when the depth of water is $0·95$ times the diameter of the circular channel.

Example 16·21. *A circular channel is of 2 metres diameter. For a given value of C and i, determine to what depth the channel should run and the angle subtended by the water surface at the centre of the channel, in order to have maximum discharge of water.*

Solution. Given : $d = 2$ m.

Depth of water for maximum discharge

We know that the depth of water in the channel for maximum discharge,

$$= 0·095 \times d = 0·95 \times 2 = 1·9 \text{ m} \quad \textbf{Ans.}$$

Angle subtended by the water surface

We also know that the angle subtended by the water surface at the centre of the channel, maximum when water flows

$$= 2\theta = 2 \times 154° = 308° \quad \textbf{Ans.}$$

Example 16·22. *A circular pipe of 1 metre diameter has a bed slope of 1 in 1500. Find the maximum discharge through the channel. Take C = 50.*

Solution. Given : $d = 1$ m or $r = 1/2 = 0·5$ m; $i = 1/1500$ and $C = 50$.

Let $\quad 2\theta =$ Total angle (in radians) subtended by the water surface at the centre of the pipe.

We know that for maximum discharge, the angle subtended by the water surface at the centre of the pipe,

$$2\theta = 308° \text{ or } \theta = 154° = 154 \times \frac{\pi}{180} = 2·688 \text{ rad}$$

$\therefore$ Area of flow, $\quad A = r^2 \left(\theta - \dfrac{\sin 2\theta}{2}\right) = (0·5)^2 \times \left(2·688 - \dfrac{\sin 308°}{2}\right) \text{ m}^2$

$$= 0·25 \left(2·688 - \frac{-\sin 52°}{2}\right) \text{ m}^2 \quad \text{...[}\because \sin(360° - \theta) = -\sin \theta\text{]}$$

$$= 0·25 \left(2·688 - \frac{-0·788}{2}\right) = 0·25 \times (2·688 + 0·394) = 0·77 \text{ m}^2$$

and perimeter, $\quad P = 2r\theta = 2 \times 0·5 \times 2·688 = 2·688$ m

$\therefore$ Hydraulic mean depth, $m = \dfrac{A}{P} = \dfrac{0·77}{2·688} = 0·286$ m and discharge through the pipe, $Q = A \cdot C \sqrt{mi} = 0.77 \times 50 \sqrt{0.286 \times \dfrac{1}{1500}}$ m³/s

$= 38.5 \times 0.0138 = 0.531$ m³/s $= 531$ litres/s   **Ans.**

16·13 Measurement of River Discharge

The measurement of river discharge is often required for various purposes. The accuracy, in the measurement of discharge, is very essential and requires a lot of skill. In order to measure the discharge of a river, or an irregular channel, we require :

1. Area of flow, and 2. Average velocity of flow.

By knowing the above mentioned data, we can conveniently find out the discharge, such that
Discharge = Area of flow × Average velocity of flow.

It is, thus, obvious that the measurement of discharge, through a river or a stream, means to find out the area of flow and the average velocity of flow.

16·14 Area of Flow

A number of methods are employed to ascertain the area of flow. But the following are important from the subject point of view :

1. Simple segments method, and
2. Simpson's rule.

16·15 Simple Segments Method

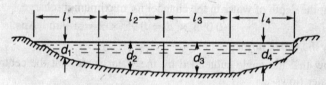

Fig. 16·17. Segments method.

In this method, the whole width of the river is divided into a number of segments as shown in Fig. 16·17.

Let $l_1, l_2, l_3 \ldots$ = Lengths of the segments, and
$d_1, d_2, d_3 \ldots$ = Mean depths of the segments.

∴ Area of flow, A = (Area of 1st segment) + (Area of 2nd segment) + ...
$= A_1 + A_2 + A_3 + \ldots$
$= l_1 d_1 + l_2 d_2 + l_3 d_3 + \ldots$

16·16. Simpson's Rule

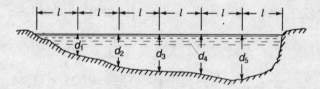

Fig. 16·18. Simpson's rule.

In this method, whole width of the river is divided into an even number of equal segments so that odd number of depths are taken at the end of each segment as shown in Fig. 16·18.

Let
l = Equal lengths of the segments, and
$d_0, d_1, d_2 \ldots$ = Depths taken at the end of segments.

Then area of flow, $A = \dfrac{l}{3} [(d_0 + d_{last}) + 2(d_1 + d_3 + d_5 \ldots) + 4(d_2 + d_4 + d_6)]$

$= \dfrac{\text{Length of segment}}{3} [(\text{First depth} + \text{Last depth})$

$+ 2 (\text{Sum of odd depths}) + 4 (\text{Sum of even depths})]$

16·17 Average Velocity of Flow

A number of methods are also employed to ascertain the average velocity of flow. But the following are important from the subject point of view :
1. Floats. 2. Pitot tube. 3. Current metre.

16·18 Floats

A simple way of measuring the velocity of flow is by means of floats. The surface velocity at any section may be easily obtained with the help of a single float. It is done by observing the time taken by the float to travel a known distance. Then the velocity is calculated by dividing the distance travelled by the float by the time taken to travel that distance. This surface velocity is then converted into an average velocity.

A better method is to use double floats or rod floats, which directly gives the average velocity of flow as discussed below :

16·19 Double Floats

A double float consists of two floats connected by a wire or a string as shown in Fig. 16·19. One of these two floats is a small wooden float, which floats on the surface. The other float is a hollow metallic sphere (heavier than water) and is suspended from the former by a wire or string of known length as shown in Fig. 16·19. Since the average velocity of flow exists at a depth of 6/10 of the total depth, therefore the length of wire or string, connecting the two floats, is such that the lower metallic float is at a depth of 6/10 of the total depth of flow.

The velocity of flow is then calculated by dividing the distance travelled by the float by the time taken to travel that distance. The double float method directly gives the value of average velocity.

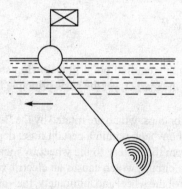

Fig. 16·19. Double float.

Fig. 16·20. Rod float.

16·20 Rod Floats

A rod float consists of a wooden rod or a hollow metallic tube weighted at the bottom, so as to keep it vertical or inclined while travelling as shown in Fig. 16·20. The length of the rod should be so adjusted that it should not touch the weeds at the bottom of the river and its top should be above the water surface as shown in Fig. 16·20.

To suit different depths, a telescopic rod may be used. Since there is every possibility of the weeds at the bottom of the river to interfere, with the rod float, therefore a section free from weeds should be chosen. The rod float method, like the double float method, directly gives us the value of average velocity.

16·21 Pitot Tube

A pitot* tube is an instrument to determine the velocity at a required point in the flowing stream. In simplest form, a pitot consists of a glass tube bent through 90° is shown in Fig. 16·21.

The lower end of the tube faces the direction of flow. The water rises up in the tube due to pressure exerted by the flowing water. By measuring the rise of water in the tube, the velocity of the water is calculated the relation :

$$v = \sqrt{2gh}$$

where g = Acceleration due to gravity, and
h = Height of water in the tube above the water surface.

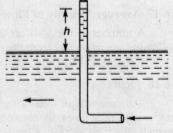

Fig. 16·21. Pitot tube.

16·22 Chemical Method for the Discharge of a River

Another method of finding out the discharge of a river or an irregular stream is by inserting uniformly a chemical solution** into the flowing water. Great care must be taken in inserting the solution equally at several places over the whole cross-section of the stream. Then a few samples of water are taken, at a lower section, where the solution has evenly mixed up the flowing water.

Let Q = Discharge of stream in m^2/s,

q = Quantity of solution inserted in m^3/s,

W = Weight of salt per cubic metre of stream water at lower section, and

w = Weight of salt per cubic metre of solution inserted.

As the weight of salt inserted per second must be equal to the weight of salt over the lower section, therefore

$$qw = QW \quad \text{or} \quad Q = \frac{w}{W} q$$

16·23 Current Meter

It is an instrument to determine the velocity of flow at required point in a flowing stream. Though there are many types of current metres available in the market, yet the basic principle for all the types of meters is the same.

A current metre consists of a wheel containing blades or cups, which are rotated by the flowing water as shown in Fig. 16·22. The number of rotations of the wheel, within a certain time, depends upon the velocity of water. An electric current is passed from the battery to the wheel by means of wire. A rotation of the wheel makes and breaks the electric circuit, which causes an electric bell to ring. Thus, by counting the ringings of the bell, the rotations of the wheel and, ultimately, the velocity of flowing water is obtained.

The current metre is suspended by means of a fine cable and lowered to the required depth. The current metre is free to move about its horizontal and vertical axis, so that it can adjust itself with the direction of the water flow.

---

* We have already discussed about pitot tube in Art. 8·17.

** Common salt solution is generally used on account of its cheapness and easily mixing with water.

Uniform Flow Through Open Channels 307

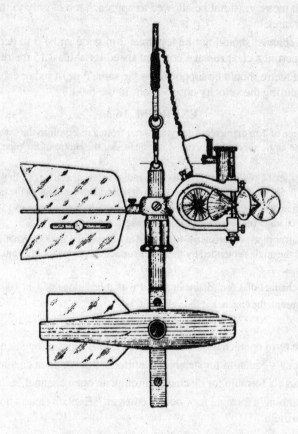

Fig. 16·22. Current metre.

16·24 Rating of Current Metres

The process of obtaining the relationship between the number of electric signals, transmitted from the current metre (or more precisely the number of revolutions of the cups of the current metre) in unit time and the velocity of water flowing past the metre is known as the calibration or rating of the current metre.

The current metre rating station consists of a long straight channel filled with water at rest. A pair of carefully levelled rails are laid on each bank of the channel. These rails form a track for an electrically driven car. The current metre is hung from the car which is driven along the track. In this way the current metre is towed through water at various selected speeds, and number of signals are recorded at the various speeds. A curve is then drawn with the help of known velocity and the signals recorded. The curve, so obtained, is known as the rating curve. Sometimes, instead of plotting a rating curve, a rating table is prepared, which directly gives the velocity of water with the corresponding number of signals.

16·25 Precautions for the Rating of Current metres

The following precautions are necessary for rating the current metres :
1. Water in the rating tank must be allowed to come completely to rest after each run and beginning of the next.

2. The current metre must not be allowed to approach too closely to the sides or bottom of the rating channel.
3. The rating channel should not be less than 2 m wide and 1·5 m deep. It should be long enough to permit a clear run at a constant speed for at least 15 metres.
4. The current metre should be supported by the same type of rod or cable, which is intended to be used during the velocity observations in the field.

EXERCISE 16·3

1. A circular pipe of 1 m diameter is carrying water from a reservoir to the town. What is the depth of water and the angle subtended by the water surface at the centre of the pipe for maximum velocity of water ? **(Ans.** 0·81 m; 257·5°)
2. A concrete pipe of circular section of 400 mm radius is laid at a slope of 1 in 850. If Chezy's constant for the pipe is 55, find the maximum velocity of water flowing through the pipe. **(Ans.** 0·93 m/s)
3. A circular channel of 600 mm diameter is discharging water at the rate of 150 litres/s. Find the bed slope of the channel for maximum velocity. Take $C = 60$. **(Ans.** 1 in 1736)
4. A circular sewer pipe of diameter 1·4 m is laid at a slope of 1 in 900. What is the depth of water in the pipe and the angle subtended by the water surface its centre for maximum discharge of water ? **(Ans.** 1·33 m; 308°)
5. A circular channel of 1·5 m diameter is laid with a bed slope of 1 in 1000. Find the maximum discharge through the channel, if the value of C is 60. **(Ans.** 2·153 m^3/s)

QUESTIONS

1. Explain the term open channel. Discuss the various types of open channels.
2. Derive Chezy's equation for steady and uniform flow in open channel.
3. What is Bazin's formula for discharge through an open channel ?
4. What is Manning's formula for Chezy's constant ? Explain clearly how does it differ from Kutter's formula.
5. Find the best form of an open channel of rectangular section of given slope and area.
6. Derive an expression for conditions of most economical section of a trapezoidal channel.
7. What are the conditions for maximum velocity and maximum discharge through a circular channel ? Prove these conditions.
8. What are the various methods adopted for taking discharge measurement in big rivers, channels and conduits ?
9. Describe with sketches the gauging of rivers by :
   (a) float method,
   (b) current metre method,
   (c) salt velocity method.
10. Draw a neat sketch of a current metre, and explain how it is calibrated.
11. What are the situations in which the following methods of flow measurement are best suited ?
    (a) Current metre, and (b) Pitot-static tube.

OBJECTIVE TYPE QUESTIONS

1. According to Chezy's formula, the discharge through an open channel is equal to
   (a) $C\sqrt{mi}$
   (b) $AC\sqrt{mi}$
   (c) $AC\sqrt{\dfrac{A}{P} \times i}$
   (d) both 'b' and 'c'

2. The discharge through an open channel according to Manning's formula is
   (a) $A.M.m^{2/3}.i^{1/2}$ (b) $A.M.m^{1/2}.i^{2/3}$ (c) $A.M.m^{3/2}.i^{1/2}$ (d) $A.M.m^{1/2}.i^{3/2}$
3. A channel is said to be of most economical cross-section, if for a given cross-sectional area and bed slope, it
   (a) gives maximum discharge
   (b) has minimum wetted perimeter
   (c) involves lesser excavation
   (d) all of the above
4. The discharge through a rectangular channel will be maximum, if its depth is
   (a) twice the width
   (b) same as its width
   (c) half of the width
   (c) one-third of the width

5. In a most economical trapezoidal section, half of the top width is equal to
   (a) bottom width of the channel
   (b) depth of the channel
   (c) sloping side of the channel
   (d) none of these
6. The velocity of water through a circular channel will be maximum, when the depth of water is .... times the diameter of the channel
   (a) 0·95 times   (b) 0·81 times   (c) 0·67 times   (d) 0·50 times

ANSWERS

1. (d)   2. (a)   3. (d)   4. (c)   5. (c)   6. (b)

17

Non-Uniform Flow Through Open Channels

1. Introduction. 2. Specific Energy of a Flowing Liquid. 3. Specific Energy Diagrams. 4. Critical Depth. 5. Critical Velocity. 6. Types of Flows. 7. Streaming Flow. 8. Critical Flow. 9. Shooting Flow. 10. Hydraulic Jump. 11. Depth of Hydraulic Jump. 12. Loss of Head due to Hydraulic Jump. 13. Venturiflume. 14. Non-modular Venturiflume. 15. Modular Venturiflume. 16. Afflux. 17. Back Water Curve. 18. Length of Back Water Curve. 19. Equation of Non-uniform Flow (Slope of Free Water Surface). 20. Characteristics of Water Curves in Non-uniform Flow. 21. Classification of Water Curves. 22. Description of Water Curves of Profiles.

17·1 Introduction

In the previous chapter, we have discussed the uniform flow through open channels (*i.e.*, a phenomenon of flow through open channels, in which the rate of flow, velocity of flow, depth of flow, area of flow and slope of bed remains constant). It may be noted that the change in any one of the above conditions causes the flow to be non-uniform. An obstruction, constructed across a channel of uniform width, will also cause the flow to be non-uniform. In this chapter, we shall discuss the effect of change in any one of the above factors.

17·2 Specific Energy of a Flowing Liquid

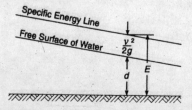

Fig. 17·1. Specific energy of a flowing liquid.

The specific energy of a flowing liquid may be defined as the energy per unit weight (say per kg) with respect to the datum, passing through the bottom of the channel as shown in Fig. 17·1.

Mathematically specific energy,

$$E = h + \frac{v^2}{2g}$$

where d = Depth of liquid flow, and
v = Velocity of the liquid.

Example 17·1. *A trapezoidal channel having bed width of 6 metres and side slope of 1 : 1 is discharging water at the rate of 8 m^3/s. To calculate the specific energy of water, if the depth of flow in the channel is 2 metres.*

Solution. Given : $b = 6$ m; Side slopes $(n) = 1$; $Q = 8$ m^3/s and $h = 2$ m.

We know that the area of flow,

$$A = d(b + nd) = 2[6 + (1 \times 2)] = 16 \text{ m}^2$$

and velocity of water, $\quad v = \dfrac{Q}{A} = \dfrac{8}{16} = 0.5$ m³/s

∴ Specific energy of water,

$$E = h + \dfrac{v^2}{2g} = 2 + \dfrac{(0.5)^2}{2 \times 9.81} = 2.013 \text{ m} \quad \text{Ans.}$$

17·3 Specific Energy Diagrams

We have seen in the previous article that the specific energy of a flowing liquid,

$$E = h + \dfrac{v^2}{2g} = E_s + E_k$$

where $\quad E_s = h =$ Static energy (also known as potential energy), and

$$E_k = \dfrac{v^2}{2g} = \text{Kinetic energy.}$$

If we have to plot the specific energy diagram for a channel, such that depth of water along Y-Y axis, and energy along X-X axis, then we may conveniently do so by first drawing two independent curves, for static energy and kinetic energy, and then by adding the ordinates of two above curves, we may get the required specific energy curve.

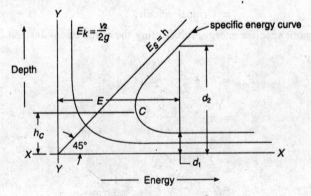

Fig. 17·2. Specific energy curve.

A little consideration will show that the curve for static energy (*i.e.*, $E_s = h$) will be a straight line, through the origin, at 45° with the horizontal as shown in Fig. 17·2. The curve for kinetic energy $\left(i.e., E_k = \dfrac{v^2}{2g}\right)$ will be a parabola as shown in Fig. 17·2.

By adding the values of these two curves, at all the points, we get the specific energy curve as shown in Fig. 17·2.

17·4 Critical Depth

We have seen in the specific energy diagram that the specific energy is minimum at C. This depth of water in a channel, corresponding to the minimum specific energy (as at C in this case) is known as critical depth. This depth can be found out by differentiating the specific energy equation and equating the same to zero. Or in other words,

$$\dfrac{dE}{dh} = 0 \quad \text{or} \quad \dfrac{d}{dh}\left(h + \dfrac{v^2}{2g}\right) = 0 \quad \ldots \left(\text{Substituting } E = h^2 + \dfrac{v^2}{2g}\right)$$

$$\frac{d}{dh}\left(h + \frac{q^2}{h^2} \times \frac{1}{2g}\right) = 0 \qquad \ldots[v = q/h \text{ where } q = \text{Discharge/unit width}]$$

$$\frac{d}{dh}\left(h + \frac{q^2}{2g} \times h^{-2}\right) = 0$$

$$\therefore \left[1 + \frac{q^2}{2g} \times \frac{(-2)}{h^3}\right] = 0$$

$$1 - \frac{q^2}{gh^3} = 0$$

or $\qquad 1 = \dfrac{q^2}{gh^3} = \dfrac{q^2}{h^2} \times \dfrac{1}{gh} = \dfrac{v^2}{gh} \qquad \ldots\left(\because v^2 = \dfrac{q^2}{h^2}\right)$

$$\therefore \qquad h = \frac{v^2}{g}$$

Since the flow is assumed to be critical, therefore critical depth,

$$h_C = \frac{v_C^2}{g} \qquad \ldots(i)$$

where $\qquad v_C$ = Critical velocity.

Now for minimum specific energy, substituting the value of h_C instead of h in the specific energy equation,

$$E_{min} = h_C + \frac{v_C^2}{2g} \qquad \ldots\left(\because E = h + \frac{v^2}{2g}\right)$$

$$= h_C + \frac{h_C \times g}{2g} \qquad \ldots\left(\because h_C = \frac{v_c^2}{2g}\right)$$

or static energy $\qquad = \dfrac{3}{2} E_{min} \qquad \ldots(ii)$

$\therefore$ Kinetic energy $\qquad = E_{min} - \dfrac{2}{3} E_{min} = \dfrac{1}{3} E_{min} \qquad \ldots(iii)$

We have seen in equation (i) that

$$h_C = \frac{v_C^2}{g} = \frac{\left(\dfrac{q}{h_C}\right)^2}{g} \qquad \ldots\left(\because v = \frac{q}{h}\right)$$

$$h_C^3 = \frac{q^2}{g} \quad \text{or} \quad h_C = \left(\frac{q^2}{g}\right)^{1/3}$$

This is the required equation for the critical depth, when the unit discharge through the channel is given.

Example 17·2 *A channel of 5 metres wide is discharging 20 cumecs of water. Determine the depth of water, when the specific energy of the flowing water is minimum.*

Solution. Given : $Q = 5$ m and $Q = 20$ m³/s.

We know that the discharge per unit width,

$$q = \frac{Q}{b} = \frac{20}{5} = 4 \text{ m}^3/\text{s}$$

Non-Uniform Flow Through Open Channels

and the depth of water when the specific energy is minimum,

$$h_C = \left[\frac{q^2}{g}\right]^{1/3} = \left[\frac{(4)^2}{9 \cdot 81}\right]^{1/3} = 1 \cdot 18 \text{ m} \quad \textbf{Ans.}$$

17·5 Types of Flows

In the previous article, we have discussed the critical depth in a channel. But, in this article, we shall discuss the importance of critical depth. As a matter fact, the critical depth as well as the normal depth of water in a channel determines the type of flow in a channel. Though there are many types of flows yet the following three types are important from the subject point of view :

1. Streaming flow.
2. Critical flow.
3. Shooting flow.

17·6 Streaming Flow

If the depth of water in the channel is greater than the critical depth, the flow is called tranquil flow or streaming flow.

17·7 Critical Flow

If the depth of water in the channel is critical, the flow is called a critical flow.

17·8 Shooting Flow

If the depth of water in the channel is less than the critical depth, the flow is called torrential flow or shooting flow.

Example 17·3. *A channel of rectangular section 8 m wide is discharging water at the rate of 12 m³/s with an average velocity of 1·2 m/s. Find the type of flow.*

Solution. Given : $b = 8$ m; $Q = 12$ m³/s and $v = 1 \cdot 2$ m/s.
We know that the discharge per unit width of the channel,

$$q = \frac{Q}{b} = \frac{12}{8} = 1 \cdot 5 \text{ m}^3/\text{s}$$

and depth of the flowing water, $h = \dfrac{q}{v} = \dfrac{1 \cdot 5}{1 \cdot 2} = 1 \cdot 25$ m ...(i)

We also know that the critical depth of water,

$$h_C = \left[\frac{q^2}{g}\right]^{1/3} = \left[\frac{(1 \cdot 5)^2}{9 \cdot 81}\right]^{1/3} = (0 \cdot 159)^{1/3} = 0 \cdot 542 \text{ m} \quad \text{...(ii)}$$

Since the depth of water (1·25 m) is more than the critical depth (0·542 m), therefore it is a streaming flow. **Ans.**

17·9 Critical Velocity

We have already discussed in Art. 17·4 that the depth of water corresponding to minimum specific energy is known as critical depth. The velocity of water, at critical depth, is known as critical velocity. Mathematically the critical depth,

$$h_C = \left(\frac{q^2}{g}\right)^{1/3}$$

where q = Discharge/unit width.

and the critical velocity corresponding to this depth,

$$v_C = \frac{q}{h_C}.$$

Example 17·4. *A cement lined rectangular channel 6 metres wide carries water at the rate of 15 cubic metres/s. Calculate the critical depth and critical velocity.*

Solution. Given : $b = 6$ m and $Q = 15$ m³/s.

Critical depth

We know that the discharge per unit width,

$$q = \frac{Q}{b} = \frac{15}{6} = 2.5 \text{ m/s}$$

and the critical depth, $h_C = \left[\frac{q^2}{g}\right]^{1/3} = \left[\frac{(2.5)^2}{9.81}\right]^{1/3} = 0.86$ m **Ans.**

Critical velocity

We also know that the critical velocity,

$$v_C = \frac{q}{h_C} = \frac{2.5}{0.86} = 2.91 \text{ m/s} \textbf{ Ans.}$$

EXERCISE 17·1

1. Find the specific energy of water, when it is flowing in a channel of 1·5 depth with a velocity of 1·25 m/s. (**Ans.** 1·58 m)
2. Water is flowing in a channel of width 3 m with side slopes 1:1. If the discharge through the channel is 4 m³/s and depth of flow is 1·4 m, find the specific energy of the water. (**Ans.** 1·42 m)
3. A rectangular channel of width 2·4 m and height 1 m is discharging water at the rate of 3·6 m³/s. What is the depth of water, when its specific energy is minimum ? (**Ans.** 0·61 m)
4. Find the type of flow in a channel of 1·5 m deep and 4·5 m wide, when it is discharging water at the rate of 7·2 m³/s with an average velocity of 1 m/s. (**Ans.** Streaming flow)
5. A rectangular channel 3·25 m wide discharges 2600 litres of water per second. Find the values of critical depth and critical velocity. (**Ans.** 0·4 m; 2 m/s)

17·10 Hydraulic Jump

We have seen in the specific energy diagram that for a given specific energy E, there are two possible depths d_1 and d_2. The depth d_1 is less than the critical depth and d_2 is greater than the critical depth.

We have also seen in Articles 17·7 and 17·9 that when the depth is less than the critical depth, the flow is said to be a shooting flow. But when this depth is greater than the critical depth, the flow is said to be a streaming flow. It has been experimentally found that the shooting flow is an unstable type of flow, and does not continue on the downstream side. The flow transforms itself to the streaming flow by increasing its depth. The rise, in water level, which occurs during the transformation of the unstable shooting flow to the stable streaming flow, is called hydraulic jump or standing wave. The place, where the hydraulic jump takes place, a lot of energy of the flowing liquid is dissipated (mainly changed into heat energy). This hydraulic jump is said to be the dissipator of the surplus energy of water. After the hydraulic jump, water flows with a greater depth, but with a less velocity. The use of this phenomenon is, sometimes, made in hydraulic structures for irrigation purposes.

17·11 Depth of Hydraulic Jump

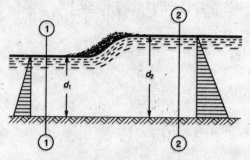

Fig. 17·3. Hydraulic jump.

Consider two sections on the upstream and downstream of a jump as shown in Fig. 17·3.

Let
    1-1 = Section on the upstream of the hydraulic jump,
    2-2 = Section on the downstream of the hydraulic jump,
    d_1 = Depth of flow at section 1-1,
    v_1 = Velocity of water at section 1-1,
    d_2, v_2 = Corresponding values at section 2-2, and
    q = Discharge per unit width
      = $d_1 v_1 = d_2 v_2$

Now consider the quantity of water between the sections 1-1 and 2-2. We know that the force on section 1-1

$$= \frac{w d_1^2}{2}$$

Similarly, force on section 2-2

$$= \frac{w d_2^2}{2}$$

The net force on the water column acting backward (because d_2 is greater than d_1)

$$= \frac{w d_2^2}{2} - \frac{w d_1^2}{2} = \frac{w}{2}(d_2^2 - d_1^2)$$

This force is responsible for change of velocity from v_1 to v_2.

We know that this force is also equal to

= Mass of water of flowing per second × Change of velocity

$$= \frac{wq}{g}(v_1 - v_2)$$

or $\quad \dfrac{w}{2}(d_2^2 - d_1^2) = \dfrac{wq}{g}(v_1 - v_2)$

---

*We know that the discharge per unit width,
$$q = \frac{Q}{b}$$
where Q = Total discharge, and
b = Width of the hydraulic jump

and total force $= wAx = w(d_1 \times 1) \times \dfrac{d_1}{2} = \dfrac{w d_1^2}{2}$

$$(d_2{}^2 - d_1{}^2) = \frac{2q}{g}(v_1 - v_2) = \frac{2q}{g}\left(\frac{q}{d_1} - \frac{q}{d_2}\right) = \frac{2q^2}{g}\left(\frac{d_2 - d_1}{d_1 d_2}\right)$$

$$(d_2 + d_1)(d_2 - d_1) = \frac{2q^2}{g \cdot d_1 d_2}(d_2 - d_1)$$

$$d_2 + d_1 = \frac{2q^2}{g \cdot d_1 d_2}$$

$$d_2{}^2 + d_1 d_2 = \frac{2q^2}{g d_1} \qquad \text{...(Multiplying by } d_2\text{)}$$

or $\quad d_2{}^2 + d_1 d_2 - \dfrac{2q^2}{g d_1} = 0$

This is a quadratic equation for d_2. Therefore

$$\therefore \quad d_2 = \frac{-d_1 \pm \sqrt{d_1{}^2 - \dfrac{4 \times 2q^2}{g d_1}}}{2}$$

$$= -\frac{d_1}{2} + \sqrt{\frac{d_1{}^2}{4} + \frac{2q^2}{g d_1}} \qquad \text{...(Taking only + sign)}$$

$$= -\frac{d_1}{2} + \sqrt{\frac{d_1{}^2}{4} + \frac{2 d_1 v_1{}^2}{g}} \qquad \text{...(Substituting } q = d_1 v_1\text{)}$$

Now the depth of the hydraulic jump or height of the standing wave is $d_2 - d_1$.

Example 17·5. *A discharge of 1000 litres/s flows along a rectangular channel 1·5 metre wide. If a standing wave is to be formed at a point, where the upstream depth is 180 mm, what would be the rise in water level ?*

Solution. Given : $q = 1000$ litres/s = 1 m³/s; $b = 1\cdot5$ m and $d_1 = 180$ mm = 0·18 m.

We know that the discharge per unit width,

$$q = \frac{Q}{b} = \frac{1}{1\cdot5} = 0\cdot67 \text{ m}^3/\text{s}$$

and depth of flow on the downstream side of the standing wave or hydraulic jump,

$$d_2 = -\frac{d_1}{2} + \sqrt{\frac{d_1{}^2}{4} + \frac{2q^2}{g d_1}} = \frac{0\cdot18}{2} \sqrt{\frac{(0\cdot18)^2}{4} + \frac{2 \times (0\cdot67)^2}{9\cdot81 \times \times 0\cdot18}}$$

$$= -0\cdot09 + 0\cdot72 = 0\cdot63 \text{ m} = 630 \text{ mm}$$

∴ Rise in water level

$$= d_2 - d_1 = 630 - 180 = 450 \text{ mm} \quad \textbf{Ans.}$$

17·12. Loss of Head due to Hydraulic Jump

The loss of head due to hydraulic jump is the difference between the total heads at sections 1-1 and 2-2. Mathematically the loss of head

$$= H_1 - H_2 = \left[d_1 + \frac{v_1{}^2}{2g}\right] - \left[d_2 + \frac{v_2{}^2}{2g}\right]$$

Note : Sometimes, the loss of head is also termed as the loss of energy.

Example 17·6. *A horizontal rectangular channel of constant breadth has a sluice opening from the bed upwards. When the sluice is partially opened, water issues at 6 m/s with a depth of 600 mm. Determine the loss of head per kN of water.*

Solution. Given : $v_1 = 6$ m/s and $d_1 = 600$ mm $= 0.6$ m.

We know that the depth of the water on the downstream side of the sluice gate,

$$d_2 = -\frac{d_1}{2} + \sqrt{\frac{d_1^2}{4} + \frac{2d_1 v_1^2}{g}} = -\frac{0.6}{2} + \sqrt{\frac{(0.6)^2}{4} + \frac{2 \times 0.6 \times (6)^2}{9.81}}$$

$$= -0.3 + 2.12 = 1.82 \text{ m}$$

Since the flow is continuous, therefore mean velocity of water on the downstream side of the sluice gate,

$$v_2 = \frac{v_1 \cdot d_1}{d_2} = \frac{6 \times 0.6}{1.82} = 1.98 \text{ m/s}$$

∴ Loss of head per kN of water

$$H_1 - H_2 = \left(d_1 + \frac{v_1^2}{2g}\right) - \left(d_2 + \frac{v_2^2}{2g}\right) = \left(0.6 + \frac{(6)^2}{2 \times 9.81}\right) - \left(1.8 + \frac{(3)^2}{2 \times 9.81}\right) \text{m}$$

$$= 2.43 - 2.26 = 0.17 \text{ m} \quad \textbf{Ans.}$$

Example 17·7. *A rectangular channel 6 metres wide discharges 1440 litres/s of water into a 6 metres wide apron, with no slope, with a mean velocity of 6 m/s. What is the height of the jump? How much energy is absorbed in the jump ?*

Solution. Given : $b = 6$ m; $Q = 1440$ litres/s $= 1.44$ m³/s and $v_1 = 6$ m/s.

Height of hydraulic jump

We know that the depth of water on the upstream side of the apron,

$$d_1 = \frac{Q}{v_1 \times b} = \frac{1.44}{6 \times 6} = 0.04 \text{ m}$$

and depth of water on the downstream side of the apron,

$$d_2 = -\frac{d_1}{2} + \sqrt{\frac{d_1^2}{4} + \frac{2d_1 v_1^2}{g}} = -\frac{0.04}{2} + \sqrt{\frac{(0.04)^2}{4} + \frac{2 \times 0.04 \times (6)^2}{9.81}} \text{ m}$$

$$= -0.02 + 0.542 = 0.522 \text{ m}$$

∴ Height of hydraulic jump

$$= d_2 - d_1 = 0.364 - 0.04 = 0.324 \text{ m} \quad \textbf{Ans.}$$

Energy absorbed in the jump

Since the flow is continuous, therefore mean velocity of water on the downstream side of the apron,

$$v_2 = \frac{v_1 \cdot d_1}{d_2} = \frac{6 \times 0.04}{0.522} = 0.46 \text{ m/s}$$

and energy absorbed in the jump,

$$E_1 - E_2 = \left(d_1 + \frac{v_1^2}{2g}\right) - \left(d_2 + \frac{v_2^2}{2g}\right) = \left(0.04 + \frac{(6)^2}{2 \times 9.81}\right) - \left(0.522 + \frac{(0.46)^2}{2 \times 9.81}\right) \text{m}$$

$$= 1.875 - 0.533 = 1.342 \text{ m} \quad \textbf{Ans.}$$

EXERCISE 17·2

1. Water flows in a rectangular channel at the rate of 3·75 m³/s per metre width. The depth of flow at a certain section is 1 m. If a hydraulic jump occurs on the downstream side of the section, find the depth of flow after the jump. **(Ans. 1·27 m)**

2. A rectangular channel 3.7 m wide conveys 7.4 cumec water at a depth of 300 mm. If a hydraulic jump occurs at a point, find the increase in water level after the jump. (Ans. 1.21 m)
3. A sluice gate discharges water into a rectangular channel with a velocity of 10 m/s with 1 m depth of flow. Find the depth of water after the hydraulic jump and the loss of total head of water.
(Ans. 4.04 m; 1.34 m)
4. Water flows at the rate of 1 m³/s along a channel of rectangular section of 1.6 m width. If a standing wave occurs at a point where upstream depth is 250 mm, find the rise in water level after the hydraulic jump. Also find the loss of head in the standing wave. (Ans. 0.203 m; 0.018 m)

17.13 Venturiflume

In every hydraulic structure (i.e., bridge, regulator etc.), which is constructed across an open channel, a few openings are left to allow water to pass. If the total width of all these openings is practically the same, as that of the channel, such a structure is called full width or unflumed structure. But, generally, the total width of such a structure is kept much less than the width of the channel to effect the economy in its construction and to increase its utility. Such a structure, whose width is less than the width of the channel, is called a flumed structure. A flumed structure (which is constructed across a channel by restricting its width) used for the measurement of the quantity of water is called venturiflume. The following two types of venturiflumes are important from the subject point of view:
 1. Non-modular venturiflume, and
 2. Modular venturiflume.

17.14 Non-modular Venturiflume

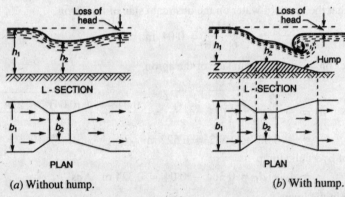

Fig. 17.4.

It is a simple venturiflume, width of the structure is reduced as shown in Fig. 17.4 (a) and (b). The restricted portion is called throat. The theory* of a non-modular venturiflume corresponds to that of the venturimeter. In a non-modular venturiflume, there is always some loss of head as shown in Fig. 17.4 (a) and (b). Now consider a venturiflume constructed in an open channel as shown in Fig. 17.4 (a).

Let
h_1 = Depth of water at section 1,
b_1 = Width of channel at section 1,
a_1 = Area of flow at section 1,
v_1 = Velocity of flow at section 1, and
h_2, b_2, a_2 and v_2 = Corresponding values at section 2.

---

*The only difference, between a venturimeter and a venturiflume, is that in a venturimeter, the flow is under pressure whereas in a venturiflume the flow is under gravity (i.e., at atmospheric pressure).

Applying Bernoulli's equation at sections 1 and 2,

$$h_1 + \frac{v_1^2}{2g} = h_2 + \frac{v_2^2}{2g} \qquad \text{...(Taking } Z_1 = Z_2\text{)}$$

or $\qquad h_1 - h_2 = \dfrac{v_2^2}{2g} - \dfrac{v_1^2}{2g} \qquad$ ...(i)

Since the discharge at sections 1 and 2 is continuous, therefore,

$$a_1 v_1 = a_2 v_2 \quad \text{or} \quad v_1 = \frac{a_2 v_2}{a_1} \quad \text{or} \quad v_1^2 = \frac{a_2^2 v_2^2}{a_1^2}$$

Substituting the value of v_1^2 in equation (i),

$$h_1 - h_2 = \frac{v_2^2}{2g} - \frac{a_2^2}{a_1^2} \times \frac{1}{2g} = \frac{v_2^2}{2g}\left(1 + \frac{a_2^2}{a_1^2}\right) = \frac{v_2^2}{2g}\left(\frac{a_1^2 - a_2^2}{a_1^2}\right)$$

Now $(h_1 - h_2)$ represents difference in the depths of water at sections 1 and 2, and is denoted by h, therefore

$$h = \frac{v_2^2}{2g}\left(\frac{a_1^2 - a_2^2}{a_1^2}\right) \quad \text{or} \quad v^2 = \left(\frac{1 a_1^2}{a_1^2} - a_2^2\right) 2gh$$

$$\therefore \qquad v = \frac{a_1}{\sqrt{a_1^2 - a_2^2}} \sqrt{2gh}$$

Now the discharge through the venturiflume,

$$Q = \text{Coefficient of venturiflume} \times a_2 v_2$$

$$= C a_2 v_2 = \frac{C a_1 a_2}{\sqrt{a_1^2 - a_2^2}} \sqrt{2gh}$$

Notes : 1. The value of C (i.e., coefficient of venturiflume) depends upon the smoothness of the bed surface and sides as well as roundness of the corners. In general, the value of C lies between 0·95 and 1·0.

2. If the venturiflume has a hump * in its throat, then the depth of water at section 1 (i.e., h_1) is measured with reference to the bed of the throat i.e., top of the hump as shown in Fig. 17·4 (b).

Example 17·8. *A rectangular channel 2·4 metres wide is provided with a venturiflume of 1·5 metre wide throat. Find the quantity of water flowing through the venturiflume, when the depth of water into upstream side is 1·2 metre and that at the throat is 0·9 metre. Take coefficient of venturiflume as 1.*

Solution. Given : $b_1 = 2·4$ m; $b_2 = 1·5$ m; $h_1 = 1·2$ m; $h_2 = 0·9$ m and $C = 1$.

We know that the area of flow in the channel,

$$a_1 = b_1 \cdot h_1 = 2·4 \times 1·2 = 2·88 \text{ m}^2$$

and the area of flow in the throat,

$$a_2 = b_2 \cdot h_2 = 1·5 \times 0·9 = 1·35 \text{ m}^2$$

---

*It is the raised portion of the throat bed, which is generally of trapezoidal corss-section. As a matter of fact, the height of hump should not exceed beyond its maximum value. Some authorities have fixed the maximum height of hump between 0·1 d_1 and 0·2 d_2, where d_1 is the depth of water in the approach channel on the upstream side of the hump. In author's opinion, the maximum height of the hump should be 0·125 d_1.

We also know that the difference of the depths of water,
$$h = 1\cdot 2 - 0\cdot 9 = 0\cdot 3 \text{ m}$$
∴ Quantity of water flowing through the venturiflume,
$$Q = \frac{Ca_1 a_2}{\sqrt{a_1^2 - a_2^2}} \times \sqrt{2gh}$$
$$= \frac{1 \times 2\cdot 88 \times 1\cdot 35}{\sqrt{(2\cdot 88)^2 - (1\cdot 35)^2}} \times \sqrt{2 \times 9\cdot 81 \times 0\cdot 3} \text{ m}^3/\text{s}$$
$$= 1\cdot 528 \times 2\cdot 426 = 3\cdot 71 \text{ m}^3/\text{s} \quad \textbf{Ans.}$$

17·15 Modular Venturiflume

It is a particular type of venturiflume, in which the width of its throat, or the depth of water at throat is decreased to such an extent that the depth of water in the throat is equal to the critical depth. The velocity of flow through the throat, corresponding to the critical depth, will also be critical. A little consideration will show that the existence of critical condition of flow in the throat will cause a standing wave on the downstream of the venturiflume. Such a venturiflume, which causes a standing wave on its downstream and have critical condition in its throat, is known as modular venturiflume or standing wave venturiflume as shown in Fig. 17·5 (a) and (b).

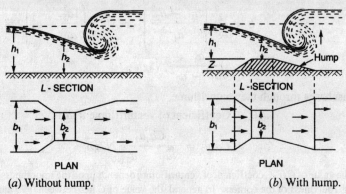

(a) Without hump.   (b) With hump.

Fig. 17·5.

In modular venturiflume, there is always some rise of water on its downstream as shown in Fig. 17·5 (a) and (b). Now consider a modular venturiflume, constructed in an open channel as shown in Fig. 17·5 (a).

Let h_1 = Depth of channel at section 1,
  b_1 = Width of channel at section 1,
  a_1 = Area of flow at section 1,
  v_1 = Velocity of flow at section 1, and
 h_2, b_2, a_2 and v_2 = Corresponding values at section 2.

∴ Specific energy at section 1,
$$Es_1 = h_1 + \frac{v_1^2}{2g}$$
Similarly,
$$Es_2 = h_2 + \frac{v_2^2}{2g}$$

Now neglecting the loss of energy in the gradually converging-portion of the flume, we have
$$Es_1 = Es_2 = h_2 + \frac{v_2^2}{2g} \qquad \ldots(i)$$

Non-Uniform Flow Through Open Channels

or
$$\frac{v_2^2}{2g} = Es_1 - h_2 \quad \text{or} \quad v_2^2 = 2g(Es_1 - h_2)$$

$$\therefore v_2 = \sqrt{2g(Es_1 - h_2)} \qquad ...(ii)$$

We know that the theoretical discharge through the venturiflume,

$$Q_{th} = a_2 v_2 = b_2 \cdot h_2 \times \sqrt{2g(Es_1 - h_2)}$$
$$= b_2 h_2 \sqrt{2g} \times \sqrt{Es_1 - h_2} \qquad ...(iii)$$
$$= b_2 \cdot \sqrt{2g} \times \sqrt{Es_1 \cdot h_2^2 - h_2^3}$$

We have already discussed that the depth of water in the throat (*i.e.*, h_2) of this venturiflume is critical. It is, therefore, obvious that the discharge through the venturiflume must be maximum. Now, for the discharge to be maximum,

$$\frac{dQ}{dh_2} = 0$$

$$\frac{d}{dh_2}\left(b \cdot \sqrt{2g} \times \sqrt{Es_1 \cdot h_2^2 - h_2^3}\right) = 0$$

$$\frac{d}{dh_2}\left(\sqrt{Es_1 \cdot h_2^2 - h_2^2}\right) = 0 \qquad ...(\text{Since } b\sqrt{2g} \text{ is constant})$$

$$\therefore Es_1 \times 2h_2 + 3h_2^2 = 0$$
$$h_2(2Es_1 - 3h_2) = 0$$
$$2Es_1 - 3h_2 = 0 \qquad ...(\text{Since } h_2 \text{ cannot be zero})$$

$$\therefore h_2 = \frac{2}{3} Es_1 \qquad .(iv)$$

We have seen in equation (*i*) that

$$Es_1 = h_2 + \frac{v_2^2}{2g}$$

$$\frac{v_2^2}{2g} = Es_1 - h_2$$

Substituting the value of h_2 from equation (*iv*) in the above equation,

$$\frac{v_2^2}{2g} = Es_1 - \frac{2}{3} Es_1 = \frac{1}{3} Es_1$$

From equation (*iii*) above, we have

$$Q_{th} = b_2 h_2 \sqrt{2g} \times \sqrt{Es_1 - h_2}$$
$$= b_2 \times \frac{2}{3} Es_1 \sqrt{2g} \times \sqrt{Es_1 - \frac{2}{3} Es_1}$$
$$= b_2 \times \frac{2}{3} Es_1 \sqrt{2g} \times \frac{1}{\sqrt{3}} (Es_1)^{1/2}$$
$$= \frac{2}{3\sqrt{3}} \sqrt{2g} \cdot b_2 (Es_1)^{3/2} = 1.70\, b_2\, Es_1^{3/2}$$

$\therefore$ Actual $\qquad Q = C \times 1.70\, b_2\, Es_1^{3/2}$

where $\qquad C = $ Coefficient of venturiflume.

Notes : 1. The value of *C* (*i.e.*, coefficient of venturiflume) depends upon the smoothness of the bed surface and sides as well roundness of the corners. In general, the value of *C* lies between 0·95 and 1·0.

2. If the venturiflume has a hump in its throat, the depth of water at section 1 (*i.e.*, h_1) is measured with reference to the bed of the throat *i.e.*, top of the hump as shown in Fig. 17·5 (*b*).

Example 17·9. *A flume has an approach channel 3 m wide and throat 1·5 m wide. Calculate the discharge, when the upstream depth is 1 m. Take C = 0·95.*

Solution. Given : $b_1 = 3$ m; $b_2 = 1·5$ m; $h_1 = 1$ m and $C = 0·95$.

First of all, let us find the discharge through the venturiflume neglecting the velocity of water in the approach channel (v_1). Therefore specific energy of water in the approach channel (neglecting v_1)

$$E_s = h_1 = 1 \text{ m}$$

and the discharge through the venturiflume,

$$Q = C \times 1·7 \times b_2 \times (Es_1)^{3/2} = (0·95 \times 1·7 \times 1·5 \times (1)^{3/2} \text{ m}^3/\text{s}$$

$$= 2·42 \text{ m}^3/\text{s}$$

∴ Velocity of water in the approach channel,

$$v_1 = \frac{Q}{b_1 \cdot h_1} = \frac{2·42}{3 \times 1} = 0·81 \text{ m/s}$$

and the specific energy of water in the approach channel (considering v_1)

$$E_s = h_1 + \frac{v_1^2}{2g} = 1 + \frac{(0·81)^2}{2 \times 9·81} = 1·03 \text{ m}$$

∴ Discharge through the venturiflume

$$Q = C \times 1·7 \times b_2 \times (Es_1)^{3/2}$$

$$= 0·95 \times 1·7 \times 1·5 \times (1·03)^{3/2} = 2·53 \text{ m}^3/\text{s} \quad \textbf{Ans.}$$

Example 17·10. *A rectangular channel of 3 m wide and laid to a slope of 0·0009 carries water at a depth of 1·5 m. Assuming Manning's N as 0·015 and flow to be uniform, calculate the maximum height of hump to produce critical depth.*

Solution. Given : $b_1 = 3$ m; $i = 0·0009$; $h_1 = 1·5$ m and $N = 0·015$.

We know that the area of flow,

$$A = b_1 \cdot h_1 = 3 \times 1·5 = 4·5 \text{ m}^2$$

and wetted perimeter, $P = 1·5 + 3 + 1·5 = 6$ m

∴ Hydraulic mean depth, $m = \dfrac{A}{P} = \dfrac{4·5}{6} = 0·75$ m

We also know that the velocity of water in the channel,

$$v_1 = \frac{1}{N} \times m^{2/3} \times i^{1/2} = \frac{1}{0·015} \times (0·75)^{2/3} \times (0·0009)^{1/2} \text{ m/s}$$

$$= 1·65 \text{ m/s}$$

∴ Specific energy of water on the channel,

$$Es_1 = h_1 + \frac{v_1^2}{2g} = 1·5 + \frac{(1·65)^2}{2 \times 9·81} = 1·64 \text{ m}$$

We know that the critical depth of water above the hump,

$$= \frac{2}{3} \times E_{s1} = \frac{2}{3} \times 1·64 = 1·09 \text{ m}$$

∴ Height of hump, $Z = 1.5 - 1.09 = 0.41$ m = 410 mm **Ans.**

17·16 Afflux

In the previous pages, while discussing the various formulae for the discharge through open channels, we have always considered a uniform flow. Sometimes, an obstruction (*e.g.*, weir or a low wall) comes across the width of the channel. Due to this obstruction, water rises up on the upstream side of the obstruction as shown in Fig. 17·6.

This increase in water level (*i.e.*, $d_2 - d_1$) is known as afflux. It has been observed that water level on the far downstream side of the obstruction again comes to the original level.

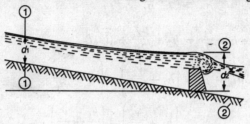

Fig. 17·6. Afflux.

17·17 Back Water Curve

In the above article, we have seen that whenever an obstruction comes across the width of the channel, the water surface on the upstream side of the obstruction, no longer remains parallel to bed; but forms a curved surface. The amount by which the water level rises is known as afflux and the curved surface (with concavity upwards) is called back water curve. The distance between sections 1-1 and 2-2 (along the bed of the channel) is known as the length of the back water curve.

where   Section 1-1 = The section from where the back water curve starts, and

Section 2-2 = The section where the back water curve ends.

It will be interesting to know that afflux or back water curve is also caused, when

1. the loss of friction is less than the bed slope, and
2. there is a decrease in the width of the channel.

17·18 Length of Back Water Curve

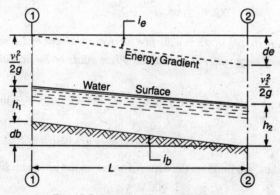

Fig. 17·7. Length of back water curve.

Consider a channel forming a back water curve. Let the length of the back water curve be between the two sections *viz.*, 1-1 and 2-2 as shown in Fig. 17·7.

Let v_1 = Velocity of water at section 1-1,

h_1 = Depth of flow at section 1-1,

v_2, h_2 = Corresponding values at section 2-2,
i_b = Slope of the channel bed,
i_c = Slope of the energy gradient, and
l = Length of the back water curve.

Applying Bernoulli's equation at sections 1-1 and 2-2,

$$d_b + h_1 + \frac{v_1^2}{2g} = h_2 + \frac{v_2^2}{2g} + d_e$$

$$(i_b \cdot l) + h_1 + \frac{v_1^2}{2g} = h_2 + \frac{v_2^2}{2g} + (i_e \cdot l)$$

or

$$l(i_b - i_e) = \left(h_2 + \frac{v_2^2}{2g}\right) - \left(h_1 - \frac{v_1^2}{2g}\right)$$

or

$$l = \frac{\left(h_2 + \frac{v_2^2}{2g}\right) - \left(\frac{h_1 - v_1^2}{2g}\right)}{i_b - i_e} = \frac{E_2 - E_1}{i_b - i_c}$$

where
E_2 = Specific energy at section 2-2, and
E_1 = Specific energy at section 1-1.

The energy gradient, i_e is determined by Manning's formula for average velocity and average hydraulic mean depth. *i.e.*,

Average velocity, $\quad v_{av} = \frac{v_1 + v_2}{2}$

and average hydraulic mean depth,

$$m_{av} = \frac{A_{av}}{P_{av}} \quad \text{or} \quad \frac{m_1 + m_2}{2}$$

Now average velocity in found out by applying Manning's formula. *i.e.*,

$$v_{av} = \frac{1}{N}(m_{av})^{2/3}(i_e)^{1/2}$$

Example 17·11. *A stream of 45 m wide has an average depth of 1·5 m. The bed slope is 1 in 2000. Find the length of back water curve caused by an afflux of 2·4 m. Take Manning's constant (N) as 0·03.*

Solution. Given : $b = 45$ m; $h_1 = 1\cdot5$ m; $i_b = 1/2000 = 0\cdot0005$ Height of afflux = 2·4 m and $N = 0\cdot03$.

Let $\quad i_e$ = Slope of the energy gradient.

From the given data, we find that the depth of water on the upstream side of the afflux,

$$h_2 = 3 + 2\cdot4 = 5\cdot4 \text{ m}$$

∴ Average depth of flow,

$$h_{av} = \frac{h_1 + h_2}{2} = \frac{3 + 5\cdot4}{2} = 4\cdot2 \text{ m}$$

First of all, let us find out the velocity of water in the stream. We know that the area of flow,

$$A = 45 \times 3 = 135 \text{ m}^2$$

and wetted perimeter, $\quad P = 3 + 45 + 3 = 51$ m

∴ Hydraulic mean depth,

$$m = \frac{A}{P} = \frac{135}{51} = 2.65 \text{ m}$$

and the velocity of water in the upstream side of the stream,

$$v_1 = \frac{1}{N} \times m^{2/3} \times i_b^{1/2} = \frac{1}{0.03} \times (2.65)^{2/3} \times \left(\frac{1}{2000}\right)^{1/2} \text{ m/s}$$

$$= 1.43 \text{ m/s}$$

Since the flow is continuous, therefore velocity of water in the downstream side,

$$v_2 = \frac{v_1 \cdot h_1}{h_2} = \frac{1.43 \times 3}{5.4} = 0.794 \text{ m/s}$$

We know that the *average velocity of flow,

$$v_{av} = \frac{v_1 \cdot h_1}{h_{av}} = \frac{1.43 \times 3}{4.2} = 1.02 \text{ m/s}$$

and the average area of flow, $\quad A_{av} = 45 \times 4.2 = 189 \text{ m}^2$

Similarly average wetted perimeter,

$$P_{av} = 4.2 + 45 + 4.2 = 53.4 \text{ m}$$

∴ Average hydraulic mean depth,

$$m_{av} = \frac{A_{av}}{P_{av}} = \frac{189}{53.4} = 3.54 \text{ m}$$

and the average velocity of water (v_{av}),

$$1.02 = \frac{1}{N} \times (m_{av})^{2/3} \times (i_e)^{1/2} = \frac{1}{0.03} \times (3.54)^{2/3} \times i_e^{1/2}$$

$$= 77.5 \sqrt{i_e}$$

∴ $\quad \sqrt{i_e} = 1.02/77.5 = 0.0132 \quad$ or $\quad i_e = 0.00017$

We also know that the specific energy of water on the upstream side,

$$Es_1 = h_1 + \frac{v_1^2}{2g} = 3 + \frac{(1.43)^2}{2 \times 9.81} = 3.1 \text{ m-kg/kN}$$

and specific energy of water on the downstream side,

$$Es_2 = h_2 + \frac{v_2^2}{2g} = 5.4 + \frac{(0.794)^2}{2 \times 9.81} = 5.43 \text{ m-kg/kN}$$

∴ Length of backwater curve,

$$l = \frac{Es_2 - Es_1}{i_b - i_e} = \frac{5.43 - 3.1}{0.0005 - 0.00017} = 7060 \text{ m} = 7.06 \text{ km} \quad \textbf{Ans.}$$

17·19 Equation of Non-uniform Flow (Slope of Free Water Surface)

Consider a non-uniform flow in an open channel between sections 1-1 and 2-2, in which the water surface** has a rising trend as shown in Fig. 17·8.

---

*It is not the average of the two velocities (i.e., v_1 and v_2). In this case, first of all the average depth of the flow is obtained (which is the average of the two depths h_1 and h_2). Now the average velocity is found out by dividing the discharge ($v_1 \times h_1$ or $v_2 \times h_2$) by average depth of flow.

** The water surface of an open channel will have a rising trend, if the loss of head is less than the bed slope.

Let
v = Velocity of water at section 1-1,
h = Depth of water at section 1-1,
*$(v - dv)$ = Velocity of water at section 2-2,
$h + dh$ = Depth of water at section 2-2,
i_b = Slope of the channel bed,
i_e = Slope of the energy gradient,
dl = Distance between sections 1-1 and 2-2,
b = Average width of the channel, and
Q = Discharge through the channel.

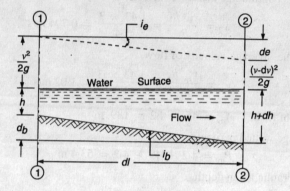

Fig. 17·8. Non-uniform flow.

Applying Bernoulli's equation at sections 1-1 and 2-2,

$$d_b + h + \frac{v^2}{2g} = (h + dh) + \frac{(v - dv)^2}{2g} + d_e$$

$$(i_b \cdot dl) + h + \frac{v^2}{2g} = (h + dh) + \frac{(v - dv)^2}{2g} + (i_e \cdot dl)$$

$$i_b \cdot dl + h + \frac{v^2}{2g} = h + dh + \frac{v^2}{2g} + \frac{dv^2}{2g} + \frac{2v \cdot dv}{2g} + i_e \cdot dl$$

∴ $\quad i_b \cdot dl = dh + \dfrac{v \cdot dv}{g} + i_e \cdot dl \qquad \ldots\left(\text{Neglecting } \dfrac{dv^2}{2g}\right)$

or $\quad i_b = \dfrac{dh}{dl} + \dfrac{v \cdot dv}{g \cdot dl} + i_e \qquad \ldots\text{(Dividing by } dl\text{)}$

$$\frac{dh}{dl} = i_b - \frac{v \cdot dv}{g \cdot dl} - i_e \qquad \ldots(i)$$

We know that the quantity of water flowing per unit width is constant, therefore

$$q = v \cdot h = \text{Constant}$$

∴ $\quad \dfrac{dq}{dl} = 0$

or $\quad \dfrac{d(vh)}{dl} = 0$

---

* Since the depth of water at section 2-2 is more than that at section 1-1, therefore, the velocity of water at section 2-2 will be less than that at section 1-1.

Non-Uniform Flow Through Open Channels

Differentiating the above equation (treating both v and h as variables),

$$\frac{v \cdot dh}{dl} + \frac{h \cdot dv}{dl} = 0$$

$$\therefore \quad \frac{dv}{dl} = -\frac{v}{h} \times \frac{dh}{dl} \qquad \ldots(ii)$$

Substituting this above value of $\frac{dv}{dl}$ in equation (i),

$$\frac{dh}{dl} = i_b - \frac{v}{g}\left(-\frac{v}{h} \times \frac{dh}{dl}\right) - i_e = i_b + \frac{v^2}{gh} \times \frac{dh}{dl} - i_e$$

or

$$\frac{dh}{dl} - \frac{v^2}{gh} \times \frac{dh}{dl} = i_b - i_e$$

$$\frac{dh}{dl}\left(1 - \frac{v^2}{gh}\right) = i_b + i_e$$

$$\therefore \quad \frac{dh}{dl} = \frac{i_b - i_e}{\left(1 - \frac{v^2}{gh}\right)}$$

Notes : A little consideration will show that above relation gives the slope* of water surface with respect of the bottom of the channel. Or in other words, it gives the variation of water depth with respect to the distance along the bottom of the channel. The value of dh/dl (i.e., zero, positive or negative) gives the following important information :

1. If dh/dl is equal to zero, it indicates that the slope of water surface is equal to the bottom slope. Or in other words, the water surface is parallel to the channel bed.

2. If dh/dl is positive, it indicates that the water surface rises in the direction of flow. The profile of water, so obtained, is called backwater curve.

3. If dh/dl is negative, it indicates that the water surface falls in the direction of flow. The profile of water, so obtained, is called downward curve.

Example 17·12. *A ractangular channel 20 m wide and having a bed slope of 0·006 is discharging water with a velocity of 1·5 m/s. The flow of channel is regulated in such a way that the slope of the energy gradient is 0·0008. Find the rate at which the depth of water will be changing at a point where water is flowing 2 m deep.*

Solution. Given : *b = 20 m; i_b = 0·006; v = 1·5 m/s; i_e = 0·0008 and h = 2 m.

We know that the rate at which the depth of water will be changing,

$$\frac{dh}{dg} = \frac{i_b - i_e}{\left(1 - \frac{v^2}{gh}\right)} = \frac{0.006 - 0.0008}{\left(1 - \frac{(1.5)^2}{9.81 \times 2}\right)} = \frac{0.0054}{0.885} = 0.0059 \text{ Ans.}$$

17·20 Characteristics of Water Curves in Non-uniform Flow

The surface curves of water are also called flow profiles or water profiles. The following depths of water effect the water profile :

1. Normal depth (y_n) It is the depth of uniform flow of water, when the discharge Q flows at a slope i_b of the bed.
2. Critical depth (y_c) It is the depth of flow of water, when the discharge Q flows as a critical flow.
3. Actual depth (y).   It is the actual depth of flow of water, when the discharge Q flows as a gradually varied flow.

---

*Superfluous data.

17·21 Classification of Water Curves

The water curves (*i.e.*, profiles) may be broadly classified into the following five types:

1. Mild slope curves ...M_1, M_2, M_3
2. Steep slope curves ...S_1, S_2, S_3
3. Critical slope curves ...C_1, C_2, C_3
4. Horizontal slope curves ...H_1, H_2, H_3
5. Adverse slope curves ...A_1, A_2, A_3

First of all, the normal depth (y_n) and the critical depth (y_c) for a given discharge is obtained. Now on the longitudinal-section of the channel, two lines parallel to the channel bottom one at a height of y_n and the other at a height of Y_c are drawn. The first line is known as normal depth line (briefly written as N.D.L.) and the second line is known as critical depth line (briefly written as C.D.L.). Now for the sake of convenience, the vertical space in the longitudinal section is divided into the following three zones:

(*a*) Zone 1 ... Space above both the lines.

(*b*) Zone 2 ... Space between both the lines.

(*c*) Zone 3 ... Space below both the lines.

In all the water curves, the letter indicates the slope and the subscript indicates the zone *e.g.*, the curve M_2 occurs in zone 2 of the mild slope. Similarly, the curve A_3 occurs in zone 3 of the adverse curve.

The following points should be clearly understood while studying the flow profiles:

1. The flow profiles approach the normal depth line tangentially. But there is an exception for the profiles on critical slopes.
2. The flow profiles approach the critical depth line perpendicularly. But there is an exception for the profiles on critical slopes.
3. All the profiles in zone 1 and 3 are backwater curve.
4. All the profiles in zone 2 are drawndown curves.
5. All the profiles in zone 3 commence from the bed of the channel.
6. The profiles C_1 and C_2 are practically horizontal.

17·22 Description of Water Curves or Profiles

A brief description of different types of flow profiles is given below. The profiles near the critical depth and channel bottom are shown by dotted lines, as at these points the streamlines are curved. And such equations of gradually varied flow are not applicable.

(*a*) Mild slope profiles

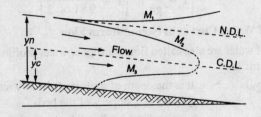

Fig. 17·9. Mild slope profiles.

A flow, in which the normal depth (y_n) is greater than the critical depth (y_c) is called streaming flow and the slope of free water surface is called mild slope or *M*-profile. There are three types of such profiles as discussed below :

1. M_1-profile. It is the most important among all the profiles and represents the back water curve. This type of profile usually occurs, when a dam of a weir is constructed across a mild long channel. In this case $y > y_n > y_c$.

2. M_2-profile. It prepresents a drawdown curve. This type of profile usually occurs, when the tail of a mild channel is submerged into a reservoir of a depth less than the normal depth. It also occurs, when the cross-section of a mild channel is subjected to a sudden enlargement. In this case $y_n > y > y_c$.

3. M_3-profile. It also represents a backwater curve. This type of profile usually occurs, when a channel after flowing below a sluice flows over a mild channel. In this case $y_n > y_c > y$.

The above profiles are shown in Fig. 17·9.

(b) **Steep slope profiles**

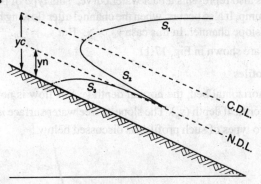

Fig. 17·10. Steep slope profiles.

A flow, in which the critical depth (y_c) is greater than the normal depth (y_n) is called a rapid flow and the slope of free water surface is called steep slope or S-profile. There are three types of such profiles as discussed below :

1. S_1-profile. It represents a backwater curve. This type of profile usually occurs, when a dam or weir is constructed across steep channel. It also occurs when the tail of a steep channel is submerged into a reservoir of a depth more than the normal depth. In this case $y > y_c > y_n$.

2. S_2-profile. It represents a drawdown curve. This type of profile usually occurs, when the steep slope of channel changes from steep to steeper. It also occurs, when the cross-section of a steep channel is subjected to a sudden enlargement. In this case $y_c > y > y_n$.

3. S_3-profile. It also represents the backwater curve. This type of profile usually occurs, when a channel after flowing below a sluice flows over a steep channel. It also occurs when the slope of the channel changes from steeper to steep. In this case $y_c > y_n > y$.

The above profiles are shown in Fig. 17·10.

(c) **Critical slope profiles**

A flow, in which the normal depth (y_n) is equal to the critical depth (y_c) is called a critical flow. And the slope of free water surface is called critical slope or C-profile. There are two types of such profiles as discussed below :

1. C_1-profile. It represents a backwater curve. This type of profile usually occurs on the critical slope portion, when the slope of the channel changes from critical to mild. In this case, $y > y_c$. But $y_c = y_n$.

2. *C_2-profile.* Since in a critical slope profile, the normal depth line and critical depth line coincide, therefore no curve is possible between these lines. However, a line coinciding with these two lines can be drawn to represent C_2 profile which will indicate a uniform critical flow. In this case $y_n = y = y_c$. Some authors do not mention the C_2 profile.

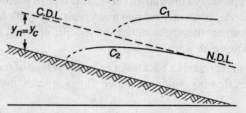

Fig. 17·11. Critical slope profiles.

3. *C_3-profile.* It is also represents a backwater curve. This type of profile, usually, occurs at the hydraulic jump. It also occurs, when the channel after flowing below a sluice gate flows over a critical slope channel. In this case $y_c > y$. But $y_c = y_n$.

The above profiles are shown in Fig. 17·11.

(d) **Horizontal slope profiles**

In a channel with horizontal bed, the normal depth (y_n) of flow is not definite and it may be either below or above the critical depth (y_n). The slope of free water surface is called horizontal slope or S-profile. There are two types of such profiles as discussed below :

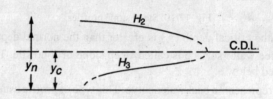

Fig. 17·12. Horizontal slope profiles.

1. *H_2-profile.* It represents a drawdown curve and is similar to M_2-profile. In this case $y_n > y > y_c$.
2. *H_3-profile.* It represents a backwater curve and is similar to M_3-profile. In this case $y_n > y_c > y$.

This above profiles are shown in Fig. 17·12.

(e) **Adverse slope profiles**

In a channel with adverse slope, the bed of channel rises in the direction of flow. As a result of this, there is no definite normal depth line, and it is assumed to be above the critical depth line. The slope of free water surface is called adverse slope or A-profile. There are two types of such profiles as discussed below :

1. *A_2-profile.* It represents a drawdown curve. This type of profile usually occurs, when the cross-section of an adverse channel is subjected to a sudden enlargement. In this case, $y_n > y > y_c$.

2. *A3-profile.* It represents a backwater curve. This type of profile usually occurs, when a channel after flowing below a sluice flows over an adverse slope. In this case $y_n > y_c > y$.

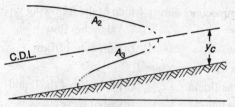

Fig. 17·13. Adverse slope profiles.

The above profiles are shown in Fig. 17·13.

EXERCISE 17·3

1. A venturiflume 0·8 m wide throat is provided in a rectangular channel of 2 m width. If the depth of water in the channel is 1 m and that in the throat is 0·75 m, find the quantity of water flowing through the venturiflume. Take coefficient of venturiflume as 0·9. **(Ans.** 1·25 m³/s)

2. A rectangular channel 4 m wide has a modular venturiflume of throat width 1·8 m. Find the discharge through the venturiflume, when the depth of water in the channel is 1·25 m. Take coefficient for the venturiflume as 0·9. **(Ans.** 11 ·02 m³/s)

3. Find the length of the backwater curve caused by an afflux of 1·5 m in a rectangular channel of width 50 m and depth 2 m. Take slope of the bed as 1 in 2000 and Manning's *N* as 0·03.

(Ans. 4·57 km)

4. Water is flowing with a velocity of 1 m/s in a rectangular channel of bed slope 1 in 1000 and 1·5 m deep. Calculate the rate of change of depth of the water surface, if slope of the energy gradient is 0·00075. **(Ans.** 0·0008)

QUESTIONS

1. Define the term 'specific energy'.
2. What are specific energy diagrams ? How are they useful in the phenomenon of flow in an open channel ?
3. Define the term critical depth as applied to the flow in an open channel.
4. Explain clearly the term critical velocity.
5. Discuss the various types of flows in an open channel.
6. What is the difference between a streaming flow and a shooting flow ? Give one example in each case.
7. What is a hydraulic jump ? Explain clearly how it is formed ?
8. What is the use of venturiflume ? Distinguish between modular and non-modular venturiflumes.
9. What do you understand by the term 'backwater curve'? Derive an equation for finding the length of the backwater curve.

OBJECTIVE TYPE QUESTIONS

1. If the depth of water is an open channel is greater than the critical depth, the flow is called
   (a) critical flow   (b) non-critical flow
   (c) streaming flow   (d) shooting flow

2. The discharge in an open channel corresponding to the critical depth is
   (a) zero
   (b) minimum
   (c) maximum
   (d) either 'a' or 'b'
3. The hydraulic jump occurs during which of the following type of flows ?
   (a) critical flow
   (b) shooting flow
   (c) streaming flow
   (d) any type of flow
4. A venturiflume is used to measure
   (a) discharge of the liquid
   (b) velocity of the liquid
   (c) pressure of the liquid
   (d) all of these
5. In a venturiflume, the flow takes place at
   (a) atmospheric pressure
   (b) gauge pressure
   (c) absolute pressure
   (d) none of the above

ANSWERS

1. (c)   2. (c)   3. (b)   4. (a)   5. (a)

18
Viscous Flow

1. Introduction. 2. Viscosity. 3. Assumptions for the Effect of Viscosity. 4. Newton's Law of Viscosity. 5. Effect of Viscosity on Motion. 6. Units of Viscosity. 7. Effect of Temperature on the Viscosity 8. Kinematic Viscosity. 9. Units of Kinematic Viscosity. 10. Effect of Temperature on Kinematic Viscosity. 11. Classification of Fluids. 12. Ideal Fluid. 13. Real Fluid. 14. Newtonian Fluid. 15. Non-Newtonian Fluid. 16. Classification of Viscous Flows. 17. Laminar Flow. 18. Turbulent Flow. 19. Critical Velocity. 20. Lower Critical Velocity. 21. Higher Critical Velocity. 22. Reynold's Experiment of Viscous Flow. 23. Reynold's Number. 24. Hagen-Poiseuille Law for Laminar Flow in Pipes. 25. Distribution of Velocity of a Flowing Liquid over a Pipe Section. 26. Loss of Head due to Friction in a Viscous Flow.

18·1 Introduction

We have already discussed in Art. 1·10 that viscosity is a property of a liquid, which controls its rate of flow. As a matter of fact, viscosity of a liquid is the property which resists (*i.e.*, retards) the motion of translation of one layer relative to other. Or in other words, viscosity is the resistance offered by a layer of the liquid to the other (in the same way as frictional resistance offered by one body to another body while sliding on it). It is, thus, obvious that a continuous supply of energy is required to overcome this resistance.

In the previous chapters, we have not taken into consideration this characteristic property of the liquids. But, in this chapter, we shall take into consideration this property also.

18·2 Viscosity

We have already discussed in the last article that viscosity of a liquid is its property which controls its rate of flow. It is fundamentally due to the cohesion between the liquid particles and is exhibited by the liquid when it is in motion.

In order to have a better understanding of this term, let us take three bottles of same size and fill them up with three different liquids, the first being some light liquid (say alcohol), second being water and third being syrup. If we invert all these three bottles at the same moment, we see that the bottle, which contains alcohol will be emptied first. After that the bottle, which contains water, will be emptied and finally the bottle, which contains syrup, will be emptied. If we analyse this phenomenon, we find that every liquid has some characteristic property known as viscosity which controls its rate of flow. A little consideration will show that a thick liquid (like a syrup) has a greater viscosity as it offers more resistance to the flow than the light liquid (like water or alcohol).

18·3 Assumptions for the Effect of Viscosity

While considering the effect of viscosity, the following two assumptions are made :

1. When a liquid is in contact with a solid boundary, the relative motion between the liquid particles (immediately adjacent to the boundary) and the solid boundary does not exist. Or in other words, if the boundary is at rest the liquid particles are also at rest. But if the boundary moves with some velocity, the liquid particles also move with the same velocity.

2. The shear stress between the two adjacent liquid layers is proportional to the rate of shear in the direction perpendicular to the motion. Or, in other words, if two adjacent layers move with a relative velocity of v, the rate of shear is v/y. The shear stress between the two liquid layers is also proportional to v/y where y is the distance between the two layers.

18·4 Newton's Law of Viscosity

It states, *"The shear stress on a layer of a fluid is directly proportional to the rate of shear strain"*. Mathematically shear stress,

$$\tau \propto \frac{v}{y} = \mu \times \frac{v}{y}$$

where v/y is the rate of shear strain (or velocity gradient) and μ is the constant of proportionality and is known as coefficient of viscosity (also known as coefficient of absolute viscosity or coefficient of dynamic viscosity).

Note : Sometimes, this equation is also written in its differential form. *i.e.*,

$$\tau = \mu \times \frac{dv}{dy}$$

18·5 Effect of Viscosity on Motion

Consider a thin layer of a liquid sandwiched between two flat parallel plates as shown in Fig. 18·1. Now let us apply a force (P) on the upper plate, which will cause it to move with respect to the lower one.

Let A = Area of the plate,

            v = Velocity of the upper plate relative to the ower one, and

            y = Thickness of the liquid layer.

We know that the force applied,

$$P = \text{Shear stress} \times \text{Area}$$
$$= \tau \cdot A = \left(\mu \times \frac{v}{y}\right) A = \mu A \times \frac{v}{y}$$

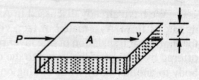

Fig. 18·1. Effect of viscosity on motion.

18·6 Units of Viscosity

We have already discussed in Art. 18·4 Newton's Law of Viscosity, where we obtained a relation,

$$\tau = \mu \times \frac{v}{y}$$

or

$$\mu = \tau \times \frac{y}{v} = \frac{F}{A} \times \frac{L^*}{\frac{L}{T}} = \frac{F}{L^2} \times \frac{L}{\frac{L}{T}} = \frac{FT}{L^2}$$

In S.I. system, there is no specific unit of viscosity. But it is used in a modified form of poise, (briefly written as P) such that

$$1 \text{ poise} = 0.1 \text{ N-s/m}^2$$

---

*For details, please refer to chapter on dimensional analysis. However, if we look at the units from the fundamentals, we find that

Length = L; Time = T; Force = F

$\therefore$ Area, $A = L \times L = L^2$

Stress $= \dfrac{\text{Force}}{\text{Area}} = \dfrac{F}{A} = \dfrac{F}{L^2}$

Length, $y = L$

Velocity, $v = \dfrac{\text{Length}}{\text{Time}} = \dfrac{L}{T}$

Viscous Flow

Sometimes, a small unit centipoise (cP) is also used, which is 1/100th of a poise.

Notes. Sometimes this unit of viscosity, which contains a unit force (F), is further analysed in terms of mass and acceleration. *i.e.*,

$$\text{Force, } F = \text{Mass} \times \text{Acceleration} = M \times \frac{L}{T^2}$$

$$\therefore \quad \mu = \frac{FT}{L^2} = \left(M \times \frac{L}{T^2}\right)\frac{T}{L^2} = \frac{M}{LT}$$

18·7 Effect of Temperature on the Viscosity

We have already discussed that the viscosity of a fluid is a measure of the internal fluid friction, which causes resistance to flow. In our daily life we see that thick liquids like coal tar, ghee, mobil oil etc., when heated become less viscous. Thus, we say that the viscosity of a fluid is a function of temperature. The following table gives the values of viscosity of water at different temperatures :

Table 18·1. Effect of temperature on the viscosity of water

Temperature of water in °C	0°	5°	10°	20°	30°	40°	60°	80°	100°
Coefficient of viscosity (μ) in poise	0·0178	0·0152	0·0131	0·0101	0·0080	0·0065	0·0048	0·0037	0·0030

Example 18·1. *A flat plate of area 1·6 m^2 is pulled with a velocity of 0·4 m/s relative to another plate located at a distance of 0·25 mm from it. Find the force required to maintain this speed, if the oil separating the plates has a viscosity of 2 poise.*

Solution. Given : $A = 1·6$ m^2; $v = 0·4$ m/s ; $y = 0·25$ mm $= 0·25 \times 10^{-3}$ m and $\mu = 2P = 0·2$ N-s/m^2

We know that the force required to maintain the speed of the plate,

$$F = \mu \cdot A \times \frac{v}{y} = 0·2 \times 1·6 \times \frac{0·4}{0·25 \times 10^{-3}} = 512 \text{ N} \quad \textbf{Ans.}$$

Example 18·2. *Two horizontal plates are placed 12·5 mm apart, the space between them being filled with oil of viscosity of 14 poise. Calculate the shear stress in the oil, if the upper plate is moved with a velocity of 2·5 m/s.*

Solution. Given : $y = 12·5$ mm $= 0·0125$ m; $\mu = 14$ P $= 1·4$ N-s/m^2 and $v = 2·5$ m/s.

We know that the shear stress in the oil,

$$\tau = \mu \times \frac{v}{y} = 1·4 \times \frac{2·5}{0·0125} = 280 \text{ N/}m^2 \quad \textbf{Ans.}$$

Example 18·3. *A plate 2·5 mm distant from a fixed plate moves at 0·8 m/s and requires a force of 40 N/m^2 to maintain this speed. Find the viscosity of the oil between the plates.*

Solution. Given : $y = 2·5$ mm $= 0·0025$ m; $v = 0·8$ m/s and $\tau = 40$ N/m^2.

Let μ = Viscosity of the oil between the plates.

We know that the force required to maintain the speed (τ),

$$40 = \mu \times \frac{v}{y} = \mu \times \frac{0·8}{0·0025} = 320 \mu$$

$$\therefore \quad \mu = 40/320 = 0·125 \text{ N-s/}m^2 = 1·25 \text{ P} \quad \textbf{Ans.}$$

18.8 Kinematic Viscosity

In many problems, it is **convenient** to use a modified form of viscosity. Thus, a new term kinematic viscosity was coined, which is the ratio of absolute viscosity in N-s/m² to the density of the liquid in kg/m³. Mathematically, kinematic viscosity ν (*nu*),

$$\nu = \frac{\mu}{\rho}$$

18.9 Units of Kinematic Viscosity

We have already discussed in the last article, a modified form of viscosity. *i.e.*, kinematic viscosity. Mathematically kinetic viscosity,

$$\nu = \frac{\mu}{\rho} = \frac{\frac{M}{LT}}{\frac{M}{L^3}} = \frac{L^2}{T} = \frac{m^2}{s}$$

In S.I. system, there is no specific unit of kinematic viscosity. But it is used in a modified form of stoke, such that

$$1 \text{ stoke} = 10^{-4} \text{ m}^2/\text{s} = \frac{1}{10000} \text{ m}^2/\text{s}$$

Sometimes, a small unit centistoke is also used which is 1/100th of a stoke.

18.10 Effect of Temperature on Kinematic Viscosity

The kinematic viscosity of a fluid, like absolute viscosity, is also a function of temperature. The following table gives the values of kinematic viscosity of water at different temperatures:

Table 18·2. Effect of temperature on kinematic viscosity

Temperature of water in °C	0°	5°	10°	20°	30°	40°	60°	80°	100°
Kinematic viscosity in stokes	0·0178	0·0152	0·0131	0·0101	0·0080	0·0065	0·0048	0·0037	0·0030

Example 18·4. *Find the kinematic viscosity of an oil in stokes, whose specific gravity is 0·95 and viscosity 0·011 poise.*

Solution. Given : Specific gravity of oil = 0·95 and $\mu = 0·011$ P = 0·0011 N-s/m²

We know that the density of the oil,

$$\rho = 0.95 \times 1000 = 950 \text{ kg/m}^3$$

and kinematic viscosity

$$\nu = \frac{\mu}{\rho} = \frac{0·0011}{950} = 0·0116 \times 10^{-4} \text{ m}^2/\text{s} = 0·0116 \text{ stoke} \quad \textbf{Ans.}$$

Example 18·5. *The space between two parallel square plates each of side 0·8 m is filled with an oil of specific gravity 0·8. If the space between the plates is 12·5 mm and the upper plate which moves with a velocity of 1·25 m/s requires a force of 51·2 N, determine (i) dynamic viscosity of the oil in poises and (ii) kinematic viscosity of oil in stokes.*

Solution. Given : Side of each square = 0·8 m; Specific gravity of oil = 0·8; y = 12·5 mm = 0·0125 m; v = 1·25 m/s and force = 51·2 N.

---

*Density, $\rho = \dfrac{\text{Mass}}{\text{Volume}} = \dfrac{M}{L^3}$

Viscous Flow

(i) Dynamic viscosity of oil in poises

Let μ = Dynamic viscosity of oil.

We know that the area of each plate,
$$A = 0.8 \times 0.8 = 0.64 \text{ m}^2$$

and the shear stress,
$$\tau = \frac{\text{Force}}{\text{Area}} = \frac{51.2}{0.64} = 80 \text{ N/m}^2$$

We also know that shear stress (τ),
$$80 = \mu \times \frac{v}{y} = \mu \times \frac{1.25}{0.0125} = 100\,\mu$$

$$\therefore \quad \mu = 80/100 = 0.8 \text{ N-s/m}^2 = 8\,P \quad \textbf{Ans.}$$

(ii) Kinematic viscosity of oil in stokes

We know that the density of oil,
$$\rho = 0.8 \times 1000 = 800 \text{ kg/m}^3$$

and kinematic viscosity,
$$\nu = \frac{\mu}{\rho} = \frac{0.8}{8000} = 10 \times 10^{-4} \times \text{m}^2/\text{s} = 10 \text{ stokes} \quad \textbf{Ans.}$$

EXERCISE 18·1

1. Two horizontal plates each of size of 1·25 m × 0·8 m are placed 8 mm apart and are filled with an oil of viscosity 15 poises. Find the force required to move the upper plate with a velocity of 0·6 m/s with respect to the lower one. **(Ans. 112·5 N)**

2. The space between two horizontal parallel plates 10 mm apart is filled with an oil of viscosity 8 poises. If the lower plate is fixed and the upper plate is moving with a velocity of 1·75 m/s, what is the shear stress of the oil? **(Ans. 140 N/m^2)**

3. Two plates are 5 mm apart from each other and the space between them is filled with an oil. If a force of 42 N/mm^2 is required to move one of these plates to maintain a speed of 1·2 m/s, determine the viscosity of the oil in poises. **(Ans. 1·75 P)**

4. A liquid has a viscosity of 5 poise and a specific gravity of 0·8. Calculate the kinematic viscosity of the liquid in stokes. **(Ans. 6·25 stokes)**

5. An oil has kinematic viscosity of 8 stokes and specific gravity of 0·9. What is its absolute viscosity in poise? **(Ans. 7·2 P)**

18·11 Classification of Fluids

The fluids may be classified into the following four types depending upon the presence of viscosity:

1. Ideal fluid,
2. Real fluid,
3. Newtonian fluid, and
4. Non-Newtonian fluid.

18·12 Ideal Fluid

A fluid, having no viscosity, is known as an ideal fluid. In actual practice, there is hardly any ideal fluid, as every fluid has some viscosity. Thus it is only a theoretical case.

18·13 Real Fluid

A fluid, having no viscosity, is known as a real fluid. In actual practice, all the fluids met with in engineering-science, are real fluids.

18·14 Newtonian Fluid

We have already discussed in Art. 18·4 Newton's law of viscosity. A fluid, which obeys this law, is termed as Newtonian fluid. Mathematically shear stress,

$$\tau = \frac{v}{y}$$ ...(where $\frac{v}{y}$ is the shear stress)

or

$$\mu = \tau \times \frac{y}{v}$$

or in other words, a fluid, whose viscosity does not change with the rate of deformation or shear strain (*i.e.*, v/y), is known as a Newtonian fluid. If we draw a graph showing shear strain as abscissa and stress as ordinate, we find that a Newtonian fluid will be represented by a straight line as shown in Fig. 18·2.

18·15 Non-Newtonian Fluid

A fluid, which does not obey Newton's law of viscosity is termed as Non-Newtonian fluid. Or in other words, a fluid, whose viscosity changes with the rate of deformation of shear strain (*i.e.*, v/y), is known as a Non–Newtonian fluid. The Non-Newtonian fluid will be represented by a curve as shown in Fig. 18·2.

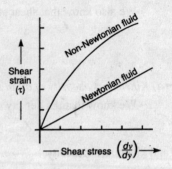

Fig. 18·2. A graph for Newtonian and Non-Newtonian fluids.

18·16 Classification of Viscous Flows

The viscous flows may be classified into the following two types depending upon the factor, whether the viscosity is dominating or not :

    1. Laminar flow.   2. Turbulent flow.

18·17 Laminar Flow

It is a flow, in which the viscosity of fluid is dominating over the inertia forces. It is more or less a theoretical flow, which rarely comes in contact with the engineers and is also known as a viscous flow.

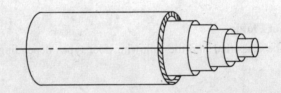

Fig. 18·3. Laminar flow.

A laminar flow can be best understood by the hypothesis that liquid moves in the form of concentric cylinders sliding one within the another as shown in Fig. 18·3. These concentric cylinders move like laminae. Such a flow, which takes place at very low velocities, is known as a laminar flow.

18·18 Turbulent Flow

It is a flow, in which the inertia force is dominating over the viscosity. It is a practical flow which comes in contact with the engineers.

Viscous Flow

In this flow the concentric cylinders diffuse or mix up with each other and the flow is a disturbed one. Such a flow, which takes place at high velocities, is known as a turbulent flow.

18·19 Critical Velocity*

It is a velocity at which the flow changes from the laminar flow to the turbulent flow. The critical velocity may be further classified into the following two types :

1. Lower critical velocity.   2. Upper critical velocity.

18·20 Lower Critical Velocity

It has been experimentally** found that when a laminar flow changes into a turbulent flow, it does not change abruptly. But it has got some transition period between the two types of flows. Thus a velocity at which the laminar flow stops, or in other words, a velocity at which the flow enters from laminar to transition period, is known as a lower critical velocity.

18·21 Upper Critical Velocity

A velocity, at which the turbulent flow starts or, in other words, a velocity at which the flow enters from transition period to turbulent flow, is known as an upper critical velocity or higher critical velocity.

18·22 Reynold's Experiment of Viscous Flow

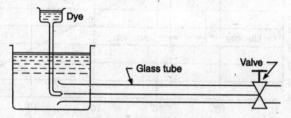

Fig. 18·4. Reynold's experiment.

Reynold's** apparatus consists of a tank, containing water and a small container containing dye. To the tank is fitted a horizontal glass tube (generally of 1·5 metre long and of 50 mm diameter) through which water can flow. The flow of water through the glass tube can be regulated by adjusting the regulating valve as shown in Fig. 18·4.

Water in the tank is first allowed to stand for several hours to allow it to come completely to rest. The outlet valve of the glass tube is then slightly opened. Then a jet of dye, having the same specific gravity as that of water, is also allowed to enter in the centre of the glass tube. It will be seen that a fine thread of the dye is carried by the flowing water as shown in Fig. 18·5 (a). The dye thread will move so steadily that it will be hardly seen to be in motion. Such a flow is known as laminar flow or streamline flow.

If we, slowly, go on increasing the velocity of water through the glass tube, we see that a stage will come, when the dye thread will start becoming irregular and then breaking up as shown in Fig. 18·5 (b). Such a velocity, at which the dye thread starts becoming irregular, is known as lower critical velocity. If we still go on increasing the velocity of water through the glass tube, we see that the length of the dye thread in the glass tube will start decreasing and ultimately a

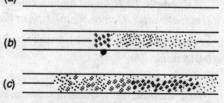

Fig. 18·5. Due thread in Reynold's experiment.

---

*For details, please refer to Art. 18.22 also.

**This experiment was first performed by Prof. Reynold Osborne in 1883.

stage will come, when we shall not be in a position to see the dye thread at all. Such a velocity, at which the whole dye thread is diffused, is known as an upper critical velocity. Beyond the upper critical velocity, the flow will be fully disturbed and such a flow is known as turbulent flow.

18·23 Reynold's Number

Prof. Reynold found that the value of critical velocity is governed by the relationship between the inertia force* and viscous* forces (*i.e.*, viscosity). He derived a ratio of these two forces and found out a dimensionless number known as Reynold's number. *i.e.*,

$$R_N = \frac{\text{Inertia forces}}{\text{Viscous forces}} = \frac{\rho v^2}{\frac{\mu v}{d}} = \frac{\rho v d}{\mu} = \frac{vd}{\nu} \qquad \ldots \left(\because \nu = \frac{\mu}{\rho} \right)$$

$$= \frac{\text{Mean velocity of liquid} \times \text{Diameter of pipe}}{\text{Kinematic viscosity of liquid}}$$

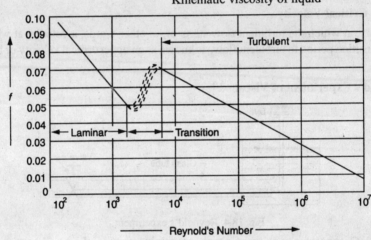

Fig. 18·6. Reynold's number

It may be noted that Reynold's number is a dimensionless* quantity which has much importance and gives us the information about the type of flow (*i.e.*, laminar or turbulent). Prof. Reynold, after carrying out a series of experiments, found that if Reynold's number for a particular flow is less than 2000, the flow is a laminar flow. But if Reynold's number is between 2000 and 2800, it is neither laminar flow nor turbulent flow. But if Reynold's number exceeds 2800 the flow is a turbulent flow. It may be noted that the value of critical velocity corresponding with $R_N = 2000$ is for a lower critical velocity and that corresponding with $R_N = 2800$ is for a higher critical velocity as shown in Fig. 18·6.

Example 18·7. *An oil having kinematic viscosity of 21·4 stokes is flowing through a pipe of 300 mm diameter. Determine the type of flow, if the discharge through the pipe is 15 litres/s.*

Solution. Given : $\nu = 21\cdot 4$ stokes $= 21\cdot 4 \times 10^{-4}$ m²/s; $d = 300$ mm $= 0\cdot 3$ m and $Q = 15$ litres/s $= 0\cdot 015$ m³/s.

---

*For details please refer to chapter on dimensionless constants.

We have seen that the Reynold's number,

$$R_N = \frac{vd}{\nu} = \frac{\frac{\text{Length}}{\text{Time}} \times \text{Length}}{\frac{\text{Area}}{\text{Time}}} = \frac{\frac{L^2}{T}}{\frac{L^2}{T}} = 1$$

Which shows that Reynold's number has no units, but is a dimensionless number.

Viscous Flow

We know that area of the pipe,
$$a = \frac{\pi}{4} \times (d)^2 = \frac{\pi}{4} \times (0.3)^2 = 0.0707 \text{ m}^2$$

and velocity of water,
$$v = \frac{Q}{a} = \frac{0.015}{0.0707} = 0.212 \text{ m/s}$$

∴ Reynold's number,
$$R_N = \frac{vd}{\nu} = \frac{0.212 \times 0.3}{21.4 \times 10^{-4}} = 29.7$$

Since Reynold's number is less than 2000, therefore the flow is laminar. **Ans.**

Example 18.8. *An oil of specific gravity of 0.95 is flowing through a pipeline of 200 mm diameter at the rate of 50 litres/s. Find the type of flow, if viscosity for the oil is 1 poise.*

Solution. Given : Specific gravity of oil = 0.85; d = 200 mm = 0.2 m; θ = 50 litres/s = 0.05 m³/s and $\mu = 1\ P = 0.1$ N-s/m².

We know that the density of oil,
$$\rho = 0.95 \times 1000 = 950 \text{ kg/m}^2$$

and area of pipeline,
$$A = \frac{\pi}{4} \times (d)^2 = \frac{\pi}{4} \times (0.2)^2 = 0.0314 \text{ m}^2$$

∴ Velocity of oil,
$$v = \frac{Q}{A} = \frac{0.05}{0.0314} = 1.59 \text{ m/s}$$

We also know that the kinematic viscosity of the oil,
$$\nu = \frac{\mu}{\rho} = \frac{0.1}{950} = 1.05 \times 10^{-4} \text{ m}^2/\text{s}$$

and Reynold's number,
$$R_N = \frac{vd}{\nu} = \frac{1.59 \times 0.2}{1.05 \times 10^{-4}} = 3029$$

Since Reynold's number is more than 2800, therefore the flow is turbulent. **Ans.**

18·24 Hagen-Poiseuille Law for Laminar Flow in Pipes

We have seen in the previous articles that some loss of head takes place in a laminar flow due to viscosity of the flowing liquid. The equation, which gives us the value of loss of head due to viscosity in a laminar flow, is known as Hagen-Poiseuille's law.

Fig. 18·7. Hagen-Poiseuille Law.

Consider a horizontal pipe of uniform diameter, in which some liquid is flowing from left to right as shown in Fig. 18·7. Let us consider two sections of this pipe.

Let
l = Length of the pipe between sections 1-1 and 2-2,
d = Diameter of the pipe,
v = Velocity of the liquid in the pipe,
w = Specific weight of the flowing liquid,
p_1 = Intensity of pressure, on liquid at section 1-1,

p_2 = Intensity of pressure, on the liquid, at section 2-2, and
H_L = Loss of head due to viscosity in the flowing liquid.

The laminar flow can be visualised as large number of concentric cylinders sliding into one another, like the tubes of a telescope (as per Art. 18·17). Now let us consider such a cylinder of radius y (in sections 1-1 and 2-2) sliding in a cylinder of radius $(y + dy)$.

We know that force acting on this cylinder of radius y

$$= \text{Area} \times \text{Difference of pressure}$$
$$= \pi y^2 (p_1 - p_2) \qquad \text{...(i)}$$

and loss of head between the sections 1-1 and 2-2,

$$H_L = \frac{p_1 - p_2}{w}$$

We also know that the shear stress

$$= \text{Shear stress per unit area} \times \text{Area of flow}$$
$$= \left(\mu \cdot \frac{dv}{dy}\right) \times (2\pi y l) \qquad \text{...(ii)}$$

Since these two forces equal (but opposite in direction), therefore equating the forces (i) and (ii),

$$\pi y^2 (p_1 - p_2) = -\mu \frac{dv}{dy} \times 2\pi y l$$

$$\pi y^2 \times w H_L = -\mu \frac{dv}{dy} \times 2\pi y l \qquad \left(\because \frac{p_1 - p_2}{w} = H_L\right)$$

$$\frac{w H_L}{2l} (y \cdot dy) = -\mu \cdot dv \qquad \text{...(iii)}$$

(Minus sign of μ indicates that v decreases as y increases)

Integrating the above equation,

$$\frac{w H_L y^2}{4l} = -\mu v + C \qquad \text{...(iv)}$$

where C is constant of integration.

We know that when $y = r$ i.e., radius of the pipe, $v = 0$.

$$\therefore \quad C = \frac{w H_L r^2}{4l}$$

Substituting this value of C in equation (iv),

$$\frac{w H_L r^2}{4l} = -\mu v + \frac{w H_L r^2}{4l}$$

or $\quad \dfrac{w H_L}{4l}(y^2 - r^2) = -\mu v$

$$\frac{w H_L}{4l}(r^2 - y^2) = \mu v \qquad \text{...(Taking minus sign outside)}$$

or $\quad v = \dfrac{w H_L}{4\mu l}(r^2 - y^2) \qquad \text{...(v)}$

A little consideration will show that the velocity is maximum at a point where $y = 0$ i.e., at the centre of the pipe. Therefore maximum velocity,

Viscous Flow

$$v_{max} = \frac{wH_L r^2}{4\mu l} \qquad \ldots(vi)$$

Now* average velocity of liquid in the pipe,

$$v = \frac{v_{max}}{2}$$

Substituting the value of v_{max} from equation (vi),

$$v = \frac{wH_L r^2}{2 \times 4\mu l}$$

$$\therefore \qquad H_L = \frac{8\mu v l}{wr^2} = \frac{32\mu v l}{wd^2} \qquad \ldots\left(\because r = \frac{d}{2}\right)$$

or pressure drop, $\qquad \dfrac{p_1 - p_1}{w} = \dfrac{32\ \mu v l}{wd^2} \qquad \ldots\left(\because H_L = \dfrac{p_1 - p_2}{w}\right)$

$$\therefore \qquad p_1 - p_2 = \frac{32\ \mu v l}{d^2}$$

Example 18·9. *In a laboratory experiment, a crude oil is flowing through a pipe of 50 mm diameter with a velocity of 1·5 m/s. During this experiment, a pressure drop of 18 kPa was recorded from two pressure gauges 8 metres apart. Find the viscosity of the flowing oil.*

Solution. Given : $d = 50$ mm $= 0.05$ m; $v = 1.5$ m/s; $(p_1 - p_2) = 18$ kPa $= 18 \times 10^3$ N/m² and length of pipe = 8 m.

Let $\qquad \mu =$ Viscosity of the flowing oil.

We know that pressure drop $(p_1 - p_2)$

$$18 \times 10^3 = \frac{32\ \mu v l}{d^2} \quad \text{or} \quad \frac{32\ \mu \times 1.5 \times 8}{(0.05)^2} = 153.6 \times 10^3\ \mu$$

$$\therefore \qquad \mu = 18/153.6 = 0.117\ \text{N-s/m}^2 = 1.17\ \text{P} \quad \textbf{Ans.}$$

Example 18·10. *An oil of mean weight of 8·8 kN/m³ flows under a head of 5 metres through 3000 metres pipe of 300 mm diameter. Due to cooling, the viscosity changes along the length and may be taken as 0·166 poise over the first 1500 metres and 0·332 poise over the second 1500 metres. Determine the flow of oil through the pipe in litres/s.*

Solution. Given : $w = 8.8$ kN/m³ $= 8.8 \times 10^3$ N/m³; $H_L = 5$ m; $l = 3000$ m; $d = 300$ mm $= 0.3$ m; $\mu_1 = 0.166$ P $= 0.0166$ N-s/m²; $l_1 = 1500$ m; $\mu_2 = 0.332$ P $= 0.0332$ N-s/m² and $l_2 = 1500$ m.

Let $\qquad v =$ Velocity of water through the pipe.

We know that loss of head due to viscosity (H_L),

$$5 = \frac{32\ \mu\ v l}{wd^2} = \frac{32 \times 0.0166 \times v \times 1500}{8.8 \times 10^3 \times (0.3)^2}$$
$$+ \frac{32 \times 0.0332 \times v \times 1500}{8.8 \times 10^3 \times (0.3)^2}$$
$$= 1.006\ v + 2.012\ v = 3.018\ v$$

$$\therefore \qquad v = 5/3.018 = 1.66\ \text{m/s}$$

---

*The average velocity, if the velocity distribution over a pipe section is parabolic, is half of the maximum velocity.

We also know that area of the pipe,

$$a = \frac{\pi}{4} \times (d^2) = \frac{\pi}{4} \times (0.3)^2 = 0.0707 \text{ m}^3$$

and flow of oil, $\quad Q = a \cdot v = 0.0707 \times 1.66 = 0.117 \text{ m}^3/\text{s} = 117$ litres/s **Ans.**

Example 18·11. *A pipeline of 300 mm diameter and 3200 m long is used to pump 50 second of an oil whose specific gravity is 0·95 and kinematic viscosity is 2·1 stokes. The centre of the pipeline at the upper end is 40 m above that at the lower end. Find the difference of pressures at the ends.*

Solution. Given : $d = 300$ mm $= 0.3$ m ; $l = 3200$ m; $Q = 50$ litres/s $= 0.05$ m^3/s; Specific gravity of oil $= 0.95$; $v = 2.1$ stokes $= 2.1 \times 10^{-4}$ m^3/s and height of the upper end $(Z_2) = 40$ m.

We know that density of oil,

$$\rho = 0.95 \times 1000 = 950 \text{ kg/m}^3$$

and specific weight of oil $\quad w = 9.81 \times 950 = 9320 \text{ N/m}^3 = 9.32 \text{ kN/m}^3$

∴ Viscosity of oil, $\quad \mu = v \cdot \rho = (2.1 \times 10^{-4}) \times 950 = 0.2 \text{ N-s/m}^2$

We also know that the area of the pipeline,

$$a = \frac{\pi}{4} \times (d)^2 = \frac{\pi}{4} \times (0.3)^2 = 0.0707 \text{ m}^2$$

and velocity of oil, $\quad v = \dfrac{Q}{a} = \dfrac{0.05}{0.0707} = 0.71$ m/s

∴ Reynold's number,

$$R_N = \frac{vd}{v} = \frac{0.71 \times 0.3}{2.1 \times 10^{-4}} = 1014$$

Since Reynold's number is less than 2000, therefore the flow is laminar. Therefore loss of head due to visocity,

$$H_L = \frac{32 \mu v l}{w d^2} = \frac{32 \times 0.2 \times 0.71 \times 3200}{9.32 \times 10^3 \times (0.3)^2} = 17.3 \text{ m}$$

Now applying Bernoulli's equation for the two ends of the pipeline. (Taking lower end of the pipe as section 1 and upper end as 2).

$$Z_1 + \frac{p_1}{w} + \frac{v_1^2}{2g} = Z_2 + \frac{p_2}{w} + \frac{v_2^2}{2g} + H_L$$

$$0 + \frac{p_1}{w} = 40 + \frac{p_2}{w} + 17.3 \qquad \ldots (\because d_1 = d_2, \text{ therefore } v_1 = v_2)$$

∴ $\quad \dfrac{p_1}{w} - \dfrac{p_2}{w} = 57.3 \quad$ or $\quad p_1 - p_2 = 57.3 \times 9.32 = 534 \text{ kN/m}^2 \quad$ **Ans.**

EXERCISE 18·2

1. Find the flow of an oil of specific gravity 0·9 and dynamic viscosity 12·5 poise flowing through a pipe of 200 mm diameter giving a discharge of 40 litres/s. **(Ans.** Laminar)
2. A 100 mm diameter pipe discharges water at the rate of 30 litres/s. Find the type of flow, if the kinematic viscosity is 1·2 stokes. **(Ans.** Turbulent)
3. An oil of specific gravity 0·91 and kinematic viscosity 0·002 m^3/s is pumped through a pipe of 150 mm diameter and 30 m long at the rate of 60 litres/s. Find the loss of head due to viscosity of the oil. **(Ans.** 29·5 m)

4. An oil having absolute viscosity of 0·8 poise flows with a velocity of 0·8 m/s through a 20 mm diameter horizontal pipe of 25 m length. Find the pressure difference between the two ends of the pipe. **(Ans. 12·8 kPa)**

18·25 Distribution of Velocity of a Flowing Liquid over a Pipe Section

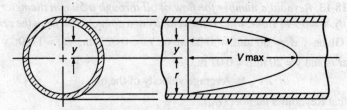

Fig. 18·8. Distribution of velocity over a pipe section.

Consider a pipe of uniform diameter, in which some liquid is flowing. Let us draw the velocity distribution diagram over the pipe section as shown in Fig. 18·8.

Let r = Radius of the pipe,
v_{max} = Maximum velocity of the liquid, and
v = Velocity of liquid at any other point at a radius y as shown in Fig. 18·8.

We have seen in Art. 18·24 that the maximum velocity of a liquid, flowing in a pipe, is at its centre. In the same article we have seen that the maximum velocity,

$$v_{max} = \frac{w H_L r^2}{4\mu l} \qquad ...(i)$$

and velocity at any section,

$$v = \frac{w H_L}{4\mu l}(r^2 - y^2) \qquad ...(ii)$$

Dividing the equation (i) by equation (ii),

$$\frac{v}{v_{max}} = \frac{\frac{wH_L}{4\mu l}(r^2 - y^2)}{\frac{w H_L}{4\mu l} \times r^2} = \frac{r^2 - y^2}{r^2} = \left(1 - \frac{y^2}{r^2}\right)$$

∴ $$v = v_{max}\left(1 - \frac{y^2}{r^2}\right)$$

Example 18·12. *Water at the rate of 65 litres/s is flowing through a pipe 200 mm diameter pipeline. Find the velocity of water at a point P, which is situated at a distance of 50 mm from the centre of the pipeline.*

Solution. Given : Q = 65 litres/s = 0·065 m³/s ; d = 200 mm = 0·2 m or r = 0·1 m and y = 50 mm = 0·05 m.

We know that the area of the pipeline,

$$a = \frac{\pi}{4} \times (d)^2 = \frac{\pi}{4} \times (0·2)^2 = 0·0314 \text{ m}^2$$

and average velocity of water in the pipeline,

$$v_{av} = \frac{Q}{a} = \frac{0·065}{0·0314} = 2·07 \text{ m/s}$$

∴ Maximum velocity of water in the pipeline,

$$v_{max} = 2 \times v_{av} = 2 \times 2·07 = 4·14 \text{ m/s}$$

We also know that the velocity of water at point P,

$$v = v_{max}\left[1 - \frac{y^2}{r^2}\right] = 4.14 \times \left[1 - \frac{(0.05)^2}{(0.1)^2}\right] = 3.11 \text{ m/s} \quad \textbf{Ans.}$$

Example 18·13. *Reynold's number for flow of oil through a 50 mm diameter pipe is 1700. If kinematic viscosity, $\nu = 0.744$ stoke, what is the velocity at a point 20 mm from the centre of the pipe?*

Solution. Given : $d = 50$ mm $= 0.05$ m or $r = 0.025$ m; $R_N = 1700$; $\nu = 0.744$ stokes $= 0.744 \times 10^{-4}$ m³/s and $y = 20$ mm $= 0.02$ m.

Let $\quad\quad\quad\quad v_{av} = $ Average velocity of the oil.

We know that Reynold's number (R_N),

$$1700 = \frac{v_{av} \cdot d}{\nu} = \frac{v_{av} \times 0.05}{0.744 \times 10^{-4}} = 672\, v_{av}$$

$\therefore\quad\quad\quad\quad v_{av} = 1700/672 = 2.53$ m/s

and maximum velocity, $\quad v_{max} = 2 \times v_{av} = 2 \times 2.53 = 5.06$ m/s.

We also know that the velocity at the point,

$$v = v_{max}\left[1 - \frac{y^2}{r^2}\right] = 5.06 \times \left[1 - \frac{(0.02)^2}{(0.025)^2}\right] = 1.82 \text{ m/s} \quad \textbf{Ans.}$$

18·26 Loss of Head due to Friction in a Viscous Flow

Consider a pipe of uniform diameter, in which a viscous flow is taking place.

Let
- $l = $ Length of the pipe,
- $d = $ Diameter of the pipe,
- $v = $ Velocity of the liquid in the pipe,
- $\mu = $ Coefficient of viscosity of the liquid,
- $\rho = $ Density of the liquid,
- $\nu = $ Kinematic viscosity of the liquid,
- $f = $ Coefficient of friction in the laminar flow, and
- $H_L = $ Loss of head in the pipe.

We have seen in Art. 18·24 that the loss of head due to viscosity,

$$H_L = \frac{32\,\mu v l}{w d^2}$$

and we have also seen in Art. 18·3 that the loss of head due to friction,

$$H_L = \frac{4 f l v^2}{2 g d} \quad\quad\quad\quad ...(ii)$$

Equating equations (i) and (ii),

$$\frac{4 f l v^2}{2 g d} = \frac{32\mu v l}{w d^2}$$

$\therefore\quad\quad\quad f = \frac{16\mu}{v d} \times \frac{g}{w} = \frac{16}{v d} \times \frac{\mu}{\rho} \quad\quad ...\left(\because \frac{w}{g} = \rho\right)$

$$= \frac{16}{v d} \times \nu \quad\quad ...\left(\because \nu = \frac{\mu}{\rho}\right)$$

or $\quad\quad\quad\quad = \frac{16}{R_N} \quad\quad ...\left(\frac{v d}{\nu} = R_N\right)$

Viscous Flow

Example 18·14. *A smooth iron pipe 200 mm in diameter carries a crude oil with a velocity of 1·5 m/s. What is the loss of head per 100 metres length of pipe? Assume the kinematic viscosity for oil as 1·0 stoke and the specific gravity as 0·9.*

Solution. Given : $d = 200$ mm $= 0·2$ m; $v = 1·5$ m/s; $l = 100$ m; $\nu = 1·0$ stoke $= 1·0 \times 10^{-4}$ m²/s and *Specific gr. of oil $= 0·9$.

We know that Reynold's number,

$$R_N = \frac{vd}{\nu} = \frac{1·5 \times 0·2}{1·0 \times 10^{-4}} = 3000$$

Since Reynold's number is more than 2800, therefore the flow is turbulent. Moreover, as the given pipe is a smooth iron pipe, therefore let us assume the coefficient of friction (f) as 0·01. We know that the loss of head,

$$h_f = \frac{4flv^2}{2gd} = \frac{4 \times 0·01 \times 100 \times (1·5)^2}{2 \times 9·81 \times 0·2} = 2·29 \text{ m} \quad \textbf{Ans.}$$

Example 18·15. *Fuel oil is pumped up in a 300 mm diameter and 1·6 kilometre long pipeline at the rate of 100 litres/s. The pipe is laid at an upgrade of 1 : 100. The specific weight of fuel oil 9 kN/m³ and its kinematic viscosity is 21·4 stokes.*

Find the power required to pump the oil.

Solution. Given : $d = 300$ mm $= 0·3$ m; $l = 1·6$ km $= 1600$ m; $Q = 100$ litres/s $= 0·1$ m³/s; $i = 1/100 = 0·01$; $w = 9$ kN/m³ and $\nu = 21·4$ stokes $= 21·4 \times 10^{-4}$ m²/s.

We know that the area of the pipeline,

$$a = \frac{\pi}{4} \times (d)^2 = \frac{\pi}{4} \times (0·3)^2 = 0·0707 \text{ m}^2$$

and velocity of oil,

$$v = \frac{Q}{a} = \frac{0·1}{0·0707} = 1·41 \text{ m/s}$$

∴ Reynold's number,

$$R_N = \frac{vd}{\nu} = \frac{1·41 \times 0·3}{21·4 \times 10^{-4}} = 197·7$$

Since Reynold's number is less than 2000, therefore the flow is laminar. We know that the coefficient of friction in a laminar flow,

$$f = \frac{16}{R_N} = \frac{16}{197·7} = 0·081$$

and the loss of head due to friction,

$$h_f = \frac{4flv^2}{2gd} = \frac{4 \times 0·081 \times 1600 \times (1·41)^2}{2 \times 9·81 \times 0·3} = 175 \text{ m}$$

∴ Total head against which the pump has to work,

$$H = h_f + (l \cdot i) = 175 + (1600 \times 0·1) = 191 \text{ m}$$

We know that the power required to pump the oil,

$$P = wQH = 9 \times 0·1 \times 191 = 171·9 \text{ kW} \quad \textbf{Ans.}$$

Example 18·16. *A crude oil of specific gravity 0·92 and absolute viscosity 1·2 poise is pumped through a pipe of 100 mm diameter laid at a slope of 1 in 100, and at the rate of 15 litres/s. What will be the difference between the pressure gauge readings, when fitted 300 m apart?*

---

*Superfluous data.

Solution. Given : Specific gravity of oil = 0·92; μ = 1·2 P = 0·12 N-s/m^2; d = 100 mm = 0·1 m; i = 1/100 = 0·01; Q = 15 litres/s = 0·015 m^3/s and l = 300 m.

We know that area of the pipe,
$$a = \frac{\pi}{4} \times (d)^2 = \frac{\pi}{4} (0\cdot 1)^2 = 7\cdot 854 \times 10^{-3} \text{ m}^2$$

and velocity of oil,
$$v = \frac{Q}{a} = \frac{0\cdot 015}{7\cdot 854 \times 10^{-3}} = 1\cdot 91 \text{ m/s}$$

We also know that the density of oil,
$$\rho = 0\cdot 92 \times 1000 = 920 \text{ kg/m}^3$$

and the kinematic viscosity,
$$\rho = \frac{\mu}{\rho} = \frac{0\cdot 12}{920} = 1\cdot 3 \times 10^{-4}$$

∴ Reynold's number,
$$R_N = \frac{vd}{\nu} = \frac{1\cdot 91 \times 0\cdot 1}{1\cdot 3 \times 10^{-4}} = 1469$$

Since Reynold's number is less than 2000, therefore the flow is laminar. We know that the coefficient of friction in a laminar flow,
$$f = \frac{16}{R_N} = \frac{16}{1469} = 0\cdot 01$$

and loss of head due to friction,
$$h_f = \frac{4flv^2}{2gd} = \frac{4 \times 0\cdot 01 \times 300 \times (1\cdot 91)^2}{2 \times 9\cdot 81 \times 0\cdot 1} = 22\cdot 3 \text{ m}$$

We know that the specific weight of the crude oil,
$$w = 9\cdot 81 \times 0\cdot 92 = 9\cdot 03 \text{ kN/m}^3$$

Now applying Bernoulli's equation at sections 1 and 2. (Taking lower point of pipe as section 1 and upper point as 2).

$$Z_1 + \frac{p_1}{w} + \frac{v_1^2}{2g} = Z_2 + \frac{p_2}{w} + \frac{v_2^2}{2g} + h_f$$

$$0 + \frac{p_1}{w} = (300 \times 0\cdot 01) + \frac{p_2}{w} + 22\cdot 3 \quad \ldots(\because d_1 = d_2, \text{ therefore } v_1 = v_2)$$

∴ $$\frac{p_1}{w} - \frac{p_2}{w} = 3 + 22\cdot 3 = 25\cdot 3$$

or $$p_1 - p_2 = w \times 25\cdot 3 = 9\cdot 03 \times 25\cdot 3 = 228\cdot 5 \text{ kN/m}^2 \quad \textbf{Ans.}$$

EXERCISE 18·3

1. In a pipeline of 320 mm diameter, water is flowing with an average velocity of 1·15 m/s. Find the velocity of water at a point 100 mm from the centre of the pipeline. **(Ans. 1·4 m/s)**
2. A fluid is flowing through a smooth pipe of length 250 m and diameter 150 mm with an average velocity of 1·75 m/s. Find the loss of head due to friction in the pipe. Take kinematic viscosity for the oil as 0·8 stoke. **(Ans. 10·4 m)**
3. An oil is flowing in a pipeline 500 m long and 250 mm diameter with a velocity of 3·5 m/s. If the kinematic viscosity of the oil is 12·5 stokes, find the type of flow. Also find the loss of head due to friction in the pipe. **(Ans. Streamline; 114·9 m)**

Viscous Flow

QUESTIONS

1. Explain clearly the term viscosity.
2. What do you understand by the term 'kinematic viscosity'? What is the relation between the absolute viscosity and kinematic viscosity?
3. What is the difference between an ideal fluid and real fluid?
4. What is meant by Newtonian and Non-Newtonian fluid?
5. Define the terms laminar flow and turbulent flow.
6. What do you understand by the term 'Reynold's number'? What useful information does it give?
7. Distinguish between lower critical velocity and higher critical velocity.
8. Derive an expression for the loss of head due to viscosity of a liquid flowing through a pipe.
9. Show that the maximum velocity of water in a pipe is at its centre.
10. From fundamentals, obtain an expression for the loss of head due to friction in a viscous flow.

OBJECTIVE TYPE QUESTIONS

1. Newton's law of viscosity for a flowing liquid gives us a relationship between its
   (a) simple stress and strain
   (b) shear stress and shear strain
   (c) temperature and volume
   (d) pressure and velocity
2. The unit of absolute viscosity is
   (a) poise
   (b) N-s/m^2
   (c) m^2/s
   (d) both 'a' and 'b'
3. A fluid having no viscosity is known as
   (a) real fluid
   (b) ideal fluid
   (c) Newtonian fluid
   (d) non-Newtonian fluid
4. The kinematic viscosity is the ratio of
   (a) viscosity to the density of the liquid
   (b) mass to the density of the liquid
   (c) viscosity to the specific weight of the liquid
   (d) density to the specific weight of the liquid
5. One stoke is equal to
   (a) 1 m^2/s
   (b) 0·1 m^2/s
   (c) 1 × 10^{-2} m^2/s
   (d) 1 × 10^{-4} m^2/s

ANSWERS

1. (b)   2. (d)   3. (c)   4. (a)   5. (d)

19
Viscous Resistance

1. Introduction. 2. Lubrication of Bearings. 3. Viscous Resistance of Oiled Bearings. 4. Viscous Resistance of Foot-step Bearings. 5. Viscous Resistance of Collar Bearings. 6. Nikuradse's Experiment on Rough Pipes. 7. Prandtl and Von Karman's Equation for Pipe Flow. 8. Coefficient of Friction for Commercial Pipes. 9. Methods for the Determination of Coefficient of Viscosity. 10. Capillary Tube Method. 11. Orifice Type Viscometer. 12. Rotating Cylinder Method. 13. Falling Sphere Method.

19·1 Introduction

In the previous chapter, we have discussed the effect of viscosity on the flow of liquids. Now in this chapter, we shall discuss the application of the theory of viscosity. Though the theory of viscosity has a number of applications, yet the viscous resistance on bearings is important from the subject point of view.

19·2 Lubrication of Bearings

The theory of viscosity has been successfully applied to the theory of lubrication of machine parts. It has been experienced that a highly viscous oil leads to a greater resistance and, thus, causes a greater power loss. On the other hand, a light oil may not be able to maintain the required film between the metal surfaces. As a result of this, one metal may come in contact with the other, which leads to the wear and tear of both the surfaces. It is, thus, obvious that the oil used for lubrication should have a correct viscosity. Since the viscosity of an oil changes with the temperature, that is why motorists use oil of different viscosities in different seasons.

In the following pages, we shall discuss the power required to overcome the viscous resistance in the following cases :

1. Viscous resistance of oiled bearings, 2. Viscous resistance of foot-step bearings, and 3. Viscous resistance of collar bearings.

19·3 Viscous Resistance of Oiled Bearings

Fig. 19·1. Oiled bearing.

A horizontal shaft revolving in an oiled bearing is always separated from the bearing by a very thin film of oil as shown in Fig. 19·1. A little consideration will show that the layer of this thin film just adjacent to the bearing will be at rest. And the layer of thin film just adjacent to the shaft, will be revolving with the same velocity as that of the shaft. The resistance offered by this thin oil film to the rotation of the shaft, is due to the viscosity of the oil. Now consider an oiled bearing* as shown in Fig. 19·1.

---

*The above bearing is, sometimes, termed as journal bearing or cylindrical bearing and the thickness of oil film t is termed and diametral clearance.

Viscous Resistance

Let D = Diameter of the shaft,
       t = Thickness of the oil film,
       l = Length of the bearing, and
       N = Speed of the shaft in r.p.m.

We know that angular velocity of the shaft,

$$\omega = \frac{\pi DN}{60} \text{ rad/s}$$

and viscous shear stress induced at the surface of the shaft,

$$\tau = \mu \times \frac{v}{y}$$

where v is the velocity of the shaft and y is the average thickness of the oil film. Since the oil film is very thin, therefore tangential velocity of the shaft may be assumed to be equal to angular velocity of the shaft and the average thickness (y) of the oil film to be equal to (t). Therefore shear stress,

$$\tau = \mu \times \frac{v}{t} = \mu \times \frac{\pi DN}{60 t} \qquad ...(i)$$

and the area of the bearing length = πDl

∴ Viscous resistance = Viscous stress × Area

$$= \mu \times \frac{\pi DN}{60 t} \times \pi Dl = \frac{\mu \pi^2 D^2 Nl}{60 t} \qquad ...(ii)$$

and the torque required to overcome the viscous resistance,

$$T = \text{Viscous resistance} \times \frac{D}{2}$$

$$= \frac{\mu \pi^2 D^2 Nl}{60 t} \times \frac{D}{2} = \frac{\mu \pi^2 D^3 Nl}{120 t}$$

We also know that the power absorbed in overcoming the viscous resistance,

$$P = T \times \omega = \frac{\mu \pi^2 D^3 Nl}{120 t} \times \frac{2\pi N}{60} \text{ W} \qquad ...\left(\because \omega = \frac{2\pi N}{60}\right)$$

$$= \frac{\mu \pi^3 D^3 N^2 l}{3600 t} \text{ W}$$

Example 19·1. *A shaft 100 mm diameter rotates at 60 r.p.m. in a 200 mm long bearing. If the surfaces are uniformly separated by a distance of 0·5 mm and linear velocity distribution in the lubricating oil having dynamic viscosity 4 centipoises, find the power absorbed is the bearing.*

Solution. Given : D = 100 mm = 0·1 m; N = 600 r.p.m.; l = 200 mm = 0·2 m; t = 0·5 mm = 0·0005 m and μ = 4 cP = 0·04 P = 0·004 N-s/m².

We know that the power absorbed in the bearing,

$$P = \frac{\mu \pi^3 D^3 N^2 l}{3600 t} = \frac{0.004 \times \pi^3 \times (0.1)^3 \times (60)^2 \times 0.2}{3600 \times 0.0005} \text{ W}$$

= 0·005 W **Ans.**

Example 19·2. *A shaft of 120 mm diameter runs in a bearing of length of 225 mm with a radial clearance of 0·3 mm at 100 r.p.m. If the power required to overcome the viscous resistance is 120 W, find the viscosity of the oil.*

Solution. Given : D = 120 mm = 0·12 m; l = 225 mm = 0·225 m; t = 0·3 mm = 0·0003 m; N = 100 r.p.m. and P = 120 W.

Let μ = Viscosity of the oil in poises.

We know that the power required to overcome the viscous resistance (P),

$$120 = \frac{\mu \pi^3 D^3 N^2 l}{3600 t} = \frac{\mu \times \pi^3 \times (0 \cdot 12)^3 \times (100)^2 \times 0 \cdot 225}{3600 \times 0 \cdot 0003} = 111 \cdot 6 \,\mu$$

$\therefore$ $\mu = 120/111 \cdot 6 = 1 \cdot 08$ N-s/m^2 = $10 \cdot 8$ P **Ans.**

19·4 Viscous of Resistance of Foot-step Bearings

Sometimes, a vertical shaft is required to rotate inside a fixed bearing. In such a case, some oil is put between the top of the bearing and bottom of the shaft as shown in Fig. 19·2. Now the viscous resistance of the foot-step bearing may be obtained by assuming the face of the shaft to be separated from the bearing surface by a thin film of the oil of uniform thickness as shown in the figure.

Let R = Radius of the shaft,
 N = Speed of the shaft in r.p.m., and
 t = Thickness of the oil film.

Now consider a thin elementary ring of the bearing surface of thickness dr and at a radius r as shown in Fig. 19·2.

$\therefore$ Area of the elementary ring
$$= 2\pi r \cdot dr$$

We know that the viscous stress induced,

$$\tau = \mu \times \frac{v}{y}$$

where v is the tangential velocity of the shaft, such that

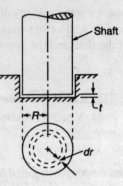

Fig. 19·2. Foot-step bearing.

$$v = \omega \cdot r = \frac{2\pi N}{60} \times r \qquad \ldots \left(\because \omega = \frac{2\pi N}{60} \right)$$

$\therefore$ Viscous stress, $\tau = \mu \times \dfrac{2\pi N r}{60 t}$

and viscous resistance $=$ Viscous stress $\times$ Area

$$= \mu \times \frac{2\pi N r}{60 t} \times 2\pi r \cdot dr = \frac{\mu \pi^2 N r^2}{15 t} dr$$

$\therefore$ Torque required to overcome the viscous resistance
$$= \text{Viscous resistance} \times \text{Radius}$$
$$= \frac{\mu \pi^2 N r^2}{15 t} dr \times r = \frac{\mu \pi^2 N r^3}{15 t} dr$$

The total viscous torque required to overcome the viscous resistance on the whole bearing may be found out by integrating the above equation between 0 and r. i.e.,

$$\therefore T = \int_0^R \frac{\mu \pi^2 N r^3}{15 t} dr = \frac{\mu \pi^2 N}{15 t} \int_0^R r^3 \cdot dr$$

$$= \frac{\mu \pi^2 N}{15 t} \left[\frac{R^4}{4} \right]_0^R$$

$$= \frac{\mu \pi^2 N R^4}{60 t}$$

Viscous Resistance

We also know that the power absorbed in overcoming the viscous resistance,

$$P = T \times \omega = \frac{\mu \pi^2 N R^4}{60 t} \times \frac{2\pi N}{60} = \frac{\mu \pi^3 N^2 R^4}{1800 t} \text{ W}$$

Example 19·3. *The lower end of a vertical shaft rests in a foot bearing. Assuming that the end of shaft and the surface of bearing are both flat and are separated by an oil film of 0·5 mm thickness, find the torque required to rotate the shaft at 750 revolutions per minute. The diameter of the shaft is 100 mm and the dynamic viscosity of oil is 1·5 poises.*

Solution. Given : $t = 0.5$ mm $= 0.0005$ m; $N = 750$ r.p.m.; Shaft diameter $= 100$ mm $= 0.1$ m or $R = 0.05$ m and $\mu = 1.5$ P $= 0.15$ N-s/m^2.

We know that the torque required to rotate the shaft,

$$T = \frac{\mu \pi^2 N R^4}{60 t} = \frac{0.15 \, \pi^2 \times 750 \times (0.05)^4}{60 \times 0.0005} = 0.231 \text{ N-m/s } \textbf{Ans.}$$

Example 19·4. *A vertical shaft of 150 mm diameter runs inside a bearing at 300 r.p.m. If the space between the lower end of the shaft and the bearing is 1 mm filled with an oil of viscosity 60 poises, determine the necessary power absorbed in overcoming the viscous resistance.*

Solution. Given : Dia. of shaft $= 150$ mm $= 0.15$ m or $R = 0.075$ m; $N = 300$ r.p.m.; $t = 1$ mm $= 0.001$ m and $\mu = 60$ P $= 6$ N-s/m^2.

We know that the power absorbed in overcoming the viscous resistance,

$$P = \frac{\mu \pi^3 N^2 R^4}{1800 t} = \frac{6 \, \pi^3 \times (300)^2 \times (0.075)^4}{1800 \times 0.001} = 294 \text{ W } \textbf{Ans.}$$

19·5 Viscous Resistance of Collar Bearings

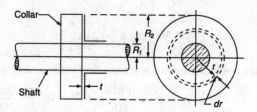

Fig. 19·3. Collar bearing.

The purpose of a collar bearing is to take up the axial thrust of a shaft. The viscous resistance of a collar bearing can be easily obtained, by assuming the face of the collar to be separated from the bearing surface by a thin film of some oil of uniform thickness as shown in Fig. 19·3.

Let
- R_1 = External radius of the collar,
- R_2 = Internal radius of the collar,
- N = Speed of the shaft in r.p.m., and
- t = Thickness of the oil film.

Now consider a thin elementary ring of the bearing surface of thickness dr and at a radius r as shown in Fig. 19·3.

∴ Area of the elementary ring

$$= 2\pi r \cdot dr$$

We know that the viscous stress induced,

$$\tau = \mu \times \frac{v}{t} \qquad \ldots(i)$$

where v is the tangential velocity of the shaft, such that

$$v = \omega r = \frac{2\pi N}{60} \times r \qquad \left(\because \omega = \frac{2\pi N}{60}\right)$$

∴ Viscous stress, $\qquad \tau = \mu \times \dfrac{2\pi Nr}{60t}$

and viscous resistance $\qquad = $ Viscous stress × Area

$$= \mu \times \frac{2\pi Nr}{60t} \times 2\pi r \cdot dr = \frac{\mu \pi^2 Nr^2}{15t} \times dr$$

∴ Torque required to overcome the viscous resistance

$$= \text{Viscous resistance} \times \text{Radius}$$

$$= \frac{\mu \pi^2 Nr^2}{15} \times dr \times r = \frac{\mu \pi^2 Nr^3}{15t} \times dr$$

The total viscous torque, required to overcome the viscous resistance on the whole collar, may be found out by integrating the above equation between R_2 and R_1. i.e.,

$$\therefore \qquad T = \int_{R_2}^{R_1} \frac{\mu \pi^2 Nr^3}{15t} \times dr$$

$$= \frac{\mu \pi^2 N}{15t} \int_{R_2}^{R_1} r^3 \cdot dr$$

$$= \frac{\mu \pi^2 N}{15t} \left[\frac{r^4}{4}\right]_{R_2}^{R_1}$$

$$= \frac{\mu \pi^2 N}{60t} \times \left(R_1^4 - R_2^4\right)$$

We also know that the power absorbed in overcoming the viscous resistance,

$$P = T \times \omega = \frac{\mu \pi^2 N}{60t}\left(R_1^4 - R_2^4\right) \times \frac{2\pi N}{60} \qquad \left(\because \omega = \frac{2\pi N}{60}\right)$$

$$= \frac{\mu \pi^3 N^2}{1800\, t} \times \left(R_1^4 - R_2^4\right) \text{ W}$$

Example 19·5. *A collar bearing has external and internal radii as 50 mm and 25 mm respectively. The oil of film, between the collar surface and the bearing, is 1·5 mm thick and has coefficient of viscosity as 15 poise. Find the power required to overcome the viscous resistance, when the shaft rotates at 600 r.p.m.*

Solution. Given : $R_1 = 50$ mm $= 0\cdot05$ m; $R_2 = 25$ mm $= 0\cdot025$ m; $t = 1\cdot5$ mm $= 0\cdot0015$ m; $\mu = 15$ P $= 1\cdot5$ N-s/m^2 and $N = 600$ r.p.m.

We know that the power required to overcome viscous resistance,

$$P = \frac{\mu \pi^3 N^2}{1800t} \times \left[R_1^4 - R_2^4\right]$$

$$= \frac{1\cdot5\, \pi^3 \times (600)^2}{1800 \times 0\cdot0015} \times \left[(0\cdot05)^4 - (0\cdot025)^4\right] \text{ W}$$

$$= (6\cdot2 \times 10^6) \times (5\cdot86 \times 10^{-6}) = 36\cdot3 \text{ W} \quad \textbf{Ans.}$$

Example 19·6. *The external and internal radii of a collar bearing are 100 mm and 75 mm respectively. The space between the collar surface and bearing is 2·5 mm and is filled with an oil. If the power lost in overcoming the viscous resistance is 23·6 W when the shaft is running at 250 r.p.m., find the viscosity of the oil.*

Solution. Given : $R_1 = 100$ mm $= 0·1$ m; $R_2 = 75$ mm $= 0·075$ m; $t = 2·5$ mm $= 0·0025$ m; $P = 23·6$ W and $N = 250$ r.p.m.

We know that the power lost in overcoming the viscous resistance (P)

$$23·6 = \frac{\mu \pi^3 N^2}{1800 t} \times \left[R_1^4 - R_2^4\right]$$

$$= \frac{\mu \pi^3 \times (250)^2}{1800 \times 0·0025} \times \left[(0·1)^4 - (0·075)^4\right] = 29·5 \,\mu$$

$$\therefore \quad \mu = 23·6/29·5 = 0·8 \text{ N-s/m}^2 = 8 \text{ P Ans.}$$

EXERCISE 19·1

1. A shaft of 150 mm length and 75 mm diameter is rotating at 100 r.p.m. coaxially within a sleeve with a clearance of 0·8 mm. Find the power absorbed in the bearing, if the viscosity of the oil is 9·6 poise. **(Ans. 6·5 W)**
2. A 100 mm diameter journal rotates axially in a bearing of length 150 mm with a radial clearance of 0·4 mm. The annular space between the journal and bearing is filled with an oil and takes 280 watt power to drive the journal at 240 r.p.m. Find the viscosity of the oil. **(Ans. 15·1 P)**
3. Find the power required to rotate a circular shaft of diameter 200 mm at 200 r.p.m. The shaft has a clearance of 1 mm from the bottom plate, which is filled with an oil of viscosity 18 poise. **(Ans. 124 W)**
4. The internal and external diameters of a collar bearing are 200 mm and 250 mm respectively. An oil of viscosity 4·5 poise is filled between the surfaces of the collar and bearing which is 1·2 mm. Find the power required to overcome the viscous resistance, when the shaft is rotating at 180 r.p.m. **(Ans. 30·1 W)**

19·6 Nikuradse's Experiment on Rough Pipes

Prof. Nikuradse conducted a series of experiments on pipes of diameters 25 mm, 50 mm and 100 mm to investigate the effect of roughness of pipe walls on the resistance to flow. The inner surfaces of these pipes were given different degrees of roughness by coating them with grains of sand of various coarseness. In these experiments, he took six different values of roughness factor (*i.e.*, r/k, where r is the radius of pipe and k is the average height of roughness projections) as 15, 30·6, 60, 126, 252 and 507. The resistance of each pipe was measured, experimentally, for various velocities of flow and resistance coefficients were obtained for various values of Reynold's number and various values of r/k. These results are plotted in Fig. 19·4.

From the above curves, we find that for smaller values of Reynold's number, the flow is laminar and follows in a straight line *AB*. But for higher values of Reynold's number, the flow passes through a transition stage and then becomes turbulent.

It will also be noticed that the turbulent flow appears to follow another straight line *CD*. The effect of roughness factor may be seen by the deviation of experimental points from the straight line *CD*. The pipes, with the roughest surface, causes the experimental points to break away at the smaller values of Reynold's number. But for smoother pipes, the experimental points coincide with the straight line *CD*, upto a larger values of Reynold's Number.

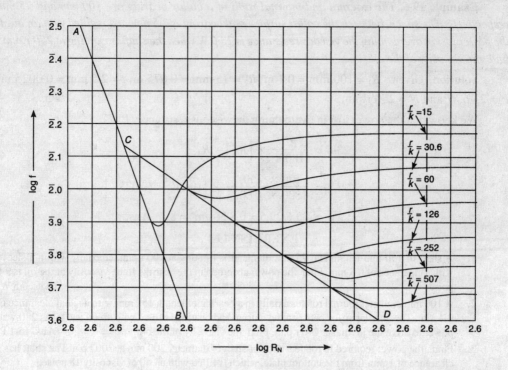

Fig. 19·4. Nikuradse's experiment.

19·7 Prandtl and Von Karman's Equation for Pipe Flow

The results of Nikuradse's experiments on rough pipes were also shown by Prandtl and Von Karman. They proved that Nikuradse's experiments may be represented in the form of equation as given below :

$$\frac{1}{\sqrt{4f}} = 2 \log_{10} \frac{r}{k} + 1 \cdot 74 \qquad \text{...(For rough pipes)}$$

and

$$\frac{1}{\sqrt{4f}} = 2 \log_{10} (R_N \sqrt{4f}) - 0 \cdot 8 \qquad \text{...(For smooth pipes)}$$

The above equations are called Prandtl-Karman resistance equations for turbulent flows. These equations are regarded as a great scientific achievement, as the experimental tests on fluid flow through pipes of all sizes and for different types of fluids agree with large values of Reynold's number.

Example 19·7. *Find the power required per km length of a pipe to maintain a flow of 450 litres/s of water through a 600 mm diameter rough pipe. Take k as 3 mm.*

Solution. Given : $l = 1$ km $= 1000$ m; $Q = 450$ litres/s $= 0.45$ m^3/s; $d = 600$ mm $= 0.6$ m or $r = 0.3$ m and $k = 3$ mm $= 0.003$ m.

Let $\qquad\qquad\qquad\qquad f = $ Coefficient of friction for the pipe.

We know that the area of pipe,

$$a = \frac{\pi}{4} \times (d)^2 = \frac{\pi}{4} \times (0 \cdot 6)^2 = 0 \cdot 2827 \text{ m}^2$$

Viscous Resistance

and velocity of water, $\quad v = \dfrac{Q}{a} = \dfrac{0.45}{0.2827} = 1.59$ m/s

We also know from the Prandtl and Von Karman's equation that

$$\dfrac{1}{\sqrt{4f}} = 2\log_{10}\left(\dfrac{r}{k}\right) + 1.74 = 2\log_{10}\left(\dfrac{0.3}{0.003}\right) + 1.74$$

$$= 2\log_{10} 100 + 1.74 = (2 \times 2) + 1.74 = 5.74$$

$$\dfrac{1}{4f} = (5.74)^2 = 32.95 \quad \text{or} \quad 4f = 1/32.95 = 0.03$$

and loss of head due to friction in the pipe

$$h_f = \dfrac{4flv^2}{2gd} = \dfrac{(0.03) \times 1000 \times (1.59)^2}{2 \times 9.81 \times 0.6} = 6.44 \text{ m}$$

and power required, $\quad P = wQh_f = 9.81 \times 0.45 \times 6.44 = 28.4$ kW **Ans.**

Example 19·8. *Find the discharge through a pipe of 100 mm diameter, if the loss of head is 1·8 m per 100 m length of the pipe. Take average height of the roughness as 0·25 mm.*

Solution. Given : $d = 100$ mm $= 0.1$ m or $r = 0.05$ m; $h_f = 1.8$ m ; $l = 100$ m and $k = 0.25$ mm $= 0.000\,25$ m.

Let $\quad v =$ Velocity of water through the pipe.

We know that the area of the pipe,

$$a = \dfrac{\pi}{4} \times (d)^2 = \dfrac{\pi}{4} \times (0.1)^2 = 7.854 \times 10^{-3} \text{ m}^2$$

We also know from the Prandtl and Von Karman's equation that

$$\dfrac{1}{\sqrt{4f}} = 2\log_{10}\left(\dfrac{r}{k}\right) + 1.74 = 2\log_{10}\left(\dfrac{0.05}{0.000\,25}\right) + 1.74$$

$$= 2\log_{10} 200 + 1.74 = (2 \times 2.3) + 1.74 = 6.34$$

∴ $\quad \dfrac{1}{4f} = (6.34)^2 = 40.2 \quad \text{or} \quad 4f = 1/40.2 = 0.025$

and loss of head (h_f), $\quad 1.8 = \dfrac{4flv^2}{2gd} = \dfrac{(0.025) \times 100\, v^2}{2 \times 9.81 \times 0.1} = 1.27\, v^2$

∴ $\quad v^2 = 1.8/1.27 = 1.42 \quad \text{or} \quad v = 1.19$ m/s

and discharge, $\quad Q = a \cdot v = (7.854 \times 10^{-3}) \times 1.19 = 0.0093$ m³/s
$\quad = 9.3$ litres/s **Ans.**

19·8 Coefficient of Friction for Commercial Pipes

In the previous article, we have discussed Nikuradse's experimental data on turbulent flow in smooth and artificially roughened pipes. But in actual practice, the commercial pipes have different wall roughness from that in the artificially roughened pipes. As a matter of fact, it is very difficult to measure the roughness of a commercial pipe. But it is comparatively easy to evaluate the roughness of a commercial pipe in terms of uniform sand roughness (k) of Nikuradse's experiments. The roughness, so obtained, is called equivalent sand grain roughness. It is equal to the roughness of an artificial roughened pipe, which gives the value of coefficient of friction (f) equal to that in the commercial pipe. The equivalent roughness of a commercial pipe is found out by experiments. The fluid is made to flow in the commercial pipe and the loss of head (h_f) is measured. The value of coefficient of friction (f) is then obtained by substituting the value of h_f in Darcy's equation.

On the basis of this, a curve was drawn by Moody, which is similar to Nikuradse's curve as shown in Fig. 19·5.

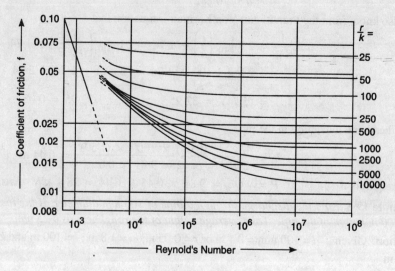

Fig. 19·5. Coefficient of friction.

The value of coefficient of friction (f) may be directly obtained from Moody's graph. The following table gives the values of equivalent sand grain roughness for different materials.

S.No.	Pipe Material	Equivalent sand grain roughness in mm
1.	Glass, drawn brass, copper	0·0015
2.	Wrought iron, steel	0·03 to 0·09
3.	Asphalted cast iron	0·06 to 0·18
4.	Galvanised iron	0·06 to 0·024
5.	Cast iron	0·12 to 0·60
6.	Concrete	0·30 to 3·00
7.	Riveted steel	0·09 to 9·00

Note. The above values correspond to the material in new and clean condition. As the pipes become older, the roughness increases due to corrosion.

19·9 Methods for the Determination of Coefficient of Viscosity

The coefficient of viscosity of a liquid may be found out, experimentally, by the following four methods :

1. Capillary tube method,
2. Orifice type viscometer,
3. Rotating cylinder method, and
4. Falling sphere method.

19·10 Capillary Tube Method

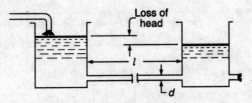

Fig. 19·6. Capillary tube method.

In this method, the liquid under test is allowed to flow from one tank to another through a capillary tube of known length (l) and diameter (d) as shown in Fig. 19·6. The rate of discharge (Q) and the loss of head (H_L) is measured at the time of performing the experiment. The velocity of flow is obtained from the relation,

$$v = \frac{Q}{a} = \frac{Q}{\frac{\pi}{4} \times (d)^2}$$

where d is the diameter of the capillary tube.

Now, the coefficient of viscosity of the flowing liquid is obtained from the relation obtained in Art. 18·24. i.e.,

$$H_L = \frac{32\,\mu v l}{w d^2}$$

Note : For accurate results, a considerable length of pipe should be used, and the fluid must be maintained at a constant temperature.

Example 19·9. *In an experiment, an oil of specific gravity 0·88 was made to flow from one tank to another through a capillary tube of 40 mm diameter 1·2 m long with a velocity of 0·2 m/s. If the viscosity of oil is 1·5 poise, find the loss of head due to viscosity of oil.*

Solution. Given : Specific gravity of oil = 0·88; d = 40 mm = 0·04 m; l = 1·2 m; v = 0·2 m/s and μ = 1·5 P = 0·15 N-s/m^2.

We know that the specific weight of the oil,

$$w = 0.88 \times 9.81 = 8.63 \text{ kN/m}^3 = 8.63 \times 10^3 \text{ N/m}^2$$

and the loss of head due to viscosity,

$$H_L = \frac{32\,\mu v l}{w d^2} = \frac{32 \times 0.15 \times 0.2 \times 1.2}{(8.63 \times 10^3) \times (0.04)^2} \text{ m}$$

$$= 0.083 \text{ m} = 83 \text{ mm} \quad \textbf{Ans.}$$

19·11 Orifice Type Viscometer

In this method, the liquid under test is allowed to flow through a short tube orifice of standard dimensions as shown in Fig. 19·7. The time taken for a given quantity (generally 50 cm^3) of the liquid to discharge through the orifice is noted. The coefficient discharge for the liquid is then found out by comparing with the coefficient of viscosity of a standard liquid or by the use of conversion factors or conversion tables.

The viscometers widely used are Redwood and Saybolt. The coefficient of viscosity (in poises) is obtained from the following relations :

$$\mu = \left(0.002\ 26\ t - \frac{1.95}{t}\right) \times \text{Sp. gr. of liquid} \qquad \text{...(Upto 100 sec)}$$

$$= \left(0.0022\ t - \frac{1.80}{t}\right) \times \text{Sp. gr. of liquid} \qquad \text{...(More than 100 sec)}$$

19·12 Rotating Cylinder Method

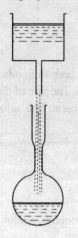

Fig. 19·7. Orifice type viscometer.

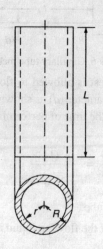

Fig. 19·8. Rotating cylinder method.

Fig. 19·9. Falling sphere method.

In this method, the liquid under test is filled within the annular space between two concentric cylinders as shown in Fig. 19·8. The *inner cylinder of radius (R) is held stationary and the outer cylinder is rotated with a uniform angular velocity (ω). A little consideration will show that this condition is like that, which we discussed in Art. 19·3. In this case,

Area of plate, $A = 2\pi R l$
Thickness of liquid, $y = R - r$
Velocity of plate, $v = \omega R$
Force applied on the outer cylinder,

$$P = \mu \cdot A \times \frac{v}{y} = \mu \times 2\pi R l \times \frac{\omega R}{y} = \frac{2\pi \mu \omega R^2 l}{y}$$

$$\frac{2\pi \mu \omega R^2 l}{(R - r)}$$

and torque, $T = P \times R = \dfrac{2\pi \mu \omega R^2 l}{y} \times R = \dfrac{2\pi \mu \omega R^3 l}{(R - r)}$

Now the coefficient of viscosity of the liquid may be found out by measuring the value of the torque (T) from the above relation.

---

*This experiment may also be conducted by holding the outer cylinder as stationary, and rotating the inner cylinder. But, in this case, there will be some effect of centrifugal force on the liquid contained in the annular space.

Example 19·10. *A 400 mm long cylinder of radius 100 mm is rotating outside a solid shaft of 98·5 mm radius. If the space between the cylinder and shaft is filled with an oil of viscosity 22·5 poises, determine the force required to rotate the cylinder at 240 r.p.m.*

Solution. Given : $l = 400$ mm $= 0.4$ m; $R = 100$ mm $= 0.1$ m; $r = 98.5$ mm $= 0.0985$ m; $\mu = 22.5$ P $= 2.25$ N-s/m^2 and $N = 240$ r.p.m.

We know that angular velocity of the shaft,

$$\omega = \frac{2\pi N}{60} = \frac{2\pi \times 240}{60} = 25.13 \text{ rad/s}$$

and force required to rotate the cylinder,

$$P = \frac{2\pi \mu \omega R^2 l}{(R - r)} = \frac{2\pi \times 2.25 \times 25.13 \times (0.1)^2 \times 0.4}{(0.1 - 0.0985)} \text{ N}$$

$$= 37.7 \text{ N} \quad \textbf{Ans.}$$

19·13 Falling Sphere Method

In this method, the liquid under test is filled in a long glass tube. A small polished steel ball is allowed to fall freely through the liquid. The falling sphere, at first, will accelerate. But as the velocity of the sphere increases, the viscous resistance to the motion will also increase. After some time, the viscous resistance will balance the pull of gravity and the sphere will move with a uniform velocity, known as terminal velocity. The time taken (t in seconds) is noted for the ball to fall through a distance (l in cm) between the two reference marks *A–A* engraved on the tube. The constant velocity of the ball :

$$v = \frac{l}{t}$$

Now the coefficient of viscosity of the liquid is obtained from the relation,

$$\mu = \frac{2r^2 g (\rho_1 - \rho_2)}{9v} \text{ poise}$$

where
$r =$ Radius of sphere,
$g =$ Acceleration due to gravity,
$\rho_1 =$ Density of the ball, and
$\rho_2 =$ Density of the liquid under test.

Note : This method gives accurate results, only if the terminal velocity is low. Moreover, no eddies should be caused by the falling ball.

Example 19·11. *In an experiment, a sphere of diameter 1·6 mm and mass density 7800 kg/m^3 was found to sink 200 mm in 21·3 seconds in an oil of density 960 kg/m^3. Find the coefficient of viscosity of the oil in poises.*

Solution. Given : Dia. of sphere $= 1.6$ mm $= 0.0016$ m or $r = 0.0008$ m; $\rho_1 = 7800$ kg/m^3; $l = 200$ mm $= 0.2$ m; $t = 21.3$ s and $\rho_2 = 960$ kg/m^3.

We know the velocity of the sphere,

$$v = \frac{l}{t} = \frac{0.2}{21.3} = 0.0094 \text{ m/s}$$

and coefficient of viscosity of the oil,

$$\mu = \frac{2r^2 g (\rho_1 - \rho_2)}{9v} = \frac{2 \times (0.0008)^2 \times 9.81 \, (7800 - 960)}{9 \times 0.0094}$$

$$= 1.02 \text{ N-s/m}^2 = 10.2 \text{ P} \quad \textbf{Ans.}$$

EXERCISE 19·2

1. A pipe of 400 mm diameter and 800 mm long is conveying water with a velocity of 1·6 m/s. Find the loss of head due to friction, if the average height of roughness is 0·5 mm. **(Ans.** 5·48 m)
2. An oil of specific gravity 0·9 and viscosity 1 poise is flowing from one tank to another through a capillary tube of 30 mm diameter and 1 m long with a velocity of 0·15 m/s. What is the loss of head due to viscosity of the oil? **(Ans.** 60 mm)
3. A cylinder of length 300 mm and diameter 180 mm is rotating outside a solid shaft with a clearance of 1 mm, which is filled with an oil of specific gravity of 16·5 poise. Find the force required to rotate the cylinder. **(Ans.** 2·27 N)

QUESTIONS

1. Explain the application of theory of viscosity to the theory of lubrication of machine parts.
2. Derive an expression for the power lost in overcoming viscous resistance in :
   (i) an oiled bearing,
   (ii) a foot-step bearing, and
   (iii) a collar bearing.
3. Describe Nikuradse's experiment on rough pipes.
4. Write Prandtl and Von Karman's equation for pipe flow.
5. Explain the methods for the determination of coefficient of viscosity.

OBJECTIVE TYPE QUESTIONS

1. Power required in overcoming the viscous resistance in case of shaft in oiled bearing is
   (a) $\dfrac{\mu \pi^3 D^3 N^2 l}{1800 t}$ (b) $\dfrac{\mu \pi^3 D^3 N^2 l}{3600 t}$ (c) $\dfrac{\mu \pi^3 N^2 R^4 l}{1800 t}$ (d) $\dfrac{\mu \pi^3 N^2 R^4 l}{3600 t}$

   where μ is the coefficient of viscosity. D is the diameter of the shaft, N angular velocity of the shaft in r.p.m., l is the length of the shaft and t is the thickness of the oil

2. In a foot-step bearing, if the speed of the shaft is doubled, then the power required to overcome the viscous resistance will be
   (a) double   (b) four times   (c) eight times   (d) sixteen times

3. Power absorbed in overcoming the viscous resistance of a collar bearing is equal to
   (a) $\dfrac{\mu \pi^3 N^2}{3600 t} \times \left(R_1^4 + R_2^4\right)$ (b) $\dfrac{\mu \pi^3 N^2}{3600 t} \times \left(R_1^4 - R_2^4\right)$
   (c) $\dfrac{\mu \pi^3 N^2}{1800 t} \times \left(R_1^4 + R_2^4\right)$ (d) $\dfrac{\mu \pi^3 N^2}{1800 t} \times \left(R_1^4 - R_2^4\right)$

   where R_1 and R_2 are the external and internal radii of the collar.

   The coefficient of viscosity (in poises) is given by the relation

   $$\mu = \left(0.0022\, t - \dfrac{1.8}{t}\right) \times \text{Sp. gr. of liquid}$$

   This equation is used to determine the coefficient of viscosity of liquids by
   (a) capillary tube method   (b) orifice type viscometer
   (c) rotating cylinder method   (d) falling sphere method

ANSWERS

1. (b)   2. (a)   3. (d)   4. (b)

20
Fluid Masses Subjected to Acceleration

1. Introduction. 2. D'Alembert's Principle. 3. Fluid Masses Subjected to Horizontal Acceleration.
4. Fluid Masses Subjected to Vertical Acceleration. 5. Fluid Masses Subjected to Acceleration along Inclined Plane.

20·1 Introduction

In the previous chapters, we have been discussing the effect of a fluid, when it is at *rest or when it moves without any outside force acting on it. But, in this chapter, we shall discuss the effect of acceleration on a fluid at rest (*i.e.*, a tank filled with liquid, being transported by a truck subjected an acceleration). In such a case, the laws of Hydrostatics, with the effect of acceleration, are applied.

20·2 D'Alembert's Principle

It states, "*A moving fluid mass may be brought to a static equilibrium position, by applying an imaginary inertia force of the same magnitude, as that of the accelerating force but in the opposite direction.*" This principle is used for changing the dynamic equilibrium of a fluid mass, into a static equilibrium. Mathematically accelerating force,

$$F = M \times \frac{dv}{dt}$$

(∵ Force = Mass × Acceleration)

∴ $$F - M \times \frac{dv}{dt} = 0$$

where M = Mass of the body,
v = Velocity of the body, and
$\frac{dv}{dt}$ = Rate of change of velocity (*i.e.*, acceleration).

The imaginary force $\left(- M \times \frac{dv}{dt}\right)$ is called the **inertia force**. Though there are many types of accelerations**, yet the following are important from the subject point of view :

1. Fluid masses subjected to horizontal acceleration,
2. Fluid masses subjected to vertical acceleration, and
3. Fluid masses subjected to acceleration along inclined plane.

Now we shall discuss all the above cases one-by-one.

---

* Strictly speaking, no body in this universe is at rest. *e.g.*, a body on the earth is said to be at rest, when it does not change its position with respect to the surrounding bodies. However, the so-called 'static body' is moving with the same velocity as that of the earth. Similarly, in a moving liquid, if the liquid particles do not move relative to each other, they are said to be in a static position. But a relative equilibrium exists between them under the action of the accelerating force.

** The fluid masses subjected to radial acceleration will be discussed in the next chapter.

[363]

20·3 Fluid Masses Subjected to Horizontal Acceleration

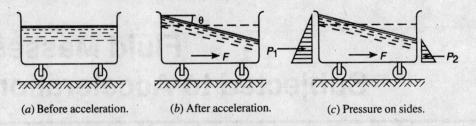

(a) Before acceleration.   (b) After acceleration.   (c) Pressure on sides.

Fig. 20·1.

Consider a tank, originally at rest and containing some liquid. We know that the liquid, at rest, maintains its surface level as shown in Fig. 20·1 (a). Now, let the tank move towards the right side with a uniform acceleration.

As the tank starts moving, we find that the liquid surface does not remain level any more. But the liquid surface falls down on the front side and rises up on the back side of the tank as shown in Fig. 20·1 (b). The static pressures on the back side and front side due to liquid are shown in Fig 20·1 (c).

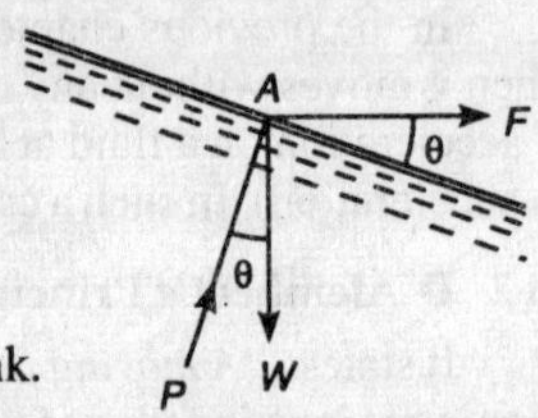

Let θ = Angle, which the liquid surface makes with the horizontal, and

a = Horizontal acceleration of the tank.

Fig. 20·2.

Now consider any particle A on the inclined liquid surface as shown in Fig. 20.2. We know that the forces acting on the liquid particle are :

1. Weight of the particle (W) acting vertically down,
2. Accelerating force (F) acting horizontally towards right, and
3. Pressure (P) exerted by the liquid particles normal to the free surface.

We know that the weight of the particle,

$$W = mg \qquad \ldots (\because \text{Force} = \text{Mass} \times \text{Acceleration})$$

where m = Mass of the liquid particle, and

g = Gravitational acceleration.

Similarly, accelerating force,

$$F = ma$$

Now resolving the forces horizontally at A,

$$P \sin \theta = F = ma \qquad \ldots(i)$$

and resolving the forces vertically at A,

$$P \cos \theta = W = mg \qquad \ldots(ii)$$

Dividing equation (i) by (ii),

$$\frac{P \sin \theta}{P \cos \theta} = \frac{ma}{mg} = \frac{a}{g}$$

or $$\tan \theta = \frac{a}{g} \qquad \ldots(iii)$$

Now consider the equilibrium of the entire mass of the liquid.

Let P_1 = Hydrostatic pressure on the back side of the tank, and

P_2 = Hydrostatic pressure on the front side of the tank.

Therefore net pressure,

$$P = P_1 - P_2$$

Now as per Newton's Second Law of Motion, this net pressure,

$$P = ma$$

or $$(P_1 - P_2) = ma \qquad \text{...(iv)}$$

Notes : 1. Since the values of tan θ are constant at all the points on the liquid surface, therefore the liquid surface will be inclined at a constant angle θ with the horizontal.

2. The inclination of the liquid surface is directly proportional to the horizontal acceleration.

Example 20·1. *An open rectangular tank 3 m long, 2·5 m wide and 1·25 m deep is completely filled with water. If the tank is moved with an acceleration of 1·5 m/s², find the slope of the free surface of water and the quantity of water which will spill out of the tank.*

Solution. Given : $l = 3$ m; $b = 2·5$ m; $d = 1·25$ m and $a = 1·5$ m/s².

Slope of the free surface of water

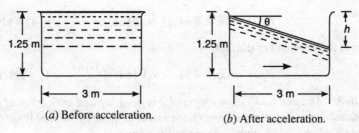

(a) Before acceleration.   (b) After acceleration.

Fig. 20·3.

Let θ = Angle which the free surface of water will make with the horizontal.

We know that $\quad \tan \theta = \dfrac{a}{g} = \dfrac{1 \cdot 5}{9 \cdot 81} = 0 \cdot 153 \quad$ or $\quad \theta = 8 \cdot 7°$ **Ans.**

Quantity of water which will spill out of the tank

We also know that the depth of water on the front side,

$$h = 3 \tan \theta = 3 \times 0 \cdot 153 = 0 \cdot 459 \text{ m}$$

∴ Quantity of water, which will spill out of the tank,

$$V = \frac{1}{2} \times 3 \times 2 \cdot 5 \times 0 \cdot 459 = 1 \cdot 72 \text{ m}^3 = 1720 \text{ litres} \quad \textbf{Ans.}$$

Example 20·2. *A rectangular tank 4 m long, 1 m wide and 3 m deep contains water to a depth of 2 m. If the tank is accelerated horizontally at 4 m/s² in the direction of its length, find the total pressure at the back and front ends of the tank.*

Solution. Given : $l = 4$ m; $b = 1$ m; $d = 3$ m; Height of water (h) = 2 m and $a = 4$ m/s².
First of all, let us find out the slope of the free surface of water due to acceleration.

Let θ = Angle which the free surface of water makes with the horizontal.

We know that $\quad \tan \theta = \dfrac{a}{g} = \dfrac{4}{9 \cdot 81} = 0 \cdot 408 \quad$ or $\quad \theta = 22 \cdot 2°$

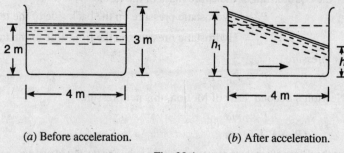

(a) Before acceleration.   (b) After acceleration.

Fig. 20·4.

From the geometry of the figure, we find that the rise of water level at the back of the tank
$$= 2 \tan \theta = 2 \times 0.408 = 0.816 \text{ m}$$
∴ Total height of water on the back side,
$$h_1 = 2 + 0.816 = 2.816 \text{ m}$$
and total height of water on the front side,
$$h_2 = 2 - 0.816 = 1.184 \text{ m}$$
We know that the total pressure at the back end of the tank,
$$P_1 = wA\bar{x} = 9.81 \ (1 \times 2.816) \times \frac{2.816}{2} = 41.4 \text{ kN Ans.}$$
and total pressure at the front end of the tank,
$$P_2 = wA\bar{x} = 9.81 \ (1 \times 1.184) \times \frac{1.184}{2} = 17.4 \text{ kN Ans.}$$

Example 20·3. *An open tank 10 m long and 2 m deep is filled with 1·5 m of oil of specific gravity 0·82. The tank is accelerated uniformly from rest to a speed of 20 m/s. What is the shortest time in which this speed may be obtained without spilling any oil?*

Solution. Given : $l = 10$ m; $d = 2$ m; Height of oil = 1·5 m;* Specific gravity of oil = 0·82; $u = 0$ and $v = 20$ m/s.

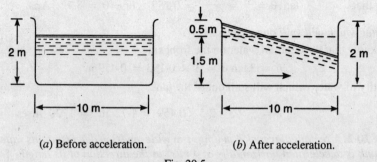

(a) Before acceleration.   (b) After acceleration.

Fig. 20·5.

Let t = Time required for the tank to be accelerated,
           θ = Angle which the surface of oil makes with the horizontal,
and a = Uniform acceleration of the tank.

---
*Superfluous data.

From the geometry of the tank, we find that when the oil is about to spill out of the tank, then

$$\tan \theta = \frac{1}{10} = \frac{a}{g} = \frac{a}{9 \cdot 81} \quad \text{or} \quad a = \frac{9 \cdot 81}{10} = 0 \cdot 981 \text{ m/s}^2$$

We know that final velocity of the tank (v),

$$20 = u + at = 0 + 0 \cdot 981 \, t$$

$$\therefore \quad t = 20/0 \cdot 981 = 20 \cdot 4 \text{ s} \quad \textbf{Ans.}$$

Example 20·4. *A closed rectangular tank 10 m long, 5 m wide and 3 m deep is completely filled with an oil of specific gravity 0.92. Find the pressure difference between the rear and front top corners of the tank, if it is moving with an acceleration of 3 m/s² in the horizontal direction.*

Solution. Given : $l = 10$ m; $b = 5$ m; $d = 3$ m; Specific gravity of oil = 0·92 and acceleration of the tank (a) = 3 m/s².

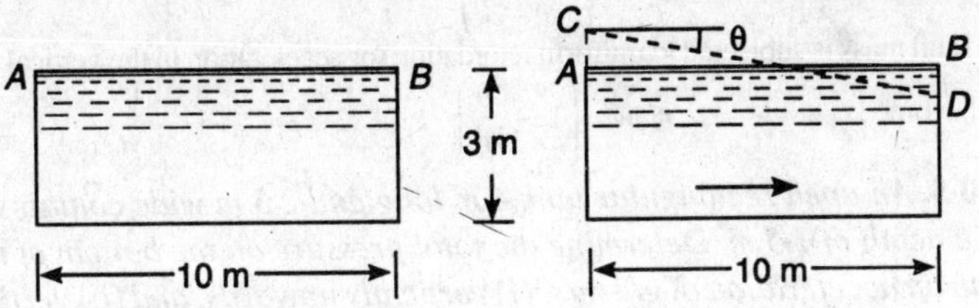

(a) Before acceleration.   (b) After acceleration.

Fig. 20·6.

We know that the specific weight of the oil,

$$w = 0 \cdot 92 \times 9 \cdot 81 = 9 \cdot 03 \text{ kN/m}^3$$

Since the top of the tank is closed, and it is completely filled with oil, therefore there cannot be any slope of the oil surface in the tank. But as a result of horizontal acceleration, the pressure will build up at the rear end A, whereas it will decrease at the front end B.

Now imagine the tank to be open at the top. As a result of the horizontal acceleration, let the oil surface be inclined at an angle θ as shown by the dotted line in Fig. 20·6 (b).

We know that $\quad \tan \theta = \dfrac{a}{g} = \dfrac{3}{9 \cdot 81} = 0 \cdot 306$

$\therefore$ Difference of levels between imaginary points C and D,

$$h = h_A - h_B = 10 \tan \theta = 10 \times 0 \cdot 306 = 3 \cdot 06 \text{ m}$$

and pressure difference between imaginary points C and D

$$p_A - p_B = wh = 9 \cdot 03 \times 3 \cdot 06 = 27 \cdot 6 \text{ kN/m}^2 = 27 \cdot 6 \text{ kPa} \quad \textbf{Ans.}$$

20·4 Fluid Mass Subjected to Vertical Acceleration

Consider a tank open at the top, containing a liquid and moving vertically upwards* with a uniform acceleration. Since the tank is subjected to an acceleration in the vertical direction only, therefore the liquid surface will remain horizontal.

Now consider a small column of the liquid of height h and area dA in the tank as shown in Fig. 20·7.

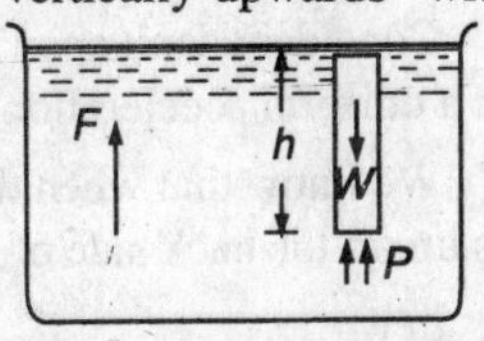

Let $\qquad p =$ Pressure due to vertical acceleration.

Fig. 20·7. Vertical acceleration.

---

*A tank, containing a liquid, kept in a rocket or an aeroplane, is the example of fluid mass subjected to vertical acceleration.

We know that the forces acting on this column are :
1. Weight of the liquid column [$W = w(h.dA)$] acting vertically downwards,
2. Acceleration force,

$$F = ma = \frac{w}{g}(h.\,dA)\,a$$

3. Pressure ($P = p.dA$) exerted by the liquid particles on the column.

Now resolving the forces vertically,

$$P = W + F$$

or

$$p.dA = w\,(h.dA) + \frac{w}{g}(h.\,dA)\,a = wh.\,dA\left(1 + \frac{a}{g}\right)$$

∴

$$p = wh\left(1 + \frac{a}{g}\right)$$

Note : If the fluid mass is subjected to uniform retardation (or acceleration in the vertical downward direction) above equation will become :

$$p = wh\left(1 - \frac{a}{g}\right)$$

Example 20·5. *An open rectangular tank 4 m long and 2·5 m wide contains an oil of specific gravity 0·85 upto a depth of 1·5 m. Determine the total pressure on the bottom of the tank, when the tank is moving with an acceleration of g/2 m/s² (i) vertically upwards, and (ii) vertically downwards.*

Solution. Given : $l = 4$ m; $b = 2·5$ m; Specific gravity of liquid $= 0·85$; $d = 1·5$ m and acceleration (a) $= g/2$.

(i) Total pressure on the bottom of the tank when it is moving vertically upwards

We know that specific weight of the oil,

$$w = 0·85 \times 9·81 = 8·34 \text{ kN/m}^3$$

and intensity of pressure on the bottom of the tank,

$$p_1 = wh\left(1 + \frac{a}{g}\right) = 8·34 \times 1·5\left(1 + \frac{g}{2g}\right) = 18·765 \text{ kN/m}^2$$

∴ Total pressure on the bottom of the tank,

$$P_1 = p_1.A = 18·765 \times (4 \times 2·5) = 187·65 \text{ kN Ans.}$$

(ii) Total pressure on the bottom of tank, when it is moving vertically downwards

We also know that the intensity of pressure on the bottom of the tank,

$$p_2 = wh\left(1 - \frac{a}{g}\right) = 8·34 \times 1·5\left(1 - \frac{g}{2g}\right) = 6·255 \text{ kN/m}^2$$

and the total pressure on the bottom of the tank,

$$P_2 = p_2.A = 6·255 \times (4 \times 2·5) = 62·55 \text{ kN Ans.}$$

20·5 Fluid Masses Subjected to Acceleration along Inclined Plane

Consider a tank open at the top, containing a liquid and moving upwards along inclined* plane with a uniform acceleration as shown in Fig. 20.8 (*a*).

We know that when the tank starts moving, the liquid surface falls down on the front side and rises up on the back side of the tank as shown in Fig. 20.8 (*b*).

Let
ϕ = Inclination of the plane with the horizontal,

---

*A wagon containing water or a tanker containing an oil and moving on mountainous regions with steep slope is the example of fluid mass subjected to acceleration along inclined plane.

θ = Angle, which the liquid surface makes with the horizontal, and
a = Acceleration of the tank.

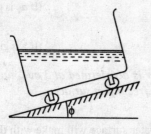
(a) Before acceleration.

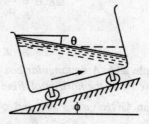
(b) After acceleration.

Fig. 20·8.

Now consider any particle A on the liquid surface. We know that the forces acting on the liquid particle are :
1. Weight of the particle (W) acting vertically down.
2. Accelerating force (F) acting on an angle φ with the horizonal.
3. Pressure (P) exerted by the liquid particles normal to the free surface.

Now resolving the accelerating force (F) horizontally and vertically, we get

$$F_H = F \cos \phi$$
and
$$F_V = F \sin \phi$$

We know that the weight of the particle,

Fig. 20·9.

$$W = mg \qquad (\because \text{Force} = \text{Mass} \times \text{Acceleration})$$

where
m = Mass of the liquid particle, and
g = Gravitational acceleration.

Similarly, the accelerating force,
$$F = ma \qquad \text{...(Where } a \text{ is the acceleration)}$$

Now resolving all the forces horizontally at A,
$$P \sin \theta = F_H = F \cos \phi = ma \cos \phi$$
or
$$P \sin \theta = ma \cos \phi \qquad (i)$$

and now resolving all the forces vertically at A,
$$P \cos \theta = F_V = F \sin \phi + W = ma \sin \phi + mg$$
or
$$P \cos \theta = ma \sin \phi + mg \qquad (ii)$$

Dividing equation (i) by (ii),

$$\frac{P \sin \theta}{P \cos \theta} = \frac{ma \cos \phi}{ma \sin \phi + mg} = \frac{a \cos + \phi}{a \sin \phi + g}$$

∴
$$\tan \theta = \frac{a_H}{a_V + g}$$

where
a_H = Horizontal component of the acceleration, and
a_V = Vertical component of the acceleration.

Note : If the fluid mass is subjected to acceleration in the downward direction, the above equation will become :

$$\tan \theta = \frac{a_H}{a_V - g} \qquad \text{...(If } a_V \text{ is greater than } g)$$

$$= \frac{a_H}{g - a_V} \qquad \text{...(If } g \text{ is greater than } a_V)$$

Example 20.6. *A rectangular box containing water is accelerated at 3 m/s² upwards on an inclined plane 30° to the horizontal. Find the slope of the free liquid surface.*

Solution. Given : $a = 3$ m/s² and $\phi = 30°$.

Let θ = Angle which the water surface will make with the horizontal.

We know that horizontal component of the acceleration,

$$a_H = a \cos 30° = 3 \times 0.866 = 2.598 \text{ m/s}^2$$

and vertical component of the acceleration,

$$a_V = a \sin 30° = 3 \times 0.5 = 1.5 \text{ m/s}^2$$

$$\therefore \quad \tan \theta = \frac{a_H}{a_V + g} = \frac{2.598}{1.5 + 9.81} = 0.2297$$

or $\theta = 12.94°$ **Ans.**

Example 20.7. *An open rectangular tank accelerates up an inclined plane having inclination of 1 vertical to 4 horizontal as shown in Fig. 20·10.*

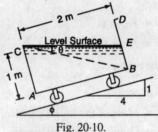

Fig. 20·10.

Find the acceleration of the tank, for the water surface to move from CE to CB as shown, where E is the mid-point of BD.

Solution. Given : $\tan \phi = \frac{1}{4} = 0.25$ (because inclination is 1 vertical to 4 horizontal) or $\phi = 14.0°$; $l = 2$ m and $d = 1$ m.

Let a = Acceleration of the tank.

We know that horizontal component of the acceleration,

$$a_H = a \cos 14° = a \times 0.9703 = 0.9703 \, a$$

and vertical component of the acceleration,

$$a_V = a \sin 14° = a \times 0.2419 = 0.2419 \, a$$

From the geometry of the figure, we find that

$$\angle ACE = 90° - 14° = 76°$$

and $\tan \angle ACB = \frac{2}{1} = 2.0$ or $\angle ACB = 63.4°$

$$\therefore \quad \theta = \angle ECB = \angle ACE - \angle ACB = 76° - 63.4° = 12.6°$$

We know that $\tan\theta = \dfrac{a_H}{a_V + g}$ or $\tan 12.6° = \dfrac{0.9703\,a}{0.2419\,a + 9.81}$

$$0.2235 = \dfrac{0.9703\,a}{0.2419\,a + 9.81}$$

$0.0541\,a + 2.1925 = 0.9703\,a$ or $0.9162\,a = 2.1925$

∴ $a = 2.1925/0.9162 = 2.4$ m/s² **Ans.**

EXERCISE 20·1

1. An open rectangular tank of 4 m long, 3 m wide and 1·8 m deep contains 1·5 m deep water. Find the slope of the free surface of water, when the tank moves on a horizontal road with an acceleration of 1 m/s². (**Ans.** 5·8°)

2. An open tank of 2 m long, 1·5 m wide and 1 m deep is completely filled with water. When the tank is moved with an acceleration of 2 m/s² on a horizontal path, find the volume of water spilled out. (**Ans.** 612 litres)

3. A rectangular tank of 5 m long 3 m wide and 2·5 m deep contains water up to a depth of 1·8 m. The tank is accelerated horizontally with an acceleration of 1·5 m/s². What will be the pressure on the back and front ends of the tank. (**Ans.** 96·3 kN; 62·6 kN)

4. A tank 2 m long and 1 m wide contains water up to a height of 1·5 m. Find the force on the bottom of the tank, when it moves vertically (*i*) upwards with an acceleration of 3 m/s² and (*ii*) downwards with an acceleration of 2 m/s². (**Ans.** 38·4 kN; 23·4 kN)

5. A rectangular box 2 m × 2 m × 2 m contains 1 m deep oil of specific gravity 0·9. If it is accelerated at 2 m/s² up an inclined plane, find the slope of the free surface of the oil. (**Ans.** 10·2°)

QUESTIONS

1. State and prove D'Alembert's principle.
2. Describe the effect of constant horizontal acceleration on a tank containing liquid.
3. Derive a relation for the inclination of the free liquid surface when accelerated horizontally.
4. Explain the effect on (*i*) the liquid surface, (*ii*) the liquid pressure on the sides, and (*iii*) the liquid surface on the bottom of the tank, when it is subjected to upward vertical acceleration and downward vertical acceleration.
5. Obtain, from fundamentals, the equation for the inclination of the free liquid surface when it is accelerated (*i*) upwards on an inclined plane, and (*ii*) downwards on an inclined plane.

OBJECTIVE TYPE QUESTIONS

1. When a tank containing liquid moves with an acceleration in the horizontal direction, then the free surface of the liquid
   (*a*) remains horizontal (*b*) falls on the front end
   (*c*) falls on the back front (*d*) becomes curved

2. An open tank containing liquid is made to move from rest with a uniform acceleration. The angle θ which the free surface of liquid makes with the horizontal is such that
   (*a*) $\tan\theta = \dfrac{a}{g}$ (*b*) $\tan\theta = \dfrac{2a}{g}$ (*c*) $\tan\theta = \dfrac{a}{2g}$ (*d*) $\dfrac{a^2}{2g}$

   where a = Horizontal acceleration of the tank, and
   g = Acceleration due to gravity.

3. A closed tank is completely filled with an oil. If it is made to move with a horizontal acceleration, then select the correct statement.
   (a) pressure at both the front and back ends will be the same
   (b) pressure at the front end will be more than that on the back end
   (c) pressure at the back end will be more than that on the front end
   (d) either 'a' or 'b'
4. An open tank containing liquid is moving with an acceleration on an inclined plane. The inclination of the free surface of the liquid will be directly proportional to
   (a) acceleration of the tank
   (b) gravitational acceleration
   (c) mass of the liquid
   (d) angle of inclined plane

ANSWERS

1. (b)   2. (a)   3. (c)   4. (a)

21

Vortex Flow

1. Introduction. 2. Forced Vortex Flow. 3. Equation of Forced Vortex Flow. 4. Height of Paraboloid of the Liquid Surface. 5. Closed Cylindrical Vessels. 6. Total Pressure on the Top and Bottom of a Closed Cylindrical Vessel Completely Filled up with a Liquid. 7. Free Vortex Flow. 8. Pressure Difference in a Free Vortex Flow.

21·1 Introduction

If we take a cylindrical vessel, containing some liquid, and start rotating it, about its vertical axis, we see that the liquid will also start revolving alongwith the vessel. After some time, we shall see that the liquid surface no longer remains level. But it has been depressed down at the axis of its rotation and has risen up near the wall of the vessel on all sides. This type of flow, in which a liquid flows continuously round a curved path about a fixed axis of rotation is called vortex flow. The vortex flows are of the following two types :
1. Forced vortex flow.   2. Free vortex flow.

21·2 Forced Vortex Flow

It is a type of vortex flow, in which the vessel, containing a liquid, is forced to rotate about the fixed vertical axis with the help of a torque. If the applied torque is removed the rotational motion will be slowly destroyed. Now consider a cylindrical vessel containing a liquid initially up to AA as shown in Fig. 21·1 (a).

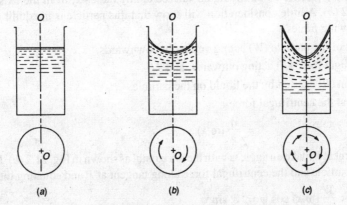

Fig. 21·1. Forced vortex flow.

Let the vessel be rotated about its vertical axis O-O. It will be noticed that the liquid surface, in the vessel, no longer remains level. But it has depressed down at the axis of its rotation and has risen up near the wall of the vessel on all sides as shown in Fig. 21.1 (b). If the vessel is revolved with an increased angular velocity, it will be noticed that the liquid has depressed down to a greater extent at its axis of rotation, and at the same time, it has also risen up to a greater height near the wall of the vessel as shown in Fig. 21·1 (c).

[373]

If we go on increasing the velocity of rotation, a stage will come when the liquid starts spilling out of the vessel. If we still go on increasing the velocity of rotation the depression of liquid, at its axis of rotation, will also go on increasing, till the axial depth of the liquid in zero.

21·3 Equation of Forced Vortex Flow

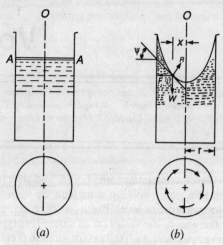

Fig. 21·2. Equation of forced vortex flow.

Consider a cylindrical vessel containing initially a liquid up to AA as shown in Fig. 21·2 (a). Now consider an instant when the vessel, containing the liquid, is revolved about its vertical axis, such that the liquid level attains a curved surface as shown in Fig. 21·2 (b).

Let $\qquad N$ = Rotation of the vessel in r.p.m.

ω = Angular velocity of the vessel in radian/s.

Now consider a particle P, on the liquid surface at any radius x, from the axis of rotation as shown in Fig. 21·2 (b). A little consideration will show that this particle is in equilibrium under the action of the following forces :

1. Weight of the particle (W) acting vertically downwards,
2. Centrifugal force (F) acting outwards, and
3. Reaction (R) exerted by the liquid on the particle.

We know that the centrifugal force

$$= \frac{W}{g}(\omega^2 x) \qquad \qquad ...(i)$$

Let the tangent at P make an angle ψ with the horizontal as shown in Fig. 21.2 (b). Now resolving the weight of the particle and the centrifugal force along tangent at P and equating the same,

$$\frac{W}{g}(\omega^2 x) \cos \psi = W \sin \psi$$

$$\therefore \qquad \tan \psi = \frac{\frac{W}{g}(\omega^2 x)}{W} = \frac{\omega^2 x}{g}$$

or $\qquad \dfrac{dy}{dx} = \dfrac{\omega^2 x}{g} \qquad \qquad \left(\because \tan \psi = \dfrac{dy}{dx}\right)$

Vortex Flow

Integrating the above equation,

$$y = \frac{\omega^2 x^2}{2g} + C$$

where C is the constant of integration. If the lowest point of the curved liquid surface is chosen as origin, then by substituting $x = 0$ and $y = 0$ in the above equation, we find that $C = 0$.

$$\therefore \quad y = \frac{\omega^2 x^2}{2g}$$

This is the equation of a parabola, which means that the surface of the liquid is a paraboloid. If the liquid particle be taken at the radius r from the axis of rotation then the above equation becomes,

$$y = \frac{\omega^2 r^2}{2g}$$

Example 21·1. *A cylinder of 100 mm diameter contains a liquid to a depth of 300 mm. Find the depth of the parabola, which the liquid surface will assume, if the cylinder is rotated about its vertical axis at 400 r.p.m.*

Solution. Given : Diameter of cylinder = 100 mm = 0·1 m or r = 0·05 m; *Depth of liquid = 300 mm = 0·3 m and N = 400 r.p.m.

We know that angular velocity of the cylinder**,

$$\omega = \frac{2\pi N}{60} = \frac{2\pi \times 400}{60} = 41·9 \text{ rad/s}$$

and depth of the parabola, $\quad y = \dfrac{\omega^2 r^2}{2g} = \dfrac{(41·9)^2 \times (0·05)^2}{2 \times 9·81} = 0·224 \text{ m} = 224 \text{ mm}$ **Ans.**

Example 21·2. *An open circular cylinder of 200 mm diameter and 1·2 m long contains water up to a height of 0·8 m. Find the speed at which the cylinder is to be rotated about its vertical axis, so that axial depth of the water becomes zero.*

Solution. Given : Diameter of cylinder = 200 mm = 0·2 m or r = 0·01 m; y = 1 m and height of water = 0·8 m.

Let $\quad \omega$ = Angular velocity of the cylinder in r.p.m.

We know that for the axial depth of the water to zero, the depth of the parabola (y) should also be 1·2 m.

We also know that depth of the parabola (y)

$$1·2 = \frac{\omega^2 r^2}{2g} = \frac{\omega^2 \times (0·1)^2}{2 \times 9·81} = 0·000\,51\, \omega^2$$

$$\therefore \quad \omega^2 = 1·2/0·000\,51 = 2353 \text{ or } \omega = 48·5 \text{ rad/s}$$

and angular speed, $\quad N = \dfrac{60\omega}{2\pi} = \dfrac{60 \times 48·5}{2\pi} = 463·1$ r.p.m. **Ans.**

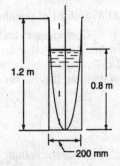

Fig. 21·3.

Example 21·3. *A cylindrical vessel of 200 mm diameter and 500 mm height is full of water. If the vessel is rotated about its vertical axis, with a speed of 450 r.p.m., find the volume of water left in the vessel.*

---

* Superfluous data.
** The angular velocity of a rotating body is given by the relation.

$$\omega = \frac{2\pi N}{60} \quad \text{or} \quad N = \frac{60\,\omega}{2\pi} \qquad \qquad \ldots (\because \omega = 2\pi \text{ radians})$$

where N is the angular speed of the rotating body in r.p.m. (*i.e.*, revolutions per minute).

Solution. Given : Diameter of cylinder = 200 mm = 0·2 m or $r = 0·1$ m ; Height of cylinder = 500 mm = 0·5 m and $N = 450$ r.p.m.

We know that angular velocity of the cylinder,
$$\omega = \frac{2\pi N}{60} = \frac{2\pi \times 450}{60} = 15\pi \text{ rad/s}$$

and height of parabola, $y = \dfrac{\omega^2 r^2}{2g} = \dfrac{(15\pi)^2 \times (0·1)^2}{2 \times 9·81}$ m

$$= 1·13 \text{ m}$$

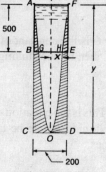

Fig. 21.14

The water surface will assume the shape as shown in Fig. 21·4. The imaginary depth of parabola (i.e., 1·13 − 0·5 = 0·63 m) and volume of water, which is left in the cylinder, is shown as the hetched volume.

Let x = Radius of the parabola at the bottom of the vessel.

We know that depth of the parabola,
$$0·63 = \frac{\omega^2 x^2}{2g} = \frac{(15\pi)^2 \times x^2}{2 \times 9·81} = 113·2\, x^2$$

∴ $\quad x^2 = 0·63/113·2 = 0·0056$ or $x = 0·075$ m

and volume of water left in the vessel

= Volume of water in cylinder ABEF
+ Volume of paraboloid GOH − Volume of paraboloid AOF

$$= \left[\frac{\pi}{4} \times (0·2)^2 \times 0·5\right] + \left[\frac{1}{2} \times \pi\, (0·075)^2 \times 0·63\right]$$

$$- \left[\frac{1}{2} \times \pi\, (0·1)^2 \times 1·13\right] \text{ m}^3$$

$$= 0·0157 + 0·0056 − 0·0178 = 0·0035 \text{ m}^3 \quad \textbf{Ans.}$$

21·4 Height of Paraboloid of the Liquid Surface

Consider a cylindrical vessel containing liquid up to AA as shown in Fig. 21·5. We have seen in Art. 21·2, that if the vessel is now revolved, the liquid surface assumes the shape of a parabola. Let the liquid surface assume the shape of parabola as shown in Fig. 21·5. Let the liquid, at its axis of rotation, depress down from AA to BB, and rise near the wall of the vessel, from AA to CC.

Let $\quad y$ = Height of the parabola,

$\quad h$ = Difference between AA and BB,

and $\quad r$ = Radius of the vessel.

Since the volume of the liquid between BB and CC, before and after the rotation remains the same, therefore

$$\pi r^2 h = \pi r^2 y - \frac{\pi r^2 y}{2} = \frac{\pi r^2 y}{2}$$

or $\quad h = \dfrac{y}{2}$

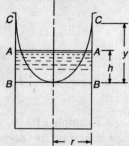

Fig. 21·5. Height of paraboloid.

This shows that the height, through which the liquid depresses down at the axis of rotation is the same through which the liquid rises up on the outer side of the cylinder.

Example 21·4 *An open cylinder of 1 m length and 0·3 m diameter contains a liquid up to a height of 0·8 m. If the cylinder is rotated about its vertical axis, find the angular velocity of the cylinder in r.p.m., so that no water spills out.*

Vortex Flow

Solution. Given : Height of cylinder = 1 m, Diameter of cylinder = 0·3 m or $r = 0·15$ m and depth of liquid = 0·8 m.

Let ω = Angular velocity of the cylinder in rad/s.
We know that for no water to spill out, the rise in water level
$$= 1 - 0·8 = 0·2 \text{ m}$$
and depth of the parabola,
$$y = 0·2 + 0·2 = 0·4 \text{ m}$$
We also know that depth of the parabola (y)
$$0·4 = \frac{\omega^2 r^2}{2g} = \frac{\omega^2 \times (0·15)^2}{2 \times 9·81} = 0·001\ 15\ \omega^2$$
$\therefore \quad \omega^2 = 0·4/0·001\ 15 = 347·8$ or $\omega = 18·6$ rad/s

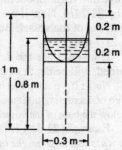

Fig. 21·6.

and angular speed, $N = \dfrac{60\ \omega}{2\pi} = \dfrac{60 \times 18·6}{2\pi} = 177·6$ r.p.m. **Ans.**

EXERCISE 21·1

1. An open circular cylinder of 200 mm diameter and 1·1 m long contains water up to a height of 600 mm. If the tank is rotated about its vertical axis at 300 r.p.m., find the depth of the parabola formed at the free surface of water. **(Ans. 0.5 m)**
2. A cylinder of 600 mm diameter is rotated about its vertical axis at 120 r.p.m. At this speed, the axial depth of the water is zero. Find the height of the cylinder. **(Ans. 724 mm)**
3. An open cylindrical vessel 120 mm diameter and 300 mm deep is filled with water up to the top. If the vessel is rotated about its vertical axis with a speed of 300 r.p.m., find the quantity of water left in the vessel. **(Ans. 0.237 m³)**
4. A 1.0 m long open cylinder of 150 mm diameter contains water up to a height of 800 mm. Find the maximum speed at which the cylinder can be rotated, about its vertical axis, so that no water spills out of the cylinder. **(Ans. 356.2 r.p.m.)**
5. A glass tube cylinder of 50 mm diameter open at the top and containing liquid rotates about its vertical axis at 600 r.p.m. What is the depression of the lowest point of the surface, below the surface of the liquid when at rest ? **(Ans. 63 mm)**

[**Hint:** Height of parabola , $y = \dfrac{\omega^2 r^2}{2g} = \dfrac{(20\pi)^2 \times (0·025)^2}{2 \times 9·81} = 126$ mm.]

21·5 Closed Cylindrical Vessels

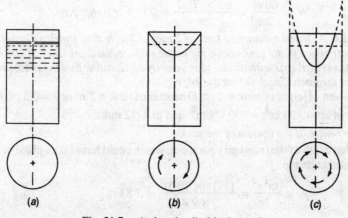

Fig. 21·7. A closed cylindrical vessel.

Consider a closed cylindrical vessel containing a liquid as shown in Fig. 21·7 (*a*). Now if we revolve this vessel about its vertical axis, the liquid surface assumes the shape of a parabola as shown in Fig. 21·7 (*b*). Now if we go on increasing the angular velocity of the vessel, the liquid will tend to spill out of the vessel. But it will not be able to do so, as the vessel is closed at its top as shown in Fig. 21·7 (*c*).

If we calculate the depth of the paraboloid, as usual, for a closed vessel also then the depth, so obtained, will be depth of the imaginary paraboloid, part of which will be in the vessel.

Note : If we still go on increasing the angular velocity of the vessel the depth of the imaginary paraboloid will go on increasing till it touches the base of the vessel. Beyond that the bottom of imaginary paraboloid will be below the base of the vessel.

Example 21·5. *A closed cylindrical vessel 150 mm in diameter is 1 m long. The vessel is filled with water up to a height of 0.7 m from bottom. Find the speed of the vessel, about its vertical axis, when the axial depth of the water is zero.*

Solution. Given : Diameter of vessel = 150 mm = 0·15 m or r = 0·075 m; Height of vessel (y)= 1 m and depth of water = 0·7 m.

Let ω = Angular velocity of the vessel in rad/s.

Since the vessel is closed, therefore, there is no change in the volume of water and air in the vessel before rotation and during rotation. We know that volume of air before rotation,

$$= \frac{\pi}{4} \times (0 \cdot 15)^2 \times 0 \cdot 3 = 0 \cdot 0053 \text{ m}^3$$

Let x = Radius of the paraboloid at the top of the vessel.

∴ Volume of paraboloid of air in the vessel during rotation

$$= \frac{1}{2} \times \pi \times x^2 \times 1 = 1 \cdot 571 \ x^2 \text{ m}^3$$

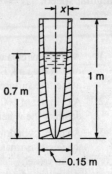

Fig. 21·8.

Since this volume is equal to the volume of air before rotation, therefore equating these two volumes,

$$0 \cdot 0053 = 1 \cdot 571 \ x^2 \text{ or } x^2 = 0 \cdot 0053 / 1 \cdot 571 = 0 \cdot 0034 \text{ or } x = 0 \cdot 058 \text{ m}$$

We know that depth of the parabola (y),

$$1 = \frac{\omega^2 x^2}{2g} = \frac{\omega^2 \times (0 \cdot 058)^2}{2 \times 9 \cdot 81} = 0 \cdot 000 \ 17 \ \omega^2$$

∴ $\omega^2 = 1/0 \cdot 000 \ 17 = 5882$ or $\omega = 76 \cdot 7$ rad/s

and angular speed $N = \frac{60 \ \omega}{2 \pi} = \frac{60 \times 76 \cdot 7}{2 \pi} = 732 \cdot 4$ r.p.m. **Ans.**

Example 21.6. *A closed cylindrical tank of 2 metres height and 1 metre diameter contains 1·5 metre water. The air space in the tank above the water surface has a pressure of 100 kPa. If the tank is rotated about its vertical axis with an angular velocity of 12 rad/s, find the pressure at the bottom of the tank:* (*i*) *at the centre, and* (*ii*) *at the edge.*

Solution. Given : Height of tank = 2 m; Diameter of tank = 1 m or r = 0·5 m; Depth of water = 1·5 m; Pressure of air = 100 kPa = 100 kN/m² and ω = 12 rad/s.

(*i*) *Pressure at the centre of the bottom of the tank*

We know that depth of the imaginary parabola, which could have taken place, if the tank would not have been closed at its top,

$$y_1 = \frac{\omega^2 r^2}{2g} = \frac{(12)^2 \times (0 \cdot 5)^2}{2 \times 9 \cdot 81} = 1 \cdot 83 \text{ m}$$

Vortex Flow

Since the vessel is closed, therefore there is no change in the volume of water and air in the tank before rotation and during rotation. We know that volume of air before rotation

$$= \frac{\pi}{4} \times (1)^2 \times 0.5 = 0.393 \text{ m}^3 \qquad ...(i)$$

Let x = Radius of the parabola at the top of the tank,
and y_2 = Depth of the actual parabola in the tank.

∴ Volume of actual paraboloid of air in the tank during rotation

$$= \frac{1}{2} \pi x^2 y_2 = \frac{\pi x^2 y_2}{2} \text{ m}^3 \qquad ...(ii)$$

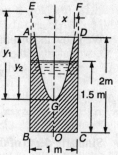

Fig. 21.9.

Since this volume is equal to the volume of air before rotation, therefore equating these two volumes,

$$\frac{\pi x^2 y_2}{2} = 0.393 \quad \text{or} \quad y_2 = \frac{0.393 \times 2}{\pi x^2} = \frac{0.25}{x^2} \qquad ...(iii)$$

We also know that depth of actual parabola in the tank,

$$y_2 = \frac{\omega^2 x^2}{2g} = \frac{(12)^2 \times x^2}{2 \times 9.81} = 7.34 x^2 \qquad ...(iv)$$

Equating these two values of y_2 from the above equations,

or $\qquad \dfrac{0.25}{x^2} = 7.34 x^2 \quad$ or $\quad x^4 = \dfrac{0.25}{7.36} = 0.034 \quad$ or $\quad x = 0.43$ m

Substituting this value of x in equation (iv),

$$y_2 = 7.34 \, x^2 = 7.34 \times (0.43)^2 = 1.36 \text{ m}$$

∴ Depth of water above the centre of the tank OG,

$$h_1 = 2 - 1.36 = 0.64 \text{ m}$$

and pressure at the centre of the bottom of the tank due to weight of the water above it

$$= wh_1 = 9.81 \times 0.64 = 6.3 \text{ kN/m}^2 = 6.3 \text{ kPa}$$

∴ Total pressure at the centre of the bottom of the tank

$$= \text{Pressure due to air + Pressure due to weight of water}$$
$$= 100 + 6.3 = 106.3 \text{ kPa } \textbf{Ans.}$$

(ii) *Pressure at the edge of the bottom of the tank*

We know that the depth of water above the edge of the bottom of the tank $(OG + y_2)$,

$$h_2 = 0.64 + 1.83 = 2.47 \text{ m}$$

and pressure at the edge of the bottom of tank due to weight of water above it

$$= wh_2 = 9.81 \times 2.47 = 24.2 \text{ kN/m}^2 = 24.2 \text{ kPa}$$

∴ Total pressure at the edge of the bottom of the tank

$$= 100 + 24.2 = 124.2 \text{ kPa } \textbf{Ans.}$$

Example 21.7. *A closed cylindrical tank of 3 metres diameter and 6 metres height, contains water to a depth of 4.5 metres. Find the area, which will be uncovered at the bottom of the tank, if the tank is rotated about its vertical axis at 12 rad/s.*

Solution. Given : Diameter of tank = 3 m or r = 1.5 m; Height of tank = 6 m; Depth of water = 4.5 m and ω = 12 rad/s.

Let x_1 = Radius of the parabola at the top of the tank, and
x_2 = Radius of the parabola at the bottom of the tank.

We know that depth of the imaginary parabola LOM,

$$y_2 = \frac{\omega^2 x_1^2}{2g} \quad \text{or} \quad x_1^2 = \frac{y_2 \times 2g}{\omega^2}$$

$$\therefore \quad x_1^2 = \frac{(y_3 + 6) \times 2g}{\omega^2} \quad \ldots (i)$$

Similarly depth of the imaginary parabola JOK,

$$y_3 = \frac{\omega^2 x_2^2}{2g}$$

or $\quad x_2^2 = \dfrac{y_3 \times 2g}{\omega^2} \quad \ldots(ii)$

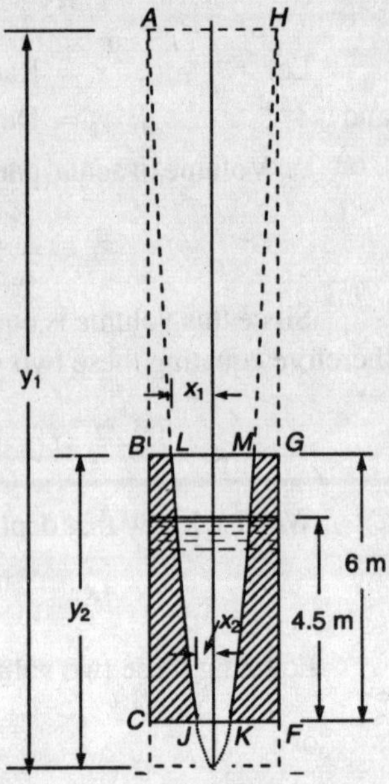

Since the vessel is closed, therefore there is no change in the volume of water and air in the tank before rotation and during rotation. We know that volume of air before rotation

$$= \frac{\pi}{4} \times (3)^2 \times 1\cdot 5 = 10\cdot 6 \text{ m}^3$$

and volume of actual paraboloid of air in the tank during rotation

= Vol. of paraboloid LOM
− Vol. of paraboloid JOK

$$= \left[\frac{1}{2}(\pi x_1^2) \times y_2\right] - \left[\frac{1}{2}(\pi x_2^2) \times y_3\right]$$

$$= \left[\frac{\pi}{2}(x_1^2)(y_3 + 6)\right] - \left[\frac{\pi}{2}(x_2^2) y_3\right] \quad \ldots(\because y_2 = y_3 + 6)$$

Fig. 21·10.

Since the two volumes of the air are equal, therefore equating these two volumes,

$$10\cdot 6 = \left[\frac{\pi}{2}(x_1^2)(y_3 + 6)\right] - \left[\frac{\pi}{2}(x_2^2) y_3\right]$$

Substituting the values of x_1^2 and x_2^2 from equations (i) and (ii) and $\omega = 12$ rad/s (given) in the above equation,

$$10\cdot 6 = \left[\frac{\pi}{2} \times \frac{(y_3 + 6) \times 2g}{(12)^2}(y_3 + 6)\right] - \left[\frac{\pi}{2} \times \frac{y_3 \times 2g}{(12)^2} \times y_3\right]$$

$$= 0\cdot 214(y_3 + 6)^2 - 0\cdot 214 y_3^2 = 0\cdot 214(y_3^2 + 36 + 12y_3 - 0\cdot 214 y_3^2)$$

$$= 2\cdot 5\, y_3 - 7\cdot 7$$

$\therefore \quad y_3 = (10\cdot 6 - 7\cdot 7)/2\cdot 57 = 1\cdot 13$ m

Now substituting this value of y_3 in equation (ii),

$$x_2^2 = \frac{y_3 \times 2g}{\omega^2} = \frac{1\cdot 13 \times 2 \times 9\cdot 81}{(12)^2} = 0\cdot 154$$

$\therefore$ Area which will be uncovered at the bottom of the tank

$$= \pi x_2^2 = \pi \times 0\cdot 154 = 0\cdot 484 \text{ m}^2 \quad \textbf{Ans.}$$

21.6 Total Pressure on the Top and Bottom of a Closed Cylindrical Vessel Completely Filled with a Liquid

Consider a closed cylindrical vessel, completely filled with some liquid, as shown in Fig. 21·11. Let the vessel be rotated about its vertical axis. The liquid surface will tend to depress down at the axis of its rotation and to rise up near the walls of the vessel. Since the vessel is closed and it is completely filled with the liquid, therefore the liquid surface cannot assume the shape of a parabola. But it will have some pressure, because it will tend to do so.

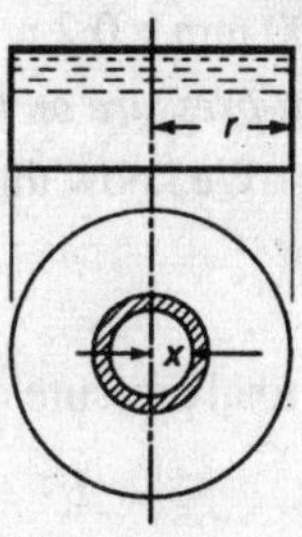

Let
w = Specific weight of liquid contained by vessel,
r = Radius of the vessel,
ω = Angular velocity of the vessel in rad/s, and
y = Height of the parabola, which could have taken place, if the cylinder would have been opened at the top and would have also been sufficiently long.

Fig. 21·11. Closed vessel completely filled with a liquid.

Now consider a ring of the liquid of thickness dx at a radius x as shown in Fig. 21.11. We know that area of the ring

$$= 2\pi x \cdot dx$$

and intensity of pressure at this ring due to imaginary liquid parabola

$$= wy = w\left(\frac{\omega^2 x^2}{2g}\right)$$

Although the pressure is horizontal, yet it will also act on the top and bottom of the vessel, as the pressure is transmitted* equally in all directions.

∴ Total pressure of the ring,

p = Intensity of pressure × Area of the ring

$$= w\left(\frac{\omega^2 x^2}{2g}\right) \times 2\pi x \cdot dx$$

The total pressure on the top of the vessel, due to centrifugal pressure may be found out by integrating the above equation between the limits 0 and r. Therefore total pressure,

$$P = \int_0^r w\left(\frac{\omega^2 x^2}{2g}\right) \times 2\pi x \cdot dx$$

$$= \frac{w\omega^2 \pi}{g} \int_0^r x^3 \, dx$$

$$= \frac{w\omega^2 \pi}{g} \left[\frac{x^4}{4}\right]_0^r$$

$$= \frac{w\omega^2 \pi r^4}{4g}$$

The total pressure, on the bottom of the vessel, will be sum of the total centrifugal pressure and the weight of the liquid in the vessel.

---

*According to Pascal's law. Please refer to Art. 2.3.

Example 21·8. *A closed cylinder of 300 mm diameter and 200 mm deep, is completely filled with water. It is rotated about its axis, which is vertical, at 240 r.p.m. Calculate the total pressure of water on the top and bottom of the cylinder.*

Solution. Given : Diameter of cylinder = 300 mm = 0.3 m or r = 0.15 m; Depth of cylinder = 200 mm = 0.2 m and ω = 240 r.p.m.

Total pressure on the top of the cylinder

We know that the angular velocity of the cylinder,

$$\omega = \frac{2\pi N}{60} = \frac{2\pi \times 240}{60} = 8\pi \text{ rad/s}$$

and total pressure on the top of the cylinder,

$$P = \frac{w\omega^2 \pi r^4}{4g} = \frac{9 \cdot 81 \times (8\pi)^2 \times \pi \times (0 \cdot 15)^4}{4 \times 9 \cdot 81} = 0 \cdot 251 \text{ kN} \quad \textbf{Ans.}$$

Total pressure on the bottom of the cylinder

We also know that weight of water in the cylinder,

$$= 9 \cdot 81 \times \pi \times (0 \cdot 15)^2 \times 0 \cdot 2 = 0 \cdot 139 \text{ kN}$$

and total pressure at the bottom of the cylinder

$$= \text{Total pressure due to rotation} + \text{Weight of water}$$
$$= 0.251 + 0.139 = 0.39 \text{ kN} \quad \textbf{Ans.}$$

21·7 Free Vortex

It is a type of flow, in which the liquid particles describe circular paths, about a fixed vertical axis, without any external force acting on the particles. The common example of a free vortex occurs when the water escapes, through the hole in the bottom of a wash basin.

Now consider an element of water having a free vortex.

Let l = Length of the water element,

dr = Thickness of the water element,

v = Tangential velocity of the water element,

dp = Change of pressure across the thickness of the water element.

A little consideration will show that the force, due to pressure, is equal to the centrifugal force acting on the water element. *i.e.*,

$$dp \times l = \frac{(wl \cdot dr) v^2}{gr} \quad \text{or} \quad \frac{dp}{w} = \frac{v^2 dr}{gr} \qquad \ldots(i)$$

We know that the total energy* of the water element,

$$E = \frac{p}{w} + \frac{v^2}{2g}$$

Differentiating, for small changes of energy,

$$dE = \frac{dp}{w} + \frac{v \, dv}{g} = \frac{v^2 dr}{gr} + \frac{v \, dv}{g} \qquad \left(\because \frac{dp}{w} = \frac{v^2 dr}{gr}\right)$$

or $\qquad \dfrac{dE}{dr} = \dfrac{v}{g}\left(\dfrac{v}{r} + \dfrac{dv}{dr}\right) \qquad \ldots(ii)$

Since in a free vortex, there is no change of energy across the streamlines, therefore equating the equation (*ii*) to zero. *i.e.*,

---

* For details, please refer to Art. 8.6.

Vortex Flow

$$\frac{v}{g}\left(\frac{v}{r}+\frac{dv}{dr}\right) = 0$$

∴ $\quad\dfrac{v}{r}+\dfrac{dv}{dr} = 0 \quad$ or $\quad \dfrac{dv}{v}+\dfrac{dr}{r} = 0$

Integrating the above equation,

$$\log_e v + \log_e r = \text{Constant}$$

∴ $\quad vr = C$

or in general $\quad v_1 r_1 = v_2 r_2 = v_3 r_3 \ldots$

or $\quad v = \dfrac{C}{r}$

where C is a constant, known as strength of the vortex. It is thus obvious, that the velocity of a particle is inversely proportional to its distance from the centre.

21·8 Pressure Difference in a Free Vortex Flow

Consider a liquid having a free vortex flow as shown in Fig. 21·12. Let the liquid move with an angular velocity of ω radians/s.

Now consider two liquid particles 1 and 2 on the liquid surface as shown in Fig. 21·12.

Let $\quad r_1$ = Radius of particle 1,

v_1 = Tangential velocity of particle 1,

p_1 = Pressure at point 1,

r_2, v_2, p_2 = Corresponding values at point 2.

Now equating the total energy at the points 1 and 2,

$$\frac{p_1}{w}+\frac{v_1^2}{2g} = \frac{p_2}{w}+\frac{v_2^2}{2g}$$

or $\quad \dfrac{p_1}{w}-\dfrac{p_2}{w} = \dfrac{v_2^2}{2g}-\dfrac{v_1^2}{2g}$

$$\frac{(p_1-p_2)}{w} = \frac{(v_2^2-v_1^2)}{2g}$$

or $\quad \dfrac{(p_2-p_1)}{w} = \dfrac{(v_1^2-v_2^2)}{2g} \quad$ ...(Taking minus sign outside)

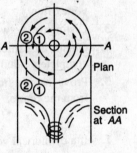

Fig. 21·12. Free vortex flow.

Example 21.9 *In a free cylindrical vortex of water, the tangential velocity and pressure at a radius of 125 mm are found to be 10 m/s and 350 kPa respectively. Calculate the pressure at a radius of 250 mm.*

Solution. Given : r_1 = 125 mm = 0·125 m; v_1 = 10 m/s; p_1 = 350 kPa = 350 kN/m² and r_2 = 250 mm = 0·25 m.

Let $\quad p_2$ = Pressure as a radius of 250 mm.

Since the flow is continuous, therefore tangential velocity at a radius of 250 mm (0·25 m),

$$v_2 = \frac{v_1 \cdot r_1}{r_2} = \frac{10 \times 0\cdot 125}{0\cdot 25} = 5 \text{ m/s}$$

We know from the pressure equation in a free vortex that

$$\frac{p_2 - p_1}{w} = \frac{v_1^2 - v_2^2}{2g} \quad \text{or} \quad \frac{p_2 - 350}{9 \cdot 81} = \frac{(10)^2 - (5)^2}{2 \times 9 \cdot 81} = 37 \cdot 5$$

∴ $p_2 = 37 \cdot 5 + 350 = 387 \cdot 5$ kN/m² = 387·5 kPa **Ans.**

EXERCISE 21·2

1. A closed cylindrical vessel of 50 mm radius and 300 mm long is filled with water to a height of 200 mm. Find the speed at which the vessel should be rotated, so that the water near the wall just touches the top of the vessel. **(Ans. 378.2 r.p.m.)**
2. A closed cylinder of 200 mm diameter and 1.2 m long contains water up to a height of 800 mm. What should be the speed of rotation of the cylinder, if the axial depth of water is zero. **(Ans. 576·2 r.p.m.)**
3. A closed cylindrical vessel of 300 mm diameter and 100 mm deep is completely filled with a liquid of specific gravity 1.2. Find the total pressure on the top of the vessel, if it is rotated about its vertical axis at 300 r.p.m. **(Ans. 6·125 kN)**
4. In a free vortex flow of water, the tangential velocity of particle was found to be 2.5 m/s. Find the difference of pressures between the water particle and the centre of the flow. **(Ans. 3·125 kPa)**

[**Hint** : Velocity at the centre in zero.]

QUESTIONS

1. What do you understand by the term vortex flow? Name the types of vortex flows.
2. Derive an expression for the depth of paraboloid, which the liquid surface assumes due to vortex flow.
3. Distinguish between forced vortex and free vortex.
4. Obtain an expression for the total pressure on the top of a closed cylindrical vessel completely filled up with some liquid, when rotated about its vertical axis.
5. Prove that in a free vortex, the tangential velocity is inversely proportional to the distance from the centre.

OBJECTIVE TYPE QUESTIONS

1. If a cylindrical vessel is rotated about its vertical axis then depth of the parabola, which the liquid surface assumes is

    (a) $\frac{\omega r}{2g}$ (b) $\frac{\omega^2 r^2}{2g}$ (c) $\frac{\omega r}{g}$ (d) $\frac{\omega r}{2g}$

2. If the angular velocity of a cylindrical vessel containing liquid is doubled, then the depth of parabola will

    (a) be halved   (b) remain the same   (c) be two times   (d) be four times

3. A cylindrical vessel containing some liquid is rotated about its vertical axis. If its angular speed is halved and radius doubled, then the depth of parabola will

    (a) be halved   (b) remain the same   (c) be four times   (d) be sixteen times

4. The total pressure on the top of a closed cylindrical vessel completely filled with water, when rotated about its vertical axis is directly proportional to

    (a) r (b) r^2 (c) r^3 (d) r^4

    where r is the radius of the vessel.

ANSWERS

1. (b)   2. (d)   3. (b)   4. (d)

22
Mechanics of Compressible Flow

1. Introduction. 2. Laws of Perfect Gases. 3. Boyle's Law. 4. Charles' Law 5. Gay-Lussac Law. 6. General Gas Equation. 7. Specific Heats of a Gas. 8. Specific Heat at a Constant Volume. 9. Specific Heat at Constant Pressure. 10. Relation Between Specific Heats. 11. Ratio of Specific Heats. 12. Isothermal Process. 13. Adiabatic Process. 14. Bulk Modulus of a Fluid.

22.1 Introduction

In the previous chapters, we have been obtaining the various basic equations of fluid flow, such as equation of continuity, Bernoulli's equation, momentum equation. At the time of deriving these equations, an important assumption was made that the fluid is incompressible. In other words, the density of the flowing liquid was assumed[*] to be constant during the process of flow. As a matter of fact, this assumption is justified for the flow of liquids, which are incompressible under ordinary conditions. But in this chapter, we shall discuss the flow of gases, which are subjected to a large change in volume and density during the process of flow. Thus are treated as compressible[**] fluids.

The laws for compressible fluids are more difficult than those for incompressible fluids. These laws may be easily understood by the combined knowledge of Thermodynamics and Fluid Mechanics. The following few basic principles of Thermodynamics should be clearly understood at this stage :

22.2 Laws of Perfect Gases

The physical properties of a gas are controlled by the following three variables :

1. Pressure exerted by the gas,
2. Volume occupied by the gas, and
3. Temperature of the gas.

The behaviour of a perfect gas, undergoing any change, in the above-mentioned variables is governed by the following laws, which have been established from the experimental results.

1. Boyle's law, 2. Charles' law, and 3. Gay-Lussac law.

22.3 Boyle's Law

It states, *"The absolute pressure of a given mass of a perfect gas varies inversely as its volume, when the temperature remains constant."*

In other words, $p \propto \dfrac{1}{V}$ or $pV = $ Constant

or $p_1 V_1 = p_2 V_2 = p_3 V_3 \ldots = $ Constant

where suffixes 1, 2, 3 .... refer to different sets of conditions.

22.4 Charles' Law

It states, *"The volume of a given mass of a perfect gas varies directly as its absolute temperature, when the pressure remains constant."*

---

[*] The reason for this assumption was to simplify the equation by reducing one variable *i.e.*, fluid density.

[**] Sometimes, if there is more than 5% change in density, the fluid is treated as compressible.

In other words, $v \propto T$ or $\dfrac{v}{T}$ = Constant

or $\dfrac{v_1}{T_1} = \dfrac{v_2}{T_2} = \dfrac{v_3}{T_3} = \ldots =$ Constant

22·5 Gay-Lussac Law

It states *"The absolute pressure of a given mass of a perfect gas varies directly as its absolute temperature, when the volume remains constant."*

In other words, $p \propto T$ or $\dfrac{p}{T}$ = Constant

or $\dfrac{p_1}{T_1} = \dfrac{p_2}{T_2} = \dfrac{p_3}{T_3} = \ldots =$ Constant

22·6 General Gas Equation

In the previous articles, we have discussed the gas laws, which give us the relation between two variables, when the third variable is kept constant. But in actual practice, all the three variables *i.e.*, pressure, volume and temperature change simultaneously. In order to deal with all practical cases, the Boyle's law and Charles' law are combined together, which give us the general gas equations in the following two types :

$$pv = C \qquad \ldots \text{(Boyle's law)}$$

and $$\dfrac{v}{T} = C \qquad \ldots \text{(Charle's law)}$$

or $$\dfrac{pv}{T} = C \qquad \ldots \text{(where } C \text{ is constant)}$$

The above equation may be written in a more useful form

$$\dfrac{p_1 v_1}{T_1} = \dfrac{p_2 v_2}{T_2} = \dfrac{p_3 v_3}{T_3} = \ldots \qquad \ldots (i)$$

The above equation may also be written as,

$$pv = mRT$$

or $$p_1 v_1 = mRT_1 \text{ and } p_2 v_2 = mRT_2 \ldots \qquad \ldots (ii)$$

where p = Absolute pressure in kP *i.e.*, kN/m^2,

v = Specific volume in m^3,

m = Mass of the gas,

R = Gas constant in kg-m/kg° K, and

T = Absolute temperature in °K..

Notes : 1. The value of gas constant (*i.e.*, R) is taken as 0·287 kJ/kg K or 287 J/kg K.
2. Sometimes, this equation is expressed as :

$$pv = RT \qquad \ldots \text{(Taking } m \text{ as unity)}$$

or $$p = \dfrac{RT}{v} = wRT = \rho \cdot g\, RT \qquad \ldots \left(\because \text{Vol.}, v = \dfrac{1}{w}\right) \ldots (\because w = \rho \cdot g)$$

Example 22·1. *Determine the volume of 10 kg mass of air at 300 kPa and 100°C. Take $R = 287$ J/kg K.*

Solution. Given : $m = 10$ kg; $p = 300$ kPa $= 300$ kN/m^2; $T = 100°C = 100 + 273 = 373$ K and $R = 287$ J/kg K $= 0.287$ kJ/kg K.

Let $\quad v$ = Volume of the air.

We know that $\quad pv = mRT \quad$ or $\quad 300\,v = 10 \times 0.287 \times 373 = 1071$

$\therefore \quad v = 1071/300 = 3.57$ m^3 **Ans.**

Example 22·2. *A gas occupies a volume of 0·105 m^3 at a temperature of 20°C and a pressure of 150 kPa absolute. Find the final temperature of the gas, if it is compressed to a pressure of 750 kPa absolute, and occupies a volume of 0·04 m^3.*

Solution. Given : $v_1 = 0.105$ m^3; $T_1 = 20°C = 20 + 273 = 293$ K; $p_1 = 150$ kPa $= 150$ kN/m^2; $p_2 = 750$ kPa $= 750$ kN/m^2 and $v_2 = 0.04$ m^3.

Let $\quad T_2$ = Final temperature of the gas.

From the general gas equation, we know that

$$\frac{p_1 v_1}{T_1} = \frac{p_2 v_2}{T_2} \quad \text{or} \quad \frac{150 \times 0.105}{293} = \frac{750 \times 0.04}{T_2}$$

$\therefore \quad T_2 = 750 \times 0.04 \times \dfrac{293}{150 \times 0.105} = 558.1$ K $= 285.1$ °C **Ans.**

22·7 Specific Heats of a Gas

The specific heat of a substance may be broadly defined as the amount of heat required to heat a unit mass of a substance through 1° rise in temperature. All the liquids and solids have only one specific heat. But a gas may have any number of specific heats (say infinite) depending upon the conditions, under which it is heated. The following two specific heats of a gas are important from the subject point of view :

1. Specific heat at constant volume, and
2. Specific heat at constant pressure.

22·8 Specific Heat at Constant Volume

In this case, as the volume of the gas remains the constant. Therefore no work is done by the gas. The whole heat energy of the gas is utilized in increasing the kinetic energy of the gas molecules. It is thus obvious, that in this process of heating the temperature and pressure of the gas will increase. The amount of heat required to raise a unit mass of the gas through 1°, when its volume remains constant, is known as specific heat at constant volume and is denoted by C_v. We know that heat added to the gas,

$$H = \text{Mass} \times \text{Sp. heat at constant volume} \times \text{Rise in temperature}$$
$$= m \cdot C_v (T_2 - T_1)$$

where T_1 and T_2 are the initial and final temperatures of the gas.

21·9 Specific Heat at Constant Pressure

In this case, as the gas is heated its temperature and pressure will increase. The volume of gas will also increase, thereby, counter-balancing any tendency for the pressure to rise. It is thus obvious, that in this process of heating, the heat is utilized for the following two purposes :

1. To raise the temperature of the gas. This heat remains within the body of the gas, and represents the increase in its internal energy. This heat is equal to $m \cdot C_v (T_2 - T_1)$.
2. To do some mechanical work during expansion. This heat is equal to $p(v_2 - v_1)$ in mechanical units and $\dfrac{p(v_2 - v_1)}{J}$ in heat units.

where v_1 and v_2 represent the initial and final volumes of the gas.

A little consideration will show, that this method of heating has a higher value of specific heat than the constant volume method. The amount of heat required to raise a unit mass of the gas through 1°, when its pressure remains constant, is known as specific heat at constant pressure, and is denoted by C_p. We know that heat added to the gas,

$$H = \text{Mass} \times \text{Sp. heat at constant pressure} \times \text{Rise in temperature}$$
$$= m \cdot C_p (T_2 - T_1)$$

where T_1 and T_2 are the initial and final temperatures of the gas.

Example 22·3. *A sample of 3 kg of an ideal gas was heated from an initial temperature of 30°C to 80°C. Find the quantity of heat added, if the gas was heated (a) at constant volume, and (b) at constant pressure. Take $C_v = 0.72$ and $C_p = 1.0$.*

Solution. Given : $m = 3$ kg; $T_1 = 30°C = 30 + 273 = 303$ K; $T_2 = 80°C = 80 + 273 = 353$ K; $C_v = 0.72$ and $C_p = 1.0$.

(a) *Quantity of heat added, if the gas was heated at constant volume*

We know that the quantity of heat added, if the gas was heated at constant volume,

$$H_1 = m \cdot C_v (T_2 - T_1) = 3 \times 0.72 \, (353 - 303) = 108 \text{ kJ Ans.}$$

(b) *Quantity of heat added, if the gas was heated at constant pressure*

We also know that the quantity of heat added, if the gas was heated at constant pressure,

$$= m \cdot C_p (T_2 - T_1) = 3 \times 1.0 \, (353 - 303) = 150 \text{ kJ Ans.}$$

22·10 Relation Between Specific Heats

It is an important relation in the field of Thermodynamics and Fluid Mechanics, and is the basic relation for all sorts of calculations. Consider a gas enclosed in a container and being heated gradually, first at a constant volume and then at constant pressure as discussed in the previous articles.

Let
- m = Mass of the gas,
- T_1 = Initial temperature of the gas,
- T_2 = Final temperature of the gas,
- C_p = Specific heat at constant pressure,
- C_v = Specific heat at constant volume,
- v_1 = Initial volume of the gas,
- v_2 = Final volume of the gas, and
- p = Constant pressure.

We know that the amount of heat supplied to the gas, when heated at constant volume,

$$= m \cdot C_v (T_2 - T_1) \qquad \ldots(i)$$

and the amount of heat supplied to the gas, when heated at constant pressure

$$= m \cdot C_p (T_2 - T_1) \qquad \ldots(ii)$$

We also know that some of the heat supplied to the gas (when heated at constant pressure) is utilized in doing some mechanical work, whereas rest of the heat remains within the body of the gas, and increases its internal energy.

∴ Heat utilized for external work

$$= p \, (v_2 - v_1)$$

and heat utilized for increase in internal energy

$$= m \cdot C_v (T_2 - T_1)$$

Mechanics of Compressible Flow

$$\therefore \quad m \cdot C_p (T_2 - T_1) = p(v_2 - v_1) + m \cdot C_v (T_2 - T_1) \quad \ldots(iii)$$

From the general gas equation, we know that

$$pv = mRT$$

If the gas is heated at constant pressure, then

$$pv_1 = mRT_1 \quad \ldots\text{(For initial condition)}$$

and $\quad pv_2 = mRT_2 \quad \ldots\text{(For final condition)}$

$$\therefore \quad p(v_2 - v_1) = mR(T_2 - T_1)$$

Now substituting the value of $p(v_2 - v_1)$ in equation (iii),

$$m \cdot C_p (T_2 - T_1) = mR(T_2 - T_1) + m C_v (T_2 - T_1)$$

$$C_p = R + C_v$$

or $\quad R = C_p - C_v$

Notes : 1. It is an important result, as it proves that the characteristic constant of a gas (R) is directly proportional to the difference of two specific heats.

2. The value of R is taken as 0·287 kJ/kg K or 287 J/kg K.

22.11 Ratio of Specific Heats

The ratio of the above two specific heats of a gas is also very important constant in the field of Thermodynamics and Fluid Mechanics. It is represented by the Greek letter gamma (γ) and is also known as adiabatic index. Such that

$$\gamma = \frac{C_p}{C_v}$$

Since C_p is always greater than C_v, therefore the value of γ is always greater than unity.

Note : The values of C_p, C_v, R and γ depend upon the type of gas and its temperature. The value of γ for air at usual temperature is taken as 1.4. The values of R and γ, for different gases of common use at 20°C are given in the following table :

Table 22·1

S.No.	Name of gas	C_p (kJ/kg K)	C_v (kJ/kg K)	$R = C_p - C_v$ (kJ/kg K)	$\gamma = \dfrac{C_p}{C_v}$
1.	Air	1.00	0.72	0.28	1.40
2.	Nitrogen	1.04	0.75	0.29	1.40
3.	Carbon dioxide	0.85	0.66	0.19	1.30
4.	Oxygen	0.91	0.65	0.26	1.40
5.	Helium	5.23	3.15	2.08	1.66
6.	Hydrogen	14.26	10.13	4.13	1.40

Example 22·4. *Find the values of gas constant and adiabatic index for methane, where values of specific heats at constant pressure and volume are 2·17 and 1·65 respectively.*

Solution. Given $C_p = 2\cdot17$ and $C_v = 1\cdot65$.

Value of gas constant for methane

We know that value of gas constant for methane,

$$R = C_p - C_v = 2\cdot17 - 1\cdot65 = 0\cdot52 \text{ kJ/kg K} \quad \text{Ans.}$$

Value of adiabatic index for methane

We also know that value of adiabatic constant for methane,

$$\gamma = \frac{C_p}{C_v} = \frac{2.17}{1.65} = 1.32 \quad \text{Ans.}$$

EXERCISE 22·1

1. Find the volume of 2 kg mass of air at 100 kPa and 310 K. Take value of the gas constant as 0·287 kJ/ kg K. (Ans. 0.178 m³)
2. A gas occupying 1·2 m³ volume at 350 K under a pressure of 500 kPa is expanded to a pressure of 350 kPa. If the air now occupies a volume of 1·5 m³, find the change in temperature. (Ans. 44 K)
3. A 1.5 kg sample of an ideal gas is cooled from a temperature of 100°C to 35°C in a process. Calculate the amount of heat lost by the gas if it is cooled at constant volume. (Ans. 97·5 kJ)
4. Determine the volumes of gas constant and adiabatic index for ammonia, if the value of its specific heat at constant pressure is 2·18 kJ/kg K and that at constant volume is 1·69 kJ/kg K. (Ans. 0·49 kJ/kg; 1·29)

22·12 Isothermal Process

A process, in which the temperature of the working substance, *i.e.*, gas remains the same during its expansion or compression, is called an isothermal process. It is thus obvious, that in an isothermal process :

1. There is no change in temperature, and
2. There is no change in internal energy.

A little consideration will show, that isothermal process is governed by Boyle's law and thus the isothermal equation of a perfect gas is :

$$pv = \text{Constant}$$

or $\quad p_1 v_1 = p_2 v_2 = p_3 v_3 \ldots\ldots$

Moreover, as there is no change in internal energy, it is thus obvious, that during the isothermal process the heat absorbed by the gas is equal to the work done by the gas. Thus work done during an isothermal expansion is given by the relation,

$$W = H = 2.3 \, p_1 v_1 \log r \qquad \ldots\left(\text{where } r = \frac{v_2}{v_1}\right)$$

Example 22·5. *A certain quantity of air has a volume of 0·4 m³ at a pressure of 500 kPa and a temperature of 80°C. It is expanded in a cylinder at a constant temperature to a pressure of 100 kPa. Determine the work done by the air during expansion.*

Solution. Given : $v_1 = 0.4 \, m^3$; $p_1 = 500 \, kPa = 500 \, kN/m^2$; *$T = 80°$ C and $p_2 = 100$ kPa $= 100 \, kN/m^2$.

Let $\qquad v_2 = $ Final volume of the air.

From the isothermal (constant temperature) expansion, we know that

$$p_1 v_1 = p_2 v_2 \quad \text{or} \quad 500 \times 0.4 = 100 \, v_2$$

∴ $\qquad v_2 = \dfrac{500 \times 0.4}{100} = 2 \, m^3$

---

*Superfluous data.

Mechanics of Compressible Flow

and expansion ratio, $\quad r = \dfrac{v_2}{v_1} = \dfrac{2}{0.4} = 5$

We also know that work done during isothermal expansion,

$$w = 2.3\, p_1 v_1 \log r = 2.3 \times 500 \times 0.4 \log 5 \text{ kJ}$$
$$= 460 \times 0.699 = 321.5 \text{ kJ } \textbf{Ans.}$$

22·13 Adiabatic Process

A process, in which the working substance, (*i.e.*, gas) neither receives nor gives out any heat to its surroundings during its expansion or compression, is called an adiabatic process. It is thus obvious, that in an adiabatic process :

1. No heat leaves or enters the gas,
2. The temperature of the gas changes, as the work done is at the cost of internal energy,
3. The change in internal energy is equal to the mechanical work done.

The adiabatic equation of a perfect gas is

$$pv^\gamma = \text{Constant} \quad \text{or} \quad p_1 v_1^\gamma = p_2 v_2^\gamma = \ldots\ldots C$$

$$\therefore \quad \dfrac{p_1}{p_2} = \left(\dfrac{v_2}{v_1}\right)^\gamma$$

Notes. The various forms of adiabatic equations are :

$$Tv^{\gamma-1} = \text{Constant} \quad \text{or} \quad T_1 v_1^{\gamma-1} = T_2 v_2^{\gamma-1}$$

$$\therefore \quad \dfrac{T_1}{T_2} = \left(\dfrac{v_2}{v_1}\right)^{\gamma-1}$$

$$\dfrac{T^\gamma}{p^{\gamma-1}} = \text{Constant} \quad \text{or} \quad \dfrac{T^\gamma}{p_1^{\gamma-1}} = \dfrac{T_2^\gamma}{p_2^{\gamma-1}}$$

$$\therefore \quad \dfrac{T_1}{T_2} = \left(\dfrac{p_1}{p_2}\right)^{\frac{\gamma-1}{\gamma}}$$

where γ is the ratio of two specific heats. *i.e.*, $\dfrac{C_p}{C_v}$.

The work done during an adiabatic expansion is given by the relation,

$$W = \dfrac{p_1 v_1 - p_2 v_2}{\gamma - 1} \qquad \ldots\text{(when } p_1, v_1, p_2 \text{ and } v_2 \text{ are given)}$$

$$= \dfrac{R(T_1 - T_2)}{\gamma - 1} \qquad \ldots\text{(when } T_1 \text{ and } T_2 \text{ are given)}$$

Note : If the adiabatic process is reversible, it is called an isentropic process. The equation for isentropic process is the same as that of adiabatic process.

Example 22·6. *A cylinder contains 10 m³ of air at 30°C and at an absolute pressure of 21 kPa. The air is compressed adiabatically to a volume of 2 m³. Determine the final pressure of the air assuming C_p/C_v as 1·4.*

Solution : Given : $v_1 = 10 \text{ m}^3$; *$T_1 = 30°$ C; $p_1 = 21$ kPa $= 21$ kN/m²; $v_2 = 2 \text{ m}^3$ and $\gamma = 1.4$.

Let $\qquad\qquad\qquad p_2 = $ Final pressure of the air.

---

* Superfluous data.

We know that for adiabatic compression,

$$\frac{p_1}{p_2} = \left(\frac{v_2}{v_1}\right)^\gamma \quad \text{or} \quad \frac{21}{p_2} = \left(\frac{2}{10}\right)^{1\cdot 4} = (0\cdot 2)^{1\cdot 4} = 0\cdot 105$$

∴ $p_2 = 21/0\cdot 105 = 200 \text{ kN/m}^2 = 200 \text{ kPa}$ **Ans.**

Example 22·7. *Find the final temperature, when a given mass of gas at 0°C is suddenly compressed to one-tenth of its initial pressure. Take γ for the gas as 1·42.*

Solution. Given : $T = 0°C = 273$ K; $p_1 = 10\ p_2$ and $\gamma = 1\cdot 42$.

Let $\qquad T_2 = $ Final temperature of the gas.

We know that for adiabatic compression,

$$\frac{T_1}{T_2} = \left(\frac{p_1}{p_2}\right)^{\frac{\gamma-1}{\gamma}} \quad \text{or} \quad \frac{273}{T_2} = \left(\frac{p_1}{10\,p_1}\right)^{\frac{1\cdot 42 - 1}{1}} = (0\cdot 1)^{0\cdot 296} = 0\cdot 506$$

∴ $T_2 = 273/0\cdot 506 = 539\cdot 5$ K $= 266\cdot 5°C$ **Ans.**

Example 22·8. *One litre of a gas at 100°C is suddenly expanded to 2 litres according to the law $pv^{1\cdot 35} = C$. Find the change in temperature of the gas.*

Solution. Given : $v_1 = 1$ litre; $T_1 = 100°C = 373$ K; $v_2 = 2$ litres and $\gamma = 1\cdot 35$.

Let $\qquad T_2 = $ Final temperature of the gas.

We know that for adiabatic expansion,

$$\frac{T_1}{T_2} = \left(\frac{v_2}{v_1}\right)^{\gamma-1} \quad \text{or} \quad \frac{373}{T_2} = \left(\frac{2}{1}\right)^{1\cdot 35 - 1} = (2)^{0\cdot 35} = 1\cdot 275$$

∴ $T_2 = 373/1\cdot 275 = 292\cdot 5$ K $= 19\cdot 5°C$

and change in temperature,

$$T = T_1 - T_2 = 100 - 19\cdot 5 = 80\cdot 5\ °C \quad \textbf{Ans.}$$

22·14 Bulk Modulus of a Fluid

The bulk modulus of a fluid is the ratio between the increase of pressure, and the volumetric strain, caused by this pressure increase. It may be noted that this ratio is applied to liquids and gases.

Let $\qquad p = $ Original pressure of the fluid,

$\qquad\qquad dp = $ Increase of pressure,

$\qquad\qquad v = $ Original volume of the fluid, and

$\qquad -dv^* = $ Change of volume.

∴ Volumetric strain $= \dfrac{\text{Change in volume}}{\text{Original volume}} = \dfrac{-dv}{v} = -\dfrac{dv}{v}$

and bulk modulus, $\qquad K = \dfrac{\text{Increase of pressure}}{\text{Volumetric strain}} = \dfrac{dp}{-\dfrac{dv}{v}}$

Note : The above equation is, sometimes, written as :

$$\frac{dp}{dv} = \frac{K}{v}$$

---

*Minus sign indicates that the volume of a fluid decreases, as the pressure increases.

Example 22·9. *In an experiment, the pressure of a liquid is increased from 65 kPa to 140 kPa. The volume of liquid was found to decrease by 1·5%. Find the bulk modulus of the liquid.*

Solution. Given : $p_1 = 65$ kPa $= 65$ kN/m^2; $p_2 = 140$ kPa $= 140$ kN/m^2 and $dv = 1.5\%$ of v $= 0.015 \, V$.

We know that change in pressure,

$$dp = p_2 - p_1 = 140 - 65 = 75 \text{ kN/m}^2$$

and volumetric strain, $\dfrac{dv}{v} = \dfrac{0.015 \, v}{v} = 0.015$

∴ Bulk modulus of the liquid,

$$K = \frac{dp}{\dfrac{dv}{v}} = \frac{75}{0.015} = 5000 \text{ kN/m}^2 = 5 \times 10^6 \text{ N/m}^2$$

$$= 5 \text{ MPa} \quad \textbf{Ans.}$$

EXERCISE 22·2

1. A sample of 0·1 m^3 of air at a pressure of 50 kPa is expanded isothermally to 0·5 m^3. Calculate the heat supplied to the air during the process. **(Ans. 80·4 kJ)**
2. A container contains a certain volume of gas at 20 kPa. Find the final pressure of the gas, when it is suddenly compressed to one-fourth of its volume. Take $\gamma = 1.4$. **(Ans. 139·3 kPa)**
3. A certain amount of gas at 150°C is expanded adiabatically from a pressure of 100 kPa, until its pressure is 40 kPa. Find the final temperature of the gas. Take ratio of specific heats of the gas as 1·4. **(Ans. 52·4 °C)**
4. Find the change in volume of 1 m^3 of water, when it is subjected to an increase of pressure of 200 kPa. Take bulk modulus of water as 2·2 GPa. **(Ans. 0·091 m^3)**

QUESTIONS

1. Explain Boyle's law, Charles' law and Gay-Lussac law.
2. Obtain a relationship between the specific heats of a gas.
3. Differentiate between 'isothermal process' and 'adiabatic process'.
4. Derive an equation for the work done in 'isothermal process' and 'adiabatic process'.
5. What do you understand by the term 'bulk modulus'? Explain it.

OBJECTIVE TYPE QUESTIONS

1. If the temperature remains constant, the volume of a given mass of a gas is inversely proportional to its pressure. This statement is known as
   (a) Boyle's law (b) Charles' law (c) Gay-Lussac law (d) Joule's law.
2. The value of gas constant (R) is
   (a) 0·287 J/kg K (b) 2·87 J/kg K (c) 28·7 J/kg K (d) 287 J/kg K.
3. The ratio of specific heats of a gas at constant pressure (C_p) to that at constant volume (C_v) is
   (a) less than unity (b) equal to unity (c) more than unity (d) either 'a' or 'b'.
4. When a gas is heated or expanded in such a way that the product of its pressure and volume remains constant, it is called
   (a) isothermal process (b) isobaric process (c) adiabatic process (d) none of these

ANSWERS

1. (a)   2. (d)   3. (c)   4. (a)

23

Compressible Flow of Fluids

1. Introduction. 2. Equation of Continuity for Flowing Gases. 3. Energy Equation for Flowing Gases. 4. Flow of Gas through an Orifice or Nozzle. 5. Relation between Pressure and Height. 6. Velocity of Sound Wave. 7. Mach Number and its Importance. 8. Mach Wave. 9. Mach Wave when the Body is Moving at a Subsonic Velocity. 10. Mach Wave when the Body is Moving at a Sonic Velocity. 11. Mach Wave when the Body is Moving at a Supersonic Velocity. 12. Stagnation Pressure. 13. Measurement of Air Speed.

23.1 Introduction

In the previous chapter, we have discussed the mechanics of compressible flow. But in this chapter, we shall discuss the applications of this mechanics in the flow of fluids.

23.2 Equation of Continuity for Flowing Gases

We have discussed in Art. 6·3 the equation of continuity, in which we assumed the flowing fluid to be incompressible. We obtained a general equation of flow on the basis of total quantity of flow. *i.e.*,

$$Q = a_1v_1 = a_2v_2 = a_3v_3 = ...$$

But in a compressible fluid flow, the mass of fluid flowing through any section remains constant. Now consider a tapering pipe through which some compressible fluid is flowing.

Let a_1 = Area of pipe at section 1,

V_1 = Velocity of fluid at section 1,

ρ_1 = Mass density of fluid at section 1,

Similarly, a_2, V_2, ρ_2 = Corresponding values at section 2, and

a_3, V_3, ρ_3 = Corresponding values at section 3.

∴ Mass of fluid passing through section 1,

$$m_1 = a_1v_1\rho_1$$

Similarly, mass of fluid passing section 2,

$$m_2 = a_2v_2\rho_2$$

and mass of fluid passing through section 3,

$$m_3 = a_3v_3\rho_3$$

In a compressible flow, the mass of the flowing fluid passing through sections 1-1, 2-2 and 3-3 is the same. Therefore

$$m_1 = m_2 = m_3 = ...$$

∴ $$a_1v_1\rho_1 = a_2v_2\rho_2 = a_3v_3\rho_3 = ...$$

or $$a_1v_1w_1 = a_2v_2w_2 = a_3v_3w_3 = ...\qquad ...(\text{where } w = \rho \cdot g)$$

23·3 Energy Equation for Flowing Gases

We have discussed in Art. 8·8 the Bernoulli's equation for liquids, in which we assumed the flowing fluid to be incompressible. But in a compressible fluid flow, the *Bernoulli's equation does not hold good, as the specific weight of the gas changes with the change in temperature and pressure. Moreover, any change in the heat appreciably changes the internal energy of the flowing fluid.

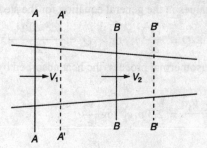

Fig. 23·1. Tapering pipe.

Consider a tapering pipe, through which some gas is flowing as shown in Fig. 23·1.

Let
- p_1 = Pressure of gas at section 1,
- T_1 = Temperature of gas at section 1,
- V_1 = Velocity of gas at section 1,
- a_1 = Area of pipe at section 1, and
- p_2, T_2, V_2, a_2 = Corresponding values at section 2.

Now consider 1 kg of gas between the two sections AA and BB move to $A'A'$ and $B'B'$ through very small lengths dl_1 and dl_2 as shown in Fig. 23·1. This movement of gas between AA and BB is equivalent to the movement of the gas between AA and $A'A'$ to BB amd $B'B'$, the remaining gas between $A'A'$ and BB being unaffected.

We know that the work done by pressure at section 1,
$$= p_1 a_1 dl_1 = p_1 v_1$$

where v_1 (equal to $a_1 dl_1$) is the volume of 1 kg of gas at section 1.

Similarly, work done by pressure at section 2.
$$= p_2 a_2 dl_2 = p_2 v_2$$

Now as per Bernoulli's equation, equating the total energy above the absolute zero temperature at the two sections and ignoring the change in altitude (being negligible) in heat units. *i.e.*,

Work done on gas + Heat absorbed by gas
= Gain in internal energy + Gain of kinetic energy

$$p_1 v_1 - p_2 v_2 + Q = E_2 - E_1 + \frac{V_2^2 - V_1^2}{2g} \qquad ...(i)$$

or
$$p_1 v_1 + E_1 + \frac{V_1^2}{2g} + Q = p_2 v_2 + E_2 + \frac{V_2^2}{2g} \qquad ...(ii)$$

It is the general equation for the steady flow of a gas through a pipe. Now we shall study this flow of a gas in the following two processes :

1. Isothermal process.
2. Adiabatic process.

---

* In a compressible flow, the Bernoulli's equation may be used in a slightly modified form. *i.e.*, the total energy above the absolute zero temperature at two sections remains the same.

1. Isothermal process

We know that in an isothermal process,
$$p_1 v_1 = p_2 v_2$$
Moreover, as there is no change in internal energy, therefore
$$E_1 = E_2$$
Now substituting these values in the general equation for the steady flow of a gas,
$$\frac{V_1^2}{2g} + Q = \frac{V_2^2}{2g} \quad \text{or} \quad Q = \frac{V_2^2 - V_1^2}{2g} \qquad \text{...(In heat units)}$$

We also know that in an isothermal process, the heat absorbed by the gas is equal to the work done. Therefore
$$\frac{V_2^2 - V_1^2}{2g} = 2 \cdot 3 \, p_1 v_1 \log \frac{v_2}{v_1} \qquad \text{...(In mechanical units)}$$
$$= 2 \cdot 3 \, p_1 v_1 \log \frac{p_1}{p_2} \qquad \text{...}(\because p_1 v_1 = p_2 v_2)$$

2. Adiabatic process

We know that in an adiabatic process, no heat leaves or enters the gas. Therefore $Q = 0$. Moreover, as the change in internal energy is equal to the mechanical work, therefore
$$E_1 - E_2 = \frac{p_1 v_1 - p_2 v_2}{(\gamma - 1)} \qquad \text{...(In heat units)}$$

Now substituting the values in the general equation (i),
$$p_1 v_1 - p_2 v_2 = -\frac{p_1 v_1 - p_2 v_2}{(\gamma - 1)} + \frac{V_2^2 - V_1^2}{2g}$$

or
$$\frac{V_2^2 - V_1^2}{2g} = p_1 v_1 - p_2 v_2 + \frac{p_1 v_1 - p_2 v_2}{\gamma - 1}$$
$$= (p_1 v_1 - p_2 v_2) \left[1 + \frac{1}{\gamma - 1} \right]$$
$$= (p_1 v_1 - p_2 v_2) \times \frac{\gamma}{\gamma - 1}$$
$$= p_1 v_1 \times \frac{\gamma}{\gamma - 1} \left[1 - \frac{p_2 v_2}{p_1 v_1} \right]$$
$$= p_1 v_1 \times \frac{\gamma}{\gamma - 1} \left[1 - \left(\frac{p_2}{p_1}\right)^{\frac{\gamma-1}{\gamma}} \right]$$
$$= RT_1 \times \frac{\gamma}{\gamma - 1} \left[1 - \left(\frac{p_2}{p_1}\right)^{\frac{\gamma-1}{\gamma}} \right] \qquad \text{...}(\because pv = RT)$$

The above isothermal and adiabatic energy equations are also called Bernoulli's equations for gases.

Compressible Flow of Fluids

Example 23·1. *The nitrogen gas flows through a duct, which has a change in its cross-section. At a section, the velocity of the gas is 40 m/s at a pressure of 80 kPa and at a temperature of 37°C. Determine the velocity of the gas at a section, where the pressure is 130 kPa and flow of gas is adiabatic in nature. Take R = 297 J/kg K and γ = 1·4.*

Solution. Given : $V_1 = 40$ m/s ; $p_1 = 80$ kPa $= 80$ kN/m²; $T_1 = 37°C = 37 + 273 = 310$ K; $p_2 = 130$ kPa $= 130$ kN/m² ; $R = 297$ J/kg K and $\gamma = 1\cdot 4$.

Let V_2 = Velocity of the gas at section 2.

We know that

$$\frac{V_2^2 - V_1^2}{2g} = RT_1 \times \frac{\gamma}{\gamma - 1}\left[1 - \left(\frac{p_2}{p_1}\right)^{\frac{\gamma-1}{\gamma}}\right]$$

$$\frac{V_2^2 - (40)^2}{2 \times 9\cdot 81} = 0\cdot 297 \times 310 \times \frac{1\cdot 4}{1\cdot 4 - 1}\left[1 - \left(\frac{130}{80}\right)^{\frac{1\cdot 4 - 1}{1\cdot 4}}\right] = -48\cdot 34$$

$V_2^2 - 1600 = -48\cdot 34 \times 19\cdot 62 = -948\cdot 4$

∴ $V_2 = -948\cdot 4 + 1600 = 651\cdot 6$ or $V_2 = 25\cdot 5$ m/s **Ans.**

23·4 Flow of Gas through an Orifice or Nozzle

The flow of a gas through an orifice or a nozzle is regarded as an isothermal process, if pressure drop of the gas is small. It is regarded as an adiabatic process, if the pressure drop is large. But in actual practice, there is always a great pressure drop of the gas, when it flows from a reservoir through an orifice or nozzle. That is why, this process is taken as an adiabatic process.

In the previous article, we have discussed the Bernoulli's equation for gases, in which we obtained the relation for adiabatic process :

$$\frac{V_2^2 - V_1^2}{2g} = p_1 v_1 \times \frac{\gamma}{\gamma - 1}\left[1 - \left(\frac{p_2}{p_1}\right)^{\frac{\gamma-1}{\gamma}}\right]$$

In this case, the velocity of gas in the reservoir $(V_1) = 0$ (as the flow commences from rest).

Let V = Velocity of the gas through the orifice or nozzle.

Substituting these values in the general equation,

$$\frac{V^2}{2g} = p_1 v_1 \times \frac{\gamma}{\gamma - 1}\left[1 - \left(\frac{p_2}{p_1}\right)^{\frac{\gamma-1}{\gamma}}\right]$$

or

$$V = \sqrt{2g\, p_1 v_1 \times \frac{\gamma}{\gamma - 1}\left[1 - \left(\frac{p_2}{p_1}\right)^{\frac{\gamma-1}{\gamma}}\right]}$$

We know that the volume of the gas flowing per second,

$$= aV = a\sqrt{2g\, p_1 v_1 \times \frac{\gamma}{\gamma-1}\left[1 - \left(\frac{p_2}{p_1}\right)^{\frac{\gamma-1}{\gamma}}\right]}$$

23·5 Relation between Pressure and Height

We have already discussed the relation between the pressure and height of an incompressible liquid in chapter 2, in which we obtained the relation :

$$p = wh \qquad \ldots(i)$$

where
p = Pressure,
w = Specific weight of the liquid, and
h = Height of the liquid.

But in the case of compressible fluids, the specific weight does not remain constant. It has also been experienced, that the pressure of air decreases with the increase of height from the surface of the earth (or technically from the mean sea level). The relation between pressure and height is obtained by considering it as an isothermal process, as the isothermal conditions exist near the surface of the earth. We know that in an isothermal process,

$$pv = p_1 v_1 = p_2 v_2$$

$$\frac{p}{w} = \frac{p_1}{w_1} = \frac{p_2}{w_2} \qquad \ldots\left(\because v \propto \frac{1}{w}\right)$$

$$\therefore \quad \frac{1}{w} = \frac{p_1}{p \cdot w_1}$$

Since the pressure of air decreases with the increase of height, therefore equation (i), in this case, is to be used as :

$$p = -wh$$

or $\qquad dp = -w \cdot dh \qquad \ldots$(In differential form)

$$\therefore \quad -dh = \frac{dp}{w} = \frac{p_1}{p \cdot w_1} \times dp$$

It is thus obvious that the change of pressure from a height h_1 to h_2 is obtained from the relation:

$$-\int_{h_1}^{h_2} dh = \int_{p_1}^{p_2} \frac{p_1}{p \cdot w_1} \times dp$$

$$= \frac{p_1}{w_1} \int_{p_1}^{p_2} \frac{1}{p} \times dp$$

$$= \frac{p_1}{w_1} \left[\log_e p\right]_{p_1}^{p_2}$$

$$-(h_2 - h_1) = \frac{p_1}{w_1} \left(\log_e \frac{p_2}{p_1}\right)$$

$$\therefore \quad \log_e \left(\frac{p_2}{p_1}\right) = -\frac{w_1}{p_1}(h_2 - h_1)$$

Compressible Flow of Fluids

Now expressing the above equation in terms of corresponding logarithm to the base 10,

$$2.3 \log\left(\frac{p_2}{p_1}\right) = -\frac{w_1}{p_1}(h_2 - h_1) = -\frac{h_2 - h_1}{RT} \quad \ldots(\because pv = RT)$$

Example 23·2. *A helicopter is travelling at a height of 1500 m above the ground level. Find the pressure of air on it, if the pressure at the ground level is 100 kPa at 20°C. Take R = 0.287 kJ/kg K.*

Solution. Given : $h_2 = 1500$ m; $h_1 = 0$ (due to ground level); $p_1 = 100$ kPa; $T_1 = 20°C = 20 + 273 = 293$ K and $R = 0.287$ kJ/kg K = 287 J/kg K.

Let $\qquad p_2$ = Pressure of air on the helicopter in kPa.

We know that

$$2.3 \log\left(\frac{p_2}{p_1}\right) = -\frac{h_2 - h_1}{RT} = \frac{1500 - 0}{287 \times 293} = -0.0178$$

$\therefore \qquad \log\left(\frac{p_2}{p_1}\right) = -\frac{0.0178}{2.3} = -0.00774$

or $\qquad \dfrac{p_2}{p_1} = $ antilog $(-0.00774) = 0.982$

$$\frac{p_2}{100} = 0.982 \quad \text{or} \quad p_2 = 0.982 \times 100 = 98.2 \text{ kPa } \textbf{Ans.}$$

23·6 Velocity of Sound Wave

It has been observed that a minor disturbance in one end of a rigid body (whose molecules are close together) is immediately transmitted to the other end. But in case of fluids, where molecules are relatively apart, each molecule has to travel through a small distance, before it collides and transmits the disturbance to the next molecule. Thus in case of fluids (liquids and gases) every molecule will first travel through a small disturbance and then transmit the disturbance. It is thus obvious, that the velocity with which the disturbance travels depends upon the distance between two molecules, which further depends upon the pressure and density of the fluid. *i.e.*, on the elastic properties of the fluid.

It will be interesting to know that every disturbance due to movement of molecules causes some difference of pressure in the medium. This pressure difference depends upon the bulk modulus and density of the material.

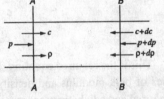

Fig. 23·2. Velocity of sound wave.

Now consider a tube of fluid, through which a pressure wave is being transmitted from left to right as shown in Fig. 23·2.

Let $\qquad a$ = Unit cross-sectional area of the tube,

v = Volume of fluid compressed by wave per second at A,

dv = Volume of fluid compressed between A and B,

c = Celerity* (*i.e.*, velocity) of pressure wave at A,

p = Pressure of the wave at A,

ρ = Density of fluid at A,

---

* The modern concept is to use the word celerity instead of velocity, which only indicates that it is the disturbance or sound that moves and not the fluid.

$(c + dc)$, $(p + dp)$, $(\rho + d\rho)$ = Corresponding values at B.

We know that volume of compressed fluid at section A,

$$v = ac = c \qquad \text{...($\because$ a is unity)},$$

and at section B, the volume of fluid compressed per second

$$= v + dv = a(c + dc)$$
$$= v + dc \qquad \text{...(Substituting $a = 1$ and $c = V$)}$$

$\therefore \qquad dv = dc \qquad \text{...(i)}$

From the equation of continuity, we also know that

$$\rho c a = (\rho + d\rho)(c + dc) a$$

or $\qquad \rho c = (\rho + d\rho)(c + dc)$

$$= \rho c + \rho . dc + c . d\rho + d\rho . dc$$
$$= \rho c + \rho . dc + c . d\rho \qquad \text{...(Neglecting $d\rho . dc$)}$$

$\therefore \qquad \rho . dc = - c . d\rho \qquad \text{...(ii)}$

We also know that

$$\text{Force} = \text{Mass} \times \text{Change of velocity}$$
$$[p - (p + dp)] a = \rho c a \times dc$$
$$- dp = \rho c . dc = \rho c . dv \qquad \text{...($\because$ $dc = dv$)}$$

$\therefore \qquad \dfrac{dp}{dv} = - \rho c \qquad \text{...(iii)}$

We have seen in Art. 22·14 that

$$\dfrac{dp}{dv} = - \dfrac{K}{v}$$

$\therefore \qquad -\dfrac{K}{v} = - \rho c$

or $\qquad \dfrac{K}{c} = \rho c \qquad \text{...($\because$ $V = c$)}$

$\therefore \qquad c = \sqrt{\dfrac{K}{\rho}} \qquad \text{...(iv)}$

This is the required equation for celerity, of a pressure wave in a fluid, which depends upon the values of bulk modulus and density of the fluid. Now we shall study this equation for celerity in isothermal and adiabatic processes.

Isothermal process

We know that in an isothermal process

$$pv = C \qquad \text{...(where C is constant)}$$

Differentiating the above equation,

$$p.dv + v.dp = 0 \quad \text{or} \quad p.dv = - v.dp$$

$\therefore \qquad p = - \dfrac{dp}{\dfrac{dv}{v}}$

We have already discussed in Art. 22·14 that $-\dfrac{dp}{\dfrac{dv}{v}}$ is bulk modulus (K) for the gas. Therefore

$$p = K$$

Now substituting the value of $p = K$ in equation (iv),

$$c = \sqrt{\dfrac{p}{\rho}} = \sqrt{R.T} \qquad \qquad ...(\because\ p = PRT)$$

This equation was originally deduced by Newton for the velocity of sound in a gas. Later on, it was found that the equation does not give correct result.

Adiabatic process

We know that in an adiabatic process,

$$pv^\gamma = C \qquad \qquad ...(\text{where } C \text{ is a constant})$$

Differentiating the above equation,

$$\gamma p^{\gamma-1} dv + v^\gamma . dp = 0$$

or $\qquad\qquad \gamma p v^{\gamma-1} . dv = -v^\gamma . dp$

$\therefore \qquad\qquad \gamma p = -\dfrac{dp}{\dfrac{dv}{v}} = K \qquad\qquad$ ...(where K is bulk modulus)

Now substituting γp instead of $p = K$ in equation (iv),

$$c = K\sqrt{\gamma RT} \quad \text{Ans.}$$

This equation was deduced by Laplace for the velocity of sound in a gas. The above equation gives the value of c as 340 m/s in atmosphere at sea level at 15°C. This value is very close to the experimental value. It is thus obvious, that isothermal process is not applicable for the velocity of sound or propagation of any disturbance through the atmosphere. The following arguments may also be given for the purpose :

(a) The temperature of air does not remain constant.

(b) For such a high velocity of sound, there is no time for any appreciable heat transfer.

Note : Sometimes the above relation is written as :

$$c = \sqrt{\gamma RT}$$

Example 23·3. *Find the speed of sound wave in dry air at sea level and at a temperature of 20° C. Take $\gamma = 1\cdot 4$ and $R = 296$ J/kg K.*

Solution. Given : $T = 20°C = 20 + 273 = 293$ K; $\gamma = 1\cdot 4$ and $R = 296$ J/kg K.

We know that velocity of sound,

$$C = \sqrt{\gamma RT} = \sqrt{1\cdot 4 \times 296 \times 293} = 348\cdot 5 \text{ m/s} \quad \text{Ans.}$$

EXERCISE 23·1

1. A gas flows through a tapering pipe. At one section, it has a velocity of 42 m/s under a pressure of 50 kPa and at 300 K. What is the velocity of the gas at a section where the pressure is 100 kPa ? Take $R = 0\cdot 287$ kJ/kg K and $\gamma = 1\cdot 4$. (Ans 21·6 m/s)
2. An army plane is flying at a height of 1200 m above the ground level. Find the pressure on the plane, when the pressure at the ground level is 80 kPa at 27° C. Take $R = 287$ J/kg K. (Ans 79·5 kPa)'
3. Find the velocity of a sound wave at a temperature of 305 K. Take $\gamma = 1\cdot 4$ and $R = 287$ J/kg K. (Ans 350·1 m/s)

23·7 Mach Number and its Importance

In the previous articles, we have discussed the relation for finding the celerity of the sound wave and the stagnation pressure. The ratio of velocity of fluid, in an undisturbed stream, to the celerity of the sound wave is known as Mach number. Mathematically Mach number,

$$M_N = \frac{V}{c}$$

The Mach number gives us an important information about the type of flow. In general, the flow of a fluid is divided into the following four types depending upon the Mach number :

1. *Subsonic flow*

   When the Mach number is less than unity, the flow is called a subsonic flow.

2. *Sonic flow*

   When the Mach number is equal to unity, the flow is called a sonic flow.

3. *Supersonic flow*

   When the Mach number is between 1 and 6, the flow is called a supersonic flow.

4. *Hypersonic flow*

   When the Mach number is more than 6, the flow is called a hypersonic flow.

23·8 Mach Wave

In the previous article, we have discussed the Mach number and its importance in finding out the type of flow (*i.e.*, subsonic flow, sonic flow, supersonic flow and hypersonic flow). As a matter of fact, the hypersonic flow is merely a theoretical case. Thus we shall discuss the effect of Mach number on the propagation of waves in a compressible flow for the first three types of flows only.

23·9 Mach Wave when the Body is Moving at a Subsonic Velocity

Consider the nose of a projectile, moving with a subsonic velocity in a fluid from left to right, at some instant to be at A. We know that the nose of the body impinging on the fluid, causes a disturbance in the form of a spherical pressure wave. This wave travels radially outwards in all directions with a sonic velocity (c) equal to the velocity of sound. After a short interval of time (say 4 seconds), let the nose of the projectile reach at B. During this time, let the movement of nose cause other pressure waves to radiate outwards, as it travels along the path AB.

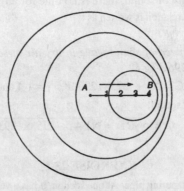

Fig. 23·3. Subsonic velocity.

A little consideration will show, that when the time t is equal to zero, the nose of the projectile is at A. The positions of the nose of the projectile, along the line at interval of 1 second as 1, 2, 3 are shown in Fig. 23·3. Let the nose of the projectile reach at B after 4 seconds. Therefore $AB = 4c$. Since the velocity of the projectile v is less than c, therefore the pressure wave is able to move ahead of the projectile. Or in other words, the nose of the projectile is always behind the wave front of the pressure

Compressible Flow of Fluids

wave, it has caused. It is thus obvious, that the projectile is always penetrating in an area of disturbed fluid. A little consideration will show, that the disturbed fluid has some effect on the fluid resistance to the motion of the projectile.

23·10 Mach Wave when the Body is Moving at a Sonic Velocity

Consider the nose of a projectile, moving with a sonic velocity in a fluid from left to right, at some instant to be at A. After a short interval time (say 4 seconds), let the nose of the projectile reach at B. During this time, let the movement of nose cause other pressure waves to radiate outwards, as it travels along the path AB.

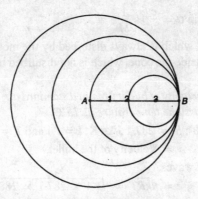

Fig. 23·4. Sonic velocity.

A little consideration will show, that when the time t is equal to zero, the nose of the projectile is at A. The positions of the nose of the projectile along the line at interval of 1 second as 1, 2, 3 are shown in Fig. 23·4. Let the nose of the projectile reach at B after 4 seconds. Therefore $AB = 4c$. Since the velocity of the projectile v is equal to c, therefore the pressure waves propagated during the projectile's movement from A to B, meet at a common point B.

It will be interesting to know that the concentration of pressure waves at B causes a great increase in the intensity of pressure at that point. It is thus obvious, that the nose of the projectile pushes this wave of intense pressure, which moves alongwith it and is known as a shock wave. It causes a great resistance to the motion of the projectile.

23·11 Mach Wave when the Body is Moving with a Supersonic Velocity

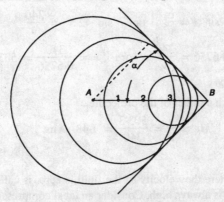

Fig. 23·5. Supersonic velocity.

Consider the nose of a projectile, moving with a supersonic velocity in a fluid from left to right, at some instant to be at A. After a short interval of time (say 4 seconds), let the nose of the projectile reach at B. During this time, let the movement of the nose cause other pressure waves to radiate outwards, as it travels along the path AB.

A little consideration will show, that when the time t is equal to zero, the nose of the projectile is at A. The positions of the nose of the projectile along the line at interval of 1 second as 1, 2, 3 are shown in Fig. 23.5. Let the nose of the projectile reach B in 4 seconds. Therefore $AB = 4c$. Since the velocity of the projectile v is more than c, therefore the pressure wave lags behind the nose of projectile. Or in other words, the nose of the projectile moves ahead of the pressure wave. The entire system of spherical pressure waves forms a conical figure, with vertex at B. The half angle α of the cone vertex is called the Mach angle. From the geometry of the figure, we find that

$$\sin \alpha = \frac{c}{v} = \frac{1}{M}$$

The area inside the cone, which is always disturbed by the motion of the projectile, is called the zone of action. The area outside the cone, which is not disturbed by the motion of the projectile, is called the zone of silence.

Example 23·4. *Find the velocity of bullet fired in a standard air, if the Mach angle is $30°$. Take $R = 287·1$ J/kg K, $k = 1·4$ and assume temperature as $15°C$.*

Solution. Given : $\alpha = 30°$; $R = 287·1$ J/kg K; $k = 1·4$ and $T = 15°C = 15 + 273 = 288$ K.

Let $\qquad v = $ Velocity of the bullet.

We know that velocity of waves,

$$c = \sqrt{kRT} = \sqrt{1·4 \times 287·1 \times 288} = 340·2 \text{ m/s}$$

and $\qquad \sin \alpha = \dfrac{c}{v} \qquad$ or $\qquad \sin 30° \dfrac{340·2}{v}$

$\therefore \qquad 0·5 = \dfrac{340·2}{v} \qquad$ or $\qquad v = \dfrac{340·2}{0·5} = 680·4$ m/s **Ans.**

Example 23·5. *Calculate the velocity and Mach number of a projectile, if the Mach angle is $42·5°$ and the temperature of air is $5°$. Take R as 297 J/kg K.*

Solution. Given : $\alpha = 42·5°$; $T = 5°C = 5 + 273 = 278$ K and $R = 297$ J/kg K.

Velocity of the projectile

Let $\qquad v = $ Velocity of the projectile.

We know that velocity of waves,

$$c = \sqrt{kRT} = \sqrt{1·4 \times 297 \times 278} = 340 \text{ m/s}$$

and $\qquad \sin \alpha = \dfrac{c}{v} \qquad$ or $\qquad \sin 42·5° = \dfrac{340}{v}$

$\therefore \qquad 0·6756 = \dfrac{340}{v} \qquad$ or $\qquad v = \dfrac{340}{0·6756} = 503·3$ m/s **Ans.**

Mach number of the projectile

We also know that Mach number of the projectile,

$$M_N = \frac{v}{c} = \frac{503·3}{340} = 1·48 \text{ \textbf{Ans.}}$$

23·12 Stagnation Pressure

A point in the flow, where the velocity of the fluid is zero, is called a stagnation point. The pressure at a stagnation point is always high. Consider an ideal compressible fluid flowing around a body as shown in Fig. 23·6. Now consider a section A in the upstream where the flow has an undisturbed stream.

Compressible Flow of Fluids

Let
 v_0 = Velocity of the fluid at A,
 ρ_0 = Density of the fluid at A,
and
 p_0 = Pressure of the fluid at A.

Now let the stagnation point be S as shown in Fig. 23·6. We know that the velocity of fluid at S will be zero.

Now let ρ = Density of the fluid at S, and
 p_s = Pressure of fluid at S.

Fig. 23·6

The stagnation pressure at the point S is given by the equation :

$$\frac{p_s - p_0}{\frac{1}{2}\rho_0 \times v_0^2} = 1 + \frac{v_0^2}{4c^2}$$

or
$$p_s = p_0 + \frac{1}{2}\rho_0 \times v_0^2 \left(1 + \frac{M^2}{4}\right) \quad \ldots \left(\because \frac{v_0}{c} = M\right)$$

where M is Mach number.

Example 23·6. *An aircraft is flying with a velocity of 200 m/s through the still air at −15°C. Find the stagnation pressure, if the mass density of the air is 1·08 kg/m³. Take pressure of the air as 80 kPa.*

Solution. Given : v_0 = 200 m/s; T = −15°C = −15 + 273 = 258 K; ρ = 1·08 kg/m³ and p_0 = 80 kPa = 80 kN/m² = 80 × 10³ N/m².

We know that celerity of sound wave,
$$c = \sqrt{\gamma RT} = \sqrt{1\cdot4 \times 287 \times 258} = 322 \text{ m/s}$$

and stagnation pressure $p_s = p_0 + \frac{1}{2} \times \rho_0 \times v_0^2 \left(1 + \frac{v_0^2}{4c^2}\right)$

$$= (80 \times 10^3) + \frac{1}{2} \times 1\cdot08 \times (200)^2 \times \left(1 + \frac{(200)^2}{4 \times (322)^2}\right) \text{ N/m}^2$$

$$= (80 \times 10^3) + (23\cdot7 \times 10^3) = 103\cdot7 \times 10^3 \text{ N/m}^2$$

$$= 103\cdot7 \text{ kPa } \textbf{Ans.}$$

Example 23·7. *An air stream has a velocity of 162 km/hour at a pressure of 10 kPa vacuum and a temperature of 47°C. Compute :*

(i) its stagnation properties, and

(ii) Mach number.

Take the atmospheric pressure = 100 kPa, R = 287 J/kg K; ρ=1.35 kg/m³ and γ = 1·4.

Solution. Given : v_0 = 162 km/hour = 45 m/s; p = 10 kPa (vacuum) = − 10 kPa; Atmospheric pressure = 100 kPa : T = 47° C = 47 + 273 = 320 K; R = 287 J/kgK; ρ = 1·35 kg/m³ and γ = 1·4.

(i) *Stagnation properties*

We know that celerity of sound wave,
$$c = \sqrt{\gamma RT} = \sqrt{1\cdot4 \times 287 \times 320} = 358\cdot6 \text{ m/s } \textbf{Ans.}$$

We also know that actual pressure of air stream,

p_0 = Atmospheric pressure − Pressure of stream in vacuum
 = 100 − 10 = 90 kPa = 90 kN/m² = 90 × 10³ N/m²

and stagnation pressure,

$$p_S = p_0 + \frac{1}{2} \rho_0 \times v_0{}^2 \left(1 + \frac{v_0{}^2}{4c^2}\right)$$

$$= (90 \times 10^3) + \frac{1}{2} \times 1\cdot 35 \times (45)^2 \times \left(1 + \frac{(45)^2}{4 \times (358\cdot 6)^2}\right) \text{ N/m}^2$$

$$= (90 \times 10^3) + (1\cdot 4 \times 10^3) = 91\cdot 4 \times 10^3 \text{ N/m}^2$$

$$= 91\cdot 4 \text{ kPa } \textbf{Ans.}$$

(ii) Mach number

We also know that Mach number,

$$M_N = \frac{v}{c} = \frac{45}{358\cdot 6} = 0\cdot 125 \textbf{ Ans.}$$

23·13 Measurement of Air Speed

The procedure for measuring the speed of air is the same as that of a body moving in air, such as an aeroplane. The general principle, used for this purpose, is that a pitot tube is used and the relative velocity of the plane to the air is obtained. In order to obtain the absolute velocity of the plane, an allowance is made for the velocity of the wind.

A metre used in an aeroplane, for the measurement of its speed, consists of a pitot static tube, which forms the mouth of the instrument at M. The other end of the tube B is connected to one of the limbs of the manometer. The inner tube is surrounded by an outer tube and an air space is provided between the two tubes. The ring of holes at A, in the outer tube, admits air in the space between the two tubes. The pressure of air is thus transmitted along the air space to its outlet at C, which is connected to the other limbs of the manometer.

The instrument is placed with its mouth facing the air flow. A little consideration will show, that the head measured by the pitot tube is only the velocity head of the flowing stream. The instrument is first calibrated, for the range over which it is to be used and the coefficient of instrument k is obtained. Now the velocity of the aeroplane is given by the relation,

$$v = k \sqrt{2gh}$$

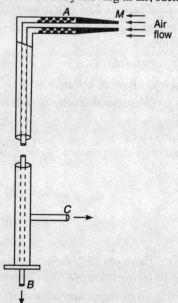

Fig. 23·7. Measurement of air speed.

Note : The speed of aeroplane is also calculated by the stagnation pressure.

EXERCISE 23.2

1. Determine the sonic speed in dry air at a temperature of 5°C. Take $\gamma = 1\cdot 4$. (**Ans.** 334·2 m/s)
2. What is the speed of the sound waves in dry air at a place where the temperature is $-5°C$? Take $R = 290$ J/kg K and $\gamma = 1\cdot 35$. (**Ans** 323·9 m/s)
3. An aeroplane is flying at 200 m/s through air at $-15°C$. Find the Mach number. (**Ans.** 0·62)
4. Find the velocity of sound waves, when a bullet is fired as a target at 27°C. Also find the velocity of bullet, if the Mach angle is 30°. Take $R = 290$ J/kg K and $\gamma = 1\cdot 4$. (**Ans.** 349 m/s; 698 m/s)

QUESTIONS

1. How will you use the Bernoulli's equation for compressible flow?
2. Obtain an expression between pressure and height.
3. Explain the effect of temperature on the velocity of sound.
4. What is Mach number? Discuss its importance.
5. Define Mach wave. Describe the Mach wave in case of subsonic speed.
6. What is stagnation pressure? Derive an expression for the same.

OBJECTIVE TYPE QUESTIONS

1. The continuity equation for flowing gases is
   (a) $a_1 V_1 = a_2 V_2$ (b) $a_1 V_1 \rho_1 = a_2 V_2 \rho_2$ (c) $a_1 V_1 w_1 = a_2 V_2 w_2$ (d) all of these
   where ρ is the mass density and w specific weight of the gas.

2. The Bernoulli's equation is applicable to the energy equation of a flowing gas in
   (a) isothermal process (b) adiabatic process
   (c) both 'a' and 'b' (d) neither 'a' nor 'b'

3. The velocity of sound wave is equal to
   (a) $\sqrt{\gamma RT}$ (b) $\sqrt{\dfrac{\gamma R}{T}}$ (c) $\sqrt{\dfrac{\gamma}{RT}}$ (d) $\sqrt{\dfrac{1}{\gamma RT}}$

4. If the Mach number of flow is unity, then the flow is
   (a) subsonic (b) sonic (c) supersonic (d) hypersonic

ANSWERS

1. (d)   2. (c)   3. (a)   4. (b)

24

Flow Around Immersed Bodies

> 1. Introduction. 2. Pressure on a Body Immersed in a Moving Liquid. 3. Newton's Law of Resistance. 4. Drag. 5. Lift. 6. Airfoil Theory. 7. Boundary Layer Theory (or Prandtl's Theory). 8. Thickness of the Boundary Layer. 9. Thickness of Boundary Layer in a Laminar Flow. 10. Thickness of Boundary Layer in a Turbulent Flow. 11. Boundary Layer Separation. 12. Prandtl's Experiment of Boundary Layer Separation. 13. Magnus Effect in a Moving Liquid. 14. Prevention of Boundary Layer Separation. 15. Prevention of Boundary Layer Separation by Providing Slots near the Leading Edge. 16. Prevention of Boundary Layer Separation by Providing Cylinder as the Leading Edge.

24·1 Introduction

We see that when a solid body is held in the path of a moving fluid and is completely immersed in it, the body will be subjected to some pressure or force. Conversely, if a body is moved with a uniform velocity through a fluid at rest, it offers some resistance to the moving body, or the body has to exert some force to maintain its steady movement. It is thus obvious, that when a submarine moves through the water or an aeroplane flies through the atmosphere, its engine must supply a sufficient force not only to run it, but also to balance the resistance offered.

24·2 Pressure on a Body Immersed in a Moving Liquid

We have discussed in the previous article that whenever a body is immersed in a moving liquid, it exerts some pressure on the immersed body. A little consideration will show, that if a plate is immersed in a liquid parallel to the flow, it will be subjected to a pressure less than that if the same plate is held immersed perpendicular to the flow.

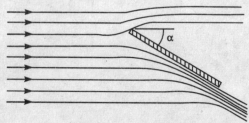

Fig. 24·1. Liquid streamlines.

It has been experimentally found that if a plate is held immersed at some angle with the direction of flow of the liquid, the streamlines of the liquid get deflected as shown in Fig. 24·1. The pressure exerted by the streamlines is the same, when the fluid is at rest and the body is moving uniformly through it; or the body is at rest and the fluid moves around it.

24·3 Newton's Law of Resistance

This law of resistance was first stated by Newton which states, *"The force exerted by a moving fluid on an immersed body is directly proportional to the rate of change of momentum due to the presence of the body."*

Mathematically the force,

$$P \propto \frac{waV^2}{g} = k \times \frac{waV^2}{g}$$

...(where k is a constant of proportionality)

where
- w = Specific weight of the fluid,
- a = Area of the body,
- V = Velocity of the fluid, and
- g = Acceleration due to gravity.

The above law of resistance is based on the following assumptions :
1. The planes of the body are completely smooth.
2. The space around the body is completely filled with the fluid.
3. The fluid has a large number of fine particles having mass but no dimension.
4. The fluid particles do not exert any influence on one another.
5. The body experiences impacts from all the particles in its path.

The above mentioned assumptions, led to a very simple formula for the constant of proportionality. Newton's calculation for a square plane, perpendicular to the direction of motion, gives the value of constant as unity. However, it was found later by experiments the value of this constant as 0·55. This figure is still less for skew planes or rounded bodies like spheres. The cause for this discrepancy is due to the following reasons:

1. The Newton's assumption takes into account only the conditions at the front of the body, while those at the sides and tail are not considered.
2. The motion of fluid particles is always influenced by its neighbouring particles. As a result of this, the path of various particles influence each other.

According to the latest theories, available with the space research organisations, the value of the constant of proportionality is taken as 0·5 for all practical purposes. Thus the force exerted,

$$P = 0 \cdot 5 \times \frac{waV^2}{g} = \frac{waV^2}{2g}$$

As per Pascal's law, the pressure exerted by the moving liquid will be at right angles to the plate. It has the following two components, which are important from the subject point of view :
1. Drag. 2. Lift.

24·4 Drag

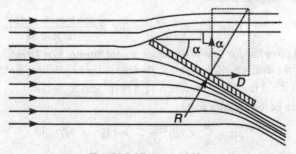

Fig. 24·2. Drag and Lift.

We have discussed in Art. 24·2 that whenever a plate is held immersed at some angle with the direction of flow of the liquid, it is subjected to some pressure. As this pressure acts at right angles to the plate, therefore it will have some component (*i*) in the direction of flow of the liquid and (*ii*) at right angles to the direction of flow of the liquid as shown in Fig. 24·2. The component of this pressure, in the direction of flow of the liquid, is known as drag.

Thus the force of drag, $P_D = \dfrac{waV^2}{2g} \times \sin \alpha = K_D \times \dfrac{waV^2}{2g}$

where K_D is a coefficient, known as drag coefficient (sometimes it is also called coefficient of drag). Its value depends upon the type of plate and the angle of inclination of the plate, which is determined experimentally.

Note : If the liquid is at rest and the immersed plate is moving with a velocity V in the liquid, then the drag may be defined as the horizontal resistance offered by the liquid to the moving plate.

24·5 Lift

We have already discussed that whenever a plate is held immersed at some angle with the direction of flow of the liquid, it is subjected to some pressure, which acts at right angles to the plate. The component of this pressure at right angle to the direction of flow of the liquid is known as lift.

Thus the force of lift, $P_L = \dfrac{waV^2}{2g} \times \cos \alpha = K_L \times \dfrac{waV^2}{2g}$

where K_L is a coefficient, known as lift coefficient (sometimes it is also called coefficient of lift). Its value also depends upon the type of plate and the angle of inclination of the plate, which is determined experimentally. The resultant force on the body,

$$R = \sqrt{P_D^2 + P_L^2}$$

Note : If the liquid is at rest and the immersed plate is moving with a velocity V in the liquid, then the lift may be defined as the vertical resistance offered by the liquid to the moving plate.

This creation of lift, on the bodies, is used for propelling sea ships and for supporting the weight of the flying aeroplanes.

Example 24·1. *A flat plate 2 m × 2 m is immersed in water flowing with a velocity of 5 m/s. Find the forces of drag and lift. Take $K_D = 0.05$ and $K_L = 0.2$.*

Solution. Given : $a = 2 \times 2 = 4$ m^2 ; $V = 5$ m/s; $K_D = 0.05$ and $k_L = 0.2$.

Force of drag

We know that force of drag,

$$P_D = K_D \times \dfrac{waV^2}{2g} = 0.05 \times \dfrac{9.81 \times 4 \times (5)^2}{2 \times 9.81} = 2.5 \text{ kN} \quad \textbf{Ans.}$$

Force of lift

We also know that force of lift,

$$P_L = K_L \times \dfrac{waV^2}{2g} = 0.2 \times \dfrac{9.81 \times 4 \times (5)^2}{2 \times 9.81} = 10.0 \text{ kN} \quad \textbf{Ans.}$$

Example 24·2. *A circular disc 3 m in diameter is held normal to a 15 m/s wind having specific weight of 11·8 N/m^3. If the drag coefficient is 0·4, find the force required to hold it.*

Solution. Given : $d = 3$ m; $V = 15$ m/s; $w = 11.8$ N/m^2 and $K_D = 0.4$.

We know that area of the circular plate,

$$a = \dfrac{\pi}{4} \times (d)^2 = \dfrac{\pi}{4} \times (3)^2 = 7.07 \text{ m}^2$$

and force required to hold the circular plate,

$$P_D = K_D \times \dfrac{waV^2}{2g} = 0.4 \times \dfrac{11.8 \times 7.07 \times (15)^2}{2 \times 9.81} = 382.7 \text{ N} \quad \textbf{Ans.}$$

Example 24·3. *A man weighing 750 N descends to the ground from an aeroplane with the help of a parachute against the resistance of air. The shape of the parachute is hemispherical of 2 m diameter. Find the velocity of the parachute with which it comes down. Take specific weight of the air as 12 N/m³.*

Solution. Given : Weight (or drag) $P_D = 750$ N; $d = 2$ m and $w = 12$ N/m³.

Let $\qquad V$ = Velocity of the parachute in m/s.

First of all, let us assume the value of drag coefficient as 0·5. We know that area of the parachute, which will be subjected to the air resistance,

$$a = \frac{\pi}{4} \times (d)^2 = \frac{\pi}{4} \times (2)^2 = 3.142 \text{ m}^2$$

and drag on the parachute (P_D),

$$750 = k_D \times \frac{waV^2}{2g} = 0.5 \times \frac{12 \times 3.142 \ V^2}{2 \times 9.81}$$

$$= 0.961 \ V^2$$

∴ $\qquad V^2 = 750/0.961 = 780.4 \quad \text{or} \quad V = 27.9 \text{ m/s} \quad \textbf{Ans.}$

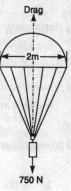

Fig. 24·3.

Example 24·4. *A truck having projected area of 6·5 square metres travelling at 72 km/hr has a total resistance of 1·6 kN. Out of this resistance, 30 per cent is due to rolling friction and 20 percent due to surface friction. The rest is due to drag. Calculate the coefficient of drag. Assume specific weight of air as 12 N/m³.*

Solution. Given : $a = 6.5$ m²; $V = 72$ km.p.h. = 20 m/s; Total resistance = 1·6 kN; Rolling friction = 30% = 0·3; Surface friction = 20% = 0·2 and $w = 12$ N/m³.

Let $\qquad k_D$ = Coefficient of drag.

We know that total friction due to rolling and surface

$$= (0.3 + 0.2) \times 1.6 = 0.8 \text{ kN}$$

and force of drag $\qquad = 1.6 - 0.8 = 0.8 \text{ kN} = 800 \text{ N}$

We also know that the force of drag (P_D),

$$800 = K_D \times \frac{waV^2}{2g} = K_D \times \frac{12 \times 6.5 \times (20)^2}{2 \times 9.81} = 1590 \ K_D$$

∴ $\qquad K_D = 800/1590 = 0.503 \quad \textbf{Ans.}$

Example 24·5. *A flat plate 1·5 m × 1·5 m moves at 45 km/hour in stationary air of specific weight 11·3 N/m³. If the coefficients of drag and lift are 0·15 and 0·75 respectively, find : (i) lift force, (ii) drag force, (iii) resultant force, and (iv) power required to keep the plate in motion.*

Solution. Given : $a = 1.5 \times 1.5 = 2.25$ m²; $V = 45$ km/hr = 12·5 m/s; $w = 11.3$ N/m³; $K_D = 0.15$ and $K_L = 0.75$.

(i) Lift force

We know that the lift force,

$$P_L = K_L \times \frac{waV^2}{2g} = 0.75 \times \frac{11.3 \times 2.25 \times (12.5)^2}{2 \times 9.81} \text{ N}$$

$$= 151.9 \text{ N} \quad \textbf{Ans.}$$

(ii) Drag force

We know that the drag force,

$$P_D = k_D \times \frac{waV^2}{2g} = 0.15 \times \frac{11.3 \times 2.25 \times (12.5)^2}{2 \times 9.81} = 30.4 \text{ N} \quad \textbf{Ans.}$$

(iii) Resultant force

We know that the resultant force,

$$R = \sqrt{P_D^2 + P_L^2} = \sqrt{(30.4)^2 + (151.9)^2} = 154.9 \text{ N} \quad \textbf{Ans.}$$

(iv) Power required to keep the plate in motion

We also know that work done in moving the plate

$$= P_D \times V = 30.4 \times 12.5 = 380 \text{ N-m/s} = 380 \text{ J/s}$$

and power required = 380 W **Ans.**

EXERCISE 24·1

1. A flat plate 1 m × 1 m moves at 6·75 m/s normal to its plane. Determine the drag on the plate, if it moves (i) in air of specific weight 11·3 N/m³ and (ii) in water. Take $K_D = 0.58$.
(**Ans.** 15·2 N; 13·2 kN)

2. A plate 5 m × 1 m is moving with a velocity of 15 m/s in air of specific weight 11·3 N/m³. Find the forces of lift and drag. Take $K_L = 0.43$ and $K_D = 0.03$. (**Ans.** 287·6 N ; 19·4 N)

3. A flying object of surface area 0·1 m² experiences a lift of 13·25 N, while moving with a velocity of 16 m/s in air of specific weight 11·5 N/m³. What is the value of coefficient of lift? (**Ans.** 0·81)

4. A plate 1 m × 1 m moves in a stationary air of specific weight 12 N/m³ at 10 m/s. If the coefficients of lift and drag are 0·75 and 0·15 respectively, determine the resultant force and power required to keep the plate in motion. (**Ans.** 46·8 N; 91·7 W)

24·6 Air Foil Theory

In the previous articles, we have discussed about the forces of drag and lift. The practical utility of these forces is derived in running sea ships, submarines and aeroplanes. We have also discussed that the value of coefficient of drag and coefficient of lift depends upon the angle of inclination of the plate with the vertical.

A little consideration will show, that in actual practice the angle of inclination depends upon the geometrical position of the body, with respect to its motion. As a matter of fact, in practical aeronautics, we are always interested in airfoils, in which the resulting force is nearly perpendicular to the direction of flow. In this case, the lift is great and the drag is small. Since the lift serves the purpose of supporting the aeroplanes, therefore more the lift the better it is. Moreover, the drag is a necessary evil, which has to be compensated for by the propeller thrust. It is thus obvious, that less the drag the better it is. It has been experimentally found that if a flat plate is inclined at about 4°, the ratio of force of lift to force of drag (*i.e.* P_L / P_D) is about 6. In order to increase this ratio, the plate is given a slight curvature or camber in its bottom face as shown in Fig. 24·4. Such a camber gives twice the ratio of these forces than the flat plate. This ratio is further increased by nicely rounding off the front end of the plate and providing a sharp edge to the tail of plate as shown in Fig. 24·4. In this way, it is possible to have lift-drag ratio of even 20 or more. The above theory is popularly known as air foil theory.

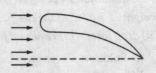

Fig. 24·4. Air foil theory

24·7 Boundary Layer Theory (or Prandtl's Theory)

The earlier scientists and engineers were of the opinion that when a body is held in the path of a moving fluid (or a body is moving through a fluid) the fluid exerts some shear force, while passing

Flow Around Immersed Bodies

over the surface of the body. This force causes some rubbing action on the surface of the body. But this conception was considered to be wrong by the scientists of the earlier twentieth century. They argued, that if there is a rubbing action of the fluid on the surface of the body, the body must show some fatigue and the sign of decay due to continuous rubbing action of the fluid on it. Prandtl was the first to publish a theory in 1904, which is now known as Boundary Layer Theory or Prandtl's Theory. According to this theory, the liquid in the vicinity of the surface of the body may be divided into the following two portions :

1. A very thin layer of the fluid, which is in the immediate contact of the body. This layer of the fluid behaves like a thin coating, as if it is fixed or glued (at its inner surface) to the boundary of the body. Since this thin layer of the fluid acts in such a way, as if its inner surface is fixed to the boundary of the body, therefore velocity of the fluid at the boundary is zero. Such a thin layer of the fluid is known as boundary layer.

    If we go away from the surface of the body, normal to the flow of the fluid (*i.e.*, across the boundary layer, the velocity of the flow will go on increasing rapidly, till at the extreme layer (*i.e.*, outer edge of the boundary layer), the velocity of flow is approximately equal to the velocity of the fluid outside the boundary layer (or more correctly the velocity of flow at the extreme edge of the velocity of flow is 0·99 times the velocity of flow of the fluid, outside the boundary layer).

2. The remaining portion of the fluid, which is outside the boundary layer. This portion has a high value of Reynold's number, because of the high velocity of flow.

24·8 Thickness of Boundary Layer

The distance from the surface of the body, to a place where the velocity of flow is 0·99 times the maximum velocity of flow, is known as thickness of the boundary layer. It is usually denoted by δ (delta).

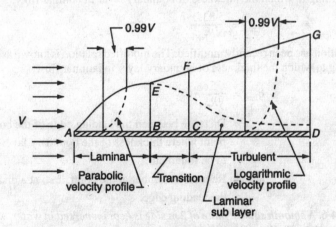

Fig. 24·5 Thickness of boundary layer

Now consider a smooth flat plate held immersed in a moving liquid as shown in Fig. 24·5.

Let V = Velocity of the liquid on the upstream side of the plate.

It has been experimentally found, that the thickness of the boundary layer is zero at the leading edge A, and increases to the trailing edge D as shown in Fig. 24·5. It has also been found that for some distance, from the leading edge, the flow is laminar. This boundary layer is also known as laminar boundary layer.

As the boundary layer continues downstream, it expands and the laminar flow ceases and the laminar boundary layer changes into transition boundary layer of very small length. As the boundary layer continues further downstream, it expands and the transition flow changes into turbulent flow and the transition boundary layer changes into turbulent boundary layer, which continues over the remaining length of the plate. However, within this turbulent boundary layer, there is a very thin film of fluid (just near the boundary of the plate) in which the flow is laminar. This layer is known as laminar sub-layer as shown in Fig. 24·5.

Now let us consider a section, where the thickness of the boundary layer is required to be found out.

Let
V = Velocity of the liquid,
ν = Kinetic viscosity of the liquid, and
x = Distance between the leading edge of the plate and the section where thickness of the boundary layer is required to be found out.

We know that in this case, the Reynold's number at a distance x from the leading edge.

$$R_{NX} = \frac{Vx}{\nu}$$

If the Reynold's number is less than 500 000, the flow is fully laminar and if it is between 500 000 and 1 000 000, the flow is in transition. But if it is more than 1 000 000 the flow is fully turbulent.

The thickness of boundary layer is different in laminar flow and turbulent flow. So we shall discuss the thickness of boundary layer in the above mentioned two types of flows.

24·9 Thickness of Boundary Layer in a Laminar Flow

Many scientists have conducted a series of experiments on the thickness of boundary layer in laminar flow and have given different relations for the same. The first relation was given by Pohlhausen, according to which the thickness of boundary layer in laminar flow,

$$\delta_{lam} = \frac{5 \cdot 835 x}{\sqrt{R_{NX}}}$$

But this relation has been recently modified. The modified relation is known as Prandtl-Blassius relation, according to which the thickness of boundary layer in laminar flow,

$$\delta_{lam} = \frac{5x}{\sqrt{R_{NX}}}$$

where
x = Distance between the leading edge of the body and the point where thickness of the boundary layer is required to be found out.
R_{NX} = Reynold's number at the point. *i.e.*, at a distance x from the leading edge.

Example 24·6. *A smooth square plate of 2 m side is kept immersed in water, which moves with a velocity of 300 mm/s. Find the thickness of the boundary layer at a distance of 0·5 m from the leading edge. Take kinematic viscosity of water as $1·0 \times 10^{-6}$ m^2/s.*

Solution. Given : Plate size = 2 m × 2 m; V = 300 mm/s = 0·3 m/s; x = 0·5 m and $\nu = 1·0 \times 10^{-6}$ m²/s.

We know that Reynold's number at a point 0·5 m from the leading edge,

$$R_{NX} = \frac{V \cdot x}{\nu} = \frac{0 \cdot 3 \times 0 \cdot 5}{1 \times 10^{-6}} = 150\ 000$$

Flow Around Immersed Bodies 415

Since the Reynold's number is less than 500 000, therefore, the flow at this section is laminar. We also know that thickness of boundary layer at the point,

$$\delta_{lam} = \frac{5x}{\sqrt{R_{NX}}} = \frac{5 \times 0.5}{\sqrt{150\ 000}} = 0.0065 \text{ m} = 6.5 \text{ mm} \quad \textbf{Ans.}$$

Example 24·7. *In an experiment, a smooth two-dimensional flat plate is exposed to a wind of velocity 90 km/hr. If laminar boundary layer exists up to a value of $R_{NX} = 2 \times 10^5$, find the maximum distance from the leading edge upto which laminar boundary layer exists and its maximum thickness. Take kinematic viscosity of air as 1.6×10^{-5} m²/s.*

Solution. Given : $V = 90$ km/hr $= 25$ m/s; $R_{NX} = 2 \times 10^5$ and $\nu = 1.6 \times 10^{-5}$ m²/s.

Maximum distance from the leading edge up to which laminar boundary layer exists

Let x = Maximum distance from the leading edge upto which laminar boundary layer exists.

We know that the Reynold's number at a distance x from the leading edge (R_{NX}),

$$2 \times 10^5 = \frac{Vx}{\nu} = \frac{25 x}{1.6 \times 10^{-5}} = \frac{15.6 x}{10^{-5}}$$

$$\therefore \quad x = \frac{(2 \times 10^5) \times (10^{-5})}{15.6} = 0.128 \text{ m} = 128 \text{ mm} \quad \textbf{Ans.}$$

Maximum thickness of the boundary layer

We also know that maximum thickness of the boundary layer,

$$\delta_{lam} = \frac{5x}{\sqrt{R_{NX}}} = \frac{5 \times 0.128}{\sqrt{2 \times 10^5}} = 0.0014 \text{ m} = 1.4 \text{ mm} \quad \textbf{Ans.}$$

24·10 Thickness of Boundary Layer in a Turbulent Flow

Though there are many relations for finding out the thickness of the boundary layer in a turbulent flow, yet the Prandtl-Blassius relation is widely used, according to which the thickness of boundary layer in a turbulent flow,

$$\delta_{tur} = \frac{0.377\ x}{(R_{NX})^{1/5}}$$

x = Distance between leading edge of the body and the section where thickness of boundary layer is required to be found out.

R_{NX} = Reynold's number at a distance x from the leading edge.

Example 24·8. *A smooth plate 1 m wide and 3 m long moves through stationary air with a velocity of 3 m/s. Determine the thickness of the boundary layer at the trailing edge. Assume ν for air as 0·15 stoke.*

Solution. Given : Plate width $= 1$ m; Plate length $= 3$ m; $V = 3$ m/s and $\nu = 0.15$ stoke $= 0.15 \times 10^{-4}$ m²/s.

We know that Reynold's number at the trailing edge (or at a distance of 3 m)

$$R_{NX} = \frac{V\ x}{\nu} = \frac{3 \times 3}{0.15 \times 10^{-4}} = 600\ 000$$

Since the Reynold's number is more than 500 000, therefore the flow near the trailing edge is turbulent. We know that thickness of the boundary layer at the trailing edge.

$$\delta_{tur} = \frac{0.377\ x}{(R_{NX})^{1/5}} = \frac{0.377 \times 3}{(600\ 000)^{1/5}} = 0.079 \text{ m} = 79 \text{ mm} \quad \textbf{Ans.}$$

24·11 Boundary Layer Separation

We have already discussed, that when a body is held immersed in a flowing liquid, a thin layer of the liquid will behave, as if it is fixed to the boundary of the body. But if the immersed body is a curved or angular one, the boundary layer does not stick to the whole surface of the body. The boundary layer leaves the surface and gets separated from it. This phenomenon is known as boundary layer separation. The point, where the boundary layer gets separated from the surface of the body, is known as point of separation.

24·12 Prandtl's Experiment of Boundary Layer Separation

Prandtl, after publishing his boundary layer theory, conducted series of experiments on boundary layer separation also. He held a cylinder in a flowing liquid and sprinkled small particles of aluminium[*] on the surface of the liquid. He allowed the liquid to flow around the cylinder, and found that the boundary layer has adhered to the surface of the cylinder throughout. The liquid also flowed in streamlines around the cylinder as shown in Fig. 24·6 (a).

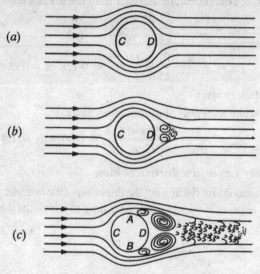

Fig. 24·6. Boundary layer separation.

After this, he went on gradually increasing the velocity of liquid. After certain velocity, he found the streamlines at D started becoming irregular. On further increasing the velocity of liquid, he found that the boundary layer has separated from the surface on both sides of D and vortices[**] have started to form as shown in Fig. 24·6 (b). He further went on increasing the velocity of liquid and found that the separation of boundary layer is taking place earlier and earlier; forming more pronounced vortices. After this, a stage came when the separation of boundary layer took place at A and B as shown in Fig. 24·6 (c).

24·13 Magnus Effect in a Moving Liquid

Consider a liquid having streamline flow from left to right. If we introduce a cylinder in the path of the streamlines, we shall see their path as shown in Fig. 24·7 (a). Now let the cylinder be rotated about its longitudinal axis. The rotating motion of the cylinder will deviate the streamlines as

---

[*]The idea, of sprinkling small particles of aluminium on the liquid surface, was that the metal particles reflect the light and thus enable the streamlines to be photographed.

[**] A flow, in which the liquid flows continuously round a curved path and about a fixed axis of rotation, is called vortex flow.

Flow Around Immersed Bodies

shown in Fig. 24·7 (b). The phenomenon of deviating the streamlines by the rotating cylinder is known as Magnus effect.

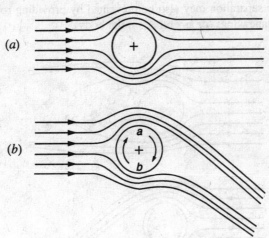

Fig. 24·7. Magnus effect in a moving liquid.

A little consideration will show, that the velocity of liquid at *a* is increased, because of the movement of the cylinder, which exerts a viscous drag on the liquid and thus increases its velocity. And the velocity of liquid at *b* is reduced, because of the viscous drag on the liquid by the moving cylinder in the opposite direction.

24·14 Prevention of Boundary Layer Separation

The separation of boundary layer in a turbulent flow may be prevented in order to have the reduced drag. Many methods have been suggested to prevent the separation of boundary layer. But the following are important from the subject point of view :

24·15 Prevention of Boundary Layer Separation by Providing Slots near the Leading Edge

The boundary layer separation may be prevented by providing slots in the wings of an aeroplane (or any other similar body) near the leading edge.

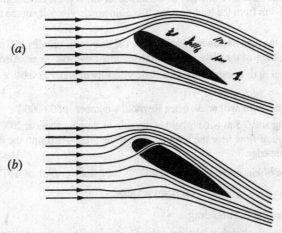

Fig. 24·8. Slot near the leading edge.

In Fig. 24·8 (a) is shown the separation of boundary layer. In Fig. 24·8 (b) is shown a slot made near the leading edge, while the remaining conditions remain unchanged. Because of the slot, the air bands pass through it and adhere on the upper surface of the wing for the whole of its length.

24·16 Prevention of Boundary Layer Separation by providing Rotating Cylinder as the Leading Edge

The boundary layer separation may also be prevented by providing rotating cylinder as the leading edge of the aeroplane wings (or any other similar body).

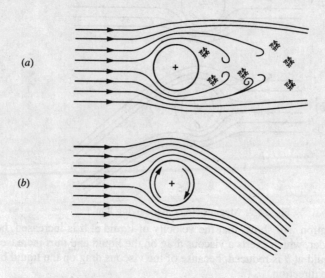

Fig. 24·9. Rotating cylinder as the leading edge.

In Fig. 24·9 (a) is shown the separation of boundary layer, when the cylinder is at rest. In Fig. 24·9 (b) is shown the rotating cylinder, while the remaining conditions remain unchanged. Because of the rotating cylinder, the Magnus effect takes place on the air bands, which adhere on the upper surface of the wing for the whole of its length.

EXERCISE 24·2

1. A circular plate moves with a velocity of 0·8 m/s in an oil. What is the thickness of boundary layer at a distance of 0·4 m from the leading edge. Take kinematic viscosity of the oil as 0·16 stoke.

    (**Ans.** 14·1 mm)

2. A plate is held in the path of flowing water which has a uniform velocity of 3 m/s. Find the distance, from the leading edge of the plate at which the laminar flow will break. Also find the thickness of the boundary layer at this point. Take kinematic viscosity of water as 0.01×10^{-4} m^3/s.

    [**Ans.** 0·167 m; 1·2 mm]

    [Hint : The laminar flow will break when Reynold's number is 500 000.]

3. A plate 1·5 m long and 0·5 m wide is held in the path of flowing water at 20°C with a velocity of 0·4 m/s. Find the type of flow near the trailing edge of the plate. Also find the thickness of boundary layer at the trailing edge. (**Ans.** Turbulent; 39·5 mm)

    [Hint : The value of kinematic viscosity at 20°C is 0·01 stoke or 0.01×10^{-4} m^3/s.]

QUESTIONS

1. Write short notes on lift and drag.
2. Explain clearly with sketches, the growth of boundary layer theory in a turbulent flow past a flat plate placed parallel to the flow.
3. What do you understand by the term thickness of boundary layer? Derive an equation for the thickness of boundary layer in a laminar flow.

4. Define the term laminar 'sub-layer'.
5. Explain clearly the phenomenon of boundary layer separation.
6. What is meant by Magnus effect?
7. Discuss the methods adopted to prevent the boundary layer separation.

OBJECTIVE TYPE QUESTIONS

1. The force exerted by a moving fluid on an immersed body is directly proportional to the rate of change of momentum due to the presence of the body. This statement is called
   (a) Newton's law of motion (b) Newton's law of viscosity
   (c) Newton's law of resistance (d) Newton's law of gravitation

2. Whenever a plate is held immersed at some angle with the direction of flow of the liquid, it is subjected to some pressure. The component of this pressure, in the direction of flow of the liquid, is called
   (a) drag   (b) lift   (c) bulk modulus   (d) none of these

3. The thickness of boundary layer in a laminar flow is
   (a) $\dfrac{5}{\sqrt{R_{NX}}}$ (b) $\dfrac{5x}{\sqrt{R_{NX}}}$ (c) $\dfrac{5x^2}{\sqrt{R_{NX}}}$ (d) $\dfrac{5\sqrt{x}}{\sqrt{R_{NX}}}$

   where x = Distance between the leading edge of the body and the point where thickness of the body is required, and
   R_{NX} = Reynold's number at a distance x from the leading edge.

4. According to Prandtl-Blassius relation, the thickness of boundary layer in a turbulent flow is
   (a) $\dfrac{0.377\,x}{(R_{NX})^{1/5}}$ (b) $\dfrac{5x}{(R_{NX})^{1/5}}$ (c) $\dfrac{5.377\,x}{(R_{NX})^{1/5}}$ (d) $\dfrac{5.835\,x}{(R_{NX})^{.15}}$

ANSWERS

1. (c)   2. (a)   3. (b)   4. (a)

25

Dimensional Analysis

1. Introduction. 2. Fundamental Dimensions. 3. Dimensional Homogeneity. 4. Uses of the Principle of Dimensional Homogeneity. 5. Methods of Dimensional Analysis. 6. Rayleigh's Method. 7. Buckingham's π-theorem. 8. Selection of Repeating Variables.

25·1 Introduction

In the recent past, there has been a rapid development in the field of Fluid Mechanics. As a result of this development, several new problems are met with by the engineers. These problems generally pertain to the design, construction as well as efficient working of various types of hydraulic structures and machines. Some of these problems may be easily solved by the mathematical analysis, whereas others may be difficult to solve due to their complex nature. The solution of such problems is considerably simplified by the use of dimensional analysis.

Strictly speaking, the dimensional analysis is a mathematical technique, which deals with the dimensions of the physical quantities involved in the phenomenon. It is based on the assumption that the phenomenon can be expressed by a dimensionally homogeneous equation, with certain variables. These days, the dimensional analysis is widely used in research work for developing design criteria and also for conducting model tests.

25·2 Fundamental Dimensions

All physical quantities are measured by comparison. This comparison is always made with respect to some arbitrarily fixed value for each independent quantity, called dimension (*e.g.,* length, mass, time, temperature etc.). Since there is no direct relationship between these dimensions, therefore they are called fundamental dimensions or fundamental quantities. Some other quantities such as area, volume, velocity, force etc., cannot be expressed in terms of fundamental dimensions and thus may be called *derived dimensions, derived quantities or secondary quantities.

There are two systems for fundamental dimensions namely *MLT* (*i.e.,* mass, length and time) and *FLT* (*i.e.,* force, length and time). The dimensional form of any quantity is independent of any system of units. Some engineers prefer to use force instead of mass as fundamental quantity, as the former is easy to measure. But in this book, we shall use the *MLT* system.

The following table gives the dimensions and units for the various physical quantities, which are important from the subject point of view :

---

e.g., Force = Mass × Acceleration = $M \times \dfrac{L}{T^2} = MLT^{-2}$

Table 25·1

S.No.	Quantity	Dimensions in terms of MLT system	Dimensions in terms of FLT system
1.	Length (l)	L	L
2.	Area (A)	L^2	L^2
3.	Volume (V)	L^3	L^3
4.	Time (t)	T	T
5.	Velocity (v)	LT^{-1}	LT^{-1}
6.	Acceleration (a)	LT^{-2}	LT^{-2}
7.	Gravitational acceleration (g)	LT^{-2}	LT^{-2}
8.	Frequency (N)	T^{-1}	T^{-1}
9.	Discharge (Q)	$L^3 T^{-1}$	$L^3 T^{-1}$
10.	Force or weight (F, W)	MLT^{-2}	F
11.	Power (P)	$ML^2 T^{-3}$	FLT^{-1}
12.	Work or Energy (E)	$ML^2 T^{-2}$	FL
13.	Pressure (p)	$ML^{-1} T^{-2}$	FL^{-2}
14.	Mass (m)	M	$FT^2 L^{-1}$
15.	Mass density (ρ)	ML^{-3}	$FT^2 L^{-4}$
16.	Specific weight (w)	$ML^{-2} T^{-2}$	FL^{-3}
17.	Dynamic viscosity (μ)	$ML^{-1} T^{-1}$	FTL^{-2}
18.	Kinematic viscosity (ν)	$L^2 T^{-1}$	$L^2 T^{-1}$
19.	Surface tension (σ)	MT^{-2}	FL^{-1}
20.	Shear stress (τ)	$ML^{-1} T^{-2}$	FL^{-2}
21.	Bulk modulus (K)	$ML^{-1} T^{-2}$	FL^{-2}

Example 25·1. *Determine the dimensions of force, pressure, power, specific weight and surface tension in M-L-T system.*

Solution. We know that the dimension for force in M-L-T system,

$$\text{Force} = \text{Mass} \times \text{Acceleration}$$
$$= M \times \frac{\text{Length}}{\text{Time}^2} = \frac{ML}{T^2} = MLT^{-2} \quad \textbf{Ans.}$$

Similarly

$$\text{pressure} = \frac{\text{Force}}{\text{Area}} = \frac{MLT^{-2}}{L^2} = ML^{-1} T^{-2} \quad \textbf{Ans.}$$

and

$$\text{power} = \frac{\text{Work done}}{\text{Time}} = \frac{\text{Force} \times \text{Distance}}{\text{Time}}$$
$$= \frac{[MLT^{-2}][L]}{T} = ML^2 T^{-3} \quad \textbf{Ans.}$$

Specific weight = $\dfrac{\text{Weight}}{\text{Volume}}$ = $\dfrac{\text{Force}}{\text{Volume}}$ ...($\because$ Weight = Force)

$$= \dfrac{MLT^{-2}}{L^3} = ML^{-2}T^{-2} \quad \textbf{Ans.}$$

and surface tension = $\dfrac{\text{Force}}{\text{Length}}$ = $\dfrac{[MLT^{-2}]}{L} = MT^{-2}$ **Ans.**

Example 25·2. *Determine the dimensions of discharge, torque and momentum in F-L-T system.*

Solution. We know that the dimension for discharge in *F-L-T* system,

$$\text{Discharge} = \dfrac{\text{Volume}}{\text{Time}} = \dfrac{L^3}{T} = L^3 T^{-1} \quad \textbf{Ans.}$$

Similarly torque = Force × Distance = $[F][L] = FL$ **Ans.**

and momentum = Mass × Velocity = $\left[\dfrac{F}{LT^{-2}}\right] \times [LT^{-1}]$

...$\left(\because \text{Mass} = \dfrac{\text{Force}}{\text{Acceleration}}\right)$

$= FT$ **Ans.**

25·3 Dimensional Homogeneity

An equation is called dimensionally homogeneous, if the fundamental dimensions have identical powers of *M-L-T* (*i.e.*, mass, length and time) on both sides. Such an equation would essentially be independent of the system of measurement. Let us consider the common equation of discharge in hydraulics,

$$Q = A \cdot V$$
$$L^3 T^{-1} = L^2 \times LT^{-1} = L^3 T^{-1}$$

We see, from the above equation that both the right and left hand sides of the equation have the same dimensions. Thus the equation is dimensionally homogeneous.

Note : Two dimensionally homogeneous equations can be multiplied or divided without effecting the homogeneity. But the two dimensionally homogeneous equations cannot be added or subtracted, as the resulting equation may or may not be dimensionally homogeneous.

Example 25·3. *Check the dimensional homogeneity of the following common equations in the field of Hydraulics.*

(i) $Q = C_d \cdot a \sqrt{2gH}$ and (ii) $v = C\sqrt{mi}$.

Solution. Given equations : $Q = C_d \cdot a \sqrt{2gH}$ and $v = C\sqrt{mi}$.

(i) *Dimensional homogeneity of the equation* $Q = C_d \cdot a \sqrt{2gH}$

Substituting the dimensions on the L.H.S. and R.H.S. of the given equation,

$$L^3 T^{-1} = [1] \times [L^2][LT^{-2} \times L]^{1/2}$$

...(The dimension of C_d, being constant is taken as 1)

$$= L^3 T^{-1}$$

Since the dimensions on both the sides of the equation are same, therefore the equation is dimensionally homogeneous. **Ans.**

Dimensional Analysis

(*ii*) *Dimensional homogeneity of the equation* $v = C\sqrt{mi}$.

Substituting the dimensions on the L.H.S. and R.H.S. of the equation,

$$LT^{-1} = [1] \times [L \times 1]^{1/2}$$

...(The dimension of C, being constant, is taken as 1)

$$= L^{1/2}$$

Since the dimensions on both the sides of the equation are not same, therefore the equation is not dimensionally homogeneous. **Ans.**

25·4 Uses of the Principle of Dimensional Homogeneity

Though the principle of dimensional homogeneity has a number of uses, yet the following are important from the subject point of view:

1. *To determine the dimension of a physical quantity*

The dimensions of any physical quantity may be easily determined with this principle. *e.g.*, the dimension of energy,

$$\text{Energy} = \text{Work} = \text{Force} \times \text{Distance}$$
$$= F \times L \qquad \text{...(in } F\text{-}L\text{-}T\text{)}$$
$$= [MLT^{-2}] \times L \qquad \text{...(}\because \text{Force, } F = MLT^{-2}\text{)}$$
$$= ML^2T^{-2}$$

2. *To check the dimensional homogeneity of an equation*

The dimensional homogeneity of an equation may be easily checked with this principle. *e.g.*, let us consider Darcy's formula for loss of head,

$$H_f = \frac{4flv^2}{2gd}$$

Now substituting the dimensions on the L.H.S. and R.H.S. of the equation,

$$\therefore \quad L = \frac{[1] \times [L] \times [LT^{-1}]^2}{[LT^{-2}][L]}$$

(The dimensions of 4 in the numerator and 2 in the denominator is not considered. The dimension of f, being constant, is taken as 1)

$$= L$$

3. *To change the coefficient of an equation while using in the other system of units*

The coefficient of an equation may be easily changed, while using the same equation in other system of units (*i.e.*, from F.P.S. to M.K.S. or *vice versa*) *e.g.*, let us consider the Manning's formula for velocity,

$$v = M \cdot m^{2/3} \cdot i^{1/2} \qquad \text{...(where } M \text{ is Manning's constant)}$$

Now substituting the dimensions on the L.H.S. and R.H.S. of the equation,

$$LT^{-1} = M[L]^{2/3}[1]^{1/2} = M[L]^{2/3}$$

...(The dimensions of i being constant, is taken as 1)

Since the dimensions of both the sides are not same, therefore the equation is dimensionally non-homogeneous.

From the above equation, we find that

$$\therefore \quad M = \frac{LT^{-1}}{L^{2/3}} = L^{1/3} T^{-1}$$

Now, in order to make the above equation applicable to F.P.S. units the value coefficient M has to be changed. We know that 1 meter = 3·281 ft, and the unit of time is the same in both the systems. Therefore the new constant

$$= L^{1/3} = (3·281)^{1/3} = 1·486$$

It is thus obvious, that the equation for F.P.S. units will be

$$v = 1·486 \, M \cdot m^{2/3} \cdot i^{1/2}$$

(d) *Dimensional analysis*

The use of this principle will be widely used in this chapter in detail.

Example 25·4. *The following equation is applicable in M.K.S. system of units.*

$$v = 50\sqrt{mi}$$

where v is the velocity of water, m hydraulic mean depth and i the longitudinal slope of the channel. Find the corresponding equation in F.P.S. system of units.

Solution. Given equation : $v = 50\sqrt{mi}$

Let us rewrite the equation as

$$v = C\sqrt{mi} \qquad \text{...(where } C = 50\text{)}$$

Now substituting the dimensions on the L.H.S. and R.H.S. of the equation,

$$LT^{-1} = C \, [L \times 1]^{1/2} \quad \text{...(The dimension of } i\text{, being constant, is taken as 1)}$$
$$= C \, [L]^{1/2}$$

Since the dimensions of both the sides are not same, therefore the equation is dimensionally non-homogeneous. From the above equation, we find that

$$C = \frac{LT^{-1}}{L^{1/2}} = L^{1/2} \, T^{-1}$$

Now in order to make the above equation applicable to F.P.S. system of units, the value of coefficient C has to be changed. We know that 1 meter = 3·281 ft, and the unit of time is the same in both the systems. Therefore the new constant

$$= L^{1/2} = (3·281)^{1/2} = 1·81$$

It is thus obvious, that the equation for F.P.S. units will be

$$v = 1·81 \times 50 \, \sqrt{mi} = 90·5 \, \sqrt{mi} \quad \textbf{Ans.}$$

25·5. Methods of Dimensional Analysis

The following two methods of dimensional analysis are important from the subject point of view :

1. Rayleigh's method, and

2. Buckingham's π-theorem.

25·6 Rayleigh's Method*

In this method, the functional relationship of some variables is expressed in the form of an exponential equation, which must be dimensionally homogeneous. *e.g.*, if Y is some function of independent variables $X_1, X_2, X_3...$ etc., then functional relationship may be written as:

$$Y = f \, [X_1, X_2, X_3...]$$

---

*This method was originally proposed by Lord Rayleigh in 1899. He used this method for determining the effect of temperature on the viscosity of gases.

Dimensional Analysis

In this equation, Y will be a dependent variable, whereas X_1, X_2, X_3... are independent variables and f is a function. A little consideration will show that from the above equation, the dependent variable is one about which information is required; whereas the independent variables are those, which govern the variation of dependent variables.

The Rayleigh's method is based on the following steps:
1. First of all, write the functional relationship with the given data.
2. Now write the equation in terms of a constant with exponents (*i.e.*, powers) a, b, c...
3. With the help of the principle of dimensional homogeneity, find out the values of a, b, c... by obtaining simultaneous equations.
4. Now substitute the values of these exponents in the main equation and simplify it.

Note : Since the main equation contains a constant, therefore multiplication or division of any function may be conveniently made.

Example 25·5. *If the capillary rise (h) depends upon the specific weight (w) surface tension (σ) of the fluid and tube radius (r), show that*

$$h = r\phi\left(\frac{\sigma}{wr^2}\right)$$

Solution. Given : Capillary rise $= h$; Specific weight of the fluid $= w$; Surface tension of the fluid $= \sigma$ and tube radius $= r$.

Let the functional relationship be,

$$h = (w, \sigma, r) = k(w^a \cdot \sigma^b \cdot r^c) \qquad \ldots(i)$$

where k is a dimensionless constant. Substituting the respective dimensions in the above equation,

$$(L) = k[(ML^{-2}T^{-2})^a \cdot (MT^{-2})^b \cdot (L)^c]$$

Now, for dimensional homogeneity, equating the powers of M, L and T on both sides of the equation,

for M, $\qquad 0 = a + b \qquad \ldots(ii)$
for L, $\qquad 1 = -2a + c \qquad \ldots(iii)$
for T, $\qquad 0 = -2a - 2b \qquad \ldots(iv)$

From equation (*ii*), we get

$$a = -b$$

Similarly, from equation (*iii*), we get

$$c = (1 + 2a) = 1 + 2(-b) = 1 - 2b$$

Now substituting the values of a, b and c in equation (*i*),

$$h = k(w^{-b} \cdot \sigma^b \cdot r^{1-2b})$$

$$= k\left(r \cdot \frac{\sigma}{wr^2}\right)^b = r\phi\left(\frac{\sigma}{wr^2}\right) \quad \textbf{Ans.}$$

Example 25·6. *Show that the resistance (R) to the motion of a sphere of diameter (D) moving with a uniform velocity (V) through a real fluid having mass density (ρ) and viscosity (μ) is given by*

$$R = \rho D^2 V^2 f\left(\frac{\mu}{\rho VD}\right)$$

Solution. Given : Resistance $= R$; Diameter of sphere $= D$; Velocity of sphere $= V$; Mass density of fluid $= \rho$ and viscosity of fluid $= \mu$.

Let the functional relationship be,
$$R = f[D, V, \rho, \mu] = k[D^a . V^b . \rho^c . \mu^d] \qquad ...(i)$$
where k is a dimensionless constant. Substituting the respective dimensions on both sides of the above equation,
$$[MLT^{-2}] = k[(L)^a (LT^{-1})^b (ML^{-3})^c (ML^{-1}T^{-1})^d]$$

Now, for dimensional homogeneity, equating the powers of M, L and T on both sides of the equation,

for M, $1 = c + d$ ...(ii)
for L, $1 = a + b - 3c - d$ ...(iii)
for T, $-2 = -b - d$ ...(iv)

From equation (ii) we get
$$c = 1 - d$$
Similarly, from equation (iv) we get
$$b = 2 - d$$
and from equation (iii) we get
$$\begin{aligned} a &= 1 - b + 3c + d \\ &= 1 - (2-d) + 3(1-d) + d \qquad \text{...(Substituting the values of } b \text{ and } c) \\ &= 1 - 2 + d + 3 - 3d + d \\ &= 2 - d \end{aligned}$$

Now substituting the values of a, b and c in equation (i),
$$R = k[D^{2-d} . V^{2-d} . \rho^{1-d} . \mu^d]$$
$$= \rho D^2 V^2 k \left(\frac{\mu}{\rho V D}\right)^d$$
$$= \rho D^2 V^2 f \left(\frac{\mu}{\rho V D}\right) \quad \text{Ans.}$$

Example 25·7. *The thrust (P) of a propeller depends upon the diameter (D), speed (V), mass density (ρ), revolutions per minute (N) and coefficient of viscosity (μ). Show that*
$$P = \rho D^2 V^2 f\left[\frac{\mu}{\rho D V} . \frac{DN}{V}\right]$$

Solution. Given : Thrust = P; Diameter of the propeller = D; Speed = V; Mass density = ρ; Revolutions = N and viscosity = μ.

Let the functional relationship be :
$$P = f[D, V, \rho, N, \mu] = k[D^a . V^b . \rho^c . N^d , \mu^e] \qquad ...(i)$$
where k is a dimensionless constant. Substituting the respective dimensions in the above equation,
$$[MLT^{-2}] = k[(L)^a (LT^{-1})^b (ML^{-3})^c (T^{-1})^d (ML^{-1}T^{-1})^e]$$

Now, for dimensional homogeneity, equating the powers of M, L and T on both sides of the equation,

for M, $1 = c + e$ ...(ii)
for L, $1 = a + b - 3c - e$ ...(iii)
for T, $-2 = -b - d - e$ ...(iv)

Dimensional Analysis

From equation (*ii*), we get
$$c = 1 - e$$
Similarly, from equation (*iv*) we get
$$b = 2 - e - d$$
and from equation (*iii*) we get
$$a = 1 - b + 3c + e$$
$$= 1 - (2 - e - d) + 3(1 - e) + e$$
...(Substituting the values of *b* and *c*)
$$= 1 - 2 + e + d + 3 - 3e + e = 2 + d - e$$

Now substituting the values of *a*, *b* and *c* in equation (*i*),
$$P = k[D^{2+d-e} \cdot V^{2-e-d} \cdot \rho^{1-e} \cdot N^d \cdot \mu^e]$$

$$= \rho D^2 V^2 k \left[\left(\frac{\mu}{\rho DV}\right)^e \cdot \left(\frac{DN}{V}\right)^d \right]$$

$$= \rho D^2 V^2 f \left[\frac{\mu}{\rho DV} \cdot \frac{DN}{V}\right] \quad \textbf{Ans.}$$

Example 25·8. *Show, by dimensional analysis, that the power P developed by a hydraulic turbine is given by :*

$$P = \rho N^3 D^5 f \left[\frac{N^2 D^2}{gH}\right]$$

where ρ is the mass density of liquid, N is the rotational speed of the turbine in r.p.m., D is the diameter of runner, H is the working head and g is the gravitational acceleration.

Solution. Given : Power = P; Mass density of liquid = ρ; Rotational speed of the turbine = N; Diameter of runner = D; Working head = H and gravitational acceleration = g.

Let the functional relationship be,
$$P = f[\rho, N, D, H, g] = k[\rho^a \cdot N^b \cdot D^c \cdot H^d \cdot g^e] \quad \text{...(i)}$$

where *k* is a dimensionless constant. Substituting the respective dimensions in the above equation,
$$[ML^2T^{-3}] = [(ML^{-3})^a \ (T^{-1})^b \ (L)^c \ (L)^d (LT^{-2})^e]$$

Now for dimensional homogeneity, equating the powers of *M*, *L* and *T* on both sides of the equation,

for *M*, $\qquad 1 = a$ ...(*ii*)
for *L*, $\qquad 2 = -3a + c + d + e$ ...(*iii*)
for *T*, $\qquad -3 = -b - 2e$ ...(*iv*)

From equation (*ii*), we get
$$a = 1$$
Similarly, from equation (*iv*), we get
$$b = 3 - 2e$$
and from equation (*iii*) we get
$$c = 2 + 3a - d - e = 2 + 3 - d - e \quad \text{...(Substituting } a = 1\text{)}$$
$$= 5 - d - e$$

Now substituting the values, of a, b and c in equation (i),

$$P = k \, [\rho^1 \, N^{3-2e} \, . \, D^{5-d-e} \, . \, g^e]$$

$$= k\left[\rho N^3 D^5 \left(\frac{g}{N^2 D}\right)^e \left(\frac{H}{D}\right)^d\right]$$

$$= \rho N^3 D^5 f\left(\frac{N^2 D}{g}\right)\left(\frac{D}{H}\right)$$

$$= \rho N^3 D^5 f\left[\frac{N^2 D^2}{gH}\right] \quad \text{Ans.}$$

Example 25·9. *The resistance (R) experienced by a partially submerged body depends upon the velocity (V), length of the body (l), viscosity of the fluid (μ), density of the fluid (ρ) and gravitational acceleration (g). Obtain a dimensionless expression for (R).*

Solution. Given : Resistance = R; Velocity = V; Length of the body = l; Viscosity of fluid = μ; Density of fluid = ρ and gravitational acceleration = g.

Let the functional relationship be,

$$R = f[V, l, \mu, \rho, g] = k\,[V^a \, . \, l^b \, . \, \mu^c \, . \, \rho^d \, . \, g^e] \qquad \ldots(i)$$

where k is a dimensionless constant. Substituting the respective dimensions in the above equation,

$$[MLT^{-2}] = k\,[(LT^{-1})^a \, (L)^b \, (ML^{-1}T^{-1})^c \, (ML^{-3})^d \, (LT^{-2})^e]$$

Now, for dimensionless homogeneity, equating the powers of M, L and T on both sides of the equation,

for M, $1 = c + d$ ...(ii)

for L, $1 = a + b - c - 3d + e$ ...(iii)

for T, $-2 = -a - c - 2e$ ...(iv)

From equation (ii) we get

$$d = 1 - c$$

Similarly, from equation (iv), we get

$$a = 2 - c - 2e$$

and from equation (iii) we get

$$b = 1 - a + c + 3d - e$$
$$= 1 - (2 - c - 2e) + c + 3(1-c) - e$$

...(Substituting the values of a and d)

$$= 1 - 2 + c + 2e + c + 3 - 3c - e$$
$$= 2 - c + e$$

Now substituting the values of d, a and b in equation (i),

$$R = k\,[V^{(2-c-2e)} . \, l^{(2-c+e)} . \, \mu^c . \, \rho^{(1-c)} . \, g^e] = k\left[(V^2 l^2 \rho)\left(\frac{\mu}{Vl\rho}\right)^c \left(\frac{lg}{V^2}\right)^e\right]$$

$$= f\left[(V^2 l^2 \rho)\left(\frac{\mu}{Vl g}\right)\left(\frac{lg}{V^2}\right)\right] \quad \text{Ans.}$$

Dimensional Analysis

Example 25·10. *Using Rayleigh's method, determine the rational formula for discharge (Q) through a sharp-edged orifice freely into the atmosphere in terms of constant head (H), diameter (d), mass density (ρ), dynamic viscosity (μ) and acceleration due to gravity (g).*

Solution. Given : Discharge = Q; Constant head = H; Diameter of orifice = d ; Mass density = ρ; Dynamic viscosity = μ and acceleration due to gravity = g.

Let the functional relationship be,

$$Q = f[H\ d, \rho, \mu, g] = k\,[H^a . d^b . \rho^c . \mu^d . g^e]$$

where k is a dimensionless constant. Substituting the respective dimensions in the above equation,

$$[L^3 T^{-1}] = k\,[(L^a)(L^b)(ML^{-3})^c (ML^{-1}T^{-1})^d (LT^{-2})^e]$$

Now, for dimensional homogeneity, equating the powers of M, L and T on both sides of the equation,

for M,	$0 = c + d$	...(ii)
for L,	$3 = a + b - 3c - d + e$	...(iii)
for T,	$-1 = -d - 2e$	...(iv)

From equation (ii) we get

$$d = -c$$

Similarly, from equation (iv) we get

$$e = \frac{1}{2} - \frac{d}{2} = \frac{1}{2} + \frac{c}{2} \qquad \ldots(\because d = -c)$$

and from equation (iii) we get

$$b = 3 - a + 3c + d - e$$
$$= 3 - a + 3c + (-c) - \left(\frac{1}{2} + \frac{c}{2}\right)$$

...(Substituting the values of d and e)

$$= 3 - a + 3c - c - \frac{1}{2} - \frac{c}{2} = \frac{5}{2} - a + \frac{3c}{2}$$

Now substituting the values of d, e and b in equation (i),

$$Q = k\left[H^a . d^{\left(\frac{5}{2} - a + \frac{3c}{2}\right)} . \rho^c . \mu^{-c} . g^{\left(\frac{1}{2} + \frac{c}{2}\right)}\right]$$

$$= k\left[\left(\frac{\rho . d^{3/2} . g^{1/2}}{\mu}\right)^c \left(\frac{H}{d}\right)^a d^{5/2} . g^{1/2}\right]$$

$$= k\left[\left(\frac{\mu}{\rho . d^{3/2} . g^{1/2}}\right)^{-c} \left(\frac{H}{d}\right)^{a-1/2} d^2 . g^{1/2} . H^{1/2}\right]$$

$$= k\left[d^2 . \sqrt{gh}\,\left(\frac{\mu}{\rho . d^{3/2} . g^{1/2}}\right)^{-c} \left(\frac{H}{d}\right)^{a-1/2}\right]$$

$$= k\left[\frac{\pi}{4} \times d^2 \times \sqrt{2gh}\,\left(\frac{\mu}{\rho . d^{3/2} . g^{1/2}}\right)^{-c} \left(\frac{H}{d}\right)^{a-1/2}\right]$$

$$= a \times \sqrt{2gh} \; f \left[\frac{\mu}{\rho \cdot d^{3/2} \cdot g^{1/2}} , \frac{H}{d} \right] \quad \text{Ans.}$$

Notes : 1. Since the main equation contains a constant (k) therefore we can multiply the equation by π.

2. A little consideration will show that in the above result, the value in the bracket (multiplied by f) stands for the coefficient of discharge (*i.e.*, C_d) in the usual formula.

EXERCISE 25·1

1. Determine the dimensions of the following in *M-L-T* and *F-L-T* systems :
   (*i*) Mass, (*ii*) Dynamic viscosity, and (*iii*) Shear stress.

   [**Ans.** M, FT^2L^{-1}; $ML^{-1}, T^{-1}, FTL^{-2}$; $ML^{-1}T^{-2}, FL^{-3}$]

2. Check the dimensional homogeneity of the following equations used in Hydraulics.
   (*a*) $\quad T = 2\pi \sqrt{\dfrac{l}{g}}$
   (*b*) $\quad Q = \dfrac{8}{15} C_d \sqrt{2g} \, \tan \dfrac{\theta}{2} (H)^{5/2}$ [**Ans.** Both homogeneous]

3. Check the dimensional homogeneity of the following equations as used in the field of Hydraulics.
   (*a*) $\quad P = wQH$
   (*b*) $\quad H_L = \dfrac{32\mu vl}{w\,d^2}$ [**Ans.** Both homogeneous]

4. A partially submerged body of length l, is towed in water of density ρ, and viscosity μ, at a velocity V. Show, by Rayleigh's method, that the total resistance R experienced by the body is given by

   $$R = \rho L^2 V^2 f\left[\left(\frac{\mu}{\rho l V}\right)\left(\frac{lg}{V^2}\right)\right]$$

   where g is the gravitational acceleration.

5. Show that the shear stress in a fluid flowing through a pipe is given by :

   $$\tau = \rho V^2 f\left[\frac{\mu}{DV}, \frac{k}{D}\right]$$

   where D is the diameter of the pipe, ρ is the mass density, V the velocity and μ the viscosity of the fluid and k is the height of the roughness projection.

6. Show that the pressure drop due to an obstruction in a pipe is given by

   $$dp = \rho V^2 f\left[\frac{DV\rho}{\mu}\right]$$

   where D is the diameter of the pipe, V the velocity, ρ the mass density and μ the dynamic viscosity of the fluid.

25·7 Buckingham's* π-Theorem

It has been observed that the Rayleigh's Method of dimensional analysis becomes cumbersome, when a large number of variables are involved. In order to overcome this difficulty, the Buckingham's method may be conveniently used. It states, *"If there are n variables in a dimensionally homogeneous equation, and if these variables contain m fundamental dimensions such as (M-L-T) they may be grouped into (n—m) non-dimensional independent π-terms."*

Mathematically, if a variable X_1 depends upon independent variables $X_2, X_3, X_4...X_n$, the functional equation may be written as :

$$X_1 = k(X_2, X_3, X_4...X_n).$$

---

*. This method was originally proposed by Edgar Buckingham in 1915.

Dimensional Analysis

The equation may be written in its general form as
$$f(X_1, X_2, X_3, X_4 \ldots X_n) = C$$
where C is a constant, and f represents some function. In this equation, there are n variables. If there are m fundamental dimensions, then according to Buckingham's π-theorem
$$f_1(\pi_1, \pi_2, \pi_3 \ldots \pi_{n-m}) = \text{Constant}.$$
The Buckingham's π-method is based on the following steps.

1. First of all, write the functional relationship with the given data.
2. Then write the equation in its general form.
3. Now choose m repeating* variables and write separate expressions for each π-term. Every π-term will contain the repeating variables and one of the remaining variables. The repeating variables are written in exponential form.
4. With the help of the principle of dimensional homogeneity find out the values of a, b, c...... by obtaining simultaneous equations.
5. Now substitute the values of these exponents in the π-terms.
6. After the π terms are determined, write the functional relation in the required form.

Note : Any π-term may be replaced by any power of it, because the power of a non-dimensional term is also non-dimensional e.g., π_1 may be replaced by $\pi_1^2, \pi_1^3, \sqrt{\pi_1}$ .....or by $2\pi_1, 3\pi_1, \pi_{1/2}$ etc.

25·8. Selection of Repeating Variables

In the previous article, we have mentioned that we should choose m repeating variables and write separate expressions for each π-term. Though there is no hard and fast rule for the selection of repeating variables, yet the following points should be borne in mind while selecting the repeating variables.

1. The variables should be such that none of them is dimensionless.
2. No two variables should have the same dimensions.
3. Independent variables should, as far as possible, be selected as repeating variables.

Notes: 1. In general, the repeating variables selected are ρ, v and l. i.e., the first representing the fluid property, the second representing the flow characteristics and the third representing the geometrical characteristics of the body.

2. In case, the example is held up, then one of the repeating variables should be changed.

Example 25·11. *Prove that the discharge over a spillway is given by the relation :*
$$Q = VD^2 f\left[\frac{\sqrt{gD}}{V}, \frac{H}{D}\right]$$
where V = Velocity of flow; D = Depth at the throat; H = Head of water and g = Acceleration due to gravity.

Solution. Given : Discharge = Q; Velocity of flow = V; Depth at the throat = D; Head of water = H and acceleration due to gravity = g.

Let the functional relationship be,
$$Q = f[V, D, H, g]$$
The above equation in its general form may be written as,
$$f(Q, V, D, H, g) = 0$$
A little consideration will show, that in the above equation, the primary dimensions are only two. Thus $m = 2$. Taking V and D as repeating variables, we get
$$\pi_1 = (V)^{a1} \cdot (D)^{b1} \cdot Q \qquad \ldots(i)$$

---

*. The most suitable variables are those, which contain flow characteristics (such as velocity, mass, density etc.).

Substituting the dimensions in both sides of the above equation,

or $\quad M^0L^0T^0 = [LT^{-1})^{b_1} \times [L]^{b_1} \times [L^3T^{-1}]$

Now for dimensional homogeneity, equating the powers of M, L and T on both sides of the equation,

For T, $\quad 0 = -a_1 - 1$ or $a_1 = -1$
For L, $\quad 0 = a_1 + b_1 + 3$
or $\quad b_1 = -3 - a_1 = -3 - (-1) = -2$

Substituting the values, of a_1 and b_1 in equation (i),

$$\pi_1 = V^{-1}.D^{-2}.Q$$

$$= \frac{Q}{VD^2} \qquad ...(ii)$$

Similarly, $\quad \pi_2 = (V)^{a_2}(D)^{b_2} H$

Substituting the dimensions in both sides of the above equation,

or $\quad M^0L^0T^0 = [LT^{-1}]^{a_2} \times [L]^{b_2} \times [L]$

Now for dimensional homogeneity, equating the powers of M, L and T on both sides of the equation,

For T, $\quad 0 = -a_2 \quad$ or $\quad a_2 = 0$
For L, $\quad 0 = a_2 + b_2 + 1 \quad$ or $\quad b_2 = -a_2 - 1 = 0 - 1 = -1$

Thus $\quad \pi_2 = V^0.D^{-1}.H$

$$= \frac{H}{D} \qquad ...(iii)$$

and $\quad \pi_3 = (V)^{a_3}(D)^{b_3} g$

Substituting the dimensions in both sides of the above equation,

or $\quad M^0L^0T^0 = [LT^{-1}]^{a_3} \times [L]^{b_3} \times [LT^{-2}]$

Now for dimensional homogeneity, equating the powers of M, L and T on both sides of the equation,

For T, $\quad 0 = -a_3 - 2 \quad$ or $\quad a_3 = -2$
For L, $\quad 0 = a_3 + b_3 + 1 \quad$ or $\quad b_3 = -a_3 - 1 = -(-2) - 1 = +1$

Thus $\quad \pi_3 = V^{-2}.D^{-1}.g = \dfrac{gD}{V^2}$

$$= \frac{\sqrt{gD}}{V} \qquad ...(iv)$$

Now the functional relationship be written as,

$$f\left[\frac{Q}{VD^2} . \frac{H}{D} . \frac{\sqrt{gD}}{V}\right] = 0$$

$$\frac{Q}{VD^2} = f\left[\frac{\sqrt{gD}}{V} . \frac{H}{D}\right]$$

$\therefore \quad Q = VD^2 f\left[\dfrac{\sqrt{gD}}{V} . \dfrac{H}{D}\right]$ **Ans.**

Example 25·12. *The resisting force (F) of a supersonic plane during flight can be considered as dependent upon the length of the aircraft (l), velocity (V), air viscosity (μ), air density (ρ) and bulk modulus of air (K). Express the functional relationship between these variables and the resisting force.*

Dimensional Analysis

Solution. Given : Resisting force = F; Length of the aircraft = l; Velocity of air craft = V; Air viscosity = μ; Air density = ρ and bulk modulus of air = K.

Let the functional relationship be,
$$F = f[l, V, \mu, \rho, K]$$
The above equation, in its general form, may be written as,
$$f(F., l, V, \mu, \rho, K) = 0$$
A little consideration will show, that in the above equation, the primary dimensions are three only. Thus $m = 3$. Taking l, V and ρ as repeating variables, we get
$$\pi_1 = (l)^{a1}.(V)^{b1}.(\rho)^{c1}.F$$
Substituting the dimension in both sides of the above equation,

$\therefore\quad M^0L^0T^0 = (L)^{a1}.(LT^{-1})^{b1}.(ML^{-3})^{c1}.(MLT^{-2})$

Now for dimensional homogeneity, equating the powers of M, L and T on both sides of the equation,

For M,	$0 = c_1 + 1$	or $c_1 = -1$
For T,	$0 = -b_1 - 2$	or $b_1 = -2$
For L,	$0 = a_1 + b_1 - 3c_1 + 1$	or $a_1 = -2$

$\therefore\quad \pi_1 = (l)^{-2}.(V)^{-2}(\rho)^{-1}.F$

$$= \frac{F}{l^2V^2\rho} \qquad ...(i)$$

Similarly $\pi_2 = (l)^{a2}.(V)^{b2}.(\rho)^{c2}.\mu$

Substituting the dimension in both sides of the above equation,

$\therefore\quad M^0L^0T^0 = (L)^{a2}.(LT^{-1})^{b2}.(ML^{-3})^{c2}.(ML^{-1}T^{-1})$

Now for dimensional homogeneity, equating the power of M, L and T on both sides of the equation,

For M,	$0 = c_2 + 1$	or $c_2 = -1$
For T,	$0 = -b_2 - 1$	or $b_2 = -1$
For L,	$0 = a_2 + b_2 - 3c_2 - 1$	or $a_2 = -1$

$\therefore\quad \pi_2 = (l)^{-1}.(V)^{-1}.(\rho)^{-1}.\mu$

$$= \frac{\mu}{lV\rho} \qquad ...(ii)$$

and $\pi_3 = (l)^{b3}.(V)^{b3}.(\rho)^{c3}.K$

Substituting the dimension in both sides of the above equation,

$\therefore\quad M^0L^0T^0 = (L)^{a3}.(LT^{-2})^{b3}.(ML^{-3})^{c3}.(ML^{-1}T^{-2})$

Now for dimensional homogeneity, equating the powers of M, L and T on both sides of the equation,

For M,	$0 = c_3 + 1$	or $c_3 = -1$
For T,	$0 = -b_3 - 2$	or $b_3 = -2$
For L,	$0 = a_3 + b_3 - 3c_3 - 1$	or $a_2 = 0$

$\therefore\quad \pi_3 = (l)^0.(V)^{-2}.(\rho)^{-1}.K$

$$= \frac{K}{V^2\rho} \qquad ...(iii)$$

Now the functional relationship may be written as :

$$f\left(\frac{F}{l^2V^2\rho} \cdot \frac{\mu}{lV\rho} \cdot \frac{K}{V^2\rho}\right) = 0$$

or $\quad F = f(l^2V^2\rho)\left(\frac{\mu}{lV\rho} \cdot \frac{K}{V^2\rho}\right)$ **Ans.**

Example 25·13. *The efficiency (η) of a fan depends upon density (ρ), dynamic viscosity (μ) of the fluid, angular velocity (ω) of rotor, diameter (D) of rotor and discharge (Q) of fluid. Explain η in terms of dimensionless parameters.*

Solution. Given : Efficiency of fan = η; Density of fluid = ρ; Dynamic viscosity of fluid = μ; Angular velocity of rotor = ω; Diameter of rotor = D and discharge of fluid = Q.

Let the functional relationship be,

$$\eta = f[\rho, \mu, \omega, D, Q]$$

The above equation, in its general form, may be written as :

$$f(\eta, \rho, \mu, \omega, D, Q) = 0$$

A little consideration will show, that in the above equation the primary dimensions are three only. Thus $m = 3$. Taking ρ, ω and D as repeating variables, we get

$$\pi_1 = (\rho)^{a1} \cdot (\omega)^{b1} \cdot (D)^{c1} \cdot \eta$$

Since efficiency (η) itself is dimensionless, therefore

$$\pi_1 = \eta \qquad \ldots(i)$$

Similarly $\quad \pi_2 = (\rho)^{a2} \cdot (\omega)^{b2} \cdot (D)^{c2} \cdot \mu$

Substituting the dimensions in both sides of the above equation,

$$\therefore \quad M^0L^0T^0 = (ML^{-3})^{a2} \cdot (T^{-1})^{b2} \cdot (L)^{c2} \cdot (ML^{-1}T^{-1})$$

Now for dimensional homogeneity, equating the powers of M, L and T on both sides of the equation,

For M,	$0 = a_2 + 1$	or	$a_2 = -1$
For T,	$0 = b_2 - 1$	or	$b_2 = -1$
For L,	$0 = -3a_2 + c_2 - 1$	or	$c_2 = -2$

Thus $\quad \pi_2 = (\rho)^{-1} \cdot (\omega)^{-1} \cdot (D)^{-2} \cdot \mu$

$$= \frac{\mu}{\rho \omega D^2} \qquad \ldots(ii)$$

and $\quad \pi_3 = (\rho)^{a_3} \cdot (\omega)^{b_3} \cdot D^{c_3} \cdot Q$

Substituting the dimensions in both sides of the above equation,

$$\therefore \quad M^0L^0T^0 = (ML^{-3})^{a_3} \cdot (T^{-1})^{b_3} (L)^{c_3} (L^3T^{-1})$$

Now for dimensional homogeneity, equating the powers of M, L and T on both sides of the equation,

For M,	$0 = a_3$	or	$a_3 = 0$
For T,	$0 = -b_3 - 1$	or	$b_3 = -1$
For L,	$0 = -3a_3 + c_3 + 3$	or	$c_3 = -3$

Thus $\quad \pi_3 = (\rho)^0 \cdot (\omega)^{-1} \cdot D^{-3} \cdot Q$

$$= \frac{Q}{\omega D^3}$$

Dimensional Analysis

Now the functional relationship may be written as,

$$f\left[\eta \cdot \frac{\mu}{\rho\omega D^2} \cdot \frac{Q}{\omega D^3}\right] = 0$$

or $\eta = f\left[\left(\frac{\mu}{\rho\omega D^2}\right), \left(\frac{Q}{\omega D^3}\right)\right]$ **Ans.**

Example 25·14. *Show that the discharge of a centrifugal pump is given by :*

$$Q = ND^3 f\left[\frac{gH}{N^2D^2}, \frac{\mu}{ND^2\rho}\right]$$

where N is the speed of the pump in r.p.m., D the diameter of the impeller, g acceleration due to gravity, H manometric head, μ viscosity of fluid and ρ the density of the fluid.

Solution. Given : Discharge = Q; Speed of the pump = N; Diameter of impeller = D; Acceleration due to gravity = g; Manometric head = H; Viscosity of fluid = μ and density of fluid = ρ.

Let the functional relationship be,

$$Q = f[N, D, g, H, \mu, \rho]$$

The above equation, in its general form, may be written as,

$$f(Q, N, D, g, H, \mu, \rho) = 0$$

A little consideration will show, that in the above equation the primary dimensions are only four. Thus $m = 4$. Taking N, D and ρ as repeating variables, we get

$$\pi_1 = (N)^{a_1} \cdot (D)^{b_1} \cdot (\rho)^{c_1} \cdot Q$$

Substituting the dimensions in both sides of the above equation,

$$M^0 L^0 T^0 = (T^{-1})^{a_1} (L)^{b_1} (ML^{-3})^{c_1} (L^3 T^{-1})$$

Now for dimensional homogeneity, equating the powers of M, L and T on both sides of the equation,

for M, $0 = c_1$ or $c_1 = 0$
for T, $0 = -a_1 - 1$ or $a_1 = -1$
for L, $0 = b_1 - 3c_1 + 3$ or $b_1 = -3$

$\therefore$ $\pi_1 = N^{-1} \cdot D^{-3} \cdot \rho^0 \cdot Q$

$$= \frac{Q}{ND^3} \qquad \qquad \ldots(i)$$

Similarly, $\pi_2 = (N)^{a_2} \cdot (D)^{b_2} \cdot (\rho)^{c_2} \cdot g$

Substituting the dimensions in both sides of the above equation,

$\therefore$ $M^0 L^0 T^0 = (T^{-1})^{a_2} \cdot (L)^{b_2} \cdot (ML^{-3})^{c_2} (LT^{-2})$

Now for dimensional homogeneity, equating the powers of M, L and T on both sides of the equation,

for M, $0 = c_2$ or $c_2 = 0$
for T, $0 = -a_2 - 2$ or $a_2 = -2$
for L, $0 = b_2 - 3c_2 + 1$ or $b_2 = -1$

Thus $\pi_2 = N^{-2} \cdot D^{-1} \cdot \rho^0 \cdot g$

$$= \frac{g}{N^2 D} \qquad \qquad \ldots(ii)$$

Substituting the dimensions in both sides of the above equation,
Similarly, $\pi_3 = (N)^{a_3} (D)^{b_3} (\rho)^{c_3} \cdot H$
Substituting the dimensions in both sides of the above equation,
$$M^0 L^0 T^0 = (T^{-1})^{a_3} \cdot (L)^{b_3} \cdot (ML^{-3})^{c_3} \cdot (L)$$
Now for dimensional homogeneity, equating the powers of M, L and T on both sides of the equation,

for M, $\quad 0 = c_3 \quad$ or $\quad c_3 = 0$

for T, $\quad 0 = -a_3 \quad$ or $\quad a_3 = 0$

for L, $\quad 0 = b_3 - 3c_3 + 1 \quad$ or $\quad b_3 = -1$

Thus, $\quad \pi_3 = N^0 \cdot D^{-1} \cdot \rho^0 \cdot H$

$$= \frac{H}{D} \qquad \ldots(iii)$$

and $\quad \pi_4 = (N)^{a_4} (D)^{b_4} (\rho)^{c_4} \cdot \mu$

Substituting the dimensions in both sides of the above equation,
$$M^0 L^0 T^0 = (T^{-1})^{a_4} \cdot (L)^{b_4} \cdot (ML^{-3})^{c_4} \cdot (ML^{-1} T^{-1})$$
Now for dimensional homogeneity, equating the powers of M, L and T on both sides of the equation,

for M, $\quad 0 = c_4 + 1 \quad$ or $\quad c_4 = -1$

for T, $\quad 0 = -a_4 - 1 \quad$ or $\quad a_4 = -1$

for L, $\quad 0 = b_4 - 3c_4 - 1 \quad$ or $\quad b_4 = -2$

Thus, $\quad \pi_4 = N^{-1} \cdot D^{-2} \cdot \rho^{-1} \cdot \mu$

$$= \frac{\mu}{ND^2 \rho} \qquad \ldots(iv)$$

Now the functional relationship may be written as,
$$f\left[\frac{Q}{ND^3}, \frac{g}{N^2 D}, \frac{H}{D}, \frac{\mu}{ND^2 \rho}\right] = 0$$

Since the product of two π-terms is dimensionless, therefore let us replace the terms π_2 and π_3 by $\dfrac{gH}{N^2 D^2}$ (i.e., multiply the terms π_2 and π_3).

$\therefore \qquad f\left[\dfrac{Q}{ND^3}, \dfrac{gH}{N^2 D^2}, \dfrac{\mu}{ND^2 \rho}\right] = 0$

or $\qquad Q = ND^3 f\left[\dfrac{gH}{N^2 D^2}, \dfrac{\mu}{ND^2 \rho}\right] \quad$ **Ans.**

Note. The expression outside the bracket may be multiplied or divided by any amount, whereas the expression inside the bracket should not be multiplied or divided.

Example 25·15. *The pressure drop (Δp) in a pipe depends upon the mean velocity of flow (v), length of pipe (l), diameter of pipe (d), viscosity of fluid (μ), average height of roughness projections on inside pipe surface (k) and mass density of fluid (ρ). By using Buckingham's π-theorem, obtain a dimensionless expression for Δp. Also show that*

$$h_f = \frac{4flv^2}{2gd}$$

Dimensional Analysis

where h_f is the head loss due to friction ($\Delta p/w$) and w is the specific weight of the fluid and f is the coefficient of friction.

Solution. Given : Pressure drop = Δp; Mean velocity of flow = v; Length of pipe = l; Diameter of pipe = d; Viscosity of fluid = μ; Average height of roughness = k and mass density of fluid = ρ.

Let the functional relationship be
$$\Delta p = f[v, l, d, \mu, k, \rho]$$
The above equation, in its general form, may be written as,
$$f_1[\Delta p, v, l, d, \mu, k, \rho] = 0$$

A little consideration will show, that in the above equation the primary dimension are only four. Thus $m = 4$. Taking v, d and ρ as the repeating variables, we get
$$\pi_1 = (v)^{a1} \cdot (d)^{b1} \cdot (\rho)^{c1} \cdot \Delta p$$
Substituting the dimensions in both sides of the above equation,
or
$$M^0L^0T^0 = [LT^{-1}]^{a1} \times [L]^{b1} \times [ML^{-3}]^{c1} \times [ML^{-1}T^{-2}]$$
Now for dimensional homogeneity, equating the powers of M, L and T on both sides of the equation,

for M, $0 = c_1 + 1$ or $c_1 = -1$
for T, $0 = -a_1 - 2$ or $a_1 = -2$
for L, $0 = a_1 + b_1 - 3c_1 - 1$ or $b = 0$

Thus, $\pi_1 = v^{-2} \cdot d^0 \cdot \rho^{-1} \cdot \Delta p$

$$= \frac{\Delta p}{\rho v^2} \qquad \qquad ...(i)$$

Similarly, $\pi_2 = (v)^{a2} \cdot (d)^{b2} \cdot (\rho)^{c2} \cdot l$

Substituting the dimensions in both sides of the above equation,
or
$$M^0L^0T^0 = (LT^{-1})^{a2} \cdot (L)^{b2} \cdot (ML^{-3})^{c2} \cdot (L)$$
Now for dimensional homogeneity, equating the powers of M, L and T on both sides of the equation,

for M, $0 = c_2$ or $c_2 = 0$
for T, $0 = -a_2$ or $a_2 = 0$
for L, $0 = a_2 + b_2 - 3c_2 + 1$ or $b_2 = -1$

Thus, $\pi_2 = v^0 \cdot d^{-1} \cdot \rho^0 \cdot l$

$$= \frac{l}{d} \qquad \qquad ...(ii)$$

Similarly, $\pi_3 = (v)^{a3} \cdot (d)^{b3} \cdot (\rho)^{c3} \cdot \mu$

Substituting the dimensions in both sides of the above equation,
or
$$M^0L^0T^0 = (LT^{-1})^{a3} \cdot (L)^{b3} \cdot (ML^{-3})^{c3} (ML^{-1}T^{-1})$$
Now for dimensional homogeneity, equating the powers of M, L and T on both sides of the equation,

for M, $0 = c_3 + 1$ or $c_3 = -1$
for T, $0 = -a_3 - 1$ or $a_3 = -1$
for L, $0 = a_3 + b_3 - 3c_3 - 1$ or $b_3 = -1$

Thus, $\pi_3 = v^{-1} \cdot d^{-1} \cdot \rho^{-1} \cdot \mu$

$$= \frac{\mu}{vd\rho} \qquad \qquad ...(iii)$$

and
$$\pi_4 = (v)^{a_4} \cdot (d)^{b_4} \cdot (\rho)^{c_4} \cdot k$$

Substituting the dimensions in both sides of the above equation,

or
$$M^0 L^0 T^0 = (LT^{-1})^{a_4} \cdot (L)^{b_4} \cdot (ML^{-3})^{c_4} (L)$$

Now for dimensional homogeneity, equating the powers of M, L and T on both sides of the equation,

for M, $\qquad 0 = c_4 \qquad$ or $\;c_4 = 0$

for T, $\qquad 0 = -a_4 \qquad$ or $\;a_4 = 0$

for L, $\qquad 0 = a_4 + b_4 - 3c_4 + 1 \qquad$ or $\;b_4 = -1$

Thus
$$\pi_4 = v^0 \cdot d^{-1} \cdot \rho^0 \cdot k$$
$$= \frac{k}{d} \qquad \ldots(iv)$$

Now the functional relationship may be written as
$$f_1\left(\frac{\Delta p}{\rho v^2}, \frac{l}{d}, \frac{\mu}{v d \rho}, \frac{k}{d}\right) = 0$$

∴
$$f_1\left(\frac{\Delta p d}{\rho v^2}, \frac{l}{d}, \frac{\mu}{v d \rho}, \frac{k}{d}\right) = 0$$

or
$$\Delta p = \frac{\rho v^2 l}{d} f_1\left(\frac{l}{d}, \frac{\mu}{v d \rho}, \frac{k}{d}\right)$$

Let
$$2f = f_1\left(\frac{l}{d}, \frac{\mu}{v d \rho}, \frac{k}{d}\right) \qquad \ldots\text{(where }f = \text{Coefficient of friction)}$$

∴
$$\Delta p = \frac{2f \rho v^2 l}{d} = \frac{4f \rho v^2 l}{2d}$$

Dividing the left hand side by w (*i.e.*, the specific weight of the fluid) and right hand side by its equivalent 'ρg' we have

$$\frac{\Delta p}{w} = \frac{4f \rho v^2 l}{2 \rho g d} \quad \text{or} \quad h_f = \frac{4f l v^2}{2gd} \quad \textbf{Ans.}$$

EXERCISE 25·2

1. Prove that shear stress (τ) in a fluid flowing through a rough pipe of diameter d with a velocity v is given by the relation :

$$\tau = \rho v^2 f\left(\frac{d v \rho}{\mu}, \frac{k}{d}\right)$$

where ρ is the density of the fluid, μ viscosity of the fluid and k average height of the roughness.

2. Show by the use of Buckingham's π-theorem that the velocity through a circular orifice is given by:

$$V = \sqrt{2gH} \; f\left[\frac{D}{H}, \frac{\mu}{\rho V H}\right]$$

where H is the head causing flow, D diameter of the orifice, μ the coefficient of viscosity, ρ the mass density and g is the gravitational acceleration.

[**Hint.** Take V, H and ρ as the repeating variables.]

Dimensional Analysis

3. Show that torque T on a shaft of diameter d, rotating at a speed N in a fluid of density ρ and viscosity μ is given by :

$$T = \rho d^5 N^2 f\left(\frac{\mu}{\rho N d^2}\right)$$

QUESTIONS

1. Explain the term 'Dimensional Analysis' and discuss its principle.
2. What is a fundamental dimension as used in the principle of Dimensional Analysis?
3. Give the dimensions of the following in M-L-T and F-L-T system.
   (i) Acceleration, (ii), Power, (iii) Kinematic viscosity, and (iv) Bulk modulus.
4. Define the term 'dimensional homogeneity'. How it is obtained?
5. Write the uses of the principle of dimensional homogeneity. Give suitable examples in support of your answer.
6. Explain the Rayleigh's method for dimensional analysis.
7. State Buckingham's π-theorem. Also discuss as to how does the Buckingham's method differs from that of Rayleigh's method.

OBJECTIVE TYPE QUESTIONS

1. The dimension of pressure in MLT system is
   (a) MLT^{-2} (b) ML^2T^2 (c) $ML^{-1}T^{-2}$ (d) $ML^{-1}T^{-3}$
2. The dimension of power in FLT system is
   (a) FL^2T (b) FLT^{-1} (c) $FL^{-2}T^{-2}$ (d) $FL^{-1}T^{-2}$
3. Two equations are said be dimensionally homogeneous, if the fundamental dimensions have identical powers of
   (a) mass   (b) length   (c) time   (d) all of these
4. The principle of dimensional homogeneity is used to
   (a) determine the dimension of a physical quantity
   (b) check the dimensional homogeneity of an equation
   (c) change the coefficient of an equation while using the other system of units
   (d) all of these

ANSWERS

1. (c)   2. (b)   3. (d)   4. (d)

26

Model Analysis
(Undistorted Models)

1. Introduction. 2. Advantages of Model Analysis. 3. Hydraulic Similarity. 4. Geometric Similarity. 5. Kinematic Similarity. 6. Dynamic Similarity. 7. Procedure for Model Analysis. 8. Selection of Suitable Scale for the Model. 9. Construction of the Model. 10. Testing of the Model. 11. Correct Prediction. 12. Classification of Models. 13. Undistorted Models. 14. Comparison of an Undistorted Model and its Prototype. 15. Velocity of Water in the Prototype for the Given Velocity of an Undistorted Model. 16. Discharge of the Prototype for the Given Discharge of an Undistorted Model. 17. Force of Resistance of the Prototype for the Given Force of Resistance of an Undistorted Model. 18. Speed of the Prototype for the Given Speed of an Undistorted Model. 19. Power Developed by the Prototype for the Given Power Developed by an Undistorted Model. 20. Time of Emptying a Prototype for the Given Time of Emptying an Undistorted Model. 21. Travelling time of the Prototype for the Given Travelling Time of an Undistorted Model.

26·1 Introduction

Since the beginning of the twentieth century, the engineers engaged on the creation or design of hydraulic structures (such as dams, spillways or large hydraulic machines) have started a new and scientific method to predict the performance of their structures and machines. This is done by preparing models and testing them in a laboratory; so as to form some opinion, about the working and behaviour of the proposed hydraulic structures, after their completion or actual installation.

The structure, of which the model is prepared, is known as prototype and the model is known as scale model or simply model.

26·2 Advantages of Model Analysis

Though there are numerous advantages of model testing, yet the following are important from the subject point of view:

1. The behaviour and working details of a hydraulic structure or a machine can be easily predicted from its model. The smooth and reliable working of a hydraulic structure or a machine can be ascertained by spending a small sum, which constitutes a negligible fraction of the total cost to be spent on the prototype.
2. If the hydraulic structure or machine is made directly, then in case of its failure, it is very difficult to change its design. Moreover, it is very costly to rebuild the same. Thus making a model and testing it in the laboratory results in the saving of human labour and material.
3. With the help of model testing, a number of alternative designs can be studied. Finally, the most economical, accurate and safe design may be selected.
4. In case, when the existing hydraulic structure is not functioning properly, then the model testing can help us in detecting and rectifying the defects.
5. Sometimes, it is difficult to design a particular portion of a complex hydraulic structure or machine. In such a case, the model testing is very essential in order to ascertain the safety and reliability of that particular portion of the prototype.

Model Analysis (Undistorted Models)

26·3 Hydraulic Similarity

If we see a photograph of a man, very carefully, we can have an idea of the proportion of various parts of his body. The photograph will also give an idea of features of each part of the man. Similarly, to know the complete working and behaviour of the prototype, from its model, there should be a complete similarity between the prototype and its scale model. This similarity is known as hydraulic similitude or hydraulic similarity. Following three types of hydraulic similarities are important from the subject point of view:

1. Geometric similarity,
2. Kinematic similarity, and
3. Dynamic similarity.

26·4 Geometric Similarity

The geometric similarity is said to exist between the model and the prototype, if both of them are identical in shape; but differ only in size. Or in other words, the geometric similarity is said to exist between the model and the prototype, if the ratios of all the corresponding linear dimensions are equal.

Let L = Length of the prototype,
B = Breadth of the prototype,
D = Depth of the prototype, and

and l, b and d = Corresponding values for the model.

Now, if the geometric similarity exists, between the prototype and the model, then the linear ratio of the prototype and the model (also called scale ratio).

$$L_r = \frac{L}{l} = \frac{B}{b} = \frac{D}{d}$$

Similarly, area ratio of the prototype and the model,

$$A_r = \left(\frac{L}{l}\right)^2 = \left(\frac{B}{b}\right)^2 = \left(\frac{D}{d}\right)^2 = (L_r)^2$$

and volume ratio of the prototype and the model,

$$V_r = \left(\frac{L}{l}\right)^3 = \left(\frac{B}{b}\right)^3 = \left(\frac{D}{d}\right)^3 = (L_r)^3$$

26·5 Kinematic Similarity

The kinematic similarity is said to exist, between the model and the prototype, if both of them have identical motions or velocities. Or in other words, the kinematic similarity is said to exist between the model and the prototype, if the ratio of the corresponding velocities at corresponding points are equal.

Let V_1 = Velocity of liquid in prototype at point 1,
V_2 = Velocity of liquid in prototype at point 2,
v_1, v_2 = Corresponding values for the model.

Now, if the kinematic similarity exists, between the prototype and the model, then the velocity ratio of the prototype to the model,

$$v_r = \frac{V_1}{v_1} = \frac{V_2}{v_2} = \frac{V_3}{v_3} = \ldots$$

26·6 Dynamic Similarity

The dynamic similarity is said to exist, between the model and the prototype, if both of them have identical forces. Or in other words, the dynamic similarity is said to exist between the model and the prototype, if the ratios of the corresponding forces acting at corresponding points are equal.

Let F_1 = Force acting in the prototype at point 1, and
F_2 = Force acting in the prototype at point 2,
f_1, f_2 = Corresponding values for the model.

Now, if the dynamic similarity exists between the prototype and the model, then the force ratio of the prototype and the model,

$$F_r = \frac{F_1}{f_1} = \frac{F_2}{f_2} = \frac{F_3}{f_3} = \ldots$$

26·7 Procedure for Model Analysis

The general procedure for the model analysis involves the following steps:
1. Selection of a suitable scale for the model.
2. Construction of the model.
3. Testing of the model.
4. Correct prediction.

26·8 Selection of Suitable Scale for the Model

It is very important point in the planning of a model. Though the selection of scale, for the model analysis, depends upon many factor, yet the following are important from the subject point of view.
1. Availability of funds.
2. Availability of time.
3. Availability of water.
4. Availability of space for accommodating the model.
5. Types of results desired.

Strictly speaking, the scale ratios adopted are different for different types of structures. But the following scale ratios are generally adopted:

1. Head works, canals, soluice gates etc. $\ldots \frac{1}{5}$ to $\frac{1}{25}$

2. Dams, spillways etc. $\ldots \frac{1}{30}$ to $\frac{1}{400}$

3. Rivers, harbours etc. $\ldots \frac{1}{100}$ to $\frac{1}{1000}$

26·9 Construction of the Model

After selecting the type and suitable material for the model, the next step is its construction very accurately according to the plan. If the model is a curved one, then a centre line is marked before its construction. The vertical heights are controlled with the help of a very accurate and precise level.

26·10 Testing of the Model

It is the most scientific and highly technical feature of the design and successful working of a hydraulic structure or machine. The following two methods are generally adopted for the testing of a model:

Model Analysis (Undistorted Models)

1. *Wind tunnel method.* In this method, a steady and uniform stream of air is provided in the laboratory, which flows around the model suspended in the stream. Though the walls interfere with the flow of air stream, yet its effect is neglected.

   In the wind tunnel, the air is set in motion by means of a fan or compressor. Sometimes the compression of air, inside the tunnel, produces an appreciable rise in temperature, which must be dissipated by a cooling arrangement.

2. *Water tunnel method.* In this method, a steady and uniform stream of water is made to flow and the model, under test, is mounted in the path of the flowing water.

   The size of water tunnel is usually expressed as the diameter of the best section. The existing water tunnels range from 500 mm to 2 m in size.

It is very essential to use very high degree instruments for the precise measurement of the hydraulic quantities in the experiments. A great care and patience is required for correctly predicting the model results.

26·11 *Correct Prediction

After obtaining the precise measurements of hydraulic quantities of a model, in the laboratory, the next step is the correct prediction of the working of its prototype. Now we shall study the correct predictions of various types of prototype in the following pages.

26·12 Classification of Models

All the hydraulic models may be broadly classified into the following two types :

1. Undistorted model, and
2. Distorted models.

26·13 Undistorted Models

A model, which is geometrically similar to the prototype (*i.e.* having geometric similarly in length, breadth, height and head of water etc.) is known as an undistorted model. The prediction of an undistorted model is comparatively easy and some of the results, obtained from the models, can be easily transferred to the prototypes as the basic condition (of geometric similarity) is satisfied. In this chapter, we shall study the undistorted models only.

26·14 Comparison of an Undistorted Model and its Prototype

We have discussed in articles 26·3 to 26·5 the different types of hydraulic similarities between the model and the prototype. If the model is to be overall similar to the prototype, then all the three similarities (*i.e.,* geometric, kinematic and dynamic) should exist even between the model and its prototype. But this is not possible in actual practice; as it is difficult to exist even two types of similarities simultaneously. In general, and undistorted model of a prototype is made keeping in view the geometric similarity only and the remaining similarities are then compared for the scale ratio (*i.e.* geometric ratio of the prototype and the model).

Though the given scale ratio provides us a wide range of data of the prototype, yet the following are important from the subject point of view :

1. Velocity of water in the prototype, for the given velocity at the corresponding point of the model.
2. Discharge of the prototype for the given discharge of the model.
3. Force of resistance on the prototype for the given force of resistance on the model.

---

*The correct prediction of a model is an important step in the study of Model Analysis. Prof. Khurmi, who is the author of this book, has given a number of methods for the correct prediction of a model. He was awarded Ph.D. Degree for these methods, which have earned an international recognition. A further research on these methods is being done in so many universities and hydraulic research centres all over the world.

4. Speed of the prototype, for the given speed of the model.
5. Power developed by the prototype for the given power developed by the model.
6. Time of emptying a prototype for the given time of emptying the model.
7. Travelling time of a prototype for the given travelling time of the model.

26·15 Velocity of Water in the Prototype for the Given Velocity in an Undistorted Model

Consider an undistorted model geometrically similar to a proposed prototype like a weir, dam, spillway etc.

Let h = Head of water over the model,
v = Velocity of water at a point in the model,
H, V = Corresponding values for the prototype.
$\dfrac{1}{s}$ = *Scale ratio of the model to the prototype.

We know that the velocity of water in the model,
$$v = c_v \sqrt{2gh} \qquad \ldots(i)$$
Similarly, velocity of water on the corresponding point in the prototype,
$$V = C_v \sqrt{2gH} \qquad \ldots(ii)$$
Dividing the equation (i) by (ii),
$$\dfrac{v}{V} = \dfrac{c_v \sqrt{2gh}}{C_v \sqrt{2gH}} = \dfrac{\sqrt{h}}{\sqrt{H}} \qquad \ldots(\text{Taking } c_v = C_v)$$
$$= \sqrt{\dfrac{h}{H}} = \sqrt{\dfrac{1}{s}} \qquad \ldots\left(\because \dfrac{h}{H} = \dfrac{1}{s}\right)$$
$$\therefore \quad V = v\sqrt{s}$$

Example 26·1. *The velocity at a point on a spillway model of a dam is 1·3 m/s for a prototype of model ratio 1:10. What is the velocity at the corresponding point in the prototype?*

Solution. Given : $v = 1·3$ m/s and $\dfrac{1}{s} = \dfrac{1}{10}$ or $s = 10$.

We know that velocity of water at the corresponding point of the prototype,
$$V = v\sqrt{s} = 1·3 \sqrt{10} = 1·3 \times 3·16 = 4·11 \text{ m/s} \quad \textbf{Ans.}$$

Example 26·2. *A 1:9 model of a barge is towed through water. What should be the towing speed of the model to simulate a speed of 6·15 m/s, if the resistance is due to waves only.*

Solution. Given : $\dfrac{1}{s} = \dfrac{1}{9}$ or $s = 9$ and $V = 6·15$ m/s.

Let v = Towing speed of the model.

We know that speed of the prototype (V),
$$6·15 = v\sqrt{s} = v\sqrt{9} = 3v$$
$$\therefore \quad v = 6·15/3 = 2·05 \text{ m/s} \quad \textbf{Ans.}$$

---

* Scale ratio exactly means the ratio of linear dimension of the model to the corresponding linear dimension of the prototype. *e.g.*, if the ratio of model to the prototype is 1:10, then

$$\text{scale ratio} = \dfrac{1}{s} = \dfrac{\text{Linear dimension of the model}}{\text{Linear dimension of the prototype}} = \dfrac{1}{10} \text{ or } s = 10$$

Example 26·3. *A model of a harbour is made to a scale of 1:150. It was observed that the waves strike the barrier with a velocity of 10 m/s. Find the velocity with which the waves in the model will strike to barrier.*

Solution. Given : $\dfrac{1}{s} = \dfrac{1}{150}$ or $s = 150$ and $V = 10$ m/s

Let $\qquad v =$ Velocity with which the waves will strike its model.

We know that the velocity with which the waves strike in the prototype (V),
$$10 = v\sqrt{s} = v\sqrt{150} = 12\cdot 25\ v$$
$$\therefore\quad v = 10/12\cdot 25 = 0\cdot 82\ \text{m/s}\quad \textbf{Ans.}$$

26·16 Discharge of the Prototype for the Given Discharge of an Undistorted Method

Consider an undistorted model geometrically similar to a prototype, like a weir, notch or spillway etc.

Let
$\qquad a =$ Area of discharge of the model,
$\qquad v =$ Actual velocity of water in the model,
$\qquad q =$ Discharge of the model,
$A, V, Q =$ Corresponding values for the prototype.
$\qquad \dfrac{1}{s} =$ Scale ratio of model to prototype.

We know that the discharge of the model,
$$q = \text{Area} \times \text{Velocity} = a \times v \qquad \ldots(i)$$

Similarly, discharge through the corresponding area of the prototype,
$$Q = A \times V \qquad \ldots(ii)$$

Dividing the equation (i) by (ii),
$$\dfrac{q}{Q} = \dfrac{a \times v}{A \times V} = \dfrac{a}{A} \times \dfrac{v}{V}$$
$$= \dfrac{1}{s^2} \times \dfrac{1}{\sqrt{s}} = \dfrac{1}{s^{2\cdot 5}}$$
$$\therefore\quad Q = q \times s^{2\cdot 5}$$

Example 26·4. *The discharge over a model, which is reduced to 1:100 in all its dimensions is 1·5 litre/s. What is the corresponding discharge in the prototype?*

Solution. Given : $\dfrac{1}{s} = \dfrac{1}{100}$ or $s = 100$ and $q = 1\cdot 5$ litres/s

We know that discharge over the prototype,
$$Q = q \times (s)^{2\cdot 5} = 1\cdot 5\ (100)^{2\cdot 5} = 150\ 000\ \text{litres/s}$$
$$= 150\ \text{m}^3/\text{s}\ \textbf{Ans.}$$

Example 26·5. *A model of spillway is constructed to a scale of 1 : 30 in a flume. The length of the spillway is 30 m. If the discharge over the spillway at the head of 6 m (depth of flow over spillway) is 443·6 cumes, calculate the corresponding head and discharge of the model required for this model study.*

Solution. Given : $\dfrac{1}{s} = \dfrac{1}{30}$ or $s = 30$; $L = 30$ m; $H = 6$ m and $Q = 443\cdot 6$ m³/s.

Head of the model

We know that head of the model,
$$h = \dfrac{H}{s} = \dfrac{6}{30} = 0\cdot 2\ \text{m} = 200\ \text{mm}\ \textbf{Ans.}$$

Discharge of the model

Let q = Discharge of the model.

We know that discharge over the spill way (Q),

$$443 \cdot 6 = q \times (s)^{2.5} = q \times (30)^{2.5} = 4930\ q$$

$\therefore\qquad q = 443 \cdot 6/4930 = 0 \cdot 09\ \text{m}^3/\text{s} = 90\ \text{litres/s}$ **Ans.**

Example 26·6. *A dam 35 m long is to discharge water at the rate of 114 m^3/s under a head of 2·7 m. Find the length of the model and head of water, if the supply available in the laboratory is 30 litres/s.*

Solution. Given : $L = 35$ m; $Q = 114\ \text{m}^3/\text{s}$; $H = 2 \cdot 7$ m and $q = 30$ litres/s $= 0 \cdot 03\ \text{m}^3/\text{s}$.

Let $\dfrac{1}{s}$ = Scale ratio of model to prototype.

We know that discharge over the prototype (Q),

$$114 = q \times (s)^{2.5} = 0 \cdot 03 \times (s)^{2.5}$$

or $\qquad (s)^{2.5} = 114/0 \cdot 03 = 3800 \quad \text{or} \quad s = 27$

$\therefore\qquad$ Length of model, $l = \dfrac{L}{s} = \dfrac{35}{27} = 1 \cdot 3$ m **Ans.**

and head of water over the model,

$$h = \dfrac{H}{s} = \dfrac{2 \cdot 7}{27} = 0 \cdot 1\ \text{m} = 100\ \text{mm} \quad \textbf{Ans.}$$

EXERCISE 26·1

1. A 1:16 model of a boat is moving in a tank with a velocity of 0·5 m/s. Find the speed at which the boat will move, if the model has similarity in wave resistance. **(Ans.** 2 m/s)

2. In a 1:100 model of a dam, the waves were found to move with a velocity of 0·25 m/s. Find the velocity of the waves in the prototype. **(Ans.** 2·5 m/s)

3. The discharge over a 1:25 model of a weir was observed to be 0·2 litre/s. Find the corresponding discharge over the weir. **(Ans.** 625 litres/s)

4. The discharge over a spillway is 50 m³/s. A model of the spillway is built with a scale of 1:25. Find the discharge over the model. **(Ans.** 16 litres/s)

5. A model of an open channel is made in a laboratory. If the actual discharge of the prototype is 102·4 m³/s and corresponding discharge over the model is 100 litres/s, find the scale of the model. **(Ans.** 1:16)

26·17 Force of Resistance of the Prototype for the Given Force of Resistance of an Undistorted Model

Consider an undistorted model geometrically similar to a prototype, like a ship, aeroplane etc.

Let $\qquad a$ = Area of model subjected to resistance,

$\qquad\qquad v$ = Velocity of the model,

$\qquad A, V$ = Corresponding values for the prototype, and

$\qquad\dfrac{1}{s}$ = Scale ratio of model to prototype.

We know that the force on the model

$$f = \text{Mass} \times \text{Acceleration}$$

$$= (\text{Volume} \times \text{Mass density}) \times \dfrac{\text{Velocity}}{\text{Time}}$$

Model Analysis (Undistorted Models)

$$= \frac{\text{Volume}}{\text{Time}} \times \text{Mass density} \times \text{Velocity}$$

$$= (\text{Area} \times \text{Velocity}) \times \text{Mass density} \times \text{Velocity}$$

$$= \text{Area} \times \text{Mass density} \times \text{Velocity}^2$$

$$= a \times \rho \times v^2 \qquad \ldots(i)$$

Similarly, force on the prototype,

$$F = A \times \rho \times V^2 \qquad \ldots(ii)$$

Dividing equation (i) by (ii),

$$\frac{f}{F} = \frac{a}{A} \times \frac{\rho}{\rho} \times \frac{v^2}{V^2} = \frac{1}{s^2} \times \frac{v^2}{(v\sqrt{s})^2} = \frac{1}{s^3}$$

or $\qquad F = f \times s^3$

Note : If the model is tested in a different fluid (*i.e.* other than the fluid, in which the prototype is to move, the corresponding value for the force will become:

$$F = f \times \frac{\rho_p}{\rho_m} \times s^3$$

where ρ_p and ρ_m are the mass densities of the fluids, in which the prototype and the model is to move.

Example 26·7. *In a sea side laboratory, the resistance measured to a ship model of 1:50 scale while moving at 0·15 m/s was 1·2 N. Calculate the corresponding resistance to the prototype.*

Solution. Given : $\frac{1}{s} = \frac{1}{50}$ or $s = 50$; $v^* = 0.15$ m/s and $f = 1.2$ N.

We know that the resistance to the prototype,

$$F = f \times (s)^3 = 1.2 \times (50)^3 = 150\ 000\ \text{N} = 150\ \text{kN Ans.}$$

Example 26·8. *The resistance on a ship is to be found out from the model tests on 1:60 model in wind tunnel. If the drag on the model was found to be 0·005 N, what will be the drag on the prototype? Take density of the air as 1·2 kg/m³.*

Solution. Given : $\frac{1}{s} = \frac{1}{60}$ or $s = 60$; $f = 0.005$ N and $\rho_m = 1.2$ kg/m³.

We know that drag on the prototype,

$$F = f \times \frac{\rho_p}{\rho_m} \times (s)^3 = 0.005 \times \frac{1030}{1.2} \times (60)^3 = 927\ 000\ \text{N}$$

$$= 927\ \text{kN Ans.}$$

Note : The mass density of sea water (ρ_p) is taken as 1030 kg/m³.

Example 26·9. *The resistance measured in fresh water of 1 to 80 model ship when moving at 1·2 m/s is 12·5 N. Calculate the corresponding velocity and resistance to the prototype in sea water.*

Solution. Given : $\frac{1}{s} = \frac{1}{80}$ or $s = 80$; $v = 1.2$ m/s and $f = 12.5$ N.

Velocity of the prototype in sea water

We know that velocity of prototype in the sea water

$$V = v\sqrt{s} = 1.2\ \sqrt{80} = 1.2 \times 8.94 = 10.7\ \text{m/s Ans.}$$

Resistance to the prototype in sea water

We also know that resistance to the prototype in sea water,

$$F = f \times \frac{\rho_p}{\rho_m} \times (s)^3 = 12.5 \times \frac{1030}{1000} \times (80)^3 = 6\ 592\ 000\ \text{N}$$

$$= 6592\ \text{kN Ans.}$$

---

*Superfluous data.

Example 26·10. *A 1/25 model of a ship, when towed in water at a velocity of 1·2 m/s, experiences a resistance of 5·0 N. The wetted area of the model is 1 m². Calculate for the prototype:*
(i) corresponding speed, and
(ii) drag, if the drag coefficient for model is 0·004 and that for prototype is 0·016.

Solution. Given : $\dfrac{1}{s} = \dfrac{1}{25}$ or $s = 25$; $v = 1\cdot2$ m/s; $f = 5\cdot0$ N; $a = 1$ m²; $k_{DM} = 0\cdot004$ and $k_{DP} = 0\cdot016$.

(i) Corresponding speed of the prototype

We know that corresponding speed of the prototype,
$$V = v\sqrt{s} = 1\cdot2\ \sqrt{25} = 1\cdot2 \times 5 = 6\cdot0 \text{ m/s} \quad \textbf{Ans.}$$

(ii) Total drag on the prototype

We know that wetted area of the prototype,
$$A = a \times s^2 = 1 \times (25)^2 = 625 \text{ m}^2$$

and drag for the model,
$$f_1 = k_{DM} \times \dfrac{wav^2}{2g} = 0\cdot004 \times \dfrac{9\cdot81 \times 1 \times (1\cdot2)^2}{2 \times 9\cdot81} \text{ kN}$$
$$= 2\cdot88 \times 10^{-3} \text{ kN} = 2\cdot88 \text{ N}$$

∴ Resistance to the prototype due to waves,
$$f_2 = f - f_1 = 5\cdot0 - 2\cdot88 = 2\cdot12 \text{ N}$$

Similarly drag for the prototype,
$$F_1 = k_{DP} \times \dfrac{wAV^2}{2g} = 0\cdot016 \times \dfrac{9\cdot81 \times 625 \times (6)^2}{2 \times 9\cdot81} = 180 \text{ kN}$$

and resistance to the prototype due to waves,
$$F_2 = f_2 \times (s)^3 = 2\cdot12 \times (25)^3 = 33\ 125 \text{ N} = 33\cdot125 \text{ kN}$$

∴ Total drag on the prototype,
$$F = F_1 + F_2 = 180 + 33\cdot125 = 213\cdot125 \text{ kN} \quad \textbf{Ans.}$$

26·18 Speed of the Prototype for the Given Speed of an Undistorted Model

Consider an undistorted model geometrically similar to a prototype, like a centrifugal pump or turbine.

Let
d = Diameter of the model impeller,
v = Tangential velocity of the model impeller,
n = Speed of the model runner,
D, V, N = Corresponding values for the prototype runner, and
$\dfrac{1}{s}$ = Scale ratio of model to prototype,

We know that tangential velocity of the model impeller,
$$v = \dfrac{\pi dn}{60} \quad \text{or} \quad n = \dfrac{60v}{\pi d} \qquad \ldots(i)$$

Similarly, speed of the prototype runner,
$$N = \dfrac{60\ V}{\pi D} \qquad \ldots(ii)$$

Model Analysis (Undistorted Models)

Divising the equation (i) by (ii),

$$\frac{n}{N} = \frac{\frac{60v}{\pi d}}{\frac{60V}{\pi D}} = \frac{v}{V} \times \frac{D}{d}$$

$$= \frac{1}{\sqrt{s}} \times s = \sqrt{s} \qquad \left(\because \frac{D}{d} = s\right)$$

or $\qquad N = \dfrac{n}{\sqrt{s}}$

Example 26·11. *A turbine model of scale 1:10 is running at 475 r.p.m. under a head of 20 metres. Find the speed of the actual turbine.*

Solution. Given : $\dfrac{1}{s} = \dfrac{1}{10}$ or $s = 10$; $n = 475$ r.p.m. and $h^* = 20$ m.

We know that speed of the prototype (or actual turbine),

$$N = \frac{n}{\sqrt{s}} = \frac{475}{\sqrt{10}} = 150 \cdot 2 \text{ r.p.m.} \quad \textbf{Ans.}$$

Example 26·12. *A pump in London sewerage disposal works is required to work at a speed of 250 r.p.m. Find the speed of the model, which has a scale ratio of 1:25.*

Solution. Given : $N = 250$ r.p.m. and $\dfrac{1}{s} = \dfrac{1}{25}$ or $s = 25$.

Let $\qquad n = $ Speed of the model.

We know that speed of the prototype (N),

$$250 = \frac{n}{\sqrt{s}} = \frac{n}{\sqrt{25}} = \frac{n}{5}$$

$\therefore \qquad n = 250 \times 5 = 1250$ r.p.m. **Ans.**

EXERCISE 26·2

1. A 1:30 ship model was found to have a resistance of 0·8 N, when it was moving in sea water. What will be the corresponding resistance to the prototype? **(Ans. 21·6 kN)**
2. The resistance measured to a ship model of 1:40 scale in fresh water was found to be 2·5 N. Find the corresponding resistance to the prototype in the sea water. **(Ans. 164·8 kN)**
3. Find the speed of a pump, if its model of scale 1 : 15 is running at 180 r·p·m. **(Ans. 46·5 r.p.m.)**
4. In a power station, a turbine is designed to run at 200 r.p.m. What is the speed of its model of scale 1:12 ? **(Ans. 692 r.p.m.)**

26·19 Power Developed by the Prototype for the Given Power Developed by an Undistorted Model

Consider an undistorted model geometrically similar to a prototype, like a turbine.

Let
- $q = $ Discharge of the model in m³/s,
- $h = $ Head of water over the model in metres,
- $p = $ Power developed by the model,
- $Q, H, P = $ Corresponding values for prototype, and
- $\dfrac{1}{s} = $ Scale ratio of the model to the prototype.

---

*Superfluous data.

We know that the power developed by the model,
$$p = wqh \qquad \ldots(i)$$
Similarly, power developed by the prototype,
$$P = wQH \qquad \ldots(ii)$$
Dividing equation (i) by (ii),
$$\frac{p}{P} = \frac{wqh}{wQH} = \frac{q}{Q} \times \frac{h}{H} = \frac{1}{s^{2.5}} \times \frac{1}{s} = \frac{1}{s^{3.5}}$$
$$P = p \times s^{3.5}$$

Notes : 1. Sometimes the model of a ship etc. is made and tested in the laboratory. In such a case, there is a difference in the specific weights of the sea water and the fresh water in the laboratory. It is thus obvious, that in such a case the two specific weights should also be considered. In such a case,
$$P = p \times \frac{w_p}{w_m} \times s^{3.5}$$

2. Similarly, if the model of the ship is tested in a wind tunnel, then the specific weight of the air in the wind tunnel should also be taken into consideration. In such a case,
$$P = p \times \frac{w_p}{w_m} \times s^{3.5}$$
where w_m is the specific weight of the air in wind tunnel.

3. The same relation can also be used for the dissipation of energy (as energy dissipated per second is power).

Example 26·13. *A Pelton wheel develops 1100 kW, while discharging 7500 litres of water per second at 300 r.p.m. Find the corresponding power of a one-ninth scale model, assuming efficiencies of the two turbines to be the same.*

Solution. Given : $P = 1100$ kW; $q = 7500$ litres/s; *$N = 300$ r.p.m. and $\frac{1}{s} = \frac{1}{9}$ or $s = 9$.

Let $\qquad p = $ Power developed by the model.

We know that power developed by the prototype (P)
$$1100 = p \times (s)^{3.5} = p \times (9)^{3.5} = 2187\, p$$
$$\therefore \qquad p = 1100/2187 = 0.5 \text{ kW} \quad \textbf{Ans.}$$

Example 26·14. *A model of ship of 1/12 size is tested in fresh water for the prediction of its performance. Find the ratio of speeds of the model to the speed of the prototype operating in sea water for geometrically similar free surface condition. Also calculate the ratio of the powers of the model with that of prototype.*

Assume mass density of fresh water as 1000 kg/m³ and that of sea water as 1030 kg/m³.

Solution. Given : $\frac{1}{s} = \frac{1}{12}$ or $s = 12$; $\rho_m = 1000$ kg/m³ and $\rho_p = 1030$ kg/m³

Ratio of speeds of the model to the speed of the prototype

Let $\qquad v = $ Speed of the model,

We know that speed of the prototype,
$$V = v\sqrt{s} = v\sqrt{12} = 3.464\, v$$
$$\therefore \qquad \frac{v}{V} = \frac{1}{3.464} = 0.289 \quad \textbf{Ans.}$$

Ratio of the powers of the model to the power of the prototype

Let $\qquad p = $ Power required by the model.

---
*Superfluous data.

Model Analysis (Undistorted Models)

We also know that power required by the prototype,

$$P = p \times \frac{\rho_p}{\rho_m} \times (s)^{3.5} = p \times \frac{1030}{1000} \times (12)^{3.5} = 6166\, p$$

$$\therefore \quad \frac{p}{P} = \frac{1}{6166} = 0.000\,16 \quad \textbf{Ans.}$$

Example 26·15. *In 1:20 model of a basin, the hydraulic jump in the model is observed to be 0·2 m. What is the height of hydraulic jump in the prototype? If the energy dissipated in the model is 0·1 kW, what is the corresponding value in the prototype?*

Solution. Given : $\dfrac{1}{s} = \dfrac{1}{20}$ or $s = 20$; $h = 0\cdot2$ m and $p = 0\cdot1$ kW

Height of hydraulic jump in the prototype

We know that the height of the hydraulic jump in the prototype,

$$H = h \times s = 0\cdot2 \times 20 = 4 \text{ m } \textbf{Ans.}$$

Energy dissipated in the prototype

We also know that energy dissipated by the prototype,

$$E = e \times s^{3.5} = 0\cdot1 \times (20)^{3.5} = 3578 \text{ kW } \textbf{Ans.}$$

26·20 Time of Emptying a Prototype for the Given Time of Emptying of an Undistorted Model.

Consider an undistorted model, geometrically similar to a prototype, like a tank or a reservoir.

Let v = Total volume of water in the model,

q = Rate of discharge of water in the model,

V, Q = Corresponding values for the prototype, and

$\dfrac{1}{s}$ = Scale ratio of the model to the prototype.

We know that time of emptying the model,

$$t = \frac{\text{Total volume}}{\text{Rate of discharge}} = \frac{v}{q} \qquad \ldots(i)$$

and time of emptying the prototype,

$$T = \frac{V}{Q} \qquad \ldots(ii)$$

Dividing equation (i) by (ii),

$$\frac{t}{T} = \frac{\frac{v}{q}}{\frac{V}{Q}} = \frac{v}{V} \times \frac{Q}{q} = \frac{1}{s^3} \times s^{2.5} = \frac{1}{\sqrt{s}}$$

$$\therefore \quad T = t \times \sqrt{s}$$

Example 26·16. *Machio dam, in Japan, was modelled with a model scale of 1/60. The prototype is an ogee spillway designed to carry a flood of 3200 m³/s. What was the discharge of the model for the designed flood in m³/s ? What time in the model is represented by one day in the prototype?*

Solution. Given : $\dfrac{1}{s} = \dfrac{1}{60}$ or $s = 60$; $Q = 3200$ m³/s and $t = 1$ day $= 24$ hours.

Discharge of the model for the designed flood

Let q = Discharge of the model for the designed flood.

We know that discharge for the prototype (Q),
$$3200 = q \times (s)^{2.5} = q \times (60)^{2.5} = 27\,885\ q$$
$$\therefore\quad q = 3200/27\,885 = 0.115\ m^3/s\ \textbf{Ans.}$$

Time represented by the model

Let $\qquad t$ = Time represented by the model.

We also know that time represented by the prototype (T),
$$24 = t \times \sqrt{s} = t \times \sqrt{60} = 7.75\ t$$
$$\therefore\quad t = 24/7.75 = 3.1\ \text{hours}\ \textbf{Ans.}$$

26·21 Travelling Time of the Prototype for the Given Travelling Time of an Undistorted Model

Consider an undistorted model geometrically similar to a prototype like ship, submarine etc., travelling in a laboratory.

Let
$\qquad l$ = Distance through which the model travels,
$\qquad v$ = Velocity of the model,
$\qquad L, V$ = Corresponding values for the prototype, and
$\qquad \dfrac{1}{s}$ = Scale ratio of the model to the prototype.

We know that time taken by the model to travel the distance,
$$t = \frac{l}{v} \qquad \ldots(i)$$
and time taken by the prototype to travel the corresponding distance.
$$T = \frac{L}{V} \qquad \ldots(ii)$$

Dividing equation (i) by (ii),
$$\frac{t}{T} = \frac{\frac{l}{v}}{\frac{L}{V}} = \frac{l}{L} \times \frac{v}{V} = \frac{1}{s} \times \sqrt{s} = \frac{1}{\sqrt{s}}$$
$$\therefore\quad T = t\sqrt{s}$$

Example 26·17. *A harbour model is made to a scale of 1:225 for studying the wave motion. If in the laboratory, a wave travels a distance in 10 seconds, what time the wave in prototype will take to cover the corresponding distance?*

Solution. Given : $\dfrac{1}{s} = \dfrac{1}{225}$ or $s = 225$ and $t = 10$ s.

We know that the time taken by the wave in the prototype to cover the correspondence distance,
$$T = t\sqrt{s} = 10 \times \sqrt{225} = 10 \times 15 = 150\ s = 2.5\ \text{min}\ \textbf{Ans.}$$

Example 26·18. *A spillway, having piers and 20 m long gates, is to be constructed to control the upstream level of a river. The discharge through each gate will be 200 m^3/s. A 1:40 scale model is to be built and tested to predict the behaviour of the prototype. Find the corresponding discharge in each gate of the model.*

If the velocity of water at a given point in the model is 2·8 m/s and the time taken for a particle to travel from the crest of the dam to a point is 1·25 second, find the time taken by a particle to travel between the corresponding points of the prototype.

Solution. Given : $L = 20$ m; $Q = 200\ m^3/s$; $\dfrac{1}{s} = \dfrac{1}{40}$ or $s = 40$; $v = 2.8$ m/s and $t = 1.25$ s.

Model Analysis (Undistorted Models)

Discharge through each gate of the model

Let q = Discharge through each gate of the model.

We know that discharge through each gate of the prototype (Q),

$$200 = q \times (s)^{2.5} = q \times (40)^{2.5} = 10\,120\, q$$

$$\therefore q = 200/10\,120 = 0.02 \text{ m}^3/\text{s} = 20 \text{ litres/s} \quad \textbf{Ans.}$$

Time taken by a particle to travel from the crest of the dam to a point on the prototype

We also know that time taken by the particle to travel from the crest of the dam to the corresponding distance in the prototype,

$$T = t \sqrt{s} = 1.25 \sqrt{40} = 1.25 \times 6.32 = 7.9 \text{ s} \quad \textbf{Ans.}$$

EXERCISE 26·3

1. If a turbine model of 1:10 scale was found to develop 0·8 kW, determine the corresponding power developed by the prototype. **(Ans. 2530 kW)**
2. In a power station, a turbine is required to develop 7500 kW, while running at 430 r.p.m. Find the power developed by its 1:12 scale model and the corresponding speed. **(Ans. 1·25 kW; 124 r.p.m.)**
3. A reservoir of scale 1:100 was emptied by a triangular notch in 1 hour. Find the time taken to empty the prototype, if the coefficients of discharge for the triangular notches in the model and prototype are the same. **(Ans. 10 hours)**
4. A 1:400 scale model of a dam reservoir is made to study the movement of waves. It was observed that a wave takes 2·5 seconds to travel a particular distance in the model. Find the time taken by the wave to travel the corresponding distance in the prototype. **(Ans. 50 s)**

QUESTIONS

1. What do you understand by the term hydraulic similarity? Name the various types of hydraulic similarities.
2. Explain the use of model in the design of hydraulic structures.
3. Explain the terms geometric similarity, kinematic similarity and dynamic similarity.
4. Give difference between the wind tunnel method and water tunnel method of testing the models.

OBJECTIVE TYPE QUESTIONS

1. The velocity of water in the prototype of an undistorted model of scale ratio (s) is
  (a) $V = v \sqrt{s}$ (b) $V = v\,(s)^{1.5}$ (c) $V = v\,(s)^{2.5}$ (d) $V = v\,(s)^{3.5}$
  where v is the velocity of water in the model.
2. In a model of scale ratio 1:25, the ratio discharges between the prototype and model is
  (a) $\dfrac{1}{25}$ (b) $\dfrac{1}{\sqrt{25}}$ (c) $(25)^{1.25}$ (d) $(25)^{2.5}$
3. If a model of scale (s) has a speed of n r.p.m., then the speed of the prototype will be
  (a) $\dfrac{n}{s}$ (b) $\dfrac{n}{\sqrt{s}}$ (c) $\dfrac{\sqrt{n}}{s}$ (d) $\dfrac{n^2}{\sqrt{s}}$
4. At a power station, a model of a turbine is made to a scale of 1 : 16. If the power developed by the model is p, then the corresponding power developed by the prototype will be
  (a) $p \times (16)^{1.5}$ (b) $p \times (16)^{2.5}$ (c) $p \times (16)^{3.5}$ (d) $p \times (16)^{4.5}$

ANSWERS

1. (a) 2. (d) 3. (b) 4. (c)

27

Model Analysis
(Distorted Models)

1. Introduction. 2. Advantages and Disadvantages of Distorted Models. 3. Prediction of Various Parameters of a Prototype from its Distorted Model. 4. Velocity of Water in the Prototype for the Given Velocity of Water of a Distorted Model. 5. Discharge of the Prototype for the Given Discharge of a Distorted Model. 6. Speed of the Prototype for the Given Speed of a Distorted Model. 7. Power Developed by the Prototype for the Given Power Developed by a Distorted Model. 8. Time of Emptying a Prototype for the Given Time of Emptying a Distorted Model.

27·1 Introduction

In the previous chapter, we have discussed the working of undistorted models (*i.e.*, in these cases, the model was geometrically similar to the prototype). But sometimes, a model does not have complete geometrical similarity with its prototype. Such a model is called distorted model. Moreover, the models of hydraulics structures, such as rivers, harbours, reservoirs etc. have very large horizontal dimensions, as compared to vertical ones. If a model of such a prototype having complete geometrical similarity is made, then the depth of water, in such a model, is so small that it cannot be accurately measured.

In order to overcome this difficulty, the models of such structures are made with different horizontal and vertical scales (*i.e.*, with greater vertical scales than the horizontal ones). A model having complete geometrical similarity with the prototype, but working under a different head of water (*i.e.*, other than the geometrically similar head of water) also behaves as a distorted model. In such models, the scale ratio of model to the prototype is taken as horizontal scale ratio and the ratio of head of water over the model to the head of water over the prototype is taken as vertical scale ratio.

As a matter of fact, the prediction of a distorted model is relatively difficult, and the results of the models being distorted cannot be easily transferred to the prototypes, as the basic condition (of geometric similarity) is not satisfied.

27·2 Advantages and Disadvantages of Distorted Models

A distorted model has the following advantages and disadvantages:

Advantages

1. The model size can be sufficiently reduced by its distortion. As a result of this, the cost of the model is considerably reduced and its operation is simplified.
2. The vertical exaggeration results in steeper slopes of water surface, which can be easily and accurately measured.
3. The Reynold's number of a model is considerably increased and surface resistance is decreased due to exaggerated water slopes. This helps in simulation of the flow conditions in the model and its prototype.

Disadvantages

1. There is an unfavourable psychological effect on the observer.
2. The behaviour of flow of a model differs in action from that of the prototype.

Model Analysis (Distorted Models)

3. The magnitude and direction of the pressures is not correctly reproduced.
4. The velocities are not correctly reproduced, as the vertical exaggeration causes distortion of lateral velocity and kinetic energy.

In spite of the above mentioned disadvantages of a distorted model, it is sometimes preferred to use a distorted model. However, by exercising an utmost care, the results of the model may be transferred to the prototype. This is, generally, done by starting from the first principles.

27·3 Prediction of Various Parameters of a Prototype from its Distorted Model

Though we can predict a wide range of parameters of a prototype from the corresponding results of its distorted model, yet the following are important from the subject point of view :

1. Velocity of water in the prototype for the given velocity at the corresponding point of the model.
2. Discharge of the prototype for the given discharge of the model.
3. Speed of the prototype for the given speed of the model.
4. Power developed by the prototype for the given power of the model.
5. Time of emptying a prototype for the given time of emptying a model.

27·4 Velocity of Water in the Prototype for the Given Velocity of Water of a Distorted Model

Consider a distorted model of prototype like a weir, dam or spillway etc.

Let h = Head of water over the model,
v = Velocity of water at some point in the model,
H, V = Corresponding values for the prototype.
$\dfrac{1}{s_H}$ = Horizontal scale ratio of the model to the prototype,
$\dfrac{1}{s_V}$ = Vertical scale ratio of the model and the prototype.

We know that the velocity of water in the model,
$$v = c_v \sqrt{2gh} \qquad \ldots(i)$$
Similarly, the velocity of water in the prototype,
$$V = C_v \sqrt{2gH} \qquad \ldots(ii)$$
Dividing equation (i) by (ii),
$$\frac{v}{V} = \frac{c_v \sqrt{2gh}}{C_v \sqrt{2gH}} = \frac{\sqrt{h}}{\sqrt{H}} \qquad \ldots(\text{Taking } c_v = C_v)$$
$$= \sqrt{\frac{h}{H}} = \sqrt{\frac{1}{s_V}} = \frac{1}{\sqrt{s_V}}$$
$$\therefore \quad V = v \sqrt{s_V}$$

Example 27·1. *A spillway model built-up to a scale of 1/10 is discharging water with a velocity of 1 m/s, under a head of 100 mm. Find the velocity of water of the prototype, if the head of water over the prototype is 5·5 metres.*

Solution. Given : $\dfrac{1}{s_H} = \dfrac{1}{10}$ or *s_H = 10; v = 1 m/s; h = 100 mm = 0·1 m and H = 5·5 m.

We know that vertical scale ratio,
$$s_V = \frac{h}{H} = \frac{0 \cdot 1}{5 \cdot 5} = 55$$

---
* Superfluous data.

and velocity of water in the prototype,
$$V = v\sqrt{s_V} = 1 \times \sqrt{55} = 7.42 \text{ m/s} \quad \textbf{Ans.}$$

Example 27·2. *A model of a reservoir is made with horizontal and vertical scale ratios of 1/100 and 1/25 respectively. If the velocity of waves in the prototype is 6 m/s, find the corresponding velocity of waves in the model.*

Solution. Given : $\dfrac{1}{s_H} = \dfrac{1}{100}$ or *$s_H = 100$; $\dfrac{1}{s_V} = \dfrac{1}{25}$ or $s_V = 25$ and $V = 6$ m/s.

Let $\qquad v = $ Velocity of waves in the model.

We know that velocity of waves in the prototype (V),
$$6 = v\sqrt{s_V} = v\sqrt{25} = 5v$$
$\therefore \qquad v = 6/5 = 1.2$ m/s   **Ans.**

27·5 Discharge of the Prototype for the Given Discharge of a Distorted Model

Consider a distorted model of a prototype like a weir, notch, spillway etc.

Let $\qquad l = $ Length of the model,

$h = $ Height of the model, through which the discharge is taking place,

$v = $ Actual velocity of water in the model,

$L, H, V = $ Corresponding values for the prototype,

$\dfrac{1}{s_H} = $ Horizontal scale ratio of the model to the prototype, and

$\dfrac{1}{s_V} = $ Vertical scale ratio of the model to the prototype.

We know that the discharge of the model,
$$q = \text{Area} \times \text{Velocity} = (\text{Length} \times \text{Height}) \times \text{Velocity}$$
$$= l \times h \times v \qquad \ldots(i)$$

Similarly, discharge of the prototype,
$$Q = L \times H \times V \qquad \ldots(ii)$$

Dividing equation (i) by (ii),
$$\frac{q}{Q} = \frac{l \times h \times v}{L \times H \times V} = \frac{l}{L} \times \frac{h}{H} \times \frac{v}{V}$$
$$= \frac{1}{s_H} \times \frac{1}{s_V} \times \frac{1}{\sqrt{s_V}} = \frac{1}{s_H} \times s_H \times s_V^{1.5}$$
$$Q = q \times s_H \times s_V^{1.5}$$

Example 27·3. *A model of weir is made to a horizontal scale of 1/40 and vertical scale 1/9. Find the discharge of the prototype, if the model is discharging 1 litre/s.*

Solution. Given : $\dfrac{1}{s_H} = \dfrac{1}{40}$ or $s_H = 40$; $\dfrac{1}{s_V} = \dfrac{1}{9}$ or $s_v = 9$ and $q = 1$ litre/s

We know that discharge of the prototype,
$$Q = q \times s_H \times s_V^{1.5} = 1 \times 40 \times (9)^{1.5} = 1080 \text{ litres/s} \quad \textbf{Ans.}$$

Example 27·4. *The discharges of a model and prototype were found to be 0·02 m^3/s and 150 m^3/s respectively. If vertical scale ratio of the model is 1:25, determine the horizontal scale ratio of the model.*

---
* Superfluous data.

Solution. Given : $q = 0.02$ m³/s; $Q = 150$ m³/s and $\dfrac{1}{s_V} = \dfrac{1}{25}$ or $s_V = 25$.

Let s_H = Horizontal scale ratio of the model.

We know that discharge of the prototype (Q),

$$150 = q \times s_H \times s_V^{1.5} = 0.02 \times s_H \times (25)^{1.5} = 2.5\, s_H$$

$\therefore \quad s_H = 150/2.5 = .60$ **Ans.**

Example 27·5. *A diversion weir 240 m long has discharging capacity of 250 m³/s under a head of 1·2 m. A model of this weir is to be constructed in laboratory where the available channel is 3 m wide and 500 m deep. Design the suitable model for the weir, if the water available in the laboratory is 25 litres/s.*

Solution. Given : $L = 240$ m; $Q = 250$ m³/s; $H = 1.2$ m; $l = 3$ m; Depth of channel = 500 mm = 0.5 m and $q = 25$ litres/s = 0.025 m³/s.

First of all, let us design an undistorted model. From given data, we find that the scale model,

$$\frac{1}{s_H} = \frac{l}{L} = \frac{3}{240} = \frac{1}{80} \text{ or } s_H = 80$$

and head of water in the model.

$$h = \frac{H}{s_H} = \frac{1.2}{80} = 0.015 \text{ m} = 15 \text{ mm}$$

We know that with 15 mm head of water, it will be difficult to take observations as the flow will be predominated by the surface tension force. Moreover, the flow with such a small head will be streamline in nature, whereas in case of the prototype the flow will be turbulent. As a result of this, we will have to exaggerate the vertical scale ratio.

Now let s_V = Vertical scale ratio of the model.

We know that discharging capacity of the weir (Q),

$$250 = q \times s_H \times s_V^{1.5} = 0.025 \times 80 \times s_V^{1.5} = 2\, s_V^{1.5}$$

$\therefore \quad s_V^{1.5} = 250/2 = 125$ or $s_V = 25$

and height of water in the model,

$$h = \frac{H}{s_V} = \frac{1.2}{25} = 0.048 \text{ m} = 48 \text{ mm} \quad \textbf{Ans.}$$

EXERCISE 27·1

1. A model of weir with horizontal and vertical scale ratio of 1:80 and 1:36 is discharging water with a velocity of 0·75 m/s. What is the velocity of water in the prototype? **(Ans. 4·5 m/s)**

2. In a laboratory, a spillway 1:40 model was found to discharge water with a velocity of 0·25 m/s under a head of 500 mm. Find the head under which the velocity of water in the prototype will be 1 m/s. **(Ans. 8 m)**

3. A model of a rectangular notch is made with horizontal and vertical scale ratios of 1:30 and 1:10·5 respectively. Find the discharge over the notch, if the discharge over the model is 1·5 litre/s. **(Ans. 1·53 m³/s)**

4. A dam model, made to 1:50 scale, is discharging water with a velocity of 0·8 m/s under a head of 400 mm. Find the velocity with which the water will flow in the dam under a head of 10 m. Also find the discharge of the dam, when the discharge of the model is 2 litres/s. **(Ans. 4 m/s; 12·5 m³/s)**

27·6 Speed of the Prototype for the Given Speed of a Distorted Model

Consider a distorted model of a prototype like a centrifugal pump or turbine.

Let
d = Diameter of the model impeller,
v = Tangential velocity of the model impeller,
n = Speed of the model runner,
D, V, N = Corresponding values for the prototype,
$\dfrac{1}{s_H}$ = Horizontal scale ratio of the model to the prototype,
$\dfrac{1}{s_V}$ = Vertical scale ratio of the model to the prototype.

We know that the tangential velocity of the model impeller,

$$\therefore \quad v = \dfrac{\pi d n}{60}$$

or $\quad n = \dfrac{60 v}{\pi d}$ ...(i)

Similarly, speed of the prototype,

$$N = \dfrac{60 V}{\pi D} \qquad ...(ii)$$

Dividing equation (i) and (ii),

$$\dfrac{n}{N} = \dfrac{\dfrac{60v}{\pi d}}{\dfrac{60V}{\pi D}} = \dfrac{v}{V} \times \dfrac{D}{d}$$

$$= \dfrac{1}{\sqrt{s_V}} \times s_H \qquad \ldots\left(\dfrac{D}{d} = s_H\right)$$

or $\quad N = \dfrac{n \sqrt{s_V}}{s_H}$

Example 27·6. *The horizontal scale of a turbine model is 1/15. If the speed of the prototype is 300 r.p.m. under a head of 10 metres, find the speed of the model r.p.m. under a head of 200 mm.*

Solution. Given : $\dfrac{1}{s_H} = \dfrac{1}{15}$ or $s_H = 15$; $N = 300$ r.p.m.; $H = 10$ m and $h = 200$ mm $= 0·2$ m.

Let n = Speed of the model in r.p.m.

We know that vertical scale ratio,

$$\dfrac{1}{s_V} = \dfrac{h}{H} = \dfrac{0·2}{10} = \dfrac{1}{50} \quad \text{or} \quad s_V = 50$$

and speed of the prototype (N),

$$300 = \dfrac{n \sqrt{s_V}}{s_H} = \dfrac{n \times \sqrt{50}}{15} = 0·471\, n$$

$\therefore \quad n = 300/0·471 = 636·9$ r.p.m.   **Ans.**

27·7 Power Developed by the Prototype for the Given Power Developed by a Distorted Model

Consider a distorted model of prototype like a turbine.

Let
q = Discharge of the model in m³/s,
h = Head of water over the model in metres,
p = Power developed by the model,

Model Analysis (Distorted Models)

Q, H, P = Corresponding values for prototype,

$\dfrac{1}{s_H}$ = Horizontal scale ratio of the model to the prototype, and

$\dfrac{1}{s_V}$ = Vertical scale ratio of the model to the prototype.

We know that the power developed by the model,
$$p = wqh \qquad \ldots(i)$$
Similarly, power developed by the prototype,
$$P = wQH \qquad \ldots(ii)$$
Dividing equation (i) by (ii),
$$\frac{p}{P} = \frac{wqh}{wQH} = \frac{q}{Q} \times \frac{h}{H}$$
$$= \left[\frac{1}{s_H} \times \frac{1}{(s_V)^{1.5}}\right] \times \left(\frac{1}{s_V}\right) = \frac{1}{s_H} \times \frac{1}{(s_V)^{2.5}}$$
$$\therefore \quad P = p \times s_H \times s_V^{2.5}$$

Example 27·8. *It is required to predict the performance of a large centrifugal pump, from that of a scale model of one-tenth diameter. The model absorbs 2 kW when pumping under a head of 7·5 metres at its best speed of 625 r.p.m. The large pump is required to pump against a head of 30 metres. What will be the working speed and power required to drive it ?*

Solution. Given : $\dfrac{1}{s_H} = \dfrac{1}{10}$ or $s_H = 10$; $p = 20$ kW; $h = 7\cdot5$ m; $n = 625$ r.p.m. and $H = 30$ m.

Working speed of the large pump

We know that vertical scale ratio,
$$\frac{1}{s_V} = \frac{h}{H} = \frac{7\cdot5}{30} = \frac{1}{4} \quad \text{or} \quad s_V = 4$$
and working speed of the large pump (*i.e.*, prototype),
$$N = \frac{n\ \sqrt{s_V}}{s_H} = \frac{625\ \sqrt{4}}{10} = 125 \text{ r.p.m. \textbf{Ans.}}$$

Power required to drive the large pump

We also know that power required to drive the large pump (*i.e.*, prototype),
$$P = p \times s_H \times s_V^{2\cdot5} = 20 \times 10 \times (4)^{2\cdot5} = 6400 \text{ kW \textbf{Ans.}}$$

Example 27·9. *A turbine model of 1:8 is tested under a head of 6 m of water. The full scale turbine is required to work under a head of 30 m of water and to run at 428 r.p.m. At what speed must the model be run?*

If the model develops 5 kW and uses 110 litres of water per second, at this speed, what power will be obtained from the full scale turbine, assuming that its efficiency is 3% better than that of the model?

Solution. Given : $\dfrac{1}{s_H} = \dfrac{1}{8}$ or $s_H = 8$; $h = 6$ m; $H = 30$ m; $N = 428$ r.p.m., $p = 5$ kW; $q = 110$ litres/s $= 0\cdot11$ m^3/s and efficiency of prototype (η_p) = 3% more that η_m.

Let $\quad P$ = Power obtained from the full scale turbine.

We know that vertical scale ratio,
$$\frac{1}{s_V} = \frac{h}{H} = \frac{6}{30} = \frac{1}{5} \quad \text{or} \quad s_V = 5$$

and discharge of the prototype,
$$Q = q \times s_H \times s_V^{1.5} = 0.11 \times 8 \times (5)^{1.5} = 9.84 \text{ m}^3/\text{s}$$
We also know that efficiency of the full scale turbine (prototype),
$$\eta_m = \frac{p}{wqh} = \frac{5}{9.81 \times 0.11 \times 6} = 0.772 = 77.2\%$$
∴ Efficiency of full scale turbine (prototype),
$$\eta_p = 77.2\% + 3\% = 80.2\% = 0.802$$
We also know the efficiency of the prototype (η_p),
$$0.802 = \frac{P}{wQH} = \frac{P}{9.81 \times 9.84 \times 30} = \frac{P}{2896}$$
∴ $P = 0.802 \times 2896 = 2325$ kW **Ans.**

27·8 Time of Emptying a Prototype for the Given Time of Emptying a Distorted Model

Consider a distorted model of a prototype, like a tank or a reservoir.

Let
- v = Total volume of water in the model,
- q = Rate of discharge of water in the model,
- V, Q = Corresponding values for the prototype,
- $\dfrac{1}{s_H}$ = Horizontal scale ratio of the model to the prototype, and
- $\dfrac{1}{s_V}$ = Vertical scale ratio of the model to the prototype.

We know that the time of emptying the model,
$$t = \frac{\text{Volume}}{\text{Rate of discharge}} = \frac{v}{q} \qquad ...(i)$$
and time of emptying the prototype,
$$T = \frac{V}{Q} \qquad ...(ii)$$
Dividing equation (i) by (ii),
$$\frac{t}{T} = \frac{\dfrac{v}{q}}{\dfrac{V}{Q}} = \frac{v}{V} \times \frac{Q}{q} = \left(\frac{1}{s_H^2} \times \frac{1}{s_V}\right) \times \left(s_H \times s_V^{1.5}\right)$$
$$= \frac{1}{s_H} \times \sqrt{s_V}$$
∴ $$T = \frac{t \cdot s_H}{\sqrt{s_V}}$$

Example 27·10. *A model of a reservoir has been built with horizontal and vertical ratios of 1:1000 and 1:400. Find the time taken to empty the reservoir, if the model is emptied in 30 seconds.*

Solution. Given : $\dfrac{1}{s_H} = \dfrac{1}{1000}$ or $s_H = 1000$; $\dfrac{1}{s_V} = \dfrac{1}{400}$ or $s_V = 400$ and $t = 30$ s.

We know that time taken to empty the reservoir,
$$T = \frac{t \cdot s_H}{\sqrt{s_V}} = \frac{30 \times 1000}{\sqrt{400}} = \frac{30\,000}{20} = 1500 \text{ s} = 25 \text{ min } \textbf{Ans.}$$

Model Analysis (Distorted Models)

Example 27·11. *A model of a weir is constructed across a reservoir with a vertical scale of 1 : 16. If the discharges of the model and prototype are 1·5 litres/s and 38·4 m³/s, find the horizontal scale of the model.*

If the reservoir is to be emptied in two days, find the time required to empty the model.

Solution. Given : $\dfrac{1}{s_v} = \dfrac{1}{16}$ or $s_V = 16$; $q = 1·5$ litres/s; $Q = 38·4$ m³/s $= 38\,400$ litres/s and
$T = 2$ days $= 2 \times 24 \times 60 \times 60 = 172\,800$ s

Horizontal scale of the model

Let s_H = Horizontal scale of the model.

We know that discharge of the prototype (Q),

$$38\,400 = q \times s_H \times s_V^{1.5} = 1·5 \times s_H \times (16)^{1.5} = 96\, s_H$$

$\therefore \qquad s_H = 38\,400/96 = 400$ **Ans.**

Time required to empty the model

Let t = Time required to empty the model.

We know that time required to empty the prototype (T),

$$172\,800 = \dfrac{t \cdot s_H}{\sqrt{s_V}} = \dfrac{t \times 400}{\sqrt{16}} = \dfrac{t \times 400}{4} = 100\, t$$

$\therefore \qquad t = 72\,800/100 = 1728$ s $= 28·8$ min **Ans.**

EXERCISE 27·2

1. A turbine model of scale ratio 1:20 has its runner speed of 500 r.p.m. under a head of 500 mm. Find the speed of the prototype runner in r.p.m. under a head of 32 m. **(Ans. 200 r.p.m.)**
2. A 1:10 model of a turbine developed 0·5 kW under a head of 2·5 m at 480 r.p.m. Find the power developed by the prototype under a head of 20 m. Also find the speed of the prototype assuming efficiencies of both the turbines to be the same. **(Ans. 905 kW; 135·8 r.p.m.)**
3. A model of a Francis turbine was tested in a laboratory under a head of 0·5 m and developed 0·2 kW. If the actual turbine develops 729 kW, find the scale ratio of the turbine. **(Ans. 1:9)**
4. A tidal model has been constructed with horizontal and vertical scale ratios of 1/7000 and 1/400 respectively. What is the period in which a tide of its natural period 12 hours 50 minutes, in the prototype can be simulated in the model? **(Ans. 2 min 12 s)**

QUESTIONS

1. What is a distorted model? How does it differ from an undistorted model?
2. Give the uses of distorted models.
3. Mention the advantages and disadvantages of distorted models.
4. How will you predict the discharge of a prototype from the discharge of a distorted model, when horizontal and vertical scale ratios are given?

OBJECTIVE TYPE QUESTIONS

1. The ratio of velocity of water in a distorted model to that of a prototype with horizontal and vertical scale ratios of s_H and s_V is

    (a) $\dfrac{s_V}{s_H}$ (b) $\dfrac{s_H^2}{\sqrt{s_V}}$ (c) $\dfrac{s_H}{\sqrt{s_V}}$ (d) $\dfrac{s_H^2}{s_V}$

2. If the discharge of a model is (q) and horizontal as well as vertical scale ratios are s_H and s_V respectively, then discharge (Q) of the prototype will be

   (a) $q \times \sqrt{s_H} \times s_V$ (b) $q \times \sqrt{s_H} \times s_V^{1.5}$ (c) $q \times s_H \times s_V^{1.5}$ (d) $q \times s_H \times s_V^{2.5}$

3. The speed of a prototype is
   (a) directly proportional to speed of the model
   (b) indirectly proportional to the horizontal scale ratio
   (c) directly proportional to the square root of the vertical scale ratio
   (d) all of these

4. In order to predict the power developed by a turbine, we made an undistorted model with horizontal and vertical scale ratios as s_H and s_V. If the power developed by the model is (p), then the power developed by the turbine will be

   (a) $p \times s_H \times s_V^{2.5}$ (b) $p \times \sqrt{s_H} \times s_V^{2.5}$ (c) $p \times s_H \times s_V^{1.5}$ (d) $p \times \sqrt{s_H} \times s_V^{1.5}$

ANSWERS
1. (b) 2. (c) 3. (d) 4. (a)

28

Non-Dimensional Constants

1. Introduction. 2. Types of Forces Present in a Moving Liquid. 3. Inertia Force. 4. Viscous Force. 5. Gravity Force. 6. Surface Tension Force. 7. Pressure Force. 8. Elastic Force. 9. Dimensional Numbers. 10. Reynold's Number. 11. Froude's Number. 12. Weber's Number. 13. Euler's Number. 14. Mach's Number or Cauchy's Number.

28·1 Introduction

In the previous two chapters, we have discussed the model analysis. In these chapters, we had focussed our discussion mainly on the comparison of a model to its prototype, without any reference to the force dominating the phenomenon. But in this chapter, we shall discuss the effect of these forces also.

28·2 Types of Forces Present in a Moving Liquid

When a liquid is in motion, some forces are always involved in the phenomenon of flow. But there are always one or two forces, which dominate the other forces, and they govern the flow of the liquid and keep it in motion. The following forces, which are present in a moving liquid, are important from the subject point of view:

1. Inertia force,
2. Viscous force,
3. Gravity force,
4. Surface tension force,
5. Pressure force, and
6. Elastic force.

28·3 Inertia Force

The inertia force (F_i) is the product of mass and acceleration of flowing liquid, and is always existing in the phenomenon of the liquid flow.

28·4 Viscous Force

The viscous force (F_v) is the product of shear stress, due to viscosity and the cross-sectional area of flow.

28·5 Gravity Force

The gravity force (F_g) is the product of mass and acceleration, due to gravity of a flowing liquid.

28·6 Surface Tension Force

The surface tension force (F_p) is the product of surface tension per unit length and length of the surface of flowing liquid.

28·7 Pressure Force

The pressure force (F_e) is the product of intensity of pressure and the area of a flowing liquid.

28·8 Elastic Force

The elastic force (F_e) is the product of the elastic stress and the area of a flowing liquid.

28·9 Dimensionless Numbers

Since the inertia force always exists in the phenomenon of the liquid flow, the conditions of dynamic similarity are always studied, by considering the ratio of the inertia force and any one of the remaining forces. Since we shall discuss only the ratio of one force to the other, therefore it will be a dimensionless constant. The following five dimensionless constants are important from the subject point of view:

1. Reynold's number,
2. Froude's number,
3. Weber's number,
4. Euler's number, and
5. Mach's number or Cauchy's number.

28·10 Reynold's Number

The Reynold's number is the ratio of inertia force to the viscous force. If viscous force is the predominating force in the model and the prototype, dynamic similarity is said to exist between the two, when the Reynold's number for the model and the prototype is the same. *e.g.*, Reynold's number,

$$R_N = \frac{\text{Inertia force}}{\text{Viscous force}}$$

$$= \frac{\text{Mass} \times \text{Acceleration}}{\text{Shear stress} \times \text{Cross -sectional area}}$$

$$= \frac{\text{Volume} \times \text{Mass Density} \times (\text{Velocity / Time})}{\text{Shear stress} \times \text{Cross-sectional area}}$$

$$= \frac{(\text{Volume / Time}) \times \text{Mass Density} \times \text{Velocity}}{\text{Shear stress} \times \text{Cross-sectional area}}$$

$$= \frac{\text{Cross-sectional area} \times \text{Velocity} \times \text{Mass density} \times \text{Velocity}}{\text{Shear stress} \times \text{Cross sectional area}}$$

$$= \frac{\text{Velocity} \times \text{Mass density} \times \text{Velocity}}{\text{Shear stress}}$$

$$= \frac{V \rho V}{\mu \left(\frac{dv}{dy}\right)} = \frac{V \rho V}{\mu \cdot \frac{V}{L}} = \frac{V\rho}{\frac{\mu}{L}} = \frac{V \cdot L}{\frac{\mu}{\rho}}$$

$$= \frac{V \cdot L}{\nu} \qquad \qquad \ldots \left(\because \frac{\mu}{\rho} = \nu\right)$$

Example 28·1. *The performance of a ship was predicted by making its model and testing it in a wind tunnel. The length of a prototype ship was 350 m and its model was 1 m long. The kinematic viscosity of air is 1·25 times that of water. The velocity of air around the model in the wind tunnel was measured as 35 m/s. Find the velocity of actual ship in the water.*

Solution. Given : $L = 350$ m; $l = 1$ m; $\nu_m = 1·25\, \nu_p$ and $\nu = 35$ m/s.

Let $\qquad\qquad\qquad V$ = Velocity of the actual ship in the water.

We know that for dynamic similarity, between the model and prototype, when viscous forces are present in the flow, the Reynold's number for both of them should be equal. *i.e.*,

$$\frac{V \cdot L}{\nu_p} = \frac{v \cdot l}{\nu_m}$$

Non-Dimensional Constants

$$\therefore \quad V = \frac{v_p}{v_m} \times \frac{l}{L} \times v = \frac{1}{1\cdot 25} = \frac{1\cdot 0}{350} \times 35 = 0\cdot 08 \text{ m/s } \textbf{Ans.}$$

28·11 Froude's Number

The Froude's number is the ratio of inertia force and gravity force. If gravity force is the predominating force in the model and prototype, dynamic similarity is said to exist between the two, when the Froude's number for the model and prototype is the same.

∴ Ratio of these two forces

$$= \frac{\text{Inertia force}}{\text{Gravity force}}$$

$$= \frac{\text{Mass} \times \text{Acceleration}}{\text{Mass} \times \text{Acceleration due to gravit}}$$

$$= \frac{(\text{Velocity/Time})}{\text{Acceleration due to gravi}}$$

$$= \frac{\frac{V}{T}}{g} = \frac{V}{Tg} = \frac{V}{\frac{L}{V} \times g} = \frac{V^2}{Lg}$$

Now the Froude's number is the square root of this ratio. *i.e.*,

$$F_N = \sqrt{\frac{V^2}{Lg}} = \frac{V}{\sqrt{Lg}}$$

Example 28·2. *A channel model 250 mm deep is discharging water with a velocity of 1·5 metre/s. Find the velocity of water in the channel 4 metres deep, if the model has dynamic similarity with its prototype and the flow is governed by Froude's law.*

Solution. Given : $h = 250$ mm $= 0\cdot 25$ m; $v = 1\cdot 5$ m/s and $H = 4$ m.

Let $\qquad V = $ Velocity of water in the channel.

We know that as the flow is governed by the Froude's law, therefore for dynamic similarity between the model and prototype, the Froude's number for both of them should be equal. *i.e.*,

$$\frac{V}{\sqrt{L \cdot g_p}} = \frac{v}{\sqrt{l \cdot g_m}}$$

$$\therefore \quad V = v \times \sqrt{\frac{L}{l}} \times \sqrt{\frac{g_p}{g_m}} = 1\cdot 5 \times \sqrt{\frac{4}{0\cdot 25}} \times 1 \text{ m/s}$$

$$\dots \left(\frac{L}{l} = \frac{H}{h} \text{ and } g_p = g_m \right)$$

$$= 6 \text{ m/s } \textbf{Ans.}$$

Example 28·3. *Water of kinematic viscosity $0\cdot 77 \times 10^{-2}$ stoke is used in the prototype. If a fluid of kinematic viscosity $0\cdot 069 \times 10^{-2}$ stoke is available for testing the model, what model scale should be adopted? Also find the velocity and discharge ratios, if viscous and gravity forces are present.*

Solution. Given : Kinematic viscosity of water (prototype), $v_p = 0\cdot 77 \times 10^{-2}$ stoke and kinematic viscosity of fluid (model), $v_m = 0\cdot 069 \times 10^{-2}$ stoke.

Model scale

Let $\qquad V = $ Velocity of water in the prototype,

$\qquad L = $ Length of the prototype, and

v and $l = $ Corresponding values for the model.

We know that as the viscous and gravity forces are present, therefore for dynamic similarity between the model and prototype, the Reynold's number and Froude's number for both of them should be equal.

i.e.,
$$\frac{V \cdot L}{v_p} = \frac{v \cdot l}{v_m} \qquad \ldots\text{(Reynold's number)}$$

or
$$\frac{v}{V} = \frac{v_m}{v_p} \times \frac{L}{l} \qquad \ldots(i)$$

and
$$\frac{V}{\sqrt{L \cdot g_p}} = \frac{v}{\sqrt{l \cdot g_m}} \qquad \ldots\text{(Froude's number)}$$

or
$$\frac{v}{V} = \frac{\sqrt{l}}{\sqrt{L}} \qquad \ldots(ii)$$

Equating equations (i) and (ii),

$$\frac{v_m}{v_p} \times \frac{L}{l} = \frac{\sqrt{l}}{\sqrt{L}} \quad \text{or} \quad \left(\frac{l}{L}\right)^{3/2} = \frac{v_m}{v_p}$$

∴
$$\frac{l}{L} = \left(\frac{v_m}{v_p}\right)^{2/3} = \left(\frac{0.069 \times 10^{-2}}{0.77 \times 10^{-2}}\right)^{2/3} = \frac{1}{5} \quad \textbf{Ans.}$$

Velocity ratio

We know that velocity ratio,

$$\frac{v}{V} = \frac{1}{\sqrt{s}} = \frac{1}{\sqrt{\frac{L}{l}}} = \frac{\sqrt{l}}{\sqrt{L}} = \frac{1}{\sqrt{5}} = 0.447 \quad \textbf{Ans.}$$

Discharge ratio

We also know that discharge ratio,

$$\frac{q}{Q} = \frac{a \times v}{A \times V} = \frac{a}{A} \times \frac{v}{V} = \left(\frac{1}{5}\right)^2 \times 0.447 = 0.018 \quad \textbf{Ans.}$$

28·12 Weber's Number

The Weber's number is the ratio of inertia force to the surface tension force. If the surface tension force is the predominating force in the model and prototype, dynamic similarity is said to exist between the two, when the Weber's number for the model and the prototype is the same.

∴ Ratio of these two forces

$$= \frac{\text{Inertia force}}{\text{Surface tension force}}$$

$$= \frac{\text{Mass} \times \text{Acceleration}}{\text{Surface tension per unit length} \times \text{Length}}$$

$$= \frac{(\text{Volume} \times \text{Mass density})(\text{Velocity} / \text{Time})}{\text{Surface tension per unit length} \times \text{Length}}$$

$$= \frac{(\text{Volume} / \text{Time}) \times \text{Mass density} \times \text{Velocity}}{\text{Surface tension per unit length} \times \text{Length}}$$

$$= \frac{\text{Cross–sectional area} \times \text{Velocity} \times (\text{Mass density} \times \text{Velocity})}{\text{Surface tension per unit length} \times \text{Length}}$$

$$= \frac{L^2 V \rho V}{\sigma L} = \frac{\rho L V^2}{\sigma}$$

Now Weber's number is the square root of this ratio. i.e.,

$$W_N = \sqrt{\frac{\rho L V^2}{\sigma}} = \sqrt{\frac{V^2}{\frac{\sigma}{\rho L}}} = \frac{V}{\sqrt{\frac{\sigma}{\rho L}}}$$

Example 28·4. *In an experiment, water was made to flow over a 600 mm long weir with a velocity of 0·8 m/s. Calculate the value of Weber's number, if the ratio of surface tension to the density of water is 2·4.*

Solution. Given : $L = 600$ mm $= 0.6$ m; $V = 0.8$ m/s and $\frac{\sigma}{\rho} = 2.4$.

We know that Weber's number,

$$W_N = \frac{V}{\sqrt{\frac{\sigma}{\rho L}}} = \frac{V}{\sqrt{\frac{\sigma}{\rho} \times \frac{1}{L}}} = \frac{0.8}{\sqrt{2.4 \times \frac{1}{0.6}}} = \frac{0.8}{\sqrt{4}} = 0.4 \text{ Ans.}$$

28·13 Euler's Number

The Euler's number is the ratio of inertia force and the pressure force. If the pressure force is the predominating force in the model and the prototype, dynamic similarity is said to exist between the two, when the Euler's number for the model and the prototype is the same.

∴ Ratio of these two forces

$$= \frac{\text{Inertia force}}{\text{Pressure Force}}$$

$$= \frac{\text{Mass} \times \text{Acceleration}}{\text{Intensity of pressure} \times \text{Area}}$$

$$= \frac{(\text{Volume} \times \text{Mass density}) \times (\text{Velocity / Time})}{\text{Intensity of pressure} \times \text{Area}}$$

$$= \frac{(\text{Volume / Time}) \times \text{Mass density} \times \text{Velocity}}{\text{Intensity of pressure} \times \text{Area}}$$

$$= \frac{(\text{Cross-sectional area}) \times \text{Velocity} \times \text{Mass density} \times \text{Velocity}}{\text{Intensity of pressure} \times \text{Area}}$$

$$= \frac{\text{Velocity} \times \text{Mass density} \times \text{Velocity}}{\text{Intensity of pressure}}$$

$$= \frac{V \rho V}{P} = \frac{V^2 \rho}{p}$$

Now Euler's number is the square root of this ratio. i.e.,

$$E_N = \frac{V^2 \rho}{p} = \sqrt{\frac{V^2}{\frac{p}{\rho}}} = \sqrt{\frac{V}{\frac{p}{\rho}}}$$

Note : The reciprocal of Eular number is called Newton number.

Example 28·5 *In a pipeline water was observed to flow with a velocity of 1 m/s. If the flow is governed by the Euler's law, find the ratio of pressure to the density of water. Take Euler's number as 2·5.*

Solution. Given : $V = 1$ m/s and $E_N = 2.5$.

Let $\dfrac{p}{\rho}$ = Ratio of pressure to the density of water.

We know that Euler's number (E_N),

$$2.5 = \dfrac{V}{\sqrt{\dfrac{p}{\rho}}} = \dfrac{1}{\sqrt{\dfrac{p}{\rho}}}$$

$\therefore \sqrt{\dfrac{p}{\rho}} = \dfrac{1}{2.5} = 0.4$ or $\dfrac{p}{\rho} = (0.4)^2 = 0.16$ **Ans.**

Example 28·6. *A spillway model is to be built to a geometrically similar scale of 1:50 across a flume of 600 mm width. If the negative pressure in the model is 2 kPa, what is the pressure in prototype? The flow is governed by Froude's law and Euler's law.*

Solution. Given : $\dfrac{1}{s} = \dfrac{1}{50}$ or $s = 50$; Flume width = 600 mm and $p = 2$ kPa.

Let P = Pressure of the prototype,
 v = Velocity of water in the model, and
 V = Corresponding velocity in the prototype.

We know that for dynamic, similarity between the model and the prototype, the Froude's number for both of them should be equal. *i.e.*

$$\dfrac{V}{\sqrt{L \cdot g_p}} = \dfrac{v}{\sqrt{l \cdot g_m}}$$

$\therefore \dfrac{V}{v} = \sqrt{\dfrac{L}{l}} \times \sqrt{\dfrac{g_p}{g_m}} = \sqrt{50}$ ...($\because \dfrac{L}{l} = s$ and $g_p = g_m$)

Now, in order to have dynamic similarity for pressure, the Euler's number for the model and the prototype should also be equal. *i.e.,*

$$\dfrac{V}{\sqrt{\dfrac{P}{\rho_p}}} = \dfrac{v}{\sqrt{\dfrac{p}{\rho_m}}}$$

$\therefore \dfrac{V}{v} = \sqrt{\dfrac{P}{p}} \times \sqrt{\dfrac{\rho_m}{\rho_p}} = \sqrt{\dfrac{P}{p}}$ ...(ii)

...($\because \rho_m = \rho_p$)

Equating equations (i) and (ii),

$\sqrt{\dfrac{P}{p}} = \sqrt{50}$ or $\dfrac{P}{p} = 50$ ...(Squaring both sides)

$\therefore P = p \times 50 = 2 \times 50 = 100$ kPa **Ans.**

28·14 Mach's Number or Cauchy's Number

The Mach's or Cauchy's number is the ratio of inertia force and the elastic force. If the elastic force is the predominating force in the model and the prototype, dynamic similarity is said to exist between the two, when the Mach's number for the model and the prototype is the same.

∴ Ratio of these two forces

$$= \dfrac{\text{Inertia force}}{\text{Elastic force}}$$

$$= \dfrac{\text{Mass} \times \text{Acceleration}}{\text{Elastic stress} \times \text{Area}}$$

Non-Dimensional Constants

$$= \frac{\text{Volume} \times \text{Mass density}) \times (\text{Velocity/Time})}{\text{Elastic stress} \times \text{Area}}$$

$$= \frac{(\text{Volume/Time}) \times \text{Mass density} \times \text{Velocity}}{\text{Elastic stress} \times \text{Area}}$$

$$= \frac{(\text{Cross-sectional area} \times \text{Velocity}) \times (\text{Mass density} \times \text{Velocity})}{\text{Elastic stress} \times \text{Area}}$$

$$= \frac{\text{Velocity} \times \text{Mass density} \times \text{Velocity}}{\text{Elastic stress}}$$

$$= \frac{V\rho V}{K} = \frac{V^2 \rho}{K}$$

Now Mach's number is the square root of this ratio. *i.e.*,

$$M_N = \frac{V^2 \rho}{K} = \sqrt{\frac{V^2}{K/\rho}} = \frac{V}{\sqrt{K/\rho}} = \frac{V}{\sqrt{k/\rho}}$$

Example 28·7. *An airfoil moves at 540 km per hour through still air at 20°C. If the ratio of elastic stress and density of air is $1·5 \times 10^{-6}$, find the Mach number.*

Solution. Given : V = 540 km/hour = 150 m/s, *T = 20°C, and $K/\rho = 1·5 \times 10^{-6}$.

We know that Mach number,

$$M_N = \frac{V}{\sqrt{K/\rho}} = \frac{150}{\sqrt{1·5 \times 10^{-6}}} = \frac{150}{1225} = 0·122 \text{ Ans.}$$

EXERCISE 28·1

1. In order to predict the performance of a 300 m long ship, its 1:50 scale ratio model was made and tested in a wind tunnel. If the velocity of air around the tunnel is 30 m/s, find the velocity of the ship in the sea water. Take kinematic viscosity of sea water and air as 0·012 stoke and 0·018 stoke respectively. **(Ans. 0·4 m/s)**
2. The velocity of water in a spillway model of a dam is 1·5 m/s for a prototype of model ratio 1:16. If the model has dynamic similarity with its prototype and the flow is governed by Froude's law, find the corresponding velocity of water in the prototype. **(Ans. 6 m/s)**
3. A model of a dam 2 m long is discharging water at the rate of 25 litres/s under a head of 100 mm. If the flow in the model and prototype is governed by the Froude's law, determine the discharge of the prototype under a head of 3·5 m. **(Ans. 181·2 m³/s)**
4. A ship moves in sea with a velocity of 18 m/s. If the ratio of elastic stress to the density of water is 100, find the Mach number for the motion. **(Ans. 0·5)**

QUESTIONS

1. Explain clearly the term 'dimensionless constant'. Name the different dimensionless constants.
2. Prove that the Reynold's number of a model and that of a prototype are equal, if the flows are similar. State the forces predominating in such a type of flow.
3. Obtain a relation for the Froude's number.
4. Define the Weber's number and explain its importance.

---
*Superfluous data.

OBJECTIVE TYPE QUESTIONS

1. The Reynold's number of a ship is
   (a) directly proportional to its velocity
   (b) directly proportional to its length
   (c) indirectly proportional to the kinematic viscosity of the fluid
   (d) all of these
2. The Weber's number depends upon the value of
   (a) surface tension of the fluid   (b) mass density of the fluid
   (c) intensity of pressure   (d) both 'a' and 'b'
3. If a flow is governed by Euler law, then we must have the value of
   (a) surface tension of the fluid   (b) intensity of pressure
   (c) bulk modulus of the fluid   (d) all of these
4. The value of bulk modulus of a fluid is required to determine
   (a) Reynold's number   (b) Fraude's number
   (c) Mach number   (d) Euler's number

ANSWERS

1. (d)   2. (d)   3. (b)   4. (c)

29

Impact of Jets

1. Introduction. 2. Force of Jet Impinging Normally on a Fixed Plate. 3. Force of Jet Impinging on an Inclined Fixed Plate. 4. Force of Jet Impinging on a Hinged Plate. 5. Force of Jet Impinging on a Moving Plate. 6. Force of Jet Impinging on a Series of Moving Vanes. 7. Force of Jet Impinging on a Fixed Curved Vane. 8. Force of Jet Impinging on a Moving Curved Vane.

29·1 Introduction

We see that whenever a jet of liquid impinges (*i.e.*, strikes) on a fixed plate, it experiences some force. As per Newton's Second Law of Motion, this force is equal to the rate of change of momentum of the jet. It has been observed that if the plate is not fixed, then the plate starts moving in the direction of the jet, because of the force.

29·2 Force of Jet Impinging Normally on a Fixed Plate

Consider a jet of water impinging normally on a fixed plate as shown in Fig. 29·1.

Let V = Velocity of the jet in m/s, and

a = Cross-sectional area of the jet in m².

∴ Mass of water flowing/s = $\dfrac{waV}{g}$ kg

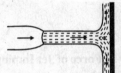

We know that the velocity of jet, in its original direction, is reduced to zero after the impact (as the plate is fixed). Therefore force exerted by the jet on the plate.

Fig. 29·1. Jet impinging normally on a fixed plate.

$$F = \text{Mass of water flowing/s} \times \text{Change of velocity}$$

$$= \dfrac{waV}{g} \times (V - 0) = \dfrac{waV^2}{g} \text{ kN}$$

where w is the specific weight of water in kN/m³.

Example 29·1. *A jet of water of 100 mm diameter impinges normally on a fixed plate with a velocity of 30 m/s. Find the force exerted on the plate.*

Solution. Given : d = 100 mm = 0·1 m and V = 30 m/s.

We know that cross-sectional area of the jet,

$$a = \dfrac{\pi}{4} \times (d)^2 = \dfrac{\pi}{4} \times (0·1)^2 = 7·854 \times 10^{-3} \text{ m}^2$$

and force exerted on the plate,

$$F = \dfrac{waV^2}{g} = \dfrac{9·81 \times (7·854 \times 10^{-3}) \times (30)^2}{9·81} \text{ kN}$$

$$= 7·07 \text{ kN Ans.}$$

Example 29·2. *A jet of water 50 mm diameter is discharging under a constant head of 70 metres. Find the force exerted by the jet on a fixed plate. Take coefficient velocity as 0·9.*

Solution. Given : $d = 50$ mm $= 0.05$ m; $H = 70$ m and $C_v = 0.9$.

We know that cross-sectional area of the jet,

$$a = \frac{\pi}{4} \times (d)^2 = \frac{\pi}{4} \times (0.05)^2 = 1.964 \times 10^{-3} \text{ m}^2$$

and velocity of the jet, $V = C_v \sqrt{2gH} = 0.9 \times \sqrt{2 \times 9.81 \times 70} = 33.4$ m/s

∴ Force exerted by the jet on the fixed plate,

$$F = \frac{waV^2}{g} = \frac{9.81 \times (1.964 \times 10^{-3}) \times (33.4)^2}{9.81} \text{ kN}$$

$$= 2.19 \text{ kN} \quad \textbf{Ans.}$$

Example 29·3. *A horizontal jet of water is issuing under an effective head of 25 m. Calculate the diameter of the jet, if the force exerted by the jet on a vertical fixed plate is 2·22 kN. Take $C_v = 1$.*

Solution. Given : $H = 25$ m; $F = 2.22$ kN and $C_v = 1$.

Let d = Diameter of the jet in metres.

We know that cross-sectional area of the jet,

$$a = \frac{\pi}{4} \times (d)^2 = 0.7854 \, d^2$$

and velocity of the jet, $V = C_v \times \sqrt{2gH} = 1 \times \sqrt{2 \times 9.81 \times 25} = 21$ m/s

We also know that force exerted by the jet (F),

$$2.22 = \frac{waV^2}{g} = \frac{9.81 \times (0.7854 \, d^2) \times (21)^2}{9.81} = 346.4 \, d^2$$

∴ $d^2 = 2.22/346.4 = 0.0064$ or $d = 0.08$ m $= 80$ mm **Ans.**

29·3 Force of Jet Impinging on an Inclined Fixed Plate

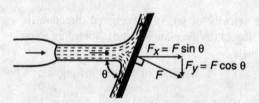

Fig. 29·2. Jet impinging on an inclined fixed plate.

Consider a jet impinging on an inclined fixed plate as shown in Fig. 29·2.

Let V = Velocity of jet in m/s,

            a = Cross-sectional area of the jet in m², and

            θ = Angle at which the plate is inclined with the jet.

We know that the force exerted by the jet in its original direction

= Mass of water flowing/s × Change of velocity

$$= \frac{waV}{g} \times (V - 0) = \frac{waV^2}{g} \text{ kN}$$

∴ Force exerted by the jet in a direction normal (*i.e.*, perpendicular) to the plate,

$$F = \frac{waV^2}{g} \times \sin \theta = \frac{waV^2 \sin \theta}{g}$$

Impact of Jets

and the force exerted by the jet in the direction of flow,

$$F_X = F \sin \theta = \frac{waV^2 \sin \theta}{g} \times \sin \theta = \frac{waV^2 \sin^2 \theta}{g}$$

Similarly, force exerted by the jet in a direction normal to flow,

$$F_Y = F \cos \theta = \frac{waV^2 \sin \theta}{g} \times \cos \theta$$

$$= \frac{2waV^2 \sin \theta \cos \theta}{2g} \qquad \text{...(Multiplying and dividing by 2)}$$

$$= \frac{waV^2 \sin 2\theta}{2g} \qquad \text{...(}\because \sin 2\theta = 2 \sin \theta \cos \theta\text{)}$$

Example 29·4. *A jet of water of 100 mm diameter, moving with a velocity of 20 m/s strikes a stationary plate. Find the force on the plate in the direction of the jet, when*

    (a) the plate is normal to the jet, and

    (b) the angle between the jet and plate is 45°.

Solution. Given : $d = 100$ mm $= 0·1$ m; $V = 20$ m/s and $\theta = 45°$.

(a) Force on the plate when it is normal to the jet

We know that cross-sectional area of the jet,

$$a = \frac{\pi}{4} \times (d)^2 = \frac{\pi}{4} \times (0·1)^2 = 7·854 \times 10^{-3} \text{ m}^2$$

and force on the plate when it is normal to the jet,

$$F = \frac{waV^2}{g} = \frac{9·81 \times (7·854 \times 10^{-3}) \times (20)^2}{9·81} = 3·142 \text{ kN } \textbf{Ans.}$$

(b) Force on the plate when the angle between the jet and plate is 45°

We also know that force on the plate when the angle between the jet and plate is 45°,

$$F = \frac{waV^2}{g} \times \sin^2 \theta = 3·142 \sin^2 45° \text{ kN}$$

$$= 3·142 \times (0·707)^2 = 1·57 \text{ kN } \textbf{Ans.}$$

Example 29·5. *A 25 mm diameter jet exerts a force of 1 kN in the direction of flow against a flat plate, which is held inclined at an angle of 30° with the axis of the stream. Find the rate of flow.*

Solution. Given : $d = 25$ mm $= 0·025$ m; $F_x = 1$ kN and $\theta = 30°$.

Let Q = Rate of flow of water.

We know that cross-sectional area of the jet,

$$a = \frac{\pi}{4} \times (d)^2 = \frac{\pi}{4} \times (0·025)^2 = 0·491 \times 10^{-3} \text{ m}^2$$

and velocity of the jet, $V = \dfrac{Q}{a} = \dfrac{Q}{0·491 \times 10^{-3}} = 2·04 \times 10^3 \, Q$ m/s

We also know that force exerted by the jet (F_x),

$$1 = \frac{waV^2}{g} \times \sin^2 \theta$$

$$= \frac{9·81 \times (0·491 \times 10^{-3}) \times (2·04 \times 10^3 \, Q)^2}{9·81} \times \sin^2 30°$$

$$= 2·04 \times 10^3 \, Q^2 \times (0·5)^2 = 0·51 \times 10^3 \, Q^2$$

∴ $Q^2 = 1/(0·51 \times 10^3) = 1·96 \times 10^{-3}$

or $Q = 0·0443$ m³/s $= 44·3$ litres/s **Ans.**

29·4 Force of Jet Impinging on a Hinged Plate

In the previous articles, we have studied that whenever a jet is impinging on a plate, it exerts some force on it. Now we shall discuss the effect of this force on a hinged plate. Now consider a plate hinged at O as shown in Fig. 29·3 (a).

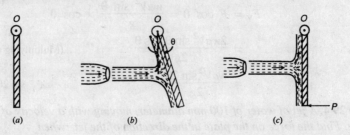

Fig. 29·3. Jet impinging on a hinged plate.

Let a jet be allowed to strike on a hinged plate. As a result of this jet strike, the plate will swing about the hinge, through some angle as shown in Fig. 29·3 (b). The angle, through which the plate will swing, may be found out by taking moments of the force of jet and the weight of plate about the hinge.

If it is required to keep the plate vertical, then some force (P) has to be applied on the plate as shown in Fig. 29·3 (c). The value of this force (P) may also be found out by taking moments about the hinge, as usual.

Note : If the jet of water strikes the plate at its centre, then the inclination of the plate, with the vertical, may be found out as discussed below.

Force exerted by the jet, $\quad F = \dfrac{waV^2}{g}$

Moment of this force about the hinge

$$= F \times \frac{d}{2} \qquad \text{...(where } d = \text{Depth of plate)}$$

and restoring moment of the plate about the hinge

$$= W \times \frac{d}{2} \sin \theta \qquad \text{... (where } W = \text{Weight of the plate)}$$

Equating these two moments,

$$W \times \frac{d}{2} \sin \theta = F \times \frac{d}{2}$$

∴ $\quad \sin \theta = \dfrac{F}{W} \qquad$ ...(i)

Now moment of the force P about the hinge
$$= P \cdot d$$

Again equating the moments of P and the force of jet,

$$P \cdot d = F \times \frac{d}{2}$$

∴ $\quad P = \dfrac{F}{2} \qquad$ ...(ii)

Example 29·6. *A jet of water 25 mm diameter strikes a flat plate normally at 30 metres/s, at a point 150 mm below the top of the plate. What force should be applied, 100 mm below the axis of the jet, in order to keep the plate vertical?*

Solution. Given : $d = 25$ mm $= 0.025$ m; $V = 30$ m/s.

Let $\qquad P =$ Force required to keep the plate vertical.

Impact of Jets

We know that cross-sectional area of the jet,

$$a = \frac{\pi}{4} \times (d)^2 = \frac{\pi}{4} \times (0.025)^2 \text{ m}^2$$
$$= 0.491 \times 10^{-3} \text{ m}^2$$

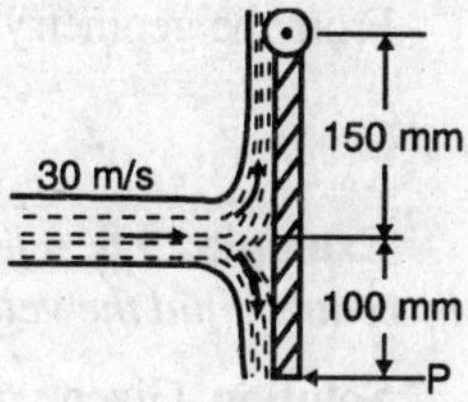

and force exerted by the jet,
$$F = \frac{waV^2}{g}$$
$$= \frac{9.81 \times (0.491 \times 10^{-3}) \times (30)^2}{9.81} \text{ kN}$$

$$= 0.442 \text{ kN} = 442 \text{ N}$$

Fig. 29·4.

Now taking moments about the hinge of the plate,
$$P \times 250 = 442 \times 150 = 66\ 300$$
∴ $$P = 66\ 300/250 = 265·2 \text{ N} \textbf{ Ans.}$$

Example 29·7. *A square plate weighing 150 N and of uniform thickness and 300 mm side is hung, so that it can swing freely about its upper horizontal edge. A horizontal jet of 20 mm diameter and having velocity of 20 m/s impinges on the plate. When the plate is vertical, the jet strikes the plate normally at its centre. Find the force, which must be applied at the lower edge of the plate, in order to keep the plate vertical.*

If the plate is allowed to swing freely, find the inclination to the vertical, which the plate will assume under the action of jet.

Solution. Given : $W = 150$ N; Side of square plate = 300 mm; $d = 20$ mm = 0·02 m; $V = 20$ m/s.

Force required to keep the plate vertical

Let $\qquad P =$ Force required to keep the plate vertical.

We know that cross-sectional area of the jet,

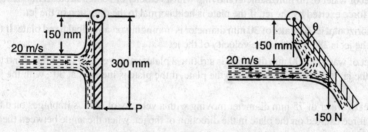

Fig. 29·5

$$a = \frac{\pi}{4} \times (d)^2 = \frac{\pi}{4} \times (0.02)^2 = 0.314 \times 10^{-3} \text{ m}^2$$

and force exerted by the jet, $F = \dfrac{waV^2}{g} = \dfrac{9.81 \times (0.314 \times 10^{-3}) \times (20)^2}{9.81} = 125·6$ N

Now taking moments about the hinge of the plate,
$$P \times 300 = 125·6 \times 150 = 18\ 840$$
∴ $$P = 18\ 840 / 300 = 62·8 \text{ N} \textbf{ Ans.}$$

Angle through which the plate will be inclined

Let $\qquad \theta =$ Angle with the vertical through which the plate will be inclined.

From the geometry of the figure, we find that

$$\sin \theta = \frac{F}{W} = \frac{125 \cdot 6}{150} = 0 \cdot 8373 \text{ or } \theta = 56 \cdot 9° \text{ Ans.}$$

Example 29·8. *A jet of water of 30 mm diameter strikes a hinged square plate weighing 100 N at its centre. Find the velocity of the jet in order to deflect the plate through an angle of 30°.*

Solution. Given : $d = 30$ mm $= 0·03$ m; $W = 100$ N and $\theta = 30°$.

Let $V = $ Velocity of the jet.

We know that cross-sectional area of the jet,

$$a = \frac{\pi}{4} \times (d)^2 = \frac{\pi}{4} \times (0·03)^2$$

$$= 0·707 \times 10^{-3} \text{ m}^2$$

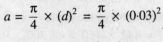

Fig. 29·6.

and force exerted by the jet, $F = \dfrac{waV^2}{g} = \dfrac{9·81 \times (0·707 \times 10^{-3}) \, V^2}{9·81}$ kN

$$= 0·707 \times 10^{-3} \, V^2 \text{ kN} = 0·707 \, V^2 \text{ N}$$

From the geometry of the figure we find that

$$\sin 30° = \frac{F}{W} \quad \text{or} \quad 0·5 = \frac{0·707 \, V^2}{100} = 0·707 \, 10^{-2} \, V^2$$

∴ $V^2 = 0·5 / (0·707 \times 10^{-2}) = 70·7$

or $V = \sqrt{70·7} = 8·41$ m/s   **Ans.**

EXERCISE 29·1

1. A jet of water of 50 mm diameter moving with a velocity of 10 m/s strikes a flat fixed plate. Calculate the force exerted by the jet, if the plate is held normal to the direction of the jet.   **[Ans. 196 N]**
2. A horizontal jet of water of 30 mm diameter is impinging on a vertical fixed plate. If the force exerted by the jet is 282·5 N, find the velocity of the jet.   **[Ans. 20 m/s]**
3. A jet of water of 100 mm diameter is striking a plate with a velocity of 35 m/s. Find the force exerted by the jet in a direction normal to the plate, if the plate is inclined at 30° with the jet.
                                                                                                         [Ans. 4·81 kN]
4. A jet of water of 75 mm diameter moving with a velocity of 15 m/s impinges on a fixed plate. Find the force exerted on the plate in the direction of the jet, when the angle between the plate and the jet is 25°.   **[Ans. 177·5 N[**
5. A horizontal jet of water of 25 mm diameter is striking a vertical plate, hinged at the top, with a velocity of 12 m/s. If the plate is allowed to swing freely, find that inclination, the plate will assume due to impact of the jet. Take weight of the plate as 180 N.   **[Ans. 23·13°]**
6. A jet of water of 50 mm diameter strikes a square plate at its centre. The plate weighing 80 N is hinged at its top. Find the velocity of the jet, which will keep the plate in equilibrium at an angle of 30° to the vertical.   **[Ans. 4·5 m/s]**

29·5 Force of Jet Impinging on a Moving Plate

Consider a jet of water impinging normally on a plate. As a result of the impact of the jet, let the plate move in the direction of the jet as shown in Fig. 29·7.

Let $V = $ Velocity of the jet in m/s,

                 $a = $ Cross-sectional area of the jet in m², and

                 $v = $ Velocity of the plate, as a result of the impact of jet in m/s.

Impact of Jets

A little consideration will show that the relative velocity of the jet with respect to the plate is equal to $(V-v)$ m/s.

For analysis purposes, it will be assumed that the plate is fixed and the jet is moving with a velocity of $(V-v)$ m/s. Therefore force exerted by the jet,

$$F = \text{Mass of water flowing per second} \times \text{Change of velocity}$$

$$= \frac{wa(V-v)}{g} \times [(V-v) - 0]$$

$$= \frac{wa(V-v)^2}{g} \text{ kN}$$

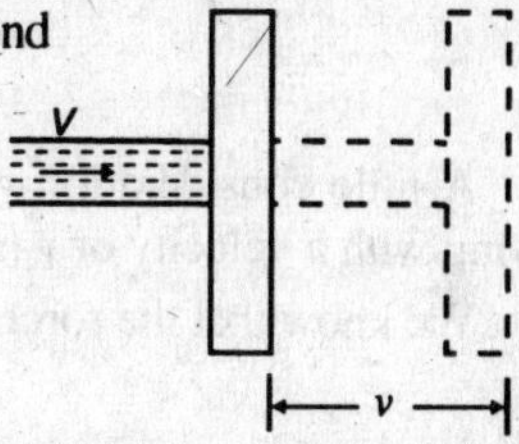

and work done by the jet $= \text{Force} \times \text{Distance}$

$$= \frac{wa(V-v)v}{g} \text{ kN-m}$$

Fig. 29·7. Jet impinging on a moving plate.

Example 29·9. *A jet of water 50 mm diameter and moving with a velocity of 26 m/s is impinging normally on a plate. Determine the pressure on the plate, when (a) it is fixed and (b) it is moving with a velocity of 10 m/s in the direction of the jet.*

Also determine the work done per second by the jet.

Solution. Given : $d = 50$ mm $= 0.05$ m; $V = 26$ m/s and $v = 10$ m/s.

Pressure on the plate when it is fixed

We know that cross-sectional area of the jet,

$$a = \frac{\pi}{4} \times (d)^2 = \frac{\pi}{4} \times (0.05)^2 = 1.964 \times 10^{-3} \text{ m}^2$$

and pressure on the plate, $P_1 = \frac{waV^2}{g} = \frac{9.81 \times (1.964 \times 10^{-3}) \times (26)^2}{9.81} = 1.33$ kN **Ans.**

Pressure on the plate when it is moving

We know that pressure on the plate when it is moving,

$$P_2 = \frac{wa(V-v)^2}{g} = \frac{9.81 \times (1.964 \times 10^{-3}) \times (26-10)^2}{9.81}$$

$$= 0.503 \text{ kN Ans.}$$

Work done by the jet

We also know that work done by the jet

$$= \text{Force} \times \text{Distance} = 0.503 \times 10 = 5.03 \text{ kN-m}$$

$$= 5.03 \text{ kJ Ans.}$$

29·6 Force of Jet Impinging on a Series of Vanes

A little consideration will show, that the case of a jet impinging on a moving plate is merely a theoretical one, which seldom arises in practice; because it requires continuously a jet following the moving plate. But in actual practice, a case similar to this arises, in which a jet of water impinges on a series of vanes, mounted on the circumference of a large wheel as shown in Fig. 29·8.

Let $V =$ Velocity of the jet in m/s,

$a =$ Cross-sectional area of the jet in m², and

$v =$ Velocity of the vanes, as a result of the jet, in m/s.

Fig. 29·8. Jet impinging on a series of vanes.

A little consideration will show, that the jet of water, after impinging on the vanes, will be moving with a velocity of v m/s.

We know that the force exerted by the jet

$$F = \text{Mass of water flowing per second} \times \text{Change of velocity}$$
$$= \frac{waV}{g} \times (V-v) = \frac{waV(V-v)}{g} \qquad \text{...(i)}$$

and work done by jet $= \text{Force} \times \text{Distance} = \dfrac{waV(V-v) \times v}{g}$ ...(ii)

$\therefore$ Work done per kN of water $= \dfrac{1}{g}(V-v) \times v$ ...(iii)

and energy of the jet water per kN of water

$$= \frac{V^2}{2g}$$

Efficiency, $\eta = \dfrac{\text{Work done per kN of water}}{\text{Energy per kN of water}}$

$$= \frac{\dfrac{1}{g}(V-v) \times v}{\dfrac{V^2}{2g}} = \frac{2(V-v) \times v}{V^2}$$

Note : The efficiency is also equal to $\dfrac{\text{Work done by the jet}}{\text{Energy of the jet}}$

Example 29·10. *A jet of water 60 mm in diameter, moving with a velocity of 20 m/s, strikes a flat vane, which is normal to the axis of the stream. Find (i) the force exerted by the jet, when the vane moves with a velocity of 8 m/s. (ii) the force exerted by the jet, if instead of one flat vane, there is a series of vanes, so arranged that each vane appears successively before the jet in the same position and always moving with a velocity of 8 m/s.*

Solution. Given : $d = 60$ mm $= 0.06$ m; $V = 20$ m/s and $v = 8$ m/s.

(i) Force exerted by the jet when a single vane is moving

We know that cross-sectional area of the jet,

$$a = \frac{\pi}{4} \times (d)^2 = \frac{\pi}{4} \times (0.06)^2 = 2.827 \times 10^{-3} \text{ m}^2$$

and force exerted by the jet,

$$F_1 = \frac{wa(V-v)^2}{g} = \frac{9.81 \times (2.827 \times 10^{-3}) \times (20-8)^2}{9.81} \text{ kN}$$

$$= 0.407 \text{ kN Ans.}$$

(ii) Force exerted by the jet when a series of vanes are moving

We also know that force exerted by the jet,

$$F_2 = \frac{waV(V-v)}{g} = \frac{9.81 \times (2.827 \times 10^{-3}) \times 20(20-8)}{9.81} \text{ kN}$$

$$= 0.678 \text{ kN Ans.}$$

Example 29·11. *A jet of water 50 mm in diameter, moving with velocity of 15 m/s, impinges on a series of vanes moving with a velocity of 6 m/s. Find (a) Force exerted by the jet, (b) Work done by the jet, and (c) Efficiency of the jet.*

Solution. Given : $d = 50$ mm $= 0.05$ m; $V = 15$ m/s and $v = 6$ m/s.

(a) Force exerted by the jet

We know that cross-sectional area of the jet,

$$a = \frac{\pi}{4} \times (d)^2 = \frac{\pi}{4} \times (0.05)^2 = 1.964 \times 10^{-3} \text{ m}^2$$

and force exerted by the jet, $F = \dfrac{waV(V-v)}{g} = \dfrac{9.81 \times (1.964 \times 10^{-3}) \times 15(15-6)}{9.8}$ kN

$$= 0.265 \text{ kN} = 265 \text{ N Ans.}$$

(b) Work done by the jet

We know that work done by the jet

$$W = \text{Force} \times \text{Distance} = 265 \times 6 = 1590 \text{ N-m} = 1590 \text{ J Ans.}$$

(c) Efficiency of the jet

We also know that efficiency of the jet,

$$\eta = \frac{2(V-v)v}{V^2} = \frac{2(15-6) \times 6}{(15)^2} = 0.48 = 48\% \text{ Ans.}$$

29·7 Force of Jet Impinging on a Fixed Curved Vane

Consider a jet of water entering and leaving a fixed curved vane tangentially as shown in Fig. 29·9.

Let V = Velocity of the jet,

a = Cross-sectional area of the jet,

α = Inlet angle of the jet, and

β = Outlet angle of the jet,

The jet, while moving through the vane, will exert some force on vane. This force may be determined by the finding out the components of the force along the normal and perpendicular to the normal of the vane.

We know that the force of the jet along normal to the vane

= Mass of water flowing per second

× Change of velocity along normal to the vane

$$= \frac{waV}{g}(V\cos\alpha + V\cos\beta) \qquad \ldots(i)$$

and force of the jet along perpendicular to the normal to the vane

$$= \frac{waV}{g}(V\sin\alpha - V\sin\beta)$$

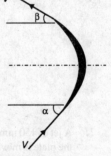

Fig. 29·9. Jet impinging on a fixed curved vane.

Note : Sometimes the total angle through which the jet is deflected is given instead of the inlet and outlet angles of the jet. In such a case, assuming $\alpha = \beta$ and $\theta = 180° - (\alpha + \beta)$, the force of the jet along X-X axis,

$$F_X = \text{Mass of water flowing per second} \times \text{Change of velocity}$$

$$= \frac{waV}{g} \times 2V \sin \frac{\theta}{2} = \frac{2waV^2}{g} \times \sin \frac{\theta}{2}.$$

Example 29·12. *Calculate the magnitude of the force exerted by a jet of cross-sectional area 2000 mm^2 and of velocity 25 m/s on a fixed smooth curved vane which deflects the jet by 120°.*

Solution. Given : $a = 2000$ mm$^2 = 0.002$ m^2; $V = 25$ m/s and $\theta = 120°$.

We know that magnitude of the force exerted by the jet,

$$F = \frac{2waV^2}{g} \times \sin \theta = \frac{2 \times 9.81 \times 0.02 \times (25)^2}{9.81} \times \sin 60° \text{ kN}$$

$$= 25 \times 0.8660 = 21.65 \text{ kN Ans.}$$

Example 29·13. *A jet of water 40 mm diameter enters a fixed curved vane with a velocity of 50 m/s at an angle 20° to the horizontal. Find the normal and tangential forces exerted by the jet, if it leaves the vane at angle 15° to the horizontal.*

Solution. Given : $d = 40$ mm $= 0.04$ m; $V = 50$ m/s and $\alpha = 20°$ and $\beta = 15°$.

Normal force exerted by the jet

We know that cross-sectional area of the jet,

$$a = \frac{\pi}{4} \times (d)^2 = \frac{\pi}{4} \times (0.04)^2 = 1.257 \times 10^{-3} \text{ m}^2$$

and normal force exerted by the jet,

$$F_n = \frac{waV}{g} (V \cos \alpha + V \cos \beta)$$

$$= \frac{9.81 \times (1.257 \times 10^{-3}) \times 50}{9.81}$$

$$\times (50 \cos 20° + 50 \cos 15°) \text{ kN}$$

$$= 0.063 \times [(50 \times 0.9397) + (50 \times 0.9659)] = 6.0 \text{ kN Ans.}$$

Tangential force exerted by the jet

We also know that tangential force exerted by the jet,

$$F_t = \frac{waV}{g} (V \sin \alpha - V \sin \beta)$$

$$= \frac{9.81 \times (1.257 \times 10^{-3}) \times 50}{9.81}$$

$$\times [(50 \sin 20°) - (50 \sin 15°)] \text{ kN}$$

$$= 0.063 \times [(50 \times 0.3420) - (50 \times 0.2588)] \text{ kN}$$

$$= 0.262 \text{ kN Ans.}$$

EXERCISE 29·2

1. A jet of 150 mm diameter is moving with a velocity of 30 m/s. Find the force exerted by the jet when the plate is moving with a velocity of 12 m/s in the direction of the jet. **[Ans. 5·73 kN]**
2. A jet of water of 100 mm diameter impinges with a velocity of 25 m/s on a plate moving with a velocity of 10 m/s in the direction of the jet. Find the force exerted by the jet.

   If the plate is now replaced with a series of vanes moving with the same velocity as that of the plate, find the force exerted by the jet on the vanes. **[Ans. 1·77 kN; 2·95 kN]**

Impact of Jets

3. A jet of 100 mm diameter, moving with a velocity of 12 m/s impinges on a series of vanes moving with a velocity of 8 m/s. Determine (i) force on the plate, (ii) work done per second, and (iii) efficiency [Ans. 377 N; 3·02 kJ; 44·4%]

4. A jet of water of 80 mm diameter moving with a velocity of 20 m/s enters tangentially a fixed curved vane and is deflected through 135°. Find the work done by the jet. [Ans. 1·42 kN]

29·8 Force of Jet Impinging on a Moving Curved Vane

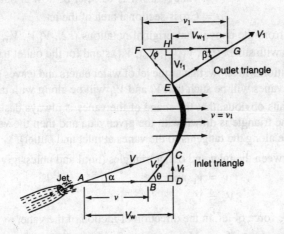

Fig. 29·10. Jet impinging or a moving curved vane.

Consider a jet of water entering and leaving a moving curved vane as shown in Fig. 20·10.

Let
- V = Velocity of the jet (AC), while entering the vane,
- V_1 = Velocity of the jet (EG) while leaving the vane,
- v_1, v_2 = Velocity of the vane (AB, FG),
- α = Angle with the direction of motion of the vane, at which the jet enters the vane,
- β = Angle, with the direction of motion of the vane, at which the jet leaves the vane,
- V_r = Relative velocity of the jet and the vane (BC) at entrance (It is the vectorial difference between V and v),
- V_{r1} = Relative velocity of the jet and the vane (EF) at exit (It is the vectorial difference between v_1 and v_1),
- θ = Angle, which V_r makes with the direction of motion of the vane at inlet (known as vane angle at inlet),
- ϕ = Angle, which V_{r1} makes with the direction of motion of the vane at outlet (known as vane angle at outlet),
- V_w = Horizontal component of V (AD, equal to $V \cos \alpha$). It is a component parallel to the direction of motion of the vane (known as velocity of whirl at inlet),
- V_{w1} = Horizontal component of V_1 (HG, equal to $V_1 \cos \beta$). It is a component parallel to the direction of motion of the vane (known as velocity of whirl at outlet),

V_f = Vertical component of V (DC, equal to $V \sin \alpha$). It is a component at right angles to the direction of motion of the vane (known as velocity of flow at inlet),

V_{f1} = Vertical component of V_1 (EH, equal to $V_1 \sin \beta$). It is a component at right angles to the direction of motion of the vane (known as velocity of flow at outlet), and

a = Cross-sectional area of the jet.

It may be seen, from the above, that original notations (*i.e.*, V, V_r, V_w, V_f) stand for the inlet triangle. The notations with suffix (*i.e.*, V_1, V_{r1}, V_{w1}, V_{f1}) stand for the outlet triangle.

A little consideration will show, that as the jet of water enters and leaves the vanes tangentially, therefore shape of the vanes will be such that V_r and V_{r1} will be along with tangents to the vanes at inlet and outlet. It is thus obvious, that the shape of the vanes is always designed according to the given data (*i.e.*, first the triangle is drawn with the given data and then the vane is drawn in such a way, that V_r and V_{r1} are along the tangents to the vanes at inlet and outlet).

The relations between the inlet and outlet triangles (until and unless given) are :

(*i*) $v = v_1$, and

(*ii*) $V_r = V_{r1}$.

We know that the force of jet, in the direction of motion of the vane,

F_X = Mass of water flowing per second × Change of velocity of whirl

$$= \frac{waV}{g} (V_w - V_{w1}) \qquad ...(i)$$

and work done in the direction of motion of the vane

$$= \text{Force} \times \text{Distance} = \frac{waV}{g} (V_w - V_{w1}) \times v \qquad ...(ii)$$

and work done per kN of water

$$= (V_w - V_{w1}) \times v \qquad ...(iii)$$

Notes : 1. It is very important to draw the correct shape of inlet and outlet triangle of velocities. Lot of patience and understanding is required for this. But, once the correct shape of the two triangles is drawn, then the example can be solved very easily.

2. The direction of V_{w1} plays an important role for finding out the work done. If v_1 is greater then ($V_{r1} \cos \phi$), the value of V_{w1} is positive. Otherwise it is negative.

3. Power developed by the vane may be found out as usual.

Example 29·14. *A jet of water, moving at 60 m/s is deflected by a vane moving at 25 m/s in a direction at 30° to the direction of the jet. The water leaves the blades normally to the motion of the vanes.*

Draw inlet and outlet triangles of velocities, and find the vane angles for no shock at entry and exit. Take relative velocity at outlet to be 0·85 of the relative velocity at inlet.

Solution. Given : $V = 60$ m/s; $v = 25$ m/s; $\alpha = 30°$ and $V_{r1} = 0.85\ V_r$.

Inlet and outlet triangles of velocities

The inlet and outlet triangles of velocities may be drawn as shown in Fig. 29·11 and as discussed below :

1. First of all, draw a horizontal line (*i.e.*, in the direction of motion of the vanes) and cut off AB equal to 25 m/s to some suitable scale, to represent the velocity of vanes (v).

Impact of Jets

2. Draw a line at an angle of 30° (*i.e.*, angle at which the jet enters the vanes) and cut off AC equal to 60 m/s to the scale to represent the velocity of jet (V).
3. Join BC, which gives the relative velocity (V_r) to the scale. Now extend AB to D, such that DC is perpendicular to AB. From the geometry of the figure, we find that the lengths AD and DC give the velocity of whirl (V_w) and velocity of flow (V_f) respectively to the scale. Measure these two lengths.
4. Now draw the curved vane, such that V_r is tangential to the vane at the inlet tip.
5. Then draw the outlet triangle of velocities with V_{r1} tangential to the vane. Now cut off (EF) *i.e.*, $V_{r1} = 0.85\ V_{r1}$.
   ... (As given)

and (EG) *i.e.*, $V_{f1} = V_f$ at right angles to v and v_1 (because it is given that the water leaves the blades normally to the motion of the blades).

6. Now take (FG) *i.e.*, $v_1 = v$ and complete the outlet triangle of velocities.

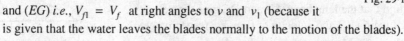

Fig. 29·11.

Vane angle at inlet

Let θ = Vane angle at inlet.

From the geometry of the inlet triangle (ACD) we find that the velocity of whirl at inlet,
$$V_w = V \cos 30° = 60 \times 0.866 = 51.96 \text{ m/s}$$

and velocity of flow at inlet, $V_f = V \sin 30° = 60 \times 0.5 = 30 \text{ m/s}$

and $$\tan \theta = \frac{V_f}{V_w - v} = \frac{30}{51.96 - 25} = 1.1128$$

or $\theta = 48.1°$ **Ans.**

Vane angle at outlet

Let ϕ = Vane angle at outlet.

From the geometry of the inlet triangle, we also find that the relative velocity of jet and vane,
$$V_r = \frac{V_f}{\sin 48.1°} = \frac{30}{0.7447} = 40.31 \text{ m/s}$$

Now in the outlet triangle (EFG) relative velocity of jet and vane,
$$V_{r1} = 0.85 \times V_r = 0.85 \times 40.31 \text{ m/s} \qquad \text{.... (given)}$$
$$= 34.26 \text{ m/s}$$

In the outlet triangle, we find that
$$\cos \phi = \frac{v_1}{V_{r1}} = \frac{25}{34.26} = 0.7297$$

∴ $\phi = 43.1°$ **Ans.**

Example 29·15. *A jet of water, having a velocity of 30 m/s impinges on a series of vanes with a velocity of 15 m/s. The jet makes an angle of 30° to the direction of motion of vanes when entering and leaves at an angle of 120°.*

Sketch velocity triangles at entrance and exit, and determine the vane angles, so that the water enters and leaves without shock.

Solution. Given : $V = 30$ m/s; $v = 15$ m/s; $\alpha = 30°$ and $\beta = 120°$.

Velocity triangles at entrance and exit

The velocity triangles at entrance and exit may be drawn as shown in Fig. 20·12 and as discussed below :

1. First of all, draw a horizontal line (*i.e.*, in the direction of motion of the vanes) and cut off *AB* equal to 15 m/s to some suitable scale, to represent the velocity of vanes (*v*).
2. Draw a line at an angle of 30° (*i.e.*, the angle at which the jet enters the vane) and cut off *AC* equal to 30 m/s to the scale to represent the velocity of jet (*V*).
3. Join *BC*, which gives the relative velocity (V_r) to the scale. Now extend *AB* to *D*, such that *DC* is perpendicular to *AB*. From the geometry of the figure, we find that the lengths of *AD* and *DC* give the velocity of whirl (V_w) and velocity of flow V_f respectively to the scale. Measure these two lengths.
4. Now draw the curved vane, such that V_r is tangential to the vane at the entrance tip.
5. Then draw a velocity triangle at exit with V_{r1} tangential to the vane. Now cut off (*EF*) *i.e.*, $V_{r1} = V_r =$ Through *F*, draw a horizontal line and cut off *FH* equal to 15 m/s to the scale to represent the velocity of vanes at exit (*i.e.*, v_1).
6. Join *EH* which will make an angle of 120° with *FH* (as the jet leaves the vane at an angle of 120°. Now draw *EG* perpendicular to *FH*.

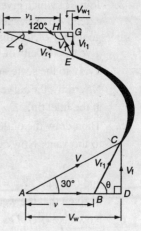

Fig. 29·12.

Vane angles

Let $\theta =$ Vane angle at inlet, and
$\phi =$ Vane angle at outlet.

From the inlet triangle, we find that the velocity of whirl at inlet,
$$V_w = V \cos 30° = 30 \times 0.866 = 25.98 \text{ m/s}$$
and velocity of flow at inlet, $V_f = V \sin 30° = 30 \times 0.5 = 15 \text{ m/s}$

∴ $$\tan \theta = \frac{V_f}{V_w - v} = \frac{15}{25.98 - 15} = 1.366$$

or $\theta = 53.8°$ **Ans.**

From the inlet triangle, we also find that the relative velocity of jet and vane,
$$V_r = \frac{V_f}{\sin 53.8°} = \frac{15}{0.8070} = 18.59 \text{ m/s}$$

From the outlet triangle, we find from the sine rule,
$$\frac{v_1}{\sin(60° - \phi)} = \frac{V_{r1}}{\sin 120°}$$

or $$\sin(60° - \phi) = \frac{v_1 \sin 120°}{V_{r1}} = \frac{15 \times \sin 60°}{18.59} = \frac{15 \times 0.866}{18.59} = 0.6988$$

∴ $(60° - \phi) = 44.3°$

or $\phi = 15.7°$ **Ans.**

Impact of Jets

Example 29·16. *A circular jet delivers water at the rate of 60 litres/s, with a velocity of 24 m/s. The jet impinges tangentially on the vane moving in the direction of the jet, with a velocity of 12 m/s. The vane is so shaped that, if stationary, it would deflect the jet through an angle 45°. Through what angle will it deflect the jet? Also find the work done/s.*

Solution. Given : Q = 60 litres/s = 0·06 m³/s; V = 24 m/s; v = 12 m/s; α = 0 (because the jet impinges tangentially on the vane moving in the direction of the jet) and ϕ = 45°.

Angle through which the jet is deflected

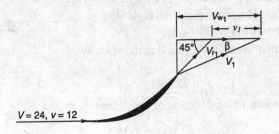

Fig. 29·13.

Since the jet of water moves in the same direction, as that of the vanes, therefore the inlet triangle will be a straight line as shown in Fig. 29·13. Thus velocity of whirl at inlet,

$$V_w = 24 \text{ m/s}$$

and relative velocity of the jet and vane,

$$V_r = 24 - 12 = 12 \text{ m/s}$$

Since the velocity of vanes at outlet (v_1) and relative velocity of the jet and vane at outlet (V_{r1}) are equal to 12 m/s, therefore the outlet triangle of velocities will be an isosceles triangle as shown in Fig. 29·13.

∴ Angle through which the jet is deflected,

$$\beta = \frac{\phi}{2} = \frac{45}{2} = 22 \cdot 5° \quad \text{Ans.}$$

Work done per second

From the outlet triangle of velocities, we find that velocity of whirl at outlet,

$$V_{w1} = 12 + 12 \cos 45° = 12 + (12 \times 0 \cdot 707) = 20 \cdot 5 \text{ m/s}$$

and work done per second

$$W = \frac{waV}{g} (V_w - V_{w1}) \times v$$

$$= \frac{9 \cdot 81 \times 0 \cdot 06}{9 \cdot 81} (24 - 20 \cdot 5) = 2 \cdot 52 \text{ kN-m} = 2 \cdot 52 \text{ kJ} \quad \text{Ans.}$$

Example 29·17. *A jet of water 100 mm in diameter, moving with a velocity of 25 m/s in the direction of the vanes, enters the vanes moving with a velocity of 12·5 m/s. If the jet leaves the vanes at an angle of 60° with the direction of motion of the vanes, find (i) force on the vanes in the direction of their motion, and (ii) work done per second.*

Solution. Given : d = 100 mm = 0·01 m; V = 25 m/s; v = 12·5 m/s; α = 0 (because the jet is moving in the direction of the vanes) and β = 60°.

(i) Force on the vanes in the direction of their motion

We know that cross-sectional area of the jet,

$$a = \frac{\pi}{4} \times (d)^2 = \frac{\pi}{4} \times (0 \cdot 1)^2 = 7 \cdot 854 \times 10^{-3} \text{ m}^2$$

Since the jet of water moves in the same direction as that of the vanes, therefore the inlet triangle will be a straight line as shown in Fig. 29·14. Thus velocity of whirl at inlet,

$$V_w = 25 \text{ m/s}$$

and relative velocity of the jet and vane,

$$V_r = 25 - 12\cdot5 = 12\cdot5 \text{ m/s}$$

Since the velocity of vane at outlet (v_1) and relative velocity of the jet and vane at outlet (V_{r1}) are equal, and the jet leaves the vanes at 60° (given), therefore the outlet triangle will be an equilateral triangle as shown in Fig. 29·14.

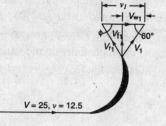

$V = 25, v = 12.5$

Thus, from the outlet triangle, we find that the velocity of whirl at outlet,

$$V_{w1} = 6\cdot25 \text{ m/s}$$

Fig. 29·14.

∴ Force on the vanes in the direction of their vanes

$$= \frac{waV}{g}(V_w - V_{w_1})$$

$$= \frac{9\cdot81 \times (7\cdot854 \times 10^{-3}) \times 25}{9\cdot81} (25 - 6\cdot25 \text{ kN}) = 3\cdot68 \text{ kN}$$

(ii) Work done per second

We also know that work done per second

$$W = \frac{waV}{g}(V_w - V_{w1}) \times v = \frac{9\cdot81 \times (7\cdot854 \times 10^{-3}) \times 25}{9\cdot81}(25 - 6\cdot25) \times 12$$

$$= 44\cdot2 \text{ kN–m/s} = 44\cdot2 \text{ kJ} \quad \textbf{Ans.}$$

Example 29·18. *A jet of water moving at 12 m/s impinges on a concave vane shaped to deflect the jet through 120° when stationary. If the vane is moving at 5 m/s, find the angle of jet so that there is no shock at outlet. What is the absolute velocity of jet at exit and the work done per kN of water? Assume that the vane is smooth.*

Solution. Given : $V = 12$ m/s; Angle through which the jet is deflected = 120°. Therefore $\phi = 180° - 120° = 60°$ and $v = 5$ m/s.

Angle of the jet at outlet

Let β = Angle of the jet at outlet.

Since no data is given for the inlet triangle of velocities, therefore the jet of water will be assumed to move in the same direction as that of the vanes. It is thus obvious, that the inlet triangle will be a straight line as shown in Fig. 29·15.

∴ Velocity of whirl at inlet,

$$V_w = V = 12 \text{ m/s}$$

and relative velocity of the jet and vane,

$$V_r = 12 - 5 = 7 \text{ m/s}$$

With the velocity of vanes at outlet (v_1) equal to 5 m/s, relative velocity ($V_{r1} = V_r$) equal to 7 m/s and vane angle at outlet (ϕ) equal to 60°, draw the outlet triangle of velocities as shown in Fig. 29·15.

From the geometry of outlet triangle of velocities, find that the velocity of flow

$$V_{f1} = V_{r1} \sin 60° = 7 \times 0\cdot866 = 6\cdot06 \text{ m/s}$$

and velocity of whirl,

$$V_{w1} = 5 - V_{r1} \cos 60° = 5 - (7 \times 0\cdot5) = 1\cdot5 \text{ m/s}$$

∴ $\tan \beta = \dfrac{V_{f1}}{V_{w1}} = \dfrac{6 \cdot 06}{1 \cdot 5} = 4 \cdot 04$

or $\beta = 76 \cdot 1°$ **Ans.**

Absolute velocity of the jet at exit

From the geometry of outlet triangle of velocities, we also find that absolute velocity of the jet at exit,

$$V_1 = \dfrac{V_{f1}}{\sin \beta} = \dfrac{6 \cdot 06}{\sin 76 \cdot 1°} = \dfrac{6 \cdot 06}{0 \cdot 9707}$$

$= 6 \cdot 24$ m/s **Ans.**

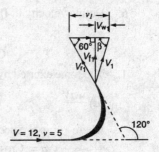

Fig. 29·15.

Work done per kN of water

We also know that work done per kN of water

$= (V_w - V_{w1}) \times v = (12 - 1 \cdot 5) \times 5 = 52 \cdot 5$ kN-m/s

$= 52 \cdot 5$ kJ/s **Ans.**

EXERCISE 29·3

1. A jet of water having a velocity of 40 m/s enters a curved vane which is moving with a velocity of 20 m/s. The jet makes an angle of 30° with the direction of motion of the vane at inlet and leaves at angle of 90° with the direction of motion at outlet. Draw the velocity triangles and determine the vane angles at inlet and outlet, so that the water enters and leaves without shock. [**Ans.** 53·8°; 36·2°]

2. A jet of water moving with a velocity of 20 m/s strikes curved vanes moving with a velocity of 10 m/s. The jet makes an angle of 20° with the direction of motion of vanes at inlet and leaves at an angle of 130° with the direction of motion of the vanes. Calculate the vane angles at inlet and outlet, so that the water has shockless entry and exit. [**Ans.** 37·9°; 6·6°]

3. A jet of water moving with a velocity of 12 m/s impinges on concave shaped vanes to deflect the jet through 120° when stationary. If the vane is moving with a velocity of 5 m/s, find the angle of jet, so that there is no shock at inlet. What is the absolute velocity of the jet at exit both in magnitude and direction? Assume the vane to be symmetrical. [**Ans.** 17·98°; 3·96 m/s; 69·1°]

[**Hint :** Since the vane is symmetrical, therefore $\theta = \phi = 30°$]

4. A 150 mm diameter jet of water moving at 30 m/s impinges on a series of vanes moving at 15 m/s in the direction of the jet and leaves at 60° with the direction of motion of the jet. Calculate (*i*) force exerted by the jet in the direction of motion of the vanes; and (*ii*) work done by the jet.
[**Ans.** 11·95 kN; 179·25 kJ]

QUESTIONS

1. What do you understand by the term 'jet of water'? Derive an expression for the force of jet on a fixed plate.

2. Show that the normal force exerted by a jet of water on an inclined plate is given by the relation,

$$F = \dfrac{waV^2 \sin \theta}{g}$$

where $a = $ Area of jet,

$V = $ Velocity of the jet, and

$\theta = $ Inclination of the plate with the jet.

3. Derive an equation between the angle, through which a hinged plate will swing, in terms of force of the jet and weight of the plate.

4. Derive an expression for the force of jet impinging on a moving plate and compare it with the force, when the same jet is impinging on a series of moving vanes.

5. Derive an expression for the force, work done and efficiency of a moving curved vane.
6. Define velocity of flow and velocity of whirl and explain their significance.

OBJECTIVE TYPE QUESTIONS

1. The force exerted by a jet of water impinging normally on a fixed plate
   (a) $\dfrac{waV}{2g}$ (b) $\dfrac{waV}{g}$ (c) $\dfrac{waV^2}{2g}$ (d) $\dfrac{waV^2}{g}$

   where w = Specific weight of water,
   a = Cross-sectional area of the jet, and
   V = Velocity of the jet.

2. The force exerted by a jet of water impinging on a fixed plate inclined at an angle θ with jet.
   (a) $\dfrac{waV}{2g} \times \sin\theta$ (b) $\dfrac{waV}{g} \times \sin\theta$ (c) $\dfrac{waV^2}{2g} \sin\theta$ (d) $\dfrac{waV^2}{g} \sin\theta$

3. The force exerted by a jet of water impinging normally with a velocity V on a plate, which moves in the direction of the jet with a velocity v is
   (a) $\dfrac{wa\,(V-v)^2}{2g}$ (b) $\dfrac{wa\,(V-v)^2}{g}$ (c) $\dfrac{wa\,(V-v)}{2g}$ (d) $\dfrac{wa\,(V-v)}{g}$

4. The efficiency of a jet of water impinging normally with a velocity V on a series of vanes moving with a velocity (v) is given by the relation
   (a) $\dfrac{2(V-v)}{V^2}$ (b) $\dfrac{(V-v)}{V^2}$ (c) $\dfrac{(v-v)^2}{2V}$ (d) $\dfrac{(V-v)^2}{V}$

ANSWERS
1. (d)   2. (c)   3. (b)   4. (a)

30

Jet Propulsion

1. Introduction. 2. Pressure of Water due to Deviated Flow. 3. Principle of Jet Propulsion. 4. Propulsion of Ships by Water Jets. 5. Propulsion of a Ships Having Inlet Orifices at Right Angles to the Direction of its Motion (i.e., Orifices Amidship). 6. Propulsion of a Ship Having Inlet Orifices Facing the Direction of Flow.

30·1 Introduction

In the previous chapter, we have discussed the jet of water impinging on the various types of plates and vanes, under different sets of conditions. As a matter of fact, it is the basic theory for the design of all types of Hydraulic Machines *i.e.*, turbine, and pumps etc. In this chapter, we shall discuss jet propulsion.

30·2 Pressure of Water due to Deviated Flow

Sometimes, a pipeline, carrying water, changes its direction from its straight path. The velocity of water flowing through pipe is also changed[*] due to change in its direction. It is thus obvious, that the deviated flow of water will cause some pressure on the pipe wall.

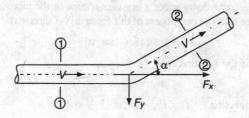

Fig. 30·1. Deviated flow

Now consider a pipeline, carrying water, and deviated from its straight path as shown in Fig. 30·1.

Let V = Velocity of water in the pipe in section 1,
 a = Area of the pipe, and
 α = Angle, through which the centre of pipe has been deviated from its straight path.

∴ Velocity of water at section 2
$$= V \cos \alpha$$
and mass of water flowing at section 1
$$= \frac{waV}{g}$$

∴ Momentum of the flowing water at section 1 in the *X-X* direction
$$= \text{Mass} \times \text{Velocity} = \frac{waV}{g} \times V = \frac{waV^2}{g}$$

---

[*]Velocity is a vector quantity, which changes because of change in direction.

Similarly, momentum of the flowing water at section 2 in the X-X direction

$$= \frac{waV}{g} \times V \cos \alpha = \frac{waV^2 \cos \alpha}{g}$$

∴ Change in momentum

$$= \frac{waV^2}{g} - \frac{waV^2 \cos \alpha}{g} = \frac{waV^2}{g}(1 - \cos \alpha)$$

We know that according to *Newton's Second Law of Motion, this force is equal to the rate of change of momentum. Therefore force exerted by the water in X-X direction,

$$F_X = \frac{waV^2}{g}(1 - \cos \alpha)$$

Similarly, it can be proved that the force exerted by the water in Y-Y direction,

$$F_Y = \frac{waV^2 \sin \alpha}{g}$$

∴ Resultant force on the bend,

$$R = \sqrt{F_X^2 + F_Y^2}$$

If the resultant force makes an angle θ with X-X direction, then

$$\tan \theta = \frac{F_Y}{F_X}$$

Note : Sometimes the water flows under a pressure p through the pipe. In such a case, force exerted because of pressure is equal to $p.a$. The component of this force in X-X direction.

$$= p.a(1 - \cos \alpha)$$

and component of this force in the Y-Y direction

$$= p.a \sin \alpha$$

∴ Total force in X-X direction $= F_X + p.a(1 - \cos \alpha)$

and total force in Y-Y direction $= F_Y + p.a \sin \alpha$

Example 30·1. *250 litre/s of water is flowing in a pipe having a diameter of 300 mm. If the pipe is bent by 135° (that is change from initial to final direction is 135°), find the magnitude and direction of resultant force on the bend.*

Solution. Given : $Q = 250$ litres/s $= 0.25$ m³/s; $d = 300$ mm $= 0.3$ m and $\alpha = 135°$.

Magnitude of the resultant force

We know that cross-sectional area of the pipe,

$$a = \frac{\pi}{4} \times (d)^2 = \frac{\pi}{4} \times (0.3)^2 = 0.0707 \text{ m}^2$$

and velocity of water, $V = \dfrac{Q}{a} = \dfrac{0.25}{0.0707} = 3.54$ m/s

∴ Force exerted by the in X-X direction,

$$F_X = \frac{waV^2}{g}(1 - \cos \alpha)$$

$$= \frac{9.81 \times 0.0707 \times (3.54)^2}{9.81} \times (1 - \cos 135°) \text{ kN}$$

---

*Newton's Second Law of Motion states "The rate of change of momentum is directly proportional to the impressed force, and takes place in the same direction in which it acts."

Jet Propulsion

$$F_X = 0.886 \times [1 - (-\cos 45°)]$$
$$= 0.886 \times (1 + 0.707) = 1.51 \text{ kN}$$

and force exerted by the water in Y-Y direction,

$$F_Y = \frac{waV^2}{g} \sin \alpha = \frac{9.81 \times 0.0707 \times (3.54)^2}{9.81} \times \sin 135°$$
$$= 0.886 \sin 45° = 0.886 \times 0.7071 = 0.626 \text{ kN} \qquad ...(ii)$$

∴ Magnitude of the resultant force,

$$R = \sqrt{F_X^2 + F_Y^2} = \sqrt{(1.51)^2 + (0.626)^2} = 1.63 \text{ kN Ans.}$$

Direction of the resultant force

Let θ = Angle, which the resultant makes with X-X direction.

We also know that $\tan \theta = \dfrac{F_Y}{F_X} = \dfrac{0.626}{1.51} = 0.4146$ or $\theta = 22.5°$ **Ans.**

Example 30·2. *A 200 mm diameter pipe carries water under a head of 20 m with a velocity of 3·5 m/s. If the axis of the pipe turns through 45°, find the magnitude and the direction of the resultant force at the bend.*

Solution. Given : $d = 200$ mm $= 0.2$ m; $h = 20$ m; $V = 3.5$ m/s and $\alpha = 45°$.

Magnitude of the resultant force at the bend

We know that cross-sectional area of the pipe,

$$a = \frac{\pi}{4} \times (d)^2 = \frac{\pi}{4} \times (0.2)^2 = 0.0314 \text{ m}^2$$

and water pressure, $p = wh = 9.81 \times 20 = 196.2 \text{ kN/m}^2$

∴ Force exerted due to water pressure,

$$F_1 = p \cdot a = 196.2 \times 0.0314 = 6.16 \text{ kN}$$

We know that component of this force in X-X direction

$$= F_1 (1 - \cos \alpha) = 6.16 \times (1 - \cos 45°) \text{ kN}$$
$$= 6.16 (1 - 0.707) = 6.16 \times 0.293 = 1.807 \text{ kN} \qquad ...(ii)$$

and component of this force in Y-Y direction

$$= F_1 \sin \alpha = 6.16 \sin 45°$$
$$= 6.16 \times 0.707 = 4.355 \text{ kN} \qquad ...(iii)$$

We also know that the force exerted by the water in X-X direction

$$= \frac{waV^2}{g} (1 - \cos \alpha)$$
$$= \frac{9.81 \times 0.0314 \times (3.5)^2}{9.81} \times (1 - \cos 45°) \text{ kN}$$
$$= 0.385 (1 - 0.707) = 0.113 \text{ kN}$$

and force exerted by the water in Y-Y direction

$$= \frac{waV^2}{g} \sin \alpha = \frac{9.81 \times 0.0314 \times (3.5)^2}{9.81} \times \sin 45°$$
$$= 0.385 \times 0.707 = 0.272 \text{ kN} \qquad ...(iv)$$

∴ Total force in X-X direction,

$$F_X = 1.807 + 0.113 = 1.92 \text{ kN}$$

and total force in Y-Y direction

$$F_Y = 4.355 + 0.272 = 4.627 \text{ kN}$$

∴ Magnitude of the resultant force at the bend,

$$R = \sqrt{F_X^2 + F_Y^2} = \sqrt{(1.92)^2 + (4.627)^2} = 5.01 \text{ kN Ans.}$$

Direction of the resultant force at the bend

Let θ = Angle, which the resultant force makes with X-X direction.

We also know that

$$\tan \theta = \frac{F_Y}{F_X} = \frac{4.627}{1.92} = 2.410 \quad \text{or} \quad \theta = 67.5° \text{ Ans.}$$

EXERCISE 30·1

1. Find the resultant force acting in a 120° bend in a pipe of cross-sectional area 0.05 m² through which water is flowing with a velocity of 5 m/s. **(Ans. 2·17 kN)**

2. Water flows through a pipe of 250 mm diameter pipe has a bend of 90°. Find the magnitude and direction of the force acting on the bend when 206 litres of water is flowing the pipe.
(Ans. 1·22 kN; 45°)

3. A 150 mm diameter pipe carries water under a head of 16 m of water with a velocity of 3 m/s. Find the magnitude and direction of the resultant Force on the bend, if the axis of the pipe turns through 75°. **(Ans. 3·5 kN; 52·5°)**

30·3 Principle of Jet Propulsion

We have discussed in Art. 29·2, that when a jet strikes a plate, is exerts some force on the plate. We have also discussed that the force,

$$F = \frac{waV^2}{g}$$

where w = Specific weight of the liquid,

a = Area of the jet, and

V = Velocity of the jet.

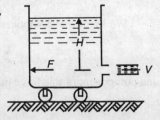

But, if such a jet is discharged from a vessel (which can move on its wheels) as shown in Fig. 30·2, then according to Newton's Third Law* of motion, the force of jet will tends to move the vessel in the opposite direction with the same force. This principle is known as the principle of jet propulsion, and is used in driving the ships and aeroplanes.

Fig. 30·2. Jet Propulsion.

30·4 Propulsion of Ships by Water Jets

The principle of jet propulsion is utilized in driving the ships through the water. The ship carries pumps, which take water from its surrounding. This water is discharged by forcing through the orifices, at the back of ship. It has been observed that efficiency of a ship depends upon the direction of the inlet orifice also. In general a ship may have:

1. Its inlet orifices at right angles to the direction of its motion, or
2. Its inlet orifices facing the direction of motion.

Notes. 1. The first way of having the orifices is commonly used, where as the second one has theoretical importance only.

2. If the given example, does not mention the direction of the inlet orifices, then their direction is assumed to be at right angles to the direction of motion.

---

*Newton's Third Law of Motion states, *"To every action there is always an equal and opposite reaction."*

30·5 Propulsion of a Ship Having Inlet Orifices at Right Angles to the Direction of its Motion (i.e., Orifices Amidship)

Consider a ship having inlet orifices at right angle to the direction of its motion as shown in Fig. 30·3.

Let
V = Velocity of the jet issuing from the ship,
a = Area of the jet,
v = Velocity of the ship.

∴ Relative velocity* of the jet and the ship,
$$V_r = V + v$$

and mass of the water flowing/s $= \dfrac{waV_r}{g}$

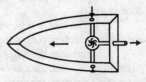

Fig. 30·3. Ship having inlet orifices at right angles to the direction of motion.

A little consideration will show, that the velocity of jet after leaving the ship will be equal to v.

∴ Force exerted on the boat
= Mass of water flowing/s × Change of velocity
$$= \dfrac{waV_r}{g} (V_r - v)$$

and work done = Force × Distance $= \dfrac{waV_r}{g} (V_r - v) v$

∴ Work done be the jet per kN of water
$$= \dfrac{1}{g} (V_r - v) v$$

We know that the energy supplied by the jet, per kN of water
$$= \dfrac{V_r^2}{2g}$$

∴ Efficiency of propulsion,

$$\eta = \dfrac{\text{Work done by the jet per kN of water}}{\text{Energy of the jet per kN of water}}$$

$$= \dfrac{\dfrac{1}{g}(V_r - v)v}{\dfrac{V_r^2}{2g}} = \dfrac{2(V_r - v)v}{V_r^2} \qquad ...(i)$$

Now the efficiency of propulsion will be maximum, when $2(V_r - v)v$ or $(V_r - v)v$ is maximum. Therefore differentiating this equation and equating the same to zero,

i.e., $\dfrac{d\eta}{dv}(V_r - v)v = 0$

$\dfrac{d\eta}{dv}(Vv_r - v^2) = 0$

---

* The relative velocity (V_r) will be equal to the vectorial difference of V and v. Since v is in the opposite direction of V, therefore their vectorial difference will be equal to $(V + v)$

∴ $V_r - 2v = 0$

or $V_r = 2v$

Now substituting this value V_r in equation (i),

$$\eta = \frac{2(2v - v)v}{(2v)^2} = \frac{2v^2}{4v^2} = \frac{1}{2} = 0.5 = 50\%$$

Notes: 1. The power required to propel the boat (*i.e.*, power of the pump) may be found out from the work done as usual.

2. In this case, it is generally mentioned that the water is taken in *amidship* (*i.e.*, at the middle of the ship) and is ejected *astern* (*i.e.*, at the stern or back of the ship).

3. Sometimes, the water is taken in through vertical pipes from the bottom of the ship instead of horizontal pipes. In such a case also, the water is ejected at the rear, through horizontal pipes.

4. Sometimes, the water is ejected out freely into the air (*i.e.*, like free orifice). But sometimes it is ejected out into the water (*i.e.*, like drowned orifice).

Example 30·3. *A boat in the water jet propulsion draws water at the rate of 0·5 m^3/s, through orifices amidship. The boat discharges water through orifices having an effective area of 0·05 m^2. If the boat travels at 5 m/s, find the propelling force of the boat.*

Solution. Given : $Q = 0.5$ m^3/s; $a = 0.05$ m^2 and $v = 5$ m/s.

We know that velocity of jet relative to water,

$$V_r = \frac{Q}{a} = \frac{0.5}{0.05} = 10 \text{ m/s}$$

and propelling force of the jet,

$$F = \frac{waV_r}{g}(V_r - v) = \frac{9.81 \times 0.05 \times 10}{9.81} \times (10 - 5) \text{ kN}$$

$$= 2.5 \text{ kN} \quad \textbf{Ans.}$$

Example 30.4. *Find the power required to propel a motor boat with jet propulsion, if it discharges 0·05 m^3 of water for second through a 0·003 m^2 hole when the boat travels at 4 m/s. Take coefficient of contraction for the jet as 0·62.*

Solution. Given : $Q = 0.05$ m^3/s; $a = 0.003$ m^2; $v = 4$ m/s and $C_c = 0.62$.

We know that velocity of the jet relative to the water,

$$V_r = \frac{Q}{a \cdot C_c} = \frac{0.05}{0.003 \times 0.62} = 26.9 \text{ m/s}$$

and work done by the engine to propel the motor boat,

$$W = \frac{waV_r}{g}(V_r - v) \times v = \frac{wQ}{g}(V_r - v) \times v$$

$$= \frac{9.81 \times 0.05}{9.81} \times (26.9 - 4) \times 4 = 4.58 \text{ k N-m/s} = 4.58 \text{ kJ/s}$$

∴ Power required to propel the motor boat,

$$P = 4.58 \text{ kW} \quad \textbf{Ans.}$$

Note : In the example the value of (a, V_r in calculations stands for (Q). *i.e.* , discharge of water.

Example 30·5. *A boat is provided with two jets, each having a cross-sectional area of 0·02 m^2. The water is drawn through orifices amidship. If the speed of the boat in the sea is 6 m/s, find the quantity of water to be pumped per second, and efficiency of the jet propulsion. The force of resistance to the motion is 4 kN.*

Jet Propulsion

Solution. Given : No. of jets = 2; Area of each jet = 0.02 m^2 or total area of two jets (a) = $2 \times 0.02 = 0.04$ m^2; $V = 6$ m/s and $F = 4$ kN.

Quantity of water to be pumped

Let V_r = Velocity of the jet relative to the water,

We know that force of resistance to the motion (F),

$$4 = \frac{w a V_r}{g}(V_r - v) = \frac{9.81 \times 0.04 \times V_r}{9.81} \times (V_r - 6)$$

$$= 0.4\, V_r^2 - 0.24 - 4 = 0$$

or $\quad V_r^2 - 6V_r - 100 = 0$

This is a quadratic equation. Therefore

$$V_r = \frac{6 \pm \sqrt{(6)^2 + 4 \times 100}}{2} = \frac{6 + 20.9}{2} = 13.45 \text{ m/s}$$

∴ Quantity of water to be pumped,

$$Q = a \cdot V_r = 0.04 \times 13.45 = 0.538 \text{ m}^3/\text{s} = 538 \text{ litres/s} \quad \textbf{Ans.}$$

Efficiency of jet propulsion

We also know that efficiency of jet propulsion,

$$\eta = \frac{2(V_r - v)v}{V_r^2} = \frac{2(13.45 - 6)6}{(13.45)^2} = 0.504 = 50.4\% \quad \textbf{Ans.}$$

Example 30·6. *A small ship is fitted up with jets of total area 0.65 m^2. The velocity through the jet is 9 m/s and speed of the ship is 18 km. p.h. in sea water. The efficiencies of the engine and pump are 85% and 65% respectively. If the water is taken amidships, determine the propelling force and the overall efficiency, assuming the pipe losses to be 10% of the kinetic energy of the jets.*

Solution. Given : $a = 0.65$ m^2 ; $V_r = 9$ m/s; $v = 18$ km.ph. $= 5$ m/s; $\eta_e = 85\% = 0.85$; $\eta_s = 65\% = 0.65$ and pipe losses = 10% of the kinetic energy of the jet.

Propelling force

We know that the propelling force,

$$F = \frac{w a V_r}{g}(V_r - v) = \frac{9.81 \times 0.65 \times 9}{9.81} \times (9 - 5) = 23.3 \text{ kN}$$

Overall efficiency

We know that kinetic energy of the jet per kN of water

$$= \frac{V_r^2}{2g} = \frac{(9)^2}{2g} = \frac{41.5}{g}$$

and pipe losses

$$= \frac{10}{100} \times \frac{41.5}{g} = \frac{4.15}{g}$$

∴ Energy available at the pump per kN of water

$$= \text{Kinetic energy of the jet} + \text{Pipe losses}$$

$$= \frac{41.5}{g} + \frac{4.15}{g} = \frac{45.65}{g} = \frac{45.65}{9.81} = 4.65$$

Since efficiencies of the engine and pump are 85% and 65% respectively, therefore total input to the engine per kN of water

$$= 4.65 \times \frac{1}{0.85} \times \frac{1}{0.65} = 8.42 \text{ kN-m}$$

and work done by the jet per kN of water

$$= \frac{1}{g}(V_r - v)v = \frac{1}{9\cdot 81}(9 - 5) \times 5 = 2\cdot 04 \text{ kN-m}$$

∴ Overall efficiency, η_0

$$= \frac{\text{Work done per kN of wate}}{\text{Total input to the engine per kN of water}}$$

$$= \frac{2\cdot 04}{8\cdot 42} = 0\cdot 242 = 24\cdot 2\% \quad \textbf{Ans.}$$

30·6. Propulsion of a Ship Having Inlet Orifices Facing the Direction of Flow

Consider a ship having inlet orifices facing the direction of flow as shown in Fig. 30·4.

Let V = Velocity of the jet issuing from the ship,
  a = Area of the jet, and
  v = Velocity of the ship.

∴ Relative velocity* of the jet and the ship,
$$V_r = V + v$$

and mass of water flowing/s $= \dfrac{waV_r}{g}$

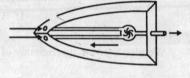

We know that the relative velocity of jet, after leaving the ship will be equal to v.

Fig. 30·4. Ship having inlet orifices facing the direction of flow.

∴ Force exerted on the boat

$$= \text{Mass of water flowing/s} \times \text{Change of velocity}$$

$$= \frac{waV_r}{g}(V_r - v)$$

and work done $= \text{Force} \times \text{Distance} = \dfrac{waV_r}{g}(V_r - v)v$

∴ Work done by the jet per kN of water

$$= \frac{1}{g}(V_r - v)v$$

We know that the energy supplied by the jet per kN of water

$$= \frac{V_r^2}{2g}$$

A little consideration will show, that when the ship moves in a steady water, the water will enter into the orifices of the ship with a velocity equal to that of the ship (*i.e.*, v). Therefore energy of the water entering into the orifices per kN of water

$$= \frac{v^2}{2g}$$

∴ Net energy supplied by the jet

$$= \left(\frac{V_r^2}{2g} - \frac{v^2}{2g}\right)$$

---

*The relative velocity V_r will be equal to the vectorial difference of V and v. Since v is in the opposite direction of V, therefore their vectorial difference will be equal to $(V + v)$.

Jet Propulsion

and efficiency of propulsion,

$$\eta = \frac{\text{Work done by the jet per kN of water}}{\text{Energy of the jet per kN of water}}$$

$$= \frac{\frac{1}{g}(V_r - v)v}{\frac{V_r^2}{2g} - \frac{v^2}{2g}} = \frac{2(V_r - v)v}{V_r^2 - v^2} = \frac{2v}{V_r + v}$$

$$\ldots [V_r^2 - v^2 = (V_r - v)(V_r + v)]$$

Note : The power required to propel the boat (*i.e.*, power of the pump) may be obtained from the work done as usual.

Example 30·7. *A boat is propelled by a jet of water of total cross-sectional area 0·025 m² and relative velocity 16 m/s. What is the propelling force of the jet, when the boat is moving with a velocity of 6 m/s? The water is drawn through the openings facing the direction of motion of the boat.*

Solution. Given : $a = 0.025$; $V_r = 16$ m/s and $v = 6$ m/s.

We know that propelling force of the jet,

$$F = \frac{waV_r}{g}(V_r - v) = \frac{9.81 \times 0.025 \times 16}{9.81} \times (16 - 6) = 4 \text{ kN}$$

Example 30·8. *A jet propelled boat is moving with the help of a jet of cross-sectional area 0·045 m² and relative velocity of 12 m/s. Find the power of the pump required to move the boat with a velocity of 27 km/hour. Assume the orifices to face the direction of the boat. Also find the efficiency of the jet propulsion.*

Solution. Given : $a = 0.045$ m²; $V_r = 12$ m/s and $v = 27$ km/hour = 7·5 m/s.

Power of the pump

We know that work done by the pump,

$$W = \frac{waV_r}{g}(V - v)v$$

$$= \frac{9.81 \times 0.045 \times 12}{9.81} \times (12 - 7.5) \times 7.5 \text{ kN-m/s}$$

$$= 2.43 \text{ kN-m/s} = 2.43 \text{ kJ/s}$$

and power of the pump $\quad P = 2.43$ kW **Ans.**

Efficiency of the jet propulsion

We also know that efficiency of the jet propulsion,

$$\eta = \frac{2v}{V_r + v} = \frac{2 \times 7.5}{12 + 7.5} = 0.667 = 66.7\% \text{ **Ans.**}$$

EXERCISE 30·2

1. A motor boat with jet propulsion draws 300 litres of water per second through orifices amidship and discharges it through an orifice of an area 0·025 m². Find the propelling force, if the boat travels at the rate of 4·5 m/s. **(Ans. 2·25 kN)**
2. A motor boat has two jets amidship each of area 6·03 m². If the boat travelling at 5 m/s experiences a resistance of 6 kN, determine the quantity of water to be pumped in litres/s. **(Ans. 768 litres/s)**
3. The water in a jet propelled boat is drawn through inlet openings facing the direction of motion of the boat. The cross-sectional area and relative velocity of the jet is 0·05 m² and 10 m/s. If the boat is moving at 4 m/s, what is the power of the pump and efficiency of jet propulsion? **(Ans. 12·0 kW; 57·1%)**

QUESTIONS

1. Derive an expression for the pressure exerted due to deviated flow.
2. If a pipe carrying water under pressure is bent through some angle, explain the effect of this bend on the pipe walls.
3. Describe how ships move in sea using 'jet propulsion.'
4. What is jet propulsion? Describe the movement of ships with this principle.
5. Obtain expression for the power required to pump the water when the ship has orifices facing the direction of flow.

OBJECTIVE TYPE QUESTIONS

1. When a pipe bends, the resultant force exerted by the flowing water is
   (a) equal to the algebrain sum of the force exerted in X-X and Y-Y directions
   (b) equal to the average of the forces exerted in X-X and Y-Y directions
   (c) more than either of the forces exerted in X-X and Y-Y directions
   (d) less than either of the forces exerted in X-X and Y-Y directions

2. A motor boat with jet propulsion draws water through orifices amidship. The propelling force of the jet is
   (a) $\dfrac{waV_r}{g}(V_r + v)$
   (b) $\dfrac{waV_r}{g}(V_r - v)$
   (c) $\dfrac{waV_r}{g}(V_r + v)^2$
   (d) $\dfrac{waV_r}{g}(V_r - v)^2$

   The terms w, a, V_r, v and g have usual meaning.

3. A jet propelled ship draws water through amidship. If the velocity of the ship is v and that of the jet relative to the water in V_r, then efficiency of propulsion is equal to
   (a) $\dfrac{2(V_r - v)v}{V_r^2}$
   (b) $\dfrac{2(V_r + v)v}{V_r^2}$
   (c) $\dfrac{(V_r - v)v}{V_r}$
   (d) $\dfrac{(V_r - v)v}{V_r}$

4. The water in a jet propelled boat is drawn through the openings facing the direction of motion of the boat. The efficiency of propulsion is given by the relation.
   (a) $\dfrac{2v}{V_r - v}$
   (b) $\dfrac{2v}{V_r + v}$
   (c) $\dfrac{V}{V_r - v}$
   (d) $\dfrac{V}{V_r + v}$

   where V_r is the velocity of jet relative to the water and v is the velocity of the jet.

ANSWERS

1. (c)   2. (b)   3. (a)   4 (b)

31

Water Wheels

1. Introduction. 2. Type of Water Wheels. 3. Overshot Water Wheel. 4. Breast Water Wheel. 5. Undershot Water Wheel 6. Poncelet Water Wheel. 7. Advantages and Disadvantages of Water Wheels. 8. Development of Water Turbines. 9. Advantages of Water Turbines. 10. Classification of Water Turbines. 11. Recent Trends in Water Power Engineering. 12. Hydroelectric Power Plant.

31·1 Introduction

The idea of using water, as a source of mechanical energy, existed since the dim ages of pre-historic times. The hydraulic energy was first converted into mechanical energy in India about 2200 years back, by passing the water through water wheels. These water wheels were originally made of wood, and are seen in technical museums even now. Such type of water wheels were taken from India to Egypt and then to European countries, and finally to America. It is believed that the water wheel was used in Europe about 500 years after its origin in India.

The systematic design of water turbines, from the water wheels, started, in 18th century. These days, lot of research is being conducted all over the world to improve upon the various existing turbines.

31·2 Types of Water Wheels

The water wheels may be broadly classified into the following three groups, depending upon the driving action of the water:

1. Wheels driven by weight of water.
2. Wheels driven partly by weight and partly by impulse of water.
3. Wheels driven entirely by impulse of water.

In general practice, a water wheel consists of a central hub and a circular wheel having a number of buckets or vanes mounted on its periphery. The water is delivered to the wheel at some convenient point on its circumference, which fills into the buckets or strikes the vanes. The following types of water wheels have been used in the olden days:

(*a*) Overshot wheel and breast wheel, and

(*b*) Undershot wheel.

31·3 Overshot Water Wheel

An overshot wheel, as the name indicates, is a water wheel in which the water enters the buckets at the top of its periphery as shown in Fig. 31·1. This wheel runs entirely by the weight of water (sometimes partly by the impulse of water).

The water, from the head race, is allowed to enter the buckets through an adjustable sluice gate. The weight of water forces the buckets downwards, and thus makes the wheel to rotate. The buckets get emptied into the tail race, as they approach the lower position. The buckets are so arranged, that the maximum water energy is utilized. Sometimes, the crown of the wheel is made slightly below the head race, as a result of which the water strikes the buckets with some initial velocity. In such case, the wheel is driven partly by weight and partly by impulse of water. The overshot wheels have the following constructional details:

Head of water, H = 10 to 25 m
Dia. of wheel, D = 3 to 20 m
No. of buckets, = 8 to 10 D
Speed of wheel, N = 4 to 8 r.p.m.
Depth of shroud = 0·5 to 1·0 m
Efficiency, η = 60 to 80%

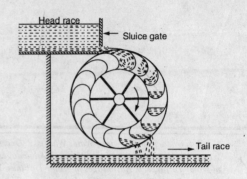

Fig. 31·1. Overshot water wheel.

The power available from an overshot water wheel is given by the relation,
$$P = wQH \text{ kW}$$

If η is the efficiency of the wheel, then actual power available,
$$P = \eta \times wQH \text{ kW}$$

and discharge of the wheel is given by the relation,
$$Q = k.b.d.v = k.b.d \times \frac{\omega D}{2}$$

where
k = Fraction of the buckets filled with water,
b = Width of the buckets,
d = Depth of shrouds,
v = Velocity of the buckets,
ω = Angular velocity of the wheel, and
D = Diameter of the wheels.

Example 31·1. *An overshot water wheel has approaching canal 1·5 m wide. The water flows in the canal with a velocity of 1·5 m/s and 200 mm deep. Determine the power available from the water wheel, if the waterfall is 20 m and the efficiency of wheel is 75%.*

Solution. Given : Canal width (b) = 1·5 m; v = 1·5 m/s; d = 200 mm = 0·2 m; H = 20 m and η = 75% = 0·75.

We know that discharge,
$$Q = a.v = (b.d)v = 1·5 \times 0·2 \times 1·5 = 0·45 \text{ m}^2$$
and power available from the water wheel,
$$P = \eta \times wQH = 0·75 \times 9·81 \times 0·45 \times 20 = 66·2 \text{ kW Ans.}$$

Example 31·2. *An overshot water wheel 5 m diameter has 500 mm deep shrouds. It is required to produce 12 kW at 5 r.p.m. Assuming the buckets to be 1/3 filled with water, find the width of the wheel when the total fall is 6 m. Take efficiency of the wheel is 70%.*

Solution. Given : D = 5 m; d = 500 mm = 0·5 m; P = 12 kW; N = 5 r.p.m.; k = 1/3; H = 6 m and η = 70% = 0·7.

Let b = Width of the wheel.
We know that angular velocity of the wheel,
$$\omega = \frac{2\pi N}{60} = \frac{2\pi \times 5}{60} = 0·524 \text{ rad/s}$$

and discharge of the wheel,
$$Q = k.b.d \times \frac{\omega D}{2} = \frac{1}{3} \times b \times 0·5 \times \frac{0·524 \times 5}{2} \text{ m}^3/\text{s}$$
$$= 0·218 \, b \text{ m}^3/\text{s}$$

We also know that power of the wheel (P),

$$12 = \eta \times wQH = 0.7 \times 9.81 \times (0.218\, b) \times 6 = 8.98\, b$$

$$\therefore \quad b = 12/8.98 = 1.34 \text{ m} \quad \textbf{Ans.}$$

31·4 Breast Water Wheel

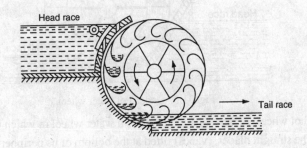

Fig. 31·2. Breast water wheel

A breast wheel, as the name indicates, is a water wheel in which the water enters the buckets at the breast height of the wheel as shown in Fig. 31·2. This wheel runs partly by weight and partly by the impulse of water.

The water, from the head race, is allowed to enter the buckets without shock through a number of passages which may be opened or closed by rock and pinion arrangement as shown in Fig. 31·2. The buckets move downwards due to weight of water, and thus make the wheel to rotate. The special feature of the wheel is that its bottom is immersed in the tail race water. As the direction of the motion of the wheel and the flow of the tail race water is the same, therefore the water while flowing further rotates the wheel. That is why, the wheel is the said to run partly by weight and partly by the impulse of water. Another special feature of this wheel is that the diameter of the wheel is more than the head of water available. A breast wheel has the following constructional details:

Head of water, H = 1 to 5 m
Dia. of wheel, D = 4 to 8 m
Speed of wheel, N = 3 to 7 r.p.m.
Depth of shroud = 300 to 600 mm
Efficiency, η = 50 to 65%

The power available from a breast water wheel is also given by the relation,

$$P = \eta \times wQH \text{ kW}$$

Example 31·3. *A breast wheel of 8 m diameter and 2 m width is working under a head of 5 m. The depth of shroud is 400 mm and the buckets move with a velocity of 1.5 m/s with 5/8 full. Calculate the power of the wheel, if its efficiency is 60%.*

Solution. Given : $D = 8$ m = $b = 2$ m; $H = 5$ m; $d = 400$ mm = 0·4 m; $v = 1·5$ m/s; $k = 5/8$ and $\eta = 60\% = 0·6$.

We know that discharge of the wheel,

$$Q = k \cdot b \cdot d \cdot v = \frac{5}{8} \times 2 \times 0·4 \times 1·5 = 0·75 \text{ m}^3/\text{s}$$

and power of the wheel,

$$P = \eta \times wQH = 0·6 \times 9·81 \times 0·75 \times 5 = 22·1 \text{ kW} \quad \textbf{Ans.}$$

31.5 Undershot Water Wheel

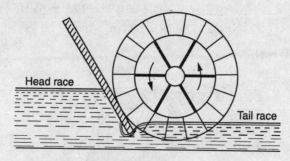

Fig. 31·3. Undershot water wheel.

An undershot wheel, as the name indicates, is a water wheel in which the water in the form of jet impinges on the straight blades (vanes) fitted at the bottom of its periphery as shown in Fig. 31·3. This wheel runs entirely by the impulse of water as the jet of water strikes on the vanes, and thus makes the wheel to rotate. The available head of water is first converted into velocity head, before the water strikes the buckets of the wheel. The undershot wheels have the following constructional details:

Head of water, $\quad H < 2$ m
Dia. of wheel, $\quad D = 2$ to $4H$
Speed of wheel, $\quad N = 2$ to 4 r.p.m.
Efficiency, $\quad \eta = 35$ to 45%

The power available from an undershot water wheel is also given by the relation,

$$P = \eta \times wQH \quad \text{kW}$$

Note : The theory of an undershot water wheel is the same as that of series of moving vanes (Art. 29·6).

Example 31·4. *An undershot wheel is working under a head of 2·5 m with a speed of 5 r.p.m. Find the diameter of the wheel, if its efficiency is 40%. Take coefficient of velocity as 0·98 and ratio of peripheral velocity of the wheel to the velocity of the jet as 0·46.*

Solution. Given : $H = 2·5$ m; $N = 5$ r.p.m ; $\eta = 40\% = 0·4$; $C_v = 0·98$ and $v = 0·6 V$ (where V is the velocity of jet).

Let $\quad D = $ Diameter of the wheel.

We know that the velocity of jet,

$$V = C_v \sqrt{2gH} = 0·98 \times \sqrt{2 \times 9·81 \times 2·5} = 6·86 \text{ m/s}$$

and peripheral velocity of wheel,

$$v = 0·46 \, V = 0·46 \times 6·86 = 3·16 \text{ m/s}$$

We also know that the peripheral velocity of the wheel (v),

$$3·16 = \frac{\pi DN}{60} = \frac{\pi \times D \times 5}{60} = 0·262 \, D$$

$\therefore \quad D = 3·16/2·62 = 1·21$ m **Ans.**

31.6 Poncelet Water Wheel

It is an improvement over the straight blade type undershot water wheel. The straight blades of the undershot wheel are replaced by curved vanes as shown in Fig. 31·4. This wheel runs entirely by the impulse of water.

Water Wheels

The vanes are curved at such an angle, that the jet of water enters the vanes tangentially (*i.e.*, without shock). The blades are made sufficiently long, so that the water does not spill over at the outlet; but actually flows down to the tail race as shown in Fig. 31·4.

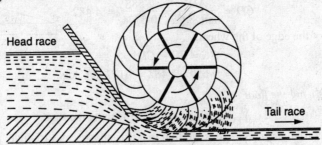

Fig. 31·4. Poncelet water wheel.

As a result of this, the efficiency of the wheel is doubled. The Poncelet wheel has the following constructional details:

Head of water, H = 2 to 3 m
Dia. of wheel, D = 2 to 4 H
Speed of wheel, N = 2 to 5 r.p.m.
Efficiency, η = 55 to 65%
Inlet angle α = 15°

The power available from a Poncelet water wheel is also given by the relation,

$$P = \eta \times wQH \text{ kW}$$

Note : The theory of a Poncelet water wheel is the same as that of moving curved vane (Art 29·8).

Example 31·5. *In a Poncelet wheel, the jet enters the curved vanes with a velocity of 7 m/s at an angle of 20° with the direction of motion of tip of the vanes. If the edge of the wheel moves with a velocity of 5·5 m/s, find the angle which the vanes make with the tangent.*

Solution. Given : $V = 7$ m/s; $\alpha = 20°$ and $v = 5·5$ m/s.

Let θ = Angle which the vanes make with the tangent.

From the inlet triangle of velocities, we find that

$BD = V \cos 20° = 7 \times 0·9397$ m/s
 $= 6·58$ m/s

and $AD = V \sin 20°$
 $= 7 \times 0·342$ m/s
 $= 2·4$ m/s

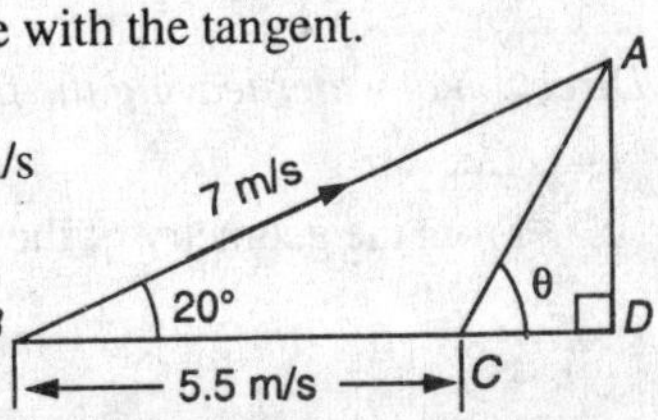

We know that $CD = BD - BC = 6·58 - 5·5$ m
 $= 1·08$ m

Fig. 31·5.

and $\tan \theta = \dfrac{AD}{CD} = \dfrac{2·4}{1·08} = 2·222$ or $\theta = 65·8°$ **Ans.**

Example 31·6. *In a Poncelet water wheel, the direction of the jet is at an angle of 15° with the tangent and the tip of the vane makes an angle of 30° with the tangent. If velocity of the jet is 10 m/s, find (i) velocity of the edge of the wheel, and (ii) velocity and direction of water leaving the float.*

Solution. Given : $\alpha = 15°$; $\theta = 30°$ and $V = 10$ m/s.

Velocity of the edge of the wheel

Let v = Velocity of the edge of the wheel.

From the inlet triangle of velocities, we find that

$BD = V \cos 15° = 10 \times 0·9659 = 9·659$ m/s

and
$$AD = V \sin 15° = 10 \times 0.2588 \text{ m/s}$$
$$= 2.588 \text{ m/s}$$

or
$$CD = \frac{AD}{\tan 30°} = \frac{2.588}{0.5774} = 4.482$$

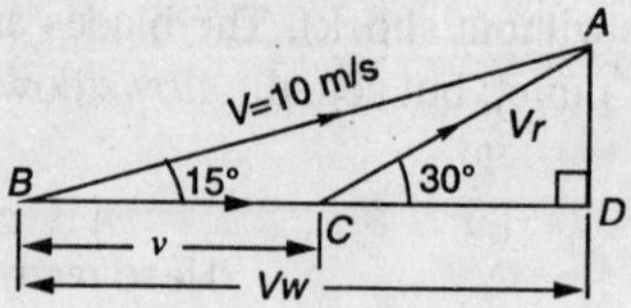

∴ Velocity of the edge of the wheel,
$$v = BC = BD - CD$$
$$= 9.659 - 4.482 = 5.177 \text{ m/s} \textbf{ Ans.}$$

Fig. 31·6.

Velocity of water leaving the float

Let V_1 = Velocity of water leaving the float.

From the inlet triangle of velocities we find that it is an isosceles triangle, as ∠ACD is twice the angle ∠ABC. Therefore relative velocity,
$$V_r = CA = BC = 5.177 \text{ m/s}$$

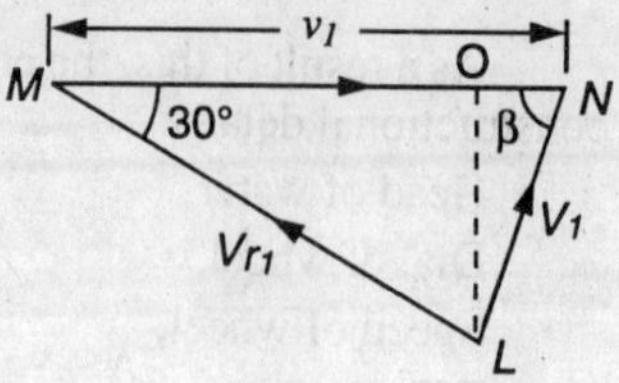

Now from the outlet triangle of velocities, we find that the relative velocity,
$$V_{r1} = V_r = LM = 5.177 \text{ m/s}$$

and velocity of edge of the wheel,
$$v_1 = v = MN = 5.177 \text{ m/s}$$

∴
$$MO = V_{r1} \cos 30° = 5.177 \times 0.866 = 4.483 \text{ m/s}$$

and
$$LO = V_{r1} \sin 30° = 5.177 \times 0.5 = 2.589 \text{ m/s}$$

Since MN (equal to 5·177) is more than MO (equal to 4·483), therefore shape of the outlet triangle will be as shown in Fig. 31·7.

Now
$$ON = MN - MO = 5.177 - 4.483 = 0.694 \text{ m/s}$$

∴ Velocity of water leaving the float,
$$V_1 = LN = \sqrt{(2.589)^2 + (0.694)^2} = 2.68 \text{ m/s} \textbf{ Ans.}$$

Direction of water leaving the float

Let β = Angle at which the water leaves the float.

From the geometry of the outlet triangle of velocities, we find that
$$\tan \beta = \frac{LO}{ON} = \frac{2.589}{0.694} = 3.73 = 75° \textbf{ Ans.}$$

31·7 Advantages and Disadvantages of Water Wheels

The water wheels have the following advantages and disadvantages :

Advantages
1. They are simple and strong in construction.
2. They are suitable even for low water heads.
3. They are cheaper.
4. They give constant efficiency, even if the discharge is not constant.

Disadvantages
1. They have slow speeds.
2. They are heavier and bigger, if compared with their capacity to produce power.
3. They are not suitable for high water heads.
4. Their speed cannot be easily regulated.

31·8 Development of Water Turbines

In the previous article, we have discussed the disadvantages of water wheels. Moreover, the hydro power was available mostly in rural and mountainous regions. As a result of this, the mills directly run by water wheels, had to be installed near the power stations. The prime movers had to run round the clock, even if some of the machines in the mills may remain idle. It was also observed that the slow moving water wheels were not suitable for all types of purposes. Lot of research was conducted by numerous scientists and engineers all over the world, to improve the working of water wheels in the 18th and 19th centuries. As a result of this research, water turbines were designed, which can operate under high head (highest head 1765 metres in Austria) and can run at higher speeds.

After the first world war, there came the evolution in the use of electric power. The use of electricity, for driving machines in industry, being very convenient to handle became very common. Side by side, with the mammoth research in electric power, high speed machines were designed. Due to rapid increase in the use of electric power, the transmission of electricity, from remote hydro power stations to the far off places, was considered necessary. Due to the above mentioned reasons, the water turbines occupied an important position among prime movers. Today, the water turbines are taken to be the back bone of electricity, which is considered to be one of the important factors for the prosperity of a nation.

31·9 Advantages of Water Turbines

The water turbines have the following advantages over water wheels or any other type of prime movers:

1. They have long life.
2. They are efficient and can be easily controlled.
3. They have outstanding ability to act as standby unit.
4. They can be made automatic controlled.
5. They can work under any head.

31·10 Classification of Water Turbines

The water turbines may be broadly classified into the following two groups:

1. Impulse or velocity turbines, and
2. Pressure or reaction turbines.

The various water turbines, belonging to both the above groups will be discussed at the appropriate place of this book.

31·11 Recent Trends in Water Power Engineering

The crying need of more and more electric energy has drawn the attention of scientists and engineers, working in the research and development programmes, to generate electricity. They have focussed their attention on:

1. Hydroelectric power,
2. Thermal power, and
3. Atomic power.

In this book, we shall discuss the hydroelectric power only, as the others are beyond the scope of this book.

31·12 Hydroelectric Power Plant

It consists of the following main components :
1. Storage reservoir,
2. Dam and its parts,
3. Water ways,
4. Water turbines and electric generators.

1. *Storage reservoir*

   The water available from the catchment area is stored in the reservoir. The capacity of reservoir should be such that the water should be available for running the turbines, for producing the desired quantity of electric power, throughout the year. A reservoir may be natural or artificial.

2. *Dams and its parts*

   A dam is constructed across a river in order to check the flow of water and impound it in the reservoir formed on the upstream side. The size and type of dam depends upon the character of river, head of water, amount of discharge etc. Its shape and other components are decided by tests on model in the laboratory.

   The dams are provided with gates for regulating the flow of water. There is also an arrangement for automatic overflow of excess water. Some means are also provided for removing silt from the reservoir just closer to the dam. It is still a problem, which the engineers working on the maintenance of dams, have not been able to solve it successfully.

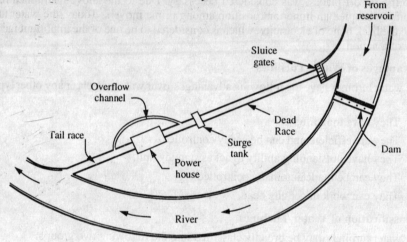

Fig. 31·8. Hydroelectric power plant.

3. *Water ways*

   A waterway is a passage through which the water is carried from the dam to the power house and then to the river. The upstream portion is known as head race and the downstream as tail race. It may consist of tunnels, canals, flumes, pipes or any other suitable arrangement. A surge tank is also provided just on the upstream of the power house to control the pressure variation and eliminate the effects of water hammer.

4. *Water turbines and electric generators*

   A place, where water turbines and electric generators are installed, is called power house. Its design is very complicated and requires lot of skill. In this book, we shall discuss the design of water turbines only, as the design of electric generator is beyond the scope of this book.

   The position of a power house is decided by considering factors such as space available, transport facilities etc. The size of a power house is decided by considering the factors such as supply height, number and size of units, type of units electrical arrangements etc. The electrical generators are directly coupled with the turbines for better efficiency.

   The turbines may be designed and laid either with their shafts horizontal or vertical. In a horizontal shaft lay out, the whole installation lies on the same floor. Thus it is very easy to carry out inspection, service or any other modification in the plant. In a vertical shaft lay out, it is convenient to connect incoming pipe and outgoing draft tube. Moreover, the generators are spaced well above the water surface, which makes their inspection, service and maintenance easier.

   The general layout of a hydroelectric power project is shown in Fig. 31·8.

Water Wheels

Note : There is a definite relation between the speed of the turbine rotor and the frequency of A.C. alternator, which is given by:

$$\frac{120f}{p} = n$$

where
f = Frequency in cycles/s,
p = No. of poles in alternators (which must be an even integral number, because they exist in pairs, and
n = Synchronous speed of the turbine rotor in r.p.m.

If the number of poles works out to be in decimal or odd number, then the next even number is selected.

EXERCISE 31·1

1. An overshot wheel is working under a head of 4 m. If 0·5 m² of water flows over the wheel in one second, find the power developed by wheel. Take efficiency of the wheel as 45%. **(Ans. 8·83 kW)**
2. An overshot wheel is working under a head of 6 m. The diameter of the wheel is 6·5 m, width 250 mm has 300 mm deep shrouds. Assuming the buckets to be 1/3 filled with water, find the power available from the wheel. Take efficiency of the wheel as 50%. **(Ans. 7·9 kW)**
3. A breast wheel of 5 m diameter 1 m wide is working under a head of 4·5 m. The depth of the shrouds is 400 mm and buckets move with a velocity of 1·2 m/s with half full. Determine the efficiency of the wheel, if it produces 5·5 kW power. **(Ans. 50·9%)**
4. In a Poncelet water wheel, the direction of jet is at angle of 30° with the tangent to the circumference. If the velocity of the jet is 4·5 m/s and that of the edge of the wheel is 2·25 m/s, determine the angle which the vane makes with the tangent. **(Ans. 53·8°)**

QUESTIONS

1. What is a 'water wheel'? Explain the various forms of water wheels.
2. Distinguish clearly between:
   (a) Overshot water and undershot water wheel.
   (b) Undershot water wheel and Poncelet water wheel.
3. Give the advantages and disadvantages of water wheels.
4. Give the advantages of water turbines.
5. What is a hydroelectric power project? Write a short note on its various parts.

OBJECTIVE TYPE QUESTIONS

1. The water wheels run by
   (a) weight of water (b) impulse of water (c) neither 'a' nor 'b' (d) both 'a' and 'b'
2. Water wheels are suitable for
   (a) low water heads (b) high water heads (c) very high water heads (d) all of these
3. The water wheels
   (a) are simple in construction   (b) are cheaper in cost
   (c) have constant efficiency   (d) all of these
4. The power developed by a water wheel is equal to
   (a) $\frac{wQH}{2}$ (b) wQH (c) $2wQH$ (d) $\frac{wQH^2}{2}$

Answers

1. (d)   2. (a)   3. (d)   4. (b)

32

Impulse Turbines

1. Introduction. 2. Pelton Wheel. 3. Nozzle. 4. Runner and Buckets. 5. Casting. 6. Braking jet. 7. Work Done by an Impulse Turbine. 8. Power Produced by an Impulse Turbine. 9. Efficiencies of an Impulse Turbine. 10. Hydraulic Efficiency. 11. Mechanical Efficiency. 12. Overall Efficiency. 13. Number of Jets for a Pelton Wheels. 14. Design of Pelton Wheels. 15. Size of Buckets of a Pelton Wheel. 16. Number of Buckets on the Periphery of a Pelton Wheel. 17. Governing of an Impulse Turbine (Pelton Wheel). 18. Other Impulse Turbines.

32·1. Introduction

An impulse turbine, as the name indicates, is a turbine which runs by the impulse of water. In an impulse turbine, the water from a dam is made to flow through a pipeline, and then through guide mechanism and finally through the nozzle. In such a process, the entire available energy of the water is converted into kinetic energy, by passing it through nozzles; which are kept close to the runner. The water enters the running wheel in the form of a jet (or jets), which impinges on the buckets, fixed to the outer periphery of the wheel.

The jet of water impinges on the buckets with a high velocity, and after flowing over the vanes, leaves with a low velocity; thus imparting energy to the runner. The pressure of water, both at entering and leaving the vanes, is atmospheric. The commonest example of an impulsive turbine is Pelton wheel, which is discussed below.

32·2. Pelton Wheel

The *Pelton wheel is an impulsive turbine used for high heads of water. It has the following main components:
1. Nozzle,
2. Runner and buckets,
3. Casing, and
4. Breaking jet.

32·3. Nozzle

It is a circular guide mechanism, which guides the water to flow at a designed direction, and also to regulate the flow of water. This water, in the form of a jet, strikes the buckets. A conical needle or spear operates inside the nozzle in an axial direction. The main purpose, of this spear, is to control or regulate the quantity of water flowing through the nozzle as shown in Fig. 32·1.

A little consideration will show, that when the spear is pushed forward into the nozzle, it reduces the area of jet. As a result of this, the quantity of water flowing through the jet is also reduced. Similarly, if the spear is pushed back out of the nozzle, it allows a greater quantity of water to flow through the jet. The movement of the spear is regulated by hand or by automatic governing arrangement depending upon the requirement. Sometimes, it is very essential to close** the nozzle suddenly. This is done with the help of spear, which may cause the pipe to burst due to sudden increase of pressure. In order to avoid such a mishap, an additional nozzle (known as bypass nozzle) is provided through which the water can pass, without striking the buckets. Sometimes, a plate (known as

---

*Named after Pelton L.A. (1829 — 1908) who was an engineer of California. He made large scale experiments and developed a turbine used for high heads and low discharge.

** It happens, when there is a sudden decrease of load on the turbine.

Impulse Turbines

deflector) is provided to the nozzle, which is used to deflect the water jet, and preventing it from striking the buckets.

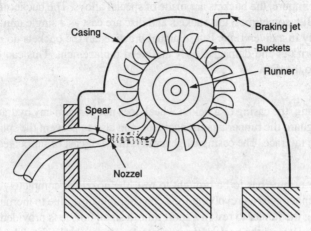

Fig. 32·1. Parts of a Pelton wheel

The nozzle is kept very close to the buckets, in order to minimise the losses due to windage.

32·4. Runner and Buckets

The runner of a Pelton wheel essentially consists of a circular disc fixed to a horizontal shaft. On the periphery of the runner, a number of buckets are fixed uniformly. A bucket resembles to a hemispherical cup or bowl with a dividing wall (known as splitter) in its centre in the radial direction of the runner as shown in Fig. 32·2.

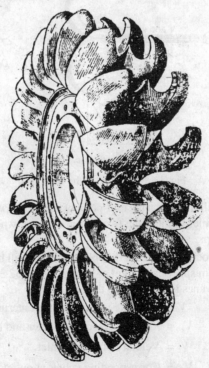

Fig. 32·2. Runner of a Pelton wheel

The surface of the buckets is made very smooth. For low heads, the buckets are made of cast iron. But for high heads, the buckets are made of bronze, stainless steel or other alloys. When the water is chemically impure, the buckets are made of special alloys. The buckets are generally bolted to the runner disc. But, sometimes, the buckets and disc are cast as a single unit. Sometimes, all the buckets wear equally in a given time. But in actual practice, all the buckets do not wear equally. A few buckets get worn out and damaged early and need replacement. This can be done only if the buckets are bolted to the disc.

32·5. Casing

Strictly speaking, the casing of a Pelton wheel does not perform any hydraulic function. But it is necessary to safeguard the runner against accident, and also to prevent the splashing of water and lead the water to the tail race. The casing is, generally, made of cast or fabricated parts.

32·6. Braking Jet

Whenever the turbine has to be brought to rest, the nozzle is completely closed. It has been observed, that the runner goes on revolving for a considerable time, due to inertia, before it comes to rest. In order to bring the runner to rest in a short time, a small nozzle is provided in such a way, that it will direct a jet of water on the back of the buckets. It acts as a brake for reducing the speed of the runner.

32·7. Work Done by an Impulse Turbine

The jet of water, issuing from the nozzle, strikes the bucket at its splitter. The splitter then splits up the jet into two parts. One part of the jet glides over the inside surface of one portion of the vane and leaves it at its extreme edge. The other part of the jet glides over the inside surface of the other portion of the vane and leaves it at its other extreme edge as shown in Fig. 32·3.

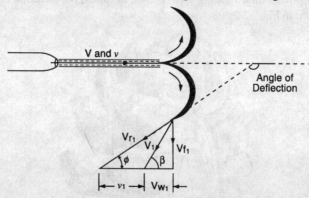

Fig. 32·3. Triangle of the velocities

A little consideration will show, that the mid-point of the bucket, where the jet strikes the splitter and gets divided, forms the inlet tip, and the two extreme edges, where the divided jet leaves the bucket, from the two outlet tips.

First of all, draw the velocity triangles at the splitter (which will be a straight line only) and *any one* of the outer tips of the hemispherical bucket as shown in Fig. 32·3. All the notations and theory of jet impinging on series of vanes is appliable in the case also.

Let
V = Absolute velocity of the entering water,
V_r = Relative velocity of water and bucket at inlet,
V_f = Velocity of flow at inlet,
V_1, V_{r1}, V_{f1} = Corresponding values at outlet. *i.e.*, at the point of leaving,

Impulse Turbines

D = Diameter of the wheel,
d = Diameter of the nozzle,
N = Revolutions of the wheel in r.p.m.
ϕ = Angle of the blade tip at outlet,
H = Total head of water, under which the wheel is working.

It will be interesting to know that inlet velocity triangle will be a straight line as shown in the figure. In this case, $\alpha = 0°$; $\theta = 0°$; $V_w = V$ and $V_r = V - v$.

As a matter of fact, the shape of the outlet velocity triangle depends upon the value of V_{w1}. If V_{w1} is in the same direction as that of the jet, its value is taken as positive. However, if V_{w1} is in the opposite (as shown in the figure) its value is taken as negative. The relation between these two velocity triangles, is

$$v_1 = v \quad \text{and} \quad V_{r1} = V_r = (V - v)$$

We know that force kN of water in the direction of motion of the jet

$$= \frac{1}{g}(V_w - V_{w1})$$

and work done $= \text{Force} \times \text{Distance} = \frac{1}{g}(V_w \cdot v - V_{w1} \cdot v_1)$

$$= \frac{1}{g}(V_w - V_{w1}) \times v \qquad \ldots(\because v_1 = v)$$

$\therefore$ Hydraulic efficiency, $\eta_h = \dfrac{\text{Work done per kN of water}}{\text{Kinetic energy of the jet}} = \dfrac{\frac{1}{g}(V_w - V_{w1}) \times v}{\dfrac{V^2}{2g}}$

$$= \frac{2(V_w - V_{w1}) \times v}{V^2}$$

Now consider a case, in which the value of V_{w1} is negative as shown in fig. Therefore work done per kN of water

$$= \frac{1}{g}[V_w - (-V_{w1})] \times v = \frac{1}{g}(V_w + V_{w1}) \times v$$

$$= \frac{V_w v}{g} + \frac{V_{w1} v}{g} = \frac{V_w v}{g} + \frac{V_{r1} \cos\phi - v)v}{g}$$

$$\ldots(\because V_{w1} = V_{r1}\cos\phi - v)$$

$$= \frac{v}{g}\{V_w + [(V - v)\cos\phi - v]\}$$

$$\ldots(\because V_{r1} = V_r = V - v)$$

$$= \frac{v}{g}(V + V\cos\phi - v) \qquad \ldots(\because V_w = V)$$

$$= \frac{v}{g}[V(1 + \cos\phi) - v(1 + \cos\phi)]$$

$$= \frac{v(V - v)(1 + \cos\phi)}{g}$$

We know that the hydraulic efficiency,

$$\eta_\eta = \frac{\text{Work done per kN of water}}{\text{Energy supplied per kN of water}}$$

$$= \frac{\frac{v(V-v)(1+\cos\phi)}{g}}{\frac{V^2}{2g}} = \frac{2v(V-v)(1+\cos\phi)}{V^2}$$

For maximum efficiency, differentiate the numerator of the above equation, with respect to v and equate it to zero (as the maximum efficiency will be, when the numerator will be maximum).

$$\frac{d}{dv}[2v(V-v)(1+\cos\phi)] = 0$$

$$\frac{d}{dv}[(2Vv - 2v^2)(1+\cos\phi)] = 0$$

$$2V - 4v = 0 \text{ or } v = \frac{V}{2}$$

It means that the velocity of the wheel, for maximum hydraulic efficiency, should be half of the jet velocity. Therefore maximum work done/kN of water

$$= \frac{v(V-v)(1+\cos\phi)}{g} = \frac{\frac{V}{2}\left(V - \frac{V}{2}\right)(1+\cos\phi)}{g}$$

...(Substituting $v = \frac{V}{2}$)

$$= \frac{V^2}{4g}(1+\cos\phi) \text{ kN}$$

∴ Maximum hydraulic efficiency,

$$\max \eta_h = \frac{\frac{V^2}{4g}(1+\cos\phi)}{\frac{V^2}{2g}} = \frac{(1+\cos\phi)}{2}$$

Notes : 1. It may be noted that the efficiency is maximum when $\cos\phi = 1$ i.e., $\phi = 180°$. But in actual practice, the jet is deflected through an angle of 160° to 165° only. Because, if the jet is made to deflect through an angle of 180°, the water discharged from one bucket, will have an impact on the bucket, in front of it.

2. In actual practice, maximum efficiency takes place when the velocity of wheel is 0·46 times the velocity of the jet (i.e., $v = 0.46V$)

3. The power generated by the turbine may be found out as usual by multiplying the discharge in m³/s with the work done per kN of water.

32·8. Power Produced by an Impulse Turbine

We have seen in the previous articles, that some work is done per kN of water, when the jet strikes the buckets of an impulse turbine. If we know the quantity of water in kN, flowing through the jets per second, and the amount of work done per second, then the power produced by the turbine may be found out with the help of velocity triangles as usual. The power produced may also be found out from the relation,

$$P = wQH \text{ kW}$$

Impulse Turbines

where
- w = specific weight of water (9·81 kN/m³),
- Q = Discharge of the turbine in m³/s, and
- H = Head of water in metres.

32·9. Efficiencies of an Impulse Turbine

In general, the term efficiency may be defined as the ratio of work done to the energy supplied. An impulse turbine has the following three types of efficiencies:
1. Hydraulic efficiency,
2. Mechanical efficiency, and
3. Overall efficiency.

32·10. Hydraulic Efficiency

It is the ratio of work done, on the wheel, to the energy of the jet. We have sen in Art. 30·7 that the hydraulic efficiency of a turbine,

$$\eta_h = \frac{2v(V-v)(1+\cos\phi)}{V^2}$$

and maximum hydraulic efficiency,

$$\max \eta_h = \frac{(1+\cos\phi)}{2}$$

32·11. Mechanical Efficiency

It has been observed that all the energy supplied to the wheel does not come out as useful work. But a part of it is dissipated in overcoming friction of bearings and other moving parts. Thus the mechanical efficiency is the ratio of actual work available at the turbine to the energy imparted to the wheel.

32·12. Overall Efficiency

It is a measure of the performance of a turbine, and is the ratio of actual power produced by the turbine to the energy actually supplied by the turbine. i.e.,

$$\eta_0 = \frac{P}{wQH}$$

Example 32·1. *A Pelton wheel develops 2000 kW under a head of 100 metres, and with an overall efficiency of 85%. Find the diameter of the nozzle, if the coefficient of velocity for the nozzle is 0·98.*

Solution. Given : $P = 2000$ kW; $H = 100$ m; $\eta_0 = 85\% = 0.85$ and $C_v = 0.98$.

Let
- d = Diameter of the nozzle, and
- Q = Discharge of the turbine.

We know that the velocity of jet,
$$V = C_v \sqrt{2gH} = 0.98 \times \sqrt{2 \times 9.81 \times 100} = 43.4 \text{ m/s}$$

and overall efficiency (η_0),

$$0.85 = \frac{P}{wQH} = \frac{2000}{9.81 \times Q \times 100} = \frac{2.04}{Q}$$

∴ $Q = 2.04/0.85 = 2.4$ m³/s

Now the total discharge of the wheel should be equal to the discharge through the jet. i.e.,

$$Q = V \times \frac{\pi}{4} \times (d)^2$$

or $\qquad 2 \cdot 4 = 43 \cdot 4 \times \dfrac{\pi}{4} \times (d)^2 = 34 \cdot 1\, d^2$

$$d^2 = 2 \cdot 4/34 \cdot 1 = 0 \cdot 0704 \text{ or } d = 0 \cdot 265 \text{ m} = 265 \text{ mm} \quad \textbf{Ans.}$$

Example 32·2. *A Pelton wheel, having semi-circular buckets and working under a head of 140 metres, is running at 600 r.p.m. The discharge through the nozzle is 500 litres/s and diameter of the wheel is 600 mm. Find:*

(a) Power available at the nozzle, and

(b) Hydraulic efficiency of the wheel, if coefficient of velocity is 0·98.

Solution. Given : $\phi = 180° - 180° = 0°$ (Because of semi-circular buckets); $H = 140$ m; $N = 600$ r.p.m.; $Q = 500$ litres/s $= 0 \cdot 5$ m^3/s; $D = 600$ mm $= 0 \cdot 6$ m and $C_v = 0 \cdot 98$.

(a) Power available at the nozzle

We know that power available at the nozzle.

$$P = wQH = 9 \cdot 81 \times 0 \cdot 5 \times 140 = 686 \cdot 7 \text{ kW} \quad \textbf{Ans.}$$

(b) Hydraulic efficiency of the wheel

We also know that velocity of the jet,

$$V = C_v \sqrt{2gH} = 0 \cdot 98 \times \sqrt{2 \times 9 \cdot 81 \times 140} = 51 \cdot 36 \text{ m/s}$$

and tangential velocity of the wheel,

$$v = \dfrac{\pi D N}{60} = \dfrac{\pi \times 0 \cdot 6 \times 600}{60} = 18 \cdot 85 \text{ m/s}$$

∴ Hydraulic efficiency of the wheel,

$$\eta_\eta = \dfrac{2v(V-v)(1+\cos\phi)}{V^2}$$

$$= \dfrac{2 \times 18 \cdot 85 \,(51 \cdot 36 - 18 \cdot 85)(1+\cos 0°)}{(51 \cdot 36)^2}$$

$$= 0 \cdot 465\,(1+1) = 0 \cdot 929 = 92 \cdot 9\% \quad \textbf{Ans.}$$

Example 32·3. *A Pelton wheel, working under a head of 500 metres, produces 13 000 kW at 430 r.p.m. If the efficiency of the wheel is 85%, determine (a) discharge of the turbine, (b) diameter of the wheel, and (c) diameter of the nozzle. Assume suitable data.*

Solution. Given : $H = 500$ m; $P = 13\,000$ kW; $N = 430$ r.p.m. and $\eta_0 = 85\% = 0 \cdot 85$

(a) Discharge of the turbine

Let $\qquad Q = $ Discharge of the turbine.

We know that overall efficiency of the Pelton wheel (η_0),

$$0 \cdot 85 = \dfrac{P}{wQH} = \dfrac{13\,000}{9 \cdot 81 \times Q \times 500} = \dfrac{2 \cdot 65}{Q}$$

∴ $\qquad Q = 2 \cdot 65/0 \cdot 85 = 3 \cdot 12 \text{ m}^3/\text{s} \quad \textbf{Ans.}$

(b) Diameter of the wheel

Let us assume coefficient of velocity $(c_v) = 0 \cdot 98$ and tangential velocity of wheel, $(v) = 0 \cdot 46\,V$ (where V is the velocity of the jet).

We know that the velocity of jet,

$$V = C_v \sqrt{2gH} = 0 \cdot 98 \sqrt{2 \times 9 \cdot 81 \times 500} = 97 \cdot 1 \text{ m/s}$$

$$v = 0 \cdot 46\,V = 0 \cdot 46 \times 97 \cdot 1 = 44 \cdot 7 \text{ m/s}$$

We also know that the tangential velocity of the wheel (v)

$$44 \cdot 7 = \dfrac{\pi D N}{60} = \dfrac{\pi D \times 430}{60} = 22 \cdot 5 D$$

∴ $\qquad D = 44 \cdot 7/22 \cdot 5 = 2 \cdot 0 \text{ m} \quad \textbf{Ans.}$

Impulse Turbines

Diameter of the nozzle

Let $\qquad d$ = Diameter of the nozzle.

The discharge through the nozzle must be equal to the discharge of the turbine. *i.e.*,

$$Q = V \times \frac{\pi}{4} \times (d)^2$$

or $\qquad 3 \cdot 12 = 97 \cdot 1 \times \frac{\pi}{4} \times (d)^2 = 76 \cdot 3 \, (d)^2$

$\therefore \qquad d^2 = 3 \cdot 12/76 \cdot 3 = 0 \cdot 041$ or $d = 0 \cdot 2$ m = 200 mm **Ans.**

Example 32·4. *In a hydroelectric scheme, the distance between high level reservoir at the top of mountains and turbine is 1·6 km and difference of their levels is 500 m. The water is brought in 4 penstocks each of diameter of 0·9 m connected to a nozzle of 200 mm diameter at the end. Find:*

(a) Power of each jet, and

(b) total power available at the reservoir, taking the value of Darcy's coefficient of friction as 0·008.

Solution. Given : l = 1.6 km = 1600 m; H = 500 m; No. of penstocks (n) = 4; Dia. of each penstock (D) 0·9 = m; Dia. of nozzle (d) = 200 mm = 0·2 m and f = 0·008.

(a) Power of each jet

We know that area of each nozzle,

$$a = \frac{\pi}{4} \times (d)^2 = \frac{\pi}{4} \times (0 \cdot 2)^2 = 0 \cdot 0314 \text{ m}^2$$

and velocity of the jet, $\qquad V = C_v \sqrt{2gH} = 0 \cdot 98 \times \sqrt{2 \times 9 \cdot 81 \times 500} = 97 \cdot 1$ m/s

...(Assuming C_v = 0·98)

∴ Discharge through each jet,

$$Q = a \cdot V = 0 \cdot 0314 \times 97 \cdot 1 = 3 \cdot 05 \text{ m}^3/\text{s}$$

and $\qquad$ power of each jet = wQH = 9·81 × 3·05 × 500 = 14 960 kW **Ans.**

(b) Total power available at the reservoir

We also know that area of the penstock,

$$A = \frac{\pi}{4} \times (D)^2 = \frac{\pi}{4} \times (0 \cdot 9)^2 = 0 \cdot 636 \text{ m}^2$$

and velocity of water in the penstock,

$$v = \frac{Q}{A} = \frac{3 \cdot 05}{0 \cdot 636} = 4 \cdot 8 \text{ m/s}$$

∴ Head lost due to friction in each penstock,

$$H_f = \frac{4flv^2}{2gd} = \frac{4 \times 0 \cdot 008 \times 1600 \times (4 \cdot 8)^2}{2 \times 9 \cdot 81 \times 0 \cdot 9} = 66 \cdot 8 \text{ m}$$

and total power available at the reservoir (in 4 penstocks)

$$P = 4 \times wQ \, (H + H_f) = 4 \times 9 \cdot 81 \times 3 \cdot 05 \, (500 + 66 \cdot 8) \text{ kW}$$
$$= 67\,836 \text{ kW} \quad \textbf{Ans.}$$

Example 32·5. *The Pykara Power House, in South India, is equipped with impulse turbines of pelton type. Each turbine delivers a maximum power of 14 250 kW, when working under a head of 900 metres and running at 600 r.p.m.*

Find the diameter of the jet, and the mean diameter of the wheel. Take overall efficiency of the turbine as 89·2%.

Solution. Given : $P = 14\,250$ kW; $H = 900$ m; $N = 600$ r.p.m. and $\eta_0 = 89.2\% = 0.892$

Diameter of the jet

Let $\quad d =$ Diameter of the jet, and
$\quad Q =$ Discharge of the turbine.

We know that overall efficiency of the turbine (η_0),

$$0.892 = \frac{P}{wQH} = \frac{14\,250}{9.81 \times Q \times 900} = \frac{1.61}{Q}$$

or $\quad Q = 1.61/0.892 = 1.8$ m³/s

and velocity of the jet, $\quad V = C_v \times \sqrt{2gH} = 0.98 \times \sqrt{2 \times 9.81 \times 900} = 130.2$ m/s

...(Assuming $C_v = 0.98$)

Now the discharge through the turbine must be equal to discharge through the jet. *i.e.*,

$$Q = V \times \frac{\pi}{4} \times (d)^2$$

or $\quad 1.8 = 130.2 \times \frac{\pi}{4} \times (d)^2 = 102.3\, d^2$

$\therefore \quad d^2 = 1.8/102.3 = 0.018$ or $d = 0.134$ m $= 134$ mm **Ans.**

Mean diameter of the wheel

Let $\quad D =$ Mean diameter of the wheel.

We know that peripheral velocity of the wheel,

$$v = 0.46 \times V = 0.46 \times 130.2 = 59.9 \text{ m/s}$$

We also know that peripheral velocity of the wheel (v),

$$59.9 = \frac{\pi DN}{60} = \frac{\pi D \times 600}{60} = 31.42 D$$

$\therefore \quad D = 59.9/31.42 = 1.91$ m **Ans.**

Example 32·6. *A Pelton wheel is required to generate 3750 kW under an effective head of 400 metres. Find the total flow in litres/second and size of the jet. Assume generator efficiency 95%, overall efficiency 80%, coefficient of velocity 0·97, speed ratio 0·46. If the jet ratio is 10, find the mean diameter of runner.*

Solution. Given : $P = 3750$ kW; $H = 400$ m; $\eta_\gamma = 95\% = 0.95\%$ $\eta_0 = 80\% = 0.8$; $C_v = 0.97$; and speed ratio $\dfrac{v}{\sqrt{2gH}} = 0.46$

Total flow of water in litres/second

Let $\quad Q =$ Total flow of water in litres/s

We know that power available,

$$P = \frac{\text{Power generated}}{\text{Generator efficiency}} = \frac{3750}{0.95} = 3947 \text{ kW}$$

and overall efficiency (η_0),

$$0.8 = \frac{P}{wQH} = \frac{3947}{9.81 \times Q \times 400} = \frac{1}{Q}$$

$\therefore \quad Q = 1/0.8 = 1.25$ m³/s $= 1250$ litres/s **Ans.**

Impulse Turbines

Size of jet

Let d = Diameter of the jet.

We know that velocity of the jet,

$$V = C_v \sqrt{2gH} = 0.97 \times \sqrt{2 \times 9.81 \times 400} = 85.9 \text{ m/s}$$

Now the total discharge of the wheel should be equal to the discharge through the jet. *i.e.*,

$$Q = V \times \frac{\pi}{4} \times (d)^2$$

or

$$1.25 = 85.9 \times \frac{\pi}{4} \times (d)^2 = 67.5\, d^2$$

∴

$$d^2 = 1.25/67.5 = 0.0185 \text{ or } d = 0.136 \text{ m} = 136 \text{ mm Ans.}$$

Example 32·7. *A Pelton wheel has a mean bucket speed of 15 m/s with a jet of water impinging with a velocity of 40 m/s and discharging 450 litres/s. If the buckets deflect the jet through an angle of 165°, find the power generated by the wheel.*

Solution. Given : $v = 15$ m/s; $V = 40$ m/s; $Q = 450$ litres/s $= 0.45$ m³/s and $\phi = 180° - 165° = 15°$ (Because the jet is deflected through 165°).

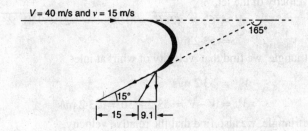

Fig. 32·4

From the inlet trianglet, we find that velocity of whirl at inlet

$$V_w = 40 \text{ m/s} \qquad \ldots(\because V = V_w)$$

and relative velocity,

$$V_r = V - v = 40 - 15 = 25 \text{ m/s}$$

From the outlet triangle, we also find that

$$V_{r1} = V_r = 25 \text{ m/s}$$

and velocity of buckets,

$$v_1 = v = 15 \text{ m/s}$$

∴ Velocity of whirl at outlet,

$$V_{w1} = v_1 - V_{r1} \cos \phi = 15 - (25 \cos 15°) = 15 - (25 \times 0.9659) \text{ m/s}$$
$$= 15 - 24.1 = -9.1 \text{ m/s}$$

...(Minus sign indicates that the direction of V_{w1} is opposite to that of V or v)

We know that work done per kN of water,

$$= \frac{v}{g}(V_w - V_{w1}) = \frac{15}{9.81}[40 - (-9.1)] = \frac{15}{9.81} \times 49.1 \text{ kN-m/s}$$
$$= 75.1 \text{ kN-m/s}$$

and total work done $= 9.81 \times 0.45 \times 75.1 = 331.5$ kN-m/s

∴ Power generated by the wheel,
$$P = 331.5 \text{ kJ/s} = 331.5 \text{ kW} \quad \textbf{Ans.}$$

Example 32·8. *A Pelton wheel has a tangential velocity of buckets of 15 m/s. The water is being supplied under a head of 150 metres at the rate of 200 litres/s. The buckets deflect the jet through an angle of 160°. If the coefficient of velocity for the nozzle is 0·98, find the power produced by the wheel and its hydraulic efficiency.*

Solution. Given : $v = 30$ m/s; $H = 150$ m; $Q = 200$ litres/s $= 0.2$ m³/s; $\phi = 180° - 160° = 20°$ (Because the jet is deflected through 160°) and $C_v = 0.98$.

Power produced by the wheel

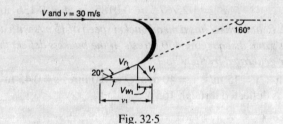

Fig. 32·5

We know that velocity of the jet,
$$V = C_v \times \sqrt{2gH} = 0.98 \times = \sqrt{2 \times 9.81 \times 150} \text{ m/s}$$
$$= 53.2 \text{ m/s}$$

From the inlet triangle, we find that velocity of whirl at inlet
$$V_w = 53.2 \text{ m/s}$$

and relative velocity, $V_r = V - v = 53.2 - 30 = 23.2$ m/s

From the outlet triangle, we also find that the relative velocity,
$$V_{r1} = V_r = 23.2 \text{ m/s}$$

and velocity of buckets, $v_1 = v = 30$ m/s

∴ Velocity of whirl at outlet,
$$V_{w1} = v_1 - V_{r1} \cos \phi = 30 - (23.2 \cos 20°) \text{ m/s}$$
$$= 30 - (23.2 \times 0.9397) = 30 - 21.8 = 8.2 \text{ m/s}$$

Since the value of $V_{v1} \cos \phi$ (28·2 cos 20° = 28·2 × 0·9397 = 21·8) in less than v_1 (25), therefore shape of the outlet triangle will be as shown in Fig. 32·5. Moreover, the velocity of whirl (V_{w1}) will be positive, as it is in the same direction as that of V_{w1}.

We know that work done per kN of water,
$$= \frac{v}{g}(V_w - V_{w1}) = \frac{30}{9.81}(53.2 - 8.2) = 137.6 \text{ kN-m/s}$$

and total work done by the water $= 9.81 \times 0.2 \times 137.6 = 270$ kN . m/s $= 270$ kJ/s

∴ Power produced by the wheel,
$$P = 270 \text{ kJ/s} = 270 \text{ kW} \quad \textbf{Ans.}$$

Hydraulic efficiency of the wheel

We also know that hydraulic efficiency of the wheel,
$$\eta_h = \frac{2(V_w - V_{w1}) \times v}{V^2} = \frac{2(53.2 - 8.2) \times 30}{(53.2)^2}$$
$$= 0.954 = 95.4\% \quad \textbf{Ans.}$$

EXERCISE 32·1

1. An impulse turbine working under a head of 100 m is required to develop 250 kW. If the overall efficiency of the turbine is 85%, find the discharge of the turbine. **(Ans. 0·3 m^3/s)**

2. A Pelton wheel has mean bucket speed of 15 m/s with a jet of water flowing under a head of 120 m. If the jet is deflected through an angle of 160° and the coefficient of velocity is 0·97, find the hydraulic efficiency of the turbine. **(Ans. 84·2%)**

3. The overall efficiency of a Pelton wheel is 86% when the power developed is 500 kW under a head of 80 m. If the coefficient of velocity for the nozzle is 0·97, find the diameter of the nozzle. **(Ans. 157 mm)**

4. A Pelton wheel of 1 meter diameter is working under a head of 150 metres. Find the speed of the runner, if the coefficient of velocity and velocity ratio is 0·98 and 0·47 respectively. **(Ans. 480 r.p.m.)**

5. A Pelton wheel is producing 1350 kW under a head of 80 metres at 300 r.p.m. Find the diameter of the wheel, if the speed ratio is 0·45. Take $C_v = 0·98$. **(Ans. 1·11 m)**

6. A Pelton wheel is working under a head of 85 m and the rate of flow of water through the jet is 800 litres/s. If the bucket speed is 14 m/s and the jet is deflected through 165°, find the power produced by the wheel and its efficiency. Take coefficient of velocity as 0·985. **(Ans. 576·8 kW; 86·5%)**

32·13. Number of Jets of a Pelton Wheel

A Pelton turbine, generally, has a single jet only. But whenever a single jet cannot develop the required power, we may have to employ more than one jets. In Fig. 32·6 show an arrangement of four jets for a Pelton wheel.

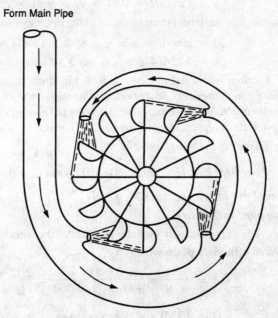

Fig. 32·6. Number of jets

In general, the maximum number of jets provided to a Pelton wheel are six. While designing the jets, care should always be taken to provide the jets at equidistant on the outer periphery of the wheel.

Sometimes, instead of providing a number of jets to a wheel (as shown in Fig. 32·6) two or three wheels are mounted on a common shaft. Such a system is known as overhung wheels.

Example 32·9. *A double overhung Pelton wheel unit is directly coupled to 1000 kW generator. Find the power developed by each wheel, if the generator efficiency is 84%.*

Solution. Given : Total output of the unit (P) = 1000 kW and η = 84% = 0·84.

We know that power generated by the Pelton wheel runner

$$= \frac{P}{\eta} = \frac{1000}{0.84} = 1190 \text{ kW}$$

Since the Pelton wheel is double overhung (*i.e.* two wheels are mounted on the runner), therefore power developed by each wheel

$$= 1190/2 = 595 \text{ kW} \quad \textbf{Ans.}$$

Example 32·10. *A Pelton wheel is supplied water under a head of 200 m through a 100 mm diameter pipes. If the quantity of water supplied to the wheel is 1.25 m^3/s. Find the number of jets. Assume $C_v = 0.97$.*

Solution. Given : H = 200 m; d = 100 mm = 0.1 m; Q = 1·25 m^3/s and C_v = 0·97.

Let n = Number of jets

We know that cross-sectional area of one pipe,

$$a = \frac{\pi}{4} \times (d)^2 = \frac{\pi}{4} \times (0.1)^2 = 7.854 \times 10^{-3} \text{ m}^2$$

and velocity of the jets,

$$V = C_v\sqrt{2gH} = 0.97 \times \sqrt{2 \times 9.81 \times 200} = 60.8 \text{ m/s}$$

Now the total discharge of the wheel must be equal to the discharge through the jets. *i.e.*,

$$1.25 = n \times V \times a = n \times 60.8 \times (7.854 \times 10^{-3}) = 0.478 \ n$$

$$\therefore n = 1.25/0.478 = 2.6 \text{ say } 3 \quad \textbf{Ans.}$$

Example 32·11. *A Pelton wheel has to develop 5000 kW under a net head of 300 m, while running at a speed of 500 r.p.m. If the coefficient of velocity for the jet = 0·97, speed ratio = 0·46 and the ratio of the jet diameter is 1/10 of wheel diameter, calculate (a) quantity of water supplied to the wheel, (b) diameter of pitch circle, (c) diameter of jets, and (d) number of jets.*

Assume overall efficiency of the wheel as 80%.

Solution. Given : P = 8000 kW; H = 300 m; N = 500 r.p.m.; C_v = 0·97; v = 0·46V; $d = \frac{D}{10}$

(where D is the diameter of the wheel) and η = 80% = 0·8.

(a) Quantity of water supplied to the wheel

Let Q = Quantity of water supplied to the wheel.

We know that overall efficiency of the wheel (η_0),

$$0.8 = \frac{P}{wQH} = \frac{8000}{9.81 \times Q \times 300} = \frac{2.72}{Q}$$

$$\therefore Q = 2.72/0.8 = 3.4 \text{ m}^3/s \quad \textbf{Ans.}$$

(b) Diameter of pitch circle

Let D = Diameter of pitch circle.

We know that the velocity of jet,

$$V = C_v\sqrt{2gH} = 0.97\sqrt{2 \times 9.81 \times 300} = 74.4 \text{ m/s}$$

and peripheral velocity, $v = 0.46 \ V = 0.46 \times 74.4 = 34.2 \text{ m/s}$

Impulse Turbines

We also know that peripheral velocity (v),

$$34 \cdot 2 = \frac{\pi DN}{60} = \frac{\pi \times D \times 500}{60} = 26 \cdot 2\, D$$

$\therefore \qquad D = 34 \cdot 2/26 \cdot 2 = 1 \cdot 3 \text{ m}$ **Ans.**

(c) Diameter of the jets

We know that diameter of the jets,

$$d = \frac{D}{10} = \frac{1 \cdot 3}{10} = 0 \cdot 13 \text{ m} = 130 \text{ mm} \quad \textbf{Ans.}$$

(d) Number of jets

Let $\qquad n$ = Number of jets.

Now total discharge of the wheel, must be equal to the discharge through the jets. *i.e.*,

$$3 \cdot 4 = n \times V \times \frac{\pi}{4} \times (d)^2 = n \times 74 \cdot 4 \times \frac{\pi}{4} \times (0 \cdot 13)^2 = 0 \cdot 99\, n$$

$$n = 3 \cdot 4/0 \cdot 99 = 3 \cdot 4 \text{ say } 4 \quad \textbf{Ans.}$$

EXERCISE 32·2

1. A double overhung impulse turbine is coupled to a 2400 kW generator. What is the power developed by each wheel, if the generator efficiency is 80%. **(Ans. 1500 kW)**
2. A Pelton wheel operating under a head of 150 m develops 2500 kW. Find the number of jets required for the wheel, if its overall efficiency is 85%. Take coefficient of velocity for the jet as 0·98 and diameter of the pipes as 120 mm. **(Ans. 4)**
3. A Pelton wheel, working under a head of 250 metres, develops 6000 kW while running at 600 r.p.m. with an overall efficiency of 90%. The ratio of jet diameter to the wheel diameter is 1/8. The coefficient of velocity for the nozzle is 0·98 and the ratio of tangential velocity of the wheel to the velocity of the wheel is 0·46. Find (*i*) rate of flow, (*ii*) diameter of the wheel, and number of jets. **(Ans. 2·45 m^3/s; 1 m; 3)**

32·14. Design of Pelton Wheels

A Pelton wheel is, generally, designed for a given head of water, power to be developed and speed of the runner. In modern design offices, a Pelton wheel is designed to find out the following data:

1. Diameter of wheel,
2. Diameter of the jet,
3. Size (*i.e.*, width and depth) of the buckets
4. No. of buckets.

While designing a Pelton wheel, if sufficient data is not available, then the following assumptions are made, which are meant for the best results:

1. Overall efficiency between 80% and 87% (preferably 85%)
2. Coefficient of velocity 0·99 (preferably 0·985)
3. Ratio of peripheral velocity to the jet velocity as 0·46.

Note. As a matter of fact, diameter of wheel and diameter of jets is obtained from the discharge condition. We have already discussed these points in our previous examples. In the following pages, we shall discuss the size of the buckets and no. of buckets.

32·15 Size of Buckets of a Pelton Wheel

In general, the buckets of a Pelton wheel have the following dimensions :
Width of the bucket $= 5 \times d$
and depth of the bucket $= 1.2 \times d$
where $d =$ Diameter of the jet.

32·16. Number of Buckets on the Periphery of a Pelton Wheel

The number of buckets, on the periphery of a Pelton wheel, is decided mainly on the following two principles, *viz.*

1. The number of buckets should be as few as possible, so that there may be as little loss, due to friction, as possible.
2. The jet of water must be fully utilised, so that no water from the jet should go waste.

Now consider a jet of water impinging on the buckets of a Pelton wheel with O as centre.

Let $R =$ Radius of the mean bucket circle,
$d =$ Diameter of the jet, and
$\alpha =$ Angle, subtended by the two adjacent buckets at the centre of the wheel.

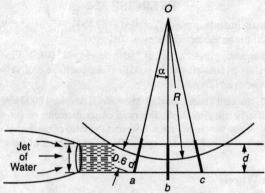

Fig 32·7. No. of buckets on the periphery

At one instant let a, b and c be the position of three adjacent buckets as shown in Fig. 32·7. Before this instant, the jet was having some impact on the bucket b also. But at this instant, the jet will be intercepted by the bucket a, which will have its impact. (This happens when the outer edge of the bucket a just touches the lower portion of the jet). It may be noted, that such a stage comes, when half of the depth of the bucket (*i.e.*, 0·6 d), projecting from the mean circumference, will just touch the lower portion of the jet. At this instant, the bucket a will move as a result of the jet impact.

When have seen in Art. 32·7 that the velocity of jet is twice the velocity of buckets. Therefore the time taken by the bucket b in travelling up to c is equal to the time taken by the last particle of water (which is just on the right side of bucket a) in travelling up to c. It is thus obvious, that the bucket b will continue to have impact till it reaches at c.

From the geometry of the figure, we find that

$$\cos \alpha = \frac{R + 0.5d}{R + 0.6d}$$

and then the number of buckets may be found out by the relation:

$$\text{Number of buckets} = \frac{360°}{\alpha}$$

Impulse Turbines

Note: It is a theoretical relation, derived for the number of buckets required for a Pelton wheel. But in actual practice, we provide the number of buckets to a Pelton wheel as half of the buckets obtained from the above equation. This unsatisfactory result has given birth to many empirical formulate. One of such formulae, which is widely used is;

$$\text{Number of buckets} = \left(\frac{D}{2d} + 15\right)$$

where
- D = Mean bucket diameter, and
- d = Diameter of the jet.

Example 32·12. *Design a Pelton wheel for a head of 350 m at a speed of 300 r.p.m. Take overall efficiency of the wheel as 85% and ratio of jet to the wheel diameter as 1/10.*

Solution. Given : $H = 350$ m; $N = 300$ r.p.m.; $\eta_0 = 85\% = 0.85$ and $\frac{d}{D} = \frac{1}{10}$ or $d = \frac{D}{10}$

...(where D is the diameter of the wheel).

1. Diameter of the wheel

Let D = Diameter of wheel.

We know that velocity of the jet,
$$V = C_v \times \sqrt{2gH} = 0.98 \times \sqrt{2 \times 9.81 \times 350} = 81.2 \text{ m/s}$$
...(Assuming $C_v = 0.98$)

and peripheral velocity of the wheel,
$$v = 0.46\ V = 0.46 \times 81.2 = 37.4 \text{ m/s}$$

We also know that peripheral velocity (v)
$$37.4 = \frac{\pi DN}{60} = \frac{\pi \times D \times 300}{60} = 15.7\ D$$

∴ $D = 37.4/15.7 = 2.4$ m **Ans.**

2. Diameter of the jet

We know that diameter of the jet,
$$d = \frac{\text{Dia. of wheel}}{10} = \frac{2.4}{10} = 0.24 \text{ m} = 240 \text{ mm} \quad \textbf{Ans.}$$

3. Width of the buckets

We know that width of the buckets
$$= 5 \times d = 5 \times 0.24 = 1.2 \text{ m} \quad \textbf{Ans.}$$

4. Depth of the buckets

We know that depth of the buckets
$$= 1.2 \times d = 1.2 \times 0.24 = 0.48 \text{ m} \quad \textbf{Ans.}$$

5. No. of buckets

We also know that the number of buckets
$$= \frac{D}{2d} + 15 = \frac{2.4}{2 \times 0.24} + 15 = 20 \quad \textbf{Ans.}$$

Example 32·13. *Design a Pelton wheel for the following data:*

- Head of water = 150 metres
- Power to be developed = 600 kW
- Speed of wheel = 360 r.p.m.

Assume, reasonably, the missing data.

Solution. Given : $H = 150$ m; $P = 600$ kW and $N = 360$ r.p.m.

1. Diameter of the wheel

Let $\qquad D$ = Diameter of the wheel.

We know that velocity of the jet,
$$V = C_v \times \sqrt{2gH} = 0.985 \times \sqrt{2 \times 9.81 \times 150} = 53.5 \text{ m/s}$$
...(Assuming $C_v = 0.985$)

and peripheral velocity, $\qquad v = 0.46\, V = 0.46 \times 53.4 = 24.6$ m/s
...(Assuming $v = 0.46V$)

We also know that peripheral velocity (v),
$$24.6 = \frac{\pi DN}{60} = \frac{\pi \times D \times 360}{60} = 18.85\, D$$
$$D = 24.6/18.85 = 1.3 \text{ m} \quad \textbf{Ans.}$$

2. Diameter of the jet

Let $\qquad d$ = Diameter of the jet.

We know that overall efficiency of the jet,
$$0.85 = \frac{P}{wQH} = \frac{600}{9.81 \times Q \times 150} = \frac{0.408}{Q}$$
...(Assuming $\eta_0 = 85\%$)

$\therefore \qquad Q = 0.408/0.85 = 0.48 \text{ m}^3/\text{s}$

Now the discharge through the wheel, must be equal to the discharge through the jet. *i.e.*,
$$Q = V \times \frac{\pi}{4} \times d^2$$

or $\qquad 0.48 \times 53.4 \times \frac{\pi}{4} \times (d)^2 = 41.9\, d^2$

$\therefore \qquad d^2 = 0.48/41.9 = 0.011 \text{ or } d = 0.105 \text{ m} = 105 \text{ mm} \quad \textbf{Ans.}$

3. Width of the buckets

We know that width of the buckets
$$= 5 \times d = 5 \times 0.105 = 525 \text{ mm} \quad \textbf{Ans.}$$

4. Depth of the buckets

We know that depth of the buckets
$$= 1.2 \times d = 1.2 \times 105 = 210 \text{ mm} \quad \textbf{Ans.}$$

5. No. of the buckets

We also know that the no. of buckets
$$= \frac{D}{2d} + 15 = \frac{1.3}{2 \times 0.105} + 15 = 21 \quad \textbf{Ans.}$$

32·17. Governing of an Impulse Turbine (Pelton Wheel)

In actual practice, load on the generator (which is coupled to an impulse turbine) is always fluctuating from time to time. This fluctuating load, on the generator, has some effect on the turbine also, because the generator is directly coupled to the turbine. A little consideration will show, that any change of load on the turbine, is sure to change its speed and rate of flow. It has been observed, that in order to have a high efficiency at different loads, a speed of the turbine must be kept constant as far as possible. The process of providing any arrangement, which will keep the speed constant and will regulate the rate of flow (according to the changing load conditions), is known as governing of

Impulse Turbines

the turbine. Though there are many methods of governing an impulse turbine, yet the servomotor method or relay cylinder method is commonly used these days which is discussed below:

The servomotor method is a mechanism consisting of the following parts as shown in Fig. 32·8.

1. Centrifugal governor,
2. Control valve,
3. Servomotor,
4. Gear pump,
5. Oil pump,
6. Spear or needle, and
7. A set of pipes, connecting oil sump with control valve, and control valve with relay cylinder.

The centrifugal governor is driven from the main shaft of the turbine, either by belt or gear arrangement. The control valve controls the direction of flow of the liquid (which is pumped by gear pump from the oil sump) either in pipe *AA* or *BB*. The servomotor or relayvalve has a piston (whose motion, towards left or right, depends upon the pressure of the liquid flowing through the pipes *AA* or *BB*) is connected to a spear or needle, which reciprocates inside the nozzle as shown in Fig. 32·8.

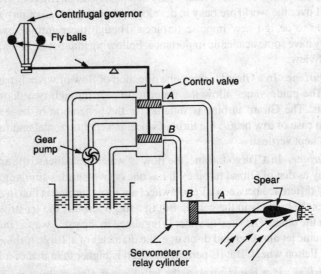

Fig. 32·8. Governing of impulse turbine

When the turbine is running at its normal speed, the positions of piston (in a servomotor or relay cylinder), control valve and fly balls of centrifugal governor will be in their normal positions as shown in the figure. The oil pumped by the gear pump, into the control valve, will come back to the oil sump as the mouths of both the pipes. *AA* and *BB* are closed by the two wings of the control valve.

Now, let the load on the turbine increase, which will decrease the speed of the turbine. This decrease in the speed of the turbine runner will also decrease the speed of centrifugal governor. As a result of this, the fly balls will come down, thus decreasing their implitude (due to decrease in centrifugal force). This coming down of the fly balls, will also bring down the sleeve, as the sleeve is connected to the central vertical bar of the centrifugal governor. This downward movement of the sleeve will raise the control valve rod (as the sleeve is connected to the control valve rod through a lever pivoted on a fulcrum). Now, a slight upwards movement of the control valve rod will open the mouth of pipe *AA* (still keeping the mouth of pipe *BB* closed). Now the oil (under pressure) will rush from the control valve to the right side of the piston in the servomotor through the pipe *AA*. This oil, under pressure, will move the piston and spear towards the left, which will open more area of the nozzle controlling the flow to the turbine. This increase in the area of flow will increase the rate of flow. As a result of this, there will be an increase in the speed of the turbine. When the speed of the runner will come up to the normal speed, fly balls will move up and the sleeve as well as the control valve rod will occupy its normal position.

It may be noted that when the load on the turbine decreases, its speed will increase. As a result of this the fly balls will go up (due to increase in centrifugal force) and sleeve will also go up. This will push the control valve downwards. This downward movement of the control valve rod will open the mouth of the pipe *BB* (still keeping the mouth of the pipe *AA* closed). Now the oil (under pressure) will rush from the control valve to the left side of the piston in servomotor through the pipe *BB*. This oil, under pressure, will move the piston and spear towards the right, which will decrease the area of the nozzle and ultimately decrease the rate of flow. This decrease in the rate of flow will decrease the speed of the turbine till the speed, once again, comes down to the normal.

32·18. Other Impulse Turbines

The Pelton wheel is considered to be the most popular impulse turbine, for high heads, and is widely used successfully all over the world. Many scientists and engineers, working in hydraulic research stations all over the world are busy in developing improved type of turbines. Some of them have also been able to design new impulse turbines. Though these turbines are not of practical importance, yet they have some academic importance. Following impulse turbines are important from the subject point of view:

1. *Girard turbine.* In a Girard turbine, the direction of flow of water is parallel to the axis of wheel. The guide vanes allow the water to impact through two diametrically opposite quadrants. The Girard turbine is suitable for the generation of large power, under low heads. In case of low heads, the turbine wheel is kept horizontal and for larger heads, the wheel is kept vertical.

2. *Turgo turbine.* In a Turgo turbine, the flow of water is parallel to the axis of wheel (in the same way as that of Girard turbine). It has one or two nozzles similar to the Pelton wheel. The only difference between a Pelton wheel are turgo turbine is that in a Pelton wheel, the jet strikes the buckets in the centre. But in a turgo turbine, the jet strikes at one end, and leaves at the other. The flow of water is regulated in the same way as that of Pelton wheel. For the same jet diameter and discharge, the diameter of a Turgo turbine is much less than that of a Pelton wheel. But its peripheral speed is higher than that of a Pelton wheel.

3. *Banki turbine.* In a Banki turbine, the jet of water after striking the buckets is made to pass through the runner. The water passing through the runner, gives some impulse to the runner. As a result of this, the velocity of water is utilised two times, which improves its efficiency.

Through the theory of utilising the velocity of water two times did not succeed in actual practice, yet this idea is utilised in steam turbines.

EXERCISE 32·3

1. A Pelton wheel working under a head of 100 metres produces 500 kW at 250 r.p.m. The overall efficiency of the wheel and coefficient of velocity for the nozzle are 80% and 0·98 respectively. If the wheel diameter is 1 metre, find

   (1) diameter of the jet   (2) width of the buckets
   (3) depth of the buckets   (4) number of buckets   (**Ans.** 137 mm; 685 mm; 164 mm; 19)

Impulse Turbines

2. A Pelton wheel is to be designed to produce 400 kW under a head of 60 m when running at 200 r.p.m. The overall efficiency of the wheel is 85% and coefficient of velocity for the jet is 0.98. Take velocity of the buckets = 0·45 times the velocity of the jet.

   (**Ans.** $d = 180$ mm; $D = 1.4$ m; Width of buckets = 900 mm; Depth of buckets = 216 mm; No. of buckets = 14)

QUESTIONS

1. What is meant by an impulse turbine?
2. Describe, withe the help of simple sketches, the working of an impulse turbine.
3. Derive an equation for the hydraulic efficiency of a Pelton wheel.
4. Show from, first principles, that the theoretical value for peripheral coefficient of a Pelton wheel is 0·5.
5. On what factors does the number of jets depend in the case of Pelton wheels?
6. What is the ratio of width of the buckets and depth of the buckets to the jet diameter?
7. By means of a neat sketch, giving complete operation, explain how the turbines are governed for constant speed.

OBJECTIVE TYPE QUESTIONS

1. A Pelton wheel is an
   - (a) axial flow impulse turbine
   - (b) inward flow impulse turbine
   - (c) outward flow impulse turbine
   - (d) all of these

2. An impulse turbine is used for
   - (a) low head of water
   - (b) medium head of water
   - (c) high head of water
   - (d) any one of these

3. The maximum hydraulic efficiency of an impulse turbine is equal to
   - (a) $\dfrac{1 + \cos \phi}{2}$
   - (b) $\dfrac{1 - \cos \phi}{2}$
   - (c) $\dfrac{1 + \sin \phi}{2}$
   - (d) $\dfrac{1 - \sin \phi}{2}$

   where ϕ is the angle of blade tip at outlet.

4. A double overhung Pelton wheel has
   - (a) two jets
   - (b) two runners
   - (c) four jets
   - (d) four runners.

5. The number of buckets on the periphery of a Pelton wheel is given by
   - (a) $\dfrac{D}{2d} + 5$
   - (b) $\dfrac{D}{2d} + 10$
   - (c) $\dfrac{D}{2d} + 15$
   - (d) $\dfrac{D}{2d} + 20$

Answers

1. (a),   2. (c),   3. (a),   4. (b),   5. (c)

33
Reaction Turbines

> *1. Introduction. 2. Main Components of a Reaction Turbine. 3. Spiral Casing. 4. Guide Mechanism. 5. Turbines Runner. 6. Draft Tube. 7. Difference between an Impulse Turbine and a Reaction Turbine. 8. Classification of Reaction Turbines. 9. Radial Flow Turbines. 10. Axial Flow Turbines. 11. Mixed Flow Turbines. 12. Inward Flow Reaction Turbines. 13. Outward Flow Reaction Turbines. 14. Discharge of a Reaction Turbine. 15. Power Produced by a Reaction Turbine. 16. Efficiencies of a Reaction Turbine. 17. Hydraulic Efficiency 18. Mechanical Efficiency. 19. Overall Efficiency. 20. Francis' Turbine. 21. Kaplan Turbine. 22. Draft Tube. 23. Types of Draft Tubes. 24. Conical Draft Tubes. 25. Elbow Draft Tubes. 26. Efficiency of a Draft Tube. 27. Other Reaction Turbines.*

33·1 Introduction

In a reaction turbine, the water enters the wheel under pressure and flows over the vanes. As the water, flowing over the vanes, is under pressure, therefore wheel of the turbine runs full and may be submerged below the tail race or may discharge into the atmosphere. The pressure head of water, while flowing over the vanes, is converted into velocity head and is finally reduced to the atmospheric pressure, before leaving the wheel.

33·2 Main Components of a Reaction Turbine

A reaction turbine has the following main components :
1. Spiral casing,
2. Guide mechanism,
3. Turbine runner, and
4. Draft tube

33·3 Spiral Casing

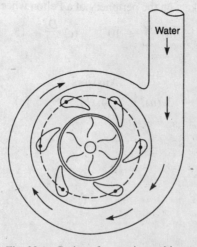

Fig. 33·1. Casing of a reaction turbine.

Reaction Turbines

The water, from a pipeline, is distributed around the guide ring in a casing. This casing is designed in such a way that its cross-sectional area goes on reducing uniformly around the circumference. The cross-sectional area is maximum at the entrance, and minimum at the tip as shown in Fig. 33·1. As a result of this, the casing will be of spiral shape. That is why, it is called a spiral casing or scroll casing.

The spiral casings are provided with inspection holes and pressure gauges. The material of a casing depends upon the head of water, under which the turbine is working as discussed below :

Concrete	...	up to 30 m
Welded rolled steel plate	...	up to 100 m
Cast steel	...	more than 100 m

33·4 Guide Mechanism

The guide vanes are fixed between two rings in the form of a wheel. This wheel is fixed in the spiral casing. The guide vanes are properly designed in order to :
1. allow the water to enter the runner without shock (This is done by keeping the relative velocity, at inlet of the runner, tangential to the vane angle).
2. allow the water to flow over them, without forming eddies.
3. allow the required quantity of water to enter the turbine. (This is done by adjusting the opening of the vanes).

All the guide vanes can rotate about their respective pivots, which are connected to the regulating ring by some mechanical means. The regulating ring is connected to the regulating shaft by means of two regulating rods. The guide vanes may be closed or opened by rotating the regulating shaft, thus allowing the required quantity of water to flow according to the need. The regulating shaft is operated by means of a governor, whose function is to govern the turbine (*i.e.*, to keep the speed constant at varying loads). The guide vanes are generally made of cast steel.

33·5 Turbine Runner

The runner of a reaction turbine consists of runner blades fixed either to a shaft or rings, depending upon the type of turbine. The blades are properly designed, in order to allow the water to enter and leave the runner without shock.

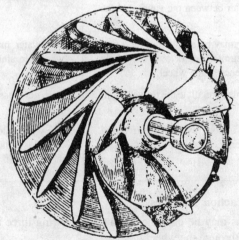

Fig. 33·2. Turbine runner.

The runner is keyed to a shaft, which may be vertical or horizontal. If the shaft is vertical, it is called a vertical turbine. Similarly, if the shaft is horizontal, it is called a horizontal turbine.

33·6 Draft Tube

The water, after passing through the runner, flows down through a tube called draft tube. It is, generally, drowned approximately 1 m below the tail race level. A draft tube has the following functions:

1. It increases the head of water by an amount equal to height of the runner outlet above the tail race.
2. It increases efficiency of the turbine.

33·7 Difference between an Impulse Turbine and a Reaction Turbine

Following are the few points of difference between a reaction turbine and an impulse turbine:

S. No.	Impulse turbine	Reaction turbine
1.	The entire available energy of the water, is first converted into kinetic energy.	The available energy, of the water, is not converted from one form to another.
2.	The water flows through the nozzles and impinges on the buckets, which are fixed to the outer periphery of the wheel.	The water is guided by the guide blades to flow over the moving vanes.
3.	The water impinges on the buckets, with kinetic energy.	The water glides over the moving vanes, with pressures energy.
4.	The pressure of the flowing water remains unchanged, and is equal to the atmospheric pressure.	The pressure of the flowing water is reduced after gliding over the vanes.
5.	It is not essential that the wheel should run full. Moreover, there should be free access of air between the vanes and the wheel.	It is essential that the wheel should always run full, and kept full of water.
6.	The water may be admitted over a part of the circumference or over the whole circumference of the wheel.	The water must be admitted over the whole circumference of the wheel.
7.	It is possible to regulate the flow without loss.	It is not possible to regulate the flow without loss.
8.	The work is done by the change in the kinetic energy of the jet.	The work is done partly by the change in the velocity head, but almost entirely by the change in pressure head.

33·8 Classification of Reaction Turbines

The reaction turbines may be classified into the following three types, depending upon the direction of flow of water through the wheel:

1. Radial flow turbines. 2. Axial flow turbines. 3. Mixed flow turbines.

Now in the following pages, we shall discuss all the above mentioned three types of reaction turbines.

Reaction Turbines

33·9 Radial Flow Turbines

In such turbines, the flow of water is radial (*i.e.*, along the radius of the wheel). The radial flow turbines may be further sub-division into the following two classes :

1. *Inward flow turbines.* In such turbines, the water enters the wheel at the outer periphery, and then flows inwards (*i.e.*, towards the centre of the wheel).
2. *Outward flow turbines.* In such turbines, the water enters at the centre of the wheel, and then flows outwards (*i.e.*, towards the outer periphery of the wheel).

33·10 Axial Flow Turbines

In such turbines, the water flows parallel to the axis of the wheel. Such turbines are also called parallel flow turbines.

33·11 Mixed Flow Turbines

These are the latest types of turbines, in which the flow is partly radial and partly axial.

33·12 Inward Flow Reaction Turbines

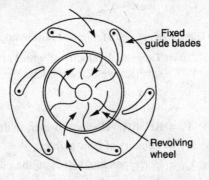

Fig. 33·3. Inward flow reaction turbine.

An inward flow reaction turbine, as the name indicates, is that reaction turbine in which the water enters the wheel at the outer periphery and then flows inwards over the vanes (*i.e.*, towards the centre of the wheel) as shown in Fig. 33·3.

An inward flow reaction turbine, in its simplest form, consists of fixed guide blades, which guide the water to enter into the revolving wheel at correct angle. *i.e.*, for the shockless entry of water. (This is done by adjusting the vane angle tangentially to the relative velocity of the water and the revolving wheel.) The water, while gliding over the vanes, exerts some force on the revolving wheel, to which the vanes are fixed. This force causes the revolving wheel to revolve.

It may be noted that whenever the load on the turbine is decreased, it causes the shaft to rotate at a higher speed. The centrifugal force, which increases due to the higher speed, tends to reduce the quantity of water flowing over the vanes, and thus the velocity of water at the entry is also reduced. It will ultimately tend to reduce the power produced by the turbine. This is the advantage of an inward flow reaction turbine, that it adjusts automatically according to the required load on the turbine. The highest efficiency is obtained, when the velocity of the leaving water is as small as possible.

Now the work done or any other detail of the turbine runner may be found out by drawing the inlet and outlet velocity triangles, as usual as shown in Fig. 33·4.

Let
V = Absolute velocity of the entering water,
D = Outer diameter of the wheel,
N = Revolutions of the wheel per minute,

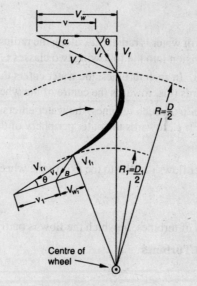

Fig. 33·4. Triangle of velocities for inward flow reaction turbine.

v = Tangential velocity of wheel at inlet (also known as spherical velocity at inlet).

$$= \frac{\pi DN}{60}$$

V_r = Relative velocity of water, to the wheel, at inlet,
V_f = Velocity of flow at inlet,
$V_1, D_1, v_1, V_{r1}, V_{f1}$ = Corresponding values at the outlet.
α = Angle, at which the water enters the wheel (also known as guide blade angle),
β = Angle, at which the water leaves the wheel,
θ = Angle of the blade tip at inlet (also known as vane angle at inlet),
ϕ = Angle of the blade tip at outlet (also known as vane angle at outlet),
H = Total head of water, under which the turbine is working,
W = Weight of the water entering the wheel in kN/s.

From the inlet triangle, we find that,

$$V_w = V \cos \alpha \quad \text{and} \quad V_f = V \sin \alpha$$

and from the outlet triangle, we find that

$$V_{w1} = V_1 \cos \beta \quad \text{and} \quad V_{f1} = V_1 \sin \beta$$

We know that the force per kN of water

$$= \frac{1}{g} \times \text{Change of velocity of whirls}$$

$$= \frac{1}{g} \left(V_w + V_{w1} \right)$$

(In this case, V_{w1} is negative, because V_{w1} is in the opposite direction as that of V_w).

Reaction Turbines

We know that work done per kN of water

= Force × Distance
= (Velocity of whirl at inlet × Tangential velocity of wheel at inlet) − (Velocity of whirl at outlet × Tangential velocity of wheel at outlet)

$$= \frac{1}{g}(V_w \cdot v - V_{w1} \cdot v_1) = \frac{V_w \cdot v}{g} - \frac{V_{w1} \cdot v_1}{g} \qquad ...(i)$$

Notes. 1. If there is no loss of energy, then

$$\frac{V_w \cdot v}{g} - \frac{V_{w1} \cdot v_1}{g} = H - \frac{V_1^2}{2g}$$

2. If the discharge of the turbines is radial, then

$$\beta = 90°; V_{w1} = 0 \text{ and } V_1 = V_{f1}$$

∴ Work done per kN of water

$$= \frac{V_w \cdot v}{g}$$

and work done

$$\frac{V_w \cdot v}{g} = H - \frac{V_1^2}{2g} = H - \frac{V_{f1}^2}{2g}$$

3. If the vanes are radial at inlet, outlet or both, then the velocity of whirl at that tip is zero.

Example 33·1. *An inward flow reaction turbine, having an external diameter of 1·5 metre runs at 400 r.p.m. The velocity of flow at inlet is 10 m/s. If the guide blade angle is 15°, find (a) absolute velocity of water, (b) velocity of whirl at inlet, (c) inlet vane angle of the runner, and (d) relative velocity at inlet.*

Solution. Given: $D = 1·5$ m; $N = 400$ r.p.m.; $V_f = 10$ m/s and $\alpha = 15°$

(a) Absolute velocity of water

From the inlet velocity triangle, we find that absolute velocity of water

$$V = \frac{V_f}{\sin 15°} = \frac{10}{0·2588} \text{ m/s}$$

= 38·64 m/s **Ans.**

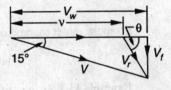

Fig. 33·5.

(b) Velocity of whirl at inlet

From the inlet velocity triangle, we find that the velocity of whirl at inlet,

$$V_w = V \cos 15° = 38·64 \times 0·9659 = 37·32 \text{ m/s}$$

(c) Inlet vane angle of the runner

Let θ = Inlet vane angle of the runner.

We know that the velocity of vane at inlet,

$$v = \frac{\pi D N}{60} = \frac{\pi \times 1·5 \times 400}{60} = 31·42 \text{ m/s}$$

and from the inlet velocity triangle, we find that

$$\tan \theta = \frac{V_f}{V_w - v} = \frac{10}{37·32 - 31·42} = 1·695 \quad \text{or} \quad \theta = 59·5° \text{ **Ans.**}$$

(d) Relative velocity at inlet

From the inlet velocity triangle, we also find that relative velocity at inlet,

$$V_r = \frac{V_f}{\sin 59·5°} = \frac{10}{0·8616} = 11·61 \text{ m/s} \text{ **Ans.**}$$

Example 33·2. *An inward flow reaction turbine has outer and inner diameters of the wheel as 1 metre and 0.5 metre respectively. The vanes are radial at inlet and the discharge is radial at outlet and the water enters the vanes at an angle of 10°. Assuming the velocity of flow to be constant and equal to 3 m/s, find (a) speed of the wheel, and (b) vane angle at outlet.*

Solution. Given : $D = 1$ m; $D_1 = 0.5$ m; $\alpha = 10°$ and $V_f = V_{f1} = 3$ m/s.

(i) Speed of the wheel

Let $\qquad N = $ Speed of the wheel.

Since the vanes are radial at inlet and outlet, therefore velocities of whirl at inlet and outlet will be zero. And the shapes of the two triangles will be as shown in Fig. 33·6.

From the inlet triangle of velocities, we find that tangential velocity of wheel at inlet,

$$v = \frac{V_f}{\tan 10°} = \frac{3}{0.1763} = 17 \text{ m/s}$$

We also know that the tangential velocity of wheel at inlet (v)

$$17 = \frac{\pi DN}{60} = \frac{\pi \times 1 \times N}{60} = 0.0524 \, N$$

$\therefore \qquad N = 17/0.0524 = 324.4$ r.p.m. **Ans.**

(ii) Vane angle at outlet

Let $\qquad \phi = $ Vane angle at outlet,

We know that the tangential velocity of wheel at outlet,

$$v_1 = \frac{\pi D_1 N}{60} = \frac{\pi \times 0.5 \times 324.4}{60} = 8.5 \text{ m/s}$$

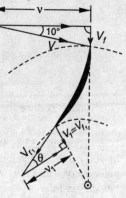

Fig. 33·6.

From the outlet triangle of velocities, we find that

$$\tan \phi = \frac{V_{f1}}{v_1} = \frac{3}{8.5} = 0.3529 \quad \text{or} \quad \phi = 19.4° \text{ **Ans.**}$$

Example 33·3. *An inward flow reaction turbine is supplied water at the rate of 600 litres/second with a velocity of flow of 6 m/s. The velocity of periphery and velocity of whirl at inlet is 24 m/s and 18 m/s respectively. Assuming the discharge to be radial at outlet, and the velocity of flow to be constant, find*

(1) vane angle at inlet, and

(2) head of water on the wheel.

Solution. Given : $Q = 600$ litres/s $= 0.6$ m³/s; $V_f = 6$ m/s; $v = 24$ m/s; $V_w = 18$ m/s and $V_{f1} = V_f$.

(1) Vane angle at inlet

Let $\qquad \theta = $ Vane angle at inlet.

From the inlet triangle of velocities, we find that

$$\tan (180° - \theta) = \frac{V_f}{v - V_w} = \frac{6}{24 - 18} = 1.0$$

$\therefore \qquad 180° - \theta = 45° \quad \text{or} \quad \theta = 135°$ **Ans.**

(2) Head of water on the wheel

Let $\qquad H = $ Head of water on the wheel.

We know that $\qquad \dfrac{V_w \cdot v}{g} = H - \dfrac{V_1^2}{2g}$

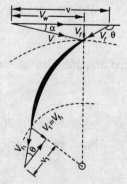

Fig. 33·7.

Reaction Turbines

or
$$\frac{18 \times 24}{9 \cdot 81} = H - \frac{V_{f1}^2}{2g} \qquad \ldots (\because V_1 = V_{f1})$$

$$44 \cdot 0 = H - \frac{V_f^2}{2g} = H - \frac{(6)^2}{2 \times 9 \cdot 81} = H - 1 \cdot 8 \qquad \ldots (\because V_{f1} = V_f)$$

∴ $H = 44 + 1 \cdot 8 = 45 \cdot 8$ m **Ans.**

Example 33·4. *An inward flow reaction turbine is working under a head of 25 metres and running at 300 revolution per minute. The velocity of periphery of the wheel is 30 m/s and velocity of flow is 4 m/s. If the hydraulic losses are 20% of the available head, and the discharge is radial, find :*

(a) *guide blade at inlet,*

(b) *wheel angle at inlet, and*

(c) *diameter of the wheel.*

Solution. Given : $H = 25$ m; $N = 300$ r.p.m., $v = 30$ m/s; $V_f = 4$ m/s and hydraulic losses = 20% of the available head = $0 \cdot 20 \times 25 = 5$ m

(a) *Guide blade angle at inlet*

Let α = Guide blade angle at inlet.

Since the discharge is radial, therefore velocity of whirl at outlet is zero.

We know that
$$\frac{V_w \cdot v}{g} = H - \frac{V_1^2}{2g} \quad \text{or} \quad \frac{V_w \times 30}{9 \cdot 81} = 25 - 5$$

$3 \cdot 06 \ V_w = 20 \quad$ or $\quad V_w = 20/3 \cdot 06 = 6 \cdot 54$ m/s

Since V_w (6·54) is less than v (30), therefore shape of the inlet triangle will be as shown in Fig. 33·8.

Now from the inlet triangle, we find

$$\tan \alpha = \frac{V_f}{V_w} = \frac{4}{6 \cdot 54} = 0 \cdot 6116$$

or $\alpha = 31 \cdot 4°$ **Ans.**

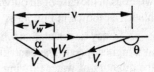

Fig. 33·8.

(b) *Wheel angle at inlet*

Let θ = Wheel angle at inlet.

From the inlet triangle of velocities, we also find that

$$\tan (180° - \theta) = \frac{V_f}{v - V_w} = \frac{4}{30 - 6 \cdot 54} = 0 \cdot 1705$$

∴ $\tan (180° - \theta) = 9 \cdot 7°$

or $\theta = 170 \cdot 3°$ **Ans.**

Diameter of the wheel

Let D = Diameter of the wheel.

We know that the velocity of periphery at inlet (v),

$$30 = \frac{\pi DN}{60} = \frac{\pi \times D \times 300}{60} = 15 \cdot 7 \ D$$

∴ $D = 30/15 \cdot 7 = 1 \cdot 91$ m **Ans.**

---

* Hydraulic losses = $\dfrac{V_1^2}{2g}$

33·13 Outward Flow Reaction Turbines

An outward flow reaction turbine, as the name indicates, is that reaction turbine, in which the water enters at the centre of the wheel and then flows outwards over the vanes (*i.e.*, towards the outer periphery of the wheel) as shown in Fig. 33·9.

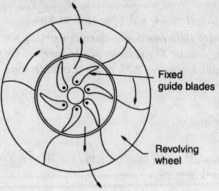

Fig. 33·9. Outward flow reaction turbine.

An outward flow reaction turbine, in its simplest form, consists of fixed guide blades, which guide the water to enter into the revolving wheel at correct angle. *i.e.*, for shockless entry of water (This is done by adjusting the vane angle tangentially to the relative velocity of water and the revolving wheel). The water, while gliding over the vanes, exerts some force on the revolving wheel to which the vanes are fixed. This force causes the revolving wheel to revolve. The only difference, between the inward flow reaction turbine and an outward flow reaction turbine, is that in case of an inward flow reaction turbine, the revolving wheel is inside the fixed guide blades as shown in Fig. 33·3; whereas in the case of an outward flow reaction turbine, the revolving wheel is outside the fixed guide blades as shown in Fig. 33·9.

It may be noted, that whenever the load on the turbine is decreased, it causes the shaft to rotate at a higher speed. The centrifugal force, which increases due to the higher speed, tends to increase the quantity of water flowing over the vanes, and thus the wheel tends to run faster and faster. It is

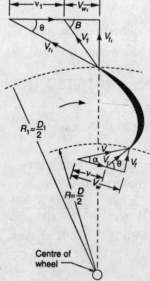

Fig. 33·10. Triangle of velocities for outward flow reaction turbine.

Reaction Turbines

the only disadvantage of an outward flow reaction turbine. Thus every outward flow reaction turbine has to be governed by a turbine governor.

All the notations for an outward reaction turbine are the same as those for inward flow reaction turbines. The inner diameter of the wheel will be denoted as D (i.e., diameter at inlet) and the outer diameter will be denoted as D_1 (i.e., diameter at outlet). All the relations, for finding out the various angles and other data will hold good for an outward turbine also.

The efficiency or the power developed by the turbine may be found out, by drawing the inlet and outlet velocity triangles as usual as shown in Fig. 33·10.

Example 33·5. *An outward flow reaction turbine has inner and outer diameters of the wheel as 1 metre and 2 metres respectively. The water enters the vanes at an angle of 20° and leaves the vanes radially. If the velocity of flow remains constant at 10 m/s and the speed of the wheel be 300 r.p.m., find the vane angles at inlet and outlet.*

Solution. Given : $D = 1$ m; $D_1 = 2$ m; $\alpha = 20°$; $V_f = V_{f1} = 10$ m/s and $N = 300$ r.p.m.

Vane angle at inlet

Let θ = Vane angle at inlet.

We know that velocity of periphery at inlet,

$$v = \frac{\pi DN}{60} = \frac{\pi \times 1 \times 300}{60} = 15.71 \text{ m/s}$$

From inlet triangle of velocities, we find that the velocity of whirl at inlet,

$$V_w = \frac{V_f}{\tan 20°} = \frac{10}{0.364} = 27.5 \text{ m/s}$$

and $\tan \theta = \dfrac{V_f}{V_w - v} = \dfrac{10}{27.5 - 15.71}$

$= 0.8482$ or $\theta = 40.3°$ **Ans.**

Vane angle at outlet

Let ϕ = Vane angle at outlet

We know that the velocity of periphery at outlet,

$$v_1 = \frac{\pi D_1 N}{60} = \frac{\pi \times 2 \times 300}{60} = 31.42 \text{ m/s}$$

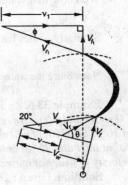

Fig. 33·11.

From the outlet triangle of velocities, we find that

$$\tan \phi = \frac{V_{f1}}{v_1} = \frac{10}{31.42} = 0.3183 \quad \text{or} \quad \phi = 17.7° \text{ Ans.}$$

EXERCISE 33·1

1. An inward flow reaction turbine having external diameter of 500 mm is running at 200 r.p.m. The guide blade angle is 30°. If the velocity of flow at inlet is 3 m/s, find the absolute velocity of water and velocity of whirl at inlet. [**Ans.** 8·77 m/s; 8·24 m/s]

2. An inward flow reaction turbine has internal and external diameters as 900 mm and 450 mm respectively. The turbine is running at 200 r.p.m. and the velocity of flow through the runner is constant and equal to 1·8 m/s. If the guide blades make an angle of 10° to the tangent at the wheel, find the relative velocity at inlet and runner blade angles. [**Ans.** 1·96 m/s; 66·5°; 20·9°]

3. The external and internal diameters of an inward flow reaction turbine are 2 m and 1 m respectively. The turbine is running at 192 r.p.m. The guide blade angle is 10° and velocity of flow at inlet and outlet is 5 m/s. Draw velocity triangles at inlet and outlet and find

    (a) Vane angles at inlet and outlet,

    (b) Absolute velocity of water leaving the guide vane. [**Ans.** 20·9°; 25·5°; 28·8 m/s]

4. An outward flow reaction turbine with outer and inner diametres as 1·2 m and 0·6 m respectively has guide blade of 25°. Determine the vane angles at inlet and outlet. Take velocity of flow constant at 4 m/s and speed of the impeller as 200 r.p.m. [**Ans.** 60·5°; 17·6°]

31·14. Discharge of a Reaction Turbine

The discharge of a reaction turbine may be found out either from the gross energy supplied to the turbine or from the actual velocity of flow at inlet or outlet as discussed below.

1. From the gross energy supplied to the turbine

Let H = Head of water supplied in metres, and
Q = Discharge through the turbine in m³/s.

Then the gross power supplied to the turbine,
$$= wQH \text{ kW}$$

2. From the velocity of flow

Let V_f = Velocity of flow at inlet,
D = Diameter of the wheel at inlet, and
b = Breadth of the wheel at inlet.

We know that water entering the wheel,
$$Q = \pi DB V_f \qquad \ldots(i)$$

Similarly, water leaving the wheel,
$$Q = \pi D_1 b_1 V_{f1} \qquad \ldots(ii)$$

Note. Since the water entering the wheel is equal to the water leaving the wheel, therefore
$$\pi Db V_f = \pi D_1 b_1 V_{f1}$$

Example 33·6. *An inward flow reaction turbine has external and internal wheel diameters as 1·0 metre and 0·5 metre respectively. The water enters the wheel with a velocity of 30 m/s at an angle of 10°. The width of wheel at inlet and outlet is 150 mm and 300 mm respectively. The vane angle is 90° at inlet and 25° at outlet. Determine : (i) tangential velocity of runner at inlet, and (ii) absolute velocity of water at outlet.*

Solution. Given : $D = 1$ m; $D_1 = 0.5$ m; $V = 30$ m/s; $\alpha = 10°$; $b = 150$ mm $= 0.15$ m; $b_1 = 300$ mm $= 0.3$ m; $\theta = 90°$ and $\phi = 25°$

(i) Tangential velocity of runner at inlet

From the inlet triangle of velocities, we find that tangential velocity of runner at inlet,
$$v = V \cos \alpha = 30 \cos 10°$$
$$= 30 \times 0.9848 = 29.5 \text{ m/s Ans.}$$

(ii) Absolute velocity of water at outlet

From the inlet triangle, we also find that the velocity of flow at inlet,
$$V_f = V \sin \alpha = 30 \sin 10°$$
$$= 30 \times 0.1736 = 5.21 \text{ m/s}$$

and tangential velocity of runner at outlet,
$$v_1 = v \times \frac{D_1}{D} = 29.5 \times \frac{0.5}{1.0} = 14.75 \text{ m/s}$$

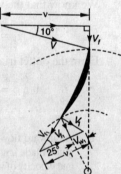

Fig. 33·12.

Since the discharge of water at inlet and outlet is equal, therefore
$$\pi Db V_f = \pi D_1 b_1 V_{f1}$$
or
$$\pi \times 1.0 \times 0.15 \times 5.21 = \pi \times 0.5 \times 0.3 \times V_{f1}$$
∴
$$V_{f1} = 0.7815/0.15 = 5.21 \text{ m/s}$$

Reaction Turbines

Now from the outlet triangle of velocities, we find that

$$\tan \phi = \frac{V_{f1}}{v_1 - V_{w1}} \quad \text{or} \quad \tan 25° = \frac{5.21}{14.75 - V_{w1}}$$

or

$$14.75 - V_{w1} = \frac{5.21}{\tan 25°} = \frac{5.21}{0.4663} = 11.17$$

$$\therefore V_{w1} = 14.75 - 11.17 = 3.58 \text{ m/s}$$

and absolute velocity of water at outlet,

$$V_1 = \sqrt{V_{f1}^2 + V_{w1}^2} = \sqrt{(5.21)^2 + (3.58)^2} = 6.32 \text{ m/s} \quad \textbf{Ans.}$$

Example 33·7. *An outward flow reaction turbine has internal and external diameters of 2·4 metres and 3·0 metres respectively. The turbine has a radial discharge of 6 m³/s, and is running at 200 r.p.m. The total head, on the turbine, is 40 metres and width of the wheel at inlet and outlet is 300 mm. Neglecting thickness of the vanes, find (i) velocity of flow at inlet, (ii) velocity of flow at outlet, and (iii) velocity of whirl at inlet.*

Solution. Given : $D = 2.4$ m; $D_1 = 3.0$ m; $Q = 6$ m³/s; $N = 200$ r.p.m.; $H = 40$ m and $b = b_1 = 300$ mm $= 0.3$ m

(i) Velocity of flow at inlet

Let V_f = Velocity of flow at inlet.

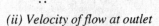

We know that the discharge through the vane at inlet (Q),

$$6 = \pi D b V_f = \pi \times 2.4 \times 0.3 \times V_f = 2.26 \, V_f$$

$$\therefore V_{f1} = 6/2.26 = 2.65 \text{ m/s} \quad \textbf{Ans.}$$

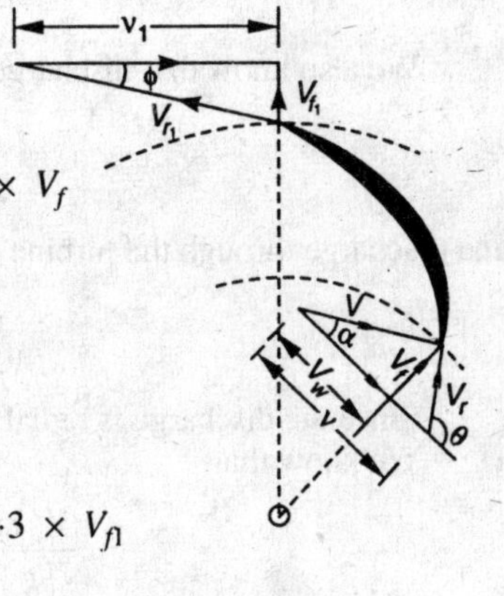

(ii) Velocity of flow at outlet

Let V_{f1} = Velocity of flow at outlet.

We know that discharge through the vanes at outlet (Q),

$$6 = \pi D_1 b_1 V_{f1} = \pi \times 3.0 \times 0.3 \times V_{f1} = 2.83 \, V_{f1}$$

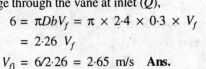

$$\therefore V_f = 6/2.83 = 2.12 \text{ m/s} \quad \textbf{Ans.}$$

Fig. 33·13.

(iii) Velocity of whirl at inlet

Let V_w = Velocity of whirl at inlet.

We know that the velocity of periphery at inlet,

$$v = \frac{\pi D N}{60} = \frac{\pi \times 2.4 \times 200}{60} = 25.13 \text{ m/s}$$

Since the discharge is radial, therefore velocity of whirl at the outlet is zero. We know that

$$\frac{V_w \cdot v}{g} = H - \frac{V_1^2}{2g} \quad \text{or} \quad \frac{V_w \times 25.13}{9.81} = 40 - \frac{(2.12)^2}{2 \times 9.81}$$

$$\ldots (\because V_1 = V_{f1})$$

$$\therefore 2.56 \, V_w = 39.8 \quad \text{or} \quad V_w = 39.8/2.56 = 15.5 \text{ m/s} \quad \textbf{Ans.}$$

Example 33·8. *An outward flow reaction turbine, running at 200 r.p.m., is supplied water at the rate of 5000 litres/s, under a head of 40 metres. The internal and external diameters of the wheel are 2 metres and 2·5 metres respectively. The wheel width at inlet and outlet is 200 mm. Assuming the discharge to be radial, determine the angles of the turbine at inlet and outlet.*

Solution. Given : $N = 200$ r.p.m.; $Q = 5000$ litres/s $= 5$ m^3/s; $H = 40$ m; $D = 2$ m; $D_1 = 2\cdot 5$ m and $b = b_1 = 200$ mm $= 0\cdot 2$ m

Guide blade angle at inlet

Let α = Guide blade angle at inlet.

We know that the velocity of periphery inlet,

$$v = \frac{\pi DN}{60} = \frac{\pi \times 2 \times 200}{60} \text{ m/s}$$

$$= 20\cdot 9 \text{ m/s}$$

and velocity of periphery at outlet,

$$v_1 = \frac{\pi D_1 N}{60} = \frac{\pi \times 2\cdot 5 \times 200}{60} \text{ m/s}$$

$$= 26\cdot 2 \text{ m/s}$$

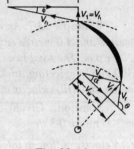

Fig. 33·14.

We also know that discharge through the turbine at inlet, (Q),

$$5 = \pi Db V_f = \pi \times 2 \times 0\cdot 2 \times V_f = 1\cdot 257\, V_f$$

∴ $V_f = 5/1\cdot 257 = 3\cdot 98$ m/s

and discharge through the turbine at outlet (Q),

$$5 = \pi D_1 b_1 V_{f1} = \pi \times 2\cdot 5 \times 0\cdot 2 \times V_{f1} = 1\cdot 571\, V_{f1}$$

∴ $V_{f1} = 5/1\cdot 571 = 3\cdot 18$ m/s

Since the discharge is radial at outlet, therefore velocity of whirl at outlet is zero.
We know that

$$\frac{V_w \cdot v}{g} = H - \frac{V_1^2}{2g} \quad \text{or} \quad \frac{V_w \times 20\cdot 9}{9\cdot 81} = 40 - \frac{(3\cdot 18)^2}{2 \times 9\cdot 81} \quad (\because V_1 = V_{f1})$$

$$2\cdot 13\, V_w = 39\cdot 5 \quad \text{or} \quad V_w = 39\cdot 5/2\cdot 13 = 18\cdot 5 \text{ m/s}$$

Since V_w (18·5) is less than v (20·9), therefore shape of the inlet triangle will be whown in Fig. 33·14. Now from the inlet triangle of velocities, we find that

$$\tan \alpha = \frac{V_f}{V_w} = \frac{3\cdot 98}{18\cdot 5} = 0\cdot 2151 \quad \text{or} \quad \alpha = 12\cdot 1° \text{ Ans.}$$

Vane angle at inlet

Let θ = Vane angle at inlet.

From the inlet triangle of velocities, we also find that

$$\tan (180° - \theta) = \frac{V_f}{v - V_w} = \frac{3\cdot 98}{20\cdot 9 - 18\cdot 5} = 1\cdot 6583$$

∴ $(180° - \theta) = 58\cdot 9°$ or $\theta = 121\cdot 9°$ **Ans.**

Vane angle at outlet

Let ϕ = Vane angle at outlet.

From the outlet triangle of velocities, we find that

$$\tan \phi = \frac{V_{f1}}{v_1} = \frac{3\cdot 18}{26\cdot 2} = 0\cdot 1214 \quad \text{or} \quad \phi = 6\cdot 92° \text{ **Ans.**}$$

Reaction Turbines

EXERCISE 33·2

1. An inward flow reaction turbine has external and internal diameters of 0·8 m and 0·4 m respectively. If the velocity of flow through the runner is constant and equal to 1·5 m/s, find the discharge through the runner and width of the turbine at inlet, if the width of the turbine at outlet is 200 mm.

   [**Ans.** 0·754 m^3/s; 400 mm]

2. An inward flow through reaction turbine has external diameter of 675 mm and corresponding width 150 mm. The similar dimensions at the inside of the runner are 500 mm and 225 mm respectively. The effective head is 21 m and velocity of flow at inlet is 3 m/s. If the guide blade angle is 12°, find the discharge through the turbine and speed of the impeller. The vanes are radial at inlet.

   [**Ans.** 0·954 m^3/s; 399·2 r.p.m.]

3. An inward flow reaction turbine, working under a head of 15 metres is running with a speed of 375 r.p.m., with a supply of 550 litres/s. The outer and inner diameters are 750 mm and 500 mm respectively. The water is discharging radially at outlet with a velocity of 3 m/s. Assuming width of the wheel to be constant, find the guide blade angle at inlet and vane angle at inlet. [**Ans.** 10·9°; 21·6°]

33·15 Power Produced by a Reaction Turbine

We have seen in the previous articles, that some work is done per kN of water, when it flows over the vanes. The power produced by the turbine is given by the relation:

$$P = wQH \text{ kW}$$

where
- w = Specific weight of water,
- Q = Discharge of the turbine in m^3/s
- H = Head of water in metres.

Note : The power produced by a reaction turbine may also be found out from the relation :

P = Quantity of water flowing over the vane is kN × Work done per kN of water

33·16 Efficiencies of a Reaction Turbine

In general, the term efficiency may be defined as the ratio of work done to the energy supplied. Following are the three types of efficiencies of a turbine :

1. Hydraulic efficiency.   2. Mechanical efficiency.   3. Overall efficiency.

33·17 Hydraulic Efficiency

It is the ratio of work done on the wheel to the head of water (or energy) actually supplied to the turbine. *i.e.*,

$$\eta_h = \frac{\text{Work done per kN of water}}{H} = \frac{\dfrac{V_w \cdot v}{g} - \dfrac{V_{w1} \cdot v_1}{g}}{H}$$

If the discharge through the wheel is radial, then the velocity of whirl at outlet is zero. *i.e.*, $V_{w1} = 0$

$$\therefore \quad \eta_h = \frac{V_w \cdot v}{gH}$$

33·18 Mechanical Efficiency

It is the ratio the actual work available at the turbine to the energy imparted to the wheel. We know that total energy imparted to the wheel (in case of radial discharge)

= Weight of water in kN × Energy imparted per kN of water

$$= wQ \times \frac{V_w \times v}{g}$$

∴ Mechanical efficiency $\eta_m = \dfrac{P}{wQ \times \dfrac{V_w \times v}{g}}$

where P = Power available at the turbine.

33·19 Overall Efficiency

It is a measure of the performance of a turbine and is the ratio of power produced by the turbine to the energy actually supplied to the turbine i.e.,

$$\eta_0 = \eta_h \times \mu_m = \dfrac{V_w \cdot v}{gH} \times \dfrac{P}{wQ \times \dfrac{V_w \times v}{g}} = \dfrac{P}{wQH}$$

where P = Power available at the turbine

Example 33·9. *An inward flow reaction turbine has tangential velocity of runner, velocity of flow and velocity whirl at inlet as 30 m/s, 3 m/s and 24 m/s respectively. Assuming the discharge to be radial at outlet and hydraulic efficiency as 78%, determine the total head on the turbine and the inlet vane angle.*

Solution. Given : $v = 30$ m/s; $V_f = 3$ m/s; $V_w = 24$ m/s and $\eta_h = 78\% = 0.78$.

Total head on the turbine

Let H = Total head on the turbine.

Since the discharge is radial at outlet, therefore velocity of whirl at outlet is zero.

We know that hydraulic efficiency (η_h)

$$0.78 = \dfrac{V_w \cdot v}{gH} = \dfrac{24 \times 30}{9.8\ H} = \dfrac{73.4}{H}$$

∴ $H = 73.4/0.78 = 94.1$ m **Ans.**

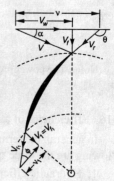

Inlet vane angle

Let θ = Inlet vane angle.

From the inlet triangle of velocities, we find that

$$\tan(180° - \theta) = \dfrac{V_f}{v - V_w} = \dfrac{3}{30 - 24} = 0.5$$

∴ $(180 - \theta) = 26.6°$ or $\theta = 153.4°$ **Ans.**

Fig. 33·15.

Example 33·10. *An inward flow reaction turbine, working under a head of 8 metres, has guide blade angle 25° and vane angle at inlet 105°. Assuming the velocity of flow to be constant and radial discharge, determine hydraulic efficiency of the turbine.*

Solution. Given : $H = 8$ m; $\alpha = 25°$; $\theta = 105°$ and $V_f = V_{f1}$

Let V = Absolute velocity of water at inlet.

From the inlet triangle of velocities, we find that velocity of whirl at inlet,

$$V_w = V \cos 25° = 0.9063\ V$$

Similarly, velocity of low at inlet,

$$V_w = V \sin 25° = 0.4226\ V$$

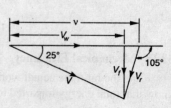

Fig. 33·16

Reaction Turbines

and tangential velocity at inlet,

$$v = V_w + \frac{V_t}{\tan 75°} = 0.9063\ V + \frac{0.4226\ V}{3.7321} = 1.0195\ V$$

Since the discharge is radial at outlet, therefore velocity of whirl at outlet is zero.

$$\therefore \quad \frac{V_w \cdot v}{g} = H - \frac{V_1^2}{2g} \quad \text{or} \quad \frac{0.9063\ V \times 1.0195\ V}{9.81} = H - \frac{V_{f1}^2}{2g}$$

$$\frac{0.924\ V^2}{9.81} = H - \frac{V_f^2}{2g} = 8 - \frac{(0.4226\ V)^2}{2 \times 9.81} \quad\quad \ldots (\because V_1 = V_{f1})$$

$$\therefore \quad \frac{1.013\ V^2}{9.81} = 8 \quad \text{or} \quad V = \sqrt{\frac{8 \times 9.81}{1\,013}} = 8.8\ \text{m/s}$$

Substituting this value of V, we find that velocity of whirl at inlet

$$V_w = 0.9063\ V = 0.9063 \times 8.8 = 7.98\ \text{m/s}$$

and tangential velocity at inlet,

$$v = 1.0195\ V = 1.0195 \times 8.8 = 8.97\ \text{m/s}$$

$\therefore$ Hydraulic efficiency,

$$\eta_h = \frac{V_w \cdot v}{gH} = \frac{7.98 \times 8.97}{9.81 \times 8} = 0.912 = 91.2\%\ \text{Ans.}$$

Example 33·11. *Show that in an inward flow reaction turbine, with radial vanes at inlet and outlet, the hydraulic efficiency is given by :*

$$\eta_h = \frac{2}{2 + \tan^2 \alpha}$$

where α is the guide blade angle. Assume the velocity of flow to be constant.

Solution. Given : Radial vanes at inlet and outlet.

From the geometry of the inlet and outlet triangles of velocities, we find that tangential velocity at inlet,

$$v = V_{w1}$$

and as the velocity of flow is constant, therefore velocity of flow at outlet,

$$V_{f1} = V_f$$

Since the discharge is radial, therefore velocity of whirl at the outlet is zero. Therefore,

$$\frac{V_w \cdot v}{g} = H - \frac{V_1^2}{2g} = H - \frac{V_{f1}^2}{2g} = H - \frac{V_f^2}{2g}$$

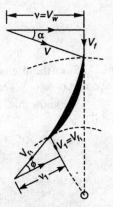

Fig. 33·17.

$$= H - \frac{(v \tan \alpha)^2}{2g} \quad\quad \ldots (\because V_f = v \tan \alpha)$$

$$\therefore \quad H = \frac{V_w \cdot v}{g} + \frac{v^2 \tan^2 \alpha}{2g} = \frac{v^2}{g} + \frac{v^2 \tan^2 \alpha}{2g} \quad\quad \ldots (\because V_w = v)$$

$$= \frac{v^2}{g}\left(1 + \frac{\tan^2 \alpha}{2}\right) = \frac{v^2}{g}\left(\frac{2 + \tan^2 \alpha}{2}\right) \quad\quad \ldots(i)$$

We also know that hydraulic efficiency.

$$\eta_h = \frac{V_w \cdot v}{gH} = \frac{v^2}{gH} \qquad \ldots (\because V_w = v)$$

Substituting the value of H from equation (i),

$$\eta_h = \frac{v^2}{g \times \dfrac{v^2}{g}\left(\dfrac{2 + \tan^2 \alpha}{2}\right)} = \frac{2}{2 + \tan^2 \alpha} \quad \textbf{Ans.}$$

Example 33·12. *An inward flow reaction turbine discharges radially and the velocity of flow is constant and equal to velocity of discharge from the suction tube. Show that the hydraulic efficiency can be expressed by :*

$$\eta_h = \frac{1}{1 - \dfrac{\dfrac{1}{2}\tan^2 \alpha}{\left(\dfrac{\tan \alpha}{\tan \theta}\right)}}$$

where α and θ are the guide and vane angles at inlet.

Solution. Given : Radial discharge at outlet; $V_f = V_{f1}$; α = Guide blade angle at inlet and θ = Vane angle at inlet.

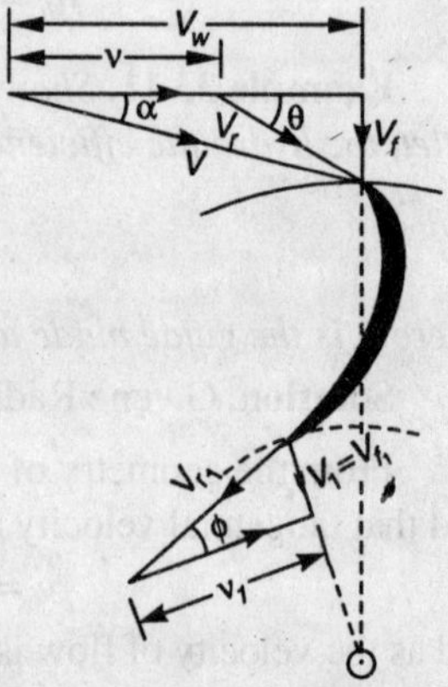

From the inlet triangle, we find that

$$V_f = V_w \tan \alpha \qquad \ldots(i)$$

and

$$v = V_w - \frac{V_f}{\tan \theta} \qquad \ldots(ii)$$

$$= V_w - \frac{V_w \tan \alpha}{\tan \theta}$$

$$= V_w \left(1 - \frac{\tan \alpha}{\tan \theta}\right) \qquad \ldots(iii)$$

Since the discharge is radial at outlet, therefore velocity of whirl at outlet is zero.

We know that

$$\frac{V_w \cdot v}{g} = H - \frac{V_1^2}{2g}$$

Fig. 33·18.

or

$$H = \frac{V_w \cdot v}{g} + \frac{V_1^2}{2g} = \frac{V_w \cdot v}{g} + \frac{V_{f1}^2}{2g} \qquad \ldots (\because V_{f1} = V_1)$$

$$= \frac{V_w \cdot v}{g} + \frac{V_f^2}{2g} \qquad \ldots (\because V_f = V_{f1})$$

Substituting the value of V_f from equation (i),

$$H = \frac{V_w \cdot v}{g} + \frac{V_w^2 \tan^2 \alpha}{2g}$$

and now substituting the value of v from equation (iii),

$$H = \frac{V_w}{g} \times V_w \left(1 - \frac{\tan \alpha}{\tan \theta}\right) + \frac{V_w^2 \tan^2 \alpha}{2g}$$

Reaction Turbines

$$= \frac{V_w^2}{g}\left(1 - \frac{\tan\alpha}{\tan\theta}\right) + \frac{V_w^2 \tan^2\alpha}{2g}$$

$$= \frac{V_w^2}{g}\left(1 - \frac{\tan\alpha}{\tan\theta} + \frac{\tan^2\alpha}{2}\right) \qquad \ldots(iv)$$

∴ Hydraulic efficiency,

$$\eta_h = \frac{V_w \cdot v}{gH} = \frac{V_w \times \left(1 - \frac{\tan\alpha}{\tan\theta}\right)}{g\left[\frac{V_w^2}{g}\left(1 - \frac{\tan\alpha}{\tan\theta} + \frac{\tan^2\alpha}{2}\right)\right]}$$

$$= \frac{1 - \frac{\tan\alpha}{\tan\theta}}{1 - \frac{\tan\alpha}{\tan\theta} + \frac{\tan^2\alpha}{2}} = \frac{1}{1 - \frac{\frac{1}{2}\tan^2\alpha}{\left(\frac{\tan\alpha}{\tan\theta}\right)}} \quad \text{Ans.}$$

Example 33·13. *Find the mean diameter and blade angles at inlet and outlet of an inward flow reaction turbine of the following particulars :*

        Output = 15 000 kW
        Speed = 300 r.p.m.
        Head = 120 metres

$$\frac{\text{Inner diameter}}{\text{Outer diameter}} = 0.6$$

Axial length of blade at inlet = 0·1 × Corresponding diameter
        Flow ratio = 0·15
Hydraulic efficiency, based on velocity triangles
        = 88 per cent
    Overall efficiency = 85 per cent

Assume radial discharge, velocity of flow constant throughout and area blocked by blade thickness as 5 per cent of area of flow.

Solution. Given : $P = 15\ 000$ kW; $N = 300$ r.p.m.; $H = 120$ m; $\frac{D_1}{D_2} = 0.6$ or $D_1 = 0.6\ D$; $b = 0.1 \times D$; $b = 0.1 \times D$; $\frac{V_f}{\sqrt{2gH}} = 0.15$ or $V_f = 0.15\ \sqrt{2gH} = 0.15 \times \sqrt{2 \times 9.81 \times 120} = 7.28$ m/s; $\eta_h = 88\% = 0.88$; $\eta_0 = 85\% = 0.85$; $V_{f1} = V_f = 7.28$ m/s and area blocked by blade thickness = 5%.

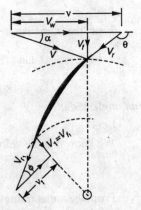

Fig. 33·19

Diameter at inlet
    Let D = Diameter at inlet, and
                            Q = Discharge through the turbine.

We know that overall efficiency (η_0)

$$0.85 = \frac{P}{wQH} = \frac{150\ 000}{9.81 \times Q \times 120} = \frac{12.7}{Q}$$

∴ $Q = 12.7/0.85 = 14.9$ m³/s

We also know that the discharge, through the turbine (Q),

$$14 \cdot 9 = \pi \, Db V_f \times 0 \cdot 95 \quad \ldots(\because 5\% \text{ area is blocked by blade thickness})$$
$$= \pi \times D \times 0 \cdot 1 \, D \times 7 \cdot 28 \times 0 \cdot 95 = 2 \cdot 17 \, D^2$$
$$\therefore \quad D_1^2 = 14 \cdot 9/2 \cdot 17 = 6 \cdot 87 \quad \text{or} \quad D = 2 \cdot 62 \text{ m} \quad \textbf{Ans.}$$

Diameter at outlet

We know that diameter at outlet,
$$D_1 = 0 \cdot 6 \, D = 0 \cdot 6 \times 2 \cdot 62 = 1 \cdot 57 \text{ m} \quad \textbf{Ans.}$$

Guide blade angle at inlet

Let α = Guide blade angle at inlet, and
V_w = Velocity of whirl at inlet.

We know that the velocity of periphery at inlet,
$$v = \frac{\pi DN}{60} = \frac{\pi \times 2 \cdot 62 \times 300}{60} = 41 \cdot 2 \text{ m/s}$$

Since the discharge is radial at outlet, therefore velocity of whirl at outlet is zero.

We know that $\eta_h = \dfrac{V_w \cdot v}{gH}$ or $0 \cdot 88 = \dfrac{V_w \times 41 \cdot 2}{9 \cdot 81 \times 120} = 0 \cdot 035 \, V_w$

$$\therefore \quad V_w = 0 \cdot 88/0 \cdot 035 = 25 \cdot 1 \text{ m/s}$$

Since V_w (25·1) is less than v (41·2), therefore the shape of the inlet triangle will be as shown in Fig. 33·18.

From the inlet triangle, we find that
$$\tan \alpha = \frac{V_f}{V_w} = \frac{7 \cdot 28}{25 \cdot 1} = 0 \cdot 2900$$

$$\therefore \quad \alpha = 16 \cdot 2° \quad \textbf{Ans.}$$

Van angle at inlet

Let θ = Vane angle at inlet.

From the inlet triangle of velocities, we also find that
$$\tan (180° - \theta) = \frac{V_f}{v - V_w} = \frac{7 \cdot 28}{41 \cdot 2 - 25 \cdot 1} = 0 \cdot 4522$$

$$\therefore \quad (180° - \theta) = 24 \cdot 3° \quad \text{or} \quad \theta = 155 \cdot 7° \quad \textbf{Ans.}$$

Van angle at outlet

Let ϕ = Vane angle at outlet.

We know that the velocity of pheriphery at outlet,
$$v_1 = \frac{\pi D_1 N}{60} = \frac{\pi \times 1 \cdot 56 \times 300}{60} = 24 \cdot 7 \text{ m/s}$$

From the outlet triangle of velocities we find that
$$\tan \phi = \frac{V_{f1}}{v_1} = \frac{7 \cdot 28}{24 \cdot 7} = 0 \cdot 2947$$

$$\therefore \quad \phi = 16 \cdot 4° \quad \textbf{Ans.}$$

Example 33·14. *Using data given below, determine the main dimensions and blade angles of an inward flow reaction turbine.*

Net head = 65 m

Reaction Turbines

Speed = 700 r.p.m.;
Power = 300 kW;
Hydraulic efficiency = 94%;
Flow ratio = 0.18

$$\frac{\text{Wheel width at inlet}}{\text{Wheel dia at inlet}} = 0.1;$$

$$\frac{\text{Inner dia}}{\text{Outer dia}} = \frac{1}{2}$$

Assume constant velocity of flow and radial discharge. Neglect area blocked by blades.

Solution. Given : $H = 65$ m; $P = 300$ kW; $\eta_h = 94\% = 0.94$; $\eta_0 = 85\% = 0.85$; $\dfrac{V_f}{\sqrt{2gH}} = 0.18$

or $V_f = 0.182 \times 9.81 \times 65 = 6.43$ m/s; $\dfrac{b}{D} = 0.1$ or $b = 0.1D$; $\dfrac{D_1}{D} = \dfrac{1}{2}$ or $D_1 = \dfrac{D}{2} = 0.5 D$;

$V_{f1} = V_f = 6.43$ m/s

Diameter of wheel at inlet

Let $\qquad D$ = Diameter of wheel at inlet, and
$\qquad\qquad Q$ = Discharge through the turbine

We know that overall efficiency (η_0),

$$0.85 = \frac{P}{wQH} = \frac{300}{9.81 \times Q \times 65} = \frac{0.47}{Q}$$

$\therefore \qquad Q = 0.47/0.85 = 0.553$ m³/s

We also know that discharge through the turbine (Q),

$$0.553 = \pi D b V_f = \pi \times D \times 0.1 D \times 6.43$$
$$= 2.02\, D^2$$

$\therefore \qquad D^2 = 0.553/2.02 = 0.274$

or $\qquad D = 0.52$ m **Ans.**

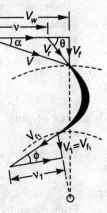

Fig. 33·20.

Diameter of wheel at outlet

We know that diameter of wheel at outlet,
$$D_1 = 0.5 D = 0.5 \times 0.52 = 0.26 \text{ m} \quad \textbf{Ans.}$$

Width of the wheel at inlet

We know that the width of wheel at inlet
$$b = 0.1 D = 0.1 \times 0.52 = 0.052 \text{ m} = 52 \text{ mm} \quad \textbf{Ans.}$$

Width of wheel at outlet

Let $\qquad b_1$ = Width of the wheel at outlet.

Since the discharge of water at inlet and outlet tips is equal, therefore
$$\pi D b V_f = \pi D_1 b_1 V_{f1}$$
$$\pi \times 0.52 \times 0.052 \times 6.43 = \pi \times 0.26 \times b_1 \times 6.43$$
$$0.546 = 5.25\, b_1 \quad \text{or} \quad b_1 = 0.546/5.25 = 0.104 = 104 \text{ mm} \quad \textbf{Ans.}$$

Guide blade angle

Let $\qquad \alpha$ = Guide blade angle, and
$\qquad\qquad V_w$ = Velocity of whirl at inlet.

We know that the velocity of pheriphery at inlet,
$$v = \frac{\pi DN}{60} = \frac{\pi \times 0.52 \times 700}{60} = 19.1 \text{ m/s}$$
Since the discharge is radial at outlet, therefore velocity of whirl at outlet is zero.
We know that hydraulic efficiency (η_h)
$$0.94 = \frac{V_w \cdot v}{gH} = \frac{V_w \times 19.06}{9.81 \times 65} = 0.03 \, V_w$$
$$\therefore \quad V_w = 0.94/0.03 = 31.3 \text{ m/s}$$

Since V_w (31·3) is more than v (19), therefore shape of inlet triangle will be as shown in Fig. 33·19.

From the inlet triangle, we find that
$$\tan \alpha = \frac{V_f}{V_w} = \frac{6.43}{31.3} = 0.2042 \text{ or } \alpha = 11.6° \quad \textbf{Ans.}$$

Vane angle at inlet

Let θ = Vane angle at inlet.

From the inlet triangle, we also find that
$$\tan \theta = \frac{V_f}{V_w - v} = \frac{6.43}{31.3 - 19.1} = 0.5270 \text{ or } \theta = 27.8° \quad \textbf{Ans.}$$

Vane angle at outlet

Let ϕ = Vane angle at outlet.

We know that the velocity of pheriphery at outlet,
$$v_1 = \frac{\pi D_1 N}{60} = \frac{\pi \times 0.26 \times 700}{60} = 9.53 \text{ m/s}$$

From the outlet triangle, we find that
$$\tan \phi = \frac{V_{f1}}{v_1} = \frac{6.43}{9.53} = 0.6747 \text{ or } \phi = 34° \textbf{ Ans.}$$

Example 33·15. *An outward flow reaction turbine has tangential velocity at its inner rim as 12 m/s, and ratio of radii is 0·8. The turbine is placed 1 metre below the water surface in the tail race and the vane angle are 90° and 20° at inlet and outlet respectively. The radial velocity of flow at inlet is 4 m/s. Neglecting frictional losses and taking velocity of outflow as radial, find :*

(a) *guide vane angle,*

(b) *velocity of flow from guides,*

(c) *total head of water and*

(d) *hydraulic efficiency.*

Solution. Given : $v = 12$ m/s; $\dfrac{r_1}{r_2} = 0.8$ or $r = 0.8 \, r_1$; $\theta = 90°$; $\phi = 20°$ and $v_f = 4$ m/s.

Guide vane angle

Let α = Guide vane angle.

We know that work done per kN of water
$$\tan \alpha = \frac{v_f}{v_w} = \frac{v_f}{v} \frac{4}{12} = 0.3333$$
or
$$\alpha = 18.4° \textbf{ Ans.}$$

Reaction Turbines

Velocity of flow from guides

From the inlet triangle of velocities, we also find that velocity of flow from the guides.

$$v = v^2 + v_f^2 = \sqrt{(12)^2 + (4)^2} \text{ m/s}$$
$$= 12.6 \text{ m/s} \quad \textbf{Ans.}$$

Total head of water

We know that tangential velocity at the outer rim,

$$v_1 = v \times \frac{r_1}{r} = 12 \times \frac{1}{0.8} = 15 \text{ m/s}$$

and from the outlet triangle of velocities, we find that velocity of flow at outlet

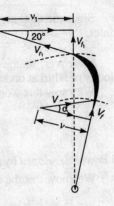

Fig. 33·21

$$v_{f1} = v_1 = v_1 \tan 20° = 15 \times 0.364 = 5.46 \text{ m/s}$$

We know that work done per kN of water

$$\frac{V_w \cdot v}{g} = \frac{12 \times 12}{9.81} = 14.7 \text{ kN-m/s}$$

∴ Total head of water at inlet

$$= \text{Energy at outlet} + \text{Work done}$$

$$= \left(\frac{V_1^2}{2g} + 1\right) + 14.68 \quad \text{...[the turbine is placed 1 m below the tail race]}$$

$$= \left[\frac{(5.46)^2}{2 \times 9.81} + 1\right] + 14.7 = 2.5 + 14.7 = 17.2 \text{ m} \quad \textbf{Ans.}$$

Hydraulic efficiency

Since the turbine is placed 1 metre below the water surface in the tail race, therefore net head of water available for the turbine,

$$H = 17.2 - 1 = 16.2 \text{ m}$$

and hydraulic efficiency,

$$\eta_h = \frac{V_w \cdot v}{gH} = \frac{12 \times 12}{9.81 \times 16.2} = 0.906 = 90.6\% \quad \textbf{Ans.}$$

Example 33·16. *Water is fed to an inward flow reaction turbine, running at 180 r.p.m. with a velocity of flow of 3 m/s. The diameter and width of the wheel at inlet in 1 metre and 135 mm respectively. Assuming the velocity of flow to be constant and radial discharge at inlet and outlet, find :*

(a) *work done per kN of water,*

(b) *power developed by the turbine,*

(c) *head of water on the turbine, and*

(d) *hydraulic efficiency of the turbine.*

Solution. Given : $N = 180$ r.p.m.; $D = 1$ m; $b = 135$ mm $= 0.135$ m and $V_f = V_{f1} = 3$ m/s

(a) *Work done per kN of water*

We know that the tangential velocity of wheel at inlet,

$$v = \frac{\pi DN}{60} = \frac{\pi \times 1 \times 180}{60} = 9.42 \text{ m/s}$$

Since the discharge is radial at inlet, therefore velocity of whirl at inlet,

$$v_w = v = 9.42 \text{ m/s}$$

Moreover, as the discharge is radial at outlet also, therefore velocity of whirl at outlet is zero.

We know that work done per kN

$$= \frac{V_w \cdot v}{g} = \frac{9.42 \times 9.42}{9.81}$$

$$= 9.05 \text{ kN-ms} \quad \textbf{Ans.}$$

Fig. 33.22.

(b) Power developed by the turbine

We know that the discharge of the turbine,

$$Q = \pi D b v_f = \pi \times 1 \times 0.135 \times 3$$

$$= 1.272 \text{ m}^3/\text{s}$$

and power developed by the turbine,

$$P = \text{Weight of water flowing per second in kN} \times \text{Work done per kN of water.}$$

$$= (9.81 \times 1.272) \times 9.05 = 112.9 \text{ kW} \quad \textbf{Ans.}$$

(c) Head of water on the turbine

Since the discharge is radial, therefore velocity of whirl at the outlet is zero.

We know that work done per kN of water

$$9.05 = H - \frac{V_1^2}{2g} = H - \frac{V_{f1}^2}{2g} \quad \ldots (\because V_1 = V_{f1})$$

$$= H - \frac{V_f^2}{2g} \quad \ldots (V_{f1} = V_f \text{ (given)})$$

$$= H - \frac{(3)^2}{2 \times 9.81} = H - 0.92$$

$$\therefore H = 9.05 + 0.92 = 9.97 \text{ m} \quad \textbf{Ans.}$$

(d) Hydraulic efficiency of the turbine

We also know that hydraulic efficiency of the turbine,

$$\eta h = \frac{V_w \cdot v}{gH} = \frac{9.42 \times 9.42}{9.81 \times 9.97} = 0.907 = 90.7\% \quad \textbf{Ans. Ans.}$$

EXERCISE 33.3

1. A reaction turbine works under a head of 120 m and at 450 r.p.m. The impeller diameter at inlet is 1.2 m and area of flow is 0.4 m². The inlet guide blade angle and vane angle is 20° and 60° respectively. Find the discharge of the turbine and power developed by it.
   [**Ans.** 5.2 m³/s; 5273.5 kW]

2. In an inward flow reaction turbine, the internal and external diameters are 0.8 m and 1.2 m respectively. The width of wheel impeller at inlet and outlet is 150 mm. The turbine is working under a head of 10 m and hydraulic efficiency is 92%. If the vane angle at outlet is 20°, find the discharge at outlet is radial and velocity of flow is 3 m/s. Also find power developed by the turbine.
   [**Ans.** 1.13 m³/s; 102 kW]

3. A radially inward flow reaction turbine working under a head of 9 metres and running at 200 r.p.m. develops 150 kW. The speed ratio and flow ratio are 0.9 and 0.3 respectively. If overall efficiency and hydraulic efficiency are 0.9 and .3 respectively, find inner diameter of the wheel width of the wheel at inlet and guide blade angle at inlet. [**Ans.** 1.14 m; 148 mm; 32.4°]

4. The external and internal diameters of an inward flow reaction turbine 700 mm and 500 mm respectively. The impeller width of the impeller at the corresponding points are 150 mm and 225

mm respectively. The effective head on the turbine is 20 m and the velocity of flow at inlet is 3·5 m/s. If the guide vane angle is 15°, find the speed of the turbine and the power developed.

[**Ans.** 356 r.p.m.; 197·2 kW]

33·20. Francis' Turbine

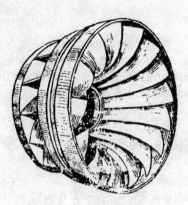

Fig. 33·23. Francis turbine runner.

The Francis' turbine is an inward flow reaction turbine, having radial discharge at outlet. It was the first turbine (inward flow reaction type) which was designed by Francis.* It is mostly used even in these days, for producing power under medium heads.

The modern Francis' turbine has a mixed flow (*i.e.,* combination of radial and axial). It has typical runner as shown in Fig. 33·24.

The height (or length) of the runner depends upon its specific speed.** A Francis turbine, having a higher specific speed, has a longer runner than that a lower specific speed. The runner of a Francis' turbine may be cast in one piece, or made of separate steel plates welded together. The runners are made of separate steel plates welded together. The runners are made of cast iron for small output, cast steel for large output and stainless steel, or other non-ferrous metal like bronze, when the water is chemically impure and there is a danger of corrosion. The blades of the runner are carefully finished.

All the relations, for finding out the various angles and other data which were used for inward flow reaction turbine will hold good for Francis turbine also.

Example 33·17. *A Francis' turbine, working under a head of 14 metres, has guide blade angle of 20° and radial vanes at inlet. The ratio of inlet and outlet diameters is 3 to 2. The velocity of flow of water, at exit, is 4 m/s. Assuming the velocity of flow to be constant, determine the peripheral velocity of water at inlet and the vane angle at outlet.*

Solution. Given : $H = 14$ m; $\alpha = 20°$; $\theta = 90°$; $D = \dfrac{3D_1}{2} = 1\cdot5\, D_1$; $v_{f1} = 4$ m/s and $v_f = v_{f1} = 4$ m/s

Pheripheral velocity of wheel at inlet

Let $\qquad\qquad\qquad v$ = Peripheral velocity of wheel at inlet.

From the inlet triangle, we find that

---

*Francis James (1815—92), was an English scientist and engineer, who went to States in 1833; where he made large scale experiments and developed the inward flow turbine, which is now known after his name. The largest Francis turbine of the world is at Krasnoyarsk (U.S.S.R.) producing 515 000 kW under a head of 95 m. Another powerful Francis' turbine is at Nohab (Sweden) producing 233 000 kW B.H.P. under a heal of 100 m.

** For details, please refer to page 836, Art. 32.7

$$\tan 20° = \frac{V_f}{v} \quad \text{or} \quad 0.3640 = \frac{4}{v} \quad \ldots(\because V_f = V_{f1} = 4\text{m/s})$$

$$\therefore \quad v = 4/0.364 = 11 \text{ m/s} \quad \textbf{Ans.}$$

Vane angle at outlet

Let $\quad \phi$ = Vane angle at outlet.

As it is a Francis turbine, therefore its discharge will be radial. Moreover, as the outer diameter of the turbine is 2/3rd of the inner diameter, therefore the peripheral velocity of wheel at outlet,

$$v_1 = \frac{2v}{3} = \frac{2}{3} \times 11 = 7.33 \text{ m/s}$$

Now from the outlet triangle, we find that

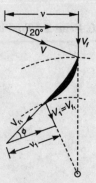

Fig. 33·24.

$$\tan \phi = \frac{V_{f1}}{v_1} = \frac{4}{7.33} = 0.5457 \quad \text{or} \quad \phi = 28.6° \quad \textbf{Ans.}$$

Example 33·18. *A Francis' 135 kW turbine, having an overall efficiency of 75%, is required to produce under a head of 9 metres and at 120 r.p.m. The velocity of periphery of the wheel and velocity of flow at inlet is $3.47 \sqrt{H}$ and $1.1 \sqrt{H}$ respectively. If the hydraulic losses in the turbine are 20% of the available energy, find:*

(a) guide blade angle at inlet,

(b) wheel vane angle at inlet, and

(c) diameter of the wheel.

Solution. Given : $\eta_0 = 75\% = 0.75$; $P = 135$ kW; $H = 9$ m; $N = 120$ r.p.m., $v = 3.47 \sqrt{H} = 3.47\sqrt{9} = 10.4$ m/s; $v_f \sqrt{H} = 3 \cdot 3.45$ m/s and hydraulic losses = 20% or hydraulic efficiency $(\eta_n) = 100\% - 20\% = 80\% = 0.8$.

Guide blade angle at inlet

Let
- α = Guide blade angle at inlet, and
- V_w = Velocity of whirl at inlet.

As it is a Francis' turbine, therefore its discharge will be radial. We know that hydraulic efficiency (η_n)

$$0.8 = \frac{V_w \cdot v}{gH} = \frac{V_w \times 10.4}{9.81 \times 9} = 0.118 \, V_w$$

$$\therefore \quad V_w = 0.8/0.118 = 6.78 \text{ m/s}$$

Since V_w (6·78) is less than v (10·41), therefore shape of the triangle will be as shown in Fig. 31·25.

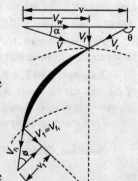

Fig. 33·25.

From the inlet triangle, we find that

$$\tan \alpha = \frac{V_f}{V_w} = \frac{3.45}{6.78} = 0.5088$$

or
$$\alpha = 27° \quad \textbf{Ans.}$$

Wheel vane angle at inlet

Let $\quad \theta$ = Wheel vane angle at inlet.

Reaction Turbines

From the inlet triangle, we also find that

$$\tan(180° - \theta) = \frac{V_f}{v - V_w} = \frac{3\cdot 45}{10\cdot 4 - 6\cdot 78} = 0\cdot 9530$$

∴ $(180° - \theta) = 43\cdot 6°$ or $\theta = 136\cdot 4°$ **Ans.**

Diameter of wheel

Let D = Diameter of the wheel.

We know that velocity of periphery of the wheel at inlet (v),

$$10\cdot 4 = \frac{\pi DN}{60} = \frac{\pi \times D \times 120}{60} = 6\cdot 283\ D$$

∴ $D = 10\cdot 4/6\cdot 283 = 1\cdot 66$ m **Ans.**

Example 33·19. *A Francis turbine is supplied 5·1 cubic metres of water per second under an average head of 20 metres. The runner diameter is 3·6 m at inlet and 2·4 m at outlet. The inlet vane angle is 120° and the turbine has radial discharge at 15 m/s with breadth of wheel constant. If the hydraulic and overall efficiencies are 90% and 80% respectively, find the power produced by the turbine and its speed.*

Solution. Given : $Q = 5\cdot 1$ m³/s; $H = 20$ m; $D = 3\cdot 6$ m; $D_1 = 2\cdot 4$ m; $\theta = 120°$; $v_1 = 15$ m/s; $b = b_1$; $\eta_h = 90\% = 0\cdot 9$ and $\eta_h = 80\% = 0\cdot 8$.

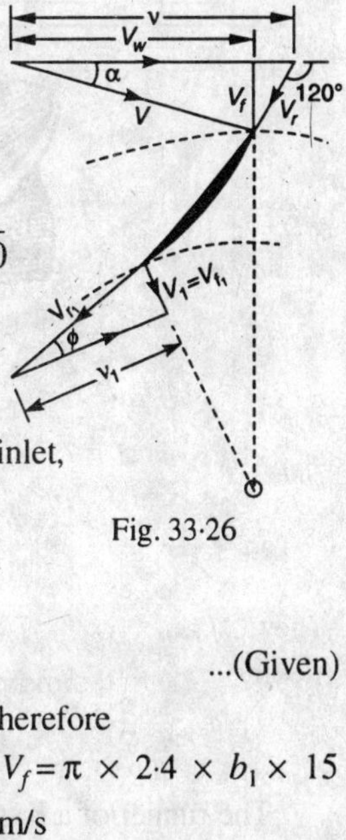

Power produced by the turbine

Let P = Power produced by the turbine.

We know that overall efficiency of the turbine (η_0)

$$0\cdot 9 = \frac{P}{uQH} = \frac{P}{9\cdot 81 \times 5\cdot 1 \times 20} = \frac{P}{1000}$$

∴ $P = 0\cdot 9 \times 1000 = 900$ kW **Ans.**

Speed of the turbine

Let N = Speed of the turbine,
    v = Tangential velocity of the wheel at inlet,
    v_w = Velocity of whirl at inlet, and
    v_f = Velocity of flow at inlet.

Fig. 33·26

Since the discharges is radial, therefore velocity of flow at outlet.

$$V_{f1} = V_1 = 15\ \text{m/sec} \qquad \text{...(Given)}$$

Moreover, as the discharge of water at inlet and outlet tips is equal, therefore

$$\pi D b V_f = \pi D_1 b_1 V_{f1} \quad \text{or} \quad \pi \times 3\cdot 6 \times b \times V_f = \pi \times 2\cdot 4 \times b_1 \times 15$$

$$3\cdot 6\ \pi V_F = 36\ \pi \quad \text{or} \quad V_f = 36/3\cdot 6 = 10\ \text{m/s}$$

Now from the inlet triangle, we find that

$$(v - V_w) = \frac{V_f}{\tan 60°} = \frac{10}{1\cdot 732} = 5\cdot 77\ \text{m/s}$$

or $V_w = v - 5\cdot 77$ m/s

Since the discharge is radial at outlet, therefore the velocity of whirl at outlet is zero. Therefore hydraulic efficiency (η_n)

$$0\cdot 9 = \frac{V_w \cdot v}{gH} = \frac{(v - 5\cdot 77)v}{9\cdot 81 \times 20} = \frac{v^2 - 5\cdot 77\ v}{196\cdot 2}$$

∴ $v^2 - 5\cdot 77\ v - 176\cdot 6 = 0$

This is a quadratic equation for v. Therefore,
$$v = \frac{5.77 \pm \sqrt{(5.77)^2 + 4 \times 176.6}}{2} = 16.5 \text{ m/s}$$

We know that tangential velocity of the wheel (v)
$$16.5 = \frac{\pi DN}{60} = \frac{\pi \times 3.6 \times N}{60} = 0.188 \, N$$

$\therefore \qquad N = 16.5/0.188 = 87.8$ r.p.m. **Ans.**

33·21. Kaplan Turbine

The Kaplan[*] turbine is an axial flow reaction turbine, in which the flow of water is parallel to the shaft. A Kaplan turbine is used where a large quantity of water is available at low heads.

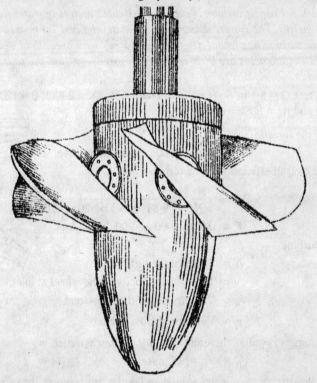

Fig. 33·27. Kaplan turbine runner.

The runner of a Kaplan turbine resembles with the propeller of a ship. That is why, a Kaplan turbine is also called a propeller turbine. The water from the scroll flows over the guide blades and then over the vanes. The water while, gliding over the vanes, exerts some force on the shaft of the turbine, which causes the shaft to revolve.

---

[*]Kaplan Victor (1815—) was a German scientist, who designed this turbine for low heads and large quantities of flow. The largest Kaplan turbine of the world is at Volga river (U.S.S.R.) which produces 130 000 kW under a head of 22.5 m, when running at 68.2 r.p.m. This turbine has a runner of diameter 9.3 m. The next largest Kaplan turbine is at Dalles which produces 93 250 kW under a head of 24.7 m. The highest head used for a Kaplan turbine is the world is at Tres Marais (Brazil). It is working under a head of 55 m and produces 75 000 kW. The turbine has a runner of diameter 8.4 m. The lowest head used for Kaplan turbine is at Vargon (Sweden). It is working under a head of 4.3 m and produces 11 500 kW. The turbine has a runner of diameter 8 m.

Reaction Turbines

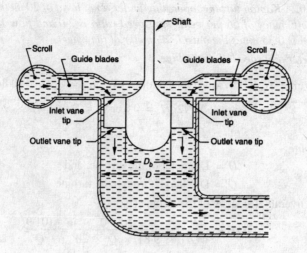

Fig. 33·28. Kaplan turbine.

The runner of a modern Kaplan turbine has the following two major differences with that of a Francis' turbine :
1. In Francis' turbine runner, the water enters radially; whereas in a Kaplan turbine runner the water strikes the blades axially.
2. In a Francis' turbine runner, the number of blades is generally between 16 and 24; whereas in a Kaplan turbine runner the number of blade is generally between 3 to 8. This reduces the frictional resistance to water.

A typical runner of a Kaplan of a Kaplan turbine is shown in Fig. 33·28. The blades of a Kaplan turbine runner can be adjusted to adjust the passage area the two blades.

The runner of a Kaplan turbine is known as boss, which is nothing but an extension of the shaft (at its lower end) as shown in Fig. 33·28 and 33·29.

Let
D = Diameter of turbine,
D_b = Diameter of the boss, and
V_f = Velocity of flow at inlet.

∴ Discharge through the turbine,

$$Q = V_f \times \frac{\pi}{4} (D^2 - D_b^2)$$

All the notations of the Kaplan turbine are the same as that of inward or outward flow reaction turbines. All the relations, for finding out the data hold good for a Kaplan turbine also.

The following table gives the ratio of boss to the outer diameter (generally termed as hub ratio) and the number of blades of a Kaplan turbine for the given head of water.

Table 33·1

Head in metres	5	20	40	50	60	70
$\frac{D_b}{D}$	0·3	0·4	0·5	0·55	0·6	0·7
No. of blades	3	4	5	6	8	10

Example 33·20. *A Kaplan turbine, operating under a net head of 20 metres, develops 20 000 kW with an overall efficiency of 66 per cent. The speed ratio is 2·0 and flow ratio is 0·6. The hub diameter of the wheel is 0·35 times the outside diameter of the wheel.*

Find the diameter and speed of the turbine.

Solution. Given : $H = 20$ m; $P = 20\ 000$ kW; $n_0 = 80\% = 0.86$; $\dfrac{v}{\sqrt{2gH}} = 2.0$ or $v = 2 \times \sqrt{2 \times 9.81 \times 20} = 39.6$ m/s; $V_f/\sqrt{2gH} = 0.6$ or $V_f = 0.6 \times \sqrt{2 \times 9.81 \times 20} = 11.9$ m/s and $D_b = 0.35\ D$.

Diameter of the turbine

Let D = Diameter of the turbine, and

                 Q = Discharge through the turbine.

We know that overall efficiency of the turbine (η_0)

$$0.86 = \frac{P}{wQH} = \frac{20000}{9.81 \times Q \times 20} = \frac{101.9}{Q}$$

∴ $Q = 101.9/0.86 = 118.5$ m³/s

and discharge through the turbine (Q)

$$118.5 = V_f \times \frac{\pi}{4}(D^2 - D_b^2) = 11.9 \times \frac{\pi}{4}[D^2 - (0.35\ D)^2] = 8.2\ D^2$$

∴ $D^2 = 118.5/8.2 = 14.45$ or $D = 3.8$ m **Ans.**

Speed of the turbine

Let N = Speed of the turbine in r.p.m.

We know that peripheral velocity of the turbine at inlet (v),

$$39.6 = \frac{\pi D N}{60} = \frac{\pi \times 3.8\ N}{60} = 0.2\ N$$

∴ $N = 39.6/0.2 = 198$ r.p.m. **Ans.**

Example 33·21. *A propeller*[*]*turbine runner has an outer diameter of 4·5 metres and an inner diameter of 2·5 metres and develops 21 000 kW when running at 140 r.p.m. under a head of 20 metres. The hydraulic efficiency is 94% and overall efficiency is 80%.*

Find discharge through the turbine, and the guide blade angle at inlet.

Solution. Given : $D = 4.5$ m; $Db = 2.5$; $P = 21\ 000$ kW; $N = 140$ r.p.m.; $H = 20$ m; $\eta_n = 94\% = 0.94$ and $\eta_0 = 80\% = 0.88$

Discharge through the turbine

Let Q = Discharge through the turbine.

We know that overall efficiency of the turbine (n0)

$$0.88 = \frac{P}{wQH} = \frac{21000}{9.81 \times Q \times 20} = \frac{107}{Q}$$

∴ $Q = 107/0.88 = 121.6$ m³/s **Ans.**

Guide blade angle at inlet

Let α = Guide blade angle at inlet,

                 Vw = Velocity of whirl at inlet, and

                 Vf = Velocity of flow at inlet.

---

[*] A propeller turbine is a Kaplan turbine.

Reaction Turbines

We know that the peripheral velocity at inlet,

$$v = \frac{\pi DN}{60} = \frac{\pi \times 4.5 \times 140}{60} = 33 \text{ m/s} \qquad \ldots(i)$$

and hydraulic efficiency (η_n)

$$0.94 = \frac{V_w \cdot v}{gH} = \frac{V_w \times 33}{9.81 \times 20} = 0.168 \, V_w$$

$$\therefore \quad V_w = 0.94/0.168 = 5.6 \text{ m/s} \qquad \ldots(ii)$$

We know that the discharge of propeller turbine, (Q),

$$121.6 = V_f \times \frac{\pi}{4}(D^2 - D_b^2) = V_f \times \frac{\pi}{4} \times [(4.5)^2 - (2.5)^2] = 11 \, V_f$$

$$\therefore \quad V_f = 121.6/11 = 11.1 \text{ m/s} \qquad \ldots(iii)$$

Now from the inlet triangle of velocities, we know that

$$\tan \alpha = \frac{V_f}{V_w} = \frac{11.1}{5.6} = 1.9821 \quad \text{or} \quad \alpha = 63.2° \textbf{ Ans.}$$

33·22. Draft Tube

It is a pipe, which connects the turbine and the tail race or outlet channel, through which the water exchausted from the runner, flows to the outlet channel. A draft tube, in addition to act as a water conduit, has the following two important functions also :

1. It enables the turbine to be placed above the tail race, so that the turbine may be inspected properly.
2. To convert the kinetic enrgy $\left(V_1^2\right)$ of the water, exhausted by the runner into pressure energy in the tube.

33·23. Types of Draft Tubes

Though there are many types of draft tubes, yet the following two types are very common these days :

1. Conical draft tubes and
2. Elbow draft tubes.

33·24. Conical Draft Tubes

Fig. 33·29. Conical draft tubes

In a conical type, the diameter of the tube gradually increases from the outlet of the runner to the channel as shown in Fig. 33·29 (a) and (b).

The conical draft tubes are commonly used in Francis' turbine. For good efficiency, the central flaring angle is kept about 8°. The conical draft tube, shown in Fig. 33·30 (b), which has a bell-mounted outlet, is best suited for inward or outward flow turbines, having helocal flow which is due to velocity of whirl at outlet of the runner. The efficiency of conical draft tubes is as large as 90%.

33·25. Elbow Draft Tubes

In elbow type, the bend of the draft tube is generally 90° and area of the tube gradually increases from the outlet of the runner to the channel as shown in Fig. 33·30 (a) and (b).

The elbow draft tubes are commonly used in Kaplan turbine. The elbow draft tube, shown in Fig. 33·30 (a), has circular sections at inlet and outlet. But the draft tube, shown in Fig. 33·30 (b) has

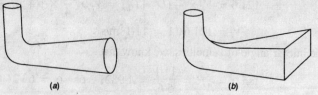

Fig. 33·31 Elbow draft tubes.

a circular section at inlet and a rectangular section at outlet. The efficiency of elbow draft tubes is generally between 60% to 70%.

33·26. Efficiency of a Draft Tube

The efficiency of a draft tube may be found out from the relation :

$$n_t = \frac{\dfrac{V_2^2}{2g} - \dfrac{V_3^2}{2g}}{\dfrac{V_2^2}{2g}} = \frac{V_2^2 - V_3^2}{V_2^2}$$

where
V_2 = Velocity of water at inlet of the draft tube, and
V_3 = Velocity of water at outlet of the draft tube.

Example 33·22. *A Kaplan turbine develops 1500 kW under a head of 6 m. The turbine is set 2·5 m above the tail water level. A vacuum gauge, inserted at the turbine outlet, records a suction head of 3·1 m. If the turbine efficiency is 85%, what will be the efficiency of draft tube, having inlet diameter of 3 m?*

Solution. Given : P = 1500 kW; H = 6 m; Height of turbine from the water level (Z) = 2·5 m; Vacuum gauge at the turbine outlet $\left(\dfrac{p_2}{w}\right)$ = 3·1 m (suction) = – 3·1 m; n_0 = 85% = 0·85 and diameter of draft tube (d_2)

Let Q = Discharge of the turbine

We know that area of the draft tube,

$$a = \frac{\pi}{4} \times (d_2)^2 = \frac{\pi}{4} \times (3)^2 = 7 \cdot 07 \text{ m}^2$$

and overall efficiency of the turbine (η_0)

$$0 \cdot 85 = \frac{P}{wQH} = \frac{1500}{9 \cdot 81 \times Q \times 6} = \frac{25 \cdot 5}{Q}$$

∴ Q = 25·5/0·85 = 30 m³/s

Reaction Turbines

and velocity of water at inlet of the draft tube,

$$V_2 = \frac{Q}{a_2} = \frac{30}{7.07} = 4.24 \text{ m/s}$$

Now applying Bernoulli's equation at points 2 and 3,

i.e.,
$$Z_2 + \frac{p_2}{w} + \frac{V_2^2}{2g} = Z_3 + \frac{p_3}{w} + \frac{V_3^2}{2g}$$

$$2.5 - 3.1 + \frac{V_2^2}{2g} = 0 + 0 + \frac{V_3^2}{2g}$$

$$\therefore \quad \frac{V_2^2}{2g} - \frac{V_3^2}{2g} = 0.6 \qquad \ldots(i)$$

We know that efficiency of the draft tube

$$\eta_t = \frac{\frac{V_2^2}{2g} - \frac{V_3^2}{2g}}{\frac{V_2^2}{2g}} = \frac{0.6}{\frac{(4.24)^2}{2 \times 9.81}} = 0.655 = 65.5\% \text{ Ans.}$$

33·27 Other Reaction Turbines

The Francis turbines and Kaplan turbines is considered to be the most popular reaction turbines for low heads, and are used successfully all over the world. Many scientists and engineers working in hydraulic reasearch stations all over the world are busy in developing improved type of turbines. Some of them have been able to design new reaction turbines. Though these turbines, like new impulse turbines are not of practical importance, yet they have some adademic importance. Following reaction turbines are important from the subject point of view :

1. *Fourneyron Turbine.* It is an outward flow reaction turbine. In this turbine an adjustable gate is provided, which can be raised can be raised or lowered by means of governor mechanism. The wheel is divided in to a number of compartments (generally four) by horizontal diaphragms so that when the turbine is workiing at part load, only the efficiency of that compartment is affected.

   A fourneyron turbine may be used for heads of 1 to 100 m with an efficiency of 75%. Its efficiency may be further increased by fitting a circular diffuser around the moving blade ring, whose function is similar to that of draft tube.

2. *Jonval Turbine.* It is inward flow reaction in with the speed regulation is obtained by cutting off the supply of water to one or more guide passages by means of a circular gate. The interesting feature of a Jonval turbine is the division of its wheel into a number of concentric compartments in such a way that each compartment forms a complete turbine in itself. This arrangement helps to cut off one or more compartments completely. Because of this arrangenment, Jonval turbine can be used for heads varying from 1 to 50 metres and for speed varying from 20 to 400 r.p.m.

3. *Thomson's Turbine.* It is an inward flow reaction turbine in which the wheel is surrounded by a large vortex chamber. The water enters at the largest part of the chamber and is guided to the moving blades. the external diameter of the wheel is generally twice the internal diameter. The flow of water is regulated by the supply to the whole of the circumferance of the wheel. The advantage of this method of regulating is that the efficiency is not very much affected. The efficiency at part load is almost equal to that at full load.

EXERCISE 33·4

1. A Francis' turbine has tangential velocity of runner, velocity of whirl and velocity of flow as 30 m/s, 24 m/s and 3 m/s respectively. The hydraulic efficiency of the turbine is 78%. If the discharge through the turbine is radial and 1000 litres/s; find (i) inlet vane angle (ii) head on the turbine and (iii) power developed by the turbine. [Ans. 27·6°; 94·2 m; 731 kW]

2. A Francis' turbine having outer diameter of 0·9 m is running at 200 r.p.m. The head of water on the turbine is 9·5 m. The velocity of flow through the runner is constant at 3 m/s. If the inlet tips of the vanes are radial and width of the runner at inlet is 150 mm, find (a) work done per kN of water, (b) hydraulic efficiency of the turbine and power produced by the turbine.

[Ans. 9·04 kN-m; 95·2%; 114·4 kW]

3. A Kaplan turbine working under a head of 5·5 metres develops 7500 kW. The speed ratio and flow ratio are 2·1 and 0·71 respectively. If the boss diameter is 1/3 of that of the runner and overall efficiency is 85%, find diameter of the runner and speed of the turbine. [Ans. 5·58 m; 75 r.p.m.]

QUESTIONS

1. What do you understand by the term reaction turbine? How does it differ from an impulse turbine?
2. Distinguish between a radial flow and an axial flow turbines.
3. Write the difference between an inward flow turbine and an outward flow turbine.
4. Show from fundamental principles that the work done in a reaction turbine
$$= \frac{V_w \cdot v}{g} \pm \frac{V_{w1} \cdot v_1}{g}$$
where V_w and V_{w1} = Velocity of water at inlet and outlet respectively.

   v and v_1 = Peripheral velocities at inlet and outlet respectively.
5. Derive an equation for the power developed by a reaction turbine.

OBJECTIVE TYPE QUESTIONS

1. In an inward flow reaction turbine, the water
   (a) flows parallel to the axis of the wheel (b) flows at right angles to the wheel
   (c) enters at the centre of the wheel and then flows towards the outer periphery of the wheel
   (d) enters the wheel at the outer periphery of the wheel and then flows towards the centre of the wheel
2. The hydraulic efficiency of a 6 reaction turbine is the ratio of
   (a) power produced by the turbine to the energy supplied to the turbine
   (b) actual work available at the turbine to the energy imparted to the wheel
   (c) work done on the wheel to the head of water (d) none of the above
3. The power produced by a reaction turbine is
   (a) directly proportional to H (b) inversely proportional to H
   (c) directional proportional to $\sqrt{H}$ (d) inversely proportional to $\sqrt{H}$
4. In a francis' turbine, runner, the number of blades are generally
   (a) 2 to 4 (b) 4 to 8 (c) 8 to 16 (d) 16 to 24

ANSWERS

1. (d) 2. (c) 3. (c) 4. (a)

34

Performance of Turbines

1. Introduction. 2. Characteristics of Turbines. 3. Unit Power. 4. Unit Speed. 5. Unit Discharge. 6. Significance of Unit Power, Unit Speed and Unit Discharge. 7. Specific Speed of a Turbine. 8. Significance of Specific Speed. 9. Selection of Turbines. 10. Selection Based on Specific Speed. 11. Selection Based on Head of Water. 12. Relation between Specific Speed and Shape of Reaction Turbine Runner. 13. Characteristic Curves of Turbine. 14. Characteristic Curves for Pelton Wheel. 15. Characteristic Curves for Francis' Turbine. 16. Cavitaiton.

34·1 Introduction

In the last two chapters (*i.e.*, in Impulse Turbines and Reaction Turbines) we have assumed that, in general, a turbine will work under a constant head, speed and output. But in actual practice, these assumptions rarely prevail. It is thus essential to review the nature of such variations, which generally take place. Though there are many types of variations, yet the following are important from the subject of point of view:

1. Keeping the discharge contant, the head of water and output may vary. In such cases, the speed should be adjusted, so that there is no appreciable change in efficiency.
2. Keeping the head of water and speed constant, the output may vary. In such cases, the discharge of the turbine should be adjusted.
3. In turbines, working under low heads, the head of water and speed may vary. Although the speed is allowed to fluctuate within narrow permissible limits, yet the head may vary up to 50%.
4. Keeping the head of water and discharge constant, the speed may vary by adjusting the load on the turbine. This is usually done in a laboratory.

34·2 Characteristics of Turbines

Sometimes, we have to compare the performances of turbines, of different outputs and speeds, working under different heads. This comparison will be much convenient, if we calculate the outputs of the turbines when the head of water is reduced to unity. *i.e.*, 1 metre. We always study the following three characteristics of a turbine under a unit head.

1. Unit power, 2. Unit speed, and 3. Unit discharge.

34·3 Unit Power

The power developed by a turbine, working under a head of 1 metre, is known as unit power. Now consider a turbine whose unit power is required to be found out.

Let
H = Head of water, under which the turbine is working
P = Power developed by the turbine under a head of water H,
Q = Discharge through the turbine under a head of water H, and
P_u = Power developed by the same turbine, under a unit head.

We know that the velocity of water (assuming C_v as unity),
$$V = \sqrt{2gH}$$

[561]

and discharge,
$$Q = aV = a\sqrt{2gH}$$
We also know that the power developed by a turbine,
$$P = wQH = w(a\sqrt{2gH})H$$
$$\propto H^{3/2}$$
$$= P_u \cdot H^{3/2}$$
or
$$P_u = \frac{P}{H^{3/2}}$$

34·4 Unit Speed

The speed of a turbine, working under a head of 1 metre, is known as unit speed. Now consider a turbine whose unit speed is required to be found out.

Let H = Head of water under, which the turbine is working,
v = Tangential velocity of the runner,
N = Speed of the turbine runner under a head of water H, and
N_u = Speed of the same turbine, under a unit head,

We know that the velocity of water (assuming C_v as unity),
$$V = \sqrt{2gH}$$
and tangential velocity of the runner,
$$v \propto \text{Velocity of water } (V)$$
$$\propto \sqrt{H} \qquad \ldots (V = C_v \sqrt{2gH})$$

We also know that the speed of the turbine runner,

or
$$N = \frac{60v}{\pi D} \qquad \ldots \left(\because v = \frac{\pi DN}{60} \right)$$
$$\propto v$$
$$\propto \sqrt{H} \qquad \ldots (\because v \propto H)$$
$$= N_u \cdot \sqrt{H}$$
$$\therefore \quad N_u = \frac{N}{\sqrt{H}}$$

34·5 Unit Discharge

The discharge of a turbine, working under a head of 1 metre, is known as unit discharge. Now consider a turbine whose unit discharge is required to be found out.

TLet H = Head of water, under which the turbine is working,
Q = Discharge through the turbine under a head of water H, and
Q_u = Discharge through the same turbine, under a unit head.

We know that the velocity of water (assuming C_v as unity),
$$V = aV = a\sqrt{2gH}$$
$$\propto \sqrt{H}$$
$$= Q_u \cdot \sqrt{H}$$
$$\therefore \quad Q_u = \frac{Q}{\sqrt{H}}$$

Performance of Turbines

Example 34·1 *A Pelton wheel develops 1750 kW under a head of 100 metres while running at 200 r.p.m. and discharging 2500 litres of water per second. Find the unit power, unit speed and unit discharge of the wheel.*

Solution. Given, $P = 1750$ kW; $H = 100$ m; $N = 200$ r.p.m. and $Q = 2500$ litres/s $= 2\cdot5$ m³/s.

Unit power of the wheel

We know that unit power of the wheel,

$$P_u = \frac{P}{H^{3/2}} = \frac{1750}{(100)^{3/2}} = \frac{1750}{1000} = 1\cdot75 \text{ kW} \quad \text{Ans.}$$

Unit speed of the wheel

We know that unit speed of the wheel,

$$N_u = \frac{N}{\sqrt{H}} = \frac{200}{\sqrt{100}} = \frac{200}{10} = 20 \text{ r.p.m.} \quad \text{Ans.}$$

Unit discharge of the wheel

We also know that unit discharge of the wheel,

$$Q_u = \frac{Q}{\sqrt{H}} = \frac{2\cdot5}{\sqrt{100}} = \frac{2\cdot5}{10} = 0\cdot25 \text{ m}^3/\text{s} \quad \text{Ans.}$$

34·6 Significance of Unit Power, Unit speed and Unit Discharge

In the last articles, we have discussed the characteristics of turbines. *i.e.*, unit power, unit speed and unit discharge. As a matter of fact, the significance of unit power, unit speed and unit discharge is of much importance in the field of Hydraulic Machines. It helps us in finding out the behaviour of a turbine, when it is put to work under different heads of water as discussed as follows:

1. *Significance of unit power*

   Let H = Head of water, under which the turbine is working.
   P = Power developed by the turbine, under head of water H, and
   P_1 = Power developed by the turbine under head of water H_1.

   We have seen in Art 34·3 that
   $$P \propto H^{3/2}$$
   $\therefore \quad P_1 = H_1^{3/2}$

   or $\quad \dfrac{P}{P_1} = \dfrac{H^{3/2}}{H_1^{3/2}}$

   $\therefore \quad P_1 = P \times \left(\dfrac{H_1}{H}\right)^{3/2} \quad \ldots(i)$

2. *Significance of unit speed*

   Let H = Head of water, under which the turbine is working,
   Q = Discharge of the turbine under a head of water H, and
   Q_1 = Discharge of the turbine under head of water H_1.

   We have seen in Art. 34·4 that
   $$N \propto \sqrt{H}$$
   $\therefore \quad N_1 \propto \sqrt{H_1}$

   or $\quad \dfrac{N}{N_1} = \dfrac{\sqrt{H}}{\sqrt{H_1}} \left(\dfrac{H}{H_1}\right)^{1/2}$

$$\therefore \quad N_1 = N \times \left(\frac{H_1}{H}\right)^{1/2} \quad \ldots(ii)$$

3. *Significance of unit discharge*

   Let H = Head of water under which the turbine is working,

   Q = Discharge of the turbine under a head of water H, and

   Q_1 = Discharge of the turbine under head of water H_1.

We have seen in Art. 34·5 that

$$Q \propto \sqrt{H}$$

$$\therefore \quad Q_1 \propto \sqrt{H_1}$$

or

$$\frac{Q}{Q_1} = \frac{\sqrt{H}}{\sqrt{H_1}} = \left(\frac{H}{H_1}\right)^{1/2}$$

$$\therefore \quad Q_1 = Q \times \left(\frac{H_1}{H}\right)^{1/2} \quad \ldots(iii)$$

Example 34·2 *An impulse turbine develops 4500 kW under a head of 200 metres. The turbine runner has a speed of 200 r.p.m. and discharges 0·8 cubic metre of water per second. If the head on the same turbine falls during summer season to 184·3 metres, find the new discharge, power and speed of the turbine.*

Solution. Given : P = 4500 kW; H = 200 m; N = 200 r.p.m.; Q = 0·8 m³/s₁ m and H_1 = 184·3 m

New discharge of the turbine

We know that new discharge of the turbine,

$$Q_1 = Q \times \left(\frac{H_1}{H}\right)^{1/2} = 0.8 \times \left(\frac{184.3}{200}\right)^{1/2} \text{ m}^3/\text{s}$$

$$= 0.8 \times 0.96 = 0.768 \text{ m}^3/\text{s} \quad \textbf{Ans.}$$

New power of the turbine

We know that new power of the turbine,

$$P_1 = P \times \left(\frac{H_1}{H_2}\right)^{3/2} = 4500 \times \left(\frac{184.3}{200}\right)^{3/2} \text{ kW}$$

$$= 4500 \times 0.88 = 3960 \text{ kW} \quad \textbf{Ans.}$$

New speed of the turbine

We also know that new speed of the turbine,

$$N_1 = N \times \left(\frac{H_1}{H_2}\right)^{1/2} = 200 \times \left(\frac{184.3}{200}\right)^{1/2} \text{ r.p.m.}$$

$$= 200 \times 0.96 = 192 \text{ r.p.m.} \quad \textbf{Ans.}$$

Example 34·3 *A reaction turbine, at best speed, produces 125 kW under a head of 64 metres. By what percent should the speed be increased for a head of 81 metres.*

Solution. Given : P = 125 kW; H = 64 m and H_1 = 81 m.

Let N = Speed of the turbine under a head of 64 metres.

We know that speed of the turbine under a head of 81 metres.

$$N_1 = N \times \left(\frac{H_1}{H_2}\right)^{1/2} = N \times \left(\frac{81}{64}\right)^{1/2} = \frac{9N}{8} \text{ r.p.m.}$$

∴ Increase in speed $= \dfrac{N_1 - N}{N} = \dfrac{\dfrac{9N}{8} - N}{N} = 0.125 = 12.5\%$ **Ans.**

EXERCISE 34·1

1. A turbine working under a head of 25 metres develops 2000 kW at 250 r.p.m. Find the unit power and unit speed of the turbine. (**Ans.** 16 kW; 50 r.p.m.)
2. A turbine develops 1000 kW under a head of 16 m at 200 r.p.m. while discharging 9 cubic metres of water per second. Find the unit power and unit discharge of the wheel. (**Ans.** 15·6 kW; 2·25 m^3/s)
3. A turbine running at 150 r.p.m. discharges 3·5 cubic metres of water per second under a head of 40 metres and produces 1000 kW. Find the normal speed and power developed by the turbine under a head of 62·5 metres. (**Ans.** 187·5 r.p.m.; 1953 kW)

34·7 Specific Speed of a Turbine

After studying the behaviour of a turbine, under unit conditions, the next step is to know the characteristics of an imaginary turbine identical[*] with the actual turbine, but reduced to such a size so as to develop a unit power under a unit head (*i.e.*, 1 kW under a head of 1 metre). This imaginary turbine is called the specific turbine and its speed is known as specific speed. Thus the specific speed of a turbine may be defined as the speed of an imaginary turbine, identical with the given turbine, which will develop a unit power under a unit head.

Let N_s = Specific speed of turbine,

D = Diameter of the turbine runner,

N = Speed of the runner in r.p.m.

v = Tangential velocity of the runner,

V = Absolute velocity of the water.

We know that the tangential velocity of the runner,

$$v \propto V$$

∴ $\propto \sqrt{2gH}$ ...(∵ $V = \sqrt{2gH}$)

$\propto \sqrt{H}$

We also know that the tangential velocity of the runner,

$$v = \dfrac{\pi DN}{60}$$

∴ $D \propto \dfrac{v}{N}$

$\propto \sqrt{H}$ ...(∵ $v \propto \sqrt{H}$)

and $D \propto \dfrac{\sqrt{H}}{N}$ ...(i)

Now let Q = Discharge through the turbine,

b = Width of the turbine runner,

V_f = Velocity of flow, and

D = Diameter of the turbine runner.

---

[*] Identical means, geometrically similar, and having same blade angles.

We know that the discharge of a turbine,
$$Q = \pi D b V_f$$
But $\quad b \propto D$
and $\quad V_f \propto \sqrt{2gH}$
$$\propto \sqrt{H}$$
$\therefore \quad Q \propto \pi D.D.\sqrt{2gH}$
$$\propto D^2\sqrt{H}$$

Substituting the value of D^2 from equation (i), in the above equation,
$$Q \propto \left(\frac{\sqrt{H}}{N}\right)^2 \times \sqrt{H}$$
$$\propto \frac{H^{3/2}}{N^2} \qquad \qquad \ldots(ii)$$

Now let $\quad P =$ Power produced by the turbine.

We know that the power,
$$P = wQH$$
$$\propto QH$$

Substituting the value of Q from equation (ii)
$$P \propto \frac{H^{3/2}}{N^2} \times H$$
$$\propto \frac{H^{5/2}}{N^2}$$

or $\quad N^2 \propto \dfrac{H^{5/2}}{P}$

$\therefore \quad N \propto \dfrac{H^{5/4}}{\sqrt{P}}$

$$= N_s \times \frac{H^{5/4}}{\sqrt{P}}$$

or $\quad N_s = \dfrac{N\sqrt{P}}{H^{5/4}}$

In the above relation for specific speed, it is useful to express P in kW, H in metres and N in r.p.m.

Example 34·3 *A reaction turbine is working under a head of 9 metres and average discharge of 11 200 litre/s for a generator speed of 200 r.p.m. What is its specific speed? Assume overall efficiency of the turbine = 92%.*

Solution. Given: $H = 9$ m; $Q = 11\,200$ litre/s $= 11·2$ m³/s; $N = 200$ r.p.m. and $\eta_0 = 92\% = 0·92$

Let $\quad P =$ Power developed by the turbine.

Perfromance of Turbines

We know that overall efficiency of the turbine (η_0),

$$0.92 = \frac{P}{wQH} = \frac{P}{9.81 \times 11.2 \times 9} = \frac{P}{988.8}$$

$\therefore$ $P = 0.92 \times 988.8 = 909.7$ kW

and specific speed of the turbine,

$$N_s = \frac{N\sqrt{P}}{H^{5/4}} = \frac{200 \times \sqrt{909.7}}{(9)^{5/4}} = \frac{6032}{15.6} = 386.7 \text{ r.p.m. Ans.}$$

Example 34·4 *A turbine develops 10 000 kW under a head of 25 metres at 135 r.p.m. What is its specific speed? What would be its normal speed and output under a head of 20 metres?*

Solution. Given : $P = 10\ 000$ kW; $H = 25$ m; $N = 135$ r.p.m.; and $H_1 = 20$ m.

Specific speed of the turbine

We know that specific speed of the turbine,

$$N_s = \frac{N\sqrt{P}}{H^{5/4}} = \frac{135 \times \sqrt{10\ 000}}{(25)^{5/4}} = \frac{13\ 500}{55.9} = 241.5 \text{ r.p.m. Ans.}$$

Normal speed and output of the turbine.

We also know that normal speed of the turbine,

$$N_1 = N \times \left(\frac{H_1}{H_2}\right)^{1/2} = 135 \times \left(\frac{20}{25}\right)^{1/2} \text{ r.p.m.}$$

$$= 135 \times 0.894 = 120.7 \text{ r.p.m. Ans.}$$

and normal output of the turbine,

$$P_1 = P \times \left(\frac{H_1}{H_2}\right)^{3/2} = 10\ 000 \times \left(\frac{20}{25}\right)^{3/2} \text{ kW}$$

$$= 10\ 000 \times 0.716 = 7160 \text{ kW Ans.}$$

Example 34·5 *One of the Kaplan turbines, installed at Ganguwal Power House (Bhakra Dam Project) is rated at 25 000 kW when working under 30 m of head at 180 r.p.m. Find the diameter of the runner, if overall efficiency of the turbine is 0·91. Assume flow ratio of 0·65 and diameter of runner hub equal to 0·3 times the external diameter of runner. Also find specific speed of the turbine.*

Solution. Given : $P = 25\ 000$ kW; $H = 30$ m; $N = 180$ r.p.m.; $\eta_0 = 0.91$; $\frac{V_f}{\sqrt{2gH}} = 0.65$

or $V_f = 0.65 \times \sqrt{2gH} = 0.65 \times \sqrt{2 \times 9.81 \times 30} = 15.8$ m/s and $D_b = 0.3D$

Diameter of the runner.

Let D = Diamter of the runner, and
Q = Discharge of the turbine.

We know that overall efficiency of the turbine (η_0),

$$0.91 = \frac{P}{wQH} = \frac{25\ 000}{9.81 \times Q \times 30} = \frac{84.9}{Q}$$

$\therefore$ $Q = 84.9/0.91 = 93.3$ m³/s

We also know that discharge of the turbine (Q),

$$93.3 = V_f \times \frac{\pi}{4}(D^2 - D_b^2) = 15.8 \times \frac{\pi}{4}\left[D^2 - (0.3D)^2\right] = 11.3D^2$$

$\therefore$ $D^2 = 93.3/11.3 = 8.26$ or $D = 2.87$ m **Ans.**

Specific speed of the turbine

We also know that specific speed of the turbine,

$$N_s = \frac{N\sqrt{P}}{H^{5/4}} = \frac{180 \times \sqrt{25\,000}}{(30)^{5/4}} = \frac{28\,460}{70 \cdot 2} = 405 \text{ r.p.m. Ans.}$$

Example 34·6 *In a hydroelectric station, the water is available under a head of 15 m at the rate of 100 m³/s. Calculate the number of turbines with a speed of 65 r.p.m. and 82% efficiency. The specific speed of the turbines is not to exceed 125 r.p.m. Also calculate the power produced by each turbine.*

Solution. Given: $H = 15$ m; Water available $= 100$ m³/s; $N = 65$ r.p.m.; $\eta_0 = 82\% = 0.82$ and $N_s = 125$ r.p.m.

Power produced by each turbine

Let P = Power produced by each turbine.

We know that specific speed of the turbine (N_s),

$$125 = \frac{N\sqrt{P}}{H^{5/4}} = \frac{65\sqrt{P}}{(15)^{5/4}} = \frac{65\sqrt{P}}{29 \cdot 5} = 2 \cdot 2\sqrt{P}$$

$\therefore \quad \sqrt{P} = 125/2 \cdot 2 = 56 \cdot 8$ or $P = 3226$ kW **Ans.**

Number of turbines

Let Q = Discharge of each turbine.

We know that efficiency of the turbine (η_0),

$$0 \cdot 82 = \frac{P}{wQH} = \frac{3226}{9 \cdot 81 \times Q \times 15} = \frac{21 \cdot 9}{Q}$$

or $\quad Q = 21 \cdot 9 / 0 \cdot 82 = 26 \cdot 7$ m³/s

$\therefore$ Number of turbines

$$= \frac{\text{Water available}}{\text{Discharge of each turbine}} = \frac{100}{26 \cdot 7} = 3 \cdot 7 \text{ say 4 Ans.}$$

34·8 Significance of Specific Speed

The significant feature of the specific speed, of a turbine, is that it is independent of the dimensions or size of the both actual and specific turbines. It is thus obvious, that all the turbines, geometrically similar, working under the same head and having the same values of speed ratio and flow ratio will have the same specific speed.

In actual practice, the conception of specific speed is of the utmost utility. The mere value of specific speed helps us in predicting the performance of a turbine, which will be discussed in the following pages.

34·9 Selection of Turbines

An engineer is often required to select the type of turbine, which he should employ for his project. It is a highly technical job and requires great experience and patience. The selection of turbine is, generally, basd upon the following two factors:

1. Selection based on the specific speed, and
2. Selection based on the head of water.

The former (*i.e.*, selection based on the specific speed) is a scientific method, and gives a precise information, whereas the later (*i.e.*, selection based on the head of water) is based on experience and observational factors only.

34·10 Selection Based on Specific Speed

First of all, specific speed of a turbine is found out, as usual. Then the type of turbine is selected. Following table shows the type of turbine to be selected, for the corresponding specific speeds.

TABLE 34·1

S. No.	Specific speed	Type of turbine
1.	8 to 30	Pelton wheel with one nozzle.
2.	30 to 50	Pelton wheel with 2 or more nozzles.
3.	50 to 250	Francis' turbine.
4.	250 to 1000	Kaplan turbine.

34·11 Selection Based on Head of Water

Following table shows the type of turbine, to be used, for the corresponding head of water.

TABLE 34·2

S. No.	Head of water in metres	Type of turbine
1.	0 to 25	Kaplan or Francis' (preferably Kaplan)
2.	25 to 50	Kaplan or Francis (preferably Francis)
3.	50 to 150	Francis
4.	150 to 250	Francis or Pelton (preferably Francis)
5.	250 to 300	Francis or Pelton (preferably Pelton)
6.	above 300	Pelton

Example 34·7 *Find the type of turbine, which should be used under a head of 150 metres to develop 1500 kW, while running at 300 r.p.m.*

Solution. Given : $H = 150$ m; $P = 1500$ kW and $N = 300$ r.p.m.

We know that specific speed of the turbine,

$$N_s = \frac{N\sqrt{P}}{H^{5/4}} = \frac{300 \times \sqrt{1500}}{(150)^{5/4}} = \frac{11\,610}{525} = 22 \cdot 1 \text{ r.p.m.}$$

Since the specific speed of the turbine lies between 8 and 30, therefore Pelton wheel with one nozzle should be used. **Ans.**

Example 34·8 *Fnd the specific speed and the type of a water turbine developing 7 000 kilowatt under a head of 20 metres when running at 100 r.p.m. Calculate its normal speed and output under 25 metres head.*

Solution. Given : $P = 7000$ kW; $H = 20$ m; $N = 100$ r.p.m. and $H_1 = 25$ m

Specific speed of the turbine and type of the turbine

We know that specific speed of the turbine,

$$N_s = \frac{N\sqrt{P}}{H^{5/4}} = \frac{100 \times \sqrt{7000}}{(20)^{5/4}} = \frac{8370}{42 \cdot 3} = 197 \cdot 8 \text{ r.p.m. \textbf{Ans.}}$$

Since the specific speed of the turbine lies between 50 and 250, therefore Francis turbine should be used. **Ans.**

Normal speed and output of the turbine

We also know that normal speed of the turbine

$$N_1 = N\left(\frac{H_1}{H}\right)^{1/2} = 100 \times \left(\frac{25}{20}\right)^{1/2} = 100 \times 1\cdot12 = 120 \text{ r.p.m. Ans.}$$

and normal output of the turbine,

$$P = P\left(\frac{H_1}{H_2}\right)^{3/2} = 7000 \times \left(\frac{25}{20}\right)^{3/2} = 7000 \times 1\cdot4 = 9800 \text{ kW Ans.}$$

Example 34·9 *A single jet Pelton wheel develops 2500 kW under a head of 70 metres. Find the maximum and minimum speeds of the turbine.*

Solution. Given : $P = 2500$ kW and $H = 70$ m

Maximum speed of turbine

Let N_1 = Maximum Speed of the wheel.

From table 34·1, we find that the maximum specific speed of a Pelton wheel with single nozzle, (N_s) is equal to 30.

We know that specific speed of the turbine (N_s),

$$30 = \frac{N_1\sqrt{P}}{H^{5/4}} = \frac{N_1\sqrt{2500}}{(70)^{5/4}} = \frac{50\,N_1}{202\cdot5} = 0\cdot247\,N_1$$

∴ $N_1 = 30/0\cdot247 = 121\cdot5$ r.p.m. **Ans.**

Minimum speed of the turbine

Let N_2 = Minimum speed of the turbine

From the same table, we also find that minimum specific speed of a Pelton wheel with single nozzle (N_s) in equal to 8.

We know that specific speed of the turbine (N_s),

$$8 = \frac{N_2\sqrt{P}}{H^{5/4}} = \frac{N_2\sqrt{2500}}{(70)^{5/4}} = \frac{50\,N_2}{202\cdot5} = 0\cdot247\,N_2$$

∴ $N_2 = 8/0\cdot247 = 32\cdot4$ r.p.m. **Ans.**

Example 34·10 *The total power generated in a hydroelectric station is 18 000 kW under a head of 16 m, while the turbines run with a speed of 192 r.p.m. Find the minimum number of turbines of the same size required in case of*

   (i) Francis' turbines with maximum specific speed of 210 r.p.m. and
   (ii) Kaplan turbines with maximum specific speed of 300 r.p.m.

Solution. Given : $P = 18\,000$ kW; $H = 16$ m and $N = 192$ r.p.m.

(i) Number of Francis turbines with maximum specific speed of 210 r.p.m.

Let P_1 = Power generated by each Francis' turbine

Performance of Turbines

We know that maximum specific speed of the Francis' turbine (N_s).

$$210 = \frac{N\sqrt{P_1}}{H^{5/4}} = \frac{192\sqrt{P_1}}{(16)^{5/4}} = \frac{192\sqrt{P_1}}{32} = 6\sqrt{P_1}$$

or $\quad \sqrt{P_1} = 210/6 = 35 \quad$ or $\quad P_1 = (35)^2 = 1225$ kW

∴ Number of Francis' turbines

$$= \frac{\text{Total power generated}}{\text{Power generated by each turbine}}$$

$$= \frac{18\,000}{1225} = 14.7 \text{ say } 15 \text{ Ans.}$$

(ii) Number of Kaplan turbines with maximum specific speed of 300 r.p.m.

Let $\quad P_2 =$ Power generated by each Kaplan turbine.

We also know that maximum specific speed of the Kaplan turbine (N_s),

$$300 = \frac{N\sqrt{P_2}}{H^{5/4}} = \frac{192\sqrt{P_2}}{(16)^{5/4}} = \frac{192\sqrt{P_2}}{32} = 6\sqrt{P_2}$$

or $\quad \sqrt{P_2} = 300/6 = 50 \quad$ or $\quad P_2 = (50)^2 = 2500$

∴ Number of Kaplan turbines

$$= \frac{\text{Total power generated}}{\text{Power generated by each turbine}}$$

$$= \frac{18\,000}{2500} = 7.2 \text{ say } 8 \text{ Ans.}$$

34·12 Relation between Specific Speed and Shape of Reaction Turbine Runner

We have already discussed in Art. 34·7 that the specific speed of a turbine,

$$N_s = \frac{N\sqrt{P}}{(H)^{5/4}}$$

Since the power produced by a turbine and the available head of water is more or less constant, for a power house, therefore specific speed is directly proportional to the speed of the turbine runner.

We have also discussed that power produced by a turbine,

$$P = wQH$$

Since the value of w is constant, therefore power produced is directly proportional to Q (*i.e.* discharge) and H (head of water). Moreover, the head of water is also fixed for every power house. It is thus obvious, that the power produced by a turbine, in a power house, depends upon the discharge. A little consideration will show that under a low head, if the same power is to be produced, then more flow is required. This can be achieved either by increasing the area of flow or velocity of water. In a reaction turbine, this is achieved by increasing the height of gates and velocity of flow.

Fig. 34·1 (*a*) to (*d*) shows the changes in the shape of blades of the turbine runner of an inward flow reaction turbine. The relative details of the inlet triangle of velocities are discussed here :

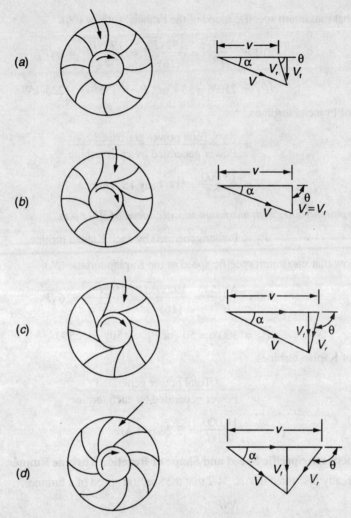

Fig. 34·1. Effect of specific speed on the shape of reaction turbine runner

1. *Fig. 34·1 (a)*

It is a general case, in which the inlet triangle of velocities is shown for a slow speed reaction turbine. In this case, the common features are:

$$N_s = 60 \text{ to } 120 \text{ r.p.m.}$$
$$\alpha = 10° \text{ to } 20°$$
$$\theta = 60° \text{ to } 90°$$

2. *Fig. 34·2 (b)*

In this case, the power developed (P) and diameter of the turbine runner (D) are the same as in the first case. Due to reduction of the available head of water, specific speed of the turbine and its discharge will increase. A little consideration will show that:

---

*The specific speed of a turbine and its discharge are inversely proportional to the available head of water, because specific speed of a turbine,

$$N_s = \frac{N\sqrt{P}}{H^{5/4}} \quad \text{and} \quad P = wQH \quad \text{or} \quad Q = \frac{P}{wH}$$

Perfromance of Turbines

(i) The reduction in the available head of water will reduce the velocity of water ($\because V = \sqrt{2gH}$).

(ii) The increase in specific speed will increase the speed of the turbine runner, and ultimately the tangential velocity of the wheel at inlet $\left(\because v = \dfrac{\pi DN}{60}\right)$

(iii) The increase in discharge of water will increase its velocities. In this case, the common features are :

$$N_s = 120 \text{ to } 180 \text{ r.p.m.}$$
$$\alpha = 20° \text{ to } 30°$$
$$\theta = 90°$$

3. Fig. 34·1 (c)

In this case, the power developed (P) and diameter of turbine runner (D) are the same as in the previous cases. Due to further reduction in the available head of water, specific speed of the turbine and its discharge will further increase. A little consideration will show, that this phenomenon will (i) reduce the velocity of water, (ii) increase the tangential velocity of wheel at inlet, and (iii) increase the velocity of flow.

The above mentioned changes will further change the shape of inlet triangle of veiocities. In this case, the common features are :

$$N_s = 180 \text{ to } 240 \text{ r.p.m.}$$
$$\alpha = 30° \text{ to } 45°$$
$$\theta = 90° \text{ to } 120°$$

4. Fig. 34·1 (d)

In this case, the power developed (P) and diameter of the turbine runner (D) are the same as in previous cases. Due to further reduction in the available head of water, specific speed of the turbine and its discharge will further increase. This phenomenon will further reduce the velocity of water, increase the tangential velocity of wheel at inlet and increase the velocity of flow. These changes will further change the shape of inlet triangle of velocities. In this case, the common features are :

$$N_s = 240 \text{ to } 300 \text{ r.p.m.}$$
$$\alpha = 45° \text{ to } 60°$$
$$\theta = 120° \text{ to } 135°$$

It will be interesting to know, that the above mentioned theory led to the development of Kaplan turbine.

34·13 Characteristics Curves of Turbines

We have discussed in the chapters 32 and 33 the various types of impulse turbines and reaction turbines. As a matter of fact, a turbine is always designed and manufactured to work under a given set of conditions (or a limited range of conditions) such as discharge, head of water, speed, power generated, efficiency etc. (at full speed or unit speed). But a turbine may have to be used under conditions, other than those for which it has been designed. It is therefore essential, that the exact behaviour of the turbine under varied conditions should be predetermined. This is represented graphically by means of curves, known as characteristic curves.

The characteristic curves are generally drawn for constant head or constant speed of the turbine runner. Sometimes, these curves are also drawn for various gate openings (briefly written as G.O.) i.e. when the gate is fully open, 0·75 open, 0·5 open etc. Though there are many types of characteristic curves, yet the following are important from subject point of view:

34·14 Characteristic Curves for Pelton Wheels

The following curves have been drawn by the various engineers, working in hydraulic research laboratories all over the globe. Though there is a little variation in the characteristic curves drawn by them, yet the most accepted and important curves are given below:

1. Characteristic curves for constant head

(a) Speed ratio versus percentage of maximum efficiency

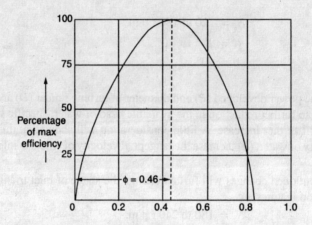

Fig. 34·2. Characteristic curve for speed ratio *vs* percentage of maximum efficiency

Fig. 34·2 shows the performance of a Pelton wheel under a constant head and discharge. It is a parabolic curve, which shows that the efficiency increases from zero, and beyond the value of $\phi = 0.46$, the efficiency decreases.

(b) Power versus efficiency

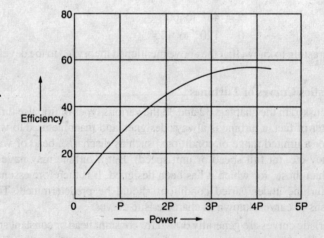

Fig. 34·3. Characteristic curve for power *vs* efficiency

Perfromance of Turbines

Fig. 34·3 shows the performance of a Pelton wheel under a constant head and speed. It is a parabolic curve, which shows that the efficiency increases with the increase in power.

2. Characteristic curves for varying gate opening

(a) Speed versus power

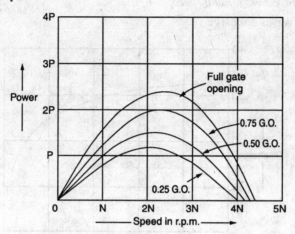

Fig. 34·4. Characteristic curves for speed vs power

Fig. 34·4 shows the performance of a Pelton wheel under a constant head. These are parabolic curves, which show that the power developed increases with the gate opening.

(b) Speed versus efficiency

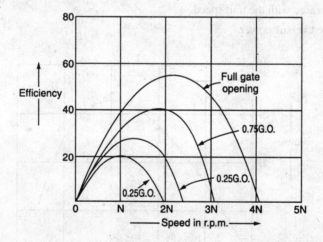

Fig. 34·5. Characteristic curves for speed vs efficiency.

Fig. 32·5 shows the performance of a Pelton wheel under a constant head. These are parabolic curves, which show that the efficiency increases with the gate opening.

34·15 Characteristic Curves of Francis' Turbines

Like characteristic curves of Pelton wheel, the following curves have also been drawn by the various engineers. In general, the characteristic curves for Francis' turbine (or any other reaction turbine) may be grouped under the following three heads:

1. Characteristic curves for unit speed,
2. Characteristic curves for speed, and
3. Characteristic curves for varying gate opening.

1. Characteristic curves for unit speed

(a) Unit speed versus discharge

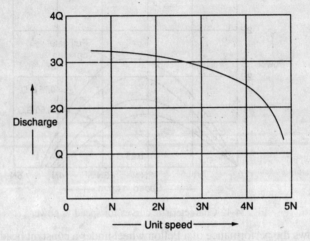

Fig. 34·6. Characteristic curves for unit speed *vs* discharge

Fig. 34·6 shows the performance of a reaction turbine. It is a parabolic curve which shows that the discharge decreases with the unit speed.

(b) Unit speed versus power

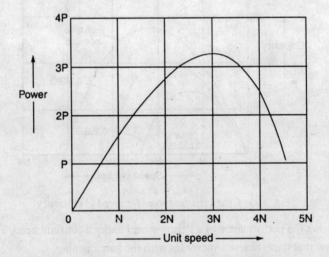

Fig. 34·7. Characteristic curves for unit speed *vs* power

Fig. 34·7 shows the performance of a reaction turbine. It is a parabolic curve, which shows that the power increases with the unit speed, and beyond a certain speed the power decreases.

(c) Unit speed versus efficiency

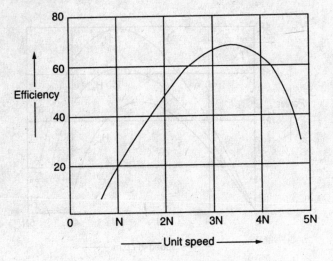

Fig. 34·8. Characteristic curves for unit speed *vs* efficiency

Fig. 34·8 shows the performance of a reaction turbine. It is a parabolic curve, which shows that the efficiency increases with the unit speed, and beyond a certain speed the efficiency decreases.

2. *Characteristic curves for speed with varying heads*

(a) Speed versus discharge

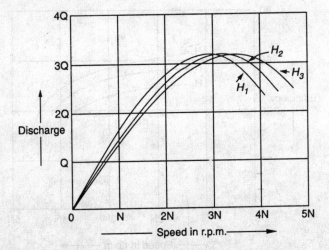

Fig. 34·9. Characteristic curves for speed *vs* discharge

Fig. 34·9 shows the performance of a Francis' turbine (or any other reaction turbine) under variable heads, but constant discharge. It is a parabolic curve, which shows that for a given head the discharge increases with the speed from zero, and beyond a certain speed the discharge decreases.

(b) Speed versus power

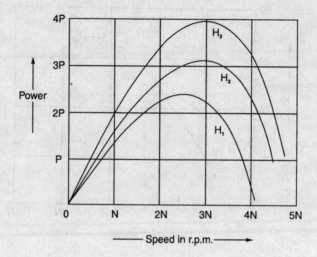

Fig. 34·10. Characteristic curves for speed *vs* power

Fig. 34·10 shows the performance of a Francis' turbine (or any other reaction turbine) under variable heads but constant discharge. It is a parabolic curve, which shows that for a given head the power increases with the speed from zero, and beyond a certain speed the power decreases.

(c) Speed versus efficiency

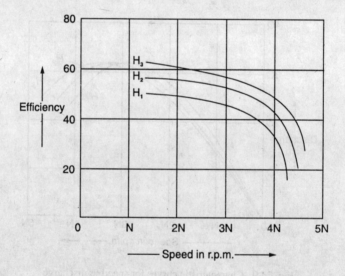

Fig. 34·11. Characteristic curves for speed *vs* efficiency

Fig. 34·11 shows the performance of a Francis' turbine (or any reaction turbine) under variable heads but constant discharge. It is a parabolic curve, which shows that for a given head the efficiency decreases with the increase in speed.

3. Characteristic curves for varying gate opening
   (a) Speed versus power

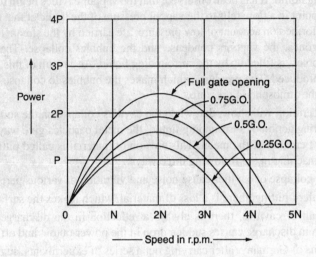

Fig. 34·12. Characteristic curves for speed *vs* power

Fig. 34·12 shows the performance of a Francis' turbine (or any other reaction turbine) under a constant head. These are parabolic curves, which show that the power developed increases with the gate opening.

   (b) Speed versus efficiency

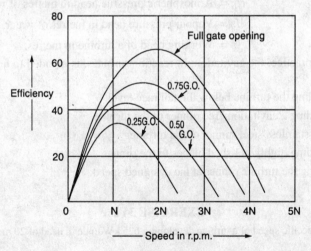

Fig. 34·13. Characteristic curves for speed *vs* efficiency

Fig. 34·13 shows the performance of a Francis' turbine (or any other reaction turbine) under a constant head. These are parabolic curves, which show that the efficiency increases with the gate opening.

34·16 Cavitation

The cavitation may be broadly defined as the formation of bubbles, filled with vapours, within the body of a moving liquid. It has been observed, that the vapour cavities begin to appear, whenever the pressure at any point in a flow falls to the vapour pressure of the liquid at that temperature. These bubbles, which are formed on account of low pressure, are carried by the stream to the zones of high pressure. In these zones, the vapours condense and the bubbles collapse. The space, previously occupied by the bubbles, is filled up by the surrounding liquid. As a result of this, some noise occurs and vibrations are produced. The pressure, which makes the bubbles to collapse, is generally of the order of 100 times the atmospheric pressure.

A little consideration will show, that when the cavities collapse on the surface of a body, due to repeated hammering action of surrounding liquid, the metal particles give way, which ultimately cause a great deal of erosion of the metal. This erosion of material is called pitting. The cavitation effects a hydraulic machine in the following three ways:

1. Irregular collapse of cavities cause noise and vibration of various parts.
2. As a result of pitting, there is a loss of material, which makes the surfaces rough.
3. As a result of cavities, there is always a reduction in the discharge of a turbine. The reduction in discharge causes sudden drop in the power output and efficiency.

Prof. D. Thoma of Germany, after carrying out a series of experiments, suggested a cavitation factor σ (sigma) to find out the zone, where a reaction turbine can work, without the effect of cavitation. The critical value of this factor is given by:

$$\sigma_{crit} = \frac{H_b - H_s}{H} = \frac{(H_a - H_v - H_s)}{H}$$

where
H_b = Barometric pressure in metres of water,
H_s = Suction pressure head in metres of water,
H_a = Atmospheric pressrue head in metres of water,
H_v = Vapour pressure head in metres of water, and
H = Working head of a turbine in metres.

But in actual practice, the cavitation, in reaction turbines, is avoided to a great extent by the following methods:

1. By installing the turbine below the tail race level.
2. By providing a cavitation-free runner of the turbine.
3. By using stainless steel runner of the turbine.
4. By providing highly polished blades to the runner.
5. By running the turbine runner at the designed speed.

EXERCISE 34·2

1. Find the specific speed of a turbine developing 625 kW under a head of 20 metres of 150 r.p.m.
   (**Ans.** 88·7 r.p.m.)
2. A turbine develops 1225 kW under a head of 64 metres while running at 120 r.p.m. Find the type of turbine suited for the project. (**Ans.** N_s = 23·2 r.p.m.; Peltan wheel with one nozzle)
3. Find the type of turbine you would employ to develop 1700 kW under a head of 150 metres at 350 r.p.m., while discharging 1·75 cubic metres of water per second.
   (**Ans.** N_s = 34·5; Peltan wheel with two nozzles).

Performance of Turbines

QUESTIONS

1. Define the terms 'unit power' 'unit speed' and 'unit discharge' with reference to a hydraulic turbines.
2. What do you understand by the terms 'specific turbine' and specific speed?
3. Derive an expression for the specific speed of a turbine in terms of power (P), head (H) and speed in r.p.m. (N).
4. Explain the significance of specific speed of water turbines.
5. Give the range of specific speed values of the Kaplan, Francis' and Pelton wheels.
6. What factors decide whether a Kaplan, Francis' or a Pelton type turbine would be used in a hydro-electric project.
7. Draw the following characteristic curves of a turbine.
    (i) Power *versus* efficiency with constant speed.
    (ii) Discharge, power and efficiency *versus* speed.
8. Draw the characteristic curves of water turbine showing rate of flow and power *versus* unit speed.

OBJECTIVE TYPE QUESTIONS

1. The value of unit power is equal to
    (a) $\dfrac{P}{H}$
    (b) $\dfrac{P}{\sqrt{H}}$
    (c) $\dfrac{P}{H^2}$
    (d) $\dfrac{P}{H^{3/2}}$

2. The value of unit speed is equal to
    (a) $\dfrac{N}{H}$
    (b) $\dfrac{N}{\sqrt{H}}$
    (c) $\dfrac{N}{H^2}$
    (d) none of these

3. The value of unit discharge is equal to
    (a) $\dfrac{Q}{H}$
    (b) $\dfrac{Q}{\sqrt{H}}$
    (c) $\dfrac{Q}{42}$
    (d) $\dfrac{Q}{H^3}$

4. The specific speed of a turbine is given by the relation
    (a) $\dfrac{NP}{H}$
    (b) $N_s = \dfrac{N\sqrt{P}}{H}$
    (c) $\dfrac{N\sqrt{P}}{H^{5/4}}$
    (d) $\dfrac{N\sqrt{P}}{H^{3/2}}$

Answers

1. (d)   2. (b)   3. (b)   4. (c)

35

Centrifugal Pumps

1. Introduction. 2. Types of Pumps. 3. Centrifugal Pump. 4. Types of Casings for the Impeller of a Centrifugal Pump. 5. Volute Casing (Spiral Casing). 6. Vortex Casing. 7. Volute Casing with Guide Blades. 8. Piping System of a Centrifugal Pump. 9. Work Done by a Centrifugal Pump. 10. Manometric Head. 11. Efficiencies of a Centrifugal Pump. 12. Manometric Efficiency. 13. Mechanical Efficiency. 14. Overall Efficiency. 15. Discharge of a Centrifugal Pump. 16. Power Required to Drive a Centrifugal Pump. 17. Increase in the Water Pressure while Flowing through the Impeller of a Centrifugal Pump. 18. Minimum Starting Speed of a Centrifugal Pump. 19. Multistage Centrifugal Pumps.

35·1 Introduction

Since the olden times, the man has been trying to find some convenient ways of lifting water to higher levels, for water supply or irrigation purposes. It is believed, that the idea of lifting water, by centrifugal force, was first given by L.D. Vinci (an Italian scientist and engineer) in the end of 16th century. Then this idea was put to experiments by French scientist and they designed centrifugal pump, having impeller and blades. At that time, the reciprocating* pumps were very popular. Then a continuous advancement of this pump has brought it to a high degree of perfection, which is used all over the world these days.

35·2 Types of Pumps

Though there are many types of pumps these days, yet the following two are important from the subject point of view:

1. Centrifugal pump, and
2. Reciprocating pump.

In this chapter, we shall discuss the centrifugal pumps only.

35·3 Centrifugal Pump

A pump, in general may be defined as a machine, when driven from some external source, lifts water or some other liquid from a lower level to a higher level. Or in other words, a pump may also be defined as a machine, which converts mechanical energy into pressure energy. The pump which raises water or a liquid from a lower level to a higher level by the action of centrifugal force, is known as a centrifugal pump.

It will be interesting to know that the action of a centrifugal pump is that of a reversed reaction turbine. In a reaction turbine, the water at high pressure, is allowed to enter the casing which gives mechanical energy at its shaft; whereas in a pump, the mechanical energy is fed into the shaft and water enters the impeller (attached to the rotating shaft) which increases the pressure energy of the out-going fluid. The water enters the impeller radially and leaves the vanes axially.

35·4 Types of Casings for the Impeller of a Centrifugal Pump

We have discussed in Art 35·3 that a centrifugal pump consists of an impeller, similar to that of a turbine, to which curved vanes are fitted. The impeller is enclosed in a water-tight casing, having a delivery pipe in one of its sides. The casing for a centrifugal pump is so designed that the kinetic

---

*Reciprocating pumps will be discussed in the next chapter.

energy of the water is converted into pressure energy before the water leaves the casing. This considerably increases the efficiency of the pump. Following are the three types of casings or chambers of centrifugal pumps :

1. Volute casing (spiral casing),
2. Vortex casing, and
3. Volute casing with guide blades.

35·5 Volute Casing (Spiral Casing)

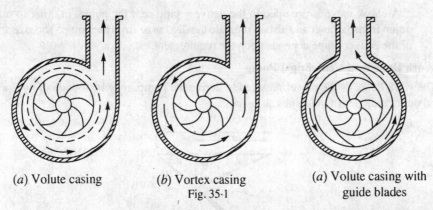

(a) Volute casing   (b) Vortex casing   (a) Volute casing with guide blades

Fig. 35·1

In a volute chamber, the impeller is surrounded by a spiral casing as shown in Fig. 35·1(a). Such a casing provides a gradual increase in the area of flow; which decreases the velocity of water, with a corresponding increase in pressure.

A considerable loss takes place due to the formation of eddies in this type of casing.

35·6 Vortex Casing

It is an improved type of a volute casing, in which the spiral casing is combined with a circular chamber as shown in Fig. 35·1 (b). In a vortex casing, the eddies are reduced to a considerable extent and an increased efficiency is obtained.

35·7 Volute Casing with Guide Blades

In this type of casing, there are guide blades surrounding the impeller as shown in Fig. 35·1 (c). These guide blades are arranged at such an angle, that the water enters without shock and forms a passage of increasing area, through which the water passes and reaches the delivery pipe.

The ring of the guide blades is called difuser and is very efficient.

35·8 Piping System of a Centrifugal Pump

The successful working of a centrifugal pump depends upon the correct selection and lay out of its piping system. An extreme care should always be taken in selecting the sizes of the pipes and their arrangement. In general, a centrifugal pump has (a) suction pipe, and (b) delivery pipe.

(a) *Suction pipe*

The suction pipe, of a centrifugal pump, plays an important role in the successful and smooth working of the pump. A poorly designed suction pipe causes insufficient net positive suction head (brief written as *NPSH), vibration, noise, water hummer, excessive wear etc. While laying the pipe, a great care should be taken to make it air tight. A strainer foot-valve is connected at the bottom of the suction pipe to avoid the entry of foreign

---

*It is an important term in pumps. For details please refer to Art. 37·9.

matter. Since the pressure at the inlet of the pump is suction (*i.e.*, negative) and its value is limited to avoid cavitation, it is therefore essential that the losses in the suction pipe should be as small as possible. For this purpose, bends in the suction pipe are avoided and its diameter is often kept larger.

Sometimes, to reduce the axial thrust, the suction pipe is branched into two parts and the liquid is allowed to enter the impeller from both sides. Such a pump is called double suction pump.

(b) *Delivery pipe*

A check valve is provided in the delivery pipe near the pump, in order to protect the pump from hammer and also to regulate the discharge from the pump. The size and length of the delivery pipe depends upon the requirement.

35·9 Work Done by a Centrifugal Pump

The work done, or the power required to derive the pump, may be found out by drawing the inlet and outlet triangles of velocities as usual.

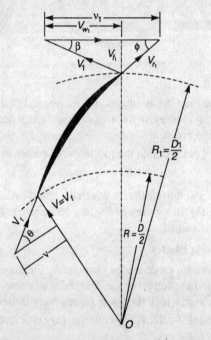

Fig. 35·2. Inlet and outlet triangle of velocities

Consider a centrifugal pump lifting water from a lower level to a higher level. Now draw the inlet and outlet triangle of velocities as shown in Fig. 35·2.

Let
- V = Absolute velocity of the entering water,
- D = Diameter of the impeller at inlet (inner diameter),
- v = Tangential velocity of impeller at inlet (Also known as peripheral velocity at inlet),
- V_r = Relative velocity of water to the wheel at inlet,
- V_f = Velocity of flow at inlet,
- $V_1, D_1, v_1, V_{r1}, V_{f1}$ = Corresponding values at the outlet,

N = Speed of the impeller in r.p.m.,
θ = Vane angle at inlet.
β = Angle at which the water leaves the impeller,
ϕ = Vane angle at outlet.

Since the water enters the impeller radially, therefore the velocity of whirl at inlet ($V_w = 0$).

∴ Work done per kN of water

$$= \frac{V_{w1} \cdot v_1}{g} \text{ kN-m}$$

where V_{w1} and v_1 are in m/s.

Example 35·1. *A centrifugal pump has external and internal impeller diameters as 600 mm and 300 mm respectively. The vane angle at inlet and outlet are 30° and 45° respectively. If the water enters the impeller at 2·5 m/s, find (a) speed of the impeller in r.p.m., (b) work done per kN of water.*

Solution. Given : $D_1 = 600$ mm $= 0.6$ m; $D = 300$ mm $= 0.3$ m; $\theta = 30°$; $\phi = 45°$ and $V = 2.5$ m/s.

(a) Speed of the impeller in r.p.m.

Let N = Speed of the impeller in r.p.m.

From the inlet triangle of velocities, we find that the tangential velocity of impeller at inlet,

$$v = \frac{V}{\tan 30°} = \frac{2.5}{0.5774} = 4.33 \text{ m/s}$$

We know that the velocity of impeller at inlet (v),

$$4.33 = \frac{\pi D N}{60} = \frac{\pi \times 0.3\, N}{60} = 0.0157\, N$$

∴ $N = 4.33/0.0157 = 275.8$ r.p.m. **Ans.**

(b) Work done per kN of water

From the outlet triangle of velocities, we find that the tangential velocity at outlet,

$$v_1 = v \times \frac{D_1}{D} = 4.33 \times \frac{0.6}{0.3} = 8.66 \text{ m/s}$$

and velocity of whirl at outlet,

$$V_{w1} = v_1 - \frac{V_{f1}}{\tan 45°} = 8.66 - \frac{2.5}{1} = 6.16 \text{ m/s}$$

...(∵ $V_{f1} = V_f = V$)

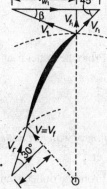

Fig. 35·3

Since the tangential velocity of impeller at outlet v_1 (8·66) is more than velocity of whirl at outlet V_{w1} (6·16), therefore shape of the outlet triangle will be as shown in Fig. 35·3. We know that work done per kN of water,

$$W = \frac{V_{w1} \cdot v_1}{g} = \frac{6.16 \times 8.66}{9.81} = 5.44 \text{ kN-m} = 5.44 \text{ kJ Ans.}$$

Example 35·2. *Calculate vane angle at the inlet of a centrifugal pump impeller having 200 mm diameter at inlet and 400 mm diameter at outlet. The impeller vanes are set back at angle of 45° to the outer rim, and the entry of the pump is radial. The pump runs at 1000 r.p.m. and the velocity of flow through the impeller is constant at 3 m/s. Also calculate the work done per kN of water and the velocity as well as direction of the water at outlet.*

Solution. Given : $D = 200$ mm $= 0.2$ m; s $D_1 = 400$ mm $= 0.4$ m; $\phi = 45°$; $N = 1000$ r.p.m. and $V_f = V_{f1} = 3$ m/s

Vane angle at inlet

Let θ = Vane angle at inlet.

We know that the tangential velocity of impeller at inlet,

$$v = \frac{\pi DN}{60} = \frac{\pi \times 0.2 \times 1000}{60}$$

$$= 10.5 \text{ m/s}$$

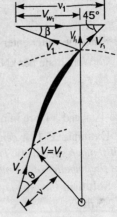

Fig. 35.4

From the inlet triangle of velocities, we find that

$$\tan \theta = \frac{V_f}{v} = \frac{3}{10.5} = 0.2857$$

$\therefore \quad \theta = 15.9°$ **Ans.**

Velocity of water at outlet

We know that the tangential velocity of impeller at outlet,

$$v_1 = \frac{\pi D_1 N}{60} = \frac{\pi \times 0.4 \times 1000}{60} = 20.9 \text{ m/s}$$

From the outlet triangle of velocities, we find that the velocity of whirl at outlet,

$$V_{w1} = v_1 - \frac{V_{f1}}{\tan 45°} = 20.9 - \frac{3}{1} = 17.9 \text{ m/s}$$

and

$$V_1 = \sqrt{V_{f1}^2 + V_{w1}^2} = \sqrt{(3)^2 + (17.9)^2} = 18.1 \text{ m/s } \textbf{Ans.}$$

Direction of water at outlet

Let β = Angle, at which the water leaves the impeller at outlet.

From the outlet triangle of velocities, we also find that

$$\tan \beta = \frac{V_{f1}}{V_{w1}} = \frac{3}{17.9} = 0.1676 \text{ or } \beta = 9.5° \textbf{ Ans.}$$

Work done per kN of water

We also know that work done per kN of water,

$$W = \frac{V_{w1} \cdot v_1}{g} = \frac{17.9 \times 20.9}{9.81} = 38.14 \text{ kN-m} = 38.14 \text{ kJ } \textbf{Ans.}$$

35.10 Manometric Head

It is an important term, in the field of centrifugal pumps, and may be defined in any one of the following four ways:

1. The manometric head is the actual head against which the pump has to work.
2. Manometric head,

$$H_m = H_s + H_{fs} + H_d + H_{fd} + \frac{V_d^2}{2g}$$

where H_s = Suction lift,

H_{fs} = Loss of head in suction pipe due to friction,

H_d = Delivery lift,

Centrifugal Pumps

H_{fd} = Loss of head in delivery pipe due to friction,

V_d = Velocity of water in the delivery pipe.

3. Manometric head,

H_m = Work done/kN of water – Losses within the impeller

$$= \frac{V_{w1} \cdot v_1}{g} - \text{Impeller losses}$$

4. Manometric head.

H_m = Energy/kN of water at outlet of impeller

– Energy/kN of water at inlet of impeller

35·11 Efficiencies of a Centrifugal Pump

A centrifugal pump has the following three types of efficiencies:
1. Manometric efficiency,
2. Mechanical efficiency, and
3. Overall efficiency.

35·12 Manometric Efficiency

It is the ratio of manometric head to the energy supplied by the impeller/kN of water. Mathematically manometric efficiency,

$$\eta_{man} = \frac{H_m}{\dfrac{V_{w1} v_1}{g}}$$

35·13 Mechanical Efficiency

It is the ratio of energy available at the impeller, to the energy given to the impeller by the prime power.

35·14 Overall Efficiency

It is the ratio of actual work done by the pump, to the energy supplied to the pump by the prime mover.

35·15 Discharge of a Centrifugal Pump

The discharge of a centrifugal may be found out by the same method as that of a reaction turbine. Now consider a centrifugal pump lifting water from a lower level to a higher level.

Let
$\quad D$ = Diameter of impeller at inlet,

$\quad V_f$ = Velocity of flow at inlet,

$\quad b$ = Width of impeller at inlet, and

D_1, b_1, V_{f1} = Corresponding values at the outlet.

Then the discharge, $\quad Q = \pi D b V_f = \pi D_1 b_1 V_{f1}$

Example 35·3. *A centrifugal pump is to discharge water at the reate of 110 litres/second at a speed of 1450 r.p.m. against a head of 23 metres. The impeller diameter is 250 mm and its width 50 mm. If the manometric efficiency is 75%, determine the vane angle at the outer periphery.*

Solution. Given : Q = 110 litres/s = 0·11 m³/s; N = 1450 r.p.m. H_m = 23 m; D_1 = 250 mm = 0·25 m; D = 50 mm = 0·05 m and η_{man} = 75% = 0·75.

Let $\quad\quad\quad\quad\quad\quad \phi$ = Vane angle at outlet,

V_{f1} = Velocity of flow at outlet, and
V_{w1} = Velocity of whirl at outlet.

We know that the tangential velocity of impeller at outlet,
$$v_1 = \frac{\pi D_1 N}{60} = \frac{\pi \times 0.25 \times 1450}{60}$$
$$= 19 \text{ m/s}$$

and discharge of the pump (Q),
$$0.11 = \pi D_1 b_1 V_{f1}$$
$$= \pi \times 0.25 \times 0.05 \times V_f$$
$$= 0.039 V_f$$
∴ $V_f = 0.11/0.039 = 2.8$ m/s.

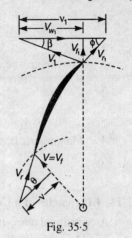
Fig. 35.5

We also know that manometric efficiency (η_{man}),
$$0.75 = \frac{H_m}{\dfrac{V_{w1} \cdot v_1}{g}} = \frac{23}{\dfrac{V_{w1} \times 19}{9.81}} = \frac{11.9}{V_{w1}}$$

∴ $V_{w1} = 11.9/0.75 = 15.9$ m/s

Now from the outlet triangle of velocities, we find that
$$\tan \phi = \frac{V_{f1}}{v_1 - V_{w1}} = \frac{2.8}{19 - 15.9} = 0.9032 \text{ or } \phi = 42.1° \textbf{ Ans.}$$

Example 35.4. *A centrifugal pump delivers water against a head of 14.5 metres while running at 1000 r.p.m. The vanes are curved back at an angle of 30° with the periphery. The impeller diameter is 300 mm and outlet width 50 mm. If manometric efficiency of the pump is 85%, find the discharge of the pump.*

Solution. Given : $H_m = 14.5$ m; $N = 1000$ r.p.m.; $\theta = 30°$; $D_1 = 300$ mm $= 0.3$ m; $b_1 = 50$ mm $= 0.05$ m and $\eta_{man} = 85\% = 0.85$,

Let V_{w1} = Velocity of whirl at outlet.

We know that tangential velocity of the impeller at outlet,
$$v_1 = \frac{\pi D_1 N}{60} = \frac{\pi \times 0.3 \times 1000}{60} \text{ m/s}$$
$$= 15.7 \text{ m/s}$$

and manometric efficiency (η_{man}),
$$0.85 = \frac{H_m}{\dfrac{V_{w1} \cdot v_1}{g}} = \frac{14.5}{\dfrac{V_{w1} \times 15.7}{9.81}}$$
$$= \frac{9.06}{V_{w1}}$$

∴ $V_{w1} = 9.06/0.85 = 10.7$ m/s

From the outlet triangle of velocities, we find that

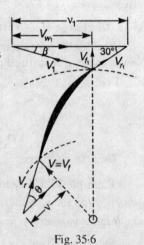
Fig. 35.6

$$\tan 30° = \frac{V_{f1}}{v_1 - V_{w1}} = \frac{V_f}{15.7 - 10.7}$$

or $\qquad 0.5774 = \dfrac{V_f}{5}$ or $V_f = 0.5774 \times 5 = 2.89$ m/s

∴ Discharge through the pump,
$$Q = \pi D_1 b_1 V_{f1} = \pi \times 0.3 \times 0.05 \times 2.89 = 1.136 \text{ m}^3/\text{s} \textbf{ Ans.}$$

Example 35·5. *A centrifugal pump discharges 7500 litres of water per minute against a total head of 25 metres when running at 660 r.p.m. The outer diameter of the impeller is 600 mm and the ratio of outer to inner diameter is 2. The area of flow, through the wheel is 0·06 m². The vanes are set back at an angle of 45°. Water enters the wheel radially and without shock. Calculate*

(a) *manometric efficiency, and*

(b) *vane angle at inlet.*

Solution. Given : $Q = 7500$ litres/min $= 0.125$ m³/s; $H_m = 25$ m; $N = 660$ r.p.m.; $D_1 = 600$ mm $= 0.6$ m; $\dfrac{D_1}{D} = 2$ or $D = \dfrac{D_1}{2} = \dfrac{0.6}{2} = 0.3$ m; Area of flow $(A) = \pi Db = 0.06$ m² and $\phi = 45°$.

(a) *Manometric effeciency*

We know that peripheral velocity at outlet,
$$v_1 = \dfrac{\pi D_1 N}{60} = \dfrac{\pi \times 0.6 \times 660}{60} \text{ m/s}$$
$$= 20.7 \text{ m/s}$$

and velocity of flow at outlet,
$$V_{f1} = \dfrac{Q}{\pi D_1 b_1} = \dfrac{0.125}{0.06} = 2.1 \text{ m/s}$$

From the inlet triangle of velocities, we find that the velocity of whirl at outlet,
$$V_{w1} = v_1 - \dfrac{V_{f1}}{\tan \phi} = 20.7 - \dfrac{2.1}{\tan 45°}$$
$$= 20.7 - \dfrac{2.1}{1} = 18.6 \text{ m/s}$$

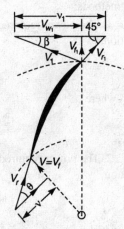

Fig. 35·7

∴ Manometric efficiency
$$\eta_{man} = \dfrac{H_m}{\dfrac{V_{w1} \cdot v_1}{g}} = \dfrac{25}{\dfrac{18.6 \times 20.7}{9.81}} = 0.637 = 63.7\% \textbf{ Ans.}$$

(b) *Vane angle at inlet*

Let $\qquad \theta =$ Vane angle at inlet.

We know that velocity of flow at inlet,
$$V_f = \dfrac{Q}{\pi Db} = \dfrac{0.125}{0.06} = 2.1 \text{ m/s} \qquad ...(\text{Given } \pi Db = 0.06 \text{ m}^2)$$

and peripheral velocity at inlet,
$$v = v_1 \times \dfrac{D}{D_1} = 20.7 \times \dfrac{0.3}{0.6} = 10.35 \text{ m/s}$$

∴ $\qquad \tan \theta = \dfrac{V_f}{v} = \dfrac{2.1}{10.35} = 0.2029$ or $\theta = 11.5°$ **Ans.**

EXERCISE 35·1

1. A centrifugal pump has external and internal diameters of 300 mm and 150 mm respectively. The vane angles of inlet and outlet are 25° and 30° and the pump runs at 1450 r.p.m. If the velocity of flow through the pump is constant, find the work done per kN of water. **(Ans. 20·2 kJ)**

2. A centrifugal pump having external and internal diameters as 750 mm and 400 mm respectively is operating at 1000 r.p.m. The vanes are curved back at 35° to the tangent at outlet. If the velocity of flow is constant at 6 m/s, find (a) vane angle at inlet, and (b) work done per kN of water.
(Ans. 16°; 123 kJ)

3. A centrifugal pump of 350 mm diameter running at 1000 r.p.m. develops a head of 18 metres. The vanes are curved back at an angle of 30° to the tangent at outlet. If the velocity of flow is constant at 2·4 m/s, find the manometric efficiency of the pump. **(Ans. 76·4%)**

4. A centrifugal pump delivers water to a height of 22 metres at a speed of 800 r.p.m. The velocity of flow is constant at a speed of 2·0 m/s and the outlet vane angle is 45°. If the pump discharges 225 litres of water per second, find the diameter of the impeller and width of the impeller at outlet.
(Ans. 375 mm; 95 mm)

35·16 Power Required to Drive a Centrifugal Pump

The power required to drive a centrifugal pump may be found out by either of the following two methods:

1. The power required to drive the pump from the manometic head may be found out by the relation,

$$P = \frac{wQH_m}{\eta_0} \text{ kW}$$

where
H_m = Manometric head of water in metres,
Q = Discharge of the pump in m³/sec,
η_0 = Overall efficiency of the pump.

2. The power required to drive the pump from the velocity traingles may be found out by the relation,

$$P = \frac{wQ \cdot V_{w1} \cdot v_1}{g} \text{ kW}$$

whirl
V_{w1} = Velocity of whirl at outlet, and
v_1 = Tangential velocity of impeller at outlet.

Example 35·6. *A centrifugal pump is requird to lift water to a total head of 40 metres at the rate of 50 litres/s. Find the power required for the pump, if its overall efficiency is 62%.*

Solution. Given : $H_m = 40$ m; $Q = 50$ litres/s = 0·05 m³/s and $\eta_0 = 62\% = 0·62$.
We know that power required to drive the pump.

$$P = \frac{wQH_m}{\eta_0} = \frac{9·81 \times 0·05 \times 40}{0·62} = 31·6 \text{ kW Ans.}$$

Example 35·7. *A centrifugal pump delivers 30 litres of water per second to a height of 18 metres through a pipe 90 metres long and of 100 mm diameter. If the overall efficiency of the pump is 75%, find the power required to drive the pump Take f 0·012.*

Solution. Given : $Q = 30$ litres/s = 0·03 m³/s; $H = 18$ m; $l = 90$ m; $d = 100$ mm = 0·1 m; $\eta_0 = 75\% = 0·75$ and $f = 0·012$.

We know that cross-sectional area of pipe,

$$a = \frac{\pi}{4} \times (d)^2 = \frac{\pi}{4} \times (0·1)^2 = 7·854 \times 10^{-3} \text{ m}^2$$

Centrifugal Pumps

and velocity of water $\quad v = \dfrac{Q}{a} = \dfrac{0.03}{7.584 \times 10^{-3}} = 3.82$ m/s

We also know that manometric head,

$$H_m = H + \text{Loss of head in pipe} + \text{Loss of head at outlet}$$

$$= 18 + \dfrac{4flv^2}{2gd} + \dfrac{v^2}{2g}$$

$$= 18 + \dfrac{4 \times 0.012 \times 90 \times (3.82)^2}{2 \times 9.81 \times 0.1} + \dfrac{(3.82)^2}{2 \times 9.81} \text{ m}$$

$$= 18 + 32.1 + 0.74 = 50.84 \text{ m}$$

and power required to drive the pump.

$$P = \dfrac{wQH_m}{\eta_0} = \dfrac{9.81 \times 0.03 \times 50.84}{0.75} = 19.9 \text{ kW Ans.}$$

Example 35·8. *A centrifugal pump, having an overall efficiency of 62%, is required to handle brine (sp. gr = 1·19) and gasoline (sp. gr. = 0·7). The discharge of each of these liquids is 50 litres/s against a net pressure of 400 kPa. Show that the same power is required for handling both the above liquids having different specific gravities.*

Also calculate the head in metres of fluid to which the brine and gasoline will be raised.

Solution. Given : $\eta_0 = 62\% = 0.62$; Specific. gravity of brince = 1·19 or specific weight of brine $(w_b) = 9.81 \times 1.19 = 11.7$ kN/m³; Specific gravity of gasoline = 0·7 or specific weight of gasoline $(w_g) = 9.81 \times 0.7 = 6.87$ kN/m³; $Q = 50$ litres/s = 0·05 m³/s and pressure = 400 kPa = 400 kN/m².

Head to which brine and gasoline will be raised

We know that head to which the brine will be raised,

$$H_b = \dfrac{p}{w_b} = \dfrac{400}{11.7} = 34.2 \text{ m Ans.}$$

and head to which the gasoline will be raised,

$$H_g = \dfrac{p}{w_g} = \dfrac{400}{6.87} = 58.2 \text{ m Ans.}$$

Power required for handling both the liquids

We know that power required to handle the brine,

$$P = \dfrac{w_b \times Q \times H_b}{\eta_0} = \dfrac{11.7 \times 0.05 \times 34.2}{0.62} = 32.3 \text{ kW}$$

and power required to handle the gasoline

$$P = \dfrac{w_g \times Q \times H_g}{\eta_0} = \dfrac{6.87 \times 0.05 \times 58.2}{0.62} = 32.3 \text{ kW}$$

From both the above results, we find that same power is required for handling both the liquids. **Ans.**

Example 35·9. *A centrifugal pump of 1·5 metre diameter runs at 210 r.p.m. and pumps 180 litres of water per second. The angle which the vane makes, at exit, with the tangent to the impeller is 25°. Assuming radial entry and velocity of flow throughout as 2·5 m/s, determine the power required to drive the pump.*

If manometric efficiency of the pump is 65 per cent, find the average lift of the pump.

Solution. Given : $D_1 = 1.5$ m; $N = 210$ r.p.m.; $Q = 180$ litres/s $= 0.18$ m³/s $\theta = 25°$; $V_f = V_{f1} = 2.5$ m/s and $\eta_{man} = 65\% = 0.65$.

Power required to drive the pump

We know that tangential velocity of the impeller, at outlet,

$$v_1 = \frac{\pi D_1 N}{60} = \frac{\pi \times 1.5 \times 210}{60}$$

$$= 16.5 \text{ m/s}$$

From the outlet velocity triangle, we find that velocity of whirl at outlet,

$$V_{w1} = v_1 - \frac{V_f}{\tan \theta}$$

$$= 16.5 - \frac{2.5}{\tan 25°} \text{ m/s}$$

$$= 16.5 - \frac{2.5}{0.4663}$$

$$= 11.1 \text{ m/s}$$

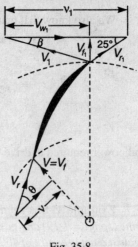

Fig. 35.8

∴ Power required to drive the pump

$$P = \frac{wQ \cdot V_{w1} \cdot v_1}{g} = \frac{9.81 \times 0.18 \times 11.1 \times 16.5}{9.81} \text{ kW}$$

$$= 33.0 \text{ kW Ans.}$$

Average lift of the pump

Let H_m = Average lift of the pump (or manometric head).

We know that manometric efficiency (η_{man}),

$$0.65 = \frac{H_m}{\dfrac{V_{w1} \cdot v_1}{g}} = \frac{H_m}{\dfrac{11.1 \times 16.5}{9.81}} = \frac{H_m}{18.7}$$

∴ $H_m = 0.65 \times 18.7 = 12.2$ m **Ans.**

Example 35·10. *A centrifugal pump delivers 50 litres of water per second against a total head of 24 metres running at 1500 r.p.m. The velocity of flow is maintained constant at 2·4 m/s and the blades are curved back at 30° to the tangent at exit. The inner diameter is half of the outer diameter. If the manometric efficiency is 80%, find:*

(a) *blade angle at inlet, and*

(b) *power required to drive the pump.*

Solution. Given : $Q = 50$ litres/s $= 0.05$ m³/s; $H_m = 24$ m; $N = 1500$ r.p.m.; $V_f = V_{f1}$ 2·4 m/s; $\phi = 30°$ $D = \dfrac{D_1}{2}$ or $\dfrac{D}{D_1} = \dfrac{1}{2}$ and $\eta_{man} = 80\% = 0.8$

(a) Blade angle at inlet

Let θ = Blade angle at inlet.

Centrifugal Pumps

We know that manometric efficiency (η_{man}),

$$0.8 = \frac{H_m}{\frac{V_{w1} \cdot v_1}{g}} = \frac{24}{\frac{V_{w1} \cdot v_1}{9.81}}$$

$$= \frac{235.4}{V_{w1} \cdot v_1}$$

$\therefore \quad V_{w1} \cdot v_1 = 235.4/0.8 = 294.3 \quad \ldots(i)$

From the outlet triangle of velocities, we find that the velocity of whirl at outlet,

$$V_{w1} = v_1 - \frac{V_{f1}}{\tan 30°} = v_1 - \frac{2.4}{0.5774}$$

$$= v_1 - 4.16$$

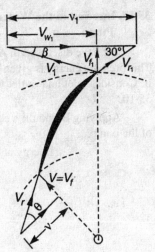

Fig. 35.9

Substituting this value of V_{w1} in equation (i),

$$(v_1 - 4.16) \times v_1 = 294.3$$

$$v_1^2 - 4.16\, v_1 = 294.3 = 0$$

This is a quadratic equation for v_1. Therefore

$$v_1 = \frac{4.16 \pm \sqrt{(4.16)^2 \pm 4 \times 294.3}}{2} = 19.3 \text{ m/s}$$

We know that peripheral valocity of impeller at inlet

$$v = \frac{D}{D_1} \times v = \frac{1}{2} \times 19.3 = 9.65 \text{ m/s}$$

From the inlet triangle of velocities, we find that

$$\tan \theta = \frac{V_f}{v} = \frac{2.4}{9.65} = 0.2487 \quad \text{or} \quad \theta = 14° \text{ Ans.}$$

(b) Power required to drive the pump

We also know that power required to drive the pump

$$P = \frac{wQ \cdot V_{w1} \cdot v_1}{g} = \frac{9.81 \times 0.05 \times (294.3)}{9.81} = 14.7 \text{ kW Ans.}$$

EXERCISE 35.2

1. A centrifugal pump delivers 60 litres of water per second to a tank situated at a height of 20 metres. If the overall efficiency of the pump is 70%, find the power required for the pump. **(Ans. 16.8 kW)**

2. A centrifugal pump having an overall efficiency of 75% is discharging 30 litres of water per second through a pipe of 150 mm diameter and 125 m long. Calculate the power required to drive the pump, if the water is lifted through a height of 25 metres. Take coefficient of friction as 0.01.
(Ans. 11.7 kW)

3. A centrifugal pump having an impeller of diameter 500 mm delivers 140 litres of water per second. The velocity of flow is 1 m/s and the vanes are curved back at outlet at 30° to the wheel tangent. If the impeller speed is 400 r.p.m., find the power to drive the pump. **(Ans. 12.9 kW)**

4. A centrifugal pump of 1.2 m diameter delivers 2000 litres of water per second against a head of 6 metres at 200 r.p.m. The vanes are curved back at an angle of 26° to the tangent at outlet and velocity of flow is constant at 2.4 m/s. Find the manometric efficiency and power required to operate the pump. **(Ans. 61.2%; 119.4 kW)**

35·17. Increase in the Water Pressure while Flowing through the Impeller of a Centrifugal Pump

We have discussed in Art. 35·3 that a pump converts mechanical energy into pressure energy. This pressure energy is given by the impeller to the water flowing through it. Consider inlet and outlet tips of a centrifugal pump as shown in Fig. 35·10.

Applying Bernoulli's equation to the inlet and outlet of the impeller of the pump,

Energy at outlet = Energy at inlet
+ Work done by the impeller

i.e.,
$$\frac{p_1}{w} + \frac{V_1^2}{2g} = \frac{p}{w} + \frac{V^2}{2g} + \frac{V_{w1} \cdot v_1}{g}$$

...(Taking $Z_1 = Z$)

$$\therefore \quad \frac{p_1}{w} - \frac{p}{w} = \frac{V^2}{2g} - \frac{V_1^2}{2g} + \frac{V_{w1} \cdot v_1}{g}$$

where $\left(\dfrac{p_1}{w} - \dfrac{p}{w}\right)$ represents the increase in the pressure of water, while flowing through the impeller, in terms of head of water.

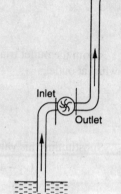

Fig. 35·10
Increase in water pressure.

Example 35·11. *From first principles, and writing down all steps of calculation, show that the theoretical pressure rise through the impeller of a centrifugal pump is given by:*

$$\frac{1}{2g}(V_f^2 + v_1^2 - V_{f1}^2 \cosec^2 \phi)$$

where
V_f = *Velocity of flow at inlet,*
V_{f1} = *Velocity of flow at outlet,*
v_1 = *Peripheral velocity of impeller at outlet, and*
ϕ = *Impeller angle at outlet.*

Solution. Given. V_f = Velocity of flow at inlet,
V_{f1} = Velocity of flow at outlet,
v_1 = Peripheral velocity of impeller at outlet, and
ϕ = Impeller angle at outlet.

Applying Bernoulli's equation at inlet and outlet of the impeller of the pump,

$$\frac{p_1}{w} + \frac{V_1^2}{2g} = \frac{p}{w} + \frac{V^2}{2g} + \frac{V_{w1} \cdot v_1}{g}$$

or
$$\frac{p_1}{w} - \frac{p}{w} = \frac{V^2}{2g} - \frac{V_1^2}{2g} + \frac{V_{w1} \cdot v_1}{g}$$

∴ Pressure rise

$$= \frac{V^2}{2g} - \frac{V_1^2}{2g} + \frac{V_{w1} \cdot v_1}{g} \qquad \ldots\left(\because \frac{p_1}{w} - \frac{p}{w} = \text{Pressure rise}\right)$$

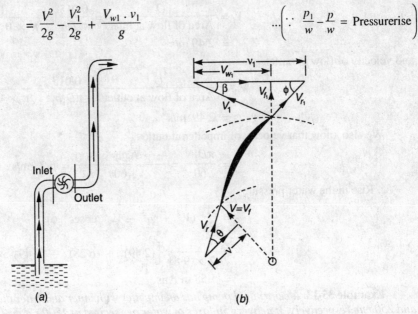

Fig. 35·11.

$$= \frac{1}{2g}\left[V^2 - V_1^2 + 2V_{w1} \cdot v_1\right]$$

$$= \frac{1}{2g}\left[V_f^2 - V_1^2 + 2V_{w1} \cdot v_1\right] \qquad \ldots(\because V = V_f)$$

$$= \frac{1}{2g}\left[V_f^2 - \left(V_{w1}^2 + V_{f1}^2\right) + 2V_{w1} \cdot v_1\right]$$

$$= \frac{1}{2g}\left[V_f^2 - V_{f1}^2 - \left(v_1 - V_{f1}\cot\phi\right)^2 + 2v_1\left(v_1 - V_{f1}\cot\phi\right)\right]$$

$(\because V_{w1} = v_1 - V_f \cot\phi)$

$$= \frac{1}{2g}\left[V_f^2 - V_{f1}^2 - \left(v_1^2 + V_{f1}^2\cot^2\phi - 2v_1 V_{f1}\cot\theta\right) + 2v_1^2 - 2v_1 V_{f1}\cot\phi\right]$$

$$= \frac{1}{2g}\left[V_f^2 - V_{f1}^2 - v_1^2 - V_{f1}^2\cot^2\phi + 2v_1 V_{f1}\cot\phi + 2v_1^2 - 2v_1 V_{f1}\cot\phi\right]$$

$$= \frac{1}{2g}\left[V_f^2 + v_1^2 - V_{f1}^2\left(1 + \cot^2\phi\right)\right]$$

$$= \frac{1}{2g}\left[V_f^2 + v_1^2 - V_{f1}^2\csc^2\phi\right] \textbf{ Ans.}$$

Example 35·12. *A centrifugal pump is discharging water at the rate of 15 litres/s at 500 r.p.m. The internal and external diameters and the impeller widths are 120 mm, 240 mm, 16 mm and 8 mm respectively. The vanes are curved back at 25° to the tangent at outlet. Find the rise in the water pressure, when it passes through the pump.*

Solution. Given : $Q = 15$ litres/s $= 0.015$ m³/s; $N = 500$ r.p.m.; $D = 120$ mm $= 0.12$ m; $D_1 = 240$ mm $= 0.24$ m; $b = 16$ mm $= 0.016$ m; $b_1 = 8$ mm $= 0.008$ m and $\phi = 25°$

We know that velocity of flow at inlet,

$$V_f = \frac{Q}{\text{Area of flow at inlet}} = \frac{0.015}{\pi D b} = \frac{0.015}{\pi \times 0.12 \times 0.016} \text{ m/s}$$
$$= 2.49 \text{ m/s}$$

and velocity of flow at outlet,

$$V_{f1} = \frac{Q}{\text{Area of flow at outlet}} = \frac{0.015}{\pi D_1 b_1} = \frac{0.015}{\pi \times 0.24 \times 0.008} \text{ m/s}$$
$$= 2.49 \text{ m/s}$$

We also know that velocity of impeller at outlet,

$$v_1 = \frac{\pi D_1 N}{60} = \frac{\pi \times 0.24 \times 500}{60} = 6.28 \text{ m/s}$$

∴ Rise in the water pressure

$$= \frac{1}{2g}(V_f^2 + v_1^2 - V_{f1}^2 \operatorname{cosec}^2 \phi)$$
$$= \frac{1}{2 \times 9.81}\left[(2.49)^2 + (6.28)^2 - (2.49)^2 \times \operatorname{cosec}^2 25°\right] \text{m}$$
$$= 1.58 \text{ m Ans.}$$

Example 35·13. *A centrifugal pump has an impeller with inner and outer diameters of 150 mm and 250 mm respectively. It delivers 50 litres of water per second at 1500 r.p.m. The velocity of flow through impeller is constant at 2·5 m/s. The blades are curved back at angle of 30° to the tangent at exit. The diameters of the suction and delivery pipes are 150 mm and 100 mm respectively. The pressure head at suction is 4 m below and that at delivery is 18 m above atmosphere. The power required to drive the pump is 18 kW. Find (i) vane angle at inlet, (ii) overall efficiency, and (iii) monometric efficiency.*

Solution. Given: $D = 150$ mm $= 0.15$ m; $D_1 = 250$ mm $= 0.25$ m; $Q = 50$ litres/s $= 0.05$ m²/s; $N = 1500$ r.p.m.; $V_f = V_{f1} = 2.5$ m/s; $\phi = 30°$; $d_s = 150$ mm $= 0.15$ m; $d_d = 100$ mm $= 0.1$ m; $\frac{p}{w} = H_{atmos} - 4 = 10.3 - 4 = 6.3$ m; $\frac{p_1}{w} = H_{atmos} + 18 = 10.3 + 18 = 28.3$ m and $P = 18$ kW

(i) Vane angle at inlet

Let θ = Vane angle at inlet.

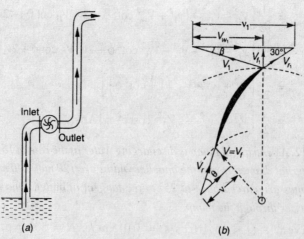

Fig. 35·12

Centrifugal Pumps

We know that tangential velocity of impeller at inlet,
$$v = \frac{\pi DN}{60} = \frac{\pi \times 0.15 \times 1500}{60} = 11.8 \text{ m/s}$$

and from inlet velocity triangle, we find that
$$\tan \theta = \frac{V_f}{v} = \frac{2.5}{11.8} = 0.2119 \quad \text{or} \quad \theta = 12° \text{ Ans.}$$

(ii) Overall efficiency

Let H_m = Manometric head of the pump.

We know that cross-sectional area of the suction pipe,
$$a_s = \frac{\pi}{4} \times (d_s)^2 = \frac{\pi}{4} \times (0.15)^2 = 17.67 \times 10^{-3} \text{ m}^2$$

and velocity of water in the suction pipe.
$$V = \frac{Q}{a} = \frac{0.05}{17.67 \times 10^{-3}} = 2.83 \text{ m/s}$$

Similarly cross-sectional area of the delivery pipe
$$a_d = \frac{\pi}{4} \times (d_d)^2 = \frac{\pi}{4} \times (0.1)^2 = 7.854 \times 10^{-3} \text{ m}^2$$

and velocity of water in the delivery pipe,
$$V_1 = \frac{Q}{a_d} = \frac{0.05}{7.854 \times 10^{-3}} = 6.37 \text{ m/s}$$

Applying Bernoulli's equation at inlet outlet of the impeller,
$$\frac{p_1}{w} + \frac{V_1^2}{2g} = \frac{p}{w} + \frac{V^2}{2g} + H_m \qquad \text{...(Assuming } Z_1 = Z\text{)}$$

or
$$H_m = \left(\frac{p_1}{w} + \frac{V_1^2}{2g}\right) - \left(\frac{p}{w} + \frac{V^2}{2g}\right)$$

$$= \left(28.3 + \frac{(6.37)^2}{2 \times 9.81}\right) - \left(6.3 + \frac{(2.83)^2}{2 \times 9.81}\right) \text{m}$$

$$= 30.4 - 6.7 = 23.7 \text{ m}$$

∴ Overall efficiency,
$$\eta_0 = \frac{wQH_m}{P} = \frac{9.81 \times 0.05 \times 23.7}{18} = 0.646 = 64.6\% \text{ Ans.}$$

(iii) Manometric efficiency

We know that peripheral velocity of the impeller at outlet,
$$v_1 = \frac{\pi D_1 N}{60} = \frac{\pi \times 0.25 \times 1500}{60} = 19.6 \text{ m/s}$$

and from the outlet triangle of velocities, we find that velocity of whirl at outlet,
$$V_{w1} = v_1 - \frac{V_{f1}}{\tan \theta} = 19.6 - \frac{2.5}{\tan 30°} = 19.6 - \frac{2.5}{0.5774} \text{ m/s}$$
$$= 15.3 \text{ m/s}$$

∴ Manometric efficiency,

$$\eta_{man} = \frac{H_m}{\dfrac{V_{w1} \cdot v_1}{g}} = \frac{23 \cdot 7}{\dfrac{15 \cdot 3 \times 19 \cdot 6}{9 \cdot 81}} = 0 \cdot 772 = 77 \cdot 2\% \text{ Ans.}$$

35·18 Minimum Starting Speed of a Centrifugal Pump

A centrifugal pump will start delivering the liquid, only when the head developed by it is equal to the manometric head. At the time of start, the liquid velocities are zero, therefore the pressure head caused by the centrifugal force

$$= \frac{v_1^2}{2g} - \frac{v^2}{2g} = \frac{v_1^2 - v^2}{2g}$$

This pressure head must give the required manometric head, *i.e.*,

$$\frac{v_1^2 - v^2}{2g} = H_m = \eta_{man} \times \frac{V_{w1} \cdot v_1}{g} \qquad \ldots \left[\because \eta_{man} = \frac{H_m}{\dfrac{V_{w1} \, v}{g}}\right]$$

Example: 35·14. *A centrifugal pump has to discharge 2000 litres of water per second and develop a total head of 22·5 metres, when the impeller rotates at 240 r.p.m. The impeller diameter is 1·5 meter and velocity of flow at outlet is 2·5 m/s. If the vanes are set back at an angle of 30° at the outlet, find*

1. *manometric efficiency of the pump, and*
2. *power required to drive the pump.*

If the inner diameter is half of the outer diameter, find the minimum speed to start pumping.

Solution. Given : $Q = 2000$ litres/s $= 2$ m³/s; $H_m = 22 \cdot 5$ m; $N = 240$ r.p.m.; $D_1 = 1 \cdot 5$ m; $V_{f1} = 2 \cdot 5$ m/s and $\phi = 30°$.

1. Manometric efficiency of the pump

We know that tangential velocity of the impeller at outlet,

$$v_1 = \frac{\pi D_1 N}{60} = \frac{\pi \times 1 \cdot 5 \times 240}{60}$$

$$= 18 \cdot 8 \text{ m/s}$$

and from outlet triangle of velocities, we find that velocity of whirl at outlet

$$V_{w1} = v_1 - \frac{V_{f1}}{\tan \phi} = 18 \cdot 8 - \frac{2 \cdot 5}{\tan 30°} \text{ m/s}$$

$$= 18 \cdot 8 - \frac{2 \cdot 5}{0 \cdot 5774} = 14 \cdot 5 \text{ m/s}$$

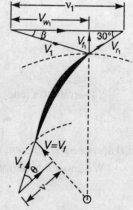

Fig. 35·13

∴ Manometric efficiency,

$$\eta_{man} = \frac{H_m}{\dfrac{V_{w1} \cdot v_1}{g}} = \frac{22 \cdot 5}{\dfrac{14 \cdot 5 \times 18 \cdot 8}{9 \cdot 81}}$$

$$= 0 \cdot 81 = 81\% \text{ Ans.}$$

35·19 Multistage Centrifugal Pumps

In the previous articles, we have seen that the head developed by a centrifugal pump is proportional to the diameter and speed of the impeller. Since there is a limitation for the diameter and speed of the impeller, therefore the head developed by a centrifugal pump is ordinarily limited to 50 metres. In some special pumps, higher heads up to 100 metres may also be developed. For still larger heads, it will be necessary to put two or more pumps in series. Liquid from one pump, is brought to the inlet of another pump, which further increases the head developed.

This type of arrangement is also possible, if we provide two or more impellers (instead of two or three pumps) and key them to the same shaft and put them in the same casing. Such a pump is called a multistage pump. For special duties, the pumps with as many as 100 stages have been manufactured.

Example 35·15. *Each impeller, of a three-stage centrifugal pump has external diameter of 375 mm and width 20 mm. The pump is discharging 3600 litres of water per minute at 900 r.p.m. The vanes are curved back at 45° to the tangent at outlet.*

If the manometric efficiency is 84%, find the manometric head generated by the pump.

Solution. Given : No. of stages = 3; D_1 = 375 mm = 0·375 m; b_1 = 20 mm = 0·02 m; Q = 3600 litres/min = 60 litres/s = 0·06 m²/s; N = 900 r.p.m.; ϕ = 45°; and η_{man} = 84% = 0·84.

Let H_m = Manometric head of each stage.

We know that tangential velocity of the impeller at outlet,

$$v_1 = \frac{\pi D_1 N}{60} = \frac{\pi \times 0.375 \times 900}{60}$$

$$= 17.7 \text{ m/s}$$

and velocity of flow at outlet, $V_{f1} = \dfrac{Q}{\pi D_1 b_1} = \dfrac{0.06}{\pi \times 0.375 \times 0.02}$

$$= 2.55 \text{ m/s}$$

From the outlet triangle of velocities, we find that the velocity of whirl at outlet,

$$V_{w1} = v_1 - \frac{V_{f1}}{\tan \phi} = 17.7 - \frac{2.55}{\tan 45°}$$

$$= 15.15 \text{ m/s}$$

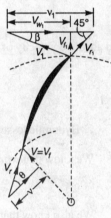

Fig. 35·14.

We also know that manometric efficiency (η_{man}),

$$0.84 = \frac{H_m}{\dfrac{V_{w1} \cdot v_1}{g}} = \frac{H_m}{\dfrac{15.15 \times 17.7}{9.81}} = \frac{H_m}{27.3}$$

or $\qquad H_m = 0.84 \times 27.3 = 22.9 \text{ m}$

∴ Total manometric head due to three stages

$$= 3 \times 22.9 = 68.7 \text{ m } \textbf{Ans.}$$

Example 35·16. *A multistage centrifugal pump is discharging 45 000 litres of water per minute against a monometric head of 60 metres. There are four equal impellers, keyed to the same shaft, which is running at 350 r.p.m. The vanes are curved back at an angle of 60° to the tangent at outer pheriphery. The velocity of flow at outlet is 0·27 times the corresponding pheripheral velocity, and the hydraulic losses in the pump are 1/3 of the velocity head at outlet of the impeller. Find*

1. Diameter of the impeller and 2. Manometric efficiency.

Solution. Given : $Q = 45\,000$ litres/min $= 750$ litres/s $= 0.75$ m³/s; Total manometric head $= 60$ m; No. of stages $= 4$ or manometric head for each stage $(H_m) = 60/4 = 15$ m; $N = 350$ r.p.m.; $\phi = 60°$; $V_{f1} = 0.27\, v_1$ and hydraulic losses $= \dfrac{1}{3} \times \dfrac{v_1^2}{2g} = \dfrac{v_1^2}{6g}$

1. Diameter of the impeller

Let $D_1 =$ Diameter of the impeller.

We know that the pheripheral velocity at outlet,

$$v_1 = \frac{\pi D_1 N}{60} = \frac{\pi \times D_1 \times 350}{60} \text{ m/s}$$

$$= 18.3\, D_1 \text{ m/s}$$

∴ Velocity of flow at outlet,

$$V_{f1} = 0.27 \times 18.3\, D_1 = 4.94\, D_1 \text{ m/s}$$

Fig. 35.15.

From the outlet triangle of velocities, we find that the velocity of whirl at outlet,

$$V_{w1} = v_1 - \frac{V_{f1}}{\tan \phi} = 18.3\, D_1 - \frac{4.94\, D_1}{\tan 60°}$$

$$= 18.3\, D_1 - \frac{4.94\, D_1}{1.732} = 15.5\, D_1 \text{ m/s}$$

and absolute velocity of water leaving the impeller,

$$V_1 = \sqrt{V_{f1}^2 + V_{w1}^2} = \sqrt{(4.94\, D_1)^2 + (15.5\, D_1)^2} \text{ m/s}$$

$$= 16.3\, D_1 \text{ m/s}$$

∴ Hydraulic losses, $= \dfrac{V_1^2}{6g} = \dfrac{(16.3\, D_1)^2}{6 \times 9.81} = 4.51\, D_1^2$ m

We know that work done by the pump per kN of water

$$W = \frac{V_{w1} \cdot v_1}{g} = \frac{15.5\, D_1 \times 18.3\, D_1}{9.81} = 28.9\, D_1^2 \text{ kN-m}$$

We also know that work done by the pump per kN of water (W),

$$28.9\, D_1^2 = \text{Manometric head} + \text{Hydraulic losses}$$

$$= 15 + 4.51\, D_1^2$$

$$24.39\, D_1^2 = 15$$

∴ $D_1^2 = 15/24.39 = 0.615$ or $D_1 = 0.784$ m **Ans.**

2. Manometric efficiency

We also know that manometric efficiency,

$$\eta_{man} = \frac{H_m}{\dfrac{V_{w1} \cdot v_1}{g}} = \frac{15}{\dfrac{(15.5 \times 0.784) \times (18.3 \times 0.784)}{9.81}}$$

$$= 0.844 = 84.4\% \text{ **Ans.**}$$

Exercise 35·3

1. A centrifugal pump is discharging water at the rate of 100 litres/s. The inetemal and external diameters of the impeller are 150 mm and 300 mm respectively. The width of the impeller at inlet and outlet are 12 mm and 6 mm. The vanes are curved bade at 45° to the tangent at outlet. Find the increase in pressure as the water passes through the impeller. **(Ans. 28·1 m)**
2. Find the minimum speed at which a centrifugal pump will start functioning against a head of 7·5 m, if the diameters of the impeller at outlet and inlet are 1 m and 0·5 m respectively. **(Ans. 267·4 r.p.m.)**
3. Determine the least no. of stages for a multistage pump to deliver 60 litres of water per second against a total head of 210 metres at 1450 r.p.m. The speed of the pump is not to exceed 670 r.p.m.
   (Ans. 5)
4. Each impeller of a two stage centrifugal pump has outer diameter of 40 mm and a width of 25 mm. The pump is discharging 60 litres of water per second at 1000 r.p.m. If the vane angle at outlet is 30°, find the total manometric head developed by the pump. Assume manometric efficiency of the pump as 80%. **(Ans. 60·2 m)**

QUESTIONS

1. What is a centrifugal pump ? On what principle does it work ?
2. What are the different types of pumps ? Explain the working principles of a centrifugal pump with sketches.
3. Name the different types of casings for the impeller of a centrifugal pump.
4. Explain the function of spiral casing for a centrifugal pump.
5. Name the different types of efficiencies of a centrifugal pump and differentiate between overall efficiency and manometric efficiency.
6. Derive an equation for the power required to drive a centrifugal pump.
7. Obtain an equation for the increase in water pressure, while flowing through the impeller of a centrifugal pump.
8. What do you understand by the term 'multistage pump'? Explain clearly the difference between a single stage and a multistage centrifugal pump.

OBJECTIVE TYPE QUESTIONS

1. In a centrifugal pump the water
   (a) enters the impeller radially and leaves the vanes axially
   (b) enters the impeller radially and leaves the vanes radially
   (c) enters the impeller axially and leaves the vanes radially
   (d) enters the impeller axially and leaves the vanes axially.
2. In the casing of a centrifugal pump, the kinetic energy of the water is converted into
   (a) potential energy   (b) pressure energy   (c) heat energy   (d) all of these
3. In a centrifugal pump the liquid enters
   (a) at the centre   (b) at the top   (c) at the bottom   (d) from sides
4. A multistage pump is used to
   (a) give high discharge   (b) produce high heads
   (c) pump viscous fluids   (d) pump chemicals

ANSWERS

1. (a) 2. (b) 3. (a) 4. (b)

36

Reciprocating Pumps

> 1. Introduction. 2. Types of Reciprocating Pumps. 3. Comparison of Centrifugal and Reciprocating Pumps. 4. Discharge of a Reciprocating Pump. 5. Slip of the Pump. 6. Negative Slip of the Pump. 7. Power Required to Drive a Reciprocating Pump. 8. Indicator Diagram of a Reciprocating Pump. 9. Variation of Pressure in the Suction and Delivery Pipes due to Acceleration of the Piston. 10. Effect of Acceleration of the Piston on the Indicator Diagram. 11. Effect of Friction in the Suction and Delivery Pipes on the Indicator Diagram. 12. Maximum Speed of the Rotating Crank. 13. Air Vessels. 14. Maximum Speed of the Rotating Crank with Air Vessels. 15. Work Done against Friction without Air Vessels. 16. Work done Against Friction with Air Vessels. 17. Work Saved against Friction by Fitting Air Vessels. 18. Flow of Water into and from the Air Vessels Fitted to the delivery Pipe of a Single Acting Reciprocating Pump. 19. Flow of Water into and from the Air Vessel Fitted to the Delivery Pipe of a Duble Acting Reciprocating Pump.

36.1 Introduction

A reciprocating pump, in its simplest form, consists of the following parts as shown in Fig. 36.1:

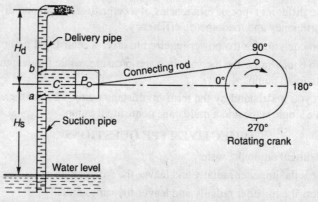

Fig. 36·1. Parts of a centrifugal pump.

1. A cylinder C, in which a piston P works. The movement of the piston is obtained by a connecting rod, which connects the piston and the rotating crank.
2. A suction pipe, connecting the source of water and the cylinder.
3. A delivery pipe, into which the water is discharged from the cylinder.
4. A suction valve a, which admits the flow from the suction pipe into the cylinder.
5. A delivery valve b, which admits the flow from the cylinder into the delivery pipe.

During the suction stroke, the piston P moves towards right (*i.e.*, from 0° to 180), thus creating vacuum in the cylinder. This vacuum causes the suction valve a to open and the water enters the cylinder. During the delivery stroke, the piston P move towards left (*i.e.*, from 180° to 360°) thus increasing pressure in the cylinder. This increase in pressure causes the suction valve a to close and delivery valve b to open, and the water is forced into the delivery pipe.

A reciprocating pump is also called a positive displacement pump, as it discharges a definite quantity of liquid during the displacement of its piston or plunger. This is why a reciprocating pump is ideally suitable for grouting operations in dam foundations.

Reciprocating Pumps

36.2 Types of Reciprocating Pumps

The reciprocating pumps may be classified as discussed below :

1. *According to action of water*
   (a) single acting pump, and
   (b) double acting pump.
2. *According to number of cylinders*
   (a) single cylinder pump,
   (b) double cylinder pump, and
   (c) triple cylinder pump etc.
3. *According to the existence of air vessels*
   (a) with air vessel, and
   (b) without air vessel.

All the above mentioned pumps will discussed, in detail, at their appropriate places in the book.

36.3 Comparison of Centrifugal and Reciprocating Pumps

Following table gives the comparison of a centrifugal pump and a reciprocating pump :

S. No.	*Centrifugal pump*	*Reciprocating pump*
1.	Simple in construction, because of less number of parts.	Complicated in construction, because of more number of parts.
2.	Total weight of the pump is less for a given discharge.	Total weight of the pump is more for a given discharge.
3.	Suitable for large discharge and smaller heads.	Suitable for less discharge and higher heads.
4.	Requires less floor area and simple foundation.	Requires more floor area and comparatively heavy foundation.
5.	Less wear and tear.	More wear and tear.
6.	Maintenance cost is less.	Maintenance cost is high.
7.	Can handle dirty water.	Cannot handle dirty water.
8.	Can run at higher speeds.	Cannot run at higher speeds.
9.	Its delivery is continuous.	Its dilivery is pulsating.
10.	No air vessels are required.	Air vessels are required.
11.	Thrust on the crankshaft is uniform.	Thrust on the crankshaft is not uniform
12.	Operation is quite simple.	Much care is required in operation.
13.	Needs priming.	Does not need priming.
14.	It has less efficiency.	It has more efficiency.

36.4 Discharge of a Reciprocating Pump

Consider a single acting reciprocating pump (*i.e.*, a pump, in which the water is acting on one side of the piston only) as shown in Fig. 36.1.

Let
- L = Length of the stroke or piston in metres,
- A = Cross-sectional area of the piston in square metres, and
- N = No. of revolutions, per minute of the crank.

∴ Discharge of water in one stroke
$$= LA$$
and discharge of the pump, $Q = \dfrac{LAN}{60}$ m³/s

If the pump is a double acting reciprocating pump (*i.e.*, a pump, in which the water is acting on both sides of the piston) the discharge is *taken to be double the discharge than that of a single acting pump. This is due to the reason that, in a double acting pump, the water is sucked on one side of the piston and delivered from the other side during the same stroke. These two processes (*i.e.*, suction on one side and delivery from the other) are reversed during the return stoke. Therefore the discharge of a double acting reciprocating pump,

$$Q = \dfrac{2LAN}{60}$$

36·5 Slip of the Pump

In the last article, we have obtained the relation for the discharge of a single acting and double acting reciprocating pumps. But in practice, the actual discharge is less than the theoretical discharge. The difference between theoretical discharge and actual discharge is known as slip of the pump. This theory is similar to that which discussed for coefficient of discharge in Art. 11·8.

36·6 Negative Slip of the Pump

Sometimes, the actual discharge of a reciprocating pump, is more than the theoretical discharge. In such cases, the coefficient of discharge will be more than unity, and the corresponding slip is known as negative slip of the pump.

This happens, when the suction pipe is long and delivery pipe is short and pump is running at high speeds. This causes the delivery valve to open before completion of the suction stroke and some water is pushed into the delivery pipe, before the piston commences its delivery stroke.

Example 36·1. *A single acting reciprocating pump has a plunger of diameter 300 mm and stroke of 200 mm. If the speed of the pump is 30 r.p.m. and it delivers 6·5 litres/sec of water, find the coefficient of discharge and the percentage slip of the pump.*

Solution. Given : D = 300 mm = 0·3 m; L = 200 mm = 0·2 m; N = 30 r.p.m. and Q_{ac} = 6·5 litres/s = 0·0065 m³/s.

Coefficient of discharge

We know that the area of the plunger,
$$A = \dfrac{\pi}{4} \times (D)^2 = \dfrac{\pi}{4} \times (0·3)^2 = 0·0707 \text{ m}^2$$

and theoretical discharge, $Q_{th} = \dfrac{LAN}{60} = \dfrac{0·2 \times 0·0707 \times 30}{60} = 0·0071$ m³/s

∴ Coefficient of discharge, $C_d = \dfrac{Q_{ac}}{Q_{th}} = \dfrac{0·0065}{0·0071} = 0·92$ **Ans.**

---

*As a matter of fact, the discharge of a double acting reciprocating pump is not double of a single acting pump, because some area is blocked by the connecting rod. However, this area being small as compared to the discharge is neglected.

Percentage slip

We also know that percentage slip

$$= \frac{Q_{th} - Q_{ac}}{Q_{th}} = \frac{0.0071 - 0.0065}{0.0071} = 0.085 = 8.5\% \text{ Ans.}$$

36.7 Power Required to Drive a Reciprocating Pump

Consider a reciprocating pump, first sucking a liquid (through the suction pipe) and then delivering the same (through the delivery pipe).

Let
H_s = Suction head of the pump in metres,
H_d = Delivery head of the pump in metres,
A = Area of piston in m²,
w = Specific weight of the liquid, and
Q = Discharge of the liquid in m³/sec.

We know that force on the piston in forward stroke

$$= w . H_s A \text{ kN}$$

and force on the piston in the backward stroke

$$= w.H_d.A \text{ kN}$$

∴ Work done by the pump $= wQ(H_s + H_d)$ kN-m

and theoretical power required to drive the pump

$$= wQ(H_s + H_d) \text{ kW}$$

Note. The actual power, required to drive the pump will be more than the theoretical, pump due to various losses.

Example 36.2. *A double acting reciprocating pump has a stroke of 300 mm and a piston of diameter 150 mm. The delivery and suction heads are 26 m and 4 m respectively including friction heads. If the pump is working at 60 r.p.m., find power required to drive the pump with 80% efficiency.*

Solution. $L = 300$ mm $= 0.3$ m; $D = 150$ mm $= 0.15$ m; $H_d = 26$ m; $H_s = 4$ m; $N = 60$ r.p.m. and $\eta = 80\% = 0.8$.

We know that area of the piston,

$$A = \frac{\pi}{4} \times (D)^2 = \frac{\pi}{4} \times (0.15)^2 = 0.0177 \text{ m}^2$$

and theoretical discharge of the double acting pump

$$Q = \frac{2LAN}{60} = \frac{2 \times 0.3 \times 0.0177 \times 60}{60} = 0.011 \text{ m}^3/s$$

∴ Theoretical power to drive the pump,

$$= wQ(H_s + H_d) = 9.81 \times 0.011 \times (4 + 26) = 3.24 \text{ kW}$$

and actual power $= \dfrac{3.24}{0.8} = 4.05$ kW **Ans.**

EXERCISE 36.1

1. A single acting reciprocating pump, having a bore of 150 mm diameter and a stroke of 300 mm length discharges 200 litres of water per minute. Neglecting losses, find (*i*) theoretical discharge in litres/min (*ii*) coefficient of discharge, and (*iii*) slip of the pump. [**Ans.** 210 litres/min; 0.95; 5.2%]

2. A single acting reciprocating pump having cylidner diameter of 150 mm and stroke 300 mm is used to raise water through a total height of 30 metres. Find the power required to drive the pump, if the crank rotates at 60 r.p.m. **(Ans. 1·56 kW)**

3. A double acting reciprocating pump of plunger diameter 100 mm and stroke of 250 mm length is discharging water into a tank fitted 20 m higher than the axis of the pump. If the pump is rotating at 45 r.p.m., find the power required to drive the pump. **(Ans. 0·57 kW)**

36·8 Indicator Diagram of a Reciprocating Pump

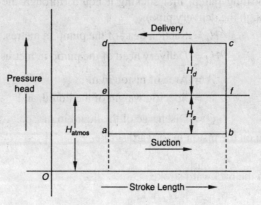

Fig. 36·2. Indicator diagram.

We have seen in the previous articles, that the pressure in the cylinder, during the suction stroke, is less than the atmospheric pressure; whereas during the delivery stroke, the pressure in the cylinder, is more than the atmospheric pressure.

Let H_s = Suction head of the pump.

H_d = Delivery head of the pump, and

L = Length of the stroke.

If we draw a diagram showing stroke length as abscissa and pressure head as ordinate, the diagram so obtained is known as an indicator diagram as shown in Fig. 36·2.

The line *ef* represents atmospheric pressure head and the line *ab* represents the pressure head during suction stroke; which is below the atmospheric pressure head by H_s. The line *cd* represents the pressure head during delivery stroke, which is above the atmospheric pressure head by H_d.

$$\therefore \text{Work done by the pump} = wQ(H_s + H_d) = \frac{wLAN}{60}(H_s + H_d)$$

It is thus obvious, that the indicator diagram, of a reciprocating pump represents the work done to some scale.

36.9 Variation of Pressure in the Suction and Delivery Pipes due to Acceleration of the Piston

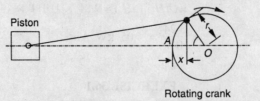

Fig. 36·3. Piston and crank.

We have seen in Art. 36·1 that the piston is connected to the rotating crank by a crank shaft. It is thus obvious that the piston will move, to and fro, with a simple harmonic motion. Therefore the

Reciprocating Pumps

velocity of the piston will not be uniform at all points. It will be zero at its extreme ends, whereas it will be maximum at its centre as shown in Fig. 36·3.

Moreover, the piston will have an acceleration at the beginning and a retradation at end of every stroke. This acceleration and retardation of the piston causes a variation of pressure in the cylinder and consequently in the suction and delivery pipes. Now consider a reciprocating pump lifting water from a sump.

Let
A = Area of the cylinder,
a = Area of the pipe,
ω = Angular velocity of the rotating crank in rad/s,
r = Radius of the rotating crank, and
l = Length of the pipe.

Let the rotating crank start from A (known as inner dead centre). After t seconds, let the angle described by the rotating crank be θ radians. Such that

$$\theta = \omega t$$

and displacement of the piston in t seconds,

$$x = r - r \cos \theta = r - r \cos \omega t$$

We know that the velocity of the piston,

$$v = \frac{dx}{dt} = \omega r \sin \omega t$$

and acceleration of the piston, $= \frac{dv}{dt} = \omega^2 r \cos \omega t$

Since the flow of water in the pipe is equal to the flow of water in the cylinder, therefore velocity of water in the pipe

$$= \frac{A}{a} \times \text{Velocity of piston} = \frac{A}{a} \times \omega r \sin \omega t$$

and acceleration of water in the pipe $= \frac{A}{a} \times \omega^2 r \cos \omega t$

Now weight of water in the pipe = wal

∴ Mass of water in the pipe $= \frac{wal}{g}$

and acceleration force = Mass × Acceleration

$$= \frac{wal}{g} \times \frac{A}{a} \times \omega^2 r \cos \omega t$$

We also know that intensity of pressure due to acceleration

$$= \frac{\text{Acceleration force}}{\text{Area}} = \frac{wl}{g} \times \frac{A}{a} \times \omega^2 r \cos \omega t$$

and acceleration pressure head, $H_a = \frac{\text{Intensity of pressure}}{\text{Sp. wt. of liquid}} = \frac{l}{g} \times \frac{A}{a} \times \omega^2 r \cos \omega t$

$$= \frac{l}{g} \times \frac{A}{a} \omega^2 r \cos \theta \qquad \ldots (\because \theta = \omega t)$$

Cor. 1. When $\theta = 0$ (*i.e.*, at the beginning of stroke),

$$H_a = \frac{l}{g} \times \frac{A}{a} \omega^2 r \qquad \ldots (\because \cos \theta = 1)$$

2. When $\theta = 90°$ (*i.e.*, at the mid of stroke)

$$H_a = 0 \qquad \ldots (\because \cos 90° = 0)$$

3. When θ = 180° (*i.e.*, at the end of stroke)

$$H_a = -\frac{l}{g} \times \frac{A}{a} \omega^2 r \qquad \qquad ...(\because \cos 180° = -1)$$

36·10 Effect of Acceleration of Piston on the Indicator Diagram

We have seen in Art. 36·8 that some acceleration pressure head is caused due to the acceleration of the piston. We have also seen that at the beginning of the suction stroke, the pressure head is below the atmospheric pressure head by $(H_s + H_a)$, where H_a is the acceleration pressure head. In the middle of the suction stroke, the pressure head is below the atmospheric pressure head by H_s (as the acceleration pressure head, $(H_a = 0$, when θ = 90°). At the end of the suction stroke, the pressure head is below atmospheric pressure head by $(H_s - H_a)$. Therefore we can modify the indicator diagram, for the suction stroke, as shown in Fig. 36·4.

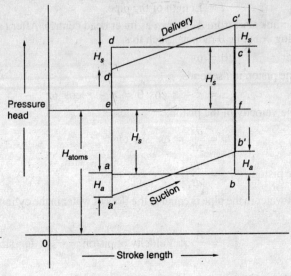

Fig. 36·4. Effect of acceleration of piston on the indicator diagram.

Similarly, at the beginning of the delivery stroke, the pressure head is above the atmospheric pressure by $(H_d + H_a)$. In the middle of the delivery stroke, the pressure head is above the atmospheric pressure head by H_d. At the end of the delivery stroke, the pressure head is above the atmospheric pressure head by $(H_d - H_a)$. Therefore we can modify the indicator diagram for delivery stroke also as shown in Fig. 36·4.

Example 36·3. *A single acting reciprocating pump, having plunger diameter 125 mm and stroke length 300 mm is drawing water from a depth of 4 metres from the axis the cylinder at 24 r.p.m. The length and diameter of suction pipe is 9 metres and 75 mm respectively. Find the pressure head on the piston at the beginning and end of the suction stroke, if the barometer reads 10·3 m of water.*

Solution. Given : D = 125 mm = 0·125 m; L = 300 mm = 0·3 m; or crank radius (r) = 0·3/2 = 0·15 m (because stroke length is equal to twice the crank radius); H_s = 4 m; N = 24 r.p.m.; l = 9 m; d = 75 mm; 0·075 m and barometer reading (H) = 10·3 m.

Pressure head at the beginning of the suction stroke

We know that area of the plunger,

$$A = \frac{\pi}{4} \times (D)^2 = \frac{\pi}{4} \times (0·125)^2 = 0·0123 \text{ m}^2$$

Similarly area of the suction pipe,

$$a = \frac{\pi}{4} \times (d)^2 = \frac{\pi}{4} \times (0·075)^2 = 0·0044 \text{ m}^2$$

Reciprocating Pumps

and angular velocity of the crank,

$$\omega = \frac{2\pi N}{60} = \frac{2\pi \times 24}{60} = 0.8 \text{ rad/s}$$

∴ Acceleration pressure head,

$$H_a = \frac{l}{g} \times \frac{A}{a} \times \omega^2 r = \frac{9}{9.81} \times \frac{0.0123}{0.0044} \times (0.8\pi)^2 \times 0.15 \text{ m}$$

$$= 2.4 \text{ m}$$

We also know that pressure head on the piston at the beginning of the stroke,

$$H_{piston} = H - (H_s + H_a) = 10.3 - (4 + 2.4) = 3.9 \text{ m Ans.}$$

Pressure head at the end of the suction stroke

We also know that the pressure head, on the piston, at the end of the stroke,

$$H_{piston} = H - (H_s - H_a) = 10.3 - (4 - 2.4) = 8.7 \text{ m Ans.}$$

36·11 Effect of Friction in the Suction and Delivery Pipes, on the Indicator Diagram

We have seen in Art. 13·3 that whenever the water is flowing through a pipe, there is always some loss of head due to friction of the pipe, which offers resistance to the flow of water. Similarly, as the water is flowing through the suction and delivery pipes, of a reciprocating pump, there will be some loss of head, due to friction, in both the pipes. Now consider a reciprocating pump lifting water from a sump.

Let
A = Area of the cylinder or bore,
d = Diameter of the pipe,
a = Area of the pipe,
ω = Angular velocity of the rotating crank in rad/s,
r = Radius of the rotatory crank,
l = Length of the pipe,
f = Coefficient of friction, and
v = Velocity of water in the pipe

We know that velocity of the piston at any instant,

$$= \omega r \sin \omega t = \omega r \sin \theta$$

∴ Velocity of water in the pipe at that instant,

$$v = \frac{A}{a} \times \omega r \sin \theta$$

We know that the loss of head due to friction,

$$H_f = \frac{4f l v^2}{2gd} = \frac{4fl}{2gd} \left(\frac{A}{a} \times \omega r \sin \theta \right)^2$$

Now we shall discuss the effect of this pipe friction on the indicator diagram at the beginning, middle and end of the stroke one by one.

1. At the beginning of the stroke, $\theta = 0$. Therefore the velocity of water in the pipe is zero, consequently there is no loss of head due to friction.
2. At the middle of the stroke, $\theta = 90°$. Therefore $\sin \theta = 1$. Therefore the loss of head due to friction,

$$H_f = \frac{4fl}{2gd} \left(\frac{A}{a} \times \omega r \right)^2$$

3. At the end of the stroke, $\theta = 180°$. Therefore the velocity of water in the pipe is zero, consequently there is no loss of head due to friction.

Now we shall study the effect of friction alongwith the effect of acceleration of piston on the indicator diagram.

In the suction stroke

1. At the beginning of the suction stroke, H_f is zero. Therefore pressure head will be below the atmospheric pressure head by $(H_s + H_a)$.
2. In the middle of the suction stroke, H_a is zero. Therefore pressure head will be below the atmospheric pressure head by $(H_a + H_f)$.
3. At the end of the suction stroke. H_f is zero. Therefore pressure head will be below the atmospheric pressure head by $(H_s - H_a)$.

In the delivery stroke

1. At the beginning of the delivery stroke, H_f is zero. Therefore pressure head will be above the atmospheric pressure by $(H_d + H_a)$.
2. In the middle of the delivery stroke, H_a is zero. Therefore pressure head will be above the atmospheric pressure head by $(H_d + H_f)$.
3. At the end of the delivery stroke, H_f is zero. Therefore pressure head will be above the atmospheric pressure head by $(H_d - H_a)$.

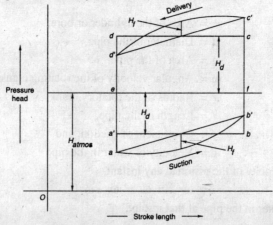

Fig. 36·5. Effect of friction in the suction and delivery pipes on the indicator diagram.

Now we can again modify the indicator diagram for suction stroke and delivery stroke as shown in Fig. 36·5.

Example 36·4. *A single acting reciprocating pump has plunger diameter of 200 mm and stroke length 300 mm. The suction pipe is 100 mm diameter and 8 metres long. The pump draws water 4 metres below the cylinder axis at 30 r.p.m. Find the pressure head on the piston :*

1. at the beginning of the suction stroke,

2. in the middle of the suction stroke, and

3. at the end of the suction stroke.

Take $f = 0·01$ and atmospheric pressure head = 10·3 metres of water.

Solution. Given : $D = 200$ mm $= 0·2$ m; $L = 300$ mm $= 0·3$ m or crank radius $(r) = 0·3/2 = 0·15$ m (because stroke length is equal to twice the crank radius); $d = 100$ mm $= 0·1$ m; $l_s = 8$ m; $H_s = 4$ m; $N = 30$ r.p.m.; $f = 0·01$ and $H = 10·3$ m.

Reciprocating Pumps

1. Pressure head at the beginning of the suction stroke

We know that area of the plunger,

$$A = \frac{\pi}{4} \times (D)^2 = \frac{\pi}{4} \times (0.2)^2 = 0.03142 \text{ m}^2$$

Similarly area of suction pipe,

$$a = \frac{\pi}{4} \times (d)^2 = \frac{\pi}{4} \times (0.1)^2 = 0.00785 \text{ m}^2$$

and angular velocity of crank, $\omega = \frac{2\pi N}{60} = \frac{2\pi \times 30}{60} = \pi$ rad/s

∴ Acceleration pressure head,

$$H_a = \frac{l}{g} \times \frac{A}{a} \times \omega^2 r = \frac{8}{9.81} \times \frac{0.03142}{0.00785} \times (\pi)^2 \times 0.15 \text{ m}$$

$$= 4.8 \text{ m}$$

We know that the pressure head on the piston at the beginning of the stroke

$$H_{piston} = H - (H_s + H_a) = 10.3 - (4 + 4.8) = 1.5 \text{ m Ans.}$$

2. Pressure head in the middle of the suction stroke

We know that the velocity of water in the suction pipe in the* middle of the stroke

$$v = \frac{A}{a} \times \omega r \sin \theta = \frac{0.03142}{0.00785} \times \pi \times 0.15 \times 1 = 1.88 \text{ m/s}$$

and loss of head due to friction in the suction pipe,

$$H_f = \frac{4flv^2}{2gd} = \frac{4 \times 0.01 \times 8 \times (1.88)^2}{2 \times 9.81 \times 0.91} = 0.6 \text{ m}$$

∴ Pressure head on the piston in the middle of the stroke,

$$H_{piston} = H - (H_s + H_f) \qquad \qquad ...(\because H_a = 0)$$

$$= 10.3 - (4 + 0.6) = 5.7 \text{ m Ans.}$$

3. Pressure head at the end of the suction stroke

We know that the pressure head on the piston at the end of the stroke,

$$H_{piston} = H - (H_s - H_a) \qquad \qquad ...(H_f = 0)$$

$$= 10.3 - (4 - 4.83) = 11.1 \text{ m Ans.}$$

Example 36.5. *A single acting reciprocating pump has a stroke of length 150 mm. The suction pipe is 7.5 metres long and the ratio of plunger diameter to the suction diameter is 4/3. The water level in the pump is 2.5 metres below the axis of the pump cylinder, and the pipe connecting the sump and pump cylinder is 75 mm diameter. If the crank is running at 75 r.p.m. find the pressure head on the piston (a) in the beginning of the suction stroke, (b) in the middle of the suction stroke, and (c) in the end of the suction stroke. Take coefficient of friction as 0.01.*

Solution. Given : $L = 150$ mm $= 0.15$ m or crank radius $(r) = 0.15/2 = 0.075$ m (because stroke length is equal to twice the crank radius); $l_s = 7.5$ m; $\frac{\text{Plunger diameter}}{\text{Suction diamete}} = \frac{4}{3}$ or $\frac{\text{Plunger area}}{\text{Suction area}}$

$= \left(\frac{4}{3}\right)^2 = \frac{16}{9}$; $H_s = 2.5$ m; $d = 75$ mm $= 0.075$ m; $N = 75$ r.p.m. and $f = 0.01$.

---

* In the middle of the stroke, $\theta = 90°$ and $\sin 90° = 1$.

(a) Pressure head in the beginning of the suction stroke

We know that the angular velocity of the crank,

$$\omega = \frac{2\pi N}{60} = \frac{2\pi \times 75}{60} = 2.5\,\pi \text{ rad/s}$$

and acceleration pressure head, $H_a = \frac{l}{g} \times \frac{A}{a} \times \omega^2 r = \frac{7.5}{9.81} \times \frac{16}{9} \times (2.5\,\pi)^2 \times 0.075$ m

$$= 6.3 \text{ m}$$

∴ Pressure head on the piston in the beginning of the stroke,

$$H_{piston} = H - (H_s + H_a) = H - (2.5 + 6.3) = H - 8.8 \text{ m}$$
$$= 8.8 \text{ m (vacuum)} \quad \textbf{Ans.}$$

(b) Pressure head in the middle of the suction stroke

We know that the velocity of water in the suction pipe in the middle of the stroke,

$$v = \frac{A}{\omega} \times \omega r \sin\theta = \frac{16}{9} \times 2.5\,\pi \times 0.075 \times 1 = 1.05 \text{ m/s}$$

and loss of head due to friction in the suction pipe,

$$H_f = \frac{4flv^2}{2gd} = \frac{4 \times 0.01 \times 7.5 \times (1.05)^2}{2 \times 9.81 \times 0.075} = 0.2 \text{ m}$$

∴ Pressure head on the piston in the middle of the stroke,

$$H_{piston} = H - (H_s + H_f) \qquad \ldots(\because H_a = 0)$$
$$= H - (2.5 + 0.2) = H - 2.7 \text{ m}$$
$$= 2.7 \text{ m (vacuum)} \quad \textbf{Ans.}$$

(c) Pressure head in the end of the suction stroke

We also know that the pressure head on the piston in the end of the stroke,

$$H_{piston} = H - (H_s - H_a) = H - (2.5 - 6.3) = H + 3.8 \text{ m}$$
$$= 3.8 \text{ m (gauge)} \quad \textbf{Ans.}$$

36·12 Maximum Speed of the Rotating Crank

It has been experienced that the continuous flow of water ceases to act (known as separation) whenever pressure in the pipe falls below the required pressure. In a reciprocating pump, the separation may take place in the suction pipe as well as delivery pipe. Now we shall discuss both the cases one by one.

(a) Separation in suction pipe

We have seen in Art. 36·11 that at the beginning of a suction stroke, the pressure head is below the atmospheric pressure head by $(H_s + H_a)$, where H_s is the suction head and H_a is the acceleration pressure head. It has been experimentally found that when this vacuum pressure head (*i.e.*, $H_s + H_a$) reaches 7·8 metres of water or 2·5 metres absolute [*i.e.*, $H - (H_s + H_a)$] the continuity of flow will stop, as the separation will take place; because the water will commence to vaporise. The head of water at which the separation takes place is known as separation head.

∴
$$H_{sep} = H - (H_s + H_a) = H - H_s - H_a$$

Since the suction head is constant for a set reciprocating pump, therefore, in order to avoid separation at the beginning of the suction stroke, the acceleration pressure head should be limited. We know that the acceleration pressure head,

$$H_a = \frac{l}{g} \times \frac{A}{a} \times \omega^2 r$$

where l is the length of the suction pipe.

Reciprocating Pumps

(b) Separation in delivery pipe

We have also seen, that the maximum pressure head in the delivery pipe is at the end of the delivery stroke is $(H + H_d - H_a)$. It has also been experimentally found that when this absolute pressure head is less than 2·5 m of water, the continuity of flow will stop and the separation will take place. The head at which the separation will take place is known as separation head and it denoted by H_{sep}.

$$\therefore \quad H_{sep} = H + H_d - H_a$$

In this case also, the relation for acceleration pressure head (as given above) is used. But the value of l is taken as the length of the delivery pipe.

It will be interesting to know that sometimes a reciprocating pump is used to deliver water through various types of delivery pipes. Two typical delivery pipes are shown in Fig. 36·6 (a) and (b). In Fig. 36·6 (a) the delivery pipe is first vertical, and then it is bent to be horizontal. In this case, the delivery head will be zero at B (i.e., at the bend) and there will be still considerable acceleration pressure head. It is thus obvious, that there is always a possibility of separation taking place at B. In Fig. 36·6 (b) the pipe is first laid horizontal, and then it is bent to be vertical. In this case, there is a little possibility of separation taking place at B; as at this point there will be considerable delivery head. It is thus desirable to lay the delivery pipe as shown in Fig. 36·6 (b) in order to avoid cavitation.

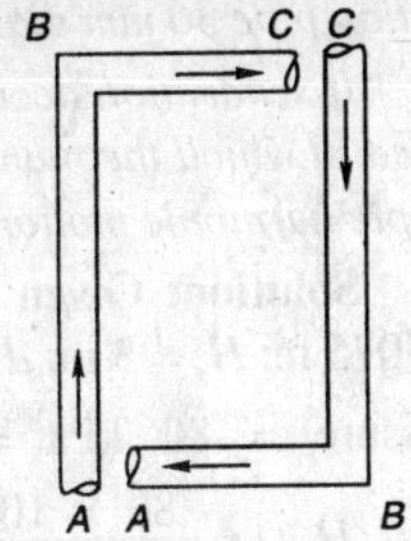

Fig. 36·6. Typical delivery pipes.

A little consideration will show, that to limit the acceleration pressure head, we have to limit the value of ω (i.e., angular velocity of the rotating crank) as all the other things are constant. Since the angular velocity of the crank,

$$\omega = \frac{2\pi N}{60}$$

therefore to limit the value of ω means to limit the value of N i.e., speed of the rotating crank. Or in other words, we may find out the maximum speed of the rotating crank in order to avoid separation.

Note : If no details of the layout of the delivery pipe is given, then the layout as shown in Fig. 36·6 (b) is assumed.

Example 36·6. *A double acting reciprocating pump runs at 90 r.p.m. The diameter and stroke are 100 mm and 250 mm respectively. The suction pipe is of 100 mm diameter and 5 m long. Calculate the maximum permissible suction lift assuming no air vessel is fitted and separation occurs at 2 m of water absolute.*

Solution. Given : $N = 90$ r.p.m.; $D = 100$ mm $= 0·1$ m; $L = 250$ mm $= 0·25$ m or crank radius $(r) = 0·25/2 = 0·125$ m (because stroke length is equal to twice the crank radius); $d = 100$ mm $= 0·1$m; $l = 5$ m and $H_{sep} = 2$ m.

Let H_s = Maximum permissible suction lift.

We know that area of the piston,

$$A = \frac{\pi}{4} \times (D)^2 = \frac{\pi}{4} \times (0·1)^2 = 0·007\ 85\ m^2$$

Similarly, area of suction pipe,

$$a = \frac{\pi}{4} \times (d)^2 = \frac{\pi}{4} \times (0·1)^2 = 0·007\ 85\ m^2$$

and angular velocity of the crank,

$$\omega = \frac{2\pi N}{60} = \frac{2\pi \times 90}{60} = 9·42\ rad/s$$

∴ Acceleration pressure head,

$$H_a = \frac{l_s}{g} \times \frac{A}{a} \times \omega^2 r = \frac{5}{9.81} \times \frac{0.007\ 85}{0.007\ 85} \times (9.42)^2 \times 0.125\ \text{m}$$

$$= 5.65\ \text{m}$$

We know that separation head (H_{sep}),

$$2 = H - H_s - H_a = 10.3 - H_s - 5.65 \quad \text{...(Assuming } H = 10.3\ \text{m)}$$

$$= 4.65 - H_s$$

∴ $H_s = 4.65 - 2 = 2.65\ \text{m}$ **Ans.**

Example 36·7. *A single acting reciprocating pump (with no air vessel) has a plunger of 80 mm diameter and a stroke of 150 mm. It draws water from a sump 3 m below the pump axis through a suction pipe 30 mm diameter and 4·5 m long.*

If separation occurs at a pressure of 80 kPa below atmospheric pressure, find the maximum speed at which the pump may be operated without separation. Assume that the plunger moves with simple harmonic motion.

Solution. Given : $D = 80$ mm $= 0.08$ m; $L = 150$ mm $= 0.15$ m; or crank radius $(r) = 0.15/2 = 0.075$ m; $H_s = 3$ m; $d = 30$ mm $= 0.03$ m; $l_s = 4.5$ and separation pressure below the atmospheric pressure $= 80$ kPa $= 80$ kN/m² or separation pressure head below the atmospheric head

$$(H - H_{sep}) = \frac{80 \times 10^3}{9.81 \times 10^3} = 8.2\ \text{m}.$$

Let $H_a =$ Acceleration pressure head,

$\omega =$ Angular velocity at which the pump may be operated without separation, and

$N =$ Maximum speed of the pump.

We know that area of the plunger,

$$A = \frac{\pi}{4} \times (D)^2 = \frac{\pi}{4} \times (0.08)^2 = 5.03 \times 10^{-3}\ \text{m}^2$$

and area of suction pipe,

$$a = \frac{\pi}{4} \times (d)^2 = \frac{\pi}{4} \times (0.03)^2 = 0.707 \times 10^{-3}\ \text{m}^2$$

We know that the separation head,

$$H_{sep} = H - H_s - H_a$$

or $H_a = H - H_{sep} - H_s = (H - H_{sep}) - H_s = 8.2 - 3 = 5.2\ \text{m}$

We also know that acceleration pressure head (H_a),

$$5.2 = \frac{l_s}{g} \times \frac{A}{a} \times \omega^2 r = \frac{4.5}{9.81} \times \frac{5.03 \times 10^{-3}}{0.707 \times 10^{-3}} \times \omega^2 \times 0.075$$

$$= 0.245\ \omega^2$$

∴ $\omega^2 = 5.2/0.245 = 21.2$ or $\omega = 4.6$ rad/s

We know that angular velocity (ω),

$$4.6 = \frac{2\pi N}{60} = 0.105\ N \quad \text{or} \quad N = \frac{4.6}{0.105} = 43.8\ \text{r.p.m.}\ \textbf{Ans.}$$

Reciprocating Pumps

Example 36·8. *A single acting reciprocating pump of 250 mm diameter and 500 mm stroke delivers water through a 100 mm diameter vertical delivery pipe to a tank situated at 15 metres above it and 30 metres horizontally from it. Find the safe speed of the pump, if separation pressure corresponds to 23 kPa, when the delivery pipe is :*

(i) vertical from the pump and then horizontal upto the tank,

(ii) horizontal from the pump and then vertical upto the tank.

Assume atmosphere pressure head as 10·3 m.

Solution. Given : $D = 250$ mm $= 0.25$ m; $L = 500$ mm $= 0.5$ m or crank radius (r) $0.5/2 = 0.25$ m; $d_d = 100$ mm $= 0.1$ m; $H_d = 15$ m; Length of vertical delivery pipe $= 15$ m; Length of horizontal delivery pipe $= 30$ m or total length of delivery pipe $(l_d) = 15 + 30 = 45$ m; Separation pressure $(p_{sep}) = 23$ kPa $= 23$ kN/m^2 or separation pressure head $(H_{sep}) = \dfrac{p_{sep}}{w} = \dfrac{23 \times 10^3}{9.81 \times 10^3} = 2.3$ m

and $H = 10.3$ m.

Safe speed for the pump, when the delivery is horizontal from the pump and then vertical upto the tank

Let $H_a =$ Acceleration pressure head,

$\omega =$ Safe speed of the pump in rad/s,

$N =$ Safe speed of the pump in r.p.m.

We know that area of piston

$$A = \frac{\pi}{4} \times (D)^2 = \frac{\pi}{4} \times (0.25)^2 = 0.0491 \text{ m}^2$$

and area of delivery pipe,

$$a = \frac{\pi}{4} \times (d_d)^2 = \frac{\pi}{4} \times (0.1)^2 = 0.00785 \text{ m}^2$$

We know that the separation head (H_{sep}),

$$2.3 = H + H_d - H = 10.3 + 15 - H_a = 25.3 - H_a$$

∴ $H_a = 25.3 - 2.3 = 23$ m

We also know that acceleration pressure head (H_a),

$$23 = \frac{l}{g} \times \frac{A}{a} \times \omega^2 r = \frac{45}{9.81} \times \frac{0.0491}{0.00785} \times \omega^2 \times 0.25$$

$$= 7.17 \, \omega^2$$

∴ $\omega^2 = 23/7.17 = 3.208$ or $\omega = 1.79$ rad/s

We know that angular velocity (ω),

$$1.79 = \frac{2\pi N}{60} = 0.105 \, N \quad \text{or} \quad N = \frac{1.79}{0.105} = 17 \text{ r.p.m.} \quad \textbf{Ans.}$$

Safe speed for the pump, when the delivery pipe is vertical from the pump and then horizontal upto the tank.

In this case, the separation will take place at the bend after the vertical pipe. Therefore we shall take delivery head as zero in the relation for separation head. Thus separation head (H_{sep}),

$$2.3 = H - H_a = 10.3 - H_a$$

or $H_a = 10.3 - 2.3 = 8.0$ m

We know that acceleration pressure head (H_a),

$$8 = \frac{l}{g} \times \frac{A}{a} \times \omega^2 r = \frac{45}{9\cdot 81} \times \frac{0\cdot 0491}{0\cdot 00785} \times \omega^2 \times 0\cdot 25$$

$$= 7\cdot 17\, \omega^2$$

∴ $\omega^2 = 8/7\cdot 13 = 1\cdot 116$ or $\omega = 1\cdot 06$ rad/s

We know that angular velocity (ω),

$$1\cdot 06 = \frac{2\pi N}{60} = 0\cdot 105\, N \quad \text{or} \quad N = \frac{1\cdot 06}{0\cdot 105} = 10\cdot 1 \text{ r.p.m.} \textbf{ Ans.}$$

EXERCISE 36·2

1. A single acting reciprocating pump has plunger of diameter 150 mm and stroke of length 400 mm. It draws water from a depth of 4·5 m through a pipe 8 m long and 80 mm diameter at 25 r.p.m. If the atmospheric pressure head is 10·3 m, find the pressure head on the piston at the beginning and end of the suction stroke. **(Ans.** 1·8 m; 9·8 m)

2. A single acting reciprocating pump has plunger of diameter 150 mm and stroke of 300 mm. The lengths of suction and delivery pipe are 6·5 m and 39 m respectively and both the pipes are of the same diameter of 75 mm. The axis of the pump is 5 m above the level of water in the sump and 33 m below the delivery water level. If the atmospheric pressure head is 10·3 m of water and coefficient of friction for both the pipes is 0·01, find the pressure head on the piston at the beginning, middle and end of the suction strokes. Take speed of the crank as 30 r.p.m. **(Ans.** 1·4 m; 4·67 m; 9·2 m)

3. The bore and stroke of a single acting reciprocating pump are 100 mm and 200 mm respectively. The suction pipe is 80 mm in diameter and 4 m long and centre of the pump in 3·5 m above the water level of the sump. Find the maximum speed at which can be run without separation taking place. Assume separation to occur at 2·5 m of water and atmospheric pressure head = 10·3 m of water.

(Ans. 78·5 r.p.m.)

4. A single acting reciprocating pump has stroke length of 300 mm and plunger diameter of 80 mm. The suction head is 3 metres and the suction pipe is 6 metres long. If the separation takes place at a pressure head of 2·5 metres, determine the maximum speed of the pump, so as to avoid separation. Take atmospheric pressure head as 10·3 metres of water. **(Ans.** 58·6 r.p.m.)

36·13 Air Vessels

An air vessel is a cast iron closed chamber, having an opening at its base, through which the water flows into the vessel or from the vessel. The vessel is filled up with compressed air.

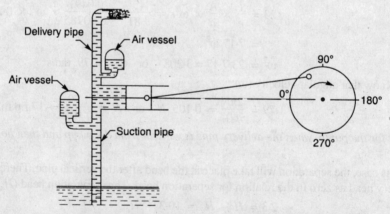

Fig. 36·7. Air vessels fitted to the suction and delivery pipes.

The air vessels are fitted to the suction pipe and delivery pipe close to the cylinder of the pump as shown in Fig. 36·7. The object of fitting the air vessels is to obtain a uniform discharge from a

Reciprocating Pumps

reciprocating pump. Consider an air vessel fitted to the delivery pipe as shown in Fig. 34·7. During the first half of the delivery stroke, the piston moves with acceleration, thus forcing the water into the delivery pipe with a velocity more than the mean velocity. The excess flow of water, flows into the air vessel thus compressing the air inside the vessel. During the second half of the delivery stroke, the piston moves with retardation thus forcing the water into the delivery pipe, with a velocity less than the mean velocity. The water, stored into the air vessel, then starts flowing into the delivery pipe, thus making up the deficiency of the flow.

Thus the discharge in the delivery pipe, beyond the air vessel, is more or less uniform. But for all practical purposes, velocity of water in the delivery pipe, beyond air vessel is taken to be uniform. Similarly, on the suction side the water first flows from the suction pipe into the air vessel (during first half of the suction stroke) and then from the air vessel to the cylinder (during the second half of the suction stroke). Thus for all practical purpose, velocity of water in the suction pipe upto the air vessel is also taken to be uniform.

Example 36·9. *A double acting pump has a cylinder of 200 mm diameter and stroke of 400 mm. The pump delivers water to a height of 10 metres through a pipe 36 metres long and 150 mm diameter at 40 r.p.m. Find the pressure in the cylinder at the beginning of the delivery stroke, if a large air vessel is fitted in the delivery pipe at the same level of the pump, but 3 metres from the cylinder. Take f = 0·008.*

Solution. Given : $D = 200$ mm $= 0.2$ m; $L = 400$ mm $= 0.4$ m or crank radius $(r) = 0.4/2 = 0.2$ m; $H_d = 10$ m; $l = 36$ m; $d_d = 150$ mm $= 0.15$ m; $N = 40$ r.p.m.; Distance between the pump level and air vessel $= 3$ m and $f = 0.008$.

We know that area of the cylinder,

$$A = \frac{\pi}{4} \times (D)^2 = \frac{\pi}{4} \times (0.2)^2 = 0.03142 \text{ m}^2$$

and area of the delivery pipe,

$$a = \frac{\pi}{4} \times (d)^2 = \frac{\pi}{4} \times (0.15)^2 = 0.01767 \text{ m}^2$$

Since the air vessel is fitted to the delivery pipe, at a distance of 3 metres from the cylinder, therefore :
1. there will be an acceleration pressure head in the delivery pipe for a length of 3 metres.
2. there will be a loss of head due to friction in the delivery pipe for a length of $36 - 3 = 33$ metres.

We know that angular velocity of the crank,

$$\omega = \frac{2\pi N}{60} = \frac{2\pi \times 40}{60} = 4.19 \text{ rad/s}$$

and acceleration pressure head,

$$H_a = \frac{l}{g} \times \frac{A}{a} \times \omega^2 r = \frac{3}{9.81} \times \frac{0.03142}{0.01767} \times (4.19)^2 \times 0.2 \text{ m}$$

$$= 1.91 \text{ m}$$

We know that the velocity of water in the delivery pipe,

$$v = \frac{Q}{a} = \frac{2LAN}{60 \times a} \qquad \text{...(}\because \text{ of double acting pump)}$$

$$= \frac{2 \times 0.4 \times 0.0314 \times 40}{60 \times 0.01767} = 0.95 \text{ m/s}$$

and loss of head due to friction in 33 m long pipe,

$$H_f = \frac{4flv^2}{2gd} = \frac{4 \times 0.008 \times 33 \times (0.95)^2}{2 \times 9.81 \times 0.15} = 0.32 \text{ m}$$

∴ Pressure head in the cylinder at the beginning of the delivery stroke,

$$= H_d + H_a + H_f = 10 + 1.91 + 0.32 = 12.23 \text{ m } \textbf{Ans.}$$

Example 36·10. *A single acting reciprocating pump is to raise a liquid of 11·8 kN/m³ through a vertical height of 11·5 metres, from 2·5 metre below pump axis to 9 metres above it. The plunger, which moves with S.H.M., has diameter 125 mm and stroke 225 mm. The suction and delivery pipes are 75 mm diameter and 3·5 metres and 13·5 metres long respectively. There is a large air vessel placed on the delivery pipe near the pump axis. But there is no air vessel on the suction pipe. If separation takes place at 90 kPa below atmospheric pressure, find*

   (a) maximum speed, with which the pump can run without separation taking place, and

   (b) power required to drive the pump, if f = 0·02.

   Neglect slip for the pump.

Solution. Given : $w = 11.8$ kN/m³; $H = 11.5$ m; $H_s = 2.5$ m; $H_d = 9$ m; $D = 125$ mm $= 0.125$ m; $L = 225$ mm $= 0.225$ m; or crank radius $(r) = 0.225/2 = 0.1125$ m; $d = 75$ mm $= 0.075$ m; $l_s = 3.5$ m; $l_d = 13.5$ m; Separation pressure below the atmospheric pressure $= 90$ kPa or separation pressure head below the atmospheric pressure head $(H - H_{sep}) = \dfrac{90 \times 10^3}{11.8 \times 10^3} = 7.6$ m and and $f = 0.02$.

(a) Maximum speed, with which the pump can run without separation taking place

   Let H_a = Acceleration pressure head,

   ω = Angular velocity at which the pump can run without separation taking place, and

   N = Maximum speed of the pump.

We know that area of plunger,

$$A = \frac{\pi}{4} \times (D)^2 = \frac{\pi}{4} \times (0.125)^2 = 0.0123 \text{ m}^3$$

and area of suction pipe,

$$a = \frac{\pi}{4} \times (d)^2 = \frac{\pi}{4} \times (0.075)^2 = 0.0044 \text{ m}^2$$

We know that the separation head,

$$H_{sep} = H - H_s - H_a$$

∴ $H_a = (H - H_{sep}) - H_s = 7.6 - 2.5 = 5.1$ m

We also know acceleration pressure head (H_a),

$$5.1 = \frac{l_s}{g} \times \frac{A}{a} \times \omega^2 r = \frac{3.5}{9.81} \times \frac{0.0123}{0.0044} \times \omega^2 \times 0.1125$$

$$= 0.112 \, \omega^2$$

∴ $\omega^2 = 5.1/0.112 = 45.5$ or $\omega = 6.75$ rad/s

We know that angular velocity (ω),

$$6.75 = \frac{2\pi N}{60} = 0.105 \, N \quad \text{or} \quad N = \frac{6.75}{0.105} = 64.3 \text{ r.p.m. } \textbf{Ans.}$$

Reciprocating Pumps

(b) Power required to drive the pump

We know that the discharge of the pump,

$$Q = \frac{LAN}{60} = \frac{0.225 \times 0.0123 \times 64.3}{60} = 0.003 \text{ m}^3/\text{s}$$

∴ Velocity of water in the delivery pipe,

$$v = \frac{Q}{a} = \frac{0.003}{0.0044} = 0.68 \text{ m/s}$$

Since there is an air vessel on the delivery pipe near the pump axis, therefore there will be no acceleration head on the delivery side. Hence we shall consider the loss of head due to friction only.

∴ Loss of head due to friction,

$$H_{fd} = \frac{4fl_d v^2}{2gd} = \frac{4 \times 0.02 \times 13.5 \times (0.68)^2}{2 \times 9.81 \times 0.075} = 0.34 \text{ m}$$

Since there is no air vessel on the suction pipe, therefore there will be loss of head due to friction. But we shall consider the acceleration head only i.e., $H_{as} = 5.1$ m.

∴ Total head, against which the pump has to work,

$$H = (H_s + H_d) + (H_{as} + H_{ad}) + (H_{fs} + H_{fd})$$
$$= (2.5 + 9) + (5.1 + 0) + (0 + 0.34) = 16.94 \text{ m}$$

and power required to drive the pump,

$$P = wQH = 9.81 \times 0.003 \times 16.94 = 0.5 \text{ kW} \quad \textbf{Ans.}$$

36·14 Maximum Speed of the Rotating Crank with Air Vessels

We have seen in Art. 36·13 that by fitting air vessels, the velocity of water in the suction pipe upto the air vessels (i.e., for a length of l_s in Fig. 36·8) is uniform. The acceleration and retardation to the velocity of water will take place in the suction pipe beyond the air vessel (i.e., for a length of l_s' in Fig. 36·8).

Similarly acceleration and retardation to the velocity of water will take place in the delivery pipe upto the air vessel (i.e., for a length of l_d'') in Fig. 36·8. The velocity of water in the delivery pipe beyond the air vessel (i.e., for a length of l_d in Fig. 36·8) is constant.

Thus for finding out the maximum speed of the crank we have to limit the separation head. i.e.,

$$H_{sep} = H - [H_s + (H_a \text{ for } l_s') + (H_f \text{ for } l_s)]$$

The constant velocity of water, in the suction or delivery pipe, may be found out by dividing the discharge of the pump by the area of the respective pipe.

Now consider a reciprocating pump fitted with air vessels on suction and delivery pipes as shown in Fig. 36·8.

Let
- L = Length of the stroke,
- A = Area of the piston,
- N = Speed of the pump in r.p.m.,
- ω = Angular velocity of the crank,
- a = Area of the pipe,
- r = Radius of the crank, and
- v = Velocity of water in the pipe.

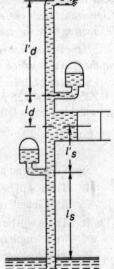

Fig. 36·8. Air vessels fitted to the pipes.

For single acting pump

We know that the discharge of a single acting pump,
$$Q = \frac{LAN}{60}$$

and velocity of water, $\quad v = \frac{Q}{a} = \frac{LAN}{60 \times a} = \frac{A}{a} \times \frac{LN}{60}$

Substituting $L = 2r$, $\omega = \frac{2\pi N}{60}$ or $N = \frac{60\omega}{2\pi}$ in the above equation,

$$v = \frac{A}{a} \times \frac{2r \times \frac{60\omega}{2\pi}}{60} = \frac{A}{a} \times \frac{\omega r}{\pi}$$

For double acting pump

We know that the discharge of a double acting pump,
$$Q = \frac{2LAN}{60}$$

and velocity of water, $\quad v = \frac{Q}{a} = \frac{2LAN}{60 \times a} = \frac{2A}{a} \times \frac{LN}{60}$

Substituting, $L = 2r$, $\omega = \frac{2\pi N}{60}$ or $N = \frac{60\omega}{2\pi}$ in the above equation,

$$v = \frac{2A}{a} \times \frac{2r \times \frac{60\omega}{2\pi}}{60} = \frac{2A}{a} \times \frac{\omega r}{\pi}$$

Example 36·11. *A single acting reciprocating pump has a plunger diameter of 250 mm and stroke length of 450 mm. The suction pipe is 125 mm diameter and 12 metres long with a suction lift of 3 metres. An air vessel is fitted to the suction pipe at a distance of 1·5 metre from the cylinder and 10·5 metres from the sump of water level.*

If the barometer reads 10·0 metres of water and separation takes place at 2·5 metres vacuum, find the speed at which the crank can operate without separation to occur. Take f = 0·01.

Solution. Given : $D = 250$ mm $= 0.25$ m; $L = 450$ mm $= 0.45$ m or crank radius (r) $= 0.45/2 = 0.225$ m; $d = 125$ mm $= 0.125$ m; $l = 12$ m; $H_s = 3$ m; $l_s' = 1.5$ m; $l_s = 10.5$ m; $H = 10$ m; $H_{sep} = 2.5$ m and $f = 0.01$.

Let $\quad \omega =$ Speed of the crank in rad/s, and
$\quad\quad\quad N =$ Speed of the crank in r.p.m.

Since there is an air vessel fitted to the suction pipe, therefore

1. There will be a loss of head due to friction in the suction pipe for a length of 10·5 metres.
2. There will be an acceleration pressure head in the suction pipe for a length of 1·5 metre.

We know that area of the plunger,
$$A = \frac{\pi}{4} \times (D)^2 = \frac{\pi}{4} \times (0.25)^2 = 0.0491 \text{ m}^2$$

and area of the suction pipe,
$$a = \frac{\pi}{4} \times (d)^2 = \frac{\pi}{4} \times (0.125)^2 = 0.0123 \text{ m}^2$$

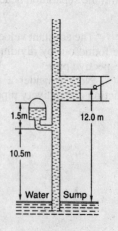

Fig. 36·9.

Reciprocating Pumps

∴ Velocity of water in the suction pipe of single acting reciprocating pump,

$$v = \frac{A}{a} \times \frac{\omega r}{\pi} = \frac{0.0491}{0.0123} \times \frac{\omega \times 0.225}{\pi} = 0.286 \; \omega \; \text{m/s}$$

We know that the loss of head due to friction,

$$H_f = \frac{4flv^2}{2gd} = \frac{4 \times 0.01 \times 10.5 \; (0.286 \, \omega)^2}{2 \times 9.81 \times 0.125} = 0.014 \; \omega^2$$

and acceleration pressure head in the suction pipe for a length of 1.5 metre,

$$H_a = \frac{l}{g} \times \frac{A}{a} \times \omega^2 r = \frac{1.5}{9.81} \times \frac{0.0491}{0.0123} \times \omega^2 \times 0.225$$

$$= 0.137 \; \omega^2 \qquad \ldots(ii)$$

We know that the separation head (H_{sep})

$$2.5 = H - (H_s + H_a + H_f) = 10 - (3 + 0.137 \, \omega^2 + 0.014 \, \omega^2)$$

$$= 7 - 0.151 \, \omega^2$$

∴ $$\omega^2 = \frac{7 - 2.5}{0.151} = 29.8 \quad \text{or} \quad \omega = 5.46 \; \text{rad/s}$$

We know that angular velocity (ω),

$$5.46 = \frac{2\pi N}{60} = 0.105 \; N \quad \text{or} \quad N = \frac{5.46}{0.105} = 52 \; \text{r.p.m.} \quad \textbf{Ans.}$$

Example 36·12. *A single-acting single-cylinder reciprocating pump is installed 3.5 metres above the water of a sump. The suction pipe is 200 mm diameter and 10 metres long. The pump cylinder is of 300 mm diameter and 500 mm stroke. Find*

(a) *the speed, at which the separation may take place at the commencement of suction stroke.*

(b) *the change in speed of the pump, if an air vessel is fitted on the suction side 2.5 metres above the pump water level.*

Assume (i) barometric head = 10.3 metres of water, (ii) f = 0.01 and, (iii) separation occurs at 2.5 metres of water absolute.

Solution. Given : H_s = 3.5 m; d = 200 mm = 0.2 m; l = 10 m; D = 300 mm = 0.3 m; L = 500 mm = 0.5 m or crank radius (r) = 0.5/2 = 0.25 m; H = 10.3 m; f = 0.01 and H_{sep} = 2.5 m.

(a) *Speed at which the separation may take place without air vessel*

Let H_a = Acceleration pressure head,

ω = Angular velocity at which the separation may take place at the commencement of the suction stroke, and

N = Speed of the crank in r.p.m.

We know that the area of pump cylinder,

$$A = \frac{\pi}{4} \times (D)^2 = \frac{\pi}{4} \times (0.3)^2 = 0.0707 \; \text{m}^2$$

and area of suction pipe,

$$a = \frac{\pi}{4} \times (d)^2 = \frac{\pi}{4} \times (0.2)^2 = 0.0314 \; \text{m}^2$$

We know that separation head,

$$H_{sep} = H - H_s - H_a$$

or $$H_a = H - H_s - H_{sep} = 10.3 - 3.5 - 2.5 = 4.3 \; \text{m}$$

We also know that acceleration pressure head (H_a),

$$4.3 = \frac{l}{g} \times \frac{A}{a} \times \omega^2 r = \frac{10}{9.81} \times \frac{0.0707}{0.0314} \times \omega^2 \times 0.25$$

$$= 0.574\,\omega^2$$

$$\therefore \quad \omega^2 = 4.3/0.574 = 7.49 \quad \text{or} \quad \omega = 2.74 \text{ rad/s}$$

We know that angular velocity (ω),

$$2.74 = \frac{2\pi N}{60} = 0.105\,N \quad \text{or} \quad N = \frac{2.74}{0.105} = 26.1 \text{ r.p.m.} \quad \textbf{A}$$

Change in the speed of the pump with air vessel

Since the air vessel is fitted in the suction pipe 2.5 metres above the sump water level, therefore

1. There will be loss of head due to friction in the suction pipe for a length $\dfrac{10 \times 2.5}{3.5} = 7.14$ m.

2. There will be an acceleration pressure head in the suction pipe for a length of $10 - 7.14 = 2.86$ m.

We know that velocity of water,

$$v = \frac{A}{a} \times \frac{\omega r}{\pi} = \frac{0.0707}{0.0314} \times \frac{\omega \times 0.25}{\pi}$$

$$= 0.18\,\omega \text{ m/s}$$

and loss of head due to friction,

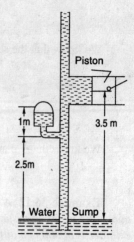

Fig. 36.10.

$$H_f = \frac{4flv^2}{2gd} = \frac{4 \times 0.01 \times 7.14\,(0.18\,\omega)^2}{2 \times 9.81 \times 0.2} = 0.002\,\omega^2$$

We also know that the acceleration pressure head in the suction pipe,

$$H_a = \frac{l}{g} \times \frac{A}{a}\,\omega^2 r = \frac{2.86}{9.81} \times \frac{0.0707}{0.0314} \times \omega^2 \times 0.25 = 0.164\,\omega^2$$

and the separation head, (H_{sep}),

$$2.5 = H - (H_s + H_a + H_f)$$

$$= 10.3 - (3.5 + 0.164\,\omega^2 + 0.002\,\omega^2) = 6.8 - 0.166\,\omega^2$$

$$\therefore \quad \omega^2 = \frac{6.8 - 2.5}{0.166} = 25.9 \quad \text{or} \quad \omega = 5.1 \text{ rad/s}$$

We know that angular velocity (ω)

$$5.1 = \frac{2\pi N}{60} = 0.105\,N \quad \text{or} \quad N = \frac{5.1}{0.105} = 48.6 \text{ r.p.m.} \quad \textbf{Ans.}$$

Note. It will be interesting to know that the angular velocity of the rotating shaft has increased from 26.1 r.p.m. to 48.5 r.p.m. by fitting air vessel.

36.15 Work Done against Friction without Air Vessels

We have seen in Art. 36.11, that at the beginning and end of a stroke, there is no loss of head due to friction. The maximum loss of head is only in the middle of the stroke. We have also seen in Art. 36.12, that in the indicator diagram, the pressure head due to friction is a parabola. Now consider a reciprocating pump lifting water from a sump.

Let A = Area of the cylinder,

Reciprocating Pumps

a = Area of the pipe,
ω = Angular velocity of the crank,
r = Radius of the crank,
l = Length of the pipe,
d = Diameter of the pipe, and
f = Coefficient of friction.

We know that in the middle of stroke speed of water in pipe,

$$v = \frac{A}{a} \times \omega r$$

and loss of head due to friction,

$$H_f = \frac{4flv^2}{2gd} = \frac{4fl}{2gd}\left(\frac{A}{a} \times \omega r\right)^2$$

We also know that area of a parabola is numerically equal to

$$= \frac{2}{3} \times \text{Base} \times \text{Height}$$

∴ Work done per stroke

$$= \frac{2}{3} \times \frac{4fl}{2gd}\left(\frac{A}{a} \omega r\right)^2$$

Note: The power required to overcome the friction of the delivery pipe will be given by the relation:

$$P = wQ \left(\frac{2}{3} \times H_f\right)$$

36·16 Work Done against Friction with Air Vessels

We have seen in Art. 36·13, that whenever an air vessel is fitted to a pipe, there is no acceleration of water, and hence there is no acceleration pressure head. Now consider a reciprocating pump lifting water from a sump.

Let
A = Area of the cylinder,
ω = Angular velocity of the crank,
l = Length of the pipe,
a = Area of the pipe,
f = Coefficient of friction, and
d = Diameter of the suction pipe.

We have seen in Art. 36·14 that the velocity of water with air vessel in the pipe,

$$v = \frac{A}{a} \times \frac{\omega r}{\pi}$$

∴ Loss of head due to friction,

$$H_f = \frac{4flv^2}{2gd} = \frac{4fl}{2gd}\left(\frac{A}{a} \times \frac{\omega r}{\pi}\right)^2$$

and work done per stroke against friction

$$= \frac{4fl}{2gd}\left(\frac{A}{a} \times \frac{\omega r}{\pi}\right)^2$$

Example 26·13. *A single acting reciprocating pump, running at 60 r.p.m., has a plunger diameter of 250 mm and a stroke of 500 mm. The delivery pipe is 100 mm diameter and 50 metres long. If the motion of the pump is simple harmonic, find the power required to overcome friction of the delivery pipe when :*

(a) no air vessel is fitted, and (b) a large air vessel is fitted at the centre line of the pump. Assume f = 0·01.

Solution. Given : $N = 60$ r.p.m.; $D = 250$ mm $= 0.25$ m; $L = 500$ mm $= 0.5$ m or crank radius $(r) = 0.5/2 = 0.25$ m; $d = 100$ mm $= 0.1$ m; $l = 50$ m and $f = 0.01$.

Power required to overcome friction of the delivery pipe, when no air vessel is fitted

We know that area of plunger,
$$A = \frac{\pi}{4} \times (D)^2 = \frac{\pi}{4} \times (0.25)^2 = 0.0491 \text{ m}^2$$

and area of delivery pipe,
$$a = \frac{\pi}{4} \times (d)^2 = \frac{\pi}{4} \times (0.1)^2 = 0.00785 \text{ m}^2$$

∴ Discharge of the pump,
$$Q = \frac{LAN}{60} = \frac{0.5 \times 0.0491 \times 60}{60} = 0.025 \text{ m}^3/\text{s}$$

We also know that angular velocity of the crank,
$$\omega = \frac{2\pi N}{60} = \frac{2\pi \times 60}{60} = 6.28 \text{ rad/s}$$

and maximum velocity of water in the delivery pipe,
$$v = \frac{A}{a} \times \omega r = \frac{0.0491}{0.00785} \times 6.28 \times 0.25 = 9.82 \text{ m/s}$$

∴ Loss of head due to friction,
$$H_f = \frac{4flv^2}{2gd} = \frac{4 \times 0.01 \times 50 \times (9.82)^2}{2 \times 9.81 \times 0.1} = 98.3 \text{ m}$$

and power required to overcome friction of the delivery pipe
$$P_1 = wQ\left(\frac{2}{3} \times H_f\right) = 9.81 \times 0.025 \times \left(\frac{2}{3} \times 98.3\right) \ 16.1 \text{ kW} \textbf{ Ans.}$$

Power required to overcome friction of the delivery pipe, when a large air vessel is fitted at the centre line of the pump

We know that average velocity of water in the delivery pipe,
$$v = \frac{Q}{a} = \frac{0.025}{0.00785} = 3.18 \text{ m/s}$$

and loss of head due to friction,
$$H_f = \frac{4flv^2}{2gd} = \frac{4 \times 0.01 \times 50 \times (3.18)^2}{2 \times 9.81 \times 0.1} = 10.3 \text{ m}$$

∴ Power required to overcome friction of the delivery pipe,
$$P = wQH_f = 9.81 \times 0.025 \times 10.3 = 2.53 \text{ kW} \textbf{ Ans.}$$

36·17 Work Saved against Friction by Fitting Air Vessels

The work saved against friction by fitting air vessel, may be found out, first by finding out the work done against friction without air vessels, and then subtracting, from it the work done against friction with air vessels.

Reciprocating Pumps

Example 36·14. *Show from first principles, that the work saved against friction in the delivery pipe of a single acting reciprocating pump, by fitting air vessel, is 84·8 %.*

Solution. Given : Single acting reciprocating pump.

Let
- A = Area of the cylinder,
- L = Length of the stroke,
- r = Radius of the crank,
- l = Length of the delivery pipe,
- d = Diameter of the delivery pipe,
- ω = Angular velocity of the crank in rad/s, and
- N = Speed of the crank in r.p.m.

We know that the velocity of water in the pipe, with air vessel,

$$v = \frac{A}{a} \times \frac{\omega r}{\pi}$$

and loss of head due to friction,

$$H_f = \frac{4flv^2}{2gd} = \frac{4fl}{2gd}\left(\frac{A}{a} \times \frac{\omega r}{\pi}\right)^2$$

∴ Work done per stroke against friction

$$W_1 = \frac{4fl}{2gd}\left(\frac{A}{a} \times \frac{\omega r}{\pi}\right)^2$$

We also know that the maximum velocity of water in the delivery pipe without air vessels,

$$v = \frac{A}{a} \times \omega r$$

and loss of head due to friction,

$$H_f = \frac{4flv^2}{2gd} = \frac{4fl}{2gd}\left(\frac{A}{a} \times \omega r\right)^2$$

We also know that work done per stroke against friction,

$$W_2 = \frac{2}{3} \times \frac{4fl}{2gd}\left(\frac{A}{a} \times \omega r\right)^2 \qquad \ldots(ii)$$

∴ Saving in work done per stroke

$$= \frac{W_2 - W_1}{W_2}$$

$$= \frac{\dfrac{2}{3} \times \dfrac{4fl}{2gd}\left(\dfrac{A}{a} \times \omega r^2\right)^2 - \dfrac{4fl}{2gd}\left(\dfrac{A}{a} \times \dfrac{\omega r}{\pi}\right)^2}{\dfrac{2}{3} \times \dfrac{4fl}{2gd}\left(\dfrac{A}{a} \times \omega r\right)^2}$$

$$= \frac{\dfrac{2}{3} - \dfrac{1}{\pi^2}}{\dfrac{2}{3}} = 0{\cdot}848 = 84{\cdot}8\% \textbf{ Ans.}$$

Example 36·15. *Find the saving, in work done, against friction in the delivery of a double acting reciprocating pump, by fitting air vessels.*

Solution. Given : Double acting reciprocating pump.

Let
A = Area of the cylinder,
L = Length of the crank,
r = Radius of the stroke,
l = Length of the delivery pipe,
d = Diameter of the delivery pipe,
a = Area of the delivery pipe,
ω = Angular velocity of the crank in rad/s, and
N = Speed of the crank in r.p.m.

We know that the velocity of water in the pipe, with air vessel,

$$v = \frac{2A}{a} \times \frac{\omega r}{\pi}$$

and loss of head due to friction,

$$H_f = \frac{4flv^2}{2gd} = \frac{4fl}{2gd}\left(\frac{2A}{a} \times \frac{\omega r}{\pi}\right)^2$$

∴ Work done per stroke against friction,

$$W_1 = \frac{4fl}{2gd}\left(\frac{2A}{a} \times \frac{\omega r}{\pi}\right)^2 \qquad ...(i)$$

We know that the maximum velocity of water in the delivery pipe, without air vessels,

$$v = \frac{A}{a} \times \omega r$$

and loss of head due to friction,

$$H_f = \frac{4flv^2}{2gd} = \frac{4fl}{2gd}\left(\frac{A}{a} \times \omega r\right)^2$$

We also know that work done per stroke against friction,

$$W_2 = \frac{2}{3} \times \frac{4fl}{2gd}\left(\frac{A}{a} \times \omega r\right)^2 \qquad ...(ii)$$

∴ Saving in work done per stroke

$$= \frac{W_2 - W_1}{W_2}$$

$$= \frac{\frac{2}{3} \times \frac{4fl}{2gd}\left(\frac{A}{a} \times \omega r\right)^2 - \frac{4fl}{2gd}\left(\frac{2A}{a} \times \frac{\omega r}{\pi}\right)^2}{\frac{2}{3} \times \frac{4fl}{2gd}\left(\frac{A}{a} \times \omega r\right)^2}$$

$$= \frac{\frac{2}{3} - \frac{4}{\pi^2}}{\frac{2}{3}} = 0 \cdot 392 = 39 \cdot 2\% \text{ Ans.}$$

36·18 Flow of Water into and from the Air Vessel Fitted to the Delivery Pipe of a Single Acting Reciprocating Pump

We have seen in Art. 36·13 that the velocity of water, in the delivery pipe, beyond the air vessel is constant. But the velocity of water from the cylinder to the air vessel is subjected to acceleration and retardation. Now we shall discuss the rate of flow of water into and from the air vessel fitted to the delivery pipe of a single reciprocating pump.

Consider a single acting reciprocating pump, fitted with air vessels on both the suction and delivery pipes. Let us consider discharge from the cylinder to the delivery pipe. We know that the velocity of water in the delivery pipe from the cylinder upto the air vessel

$$= \frac{A}{a} \times \omega r \sin \theta$$

∴ Discharge from the cylinder,

$$= a \times \frac{A}{a} \times \omega r \sin \theta = A\omega r \sin \theta \qquad \ldots(i)$$

We also know that the velocity of water in a single acting reciprocating pump beyond the air vessel,

$$= \frac{A}{a} \times \frac{\omega r}{\pi}$$

∴ Discharge in the delivery pipe beyond the air vessel

$$= a \times \frac{A}{a} \times \frac{\omega r}{\pi} = \frac{A\omega r}{\pi} \qquad \ldots(ii)$$

A little consideration will show, that the difference of the above two discharges will be the discharge, into or form the air vessel.

∴ Discharge into the air vessel,

Q = Discharge from the cylinder − Discharge beyond air vessel

$$= A\omega r \sin \theta - \frac{A\omega r}{\pi}$$

$$= A\omega r \left(\sin \theta - \frac{1}{\pi} \right) \qquad \ldots(iii)$$

If the above equation works out to be +ve, it means that the discharge is taking place into the air vessel. But if this equation works out to be −ve, it means that the discharge is taking place from the air vessel.

Note : If we consider the flow, into or from the air vessel fitted to the suction pipe, then the above condition is reserved. *i.e.*, if the equation (*iii*) above works out to be +ve the discharge is taking place from the air vessel. But if this equation works out to be −ve, the discharge is taking place into the air vessel.

36·19 Flow of Water, into and from the Air Vessel Fitted to the Delivery Pipe of a Double Acting Reciprocating Pump

Consider a double acting reciprocating pump, fitted with air vessels on both the suction and delivery pipes. Let us consider the discharge from the cylinder to the delivery pipe.

We know that the velocity of water in the delivery pipe from the cylinder up to the air vessel

$$= \frac{A}{a} \omega r \sin \theta$$

∴ Discharge from the cylinder

$$= a \times \frac{A}{a} \times \omega r \sin \theta = A\omega r \sin \theta \qquad \ldots(i)$$

We know that the velocity of water in a double acting reciprocating pump beyond the air vessel

$$= \frac{2A}{a} \times \frac{\omega r}{\pi}$$

∴ Discharge in the delivery pipe beyond the air vessel

$$= a \times \frac{2A}{a} \times \frac{\omega r}{\pi} = \frac{2A\omega r}{\pi} \qquad \ldots(ii)$$

A little consideration will show, that the difference of the above two discharges will be the discharge into or form the air vessel.

∴ Discharge into the air vessel,

$$Q = \text{Discharge from the cylinder}$$
$$\quad - \text{Discharge beyond the air vessel}$$
$$= A\omega r \sin\theta - \frac{2A\omega r}{\pi}$$
$$= A\omega r \left(\sin\theta - \frac{2}{\pi}\right) \qquad \ldots(iii)$$

If the above equation works out to be +ve, it means that the discharge is taking place into the air vessel. But if this equation works out to be –ve, it means that the discharge is taking place from the air vessel.

Note. If we consider the flow into or from the air vessel fitted to the suction pipe, then the above condition is reversed. *i.e.*, if the equation (*iii*) above works out to be +ve, the discharge is taking place from the air vessel. But if this equation work out to be –ve, the discharge is taking place into the air vessel.

Example 36·16. *A double acting reciprocating pump has a bore of 175 mm and a stroke of 350 mm. The suction pipe has a diameter of 150 mm and is fitted with an air vessel. Determine the crank angle, at which there is no flow of water to or from the vessel.*

Assume the crank speed as 150 r.p.m. and plunger has simple harmonic motion.

Solution. Given : D = 175 mm = 0·175 m; L = 350 mm = 0·35 m or crank radius (r) = 0·35/2 = 0·175 m and N = 150 r.p.m.

We know that area of bore,

$$A = \frac{\pi}{4} \times (D)^2 = \frac{\pi}{4} \times (0\cdot175)^2 = 0\cdot024 \text{ m}^2$$

and angular velocity of the crank,

$$\omega = \frac{2\pi N}{60} = \frac{2\pi \times 150}{60} = 5\pi \text{ rad/s}$$

We also know that discharge from the sump upto the vessel

$$= \frac{2A\omega r}{\pi} = \frac{2 \times 0\cdot024 \times (5\pi) \times 0\cdot175}{\pi} = 0\cdot042 \text{ m}^3/\text{s} \qquad \ldots(i)$$

and discharge beyond the air vessel

$$= A\omega r \sin\theta = 0\cdot024 \times (5\pi) \times 0\cdot175 \sin\theta$$
$$= 0\cdot066 \sin\theta \text{ m}^3/\text{s} \qquad \ldots(ii)$$

*For no flow of water to or from the air vessel, the above two discharges should be equal. Thus equating the above two discharges,

$$0.666 \sin \theta = 0.042$$

or
$$\sin \theta = 0.042/0.066 = 0.6364$$

$$\therefore \quad \theta = 39.5° \quad \text{or} \quad 140.5° \quad \textbf{Ans.}$$

Example 36·17. *A 100 mm diameter suction pipe of a double acting reciprocating pump having 200 mm bore and 400 mm stroke is fitted with an air vessel. The pump runs at 120 r.p.m. and it may be assumed that the piston makes simple harmonic motion. Calculate the rate of flow into or from the air vessel in litres/s, when crank makes 30°, 90° and 120° with the inner dead centre.*

Solution. Given : $d_s = 100$ mm $= 0.1$ m; $D = 200$ mm $= 0.2$ m; $L = 400$ mm $= 0.4$ m or crank radius $(r) = 0.4/2 = 0.2$ m; and $N = 120$ r.p.m.

Rate of flow of water when the crank makes an angle of 30°

We know that area of bore,

$$A = \frac{\pi}{4} \times (D)^2 = \frac{\pi}{4} \times (0.2)^2 = 0.0314 \text{ m}^2$$

and angular velocity of the crank,

$$\omega = \frac{2\pi N}{60} = \frac{2\pi \times 120}{60} = 4\pi \text{ rad/s}$$

∴ Rate of flow of water, when $\theta = 30°$,

$$Q = A\omega r \left(\sin \theta - \frac{2}{\pi} \right)$$

$$= 0.0314 \times 4\pi \times 0.2 \left(\sin 30° - \frac{2}{\pi} \right) \text{ m}^3/\text{s}$$

$$= 0.0789 \, (0.5 - 0.637) \text{ m}^3/\text{s}$$

$$= -0.0108 \text{ m}^3/\text{s} = -10.8 \text{ litres/s} \quad \textbf{Ans.}$$

Since the result work out to be −ve and we are considering flow in the suction pipe, therefore the discharge is taking place into the air vessel. **Ans.**

Flow of water when the crank makes an angle of 90°

We know that the rate of flow of water, when $\theta = 90°$

$$Q = A\omega r \left(\sin \theta - \frac{2}{\pi} \right)$$

$$= 0.0314 \times 4\pi \times 0.2 \left(\sin 90° - \frac{2}{\pi} \right) \text{ m}^3/\text{s}$$

---

*This may also be found from the relation of discharge to or from the air vessel. *i.e.*,

$$Q = A\omega r \left(\sin \theta - \frac{2}{\pi} \right)$$

For no flow of water to or from the air vessel, let us equate this equation to zero.

i.e.,
$$A\omega r \left(\sin \theta - \frac{2}{\pi} \right) = 0$$

$$\sin \theta - \frac{2}{\pi} = 0 \qquad \text{...(Dividing both sides by } A\omega r)$$

$$\sin \theta = \frac{2}{\pi} = 0.6364$$

$$\therefore \quad \theta = 39.5° \quad \text{or} \quad 140.5° \quad \textbf{Ans.}$$

$$Q = 0.0789 \ (1.0 - 0.637) \ m^3/s$$
$$= + 0.0286 \ m^3/s = + 28.6 \ \text{litres/s}$$

Since the result works out to be +ve and we are considering flow in the suction pipe, therefore the discharge is taking place from the air vessel. **Ans.**

Rate of flow of water when the crank makes an angle of 120°

We also know that the rate of flow of water, when $\theta = 120°$,

$$Q = A\omega r \left(\sin \theta - \frac{2}{\pi} \right)$$

$$= 0.0314 \times 4\pi \times 0.2 \left(\sin 120° - \frac{2}{\pi} \right) m^3/s$$

$$= 0.0789 \ (0.866 - 0.637) \ m^3/s$$

$$= + 0.0181 \ m^3/s = + 18.1 \ \text{litres/s}$$

Since the result works out to be +ve and we are considering flow in the suction pipe, therefore the discharge is taking place from the air vessel. **Ans.**

EXERCISE 36.3

1. A double acting reciprocating pump, having piston diameter 150 mm and stroke length 450 mm, draws water from a sump 3.6 metres below it through a 100 mm diameter and 6 metres long pipe. A large air vessel is fitted on the suction pipe close to the pump. Find the maximum speed of the pump, in order to avoid separation, which takes place at a pressure head of 2.5 metres, if the barometer reads 10.3 metres. Take $4f = 0.025$. **(Ans. 107.6 r.p.m.)**
2. A single acting reciprocating pump has a plunger of 375 mm diameter and stroke of 600 mm. The delivery pipe is 90 metres long and 150 mm diameter. Find the power saved by installing an air vessel near the pump, if the pump runs at 50 r.p.m. Take $f = 0.008$. **(Ans. 31.5 kW)**
3. A double acting reciprocating pump has an air vessel fitted on the suction pipe. The plunger is 150 mm diameter and 300 mm long. The suction pipe is 8 metres long and 100 mm diameter. Determine the rate of flow into or from the air vessel at crank positions of 30°, 90° and 120° from the inner dead centre. Take speed of the pump as 120 r.p.m. **[Ans. 4.33 litres/s (into the air vessel); 12.54 litres/s (from air vessel); 8.04 litres/s (from air vessel)]**

QUESTIONS

1. Why is a reciprocating pump called a positive displacement pump ?
2. Distinguish between coefficient of discharge and slip of a reciprocating pump.
3. Explain the working principle of reciprocating pump with sketches.
4. Show that the maximum acceleration head, in a reciprocating pump, without air vessel is given by the relation :

$$H_a = \frac{l}{g} \times \frac{A}{a} \times \omega^2 r$$

where
l = Length of pipe,
$\frac{A}{a}$ = Ratio of cylinder area to pipe area, and
ω = Angular velocity of crank shaft.

5. Define 'separation' in a reciprocating pump, and explain how it can be avoided ?
6. Explain the function of air vessel in a reciprocating pump.

OBJECTIVE TYPE QUESTIONS

1. A reciprocating pump is suitable for
   (a) less discharge
   (b) more discharge
   (c) higher heads
   (d) both 'a' and 'c'
2. The slip of a reciprocating pump is
   (a) ratio of actual discharge to theoretical discharge
   (b) product of actual discharge and theoretical discharge
   (c) difference of theoretical discharge and actual discharge
   (d) sum of theoretical discharge and actual discharge
3. Air vessels are used in a reciprocating pump to
   (a) smoothen the flow
   (b) increase the flow
   (c) reduce acceleration head
   (d) all of these
4. By fitting air vessel in the reciprocating pump, there is always some saving in power. This saving in a single acting reciprocating pump is
   (a) 39·2%   (b) 48·8%   (c) 84·8%   (d) 88·4%

ANSWER

1. (d)   2. (c)   3. (c)   4. (c)

37

Performance of Pumps

1. Introduction 2. Variation in Speed and Diameter of a Centrifugal Pump. 3. Effect of Variation in Speed. 4. Effect of Variation in Diameter. 5. Specific Speed of a Centrifugal Pump. 6. Selection of Centrifugal Pumps Based on Specific Speed. 7. Suction Head. 8. Vapour Pressure. 9. Net Positive Suction Head. (NPSH) 10. Cavitation in Centrifugal Pumps. 11. Multiple Cylinder Reciprocating Pumps. 12. Double Cylinder Pump. 13. Triple Cylinder Pump.

37·1 Introduction

In the last two chapters *i.e.*, (Centrifugal Pumps and Reciprocating Pumps), we have assumed that the pump will be required to lift a constant quantity of water to a constant height. But in actual practice, these assumptions rarely prevail. It is thus essential to know the working of the pump under changed conditions also.

37·2 Variation in Speed and Diameter of a Centrifugal Pump

Sometimes, there is a minor[*] change in the requirement of the head of water or discharge of a pump from its designed head of water or discharge. In such a case, a slight adjustment in the pump is made to suit the new set of conditions. This is done either :

1. By varying speed of the pump impeller, or
2. By changing the diameter of the pump impeller.

Now we shall study the effect of there two variations on the discharge, head of water, and the power required to drive the pump.

37·3 Effect of Variation in Speed

Consider a centrifugal pump, whose speed is changed to suit the new set of conditions.

Let N = Designed speed in r.p.m.,

Q = Discharge of pump with the designed speed of N r.p.m.,

H = Head of water with the designed speed of N r.p.m.,

P = Power required to drive the pump with the designed speed of N r.p.m.,

N_1 = New speed to suit the changed set of conditions, and

Q_1, H_1, P_1 = Corresponding values with the new speed of N_1 r.p.m.

A little consideration will show, that when the speed of the impeller is changed from N to N_1, shape of the velocity triangle will remain the same (*i.e.*, various angles will remain the same). But the values of the velocities will change proportionately. We know that the tangential velocity of the impeller at inlet,

$$v = \frac{\pi DN}{60}$$

$$\propto N$$

---

[*]If the change is a major one, then a new pump is designed and manufactured with altogether different dimensions.

Performance of Pumps

Similarly $v_1 \propto N$

and $V_{w1} \propto N$ ...($\because V_{w1} \propto v_1$)

We also know that velocity of flow,
$$V_f \propto v$$
$$\propto N \qquad \text{...}(\because v \propto N)$$

But discharge, $Q = \pi D b V_f$
$$\propto V_f$$
$$\propto N \qquad \text{...}(\because V_f \propto N)$$

Similarly $Q_1 \propto N_1$

$\therefore \quad \dfrac{Q}{Q_1} = \dfrac{N}{N_1}$ ...(i)

It is thus obvious that the discharge of a centrifugal pump is proportional to the speed of its impeller.

We know that the head of water,
$$H = \eta \times \dfrac{V_{w1} \cdot v_1}{g}$$
$$\propto V_{w1} \cdot v_1$$
$$\propto N \cdot N \qquad \text{...}(\because V_{w1} \propto N \text{ and } v_1 \propto N)$$
$$\propto N^2$$

Similarly $H_1 \propto N_1^2$

$\therefore \quad \dfrac{H}{H_1} = \dfrac{N^2}{N_1^2} = \left(\dfrac{N}{N_1}\right)^2$ ...(ii)

It is thus obvious, that the head of a centrifugal pump is proportional to the square of the speed of its impeller.

We also know that the power required to drive a pump,
$$P = \dfrac{wQV_{w1} \cdot v_1}{g}$$
$$\propto Q \cdot V_{w1} \cdot v_1$$
$$\propto N \cdot N \cdot N \propto N^3$$

Similarly $P_1 \propto N_1^3$

$\therefore \quad \dfrac{P}{P_1} = \dfrac{N^3}{N_1^3} = \left(\dfrac{N}{N_1}\right)^3$ ...(iii)

It is thus obvious, that the power required to drive a centrifugal pump is proportional to the cube of the speed of its impeller.

It may be noted that the hydraulic losses vary, more or less, uniformly with the speed of the impeller. Thus the hydraulic efficiency practically, remains unchanged with the change of impeller speed. But mechanical losses are relatively small at higher speeds. Thus the mechanical efficiency will slightly increase with the increase in the impeller speed.

Example 37·1. *A centrifugal pump delivers 30 litres of water per second against a head of 12 metres and running at 1200 r.p.m. requires 6 kW power. Determine the discharge, head of the pump and power required, if the pump runs at 1500 r.p.m.*

Solution. Given : Given $Q = 30$ litres/s; $H = 12$ m; $N = 1200$ r.p.m. and $N_1 = 1500$ r.p.m.

Discharge under new speed

Let $\qquad Q_1 =$ Discharge under the new speed.

We know that $\qquad \dfrac{Q}{Q_1} = \dfrac{N}{N_1}$ or $\dfrac{30}{Q_1} = \dfrac{1200}{1500} = 0.8$

$\therefore \qquad Q = 30/0.8 = 37.5$ litres/s **Ans.**

Head of the pump under the new speed

Let $\qquad H_1 =$ Head under the new speed.

We know that $\qquad \dfrac{H}{H_1} = \left(\dfrac{N}{N_1}\right)^2$ or $\dfrac{12}{H_1} = \left(\dfrac{1200}{1500}\right)^2 = 0.64$

$\therefore \qquad H_1 = 12/0.64 = 18.75$ m **Ans.**

Power required to drive the pump under the new speed

Let $\qquad P_1 =$ Power required under the new speed.

We know that $\qquad \dfrac{P}{P_1} = \left(\dfrac{N}{N_1}\right)^3$ or $\dfrac{6}{P_1} = \left(\dfrac{1200}{1500}\right)^3 = 0.512$

$\therefore \qquad P_1 = 6/0.512 = 11.7$ kW **Ans.**

37.4 Effect of Variation in Diameter

We have discussed in Art. 37.2 that a minor change in the requirement of the head of water or discharge of a pump is made either by varying speed or diameter of the pump impeller. It has been experienced that the former (*i.e.*, varying the speed of the pump impeller) is not possible, because the pump impeller is driven by motor, whose speed is fixed. It is thus obvious, that in majority of the cases, the diameter of the pump impeller is enlarged or reduced, whenever the head of water or discharged is to be increased or decreased. It is done either by changing the blades of the impeller or fixing rings to its outside diameter.

Now consider a centrifugal pump, whose diameter is changed to suit the new set of conditions.

Let $\qquad D =$ Outside diameter of the pump,

$\qquad Q =$ Discharge of the pump with diameter D,

$\qquad H =$ Head of water with diameter D,

$\qquad P =$ Power required to drive the pump with diameter D.

$\qquad D_1 =$ New outside diameter to suit the changed requirement, and

$\qquad Q_1, H_1, P_1 =$ Corresponding values with diameter D_1.

A little consideration will show, that when the diameter of the impeller is changed from D to D_1, the shape of the velocity triangle will remain the same (*i.e.*, the various angles will remain the same). But the values of the velocities will change proportionately. We know that the tangential velocity of the impeller,

$$v = \dfrac{\pi DN}{60} \propto D$$

Similarly, velocity of flow, $V_f \propto v \propto D$ $\qquad\qquad\qquad$ ...($\because D \propto v$)

But discharge, $\qquad Q = \pi D b V_f \propto D \cdot V_f$

$\qquad\qquad\qquad \propto D \cdot D \propto D^2$ $\qquad\qquad\qquad$ ...($\because D \propto V_f$)

Performance of Pumps

Similarly
$$Q_1 \propto D_1^2$$

$$\therefore \quad \frac{Q}{Q_1} = \frac{D^2}{D_1^2} = \left(\frac{D}{D_1}\right)^2 \qquad ...(i)$$

It is thus obvious, that the discharge of a centrifugal pump is proportional to the square of the diameter of its impeller. We know that head of water,

$$H = \eta \times \frac{V_{w1} \cdot v1}{g} \propto V_{w1} \cdot v_{s1} \propto D \cdot D \propto D^2$$

Similarly
$$H_1 \propto D_1^2$$

$$\therefore \quad \frac{H}{H_1} = \frac{D^2}{D_1^2} = \left(\frac{D}{D_1}\right)^2 \qquad ...(ii)$$

It is thus obvious, that the head of water of a centrifugal pump is also proportional to the square of the diameter of its impeller. We also know that the power required to drive a pump,

$$P = \frac{w \cdot Q V_{w1} \cdot v1}{g}$$

$$\propto Q \cdot V_{w1} \cdot v1 \propto D^2 \cdot D \cdot D \propto D^4$$

Similarly
$$P_1 \propto D_1^4$$

$$\therefore \quad \frac{P}{P_1} = \frac{D^4}{D_1^4} = \left(\frac{D}{D_1}\right)^4 \qquad ...(iii)$$

It is thus obvious, that the power required to drive a centrifugal pump is proportional to the fourth power of the diameter of its impeller.

Example 37·2. *A centrifugal pump was built to supply water against a head of 22·5 metres. But later on it was required to supply the required quantity of water against a head of 20 metres. Find the necessary reduction in the impeller diameter, if it is planned to reduce the original diameter of 300 mm without reducing the speed of impeller.*

Solution. Given : $H = 22\cdot 5$ m; $H_1 = 20$ m and $D = 300$ mm $= 0\cdot 3$ m.

Let $\qquad D_1 =$ New impeller diameter.

We know that $\qquad \dfrac{H}{H_1} = \left(\dfrac{D}{D_1}\right)^2 \quad$ or $\quad \dfrac{22\cdot 5}{20} = \left(\dfrac{0\cdot 3}{D_1}\right)^2 = \dfrac{0\cdot 09}{D_1^2}$

$$\therefore \quad D_1^2 = \frac{0\cdot 09 \times 20}{22\cdot 5} = 0\cdot 08 \text{ or } D_1 = 0\cdot 283 \text{ m} = 283 \text{ mm Ans.}$$

Note. The above example may also be solved without changing the original impeller diameter into metres *i.e.*, keeping the value of D as 300 mm as discussed below :

$$\frac{H}{H_1} = \left(\frac{D}{D_1}\right)^2 \text{ or } \frac{22\cdot 5}{20} = \left(\frac{300}{D_1}\right)^2 = \frac{90\,000}{D_1^2}$$

$$D_1^2 = \frac{90\,000 \times 20}{22\cdot 5} = 80\,000 \text{ or } D = 283 \text{ mm}$$

37·5 Specific Speed of a Centrifugal Pump

The specific speed of a centrifugal pump may be defined as the speed if an imaginary pump, identical with the given pump, which will discharge one litre of water, while it is being raised through a head of one metre.

Let
- N_S = Specific speed of the pump,
- D_1 = Diameter of the impeller at outlet,
- N = Speed of the impeller in r.p.m.,
- v_1 = Tangential velocity of impeller at outlet, and
- H = Lift of the pump in metres.

We know that the tangential velocity of the impeller,
$$v \propto \sqrt{H}$$

and
$$v = \frac{\pi D_1 N}{60}$$

$\therefore \quad D_1 N \propto v \propto \sqrt{H}$

$\therefore \quad D_1 \propto \dfrac{\sqrt{H}}{N}$ ...(i)

Now let
- Q = Discharge of the pump m³/s
- b_1 = Width of the impeller at outlet, and
- V_{f1} = Velocity of flow at outlet.

We know that the discharge
$$Q = \pi D_1 b_1 V_{f1}$$

But
$b_1 \propto D_1$
$V_{f1} \propto \sqrt{H}$

$\therefore \quad Q \propto D_1^2 \sqrt{H}$

$\propto \left(\dfrac{\sqrt{H}}{N}\right)^2 \sqrt{H} \qquad \qquad \ldots\left(\because D \propto \dfrac{\sqrt{H}}{N}\right)$

$\propto \dfrac{H^{3/2}}{N^2}$

$\therefore \quad N^2 \propto \dfrac{H^{3/2}}{Q}$

or
$$N \propto \frac{H^{3/4}}{\sqrt{Q}} = \frac{N_S \cdot H^{3/4}}{\sqrt{Q}} \qquad \text{...(where } N_s \text{ is constant of proportionality)}$$

$\therefore \quad N_S = \dfrac{N \sqrt{Q}}{H^{3/4}}$

Example 37·3. *Find the specific speed of a centrifugal pump, delivering 750 litres of water per second against a head of 15 metres at 725 r.p.m.*

Solution. Given : $Q = 750$ litres/s $= 0.75$ m³/s; $H = 15$ m and $N = 725$ r.p.m.

We know that specific speed of the pump,
$$N_S = \frac{N\sqrt{Q}}{H^{3/4}} = \frac{725 \sqrt{0.75}}{(15)^{3/4}} = 82.4 \text{ r.p.m.} \quad \textbf{Ans.}$$

Performance of Pumps

Example 37·4. *A multi-stage centrifugal pump is required to lift 9 000 litres of water per minute from a mine, the total head including friction being 75 metres. If the speed of the pump is 1200 r.p.m., find the least number of stages, if the specific speed per stage is not to be less than 60 r.p.m.*

Solution. Given : $Q = 9000$ litres/min $= 150$ litres/s $= 0.15$ m^3/s; Total head $= 75$ m; $N = 1200$ r.p.m. and $N_S = 60$ r.p.m.

Let $\qquad H = $ Head of water per stage.

We know that specific speed (N_S),

$$60 = \frac{N\sqrt{Q}}{H^{3/4}} = \frac{1200\sqrt{0.15}}{H^{3/4}} = \frac{464.4}{H^{3/4}}$$

$\therefore \qquad H^{3/4} = 464.4/60 = 7.74 \quad \text{or} \quad H = 15.3$ m

We also know that number of stages

$$= \frac{\text{Total head of water}}{\text{Head od water per stage}} = \frac{75}{15.3} = 4.9 \text{ say 5 } \textbf{Ans.}$$

Example 37·5. *A single stage centrifugal pump with impeller diameter of 300 mm rotates at 2000 r.p.m. and lifts 3 m^3 of water per minute to a height of 30 metres with an efficiency of 75%. Find the number of stages and the diameter of each impeller of a similar multistage pump to lift 4·5 m^3 of water per minute to a height of 130 m when rotating at 1500 r.p.m.*

Solution. Given : $D = 300$ mm $= 0.3$ m; $N = 2000$ r.p.m.; $Q = 3$ m^3/min $= 0.05$ m^3/s; Height through which water is lifted in one stage $= 30$ m; $\eta = 75\% = 0.75$; $Q_1 = 4.5$ m^3/min $= 0.075$ m^3/s; Total height through which water is to be lifted $= 130$ m and $N_1 = 1500$ r.p.m.

Number of stages

First of all, consider the first pump. We know that actual head against which the first pump has to work

$$H = 30/0.75 = 40 \text{ m}$$

and specific speed of the pump, $N_S = \dfrac{N\sqrt{Q}}{H^{3/4}} = \dfrac{2000\sqrt{0.05}}{(40)^{3/4}} = 28.1$ r.p.m.

Now consider the second pump. Since both the pumps are similar, therefore their specific speeds must be equal.

Let $\qquad H_1 = $ Height through which the pump can lift water.

We know that specific speed of the second pump (N_S)

$$28.1 = \frac{N_1\sqrt{Q_1}}{H_1^{3/4}} = \frac{1500\sqrt{0.075}}{H_1^{3/4}} = \frac{410.8}{H_1^{3/4}}$$

or $\qquad H_1^{3/4} = 410.8/28.1 = 14.62 \quad \text{or} \quad H_1 = 35.7$ m

$\therefore$ Actual height to which the water can be lifted in one stage

$$= 35.7 \times 0.75 = 26.8 \text{ m}$$

and number of stages $= \dfrac{\text{Total height through which water is to be lifted}}{\text{Actual height of one stage}}$

$$= \frac{130}{26.8} = 4.85 \text{ say 5 } \textbf{Ans.}$$

Diameter of each impeller

Let $\qquad D_1 = $ Diameter of each impeller.

We know that $\dfrac{H}{H_1} = \left(\dfrac{D}{D_1}\right)^2$ or $\dfrac{30}{35\cdot 7} = \left(\dfrac{0\cdot 3}{D_1}\right)^2 = \dfrac{0\cdot 09}{D_1{}^2}$

$\therefore \quad D_1{}^2 = \dfrac{0\cdot 09 \times 35\cdot 7}{30} = 0\cdot 107$ or $D_1 = 0\cdot 327$ m $= 327$ mm **Ans.**

37·6 Selection of Centrifugal Pumps Based on Specific Speed

The specific speed of a centrifugal pump, like that of a turbine, helps us in selecting the type of centrifugal pump. Following table gives the type of centrifugal pump for the corresponding specific speed.

Table 37·1

S. No.	Specific speed	Type of centrifugal pump
1	10 to 30	Slow speed pump, with radial flow at outlet.
2	30 to 50	Medium speed pump, with radial flow at outlet.
3	50 to 80	High speed pump, with radial flow at outlet.
4	80 to 160	High speed pump, with mixed flow at outlet.
5	160 to 500	High speed pump, with axial flow at outlet.
6	Above 500	Very high speed pump.

Example 37·6. *A centrifugal pump delivers 120 litres of water per second against a head of 85 metres at 900 r.p.m. Find the specific speed of the pump. What type of impeller would you select for the pump ?*

Solution. Given. $Q = 120$ litres/s $= 0\cdot 12$ m³/s; $H = 85$ m and $N = 900$ r.p.m.

Specific speed of the pump

We know that specific speed of the pump,

$$N_S = \dfrac{N\sqrt{Q}}{H^{3/4}} = \dfrac{900\sqrt{0\cdot 12}}{(85)^{3/4}} = 11\cdot 1 \text{ r.p.m.} \quad \textbf{Ans.}$$

Type of impeller

Since the specific speed of the pump is 11·1 r.p.m., therefore slow speed centrifugal pump with radial flow should be selected. **Ans.**

37·7 Suction Head

In the previous articles, we have discussed the term suction head. As a matter of fact, it is of utmost importance for the smooth and efficient working of a centrifugal pump.

Strictly speaking, a pump (centrifugal or reciprocating) lifts water from a reservoir because of atmospheric pressure acting on the surface of water. The pump reduces the pressure in the casing, to such an extent, that the atmospheric pressure forces up water in the suction pipe. A little consideration will show, that as the pump cannot produce a pressure below the vapour pressure of the liquid, therefore the limiting pressure difference is the atmospheric pressure minus the vapour pressure. This available pressure difference is responsible for lifting the water in the suction pipe. A little consideration will show, that this pressure difference should be capable enough :

1. to lift water through the suction head (H_s)

Performance of Pumps

2. to overcome frictional loss in the suction pipe (H_{fs}), and
3. to produce velocity head $\left(\dfrac{v_s^2}{2g}\right)$.

Now consider a pump lifting water from a sump.

Let
p_a = Atmospheric pressure in kPa (*i.e.*, kN/m^2)
p_v = Vapour pressure in kPa (*i.e.*, kN/m^2)
H_a = Atmospheric pressure in metres,
H_v = Vapour pressure in metres, and
w = Specific weight of liquid.

We know that
$$\dfrac{p_a}{w} - \dfrac{p_v}{w} = H_a + H_v$$

$$\dfrac{p_a - p_v}{w} = H_s + H_{fs} + \dfrac{v_s^2}{2g}$$

or
$$H_s = \dfrac{p_a - p_v}{w} - H_{fs} - \dfrac{v_s^2}{2g} \qquad \text{... (In terms of } p_a \text{ and } p_v)$$

$$= H_a - H_v - H_{fs} - \dfrac{v_s^2}{2g} \qquad \text{...(in terms of } H_a \text{ and } H_v)$$

where H_s is the suction head. In actual practice, the value of H_s is not kept equal to that obtained from the above relation. But it is generally kept from 5 to 6 metres only.

37·8 Vapour Pressure

The vapour pressure of a liquid may be defined as the pressure at which the liquid will transform into vapour at the given temperature. A little consideration will show, that the vapour pressure is a function of temperature. Higher the temperature, higher will be the vapour pressure. As a matter of fact, the pressure at any point should not fall below the vapour in any pumping system. The vapour bubbles when collapse result in the corrosion of the suction pipe and other parts.

37·9 Net Positive Suction Head (NPSH)

It is a commercial term used by the pump manufacturers, and indicates the suction head which the pump impeller can produce. We have already discussed in the last articles, the suction head and vapour pressure. There we derived a relation for the suction head (or more accurately the limiting suction head), such that

$$H_s = \dfrac{p_a - p_v}{w} - H_{fs} - \dfrac{v_s^2}{2g} \qquad \text{...(in terms of } p_a \text{ and } p_v)$$

$$= H_a - H_v - H_{fs} - \dfrac{v_s^2}{2g} \qquad \text{...(in terms of } H_a \text{ and } H_v)$$

The right hand side of the above equations represents the suction head or net positive suction head.

Note. For any pump installation, a distinction is generally made between the required NPSH and the available NPSH. The required NPSH varies with the pump design, its speed, capacity etc. The available NPSH depends upon the site conditions and the available equipment. In order to have a smooth working of the pump, the available NPSH should be more (or equal) to the required NPSH.

37·10 Cavitation in Centrifugal Pumps

We have already discussed in Art. 34·16 the phenomenon of cavitation. If a centrifugal pump, is working at high suction head or high vapour pressure, cavitation is formed. It damages the impeller.

The cavitation can be eliminated (or minimised) by the following precautions :
1. The liquid temperature should be as low as possible to keep the vapour pressure down and to obtain an increased NPSH.
2. The velocity of liquid in the suction pipe should be as low as practicable.
3. As far as possible, the sharp bends in the suction pipe should be avoided to reduce loss of head.

37·11 Multiple Cylinder Reciprocating Pumps

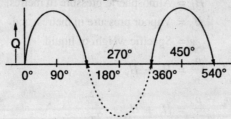

Fig. 37·1. Flow pattern in a reciprocating pump.

We have already discussed in the last chapter that in a single acting reciprocating pump, the discharge takes place only during the delivery stroke. Even the rate of flow, during the delivery stroke, is not constant and fluctuates as a sine curve as shown in Fig. 37·1. Since the velocity of water is proportional to the velocity of crank, therefore the figure represents the velocity of water to some scale.

In order to have a continuous flow (preferably a uniform flow too) multiple cylinder pumps are used. Though there are many types of multiple cylinder pumps, yet the following are important from the subject point of view.
1. Double cylinder pump, and
2. Triple cylinder pump.

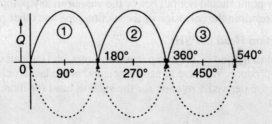

Fig. 37·2. Double cylinder pump.

37·12 Double Cylinder Pump

A double cylinder pump is a pump which has two cylinders in which the pistons are moving simultaneously. The motion of the two pistons is so arranged that during each stroke, there is a delivery stroke in one cylinder and suction stroke in the other at the same time as shown in Fig. 37·2. By this arrangement, we get water continuously from the delivery pipe as shown by the resultant of two sin curves at a phase difference of 180° as shown in Fig. 37·2.

37·13 Triple Cylinder Pump

A triple cylinder pump is a pump which has three cylinders in which the pistons are moving simultaneously. The motion of the three pistons is arranged at a phase difference of 120°. The velocity of water or discharge received by a triple cylinder pump is fairly uniform as shown in Fig. 37·3.

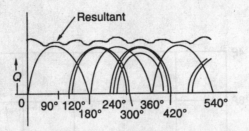

Fig. 37·3. Triple cylinder pump.

37·14 Priming of a Centrifugal Pump

We have already discussed that the pressure developed by the impeller of a centrifugal pump, is proportional to the density of the fluid in the impeller. It is thus obvious, that if impeller is running in air, it will produce only a negligible pressure, which may not suck water, from its source, through the suction pipe. To avoid this, the pump is first primed. *i.e.*, filled up with water.

For doing so, first of all the suction pipe and the impeller is completely filled with water. The delivery valve is closed and the pump is started. The rotating impeller pushes the water in the delivery pipe, opens the delivery valve and sucks water through the suction pipe.

37·15 Characteristic Curves of Centrifugal Pumps

As a matter of fact a centrifugal pump, like a turbine is designed and manufactured to work under a given set of conditions (or a limited range of conditions) such as discharge, speed, power required, head of water, efficiency etc. But a pump may have to be used under conditions, other than those for which it has been designed. It is, therefore, essential that the exact behaviour of the pump under varied conditions should be predetermined. This is represented graphically by means of curves, known as characteristic curves. Though there are many types of characteristic curves, yet the following are important from the subject point of view :

1. Characteristic curves for speed, and
2. Characteristic curves for discharge with varying speeds.

1. *Characteristic curves for speed*

   Speed versus discharge

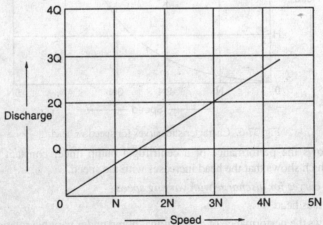

Fig. 37·4. Characteristic curve for speed *vs* discharge.

Fig. 37·4 shows the performance of a centrifugal pump under constant head. It is a straight line, which shows that the discharge increases with the speed.

Speed versus power

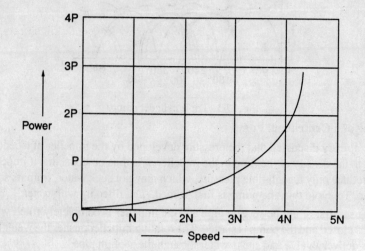

Fig. 37·5. Characteristic curve for speed *vs* power.

Fig. 37·5 shows the performance of a centrifugal pump under a constant head and discharge. It is a parabolic curve, which shows that the power increases with the speed.

Speed versus head

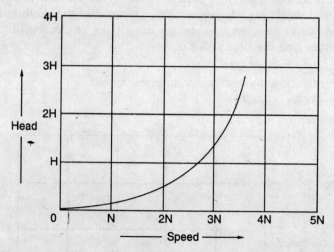

Fig. 37·6. Characteristic curves for speed *vs* head.

Fig. 37·6 shows the performance of a centrifugal pump under constant discharge. It is a parabolic, curve which shows that the head increases with the speed.

2. *Characteristic curves for discharge with varying speed*

Discharge versus head

Fig. 37·7 shows the performance of a centrifugal pump under variable rotational speeds. It is a parabolic curve, which shows that for a given rotational speed, the manometric head decreases with the discharge.

Performance of Pumps

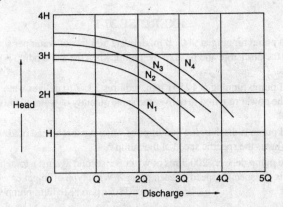

Fig. 37·7. Characteristic curves for discharge *vs* head.

Discharge versus power

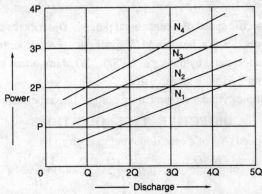

Fig. 37·8. Characteristic curves for discharge *vs* power.

Fig. 37·8 shows the performance of a centrifugal pump under variable rotational speeds. It is almost a straight curve, which shows that for a given rotational speed, the power increases with the discharge.

Discharge versus efficiency

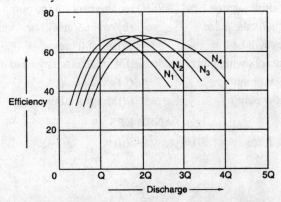

Fig. 37·9. Characteristic curves for discharge *vs* efficiency.

Fig. 37·9 shows the performance of a centrifugal pump under variable rotational speeds. It is a parabolic curve, which shows that for a given rotational speed, the efficiency increases with the discharge. And beyond a certain discharge, the efficiency decreases.

EXERCISE 37·1

1. A centrifugal pump running at 900 r.p.m. delivers 50 litres of water per second against a head of 10 metres. Find the discharge and head of the pump, when the same pump is running at 1000 r.p.m.
   (Ans. 55·6 litres/s; 12·3 m)

2. A centrifugal pump running at 1250 r.p.m. requires 3 kW to deliver water against a certain head of water. Find the power required to deliver the same quantity of water to the same head, while running at 1000 r.p.m.
   (Ans. 1·54 kW)

3. A centrifugal pump is discharging 500 litres of water against a head of 200 metres, while running at 900 r.p.m. What is the specific speed of the pump?
   (Ans. 67·3 r.p.m.)

4. A three stage pump delivers 200 litres of water per second against a total head of 50 metres at 1450 r.p.m. What is the specific speed of each pump? What type of impeller would you use?
   (Ans. 34·5 r.p.m.; Medium pump with radial flow at outlet)

QUESTIONS

1. Explain the effect of change of speed or diameter of the pump impeller on the discharge of the pump.
2. What is the specific speed of a centrifugal pump? Describe its uses.
3. Define the terms 'Suction head', and 'Vapour pressure'. Describe their importance.
4. What do you understand by the term 'NPSH'? What important value does it give?
5. What is priming? Explain its necessity.
6. What is a multiple cylinder reciprocating pump? Describe the speciality of such a pump.

OBJECTIVE TYPE QUESTIONS

1. The specific speed (N_s) of a centrifugal pump is given by
   (a) $\dfrac{N\sqrt{Q}}{H^{5/4}}$ (b) $\dfrac{N\sqrt{Q}}{H}$ (c) $\dfrac{N\sqrt{Q}}{H^{3/4}}$ (d) $\dfrac{N\sqrt{Q}}{H^{2/3}}$

2. The type of centrifugal pump preferred for a specific speed of 40 r.p.m. is
   (a) slow speed pump with radial flow at outlet
   (b) medium speed pump with radial flow at outlet
   (c) high speed pump with radial flow at outlet
   (d) high speed pump with axial flow at outlet

3. If the net positive suction head (NPSH) requirement for the pump is not satisfied, then
   (a) no flow will take place   (b) cavitation will be formed
   (c) efficiency will be low   (d) all of these

4. Which of the following pump is suitable for small discharge and high heads?
   (a) centrifugal pump   (b) axial pump
   (c) mixed flow pump   (d) reciprocating pump

ANSWERS

1. (c)   2. (b)   3. (b)   4. (d)

38

Pumping Devices

1. Introduction. 2. Hydraulic Ram. 3. Air lift Pump. 4. Jet Pump. 5. Rotary Pump. 6. External Gear Pump. 7. Internal Gear Pump. 8. Lobe Pump. 9. Vane Pump.

38·1 Introduction

In the previous chapter, we have discussed centrifugal and reciprocating pumps. But, sometimes, in certain industrial plants and power stations, oil and other liquids have to be handled, under different conditions. Moreover, instead of conventional devices like electric motor, petrol or diesel engines, other sources of power such as falling water, compressed air or steam is also utilized for lifting the liquids. Though there are numerous pumping devices, yet the following are important from the subject point of view :

38·2 Hydraulic Ram

This is an automatic machine, which can lift a small quantity of water to a greater height, when a large quantity of water is available at a smaller height.

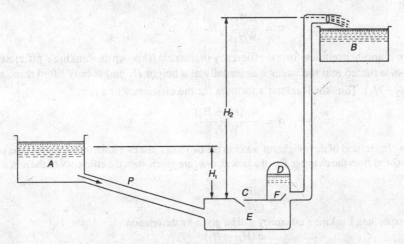

Fig. 38·1. Hydraulic ram.

Fig. 38·1 shows a diagrammatic view of a hydraulic ram, in which water is available from source A at a height H_1. By means of the hydraulic ram, a small quantity of this water is raised through a height H_2 into tank B.

The water starts flowing from the tank A to the chamber E through the pipe P. The waste-value C is open and the water flows out through it. As the speed of the water in the pipe P increases, the dynamic pressure on the waste-valve C increase, until it is greater than the weight of the valve lid, which will suddenly close the valve lid. This sudden closure of the valve brings the water in the pipe and chamber suddenly to rest, which will increase the pressure in the chamber E. This increase of pressure, in the chamber E, lifts the valve F and some water flows into the air vessel D, which will compress the air in the vessel, causing its pressure to increase. This increased air pressure in the vessel D closes the valve F and forces up the water into the tank B.

When the momentum of the water in the chamber E is destroyed, the waste-valve C opens which causes the flow of water from the tank A to recommence.

Let
- W = Weight of the water flowing from tank A into the chamber E,
- w = Weight of the water flowing from the chamber E into the tank B,
- H_1 = Height of water in the tank A, above the chamber E, and
- H_2 = Height of water in the tank B, above the chamber E.

We know that nergy supplied by the tank A
$$= WH_1 \qquad \ldots(i)$$
energy supplied to the tank $B = wH_2 \qquad \ldots(ii)$

Equating equations (*i*) and (*ii*),
$$WH_1 = wH_2$$
$$\therefore \quad w = \frac{H_1}{H_2} \times W$$

If losses are taken into account, then efficiency of the ram (known as D' Aubuisson's efficiency),
$$\eta = \frac{wH_2}{WH_1}$$

There is another relation for the efficiency of the ram ((known as Rankine's efficiency). In this relation, it is assumed that the water was initially at a height H_1 and is only lifted through a height equal to $(H_2 - H_1)$. Thus the Rankine's formula for the efficiency of a ram,
$$\eta = \frac{w \ (H_2 - H_1)}{WH_1}$$

Notes : Ig, instead of the weights of water in the two tanks, the two discharge (*i.e.*, from the tank A to the chamber, as Q and from the chamber E to the tank B, as q) are given, then the efficiency of the ram will be given by the relation,
$$\eta = \frac{qH_2}{QH_1}.$$
and the corresponding Rankine's efficiency will be given by the relation,
$$\eta = \frac{q \ (H_2 - H_1)}{QH_1}$$

Example 38·1. *A hydraulic ram, suitable for demonstration, in a research laboratory has the following data :*

Supply head	= 2·5 metres
Total water supplied	= 1000 litres/min
Discharge	= 100 litres/min
Delivery head	= 15 metres

Determine the Rankine's efficiency.

Solution. Given : $H_1 = 2\cdot5$ m; $Q = 1000$ litres/min; $q = 100$ litres/min; $H_2 = 15$ m.

We know that Rankine's efficiency,
$$\eta = \frac{q(H_2 - H_1)}{QH_1} = \frac{100 \times (15 - 2.5)}{1000 \times 2.5} = 0.50 = 50\% \text{ Ans.}$$

Example 38·2 *A hydraulic ram utilizes water under a head of 3 metres and delivers against an effective head of 30 metres. If the ratio of water raised to the water wasted by the ram is 1 : 15, calculate the D'Aubuisson's and Rankine's efficiencies of the ram.*

Solution. Given : $H_1 = 3$ m; $H_2 = 30$ m and ratio of water raised to the water wasted = 1 : 15.

D' Aubuisson's efficiency

Let the quantity of water raised by the ram,
$$q = 1 \text{ m}^3$$
Since the ratio of water raised to the water wasted is 1 : 15, therefore, quantity of water wasted = 15 m³. Or in other words total quantity of water delivered,
$$Q = 1 + 15 = 16 \text{ m}^3$$
We know that D'Aubuisson's efficiency,
$$\eta = \frac{qH_2}{QH_1} = \frac{1 \times 30}{16 \times 3} = 0.625 = 62.5\% \text{ Ans.}$$

Rankine's efficiency

We also know that Rankine's efficiency,
$$\eta = \frac{q(H_2 - H_1)}{QH_1} = \frac{1(30 - 3)}{16 \times 3} = 0.5625 = 56.25\% \text{ Ans.}$$

38·3 Air Lift Pump

It is successfully used for lifting water form deep wells or against high heads. An air lift pump, in its simplest form, consists of an open vertical pipe with its lower end submerged into the liquid to be raised. The upper end of the pipe leads to the required height. The compressed air is introduced at the bottom of the pipe through a nozzle as shown in Fig. 38·2.

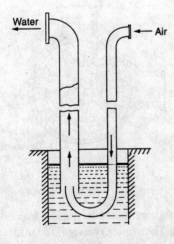

Fig. 38·2. Air lift pump.

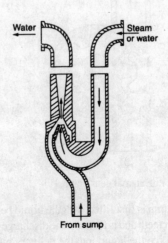

Fig. 38·3. Jet Pump.

The efficiency of this pump is very low (about 25%). This is due to the loss of energy of the compressed air while mixing with the liquid. An air jet has the following advantages :

1. It is quite simple in design.
2. Since there are no moving parts, therefore, there is no wear and tear.
3. There is no problem of lubrication.
4. The initial as well as maintenance cost is very small.

38·4 Jet Pump

It is successfully used for lifting water to the boilers. A jet pump, in its simplest form, consists of a pipe having a convergent end at its bottom. The upper end of the pipe leads to the required. height. Now steam (or sometimes water) under a high pressure is introduced through a nozzle as shown in Fig. 38·3. The pressure energy of the steam is converted into kinetic energy, as it passes through the nozzle. As a result of this, the pressure in the convergent portion of the pipe is considerably reduced and water is sucked into the pipe from the sump. The sucked water after coming in contact with the jet, is carried into the delivery pipe.

Here the kinetic head of the water steam is converted into pressure head, which forces the water into the delivery pipe.

38·5 Rotary Pumps

A rotary pump resembles a centrifugal pump in its outward appearance. But it differs in action. It combines the advantages of both the centrifugal and reciprocating pumps (*i.e.*, constant discharge of a centrifugal pump, and positive displacement of a reciprocating pump). The following rotary pumps are important from the subject point of view :

1. External gear pump, 2. Internal gear pump, 3. Lobe pump, and 4. Vane pump.

38·6 External Gear Pump

An external gear pump, in its simplest form, consists of two identical intermeshing spur wheels *A* and *B* working with a fine clearance inside inside the casing. The wheels are so designed, that they form a fluid tight joint at the point of contact as shown in Fig. 38·4. One of the wheels is keyed to the driving shaft, and the other revolves as a driven wheel.

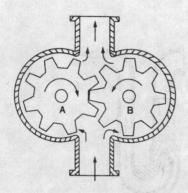

Fig. 38.4. External gear pump.

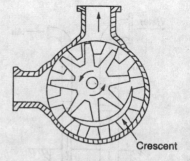

Fig. 38.5. Internal gear pump.

The pump is first filled with the liquid before it is starts. As the gears rotate, the liquid is trapped in between their teeth and is flown to the discharge end round the casing. The rotating gears build up sufficient pressure to force the liquid into the delivery pipe. A little consideration will show, that each tooth of the gear acts like a piston or plunger of a reciprocating pump to force liquid into the discharge pipe.

38·7 Internal Gear Pump

As internal gear pump, in its simplest form, consists of two spur wheels intermeshing internally. the wheels are so designed, that on one side, they form a fluid tight joint at the point of contact, and a space for crescent on the other as shown in Fig. 38·5.

A crescent shaped partition is provided between the two wheels to act as a seal between suction and discharge. This is done by placing the inner wheel eccentrically in the outer wheel. The inner wheel is keyed to the driving shaft and the outer revolves as a driven wheel.

The pump is first filled with the liquid before it is started. As the wheels rotate, the teeth come out of the mesh, near the suction end. As a result of this, the pace between the two wheels increases and the liquid flows into this space. As the wheels continue to rotate, the liquid is trapped, between the teeth and crescent of both the wheels and flown to the discharge end. A little consideration will show at each tooth of the gear, like external gear pump, acts like a piston or plunger of a reciprocating pump to force the liquid inside the discharge pipe.

38·8 Lobe Pump

A lobe type pump resembles with a gear pump in action. There are many designs of lobe type rotary pumps. But the wheels have usually two or three lobes, and sometimes even more. In all cases, the action remains the same. In Fig. 38·6 shown two types of lobe pumps.

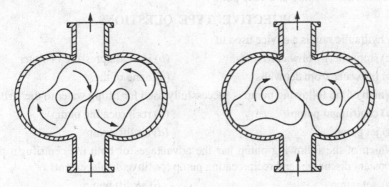

Fig. 38.6. Lobe pumps.

The lobes are so designed, that they from a fluid tight joint at the point of contact. As the lobes rotate, the liquid is trapped in the pockets formed between the lobes and casing. The lobes build up sufficient pressure to force the liquid into delivery pipe. The only drawback, in a lobe type, is that its discharge is not so constant as that of a gear pump.

38·9 Vane Pump

A vane pump in its simplest form, consists of a disc rotating eccentrically in the pump casing. The disc has a number of slots (generally 4 to 8) containing vanes, which are free to slide radially into the vanes. When the rotor rotates the disc, the vanes are pressed against the casing due to centrifugal force, and forms a liquid tight seal. As the dics rotates, the liquid is trapped in the pockets formed between the vanes and the casing. The vanes build up sufficient pressure to force the liquid into the delivery pipe.

In some designs the springs are used to press the vanes against the casing. But in some more designs, swinging vanes are used. In this type, the vanes are hinged, which swing out as the disc rotates due to centrifugal force.

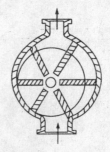

Fig. 38.7. Vane pump.

EXERCISE 38·1

1. A hydraulic ram model has the following data :
   Supply head = 1 m
   Delivery head = 6 m
   Total water available = 1 500 litres/s
   Discharge = 100 litres/s
   Find the efficiencies of the ram. [**Ans.** D' Aubuisson's $\eta = 40\%$; Rankine's $\eta = 33\%$]

2. A hydraulic ram is supplied water under a head of 2 metres and is to deliver against an effective head of 16·8 metres. If the ratio of water used to the water supplied is 16, find the efficiencies of the ram.
   [**Ans.** D' Aubuisson's $\eta = 49.4\%$; Rankine's $\eta = 46.25\%$]

QUESTIONS

1. Explain, with a neat sketch, the working of hydraulic ram and define its efficiency.
2. What is an air lift pump ? Where it is used ?
3. Give the advantages of air lift pump.
4. Distinguish between an internal gear pump and external gear pump.
5. Describe lobe pump. On what principle does it work ?
6. Explain the working of a vane pump.

OBJECTIVE TYPE QUESTIONS

1. A hydraulic ram is a device used to
   (a) store energy of water
   (b) increase pressure of water
   (c) lift water from depwells
   (d) none of these

2. Which of the following pump is successfully used for lifting water to the turbines
   (a) centrifugal pump
   (b) reciprocating pump
   (c) jet pump
   (d) air lift pump

3. Which of the following pump has the advantages of both the centrifugal pumps (*i.e.*, constant discharge) and reciprocating pump (positive displacement) ?
   (a) jet pump
   (b) air lift pump
   (c) rotary pump
   (d) all of these

ANSWERS

1. (d)   2. (c)   3. (c)

39

Hydraulic Systems

1. Introduction. 2. Hydraulic Press. 3. Hydraulic Accumulator. 4. Hydraulic Intensifier 5. Hydraulic Crane. 6. Hydraulic Lift. 7. Direct Acting Hydraulic Lift. 8. Suspend Hydraulic Lift. 9. Hydraulic Coupling. 10. Hydraulic Torque Convertor.

39·1 Introduction

In the previous chapters, we have discussed the various types of machines, in which the water moves bodily from one point to another. In some of the machines (*i.e.*, turbines) the potential or kinetic energy of water is converted into mechanical energy. But in other machines (*i.e.*, pumps) mechanical energy is converted into potential or kinetic energy.

In this chapter we shall discuss such hydraulic machines in which the liquid (water or oil) acts as a medium of transmission of power or pressure based on the principles of hydrostatics and Hydraulics.

Though there are numerous devices, based on the principle of Hydrostatics and Hydraulics, yet the following are important from the subject point of view :

1. Hydraulic press,
2. Hydraulic accumulator,
3. Hydraulic intensifier,
4. Hydraulic crane, and
5. Hydraulic lift.

39·2 Hydraulic Press

It is a device, by which we can lift a larger load, by the application of a comparatively much smaller force.

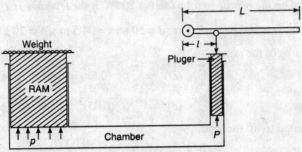

Fig. 39.1. Hydraulic press.

A hydraulic press in its simplest from, consists of two cylinders, one larger and the other smaller, connected to a chamber containing some liquid. The larger cylinder contains a ram and the smaller one a plunger as shown in Fig. 39·1.

A smaller force P acts on the plunger, in the downward direction, which presses the liquid below it. This pressure is transmitted, equally, in all directions and raises the ram. The heavier load, placed on the ram, is then lifted up.

Let
A = Area of the ram,
a = Area of the plunger,
p = Intensity of pressure, and
W = Weight lifted by the ram.

Since the intensity of pressure in the chamber, is the same in all the directions, therefore pressure at the bottom of the plunger,

$$P = pa \quad \text{or} \quad p = \frac{P}{a} \qquad \ldots (i)$$

and pressure at the bottom of the ram,

$$W = pA \quad \text{or} \quad p = \frac{W}{A} \qquad \ldots (ii)$$

Equating equations (i) and (ii),

$$\frac{P}{a} = \frac{W}{A}$$

$$\therefore \quad W = P \times \frac{A}{a} \qquad \ldots \text{(For } W\text{)}$$

or
$$P = W \times \frac{a}{A} \qquad \ldots \text{(For } P\text{)}$$

In losses are taken into account, then the efficiency of the hydraulic press,

$$\eta = \frac{W/A}{P/a} = \frac{W}{P} \times \frac{a}{A}$$

The term $\left(\frac{W}{P}\right)$ i.e., (ratio of load lifted to the effort applied) is known as mechanical advantage.

It is increased by applying a force (F) on the plunger by mass of a lever as shown in the figure. In this case

$$P = W \times \frac{a}{A} \times \frac{l}{L}$$

Note. The term L/l is known as leverage of the hydraulic press.

Example 39·1. *A hydraulic press has a ram of 500 mm and a plunger of 50 mm diameter. What force is required on the handle to lift a load of 20 kN, if the leverage is 1 : 10 ?*

Solution. Given : D = 500 mm = 0·5 m ; d = 50 mm = 0·05 m ; W = 20 kN and $\frac{l}{L} = \frac{1}{10}$.

We know that area of the ram,

$$A = \frac{\pi}{4} \times (D)^2 = \frac{\pi}{4} \times (0.5)^2 = 0.0625\ \pi\ m^2$$

and area of the flunge,
$$a = \frac{\pi}{4} \times (d)^2 = \frac{\pi}{4} \times (0.05)^2 = 0.000\ 625\ \pi\ m^2$$

$\therefore$ Force required to lift the load,

$$P = W \times \frac{a}{A} \times \frac{l}{L} = 20 \times \frac{0.000\ 625\ \pi}{0.0625\ \pi} \times \frac{1}{10}\ kN$$

$$= 0.02\ kN = 20\ N \quad \textbf{Ans.}$$

Example 39·2. *The ram and plunger of a hydraulic press are 150 mm and 25 mm in diameter respectively. With a plunger stroke of 300 mm, the press is able to lift a load of 10 kN through 1·6 metre in a 30 seconds. What is the (i) load on the plunger and (ii) power required to drive the plunger?*

Solution. Given: $D = 150$ mm $= 0.15$ m; $d = 25$ mm $= 0.025$ m; Plunger stroke $= 300$ mm $= 0.3$ m; $W = 10$ kN; $H = 1.6$ m and $t = 30$ s.

(i) Load on the plunger

We know that area of ram, $A = \dfrac{\pi}{4} \times (D)^2 = \dfrac{\pi}{4} \times (0.15)^2 = 0.0177$ m^2

and area of plunger, $a = \dfrac{\pi}{4} \times (d)^2 = \dfrac{\pi}{4} \times (0.025)^2 = 0.000\,491$ m^2

∴ Load on the plunger, $P = W \times \dfrac{a}{A} = 10 \times \dfrac{0.000\,491}{0.0177} = 0.277$ kN $= 277$ N **Ans.**

(ii) Power required to drive the plunger

We know that work done by the hydraulic press in one second

$$= \dfrac{WH}{t} = \dfrac{9.81 \times 1.6}{30} = 0.52 \text{ kN-m}$$

and power required to drive the plunger

$$= 0.52 \text{ kW} \quad \textbf{Ans.}$$

(iii) No. of strokes done by the pump

We also know that no. of strokes done by the pump,

$$= \dfrac{\text{Volume of water displaced in the cylinder}}{\text{Volume of water pumped by plunger in one stroke}}$$

$$= \dfrac{\text{Area of cylinder} \times \text{Height through which the load is lifted}}{\text{Area of plunger} \times \text{stroke length of the plunger}}$$

$$= \dfrac{0.0177 \times 1.6}{0.000\,491 \times 0.3} = 192 \quad \textbf{Ans.}$$

39.3 Hydraulic Accumulator

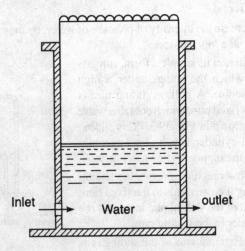

Fig. 39.2. Hydraulic accumulator.

It is a device, to store pressure energy, which may be supplied to a machine later on. A hydraulic accumulator, in its simplest form, consists of a vertical cylinder containing a sliding ram as shown in Fig. 39.2.

The sliding ram is loaded with weights. When the water, under pressure, enters the accumulator cylinder through the inlet valve, it lifts the loaded ram until the cylinder is full of water. At this stage, the accumulator has stored it maximum amount of energy, which is also known as capacity of the accumulator.

Let
A = Area of the cylinder,
H = Lift of the cylinder, and
p = Intensity of water pressure supplied by the accumulator,

Then the capacity of accumulator = pAH

Example 39·3. *An accumulator has a ram of area 2 square metres and a lift of 10 metres. Find capacity of the accumulator, if the water is supplied at a pressure of 150 kPa.*

Solution. Given : $A = 2$ m^2 ; $H = 10$ m and $p = 150$ kPa = 150 N/m^2.
We know that capacity of the accumulator
$$= pAH = 150 \times 2 \times 10 = 3000 \text{ kN-m} \quad \textbf{Ans.}$$

Example 39·4. *The displacement volume of an accumulator is 4 litres of water and diameter of its plunger is 375 mm. Find the length of the accumulator stroke.*

Solution. Given : Displacement volume = 4 litres = 0·004 m^3 and diameter of the plunger $(d) = 375$ mm = 0·375 m.
We know that area of the plunger
$$= \frac{\pi}{4} \times (d)^2 = \frac{\pi}{4} \times (0.375)^2 = 0.11 \text{ m}^2$$

and length of the accumulator stroke
$$= \frac{\text{Displacement volume}}{\text{Area of the plunger}} = \frac{0.004}{0.11} \text{ m}$$
$$= 0.0364 \text{ m} = 36.4 \text{ mm} \quad \textbf{Ans.}$$

39·4 Hydraulic Intensifier

It is a device to increase the intensity of pressure of water, by means of energy available from a larger quantity of water at a low pressure.

A hydraulic intensifier, in its simplest form, consists of a fixed ram through which the water, under a high pressure, flows to the machine. A hollow sliding ram is mounted externally on this fixed ram, which contains water under a high pressure as shown in Fig. 39·3. This sliding ram is encased in a fixed cylinder which contains water under a low pressure from the supply as shown in Fig. 39·3.

The water under a low pressure, presses the sliding ram on the top, thus forcing it downwards to the fixed ram. This downward movement of the sliding ram increases the intensity of pressure of water in the sliding ram.

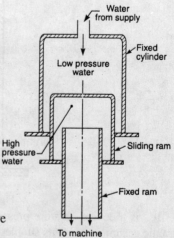

Fig. 39·3. Hydraulic intensifier.

Let
A = External area of the sliding ram,
a = Area of the end fixed ram,
p = Intensity of pressure (of low pressure water) in the fixed cylinder, and
P = Intensity of pressure (of high pressure water) in the sliding ram.

Hydraulic Systems

We know that total upward force

$$= \text{Intensity of pressure} \times \text{Area} = p \times a \qquad ...(i)$$

Similarly, total downward force

$$= p \times A \qquad ...(ii)$$

Since the two forces are equal, therefore equating equations (i) and (ii),

$$P \times a = p \times A$$

$$\therefore P = \frac{A}{a} \times p$$

Note: Sometimes, there is a loss due to friction (k) at each packing. In such a case, the pressure of water in the sliding ram,

$$P = \frac{A}{a} \times p (1 - k)^2$$

Example 39·5. *Find the pressure in a small cylinder of an intensifier, if the pressure in the larger cylinder is 5·4 MPa. The diameters of the smaller and large plungers of the intensifier are 100 mm and 300 mm respectively.*

Solution. Given: $P = 5·4$ MPa $= 5·4 \times 10^6$ N/m²; $d = 100$ mm $= 0·1$ m and $D = 300$ mm $= 0·3$ m.

Let $p =$ Pressure in the small cylinder.

We know that area of smaller cylinder,

$$a = \frac{\pi}{4} \times (d)^2 = \frac{\pi}{4} \times (0·1)^2 = 0·0025 \pi \text{ m}^2$$

and area of larger cylinder,

$$A = \frac{\pi}{4} \times (D)^2 = \frac{\pi}{4} \times (0·3)^2 = 0·0225 \pi \text{ m}^2$$

We also know that pressure in the larger cylinder (p),

$$5·4 \times 10^6 = \frac{A}{a} \times p = \frac{0·0225 \pi}{0·0025 \pi} \times p = 9p$$

$$\therefore p = \frac{5·4 \times 10^6}{9} = 0·6 \times 10^6 \text{ N/m}^2 = 0·6 \text{ MPa} \quad \textbf{Ans.}$$

Example 39·6. *An intensifier has a 150 mm diameter and a sliding cylinder of 900 mm diameter. If there is 10% loss due to friction in each packing and the supply pressure is 500 kPa, find the water pressure leaving the cylinder.*

Solution. Given: $d = 150$ mm $= 0·15$ m; $D = 900$ m $= 0·9$ m; $k = 10\% = 0·1$ and $p = 500$ kPa $= 500 \times 10^3$ N/m²

We know that area of intensifier,

$$a = \frac{\pi}{4} \times (d)^2 = \frac{\pi}{4} \times (0·15)^2 = 0·0177 \text{ m}^2$$

and area of sliding cylinder,

$$A = \frac{\pi}{4} \times (D)^2 = \frac{\pi}{4} \times (0·9)^2 = 0·6362 \text{ m}^2$$

∴ Water pressure leaving the cylinder,

$$p = \frac{A}{a} \times p \times [1 - k^2]$$

$$= \frac{0·6362}{0·0177} \times (500 \times 10^3) \times [1 - (0·1)^2] \text{ N/m}^2$$

$$p = 17790 \times 10^3 \text{ N/m}^2 = 17·79 \times 10^6 \text{ N/m}^2$$

$$= 17·79 \text{ MPa} \quad \textbf{Ans.}$$

39·5 Hydraulic Crane

It is a device for raising or transferring heavy loads (up to 250 tonnes) and is widely used on dock sidings, warehouses or workshops.

A hydraulic crane, in its simplest from, consists of a vertical post, tie and jib (*i.e.*, basic requirements of a crane) having guide pulleys. Near the foot of the vertical crane post, is provided a jigger. The jigger consists of a fixed cylinder, having a pulley block and containing a sliding ram. One end of the ram is in contact with water and the other carries a pulley block. A chain or wire rope, one end of which is fixed, is taken round all the pulleys of the two blocks, through the vertical post and finally over the guide pulleys. A hook is attached to the other end of the rope for handling the load as shown in Fig. 39·4. There is a pipe connection for supplying water under a high pressure to the fixed cylinder of the jigger as shown in Fig. 39·4.

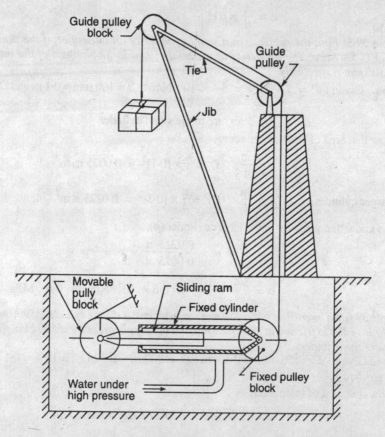

Fig. 39.4. Hydraulic crane.

The load to be lifted, is suspended on the free end of the wire rope. The water, under a high pressure, is admitted into the cylinder of the jigger. This water forces the sliding ram to move towards the left. This outward movement of the sliding ram, makes the pulley block to move outwards. Due to the increased distance between the two pulley blocks, the wire rope is pulled and the load is lifted up. If the load is required to be shifted, the vertical post, of the crane, can be rotated through the desired horizontal angle.

39·6 Hydraulic Lift

It is a device to lift or bring down loads or passengers, from one floor to another in a multi-storeyed building. Following are the two types of hydraulic lifts :
1. Direct acting hydraulic lift, and
2. Suspended hydraulic lift.

39·7 Direct Acting Hydraulic Lift

A direct acting hydraulic lift, in its simplest form, consists of a cage (for placing the loads or standing passengers) firmly secured to the top of a vertical ram, sliding in a fixed cyclinder. There is a pipe connection for supplying water, under a high pressure to the fixed cylinder as shown in Fig. 39·5.

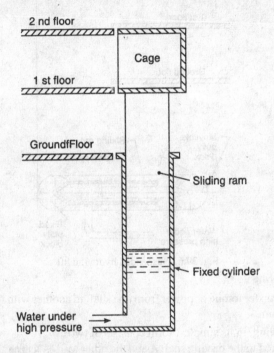

Fig. 39·5. Direct acting hydraulic lift.

The load, to be lifted, is placed in the cage of lift. The water under high pressure, is made to enter in the fixed cylinder, which moves up the sliding ram and the cage fixed to it.

39·8 Suspended Hydraulic Lift

A suspended hydraulic lift, in the simplest form, consists of a cage (for placing the loads or standing of passengers) which is suspended from a wire rope. The hydraulic lift obtains its motion from the jigger, in the same way as the hydraulic crane (Art. 39·5). Near the foot of the cage hole a jigger is provided. The jigger consists of a fixed cylinder, having a pulley-block, and containing a sliding ram. One end of the ram is in contact with the water and the other carries a pulley-block. A wire rope, one end of which is fixed, is taken round all the pulleys of the two blocks and finally over the guide pulleys. The cage is suspended from the other end of the rope. There is a pipe connection for supplying water under a high pressure to the fixed cylinder of the jigger as shown in Fig. 39·6.

The load to be lifted, is placed in the cage. The water under a high pressure, is admitted into the cylinder of the jigger. This water forces the sliding ram to move towards the left. This outward movement of the sliding ram, makes the pulley-block to move outwards. Due to increased distance, between the two pulley-blocks, the wire rope is pulled and the cage is lifted up.

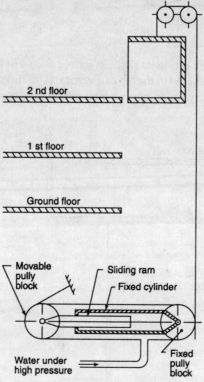

Fig. 39.6. Suspended hydraulic lift.

39.9 Hydraulic Coupling

It is a device to transfer torque or power from one shaft to another with the help of an oil, which acts as a working medium.

A hydraulic coupling, in its simplest form consists of two identical halves, one fixed to the driving shaft A and the other to the driven shaft B. Both these havles are housed in a common casing fitted with oil, which is used as a working medium. The half, which if fixed to the driving shaft A is in the of a pump impeller, which imparts energy to the oil. The other half which is fixed to the driven shaft B is in the form of a reaction turbine runner as shown in Fig. 39.7. It receives energy from the oil.

As the driving shaft A starts rotating, the oil in the casing is forced out through the periphery of the pump impeller. This oil enters the reaction turbine runner and makes it to rotate. The oil flows back into the pump impeller from the turbine runner. In actual practice, the speed of shaft B is always less than that of shaft A, by 2 to 4 per cent. This slip is due to friction and turbulence loss in the impeller and the runner passages.

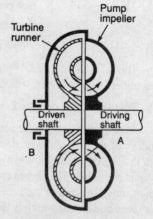

Fig. 39.7. Hydraulic coupling.

Hydraulic Systems

The efficiency of a hydraulic coupling may be defined as the ratio of power output at shaft B to the power input at shaft A. Mathematically

$$\eta = \frac{P_B}{P_A} = \frac{N_B}{N_A}$$

and slip,

$$s = 1-\eta = 1 - \frac{N_B}{N_A}$$

39·10 Hydraulic Torque Converter

A hydraulic torque converter is an improved form of hydraulic coupling, in which the torque (or in other words, speed) at the driven shaft may be increased or decreased. It is achieved by providing a third member between the pump impeller and reaction turbine runner, known as a guide ring or a reactionary member. The guide ring can give a gear ratio of 5 : 1 with an efficiency of 85% to 90 per cent.

The guide ring consists of a series of fixed guide vanes, whose function is to change the direction of the oil. As a result of this, the oil reacts upon the turbine runner, which multiplies the torque of the turbine runner.

EXERCISE 39·1

1. A hydraulic press has a ram of 800 mm diameter and a plunger of 40 mm diameter. Find the force required to lift a load of 50 kN on the ram. [Ans. 125 N]
2. A hydraulic press has a ram of 100 mm diameter and a plunger of 12·5 mm diameter. The machine is lifting a load of 60 kN through a distance of 1·2 metre. Find the number of stroks required by the plunger, if the stroke of the plunger is 300 mm. [Ans. 256]
3. A hydraulic accumulator has sliding ram of 1·2 m diameter and 5 m lift. If the water supplied to the accumulator is at a pressure of 100 kPa, what is the capacity of the accumulator? [Ans 565·5]
4. The diameters of fixed and sliding rams of a hydraulic intensifier are 100 mm and 300 mm respectively. Find the intensity of pressure of the water leaving the intensifier, if the water supplied to the intensifier is under a pressure of 200 kPa. [Ans 1·8 MPa]

QUESTIONS

1. Explain, with the help of a neat sketch, the working and principle of hydraulic press.
2. Draw a neat sketch, and explain the working of a hydraulic accumulator.
3. Define the capacity of a hydraulic accumulator.
4. Draw a neat sketch of a hydraulic intensifier and explain its working principles.
5. Explain the working of a hydraulic crane with diagram.
6. Describe the working of lift.
7. Describe the working of a hydraulic coupling.
8. Give the uses of hydraulic torque convertor.

OBJECTIVE TYPE QUESTIONS

1. A hydraulic press is a device used to
   (a) store energy which may be supplied to a machine at a later stage
   (b) increase the intensity of pressure from large quantity of water at a low pressure
   (c) lift a large load by the application of a comparatively much smaller force
   (d) all of these

2. The efficiency of a hydraulic press is given by

   (a) $\dfrac{W}{P} \times \dfrac{A}{a}$ (b) $\dfrac{W}{P} \times \dfrac{a}{A}$ (c) $\dfrac{P}{W} \times \dfrac{a}{A}$ (d) $\dfrac{P}{W} \times \dfrac{A}{a}$

   where W = Weight lifted by the ram,
   P = Force applied on the plunger,
   A = Area of the ram, and
   a = Area of the plunger.

3. Which of the following type of energy is stored in an accumulator?
   (a) pressure energy   (b) kinetic energy
   (c) potential energy   (d) strain energy

4. The capacity of a hydraulic accumulator is specified as the
   (a) minimum energy it can store
   (b) maximum energy it can store
   (c) average of maximum and minimum energy it can store
   (d) product of maximum and minimum energy it can store

ANSWERS

1. (c)   2. (b)   3. (a)   4. (b)

Index

A

Absolute pressure, 16
Accumulator, Hydraulic, 653
Adiabatic process, 391
Advantages of distorted model, 454
     — model analysis, 440
     — syphon spillway, 229
     — triangular notch over a rectangular notch, 197
     — water turbines, 505
     — water wheels, 504
Adverse slope profiles, 330
Air foil theory, 412
Air lift pump, 647
Air vessels, 616
Analytical method for metacentric height, 90
Anicut, 227
Applications of Bernoulli's equation, 126
     —hydrostatics, 69
     —weirs, 227
Approach, Velocity of, 217
Archimedies' principle, 87
Area of flow, 304
Assumptions for the effect of viscosity, 333
Atmospheric pressure, 15
Average velocity of flow, 305
Axial flow turbines, 531

B

Backwater curve, 323
Banki turbine, 526
Barrage, 227
Bazin's formula for discharge over a rectangular weir, 214
     —through open channels, 287
Beginning and development of Hydraulics, 1
Bell mouthpiece, 272
Bernoulli's equation, 118
     —Limitations of 121
     —Practical applications of, 126
Borda's mouthpiece, 182
Boundary layer separation, 416
     — theory, 412
Bourdon's tube pressure gauge, 29
Boyle's law, 385

Braking jet, 510
Branched pipes, Discharge through, 258, 264
Breast water wheel, 501
Bridge openings, 228
Broad-crested weir, 221
Buckets of a Pelton wheel, 509
Bulk modulus of a fluid, 392
Buoyancy, 87

C

Capillarity of water, 8
Capillary tube method, 359
Casing of a Pelton wheel, 510
Cauchy's number, 467
Cavitation, 580
     — in centrifugal pumps, 639
Centre of buoyancy, 87
Centre of pressure, 42
     —of a composite section, 54
     —of an inclined immersed surface, 49
     —of a vertically immersed surface, 43
Channels of most economical cross section, 293
Characteristic curves for centrifugal pumps, 641
     — Francis' turbine, 575
     — Pelton wheel, 574
     — turbines, 573
Characteristics of turbines, 561
     — water curves in non-uniform flow, 327
Charle's law, 385
Chemical method for discharge through a river, 306
Chezy's formula for loss of head in pipes, 235
     —discharge through an open channel, 284
Cippoletti weir, 318
Classification of fluids, 337
     — models, 443
     — reaction turbines, 530
     —viscous flows, 338
     —water turbines, 505
     — water curves, 328
Clinging nappe, 227
Closed cylindrical vessel, 377

Coefficient of contraction, 139
     —discharge, 140
     —resistance, 140
     —velocity, 139
Coefficient of friction for commercial pipes, 357

Comparison of a model and its prototype, 443, 444, 445, 446, 448, 449, 451, 452
Components of a reaction turbine, 528
Composition of forces, 10
Compressibility of water, 7
Compressible flow, 112
Compound pipe, Discharge through, 253
Conditions for equilibrium of a floating body, 92
     —maximum discharge through a channel of circular section, 302
     —maximum velocity through a channel for circular section, 300
     —rectangular section, 293
     —trapezoidal section, 295
Condition for stability of a dam, 83
Conical buoys floating in liquid, 99
     — draft tubes, 557
Conservation of energy, Law of, 11
Construction of the model, 442
Correct prediction of the model, 443
Crane, Hydraulic, 656
Critical depth, 311
     —flow, 313
     —slope profiles, 329
     —velocity, 313, 339
Current meter, 306
     —Rating of, 307

D

D' Alembert's principle, 363
Darcy's formula for loss of head in pipes, 232
Dead weight pressure gauge, 30
Density of water, 6
Depressed nappe, 227
Depth of hydraulic jump, 315

[661]

Description of water curves or profiles, 328
Design of Pelton wheels, 521
Determination of velocity of approach, 218
Development of hydraulic turbines, 505
Diameter of the nozzle for maximum transmission of power, 275
Diaphragm pressure gauge, 30
Difference between impulse turbine and reaction turbine, 530
Differential manometer, 25, 28
Dimensional homogeneity, 422
Dimensionless number, 463
Direct acting hydraulic lift, 657
Disadvantages of distorted model, 454
— water wheels, 504
Discharge of a centrifugal pump, 587
— reaction turbine, 538
— reciprocating pump, 603
Discharge of the prototype for the given discharge of the model, 445, 456
Discharge over a broad crested weir, 221
— Cippoletti weir, 215
— drowned weir, 225
— narrow crested weir, 220
— ogee weir, 224
— rectangular notch, 194
— rectangular weir, 211, 213
— sharp crested weir, 223
— stepped notch, 199
— submerged weir, 225
— trapezoidal notch, 198
— triangular notch, 196
Discharge through branched pipes, 258, 264
— Borda's mouthpipe, 182
— circular channel, 291
— compound pipe (ie., pipe in series), 253
— convergent-divergent mouthpiece, 186
— convergent mouthpiece, 185
— drowned orifice, 147, 148, 149

— rectangular orifice, 144, 145
— mouthpiece, 180, 181, 182, 185, 186
— pipes in parallel, 256
— syphon pipes, 265
— venturimeter, 127
Distribution of velocity of a flowing liquid on a pipe section, 345
Draft tube, 530, 557
Drag, 409
Drowned orifice, 147, 148
— weir, 225
Double cylinder reciprocating pump, 640
— floats, 305
Dynamic similarity, 442

E

Effect of acceleration of the piston on the indicator diagram, 608
— end contractions over a rectangular weir, 213
— friction in the suction and delivery pipes on the indicator diagram, 609
— pipe elasticity on hammer blow, 280
— temperature in the viscosity, 335, 336
— variation in diameter, 634
— variation in speed, 632
— viscosity on motion, 334
Effect on discharge over a notch, due to error in the measurement of head, 205
— over a rectangular notch, 205
— over a triangular notch, 207
Efficiencies of a centrifugal pump, 587
— draft tubes, 558
— impulse turbines, 513
— power transmission through a nozzle, 273
— reaction turbine, 541
Elastic force, 463
Elbow draft tubes, 558
Energy, 11
— equation for flowing gases, 395
— of a liquid in motion, 116

Equation of continuity, 107, 394
— of forced vortex flow, 374
— non-uniform flow, 325
Equipotential lines, 110
Equivalent size of a pipe, 255
Euler's equation for motion, 119
— number, 466
Experimental method for hydraulic coefficients, 141
— metacentric height, 102
External gear pump, 648

F

Falling sphere method, 360
Filament lines, 110
Floating bodies anchored at base, 98
Floats, 305
Flow of gas through an orifice or nozzle, 397
— liquid from one vessel into another under pressure, 149
— water into and from the air vessels fitted to the delivery pipe of a reciprocating pump, 627
Flow through syphon pipes, 265
Flow net, 110
Fluid masses subjected to acceleration along inclined plane, 368
— horizontal acceleration, 364
— vertical acceleration, 367
Force, 9
Force of a jet impinging normally on a fixed plate, 471
— on a fixed curved vane, 479
— on a hinged plate, 474
— on an inclined fixed plate, 472
— on a moving curved vane, 481
— on a moving plate, 476
— on a series of vanes, 477
Force of resistance of the prototype for the given force of resistance of a model, 446
Forced vortex flow, 373
Francis' formula for discharge over a rectangular weir, 213
— turbine, 551
Free nappe, 227

Index

Froude's number, 464
Fundamental dimensions, 420
— units, 3
Furneyron turbine, 559

G

Gauge pressure, 16
Gay-Lussac law, 386
General gas equation, 386
Geometrical similarity, 441
Girard turbine, 526
Govering of an impulse turbine, 524
Graphical representation of pressure head and velocity head, 239
Gravity force, 463
Guide mechanism, 529

H

Hagen-Poiseulle law for laminar flow in pipes, 341
Height of paraboloid of the liquid surface, 376
Higher critical velocity, 339
Horizontal slope profiles, 330
Hydraulic coefficients, 139
— Experimental determination of, 141
Hydraulic accumulator, 653
— coupling, 658
— crane, 656
— gradient line, 239
— intesifier, 654
— jump, 314
— lift, 657
— press, 651
— ram, 645
— similarity, 441
— torque convertor, 659
Hydraulic efficiency of an impulse turbine, 513
— reaction turbine, 541
Hydroelectric power plant, 505

I

Ideal fluid, 337
Impulse turbine, Govering of, 524
Inclined venturimeter, 131
Incompressible flow, 112
Increase in the water pressure while flowing through the impeller of a centrifugal pump, 594

Indicator diagram of a reciprocating pump, 606, 608, 609
Inertia force, 462
Intensifier, Hydraulic, 654
Internal gear pump, 649
International system of units, 3
Inverted differential manometer, 28
Irrotational flow, 112
Inward flow reaction turbine, 531
Isothermal process, 390

J

Jet of water, 138
Jet pump, 648
Jonval turbine, 559

K

Kaplan turbine, 554
Kilogram, 4
Kinematic similarity, 441
— viscosity, 336
Kinetic energy of a liquid particle in motion, 116
Kutter's formula for discharge through open channels, 288

L

Laminar flow, 338
Law of conservation of energy, 11
Laws of motion, 10
— perfrct gases, 385
Length of backward curve, 323
Lift, 410, 657
Limitations of Bernoulli's equation, 121
Liquids and their properties, 6
Lobe pump, 649
Lock gates, Water pressure on, 72
Loss of head due to hydraulic jump, 316
Loss of head due to friction in a viscous flow, 346
Loss of head in pipes, 231
— Chezy's formula, 235
— Darcy's formula, 232
Loss of head of a liquid flowing in a pipe, 173
— due to entrance in a pipe, 178
— due to exit in a pipe, 178

— due to obstruction in a pipe, 179
— due to sudden contraction, 175
— due to sudden enlargement, 173
Lower critical velocity, 339
Lubrication of bearings, 350

M

Mach number, 467
— its importance, 402
— wave, 403, 404
Main components of a reaction turbine, 528
Manning's formula for discharge through open channels, 289
Manometer, 17
Manometric efficiency of a centrifugal pump, 587
— head of a centrifugal pump, 586
Maximum length of a vertical floating body, 95
— speed of the rotating crank of a reciprocating pump, 612, 619
Measurement of air speed, 406
— river discharge, 304
— fluid pressure, 16
Magnus effect in a moving liquid, 416
Mechanical efficiency of a centrifugal pump, 587
— impulse turbine, 513
— reaction turbine, 541
Mechanical gauges, 29
Metacentric height, 89
— Analytical method for, 90
— Experimental method for, 102
Methods of dimensional analysis, 424, 430
— determination of coefficients of viscosity, 358
Metre, 4
Micromanometer, 21
Mild slope profiles, 328
Minimum starting speed of a centrifugal pump, 598
Mixed flow turbines, 531
Modular venturiflume, 320
Mouthpiece running free, 183

—running full, 184
Multiple cylinder reciprocating pump, 640
Multistage centrifugal pump, 599

N

Narrow cred weir, 220
Negative slip of the pump, 604
Net positive suction head (NPSH), 639
Neutral equilibrium of a floating body, 93
Newtonian fluid, 338
Newton's laws of motion, 10
— resistance, 408
—viscosity, 334
Nikuradse's experiment in rough pipes, 355
No. of buckets on the periphery of a Pelton wheel, 522
— jets for a Pelton wheel, 519
Non-modular venturiflume, 318
Non-Newtonian fluid, 338
Non-uniform flow, 111
Nozzle of a Pelton wheel, 508

O

Ogee weir, 224
One-dimensional flow, 113
Orifice meter, 134
Orifice viscometer, 359
Other impulse turbines, 526
— reaction turbines, 559
Outward flow reaction turbines, 536
Overall efficiency of a centrifugal pump, 587
— impulse turbine, 513
— reaction turbine, 542
Overshot water wheel, 499

P

Parallelogram law of forces, 10
Pascal's law, 14
Path lines, 110
Pelton wheel, 508
— Design of, 521
Piezometer tube, 16
Pilot tube, 135, 306
Piping system of a centrifugal pump, 583
Polygon law of forces, 10
Poncelet water wheel, 502

Potential energy of a liquid in motion, 116
—lines, 110
Power, 10
— developed by a prototype for the power developed by a model, 449, 458
— produced by an impulse turbine, 512
— reaction turbine, 541
— required to drive a centrifugal pump, 590
— reciprocating pump, 605
Practical applications of Bernoulli's equations, 126
—hydrostatics, 69
—weirs, 277
Prandtl and Von Karman's equation for pipe flow, 356
— experiment of boundary layer separation, 412
Precautions for rating of current meters, 307
Prediction of a model, 443, 45
Press, Hydraulic, 651
Presentation of units and their values, 4
Pressure diagrams, 64
— difference in a free vortex flow, 383
—energy of a liquid particle in motion, 116
— force, 462
—head, 13
—due to one kind of liquid on one side, 65, 68
Pressure in a mouthpiece, 187, 189, 190, 191
Pressure of water due to deviated flow, 489
Pressure on a body immersed in a moving liquid, 408
—curved surface, 58
Prevention of boundary layer separation, 417, 418
Priming of a centrifugal pump, 641
Principle, Archimedes', 87
— of jet propulsion, 492
Procedure for model analysis, 442
Proof of Bernoulli's equation, 118
—Pascal's law, 14
Properties of liquid, 6
Propulsion of ships, 492, 493, 496

R

Radial flow turbines, 531
Raised weir, 227
Ram, Hydraulic, 645
Rate of discharge, 107
Rating of current metres, 307
Ratio of specific heats, 389
Rayleigh's method for dimensional homogeneity, 424
Real fluid, 339
Recent trends in water power engineering, 505
Rectangular notch, Discharge over, 144, 145
Relation between pressure and height, 398
— specific heats, 386
— specific speed and shape of reaction turbine runner, 571
Resultant force, 9
Reynold's experiment of viscous flow, 339
Reynold's number, 340, 463
Rod floats, 305
Rotating cylinder method, 360
Rotational flow, 112
Runner of Pelton wheel, 509

S

Scalars and vectors, 11
Second, 4
Selection of centrifugal pump, 638
— repeating variables, 431
— suitable scale for model, 442
— turbines, 568, 569
Sharp crested weir, 223
Shooting flow, 313
S. I. units, 3
Significance of specific speed, 568
— unit power, unit speed and unit discharge, 563
Simple manometer, 17
—segment method, 304
Simpson's rule, 304
Siphon spillway, 229
—Advantages of, 229
Size of a buckets of a Pelton wheel, 522
Slip of pump, 604
Slope of free surface of water, 325

Index

Sluice gates, Water pressure on, 70
Specific energy of a flowing liquid, 310
— diagrams, 311
— gravity of water, 7
— weight of water, 6
Specific heats of a gas, 387
Specific speed of a centrifugal pump, 635
— of a turbine, 565
— shape of reaction turbine runner, 571
Speed of the prototype for the given speed of a model, 448, 457
Spiral casing for a centrifugal pump, 583
— reaction turbine, 528
Stable equilibrium of a floating body, 92
Stagnation pressure, 404
Steady flow, 112
Steep slope profile, 329
Stepped notch, 199
Streaklines, 110
Streaming flow, 313
Stream function, 113
Streamline flow, 112
Streamlines, 110
Submerged orifice, 147
— weir, 225
Suction head, 638
Surface tension force, 462
Surge tanks, 282
Suspended hydraulic pump, 657
Syphon pipes, 265
— spillway, 229

T

Testing of the model, 442
Theorem, Bernoulli's, 118
Thompson's turbine, 559
Thickness of boundary layer, 413, 414, 415
Three dimensional flow, 113
— a tank through an orifice at its bottom, 153
— through a long pipe, 244
— over a rectangular notch, 201
— over a triangular notch, 203
Time of emptying a hemispherical tank through an orifice at its bottom, 156
— circular horizontal tank through an orifice at the bottom, 158
— prototype for the given time of emptying a model, 451, 460
— tank of variable cross-section through an orifice, 161
— tank through two orifices, 163
Time of flow of liquid from, one vessel into another, 166
— through a long pipe, 247
Time of rolling (oscillation) of a floating body, 104
Total energy line, 239
— energy of liquid particle in motion, 117
Total pressure, 37
— head of a liquid particle in motion, 117
— on a horizontal immersed surface, 37
— on an anclined surface, 40
— on the top and bottom of a closed cylindrical vessel completely filled with water, 381
— on a vertically immersed surface, 38
Transmission of power through nozzle, 272
— pipes, 241
Trapezoidal notch, 198
Travelling time of a prototype for the given travelling time of a model, 452
Triangular notch, 196
Triangle law of forces, 10
Tripple cylinder reciprocating pump, 640
Tube gauges, 16
Turbine runner, 529
Turbulent flow, 112, 338
Turgo turbine, 526
Two-dimensional flow, 113
Types of casings for the impeller of a centrifugal pump, 582
— draft tubes, 557
— flows in cannel, 313
— flow lines, 110
— flows in a pipe, 111
— forces present in a moving liquid, 462
— mouthpieces, 172
— notches, 194
— orifices, 138
— pumps, 582
— reciprocating pump, 603
— water wheels, 499
— weir, 210, 220

U

Undesshot water wheel, 502
Undistorted models, 443
Uniform flow, 111
Unit discharge, 562
— power, 561
— speed, 562
Units of kinematic viscosity, 336
— viscosity, 334
Unstable equilibrium of a floating body, 92
Unsteady flow, 112
Upper critical velocity, 339
Useful data, 9
Use of flow nets, 110
— nozzles, 278
Uses of principle of dimensional homogeneity, 423

V

Values of Chezy's constant in the formula for discharge through an open channel, 287
Vane pump, 649
Vapour pressure, 639
Vectors and scalars, 11
Velocity of approach, 217, 218
— sound wave, 399
— water through a nozzle, 276
Velocity of water in the prototype for the given velocity of water in the model, 444, 455
Vena contract, 139
Ventilation of rectangular weirs, 226
Venturiflume, 318
Venturimeter, 126, 131
— Discharge through, 189
Viscous force, 462
Viscosity, 9, 333, 358
Viscous flow, 373
Viscous resistance of collar bearings, 353

—foot step bearings, 352
—oiled bearings, 350
Variation of pressure in the suction and delivery pipe due to acceleration of the piston of a reciprocating pump, 606
— speed and diameter of a centrifugal pump, 632

W

Water hammer, 278
—when valve is gradually closed, 278
—when valve is suddenly closed, 279
Water pressure on sluice gates, 70
—one lock gates, 72
—on masonry dams, 78
—on manonry walls, 76
—on rectangular dams, 78
—on trapezoidal dams, 79
Weber's number, 465
Weight, 14
Work, 10
Work done against friction in a reciprocating pump, 622, 623
Work done by a centrifugal pump, 584
— impulse turbine, 510
Work saved against friction by fitting air vessels in a reciprocating pump, 624

Attention: Readers

We request you for your frank assessment regarding some of the aspects of the book given as under:

10 026   A Textbook of Hydraulics, Fluid Mechanics and Hydraulic Machines
            R.S. Khurmi   Reprint 2007

Please fill up the given space in neat capital letters.

(i)   What topics of your syllabus that are important from your examination point of view are not covered in the book?
....
....
....
....
....

(ii)   What are the chapters wherein the treatment of the subject matter is not systematic or organised or updated?
....
....
....
....
....

(iii)   Have you come across misprints/mistakes/factual inaccuracies in the book? Please specify the chapters and the page numbers.
....
....
....
....
....

(iv)   What other books on the same subject have you found/heard better than the present book? Please specify in terms of price and quality.
....
....
....
....
....
....
....

(v) Further suggestions and comments for the improvement of the book:
....
....
....
....
....

(PS: If need be please attach a separate sheet to write your views.)

Other Details:

(i) Who recommended you the book? (Please tick in the box near the option relevant to you.)
☐ Teacher ☐ Friends ☐ Bookseller

(ii) Name of the recommending teacher, his designation and address:
....
....
....

(iii) Name and address of the bookseller you purchased the book from:
....
....
....

(iv) Name and address of your institution (Please mention the University or Board, as the case may be)
....
....
....

(v) Your name and complete postal address:
....
....
....

(vi) Course you are enrolled in:
....
....

The best assessment will be awarded each month. The award will be in the form of our publications as decided by the Editorial Board.

Please mail the filled up coupon at your earliest to:
The EDP Department (FB)
S. CHAND & COMPANY LTD.,
Post Box No. 5733, Ram Nagar, New Delhi 110 055

IMPORTANT BOOKS ON MECHANICAL ENGINEERING

HEAT AND MASS TRANSFER
R.K. Rajput

The book is meant for the students preparing for engineering undergraduate, AMIE, U.P.S.C. and other competitive examinations.

CONTENTS: Basic Concepts • **Part I: Heat Transfer by conduction** • Conduction-Steady State one dimension • Conduction-Steady State Two Dimensions and three dimensions Conduction-Unsteady State (Transient) **Part-II: Heat Transfer by Convection** • Thermal Radiation-basic Relations • **Part III: Radiation Exchange** • Between Surfaces **Part-IV: Mass Transfer** • **Part-V: Objective Type Questions Bank** • Index

10 202   1st Edn. 1998   ISBN:81-219-1777-8   pp. 580

FLUID MECHANICS & HYDRAULIC MACHINES
R.K. Rajput

The book contains comprehensive treatment of the subject-matter in simple & lucid language, supported by self-explaining figures and a large number of worked out examples. At the end of each chapter, there are (i) Highlights (ii) Objective Type Questions (iii) Theoretical Questions and (iv) Unsolved Problems. The book is specially suited to engineering undergraduate, AMIE (section B) and other competitive examinations.

CONTENTS: Part I: Fluid Mechanics— Properties of Fluids • Pressure Measuremem • Hydrostatic Forces on Surfaces • Buoyancy and Floatation • Fluid Kinematics • Fluid Dynamics • Dimensional & Model Analysis • Flow through Orifices and Mouthpieces • Flow Over Notches & Weirs • Laminar Flow • Turbulent Flow in Pipes • Flow Through Pipes • Boundary Layer Theory • Flow Around Submerged Bodies Drag and Lift • Compressible Flow • Flow in Open Channels • **Part II: Hydraulic Machines** -Impact of Free Jets • Hydraulic Turbines • Centrifugal Pumps • Reciprocating Pumps • Misc. Hydraulic Machines • Water Power Development

10 185   1st Edn. 1998   ISBN:81-219-1666-6   pp. 1088

A TEXTBOOK OF FLUID MECHANICS
R.K. Rajput

The book covers comprehensively the subject of "Fluid Mechanics" in 16 chapters. Each chapter starts with needed text, supported by simple and self-explanatory figures, and a large number of worked out examples, including typical ones, suited for Competitive Examinations. At the end of each chapter, there are (i) Highlights (ii) Objective Type Questions (iii) Theoretical Questions and (iv) Unsolved Problems. The book is specially suited to students appearing for Engineering Undergraduate Examination, Section B of AMIE Examination and Diploma Examinations.

CONTENTS: Properties of Fluids • Pressure Measurement • Hydrostatic Forces on Surfaces • Buoyancy and Floatation • Fluid Kinematics • Fluid Dynamics • Dimensional and Model Analysis • Flow through Orifices and Mouthpieces • Flow over Notches and Weirs • Laminar Flow • Turbulent Flow • Flow Through Pipes • Boundary Layer Theory • Flow Around Submerged Bodies -Drag and Lift • Compressible Flow • Flow in Open Channels • Index

10 192   1st Edn. 1998   ISBN:81-219-1667-4   pp. 784

A TEXTBOOK OF HYDRAULIC MACHINES
(Fluid Power Engineering)
R.K. Rajput

The book covers comprehensively the subject of "Hydraulic Machines" (Fluid Power Engineering). All the chapters start with the much needed text, supported by simple and self-explanatory figures, and a large number of worked out examples, including typical one, suited for competitive examinations. At the end of each chapter there are (i) Highlights (ii) Objective Type Questions (iii) Theoretical Questions and (iv) Unsolved Problems. Thus the book is a comprehensive treatise on the subject and is specially suited to students appearing for Engineering Undergraduate Examination, Section B of AMIE (India) Examination and other Competitive Examinations.

CONTENTS: Impact of Free Jets • Hydraulic Turbines • Centrifugal Pumps • Reciprocating Pumps • Miscellaneous Hydraulic Machines • Water Power Development • Index

| 10 194 | 1st Edn. 1998 | ISBN:81-219-1668-2 | pp. 320 |

A TEXTBOOK OF HYDRAULICS

R.K. Rajput

This treatise contains comprehensive treatment of the subject-matter in simple and lucid language and envelops a large number of solved problems properly graded including typical examples from examination point of view.

CONTENTS: Properties of Fluids • Pressure Measurement • Hydrostatic Forces on Surfaces • Buoyancy and Floatation • Fluid Kinematics • Fluid Dynamics • Dimensional and Model Analysis • Flow Through Orifices and Mouthpieces • Flow over Notches and Weirs • Laminar Flow • Flow Through Pipes • Flow Around Submerged Bodies-Drag and Lift • Flow in Open Channels • Impact of Free Jets • Hydraulic Turbines • Centrifugal Pumps • Reciprocating Pumps • Miscellaneous Hydraulic Machines • Experiments • Index

| 10 198 | 1st Edn. 1998 | ISBN:81-219-1731-X | pp. 546 |

A TEXTBOOK OF ENGINEERING MECHANICS
(Applied Mechanics)

R. S. Khurmi

A large number of worked examples have been given in a systematic manner and logical sequence to assist the student to understand the text of the subject. At the end of each chapter highlights and exercises have been added.

CONTENTS: Introduction • Composition and Resolution of Forces •Moments and their Applications • Parallel Forces and Couples • Equilibrium of Forces • Centre of Gravity • Moment of Inertia • Friction • Principles of Lifting Machines • Simple Lifting Machines • Support Reactions • Analysis of Perfect Frames (Analytical Method & Graphical Method) • Equilibrium of Strings • Virtual Work • Plane Motion • Motion Under Variable Acceleration • Relative Velocity • Projectiles • Motion of Rotation • Combined Motion of Rotation and Translation • Simple Harmonic Motion • Laws of Motion • Motion of Connected Bodies • Helical Springs and Pendulums • Collision of Elastic Bodies • Motion Along a Circular Path • Balancing of Rotating Masses • Work, Power and Energy • Kinetics of Motion of Rotation • Motion of Vehicles • Transmission of Power by Belts and Ropes • Gear Trains • Hydrostatics • Equilibrium of Floating Bodies• Index

| 10 023 | 8th Edn. Rep. 1998 | ISBN:81-219-0651-2 | pp. 848 |

APPLIED MECHANICS AND STRENGTH OF MATERIALS

R.S. Khurmi

This book is suitable for B.Sc. Engg., A.M.I.E. and Diploma courses. A large number of worked out examples, highlights and illustrations have been given.

CONTENTS: Introduction • Composition and Resolution of Forces • Moments and their Application • Parallel Forces and Couples • Equilibrium of Forces • Centre of Gravity • Moment of Inertia • Principles of Friction • Applications of Friction •Principles of Lifting Machines • Simple Lifting Machines • Linear Motion • Circular Motion • Projectiles • Laws of Motion • Work, Power and Energy • Simple Stresses and Strains • Thermal Stresses and Strains • Elastic Constants • Strain Energy and Impact Loading • Bending Moment and Shear Force • Bending Stresses in Beams • Shearing Stresses in Beams • Deflection of Beams • Deflections of Cantilevers • Torsion of Circular Shafts • Riveted Joints • Thin Cylindrical and Spherical Shells • Analysis of Perfect Frames (Analytical Method & Graphical Method)

| 10 025 | 13th Edn. Rep. 1997 | ISBN:81-219-1077-3 | pp. 580 |

A TEXTBOOK OF STRUCTURAL MECHANICS
R.S. Khurmi
10 187   13th Edn. 1997   ISBN:81-219-1642-9   pp. 352

A TEXTBOOK OF APPLIED MECHANICS
R.S. Khurmi
10 191   13th Edn. 1997   IBN:81-219-1643-7   pp. 352

STRENGTH OF MATERIALS
R.S. Khurmi

This student-oriented book has been widely adopted by undergraduates in Engineering throughout India. Subject-matter has been amply supported with illustrations and solved, unsolved and well-graded examples.

CONTENTS: Introduction • Principles of Simple Stresses and Strains • Thermal Stresses and Strains • Elastic Constants • Principal Stresses and Strains • Strain Energy and Impact Loading • Centre of Gravity • Moment of Intertia • Analysis of Perfect Frames (Analytical Method & Graphical Method) • Shear Force and Bending Moment of Beams • Bending Stresses in Beams •Shearing Stresses in Beams • Deflection of Beams • Deflection of Cantilevers • Deflection by Conjugate Beam Method • Propped Cantilevers and Beams • Fixed Beams • Theorem of Three Moments • Moment Distribution Method • Torsion of Circular Shafts • Springs • Riveted Joints • Welded Joints • Thin and Thick Cylindrical and Spherical Shells • Direct and Bending Stresses • Dams and Retaining Walls • Columns and Struts • Introduction to Reinforced Concrete • Mechanical Properties of Materials • Index

10 024   21st Edn. Rep.1998   ISBN:81-219-0533-8   pp. 1000
10 156   ISBN:81-219-0898-1   Hardbound   pp. 1000

STRENGTH OF MATERIALS
R.K. Rajput

The book deals the subject of Strength of Materials exhaustively in a lucid, direct and easily understandable style. It contains a large number of worked out simple, problems arranged in a scientific, and a graded manner to enable the students to grasp the subject effectively from the examination point of view.

Another salient feature of the book is "Experiments" at the end of the chapters to enable the students to have an access to the practical aspects of the subject.

CONTENTS: Simple Stresses and Strains • Principal Stresses and Strains • Centroid and Moment of Inertia • Bending Stresses • Combined Direct and BendingStresses • Shear Stresses in Beam • Thin Shells • Thick Shells • Riveted and Welded Joints • Torsion of Circular and Non-Circular Shafts • Springs • Strain Energy and Deflection due to Shear & Bending • Columns and Struts • Analysis of Framed Structures • Theories of Failure • Rotating Discs and Cylinders • Bending of Curved Bars • Unsymmetrical Bending • Material Testing Experiments • Index

10 174   1st Edn. 1996   ISBN:81-219-1381-0   pp.1056

A TEXTBOOK OF THERMAL ENGINEERING
R.S. Khurmi & J.K.Gupta,

The entire text of the book has been presented in SI units and arranged in a systematic manner. The text has been profusely illustrated by incorporating a number of solved, unsolved and well-graded examples.

CONTENTS: Introduction • Properties of Perfect Gases • Thermodynamic Processes of Perfect Gases • Entropy of Perfect Gases • Thermodynamic Air Cycles • Formation and Properties of Steam • Entropy of Steam • Thermodynamic Processes of Vapour • Thermodynamic Vapour Cycles • Fuels • Combustion of Fuels • Steam Boilers • Boiler Mountings and Accessories • Performance of Steam Boilers • Boiler Draught • Simple Steam Engines • Compound Steam Engines • Performance of Steam Engines • Steam Condensers • Steam Nozzles • Impulse Turbines • Reaction Turbines

• Performance of Steam Turbines • Modern Steam Turbines • Internal Combustion Engines • Testing of Internal Combustion Engines • Reciprocating Air Compressors • Air Motors • Gas Turbines • Performance of Gas Turbines • Introduction to Heat Transfer • Air Refrigeration Systems • Vapour Compression Refrigeration • Psychrometry Air Conditioning Systems • Index

10 172   14th Edn. 1997   ISBN:81-219-1381-0   pp. 846

A TEXTBOOK OF HYDRAULICS
R.S. Khurmi

This revised edition continues to cater to the needs of the undergraduate students of Engineering. Latest research works conducted in various countries have been incorporated.

CONTENTS: Introduction • Fluid Pressure and its Measurement • Hydrostatics • Applications of Hydrostatics • Equilibrium of Floating Bodies • Hydrokinematics • Bernoulli's Equation and its Applications • Flow Through Orifices (Measurement of Discharge) • Flow Through Orifices (Measurement of Time) • Flow Through Mouthpieces • Flow Over Notches • Flow Over Weirs • Flow Through Simple Pipes • Flow Through Compound Pipes • Flow Through Nozzles • Uniform Flow Through Open Channels • Non-Uniform Flow Through Open Channels • Viscous Flow • Viscous Resistance • Impact of Sets • Hydraulic Turbines • Hydraulic Pumps • Pumping Devices • Hydraulic System • Index

10 027   19th Edn. Rep. 1998   ISBN:81-219-0135-9   pp.424

A TEXTBOOK OF HYDRAULICS, FLUID MECHANICS AND HYDRAULIC MACHINES
R.S. Khurmi

About 50 new examples have been included in this revised edition. The subject matter has been amply illustrated by incorporating a large number of solved and well-graded examples.

CONTENTS: Introduction • Fluid Pressure and its Measurement • Hydrostatics • Applications of Hydrostatics • Equilibrium of Floating Bodies • Hydro-kinematics • Bernoulli's Equation and its Applications • Flow through Orifices (Measurement of Discharge & time • Flow through Mouthpieces • Flow over Notches • Flow over Weirs • Flow through Simple Pipes • Flow through Compound Pipes • Flow through Nozzles • Uniform Flow through Open Channels • Non-uniform flow through Open Channels • Viscous Flow • Viscous Resistance Fluid Masses Subjected to Acceleration • Vortex Flow • Mechanics of Compressible Flow • Compressible Flow of Fluids • Flow Around Immersed Bodies • Dimensional Analysis • Model Analysis (Undistorted Models and Distorted Models) • Non-Dimensional Constants • Impact of Jets • Jet Propulsion • Water Wheels • Impulse Turbines • Reaction Turbines • Performance of Turbines • Centrifugal Pumps • Reciprocating Pumps • Performance of Pumps • Pumping Devices • Hydraulic Systems • Index

10 026   19th Edn. 1998   ISBN:81-219-0162-6   pp. 656

THEORY OF MACHINES
R.S. Khurmi & J.K. Gupta

This book is written for the students of Degree, Diploma and A.M.I.E. courses. The subject matter has been amply illustrated by incorporating a large number of solved, unsolved well-graded examples.

CONTENTS: Introduction • Kinematics of Motion • Kinetics of Motion • Simple Harmonic Motion • Simple Mechanisms • Velocity in Mechanisms (Instantaneous Centre Method) • Velocity in Mechanisms (Relative Velocity Method) • Acceleration in Mechanisms • Mechanisms with Lower Pairs • Friction • Belt, Rope and Chain Drives • Toothed Gearing • Gear Trains • Gyroscopic Couple and Precessional Motion • Inertia Forces in Reciprocating Parts • Turning Moment Diagrams and Flywheel • Steam Engine Valves and Reversing Gears • Governors • Brakes and Dynamometers • Cams • Balancing of Rotating Masses • Balancing of reciprocating Masses • Longitudinal and Transverse Vibrations • Torsional Vibrations • Computer Aided Analysis and Synthesis of Mechanisms • Index

10 013   12th Edn. Rep. 1998   ISBN:81-219-0132-4   pp. 1010